Water and Aqueous Solutions

Structure, Thermodynamics, and Transport Processes

WATER AND AQUEOUS SOLUTIONS

Structure, Thermodynamics, and Transport Processes

Edited by

R. A. HORNE

JBF Scientific Corporation
Burlington, Massachusetts

WILEY-INTERSCIENCE

a Division of John Wiley & Sons, Inc.
New York · London · Sydney · Toronto

Preface

In assembling the chapters for this book I have tried to accomplish two things, which, in the growing awareness of the incredible complexity of water and its solutions, have been somewhat neglected. First, by including discussions of seawater, cell water, and the state of water in living cells and tissues, I have tried to remind the reader of the importance of water structure in addition to its great scientific fascination. Second, we have become so impressed with the abnormality of this liquid that we sometimes neglect the points of comparison with its close sister—ice—and with other solvent systems. Such comparisons can, I am sure, contribute significantly to our understanding; therefore in addition to chapters specifically about water and its solutions I have included some chapters on related systems such as ice and fused salts.

The editor of a collected volume such as this has mainly his contributors to thank. I particularly wish to thank those who were prompt and who, for this virtue, have suffered with me the nervous worry that delays might outdate their work. And I cannot help but thank even those who were last to mail their manuscripts—it was such a relief to get them!

R. A. HORNE

Boston, Massachusetts
February 1971

Contents

Water and Aqueous Solutions

Structure, Thermodynamics, and Transport Processes

Introduction

R. A. Horne,
JBF Scientific Corp.
Burlington,
Massachusetts

Many modern scientists question almost belligerently the relevancy of the history of science. For my own part, however, I have preferred to find a profound if somewhat ill-defined significance in the fact that the central concern of the first scientist, Thales of Miletus (ca. 640–546 B.C.), was water. Thales held that "Water is best," that all things are composed of water,[1] and he based his whole cosmology on that one primal substance. Sir Isaac Newton, an avid alchemist, took water as his central principle,[2] and the notion of the primacy of water later reappeared, thinly disguised in Prout's hypothesis.

In our own century, between wars, the investigation of aqueous electrolytic solutions was diligently pursued, and the principles laid down were later enshrined in the two classic texts of Robinson and Stokes[3] and Harned and Owen;[4] the vast wealth of accumulated experimental data on the properties of water itself was summarized by Dorsey.[5] After World War II, however, except for the fields of ion exchange and solvent extraction, there seemed to be far more fascinating substances for chemists to examine than ordinary water and its solutions; this decline in interest occurred because of the explosion in organic chemistry, the renaissance in inorganic chemistry, and the rapid advances in spectroscopy, not to mention radiochemistry and solid-state chemistry. Furthermore, it had become clear that other than infinitesimally dilute aqueous electrolytic solutions represent systems of a very discouraging level of difficulty.

That water is by no means a simple liquid was discerned in the nineteenth century; but modern theories of the structure of liquid water may be said to begin with the magnificent paper by Bernal and Fowler on the interpretation of the then fresh X-ray data on liquid water, which appeared in the first volume of the *Journal of Physical Chemistry*.[6] Over the years this paper has been followed by a virtual avalanche of theories on the structure of liquid water and on the effects of solutes on that structure.[7] But throughout the

1

1940s and early 1950s these advances appeared to be oriented more toward the physicist than toward the chemist. From the vantage point of the more prestigious academic research laboratories, the physical chemistry of aqueous electrolytic solutions may well have seemed, if not an exhausted field, at least a field that had reached the point of diminishing returns in the face of enormous complexities. Among major American universities, Yale alone maintained a strength in the field of water and aqueous solutions in the persons of Raymond Fuoss and Lars Onsager.

In 1964 events took a spectacular turn for the better. In October of that year the Office of Naval Research, the National Aeronautics and Space Administration, and the New York Academy of Science jointly sponsored a conference on "The Role of Water in Biologic Systems." It was easily the most lively, productive, and stimulating scientific meeting that I and many of the other participants had ever attended. Those present included physical chemists, molecular biologists, and hospital-based physiologists. There was vigorous and creative rapport throughout, although not always agreement. It became apparent to the chemists present that, while their discipline had slept, a scientific revolution had occurred. A new field now occupied the center of the vanguard of science, having swept past solid-state physics, inorganic chemistry, atomic energy, and that field comprised the structure and properties of liquid water and of aqueous solutions.

Just as nuclear physics had been the pathfinder science in the first half of our century, molecular biology leads the second. And the biologists had discovered water! Previously they had looked upon water merely as a liquid that filled up the "holes" in living cells and tissues. But then they realized that this most peculiar of solvents plays a major, possibly even the predominant, role in the determination of biomacromolecular conformation and changes of conformation, and that it is also central in a wide spectrum of biological processes, in particular those involving chemical mass transport through biomembranes.

Subsequently, in the United States the renewal of interest in water and aqueous solutions received strong support (including monetary support) from quite a different direction. One of the first areas in which the United States began to feel the pressure of overpopulation and uncontrolled industrial expansion was that of water resources. In order to establish the necessary desalination technology, the federal Office of Saline Water (OSW) recognized the need to support fundamental water research on a scale hitherto unequaled. As a consequence of the neglect of solution chemistry by many of the major universities, the quality of scientific talent ran somewhat behind the funding monies that became available. Nevertheless OSW was able to sponsor work that resulted directly in a very substantial body of fresh and reliable experimental data on the properties of water and aqueous solution.

Inasmuch as the oceans of this planet are a moderately concentrated, mixed, aqueous electrolytic solution, one might expect still a third current of interest in water and its solutions to originate from chemical oceanography. At the present time, however, this expectation is largely unfulfilled. A minority of the younger chemical oceanographers have begun to concern themselves with the more fundamental physical–chemical properties of the marine environment, but the majority still work with traditional chemical analyses, particularly of trace elements, even though they realize that the oceans contain 300,000,000 times more water than the most important trace element (iron). I have tried to remedy this situation by the preparation of a textbook on the chemistry of the marine environment that gives strong emphasis to the structure and properties of water and aqueous solutions,[8] but the modernization of a whole vast, sprawling science clearly cannot be done single-handed, especially by a person as humbly equipped as I am. If only chemical oceanography could attract *and keep* superior students, the level of sophistication of the discipline could be upgraded to the level of the exact physical sciences, and this would necessarily involve detailed concern with water.

At this point it should be noted that the United States Navy, which has long recognized the importance of the medium in which it must work, has been prominent in supporting studies of water. Mention might be made here of a particular study on the effect of pressure on the dissociation constant of magnesium sulfate.[9] In a sense this subject was an ideal one. Not only did the study improve our understanding of a fundamental equilibrium phenomenon, but the particular equilibrium is an important one in seawater; further still, it is an equilibrium of immediate practical concern to the Navy, since it is responsible for the abnormal absorption of acoustic energy in the seas. This research was supported by the Office of Naval Research, which, perhaps more than any other government agency, has distinguished itself by the superior quality of the fundamental research it has sponsored, but which, alas, because of changing politics, is now finding it increasingly difficult to continue its excellent work.

Although the prospects for chemical oceanography do not appear good, those for the environmental sciences generally are considerably brighter. Water is the principal carrier and repository of pollution. Concern with the horrendous problems associated with the deterioration of our environment must inevitably result in new and widespread assaults on the fundamental scientific problems of water physical chemistry. Fortunately, this concern is shared by the public (and reflected through its government in the form of financial support) and by the more intelligent students, who, by virtue of their superior talents, are also particularly sensitive to the sociological problems that modern technology has created. A better understanding of these problems, clearly, is a necessary precursor to systematic technological solutions.[10]

But to continue tracing the progress of the state of our knowledge of the structure of liquid water and aqueous solutions, we should note that theories of water structures are still produced in abundance. As Franks and Ives[11] succinctly phrased it, "increased efforts in recent years have provided an embarrassment of partially successful alternative models...." Among these many theories, my personal favorite has been the "flickering cluster" model of Frank and Wen[12] as developed in a semiquantitative way by Nemethy and Scheraga.[13] Two important advantages of this model, I continue to feel, are that it does not assign a specific spacial lattice to the hydrogen-bonded regions and that it speaks of "ice-like-ness" and "free" water only with extreme caution. I suspect that others shared my disappointment when Pauling's fascinating self-clathrate model[14] failed to conform with the X-ray facts of life; and my discouragement at what seemed to be negative progress in a fast-growing glut of measurements and theories perhaps reached its nadir when Falk and Ford[15] pointed out with ill-disguised indignation that published estimates of the percentage of broken H-bonds in liquid water at 0°C ranged from 3 to 73%.

More recently, however, the situation seems to be improving, and Frank appears to share this optimism; "The acquisition of new kinds of data," he writes, "... will permit us, at least in principle, to draw firmer conclusions than formerly, so that after many years of speculating about what water *might* be like—we can now begin to ask what water *must* be like if certain pieces of data are, simultaneously, to be accepted as reliable."[16]

Reflecting upon the controversy and confusion, it becomes increasingly clear that certain definite, important strides forward have been made. As examples I might mention four problems that either have been solved or now appear well along the road to resolution. The concept of Samoilov[7c] and others of solute hydration in terms of relative solvent residence times forms a badly needed supplement to the very inadequate concept of "hydration number." The solution to the long mysterious problem of "kinks" or anomalous deflections in the temperature dependence of water and solution properties at last appears to be attributable as a characteristic of near-surface or vicinal water rather than the bulk phase.[17]

Although the last volley is still far from being fired, the end of the battle between the proponents of uniformist models and the advocates of mixture models seems to be in sight, thanks to the cogency of Raman spectroscopy[18] and other powerful instrumental techniques. Finally, the announcement of the discovery of ortho-water, a new form of liquid water alleged to be formed catalytically by silica surfaces, has excited the imaginations of scientists and galvanized their energies to such a degree that we can confidently expect that the question of whether this substance is indeed a new form or simply vicinal water with various levels of impurities will be resolved in the near future.

Aqueous systems are so exceedingly complex that our knowledge of them will surely forever remain incomplete. Nevertheless, because of the impetus being given to water research by molecular biology and the impetus it must inevitably receive from the environmental sciences, not to mention that impetus which flows from the insatiability of human curiosity, I think we can reasonably expect that water will reveal many of its secrets in our lifetime; thus we shall eventually possess some keys to the most profound mysteries of the physical order that surrounds *and* constitutes us.

So many specialized papers and monographs on water structure have appeared in recent years that I thought it might be useful in this volume to talk around the subject. Then, too, such emphasis has been given to the eccentricities of water that there has been some tendency to neglect the insights that can be gained by comparison of water with more "normal" liquids. Consequently the first part of this volume is devoted to systems other than liquid water, starting with the structure and transport and interfacial properties of ice, inasmuch as the solid has served as the point of departure for many modern models of the structure of the liquid. These chapters are followed by one on nonaqueous solutions, intended to give some idea of the baseline "normal" behavior upon which, presumably, the peculiarities of water are superimposed. Pure water and fused salts constitute the two extreme ends of a spectrum in which the aqueous systems of the real world lie. Although many solution chemists have taken pains to inform themselves about the former, I suspect that they have paid much less attention to the latter path of potential insight.

Speaking of the real world, the obvious is often the most difficult to propagandize. It should not be necessary to insist again how important aqueous electrolytic solutions are—they are the most important type of physical–chemical system in our world—but the emphasis of chemistry curricula on the solid and gaseous states and their continued neglect of the liquid state, especially aqueous systems, would seem to indicate that the message is still not grasped. For the benefit of those who have not seen the light, and for the congratulation of those who have, two chapters are included, one describing the aqueous system that dominates our physical world–seawater and one devoted to the aqueous systems that are ourselves.

The chapters on the structure of liquid water itself are intended to review and evaluate current views, to interject some fresh ideas, and to place some emphasis on the phenomenon responsible for water's peculiar properties—hydrogen bonding.

I am not sure that existing information on the thermodynamics of aqueous solutions has been exploited as fully as it should be, and for this reason chapters are included dealing with that subject and with partial molal volumes. The complexity of the hydration envelope of solutes renders difficult the interpretation of partial molal volumes, but despite these obstacles I feel that consideration of this parameter holds considerable promise; in particular I am convinced that the volume changes accompanying certain transformations involving biomacromolecules will prove to be very useful in determining the nature of the hydration sheath of those substances.

I have already mentioned the utility of the view that interprets ionic hydration in terms of residence times rather than the more familiar "hydration numbers," and the subject of hydration forms the basis of the next grouping of chapters. As for the tetralkyl ammonium salts, I do not think that their importance, as representatives of a second type of solute hydration and also as forming a continuous series connecting the two types of hydration, has been adequately appreciated, and I hope the detailed review given here will help to remedy this situation. Even although our knowledge of the hydration of relatively simple systems is still imperfect, it is always fascinating, and not too terribly premature, to try to apply our bits and snatches of knowledge to that most important and complex situation, the state of water in living cells and tissues.

After so much concern with structure, I think it fitting that the book should conclude with some chapters devoted to process, although, of course, form and function are inseparable and can never be independent of each other. Of transport processes in aqueous electrolytic solutions, perhaps the first that comes to mind and the one that has certainly received the most experimental and theoretical attention, is the electrical conductivity resulting from the movement of ionic solutes. The accepted theories of electrical conduction bear little relation to modern ideas of the structure of aqueous solutions. It is insufficient to reply that they can be made to fit the dilute solution data. Fresh thinking, new hypotheses, are needed, and I am happy to include a presentation of one such possible alternative approach. Two points of leverage that may be brought to bear on the resolution of solution structure and the mechanism of transport processes in solution are temperature and pressure; hence it is appropriate for the book to end with a review of aqueous solutions under environmental extremes of temperatures and pressure.

REFERENCES

1. Aristotle, *De Caelo*, 294a28.
2. T. S. Kuhn, *Isis*, **43**, 12 (1952).
3. R. A. Robinson and R. H. Stokes, 2nd ed., Butterworths, London, 1959.
4. H. S. Harned and B. B. Owen, *The Physical Chemistry of Electrolytic Solutions*, 3rd ed., Reinhold, New York, 1958.
5. N. E. Dorsey, *Properties of Ordinary Water Substance*, Reinhold, New York, 1940.

6. J. D. Bernal and R. H. Fowler, *J. Amer. Phys.*, **1**, 515 (1933).

7. Earlier theories of water structures are revealed in H. M. Chadwell, *Chem. Rev.* **4**, 375 (1929). For recent reviews, see (a) J. L. Kavanau, *Water Solute-Water Interactions*, Holden-Day, San Francisco, 1964; (b) W. Luck, *Fortschr. Chem. Forsch.*, **4**, 653 (1964); (c) O. Y. Samoilov, *The Structure of Aqueous Electrolytic Solutions and the Hydration of Ions*, Consultants Bureau, New York, New York, 1965; (d) E. Wicke, *Angew. Chem.*, **5**, 106, 122 (1966); (e) W. Drost-Hansen, in W. Stumm, Ed., *Equilibrium Concepts in Natural Water Systems*, Advan. Chem. Ser. No. 67, Amer. Chem. Soc., Washington, D.C., 1967; (f) F. Franks, *Chem. Ind. (London)*, 560 (May, 1968); (g) R. A. Horne, *Surv. Progr. Chem.*, **4**, 1 (1968); (h) D. Eisenberg and W. Kauzmann, *The Structure and Properties of Water*, Oxford University Press, 1969.

8. R. A. Horne, *Marine Chemistry*, Wiley-Interscience, New York, 1969.

9. F. H. Fisher, *J. Phys. Chem.*, **66**, 1607 (1962).

10. As evidence of awareness and progress in this direction, see books such as, W. Stumm, Ed., *Equilibrium Concepts in Natural Water Systems*, Advan. Chem. Ser. No. 67, Amer. Chem. Soc., Washington, D.C., 1967.

11. F. Franks and D. J. G. Ives, *Quart. Rev. (London)*, **20**, 1 (1966).

12. H. S. Frank and W. Y. Wen, *Discussions Faraday Soc.*, **24**, 133 (1957).

13. G. Nemethy and H. A. Scheraga, *J. Chem. Phys.*, **36**, 3382 (1962).

14. L. Pauling, *Science*, **134**, 15 (1961).

15. M. Falk and T. A. Ford, *Can. J. Chem.*, **44**, 1699 (1966).

16. H. S. Frank, *Science*, **169**, 635 (1970).

17. W. Drost-Hansen, *Chem. Phys. Lett.*, **2**, 647 (1968).

18. G. C. Walrafen, *J. Chem. Phys.*, **48**, 244 (1968).

19. B. V. Derjaguin, *Discussions Faraday Soc.*, **42**, 109 (1966).

I   Structure of the Ices

B. Kamb, California Institute of
Technology, Pasadena, California *

1   Significance of Ice Polymorphism

The relation of the structure of water to the structure of ice has been pointed out many times. Less frequently has it been realized that the structural relation between the liquid and the solid involves not only ordinary ice I, but also the other ice phases. Long ago Tammann[1] suggested that there should be as many different "kinds" of water as there are different phases of ice. At that time, nothing was known about the structure of these phases, and in thinking of different "kinds" of water, Tammann had in mind a molecular model of monomers, dimers, trimers, and so on. The idea of the presence of such polymers in water has cropped up numerous times since 1926, most recently in theories for the supposed anomalous phase of water.[2] By the time of Bernal and Fowler's classic paper on water,[3] the structure of ordinary ice I had become known,[4] and on this basis Bernal and Fowler were able to recognize the important role of hydrogen bonding in the structure and properties of water. Starting from the known structural analogy between ice I and the tridymite polymorph of SiO_2, Bernal and Fowler explained the increase in density on melting of ice to water by proposing an analogy between water and quartz, the density increase on going from ice I to water being the same as from tridymite to quartz. This proposal introduced the concept of hydrogen-bond bending in water, since the Si–O–Si linkage in quartz is bent through an angle of about 35 degrees. Bernal and Fowler pointed out that, if this idea were correct, there should occur among the dense polymorphs of ice a structure analogous to that of quartz.

It is, of course, widely recognized that a basic structural relation exists between the two types of condensed phases, liquid and solid, and that the structural change in melting is primarily a loss of crystalline long-range order, accompanied by a great increase in molecular or atomic mobility.[5] The determination of melting curves, under pressure, for substances showing

* Division of Geological Sciences, Contribution No. 1909.

9

high-pressure polymorphism has led recently to recognition of the important fact that the liquid phase is able to anticipate structurally the features of dense polymorphs at pressures up to several kilobars below the corresponding solid–solid phase transitions.[6] In effect, the labile structure of the liquid incorporates features of atomic arrangement from a range of polymorphs accessible at reasonable energies. When the dense contribution from a high-pressure polymorph is sufficiently abundant, the change of volume on melting is negative over a range of pressures below the solid–solid phase transition. From this point of view, the density increase on melting of ordinary ice is directly traceable to the fact that, above 2 kbar, ice shows a rich high-pressure polymorphism. The structural change responsible for the increase in density is basically the same as what occurs in the transition from ice I to the dense ice phases.

Although ice polymorphism thus has an important bearing on water structure, this in no way precludes the possibility that some of the significant structural features of liquid water are not attributable to the dense ice phases. By comparison with the crystalline solids, some type of major structural breakdown must be incorporated in liquid water, as it is in any liquid. The relevance of ice polymorphism is that, contrary to what has usually been assumed in discussions of water structure, the structural breakdown does not necessarily correspond to features of denser molecular packing and higher coordination in liquid water, any more than it does in liquids generally, where an expansion of roughly 10% typically accompanies the structural breakdown.

This chapter discusses pertinent structural aspects of the ice phases but does not develop in any detail their implications for the structure and properties of liquid water. Perhaps because the dense ice structures have only recently become known, their implications for liquid water have not yet been thoroughly investigated; a preliminary discussion, has, however, been given.[7]

2   The Ice Phases

Altogether we can recognize 13 distinct ice phases, if we include among the distinctions certain temperature-dependent order–disorder phenomena. The phases are listed in Table 1. An extensive bibliography of the sources for the information in the table has been given elsewhere,[8] and is therefore omitted here.

As far as known, ordinary ice I is the only phase of ice stable at low pressures. Ice Ic, sometimes called cubic ice, is formed by condensation of water vapor below −110°C and also by inversion of quenched high-pressure ices. At about −90°C, ice Ic inverts to ice I, probably with the release of a small amount of heat, which shows that ice Ic is metastable with respect to ice I at

Table 1   The Ice Phases

| | | | Limits of Temperature range | | |
Phase	Density g cm^{-3}	Pressure kb	Lower °C	Upper °C	Transition at upper limit
I	0.92	0	−273°	0°	melts
Ic	0.92	0	−150° (?)	−90	inv. to I
vit.	0.94	0	−273° (?)	−120°	inv. to Ic
II	1.21	3	−273°	−30°	trans. to III
III	1.15	3	−90°	−20°	melts
IV	1.29	5	−40° (?)	−25°	melts
V	1.28	5	−150° (?)	−10°	melts
VI	1.38	14	−150° (?)	+50°	melts
VII	1.57	25	+2	+110°	melts
VIII	1.63	25	−273°	+2°	trans. to VII
IX	1.19	3	−273°	−105°	trans. to III
X*	~1.4	−273°	−273°	−150° (?)	trans. to VI
XI*	~1.3	5	−273°	−150° (?)	trans. to V

Sources of data are given in Ref. 8.
* Provisional designation.

that temperature. The vitreous form of ice (vit. in Table 1) has been made by condensation of water vapor below −150°C, but it is impossible, as far as we know, to convert liquid water to a vitreous form by rapid cooling.

The high-pressure polymorphism of ice starts at about 2 kbar. Representative pressures near the middle of the stability fields for each of the phases are given in Table 1. Also given are the upper and lower temperature limits of stability at the pressures listed; in some cases, as indicated, the upper limit represents melting, in others a solid–solid phase transition. The densities given for the different phases are those measured or estimated at the pressures listed; the densities at atmospheric pressure, which would be pertinent to water structures, are slightly lower owing to decompression. All the high-pressure phases listed in Table 1 are stable under the conditions indicated, except for ice IV, whose existence is at best metastable and transitory and has been in doubt, but is supported by recent experiments using special nucleating agents.[9]

3   Structural Information

Elucidation of the ice structures has been accomplished by means of X-ray and neutron diffraction, infrared and Raman spectroscopy, and measurement of dielectric properties. The relative entropies, determined from the H_2O phase diagram, are also informative. Except in the case of ice I, somewhat special experimental methods have been required because of the high pressures or low temperatures at which the ice phases are formed. These phases can be retained metastably at atmospheric pressure when cooled to liquid nitrogen temperature, and most of the detailed structural information for the high-pressure phases has been obtained with the help of this property. Some structural information has also been obtained at high pressure, however. A discussion of experimental methods and problems, together with a detailed bibliography of the original papers, is given in the review article cited earlier.[8]

4   Coordination and Bonding

The well-known ice I structure, in which each water molecule is hydrogen bonded to four others in nearly perfect tetrahedral coordination, serves as a basis for the concept of water as a tetrahedral molecule. In this concept, the hydrogen bond is formed primarily by the electrostatic attraction between a proton of one molecule and an unshared electron pair of another, the tetrahedral bond directionality being a consequence of an approach to sp^3 orbital hybridization in the electronic structure of the water molecule. This theme is repeated, or in fact improved upon, in the structure of ice Ic, where the water molecule coordination is constrained to be perfectly tetrahedral by the cubic symmetry of the crystal. (By contrast, the nearly perfect tetrahedral coordination in ice I is achieved by unconstrained adjustment of the a and c axes of the hexagonal crystal and of the coordinate that fixes the position of the oxygen atom within the unit cell.) The hydrogen-bond network in ice Ic, illustrated in Figure 1, is related to that in ice I in the same way that the bond networks are related in the cristobalite and tridymite forms of SiO_2, that is, by a simple change in stacking sequence similar to cubic and hexagonal closest packing. With regard to nearest-neighbor coordination, which is the feature most clearly expressed in liquid structure, the ice I and Ic structures are essentially identical. Evidently the differences in nonnearest-neighbor coordination are responsible for the difference in stability of the two phases, but the difference in energy is so small that its measurement remains uncertain. However, the nonnearest-neighbor coordinations have significance for liquid structure models, inasmuch as they determine the size and shape of the void spaces in the rather open tetrahedral framework, which may contain interstitial water molecules, according to some theories for liquid water.[10]

In spite of its obvious pertinence to liquid water, very little is known about the vitreous form of ice. Its structure appears to be closely related to ice Ic, on the basis of close relationships in the X-ray diffraction patterns and near equality of densities.[27] In fact some workers consider that this form is not vitreous at all, but is simply very fine-grained ice Ic;[11] however, as the grain size becomes very small, this distinction becomes moot.

Although the first X-ray[12] and infrared[13] studies of the dense forms of ice suggested that normal hydrogen bonding had been eliminated, and even that the structure of the water molecule had been broken down and replaced with an arrangement of H^+ and O^{2-} ions,[12] later work shows the situation to be quite different.[7,8] All the forms of ice are hydrogen bonded, although the strength of the hydrogen bonding generally decreases with increasing density, and in none of the dense forms is the bonding as strong as it is in ice I. In all the ice phases, the water molecules maintain their individuality almost unchanged from their configuration in water vapor, and each molecule forms four hydrogen bonds, making a bond framework that traverses the entire crystal. Thus, although we can consider the water molecules to be polymerized by hydrogen bonding,[14] the early idea that the different forms of ice correspond to different water polymers[1] is not really valid, in the sense that different discrete, tightly bonded groups of molecules cannot be singled out in the structures. The only groups of this type definitely known are the aquohydrogen ions $H_5O_2^+$, $H_7O_3^+$, and so on, which occur in certain hydrates.[15] In the ice phases, we are dealing instead with various three-dimensional bond frameworks, in which no discrete groupings of this type occur.

The bonding coordination in the dense ice phases is tetrahedral in the sense that each molecule forms four hydrogen bonds of approximately equal strength, but the arrangement of the bonded neighbors is more or less distorted from the ideal tetrahedral geometry, by contrast with ices I and Ic. This distortion represents a bending of the hydrogen bonds, because of the unchanging geometry of the individual water molecules; the hydrogen-bonding interactions are not strong enough to cause any significant rehybridization that would allow a change in the H–O–H angle or in the spatial distribution of the unshared electron pairs. The protons of a water molecule in distorted tetrahedral coordination must in general lie off the O···O centerlines to the neighbors, and the acceptor electron pairs, tetrahedrally disposed relative to the H–O–H angle, cannot in general be directed either exactly toward protons of neighboring molecules or along the O···O centerlines. Force constants for these different aspects of hydrogen bond bending determine the equilibrium orientation that a water molecule assumes relative to its four neighbors in distorted tetrahedral coordination. Observed orientations indicate that the donor relation (alignment of O–H in relation to the O···O line) has a more important effect than the acceptor relation (alignment of the lone pair directions), but actual force constants for the different aspects of hydrogen bond bending have not yet been determined. A simple overall measure of bond bending, which overlooks these detailed aspects, is the mean square deviation

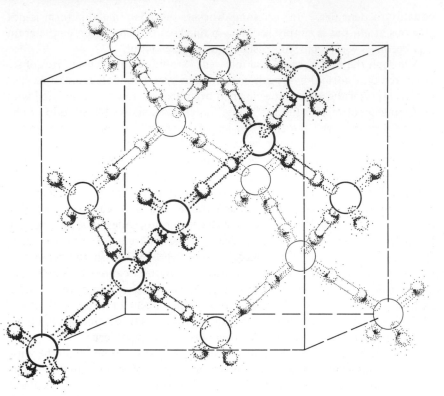

Figure 1.   Molecular arrangement in ice Ic. Large solid balls represent oxygen atoms, and fuzzy small balls hydrogen atoms statistically distributed over the positions shown, so as to link adjacent oxygen atoms by asymmetric hydrogen bonds (shown dotted). Two structures depicted (left and right) are identical and differ only in the choice of the origin point for the cubic unit cells (shown dashed). Water molecules at four of the corners of the cube on the left are omitted, because they form H-bonds only to molecules outside the cell shown.

of the O···O···O bond angles from tetrahedral (109.5°). This quantity is given for each of the ice phases in Table 2.

The extent of bond bending, as indicated in Table 2, increases progressively with increased density (except for ices VII and VIII). The distortions from ideal tetrahedral coordination around each water molecule make possible increased densities by allowing the accommodation of neighbors in the distance range 3 to 4 Å, too far away to be hydrogen-bonded directly to the central molecule, but close enough to interact significantly with it by dispersion forces. In ice I, the next-nearest neighbors are at the relatively large distance 4.5 Å and are second neighbors as counted outward along the bond

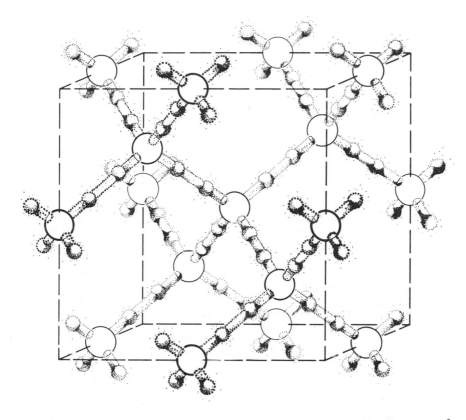

network. In the denser ice structures, by contrast, the neighbors at 3 to 4 Å are actually fourth or more distant neighbors as counted along the bond network. This results from a "doubling back" feature of the network, and allows the near neighbors to be accommodated with a minimum of bond bending.

The bond-bending mechanism of densification is offset to some extent by a lengthening of the hydrogen bonds in the denser ice phases, as shown by the mean bond lengths listed in Table 2. The lengthening appears to be coupled to the bond bending and reflects weakening of the bonds by bending. The weakening is demonstrated also in infrared spectra of the ice phases, the O–H stretching frequencies showing a systematic increase as the bonds lengthen. This effect is shown by the mean stretching frequencies listed for the ice phases in Table 2.

Although hydrogen-bond bending is abundant in the dense forms of ice, the particular pattern of bond bending represented by the original proposal of Bernal and Fowler[3] is not actually realized, because none of the ice phases is in fact a structural analog of quartz. The failure of a quartz analog to occur

Table 2   Structural Features of the Ices

Phase	Bond Frameworks[b]	Bond Bending[c] $(deg^2 \times 10^3)$	Bond Length[d] (Å)	O–H Frequency[e] (cm^{-1})	Protons[f]
I	1	0	2.75	3277	Disorder
Ic	1	0	2.75	3277	Disorder
II	1	0.29	2.80	3350	Order
III	1	(0.27)	(2.78)	—	Disorder
V	1	(0.34)	(2.80)	—	Disorder
VI	2	(0.53)	(2.81)	—	Disorder
VII	2	0	(2.96)	—	Disorder
VIII	2	0.01	2.96	3442	Order
IX	1	0.27	2.78	3321	Order
X[a]	2	0.53	2.81	3356	Partial order
XI[a]	1	0.34	2.80	3349	Partial order

From Refs. 7 and 8.
[a] Provisional designation.
[b] Number of independent hydrogen-bond frameworks.
[c] Mean square deviation of O···O···O bond angles from tetrahedral.
[d] Mean O···O distance for hydrogen bonds, at 77°K and 1 atm; for ices III–VI, the values are assumed the same as in the corresponding low-temperature (ordered) structures.
[e] Mean O–H stretching frequency from infrared spectra, or, for ices VIII and IX, from Raman spectra, at 77°K and 1 atm. Data from Ref. 28.
[f] State of water-molecule orientational order or disorder.

doubtless results from the differences between the bond-bending energetics of hydrogen bonds and Si–O–Si linkages, and also from differences between the packing requirements that arise when the relatively large oxygen atoms lie at the tetrahedral centers, as in H_2O, and when they lie at or near the centers of the bond linkages, as in SiO_2. For these reasons, and also because silicon is able to achieve 6-coordination by oxygen under high pressure, the general structural analogy between SiO_2 and H_2O is decidedly incomplete, but it continues nevertheless to be useful for the less dense polymorphs.[16]

Typical features of the bent hydrogen-bond networks in the dense ice phases are illustrated by ice V, shown in Figure 2. In this structure there are 4-rings of water molecules, which are rather more compact groups than the 6-rings that occur in ices I and Ic. Such 4-rings have been proposed as water tetramers.[17] In ice V, however, the 4-rings are fully hydrogen-bonded to the rest of the framework, and the hydrogen-bond lengths in the 4-rings are somewhat longer than in the rest of the framework, so that the 4-rings cannot be considered as relatively tightly bonded units and thus are not discrete tetramers. Although the distortion of the tetrahedral coordination can to some extent be judged from the illustration, a full demonstration requires three-

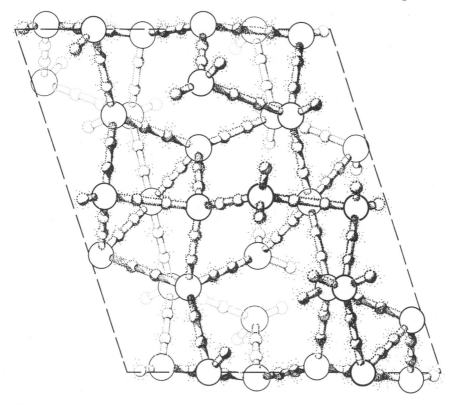

Figure 2.   Molecular arrangement in ice V, shown with the same conventions as in Figure 1
The unit cell is monoclinic and viewed along the *b* axis.

dimensional diagrams or tables of bond angles. The O···O···O bond angles
for a representative water molecule in ice V are shown in Figure 3. Also
shown are the hydrogen-bond lengths; the general increase over the length
2.75Å in ices I and Ic is evident. A nonbonded near neighbor made possible by the tetrahedral distortion is the molecule at distance 3.28 Å shown in
Figure 3. This molecule is five molecules distant from the central molecule
as counted outward along the bond network, which can be done by referring
back to Figure 2.

   In ices VI, VII, and VIII a new structural feature appears, which is not
present in ice structures of lower density. The structures consist of two interpenetrating but not interconnected hydrogen-bond networks. I call this
structural feature self-clathration, because of its analogy with clathration in
crystalline hydrates,[18] where icelike hydrogen-bonded water frameworks

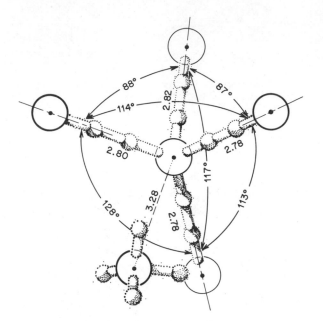

Figure 3. O···O bond distances (in angstroms) and O···O···O bond angles for a representative water molecule in ice V. To identify the same molecule in Figure 2, note that the oxygen atom at the top here is the one in the upper right-hand corner of Figure 2. Here and in Figure 2 the statistically distributed protons are shown for simplicity on the O···O centerlines, but in reality lie somewhat off to the side, because the O···O···O angles do not match the H–O–H angle of 104.5°. The water molecule shown at distance 3.28 Å is not H-bonded to the central molecule.

enclose guest molecules; in self-clathration, the "guest" molecules are water, forming a second hydrogen-bonded framework identical to the first. Self-clathration seems to be favored as a way of obtaining high packing densities of water molecules without requiring excessive hydrogen-bond bending.

In ice VII, which is illustrated in Figure 4, the bond bending is in fact reduced to nil. The bonding coordination is again perfectly tetrahedral, as in ice Ic. This is achieved by allowing the total nearest-neighbor coordination of each water molecule to rise to 8. The eight neighbors lie at the corners of a cube around each molecule, and of these eight, four, in tetrahedral orientation around the central molecule, are hydrogen-bonded to it; the remaining four, also tetrahedrally arranged, are not hydrogen-bonded to the central molecule, although they are constrained by the structural geometry to lie at the same distance as the bonded neighbors. Nonbonded water molecules at this distance are close enough to repel one another, and as a result the coordination cube is forced to expand, stretching the hydrogen bonds until equilibrium is achieved between tension of the stretched bonds and repulsion of the nonbonded neighbors.[19] The resulting O···O distances, 2.96 Å, represent the longest and weakest hydrogen bonds occurring in any of the ice phases. The bond weakening in this case is caused entirely by forced bond stretching, and not by bending. The hydrogen bonds in ice VII actually form two separate, inter-penetrating networks, each identical to the network in ice Ic. This is, again, an example of the self-clathrate feature, but it is somewhat less obvious since the nonbonded O···O distances, which represent contacts between the two separate bond frameworks, are just as short as the hydrogen bonds within the frameworks. However, when the two frameworks are dissected out of the structure, as shown in Figure 4, the self-clathrate feature becomes evident.

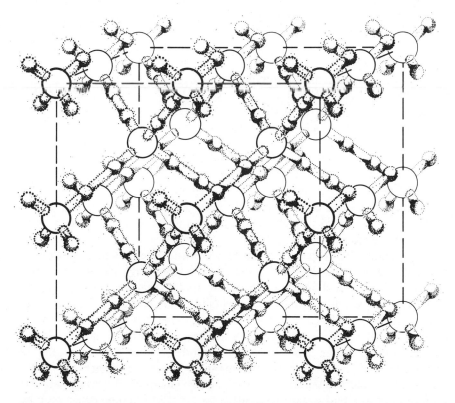

Figure 4. Molecular arrangement in ice VII, shown with the conventions of Figure 1. The ice VII structure is obtained by combining into a single cube the contents of the two cubes shown separately in Figure 1. This permits a dense molecular packing without breakage of hydrogen bonds.

In ice VIII a distortion occurs from the structure of ice VII, such that the hydrogen bonds retain their length of 2.96 Å, whereas the four nonbonded O···O contacts per molecule separate into two at a distance of 3.15 Å, and two at 2.80 Å. Here we have a remarkable situation in which some of the nonbonded O···O distances are distinctly shorter than the hydrogen bonds. The cause of this is explained later.

Although it is true that the bonding coordination in all the ice phases is tetrahedral (either perfect or distorted), the examples of ices VII and VIII show that the total near-neighbor coordination, bonded and nonbonded, can rise to six, eight, or even nine, if in ice VIII we count a neighbor at 3.19 Å as part of the near-neighbor coordination group. This situation is reminiscent of the fivefold coordination of water molecules in certain rare hydrates.[20] Although a closest-packed ice structure, with total coordination 12, is theoretically possible,[21] it has not actually been realized and probably is not the stable structure under any conditions of temperature or pressure, because of unfavorable energy and entropy.[19]

In spite of these complications of total coordination, the fact remains that each water molecule in the ice phases is fully hydrogen bonded in that it forms four bonds, to two of which it donates protons and in two of which it accepts protons from other molecules. No water molecules are unbonded, either partially or completely. In particular, there are no ice structures in which unbonded water molecules occupy cavities in a hydrogen-bonded framework. This has special significance in relation to liquid water, because several models for water structure involve such unbonded interstitial molecules as an important feature.[10] Since the stabilizing and destabilizing aspects of such a structural feature (high rotational entropy, high bond-breakage energy) would be just as effective in a crystalline solid as in a liquid, the absence of any solid phases with this type of structure argues against its importance as the primary structural feature of the liquid.[22]

Nevertheless, the properties of liquid water indicate that there must be a substantial amount ($\sim 20\%$) of hydrogen-bond breakage in the structure.[7] The type of bond breakage that can effect rapid relaxation of the hydrogen-bonded structure, allowing the observed fluidity and short dielectric relaxation time of the liquid, has the form of incompleteness in the hydrogen-bond framework, which occurs wherever a water molecule of the framework is not able to complete four hydrogen bonds. Bond-breakage defects (L and D defects) do occur in low concentration in the ice structures and are responsible for dielectric and elastic relaxation in the ice phases.[23] However, it seems unlikely that in liquid water these particular types of bonding defects can be mainly responsible for the relaxation phenomena, which are many orders of magnitude more rapid than in the solids.

Probably a more fundamental breakdown in the bond network is required, in which not only are there local imbalances in availability of the needed donor protons and acceptor electron pairs, as in the L and D defects, but also there are large departures or distortions from tetrahedral coordination of individual molecules, so that some protons and some acceptors are left "dangling," out of range of any possible hydrogen-bonding interactions. Bifurcated hydrogen bonds perhaps represent an approach to a local situation of this kind. Although distortions from ideal tetrahedral coordination are large in some of the ice structures, they do not become severe enough to lead to bond-breakage effects of the above types, hence the ice structures cannot provide direct insight into the bond-breakage aspects of liquid water structure.

5 Order and Disorder

The presence of discrete water molecules in the ice phases is associated with the asymmetry of the hydrogen bonds, such that the proton in each bond lies closer to one oxygen atom than the other. The hydrogen-bond asymmetry is always found for O···O distances longer than about 2.5 Å, the proton in the bond lying about 1.0 Å from the oxygen atom at one end. Ions such as H_3O^+ and OH^- occur only in very low concentration in the ice phases; normally each oxygen atom has only two nearby protons, forming a discrete water molecule. Because of tetrahedral coordination, there are six ways of orienting the molecule to form hydrogen bonds to the four neighbors. If the coordination is perfectly tetrahedral, these six orientations are equivalent by symmetry and result in exactly the same bonding energy, hence they occur statistically with equal probability. This is the phenomenon of proton disorder, or water-molecule orientation disorder, which was discovered originally in ice I[24] and is illustrated in Figure 5.

In ice I the departure from ideal tetrahedral coordination is not important enough to interfere with the statistical occurrence of the different molecular orientations. The gain in configurational entropy that accompanies disordering the protons (0.81 eu) is large enough that, even with the much larger departures from ideal tetrahedral coordination that occur in the dense ice phases, the effects of proton disorder are still strongly observed in the physical properties of these phases at higher temperatures, near melting.[25]

The hydrogen-bond energy, however, must be affected by bond-bending factors such as lack of a match between the H–O–H angle of the water molecule and the O···O···O coordination angle into which the molecule donates its protons. Owing to the relatively high force constant for H–O–H bending, this angle is relatively little altered in ice phases and hydrates from its value in the free water molecule. A water molecule oriented so that it donates its protons into an O···O···O angle differing substantially from tetrahedral (or from

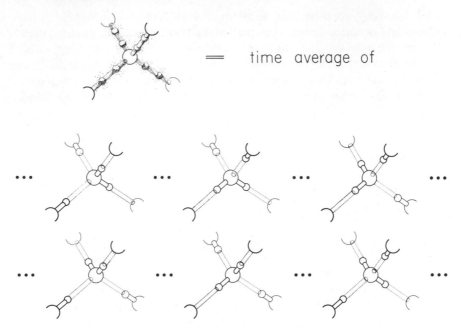

= time average of

Figure 5. Water-molecule orientation disorder (proton disorder), depicted for a particular water molecule in ice Ic. This clarifies the symbolic representation of the statistical proton arrangement that is used in Figures 1–4. At high temperatures, each water molecule flips around in the course of time among the six different orientations as shown. At low temperatures, however, one of the six orientations becomes frozen in at each site. In ices I and Ic this occurs in a random way, so that spatial average over crystallographically equivalent water molecules at low temperature is the same as a time average over a single molecule at high temperature.

104.5°, the angle in the free molecule) should therefore experience a definite bond-energy destabilization, and this orientation should tend to be avoided. The extent to which such orientation avoidance occurs statistically in the ice phases at higher temperatures is not known, but at low temperatures a definite pattern of proton order develops in all cases except ices I and Ic. Even in ice I, there is indirect evidence that some proton order develops in response to elastic strain.[26]

The transformation of ice VII to ice VIII, discussed earlier, is an example of a proton-ordering transition. Such transitions are responsible for several of the distinctions between ice phases listed in Table 1. In most of these transitions, the development of proton order can be traced more or less directly to the existing distortions from tetrahedral coordination in the basic structure;

the favored water molecule orientations are those that occupy O···O···O angles as near to 104.5° as possible. In ice VII, the coordination symmetry is perfectly tetrahedral, and a state of proton order can develop only if a distortion from this ideal symmetry occurs. This is the source of the structural distortion that takes place in going from ice VII to ice VIII. The distortion causes the O···O···O angles that become occupied preferentially in ice VIII to decrease from 109.5 to 107.9°.

The proton disordering that occurs, conversely, on warming the ordered ice phases from low temperature is in some ways analogous to a "melting" of the proton arrangement within an essentially fixed framework provided by the oxygen arrangement. The disordering transitions are like melting in that they do not permit any appreciable superheating. By contrast, it is possible, at least in some of the phases, to quench in a state of complete or partial disorder by rapid cooling.

It stands to reason that the type of proton disorder that occurs in the bond networks of the ice phases at higher temperatures is present also in liquid water, and that proton-ordering effects do not play a significant role in the liquid structure. Although proton disorder is decisive in allowing dielectric and other types of relaxation in the ice structures, through the migration of the L and D bond defects, the much faster relaxation in liquid water probably represents a fundamentally different mechanism, in which the water molecules are reoriented more quickly in the course of breakdown and restructuring of the entire bond network than by reorientation within an unchanging network, as in the ices. Thus although the proton order–disorder phenomena in the ice structures are not of such direct interest in relation to liquid water, nevertheless they have indirect significance, because they provide some definite information about the energetics of bond bending in hydrogen-bond networks, which must govern the type and extent of bond bending that can occur in liquid water.[7]

REFERENCES

1. G. Tammann, *Z. Anorg. Allgem. Chem.*, **158**, 1 (1926).
2. E. Wicke, *Angew. Chem. (Int. Ed.)*, **5**, 109 (1966); R. W. Bolander, J. L. Kassner, Jr., and J. T. Zung, *Nature*, **221**, 1233 (1969); J. Donohue, *Science*, **166**, 1000 (1969); C. T. O'Konski, *Science*, **168**, 1089 (1970).
3. J. D. Bernal and R. H. Fowler, *J. Chem. Phys.*, **1**, 515 (1933).
4. W. H. Barnes, *Proc. Roy. Soc.*, **125**, 670 (1928).
5. A. R. Ubbelohde, *Melting and Crystal Structure*, Oxford University Press, New York, 1965.
6. G. C. Kennedy, A. Jayaraman, and R. C. Newton, *Phys. Rev.*, **126**, 1363 (1962); W. Klement and A. Jayaraman, *Prog. Solid State Chem.*, **3**, 289 (1966); E. Rapoport, *J. Chem. Phys.* **46**, 2891 (1967).

7. B. Kamb, in *Structural Chemistry and Molecular Biology*, A. Rich and N. Davidson, Eds., Freeman, San Francisco, 1968, p. 507.

8. B. Kamb, *Trans. Amer. Cryst. Assoc.*, **5**, 61 (1969).

9. L. F. Evans, *J. Appl. Phys.*, **38**, 4930 (1968).

10. A. H. Narten and H. A. Levy, *Science*, **165**, 447 (1969).

11. M. Kumai, *J. Glaciol.*, **7**, 95 (1968).

12. R. L. McFarlan, *J. Chem. Phys.*, **4**, 60, 253 (1936).

13. E. R. Lippincott, C. E. Weir, and A. Van Valkenburg, *J. Chem. Phys.*, **32**, 612 (1960).

14. R. A. Horne, *Science*, **168**, 151 (1970).

15. J. M. Williams, *Inorg. Nucl. Chem. Lett.*, **3**, 297 (1967); J. Lundgren and J. Olovsson, *Acta Cryst.*, **23**, 966, 971 (1967); J. Lundgren and J. Olovsson, *J. Chem. Phys.*, **49**, 1068 (1968) J. Olovsson, *J. Chem. Phys.*, **49**, 1063 1968.

16. B. Kamb, *Science*, **148**, 232 (1965).

17. L. V. Bellamy, A. R. Osborn, E. R. Lippincott, and A. R. Bandy, *Chem. and Ind.* (Lond.) **1969**, 686.

18. G. A. Jeffrey, and R. K. McMullen, *Prog. Inorg. Chem.*, **8**, 43 (1967).

19. B. Kamb, *J. Chem. Phys.*, **43**, 3917 (1965).

20. W. C. Hamilton and J. A. Ibers, *The Hydrogen Bond in Solids*, Benjamin, New York, 1969, p. 207.

21. B. Kamb and B. L. Davis, *Proc. Nat. Acad. Sci. U.S.*, **52**, 1433 (1964).

22. B. Kamb, *Science*, **167**, 1520 (1970).

23. N. H. Fletcher, *The Chemical Physics of Ice*, Cambridge University Press, Cambridge, 1970.

24. L. Pauling, *J. Amer. Chem. Soc.*, **57**, 2680 (1935).

25. B. Kamb, *Acta Cryst.*, **17**, 1446 (1964); G. J. Wilson, R. K. Chan, D. W. Davidson, and E. Whalley, *J. Chem. Phys.*, **43**, 2384 (1965); E. Whalley, D. W. Davidson, and J. B. R. Heath, *J. Chem. Phys.*, **43**, 2384 (1966).

26. R. Bass, *Zeitschr. Phys.*, **153**, 16 (1958).

27. J. A. Ghormley and C. J. Hochanadel, *Science*, **171**, 62 (1971).

28. J. E. Bertie and E. Whalley, *J. Chem. Phys.*, **40**, 1637 (1964); J. E. Bertie, H. J. Labbe, and E. Whalley, *J. Chem. Phys.*, **49**, 2141 (1968); J. P. Marckmann and E. Whalley, *J. Chem. Phys.*, **32**, 612 (1960); M. J. Taylor and E. Whalley, *J. Chem. Phys.*, **40**, 1660 (1964).

2   Transport Properties of Ice

*C. Jaccard, Institut de physique de l'Université de
Neuchâtel, Neuchâtel, Switzerland*

1   Introduction

The study of the physical properties of ice can be traced back to the middle
of the nineteenth century when Faraday observed various electrical effects.
But even though ice is a very common substance, most of its properties have
been investigated only during the last two decades. Compared to water
chemistry or to semiconductor physics, our current knowledge of ice is still
fragmentary, and several of the basic problems remain unsolved.

Interest in ice dynamics derives from the character of ice as a hydrogen-
bonded molecular crystal, together with its almost purely protonic conduction. Certain models can then be borrowed from the chemistry of aqueous
solution, others from the physics of semiconductors or insulators, but each
must be adapted to the particular situation. As in many other cases, a theoretical description from first principles is hopeless. A good part of the transport properties can be explained thanks to the lattice model evolved from
X-ray and neutron diffraction experiments, but the picture obtained thus far
rests also on concepts such as structural and chemical defects, which play an
essential part in modern solid-state physics. In this respect, the microscopic
picture lags behind the phenomenological description and a good proportion
of future efforts must be devoted to elucidating it. On the experimental side,
the foundation has been considerably widened by recent improvement of
existing techniques (e.g., crystal growth and nonblocking electrodes) and also
by new methods such as magnetic resonance and isotope tracers.

The main part of this chapter is devoted to electrical conductivity and to
the responsible defects. This transport mechanism is conditioned by the
crystal structure in the most sensitive way. The diffusion properties are the
object of the second part of the chapter. The electrically active defects give
rise to the well-known chemical and thermal potential differences, the study
of which contributes significantly to the picture of general conduction, whereas molecular diffusion is related to the neutral point defects. The last section
discusses briefly macroscopic mass transport by plastic deformation, for

which a connection with electrically active defects has been discovered recently. Thermal conductivity is left aside, since it has not yet raised problems peculiar to ice.

2 Electrical Properties

Early electrostatic experiments indicate that a charge transfer is possible through ice, and this implies a finite electrical conductivity. The magnitude of such a charge transfer was determined for the first time at the turn of the century (Johnstone, 1912); it depends upon the temperature according to an Arrhenius law and amounts to $10^{-10}/(\Omega)(cm)$ at $-10°C$. The first question that arises pertains to the nature of the charge carriers: electrons or ions. It has been settled by electrolysis experiments performed on impure ice and on pure ice (Workman et al., 1954; Decroly et al., 1957). Within an experimental error of the order of one percentage point, the amount of hydrogen gas evolved at the cathode corresponds to the electric charge transferred through the crystal, showing the ionic character of the conductivity (at least at high temperature).

Although the first conductivity measurements had been performed carefully, some doubts persisted (and still persist!) about their reliability, for the following reasons. As in many nonmetallic conductors, the measurement of direct current conductivity involves several sources of error, which implies the need for an appropriate design of the experiment. The most prominent source of error is due to electrode polarization: the mobile electric charges, which are in the form of the metallic electrons, have to be transferred to and from ions at the metal–ice interface. This process is generally far from completion—a significant part of the ions cannot discharge even under high potential differences. The electrodes then have a so-called blocking character, almost perfect when they consist of evaporated gold. The space charge that builds up gives rise to a current–voltage characteristic that does not reveal the properties of the bulk, but rather those of the interfaces. To obviate this, several methods have been devised which yield indeed roughly the same magnitude for the conductivity at $-10°C$, but different activation energies (Table 1).

A convenient source and sink for protons can be made by massive palladium electrodes, partly saturated with hydrogen (Bradley, 1957). A refinement consists in a palladium black layer on a massive metallic electrode, but separated from the ice by a palladium foil with a thickness of 200 Å (Bullemer et al., 1969). Another trick, used in the semiconductor technology, is to enhance considerably the conductivity of the layer near the electrode ("sandwich" electrode with HF doping) (Jaccard, 1959; Gross, 1962). The same effect is

Table 1   Direct Current Conductivity of Pure Ice at $-10°C$

Conductivity $(\Omega^{-1} cm^{-1})$	Activation Energy (kcal/mole)	Crystallinity	Electrodes	Volume of Activation $(cm^3/mole)$	Reference
H_2O 1.1×10^{-9}	19.5	Polycrystalline	Solid (potential probes)		Johnstone (1912)
1.4×10^{-9}	12.2	Polycrystalline	Solid		Bradley (1957)
1.0×10^{-9}	14.0	Monocrystalline	Sandwich		Jaccard (1959)
1.0×10^{-9}	11.0	Polycrystalline	Liquid (potential probes)		Eigen et al. (1964)
1.1×10^{-10}	8.0	Monocrystalline	Au/Pd (potential probes guard rings)		Bullemer et al. (1969)
		Polycrystalline	Solid	$-11 + 3$	Chan et al. (1965)
D_2O 3.6×10^{-11}	13.1	Polycrystalline	Liquid (potential probes)		Eigen et al. (1964)

obtained with ion-exchange membranes (Durand et al., 1967). The standard method of interjacent potential probes has also been used by many investigators (Johnstone, 1912; Eigen et al., 1964; Bullemer et al., 1966). In an ingenious device, a long U-shaped ice sample has strong temperature gradients at its ends, so that a water layer bridges the gap between platinum current electrodes and the ice, thus preventing almost completely electrode polarization (Eigen et al., 1964). Unfortunately, this method is restricted to a temperature range near the melting point. Another way of controlling electrode behavior is to introduce deliberately an insulating layer between metal and ice, making the electrode perfectly blocking (Mounier and Sixou, 1969) and to analyze the resulting Maxwell–Wagner system with a low-frequency a-c method.

The activation energies obtained thus far seem to converge toward the value given by Eigen, who took the maximum care to eliminate unwanted effects. However, later investigations of the surface conductivity revealed that it can, under certain circumstances, significantly bypass the bulk conduction and induce a sensitive potential distortion that falsifies even potential probe measurements with ohmic electrodes (Riehl et al., 1966a; Jaccard, 1967). The only way to cancel this disturbance is to choose an appropriate geometry and to mount around the current and the probe electrodes guard rings that follow their potential by means of a servosystem (Bullemer et al., 1969). The resulting conductivity, which may then be attributed unequivocally to the bulk, is lower by a factor of 10 ($1.1 \times 10^{-10}/(\Omega)(cm)$ at $-10°C$), and the activation energy is smaller by 3 kcal/mole (8.0 ± 0.5 kcal/mole). According to our present knowledge, these values may be taken as the most reliable.

The next question raised by the conductivity pertains to its composite nature: it is the product of the carriers' concentration and their mobility (Table 2). A direct determination of each of the factors has not yet been performed for pure ice, but an indirect estimate can be drawn from the

Table 2   Properties of Charge Carriers in Pure Ice at $-10°C$

Properties	H_2O	D_2O	Unit	Reference
		Ions		
Concentration	1.4×10^{-10}	0.32×10^{-10}	mole/liter	Eigen et al. (1964)
	5×10^{-10}			Bullemer et al. (1969)
Dissociation rate	3.2×10^{-9}	2.7×10^{-11}	sec^{-1}	Eigen et al. (1964)
	1.1×10^{-9}			Bullemer et al. (1969)
Pair formation energy	22.5	25	kcal/mole	Eigen et al. (1964)
	17.5			Bullemer et al. (1969)
Mobility	7.5×10^{-2}	1.2×10^{-2}	$cm^2/V\text{-}sec$	Eigen et al. (1964)
(saturation current)	2.5×10^{-3}			Bullemer et al. (1969)
(Hall effect, $-8°C$)	1.4			Bullemer et al. (1969)
(Space charge, $-60°C$)	5			Engelhardt et al. (1969)
		Bjerrum Defects		
Concentration	1×10^{-5}		mole/liter	Jaccard (1959)
Pair formation energy	15.5		kcal/mole	Jaccard (1959)
Mobility	2×10^{-4}		$cm^2/V\text{-}sec$	Jaccard (1959)
Activation energy	5.3		kcal/mole	Jaccard (1959)

saturation current. If the field applied to a sample is large enough to extract all the charge carriers as soon as they are produced by chemical dissociation, the current reaches a saturation value that is no longer proportional to the applied potential. In first approximation it is a constant, proportional to the volume and to the dissociation constant, and thus it yields this parameter in an unambiguous way. But since the field must be large, the process of dissociation is slightly enhanced during its first stage, while the ions are still within atomic separation. Combining the values of the saturation current and of the field dissociation yields the equilibrium concentration of the charge carriers. First performed by Eigen et al. (1964), this experiment indicated a rather low dissociation constant of $3.2 \times 10^{-9}/sec$ at $10°C$, with an activation energy of 22.5 kcal/mole and an ionic concentration of 1.4×10^{-10} mole/liter. This implies a mobility of 7.5×10^{-2} cm²/V-sec, several orders of magnitude higher than the values expected for ions at this temperature. This suggests a particular transfer process for the proton, especially as the activation energy vanishes (see Section 4). The experiment has been resumed by the Munich school, with the guard-ring system eliminating the surface contribution (Bullemer et al., 1969). The dissociation rate is smaller by a factor of 3, and its activation energy is only 17.5 kcal/mole. Combining with the conductivity, this yields only 2.5×10^{-3} cm^2/V-sec for the mobility which is still a large value for ions. Analysis of temperature dependences seems to indicate a small negative activation energy; that is, the mobility increases slightly with decreasing temperature.

Information on mobility has also been obtained from proton injection experiments. A palladium anode behaves in an ohmic way at low field strength; but above a critical value, the current becomes proportional to the square of the applied potential, indicating a space charge limited current of the field-injected protons (Engelhardt et al., 1969). The peculiar temperature

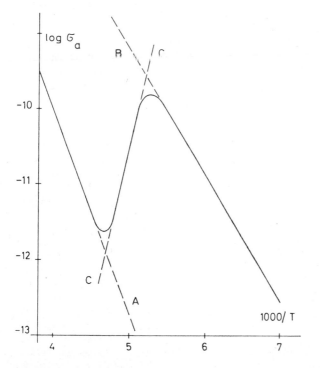

Figure 1. Apparent conductivity of pure ice with proton injection from a palladium electrode (average field 100 V/cm): curves A and B—normal conductivity of the ions, and of the Bjerrum defects respectively; curve C—injection ions. After Englehardt et al., 1969.

dependence of the current at constant voltage is shown schematically on Figure 1 and it is interpreted in the following way. Down to $-60°C$, the injected current is masked by the intrinsic one, which decreases greatly with temperature. Below $-60°C$, the injected protons prevail and the positive slope of the curve is caused by a rise of their mobility. In the third range, below $-90°C$, the fall of the apparent mobility is the result of a mechanism that is described later (see Sections 3 and 4). The mobility can be calculated from the law expressing the space-charge limited current. However, this law predicts a dependence of the third power of the sample length, which has not been tested. It seems therefore premature to grant a quantitative significance to this curve, because the injection mechanism itself might also vary with temperature. Anyhow, these results are compatible with a mobility at least of the order of the one determined from the conductivity.

A different method, largely used in semiconductor research, is to measure the Hall effect. This presents many difficulties in ice, because of its small magnitude and also because of the occurrence of electrode polarization and superficial short circuit. The Hall mobility obtained by the Munich school (Bullemer and Riehl, 1966b) amounts to $1.4 \text{ cm}^2/\text{V-sec}$ near $-10°C$. The discrepancy with the other determination occurs because the carriers can respond to the magnetic field only while they are in motion. This would mean that only 0.2% of the carriers can be considered to be freely moving instantly. This has to be brought together with the correlated motions of the protons, which is treated later (Section 4).

Table 3   Alternating Current Behavior of Pure Ice at $-10°C$

Property	H_2O	D_2O	Unit	Reference
Static dielectric constant	95 112[a] 96[b] 100	92		Auty et al. (1952) Humbel et al. (1953) Steinemann (1957)
High-frequency dielectric constant	3.17 3.08 3	(24 GHz) 3.03		Lamb et al. (1949) Auty et al. (1952) Humbel et al. (1953)
Relaxation time	6.0×10^{-5} 6.0×10^{-5}	9.15×10^{-5}	sec	Auty et al. (1952) Humbel et al. (1953)
Volume of activation	2.9 3.8		cm^3/mole	Chan et al. (1965) Gränicher (1969b)
High-frequency conductivity	1.5×10^{-7}		$\Omega^{-1} \text{ cm}^{-1}$	Steinemann (1957)

[a] Parallel to c-axis.
[b] Perpendicular to c-axis.

Investigations of the a-c properties, particularly of the dielectric constant, were started in the 1920s and led to the discovery of the large value of the static dielectric permittivity ε_s and of its strong dependence upon frequency and temperature (Table 3 and Figure 2). The first explorations were based on a Maxwell–Wagner mechanism, but reliable measurements performed during the 1940s (Auty and Cole, 1952; Humbel, Jona, and Scherrer, 1953) indicated a relaxation mechanism of the Debye type, with a single relaxation time τ of 5×10^{-5} sec at $-10°C$ obeying an Arrhenius law with an activation energy of 13 kcal/mole. The frequency dependence is expressed by the formula

(1)
$$\varepsilon - \varepsilon_\infty = \frac{\varepsilon_s - \varepsilon_\infty}{1 + i\omega\tau}$$

If the imaginary part is plotted as a function of the real part with the frequency as a parameter (Cole plot), a half-circle results with its center on the horizontal axis. In the low-frequency range, the dielectric constant is real and tends toward an anisotropic value of 112 parallel and 96 perpendicular to the hexagonal c-axis varying approximately as $1/T$. In the high-frequency range (up to the GHz) it has the constant value of $\varepsilon_\infty = 3.2$ (Lamb and Turney, 1949).

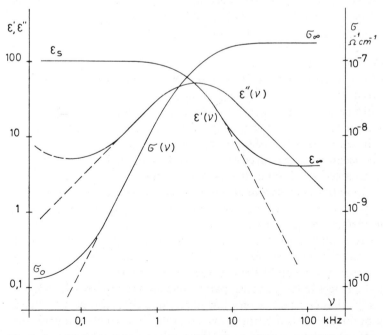

Figure 2. Alternating-current properties of pure ice at $-10°C$ in log–log plot. After Humbel et al., 1953, and Bullemer et al., 1969.

As in the case of the d-c conductivity, the sample purity is critical for reproducibility. The d-c conductivity has to be taken into account carefully; otherwise it will bias the results, as it has been stressed by Gränicher (1969a). Moreover, with partly blocking electrodes, a low-frequency space charge relaxation occurs and falsifies the extrapolated static dielectric constant (A. Steinemann, 1957). It has to be pointed out that, to our knowledge, no measurements have been performed in a way that eliminates the possibility of a surface contribution. As in the case of d-c conduction, this could require some adjustments for the hitherto accepted values of the activation energy.

The high value of the static permittivity has been attributed by Debye to the reorientation of the molecular dipole moments (in the vapor: 1.84×10^{-18} esu): to attain the polarization induced by a field of 100 V/cm, only one of 10^7 molecules has to be turned. The defect mechanism responsible for the dynamic response was not discovered until the 1950s (see Section 3), but Debye's estimate of the order of magnitude was correct—it shows that ice, like most other crystals, owes its properties to a minute concentration of very efficient agents.

The pressure dependence of some electrical parameters of polycrystals has been investigated by Chan, Davidson, and Whalley (1965). At $-23.4°C$ they observe an increase of the natural logarithm of the dielectric permittivity amounting to $(14 \pm 3) \times 10^{-6}$/bar. Starting from the Kirkwood equation for a condensed phase of freely orientable polar molecules, they show that this value can be attributed to the increase in density and to its effect on the reaction field, nothing remaining for pressure dependence of the molecular polarizability and of the dipole moment. The activation volume is found to be positive for the relaxation time (2.9 ± 0.3 cm^3/mole) and negative for the d-c conductivity (-11 ± 3 cm^3/mole). Gränicher (1969b) observe a larger activation volume of about 3.8 cm^3/mole for a mono-crystal having its c-axis making an angle of 68 degrees with the electric field. The structural change implied by these results will be discussed later.

The low concentrations involved in a-c and d-c properties alike suggest a marked sensitivity of ice to impurities. Although an efficient doping is easy to produce, the quantitative determination of the concentration has to be done very carefully because of the impurity segregation during the growth. The segregation coefficients are very sensitive to the growth parameter and can reach very high values up to 10^3. Crystal "engineering" is therefore difficult, and in most cases the concentration has to be measured by a destructive method after the completion of the electrical experiment.

The largest changes in the electrical parameters are induced by hydrofluoric acid. Its dissociation constant is very low in ice, maybe as low as 10^{-13} near the melting point (Jaccard, 1959); the mass action law predicts a square root dependence of the acid concentration for the conductivities, confirmed experimentally, and the activation energy is lowered to 7.5 kcal/mole. A peculiar

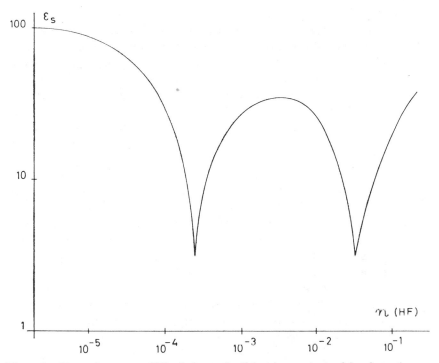

Figure 3.   Dependence at $-3°C$ of the static dielectric constant of ice from the concentration $n(HF)$ of hydrofluoric acid (mole/liter); log–log plot, after Steinemann, 1957.

effect occurs for the static dielectric permittivity (A. Steinemann, 1957). This parameter shows well-marked minima of the order of unity for two acid concentrations (Figure 3). This behavior is characteristic for the interaction between two different types of defects, presented in Section 3.

Other halogenic acids (HCl, HBr, and HI) and nitric acid in concentrations up to 10^{-2} mole/liter, have the same influence: in a log–log plot of d-c conductivity versus concentration, they yield points lying roughly on the same line with slope $\frac{1}{2}$ (Levi and Arias, 1964). These results have been confirmed by other authors (e.g., Gross, 1962) and hold for some alkali halides too. The similarity extends for the halogenic acids even to the a-c properties (Young and Salomon, 1968). Investigations on salts have been concentrated on ammonium fluoride because of its ability to form mixed crystals with the ice (Zaromb and Brill, 1956). This can be easily understood by the substitution in the lattice of one NH_4F molecule for two H_2O molecules, which does not alter the total number of protons. The effect on the d-c conductivity depends only upon the concentration ratio $(F)/(NH_4)$ (Levi et al., 1963). The conductivity is the same as for pure ice if the ratio is near unity, but in other cases it increases by a few orders of magnitude. The relaxation time is unaffected near $-10°C$, but its activation energy is depressed to 2.5 kcal/mole, depending upon the growth conditions, which, in certain cases, favor a nonstoichiometric incorporation of the ions (Dengel et al., 1966; Levi and Lubart, 1968; Brill et al., 1957). In an extended concentration range (0.002 to 0.7%) the pre-exponential factor is of the form constant $\times$ exp (activation energy/550 cal), whereas the static permittivity seems to keep its pure ice value. However, the circles of the Cole plots do not have their center on the horizontal axis, but they are slightly depressed; this indicates a mechanism with a relaxation time spectrum that must be therefore different from the mechanism observed in pure ice or in HF-doped ice, characterized by a single relaxation time.

Most of the measurements of dielectric relaxation have been performed at temperatures between the melting point and about $-60°C$, to keep the relaxation time small enough to be measured with standard a-c bridges (e.g., $\tau = 10$ msec at $-60°C$ for pure ice). If the ice is heavily doped, these measurements can be extended to liquid nitrogen temperature, showing a Debye dispersion below $-100°C$ with a spectrum for the relaxation time (Ida et al., 1966). However, these results must be considered with caution, because the high doping concentration certainly introduces structural changes, probably partly amorphous domains, and a comparison with high-temperature measurements does not seem to be justified.

3   Electrical Conduction Mechanism

The experiments mentioned thus far allow a direct phenomenological description of the conduction mechanism: a Debye relaxation for the polarization, in parallel with a finite d-c conductivity. For deeper insight, one must take into account the microscopic structure of the lattice, building up a reasonable model for its response to an electric field (Jaccard, 1959 and 1964; Onsager and Dupuis, 1960 and 1962).

As shown in Chapter 1, the structure of ice is characterized by the tetrahedral coordination of the oxygen atoms by means of the hydrogen bonds of length $r_{OO} = 2.76$ Å. The tetrahedra are piled up either in a hexagonal close-packing structure (in the usual hexagonal form of ice) or in a cubic close-packing structure (in the low-temperature diamondlike modification). The protons obey the rules postulated by Bernal and Fowler (1933): (a) they lie on the straight lines connecting neighboring oxygen atoms, (b) there is only one proton on each bond, and (c) every oxygen has two protons as nearest neighbors in order to preserve the molecule identity. The hypothesis that the protonic distribution is disordered, explaining the zero point entropy (Pauling, 1935), has been confirmed by neutron diffraction experiments on D_2O (Wollan et al., 1949; Peterson et al., 1957).

The model as it stands suffices to explain the high-frequency dielectric permittivity of 3.2 with an electronic contribution ($n^2 = 1.7$) and a component of 1.5 resulting from molecular deformation and rotational distortion in the lattice. However, the large value of the static permittivity can only be accounted for by a dipolar reorientation mechanism. Certainly the water molecules bear in solids a dipole moment of the same order of magnitude as in gas, and they have to turn from one equilibrium position into another, thereby changing the protonic configuration. Realizing that a displacement of the protons along a closed ring would require a prohibitive energy, Bjerrum (1951) postulated the existence of two complementary mobile defects, which turn the water molecules one after another; that is, with a relatively small activation energy at each step. These defects, resulting from a violation of the second Bernal–Fowler rule, consist of a bond occupied by two protons (O–H···H–O) or in an empty bond (O···O). They have been called D and L (from the German, *doppelt besetzte, leere Bindung*) (Figure 4). They come up from thermal fluctuations in the lattice, the proton of a normal water molecule jumping on an adjacent bond, thus giving rise to the formation of a pair D, L. Each of its members then diffuses for itself by successively turning the protons around the oxygen atoms (within the molecules) by a

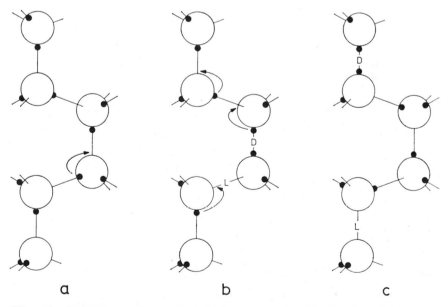

Figure 4. Formation and diffusion of Bjerrum defects: (*a*) unperturbed lattice; (*b*) formation; (*c*) separation of a pair D/L.

thermally activated process. The defect state migrates in the lattice from bond to bond, bearing an effective charge $e_i = e_D, e_L$. This charge, multiplied by the elementary step vector of the defect state, is equal to the sum of all the neighboring charge elements (nuclear and electronic) times their displacement:

$$e_i \mathbf{r}_{step} = \int \rho(\mathbf{r}') \, \Delta \mathbf{r}(\mathbf{r}') \, d^3 r'$$

The value defined this way is, of course, different from the protonic charge e, but because of its dynamic meaning it is convenient to express in a symmetrical way the properties of the defects, as we shall see below. Since the passage along a lattice chain of a D and an L defect leaves the lattice in its primitive state, the two defects must have charges of the same magnitude e_B, but with opposite signs, the positive charge belonging to the D (proton excess) and the negative one to the L (proton deficiency): $e_D = -e_L = e_B$. In the equilibrium state, without field, the configuration entropy reaches its maximum value and the Bjerrum defects perform a random motion which does not modify the average disorder. When a small electric field is applied, a coherent drift is superposed to the random motion, lowering the energy of the system proportionally to the average drift length x. But this drift increases the order in the proton configuration; that is, the entropy decreases with x^2 from its maximum

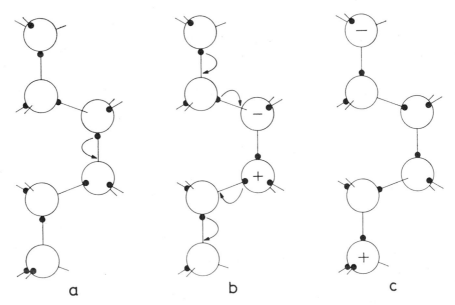

Figure 5. Formation and diffusion of ionic defects: (a) unperturbed lattice; (b) formation; (c) separation of a pair $H_3O^{+\prime}OH^-$.

value until it balances the gain in electric energy (minimum of the free enthalpy). A stable driftless situation is established in which the configurational polarization (characterizing the order) reaches a value proportional to the applied field. Since the net force acting on the defects has a restoring component proportional to the polarization but opposite to the field, and since the time derivative of the polarization is proportional to the defect current, the system dynamics is described by a linear differential equation of the first order and the response is an exponential function of the time for a transient and a Debye function of the frequency for a periodic field.

The Bjerrum defects alone thus suffice to account for the Debye relaxation, but not for the d-c conductivity. As a matter of fact, a D crossing the whole crystal transports with itself only the defect state, but no proton, since these are bound to stay within the same molecules. To obtain a net proton transport (i.e., a finite d-c conductivity), we have to postulate the existence of complementary defects able to shift the protons between the molecules. We then assume that the third Bernal–Fowler rule can also be violated and that H_3O^+ and OH^- ions exist in ice as they do in water. With sufficient thermal activation, they form pairs by shifting a proton from one molecule along the bond to the next one; each species migrates further in the lattice by a subsequent proton shift (Figure 5). Their effective electric charges are defined as for the Bjerrum defects and the signs are obvious: $e_+ = -e_- = e_I$. The passage along the same lattice chain of one Bjerrum defect and one ion is now equivalent to the net transfer of one protonic charge, whence the relation:

$$(2) \qquad\qquad e_B + e_I = e$$

But the complementary one of the two defect types also appears in the following process, which is responsible for the distinctive electric behavior (Figure 6). If a positive ion moves along a certain chain in the lattice, say upward, then after its passage all the protons sit on the upper ends of the bonds. No second positive ion can use the same way in the same direction, unless a D defect, for instance, climbs up the same chain and, by turning the protons around the molecules, places them in the lower bond ends again. The same occurs if an L goes down or if an OH^- goes up the chain. A path used by a certain defect has to be reactivated by one of the same type but of opposite sign, or by one of the other type but of the same sign, if the original proton configuration has to be restored. This shows that a significant parameter describing the polarization changes is the vector

$$(3) \qquad\qquad \Omega = \int (\mathbf{j}_+ - \mathbf{j}_- - \mathbf{j}_D + \mathbf{j}_L)\, dt$$

which is connected in first approximation with the entropy by

$$(4) \qquad\qquad \Delta S \sim -\Omega^2$$

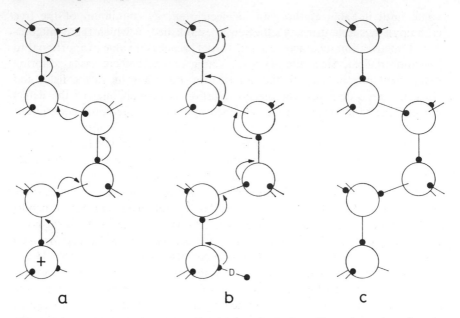

a   b   c

Figure 6. Reactivation of a chain: (*a*) initial configuration; (*b*) configuration after the positive ion has moved upward; (*c*) configuration after the D defect has moved upward (identical to the initial configuration).

The entropy production per unit time and volume includes then two terms (in a homogeneous crystal under isothermal and isobaric conditions): one expressing the Joule heat and one contribution from the configurational polarization:

$$(5) \qquad T\dot{S} = \mathbf{F} \cdot \mathbf{I} - \phi \mathbf{\Omega} \cdot \dot{\mathbf{\Omega}}$$

A detailed calculation (Jaccard, 1964) gives a value of $(16/\sqrt{3})kTr_{00}$ for the coefficient ϕ.

The currents $\mathbf{j_I}$ and $\mathbf{j_B}$ defined as

$$(6) \qquad \mathbf{j_I} = \mathbf{j_+} - \mathbf{j_-} \qquad \mathbf{j_B} = \mathbf{j_D} - \mathbf{j_L}$$

are the primary components of the electric current:

$$\mathbf{I} = e_I \mathbf{j_I} + e_B \mathbf{j_B}$$

This expression then takes the form

$$(7) \qquad T\dot{S} = \mathbf{j_I}(e_I \mathbf{F} - \phi \mathbf{\Omega}) + \mathbf{j_B}(e_B \mathbf{F} + \phi \mathbf{\Omega})$$

Neglecting crossterms (which do not alter the essence of the mechanism but only complicate the formalism), we can write for the proportionality between the currents and the forces

(8)
$$j_I = \frac{\sigma_I}{e_I^2}(e_I F - \phi\Omega) + \cdots$$

$$j_B = \frac{\sigma_B}{e_B^2}(e_B F + \phi\Omega) + \cdots$$

The coupling between the two defect types is contained in Ω, which takes the form in the periodic case

(9)
$$\Omega = \frac{j_I - j_B}{i\omega}$$

The meaning of the coefficients σ_I and σ_B is straightforward: at high frequencies, Ω vanishes and the σ's are just the partial electric conductivities of the ions and of the Bjerrum defects. In this case they move only locally and independently of each other, and the total conductivity is the sum of the partial conductivities

(10)
$$\sigma_\infty = \sigma_I + \sigma_B$$

This formula expresses the fact that the ionic and the Bjerrum conduction mechanisms are acting in parallel. In the static case, the system above yields for the d-c conductivity σ_0:

(11)
$$\frac{e^2}{\sigma_0} = \frac{e_I^2}{\sigma_I} + \frac{e_B^2}{\sigma_B}$$

meaning that the conduction mechanisms are acting in series, because the lattice chains have to be used alternatively by ions and by Bjerrum defects.

For an arbitrary frequency, the conductivity takes the form

(12)
$$\sigma - \sigma_0 = \frac{e_P^2/\phi\tau}{1 + (i\omega\tau)^{-1}}$$

where

(13)
$$\frac{1}{\tau} = \phi\left(\frac{\sigma_I}{e_I^2} + \frac{\sigma_B}{e_B^2}\right)$$

(14)
$$e_P = \frac{\sigma_I/e_I - \sigma_B/e_B}{\sigma_I/e_I^2 + \sigma_B/e_B^2}$$

Because of the relation

(15)
$$\sigma - \sigma_0 = i\omega\varepsilon_0(\varepsilon - \varepsilon_\infty)$$

the dielectric permittivity takes the Debye form

(16)
$$\varepsilon - \varepsilon_\infty = \frac{\varepsilon_s - \varepsilon_\infty}{1 + i\omega\tau}$$

with

$$(17) \qquad\qquad \varepsilon_0(\varepsilon_s - \varepsilon_\infty) = \frac{e_P{}^2}{\phi}$$

where ε_0 is the dielectric permittivity of vacuum.

The transport numbers t_i defined by

$$(18) \qquad\qquad t_i \mathbf{I} = e_i \mathbf{j}_i \qquad i = +, -, \mathrm{D, L}$$

then have the following values in a d-c process:

$$(19) \qquad t_\pm = \left(\frac{\sigma_\pm}{\sigma_0}\right)\left(1 - \frac{e_P}{e_I}\right) \qquad t_{\mathrm{DL}} = \left(\frac{\sigma_{\mathrm{DL}}}{\sigma_\mathrm{D}}\right)\left(1 + \frac{e_P}{e_\mathrm{B}}\right)$$

and they obey the relations

$$(20) \qquad t_+ + t_- = t_\mathrm{I} = \frac{e_\mathrm{I}}{e} \qquad t_\mathrm{D} + t_\mathrm{L} = t_\mathrm{B} = \frac{e_\mathrm{B}}{e} \qquad t_\mathrm{I} + t_\mathrm{B} = 1$$

A physical interpretation remains to be given for the charge e_P, remembering that the effective charges $\pm e_\mathrm{I}$ and $\pm e_\mathrm{B}$ are those that respond to the electric field $\mathbf{F}$ and, according to the Poisson equation, are conversely the sources of $\mathbf{F}$. A comparison of the configurational polarization $\mathbf{P}_c$ and of the configuration vector $\mathbf{\Omega}$ shows that they are proportional, but with e_P as a coefficient:

$$(21) \qquad\qquad \mathbf{P}_c = \varepsilon_0(\varepsilon - \varepsilon_\infty)\mathbf{F} = \frac{e_P{}^2}{\phi}(1 + i\omega\tau)^{-1}\mathbf{F}$$

$$(22) \qquad\qquad \mathbf{\Omega} = \frac{\mathbf{j}_\mathrm{I} - \mathbf{j}_\mathrm{B}}{i\omega} = \frac{e_P}{\phi}(1 + i\omega\tau)^{-1}\mathbf{F}$$

$$(23) \qquad\qquad \mathbf{P}_c = e_P\mathbf{\Omega}$$

Introducing for convenience the parameter η with the value -1 for positive ions and L defects and $+1$ in the other cases, we can write for the configurational component of the force acting on the defects:

$$(24) \qquad\qquad (\mathbf{X})_c = \eta\phi\mathbf{\Omega} = \frac{\eta e_P \mathbf{P}_c}{\varepsilon_0(\varepsilon_s - \varepsilon_\infty)}$$

This shows that ηe_P is a charge attributed to each defect, characterizing its response to the polarization $\mathbf{P}_c$. On the other hand, we have in the case of chemical equilibrium for the concentrations n_i

$$(25) \qquad\qquad \mathrm{div}\,(\mathbf{j}_\mathrm{I} - \mathbf{j}_\mathrm{B}) = -\frac{\partial}{\partial t}(n_+ - n_- - n_\mathrm{D} + n_\mathrm{L})$$

and consequently

$$(26) \qquad\qquad \mathrm{div}\,\mathbf{P}_c = e_P\,\mathrm{div}\,\mathbf{\Omega} = e_P(-n_+ + n_- + n_\mathrm{D} - n_\mathrm{L})_0{}^t$$

The charges ηe_P must thus be considered as the sources of the polarization $\mathbf{P}_c$, in complete symmetry with the electric charges e_i. As Onsager has pointed out, e_P can be interpreted as a "quantum" of polarization charge because it is directly related to the smallest amount of configurational polarization which can be produced in the lattice. According to its definition (14) e_P is a constant only for a given crystal, because it is a function of the dynamic parameters of the lattice. If the partial ionic conductivity σ_I greatly outweighs σ_B, a case in which the ions are called majority carriers, then the value of e_P is e_I. In the opposite case, when the Bjerrum defects are majority carriers, its value is $-e_B$. If the concentrations and mobilities of both defect types are balanced, e_P can vanish, and with it the differences $\varepsilon_s - \varepsilon_\infty$ and $\sigma_\infty - \sigma_0$. This peculiarity explains the behavior of the dielectric permittivity of HF-doped ice as a function of the acid concentration discovered by A. Steinemann (1957).

The experiments referred to in Section 2 indicate that acid molecules dissociate in ice as they do in water, but to a much smaller extent, with the ionic concentration proportional to the square root of the acid concentration. On the other hand, let us assume that the HF molecule is substituted in the lattice for a regular water molecule; the molecular size does not cause any difficulties, and since HF is known to exist as hydrogen-bonded polymers, it should be able to bind in the same way the four neighboring oxygen atoms (Truby, 1955). But an HF molecule brings only one proton to a place that has normally two; therefore, each HF molecule introduces into the lattice, without energy expenditure, an L defect that is only weakly bound to the parent molecule. At high temperature, the L defect concentration should then rise proportionally to the first power of the acid concentration and should predominate over the square root dependent ions. Figure 3, then, must be interpreted in the following way: the L are the majority carriers in the high-concentration range; toward the left the majority character passes over to the ions, and again to the Bjerrum defects, whenever the dielectric permittivity shows a minimum. In pure ice, the polarization is then governed by the Bjerrum defects, as well as by the high-frequency conductivity and the relaxation time, but the static conductivity is due to the ions. Remembering that $e_{P,\text{pure}} = -e_B$, the experimental results introduced in (17) give the following values for the effective charges:

parallel to the c-axis $\qquad \dfrac{e_B}{e} = 0.55, \quad \dfrac{e_I}{e} = 0.45$

perpendicular to the c-axis $\qquad \dfrac{e_B}{e} = 0.51, \quad \dfrac{e_I}{e} = 0.49$

The interpretation given thus far is confirmed by the measurements of the pressure dependence by Chan et al. (1965). The formation of Bjerrum pairs is bound to some accumulation of charges of the same sign between the molecules. This holds true even if the defects have a more complicated structure than the simple models above—even, for example, if the defect state is distributed in a volume including several molecules. This interjacent charge concentration implies repulsive forces between the molecules and, as a consequence, a local dilatation of the lattice. During the movement of the defect, the elastic strain may be released partially, but not enough to compensate for the increase in volume when a defect pair is created. The opposite occurs in the case of ions, because the charge is then concentrated around a molecule; it produces a strong central field on the neighbors which pulls them nearer by electrostriction. Since the activation volume is positive in pure ice for the relaxation time, this phenomenon results from the Bjerrum defects, in agreement with the other considerations. Of the negative volume for the d-c conductivity, Chan et al. attribute about -9 cm^3/mole to the ion pair formation and -6 cm^3/mole to the proton transfer.

The defects interact with one another by means of the electrostatic potential and the configurational polarization. The attraction between positive and negative ions or between D and L has to be taken into account for the chemical recombination coefficient. But the electric attraction can also give rise to defect associations such as L + positive ion or D + negative ion. These pairs, called sessile ions by Onsager, bear a double polarization charge and a small uncompensated electric charge; they are unable to migrate without dissociating (Figure 7). Since the dissociation field effect involves Coulomb

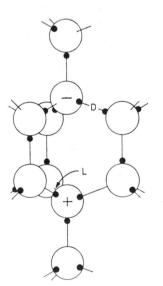

Figure 7. Defect associations H_3O^+/L and OH^-/D (sessile defects). The only degree of freedom left is a rotation of the ion between three equivalent positions.

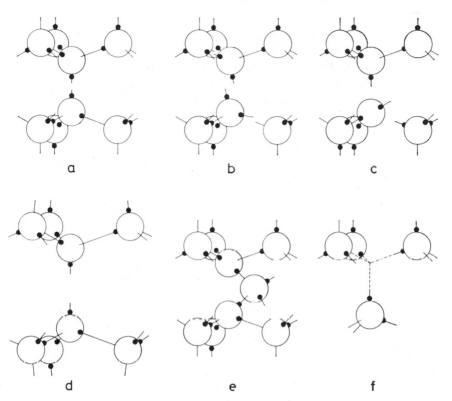

Figure 8. Models of D defects: (*a*) rigid lattice model; (*b*) rotated model (after Cohan et al., 1962); (*c*) X defect (after Dunitz, 1963); (*d*) relaxed lattice model (after Eisenberg and Coulson, 1963); (*e*) DI interstitial model (Haas, 1962); (*f*) DV vacancy model (Onsager and Runnels, 1963; Kopp, 1966).

constants for the field and for the polarization, Onsager showed from Eigen's results that the sessile ions are outnumbered by the regular ones; since they do not move, their role in the electric conduction mechanism is insignificant anyway.

The interaction between the defects leads also to the formation around each of them of a Debye–Hückel atmosphere of the majority defects, which screens off the long-range fields (Onsager, 1960). It modifies the solvation energy of the defects and their equilibrium concentration, but the inaccuracy of the experiments had not yet permitted us to see these effects.

Water molecule vacancies (Schottky defects) should also be present near the melting point in a number comparable to that of the Bjerrum defects. But the protonic configuration does not change while they migrate, and bare vacancies are thus electrically inactive. However, since the neighboring molecules have bonds dangling in the vacancy, they can turn more easily than regular fourfold coordinated molecules. The vacancies must thus be considered as sources for the L's and as traps for the D's. The resulting association has been called "Bjerrum defect diluted in a vacancy" by Onsager and Runnels (1963), or "invested vacancy" by Kopp (1966) (Figure 8f). The effect of their movement on the protonic configuration is the same as that which would be produced by the investing defect alone, and such vacancies are therefore electrically active. They might compete efficiently in the conduction mechanism, but with our present knowledge, it is difficult to assess to what extent this may be true.

4   Structure and Dynamical Properties of the Defects

The simple representation given here for the defects should be considered simply as a "topological" approximation, accounting only for the interaction of the defects with the proton configuration, for what has been called by Onsager the "proton bookkeeping." Although the picture of the ions is qualitatively valid, the bond angles and lengths are expected to be different from the values in the liquid or the vapor. But these adjustments are of minor importance as far as the existence of the ions is concerned; this existence is unquestionable, especially, since they are found in other water-containing crystals such as in gallium sulfate (Kydon et al., 1968).

The same is not true for Bjerrum's D defect: in the simple model, both protons are within a distance of 0.8 Å. The corresponding electrostatic repulsion energy ranges about 60 kcal/mole, three times as much as can be determined from the dielectric data (Figure 8). Cohan et al. (1962 and 1964) have suggested that the defect escapes this electrostatic repulsion by turning the two neighboring molecules around their normal bond by 20 degrees to increase the interprotonic distance. Dunitz (1963) has proposed that the excess proton stays halfway between its two regular positions, yielding what he has called an X defect. Both mechanisms reduce the energy to its experimental value. On the other hand, it has been shown by Eisenberg and Coulson (1963) that a still larger reduction occurs if the lattice relaxes elastically around the defect, the bond length increasing by 0.5 Å without disalignment. Because of the softness of the lattice, the strain and electrostatic energy in the D defect is estimated to be around 5 kcal/mole. It seems plausible that both processes, mechanical relaxation and molecular rotation, might work together and not only near a single bond, but spreading the defect on several bonds and molecules, possibly with strong deformations of the electronic wave functions. Although the L defect does not appear to present the same difficulty, its structure is by no means trivial, because it certainly implies some distortion of the lattice and of the electron distribution. Unfortunately, no experimental data have yet given more quantitative information on details of the defect structure. The small concentration makes spectroscopic detection very difficult, if not impossible, but there is some hope that other methods, such as paramagnetic resonance of suitably chosen probes or investigations of the high-pressure phases, might throw some light on this subject in the future.

The transfer mechanism bears a definite relation to the defect structure. The simple model shows that defect migration is performed by stepwise proton transfer, either along the bonds or around the molecules. The Bjerrum defects show a strong temperature dependence of the Arrhenius type not only for their partial conductivity, but also for the mobility ($\sim 10^{-4}$ cm^2/V-sec at $-10°$C), the activation energy being some tenths of an electron volt. This means that the proton transfer does not occur directly from the ground state, but that a certain thermal activation into higher lying states is necessary. The classical model of a particle jumping over a potential barrier applies in first approximation, with the activation energy of the mobility as the barrier height. Moreover, since the experiments with HF-doped ice do not show a decrease of high-frequency conductivity with acid concentration, the L defects should be at least as mobile as the D defects.

As Gosar has pointed out, however, there is some indication that the single particle model has to be replaced by a mechanism involving the cooperative motion of several protons. The relaxation frequency is equal to

(27)
$$\frac{1}{\tau} = \frac{3}{4} \left(\frac{n_B}{n_N}\right) p_B$$

where n are the concentrations and p_B is the average transfer frequency; the preexponential factor of p_B is of the order of 2×10^{15}/sec.

Since the librational frequency of the molecules (~ 800 cm^{-1}) is higher than the thermal frequency kT/h, the preexponential factor should have the form: $(kT/h) \exp (\Delta S/k)$, where ΔS is the change in entropy when a defect is excited to a level at the height of the potential barrier. But this entropy change cannot be accounted for, because the modified bending and stretching modes are too far above kT to be effective, and the protonic motion is faster than the vibrational period of the lattice. Gosar suggests therefore an energy transfer process between neighboring protons, as for Frenkel excitons. The presence of the electric charge of a Bjerrum defect polarizes this transfer, so that an excited proton, several molecules apart from the defect, may turn in advance and initiate a dominolike chain reaction. The net result is the displacement of the defect on several sites in a single step. The mechanism is somewhat analogous to the multiple collision in radiation damage. Since the dielectric relaxation frequency is proportional to the diffusivity of the Bjerrum defects, and hence to the square of the step length, it will also be proportional to the square of the number z of protons taking part in the elementary collective motion. A value of z about 20 would then yield a correct magnitude for the preexponential factor. This explanation is supported by estimates of the involved parameters; the principle is probably correct, but more experimental evidence is needed to get a reliable picture.

Although experimental determinations of the ionic mobility disagree, the order of magnitude of this parameter is large enough to suggest a significant deviation from a purely classical process. This seems particularly plausible because saturation current experiments indicate a very small, even negative, activation energy. Engelhardt's proton injection measurements (Figure 1) seem to exhibit an increase of the mobility with decreasing temperature, at least between -60 and $-90°C$. (Below this temperature, the apparent drop must certainly be attributed to the lack of intrinsic Bjerrum defects to reactivate the lattice chains used by the injected protons.) The potential in which the regular protons move is very asymmetrical along the bond, keeping the proton localized near a particular oxygen atom according to Bernal–Fowler rules. A shift of such a proton is indeed equivalent to the formation of an ion pair, and it involves energy of the order of 1 eV. But any proton belonging to a hydronium ion is in a symmetrical double-well potential, because it moves along its bond between two neutral atoms. The same occurs for a proton from a hydroxyl ion, which then moves between two negative ions. Since the electrostatic interaction associated with this motion might increase the well depth, an explanation exists for the possibly smaller mobility of the OH^-. The proton shifts by a quantum mechanical tunnel effect through the central barrier, with a frequency strongly depending upon the barrier height and width.

A theoretical study of the potential has been done by Weissmann and Cohan (1965) by an SCF–MO–LCAO method. They find indeed a lower middle barrier for the positive ion than for the negative one, but a simple tunnel effect transfer accounts for the observed mobility only if the interoxygen distance is reduced from 2.76 to 2.58 Å. Investigating several types of hydrogen bonds, Itoh (1967) obtained a tunneling frequency of about 10^{11}/sec for the $O—H\cdots O$ system, increasing by a decade if the bond length is shortened below 2.65 Å. These considerations suggest a strong coupling between the proton transfer and the phonons. For these, the frequency spectrum is known mainly from optical absorption measurements (Bertie and Whalley, 1964 and 1967; Whalley, 1969) and shows several bands centered around 200 cm^{-1} for the optical lattice modes, which are mainly responsible for the O–O bond stretching. The modes corresponding to intramolecular motions lie at higher frequencies: the molecular libration around 800 cm^{-1}, the symmetrical bending vibrations around 1600 cm^{-1}, and the bond-stretching oscillations between 2000 and 3600 cm^{-1}. In a semiclassical approach, Kim and Schmidt (1967) calculated the mobility, but they considered at first only the interaction with the 800 cm^{-1} librational mode. The obtained value is too large by more than two decades, and they had to introduce the electrostatic interaction between the ion and the molecular dipoles to reduce the mobility to a reasonable value.

Quantum mechanical calculations have been performed at first by Gosar and Pintar (1964), who considered a simplified model of the ice lattice: a linear chain of numbered molecules with double-well potentials in between, a state $|n\rangle$ being defined as an excess proton on the nth molecule. Since the tunnel effect introduces a transition probability, these nonstationary states are replaced by stationary ones in the form of waves. The possible energy levels are then concentrated in a band that is narrow compared with the range of the acoustic phonon frequencies. Therefore, conservation of energy and momentum requires consideration of processes that involve at least two phonons. The proton–phonon interaction couples the ground proton state with the excited states in the potential well, and the Boltzmann equation yields a lower limit for the relaxation time τ of 3×10^{-12} sec, and a mean free path of three lattice constants. Although the bandwidth is small (2×10^{-3} eV), it is well defined because it is 10 times larger than the ionic level spread $\hbar/\tau$. The theoretical temperature dependence of the mobility as it relates to the Debye temperature is of the forms T^{-3} and T^{-10} at high and low temperatures, respectively.

A more refined quantum mechanical description has been given by Fischer and Hofacker (1969). Considering also the linear chain approximation in the second quantization formalism, the localized states are expressed by delocalized wave states. Building in the strong interaction of the 200 cm^{-1} O–O stretching vibration, new states are created which represent the ions accompanied by the elastic relaxation of the lattice in their neighborhood. The effect of this relaxation is to depress and shrink the ground state energy, that is, to trap the defect at a certain lattice site with Gosar's interaction coupling the ground with the excited local states. Another interaction is introduced expressing the influence of the phonons on the tunneling between neighboring sites, and the resulting conductivity is governed at high temperature essentially by proton–phonon scattering within the ground band, showing, furthermore, an Arrhenius dependence $\exp(-E_a/kT)$ with a positive activation energy. This corresponds to a diffusive process with random jumps from site to site. Nevertheless, this has to be corrected in ice because of the strong coupling frequencies (which are large with respect to kT) and because of the long time correlations. Gosar's interband coupling and the tunnel–phonon coupling determine the low-temperature mobility, but for the latter mechanism, the relaxation frequency of the form T^5 should prevail over the former (which is in T^9), so that the overall mobility depends upon the temperature according to T^{-6}. The calculation has been done in view of the hydronium ion state, which has the best chance of propagating through the ground band. However, it also applies to the Bjerrum defects, for which an excitonlike transfer in the excited state must be expected.

It would be premature to interpret more details of the theory because of lack of experimental material, but it can be noted here that temperature dependence between -60 and $-90°C$ for Engelhardt's injected current is much too strong to be explained by a negative power of T, even with an exponent of the order of 10; the injection efficiency should therefore be very sensitive to temperature changes.

5   Diffusion of Electrically Active Defects and Related Phenomena

The coherent motion of the ions and Bjerrum defects can also occur under a force of another type, such as a gradient of the chemical potential. In this case a diffusion process may transport charge and mass, but the latter is very difficult to measure. Proton self-diffusion takes place in a sample that is heavily doped on one side with tagged protons (e.g., tritons), the random motion of the naturally occurring ions and Bjerrum defects imparting to the tritons a weak diffusivity. However, as in the case of d-c conductivity, both defect mechanisms act in series, and it is easy to show that the bottleneck is caused by the ions, at least in pure ice. The ratio of the waiting times for a given triton between jumps around the molecule and along the bond is of the order of the ratio of the partial conductivities. An estimate of the resulting diffusivity then gives values much smaller than those produced by molecular diffusion (see Section 6), thus making it possible to discard this effect.

The situation is quite different if the charge is transported together with the mass in a net defect flow, because a small charge accumulation is easily measurable as a potential difference. In a first method, an impurity gradient is used to initiate the diffusive movement of the defects, which then comes to a stop when the induced electric force is equal to the thermodynamic force. The latter has to be introduced in the entropy production as the negative gradient of the free enthalpy $-$ grad g_k (Jaccard, 1964). Taking the 0 level of this quantity for the regular bonds and molecules, the dissociation reactions producing the defects imply then

$$(28) \qquad g_+ + g_- = 0; \qquad g_D + g_L = 0$$

Together with the relations existing between the effective charges e_k and between the d-c transport numbers t_k, the electrochemical potential takes the form

(29)
$$U_c = -\sum_k \left(\frac{t_k}{e_k}\right) g_k = -\frac{g_+ + g_D}{e}$$

Since the defect concentration is low, the following formula for ideal solutions may be used:

(30)
$$g(p, T, c) = g°(p, T) + kT \log c$$

In ice slightly doped with HF, the ionic concentration varies with the square root of the acid concentration, but the number of D and L defects keeps its pure ice value. There, the slope of the potential difference versus the logarithm of the HF concentration should then be $-kT/2e = -11$ mV.

The measurements, performed near $-20°$C by Brownscombe and Mason, yield a different value of -17 ± 5 mV. A possible explanation for this discrepancy can be found in the concentration dependence of the metal–ice contact potential (Jaccard, 1969). Using a simple semiconductor model for the electrons in which the donor concentration is assumed to be proportional to the square root of the HF concentration (e.g., the fluorine ions act as donors), an additional term of -4.5 mV is found, bringing the theoretical value within the experimental range.

In a second method, the driving force is provided by a thermal gradient $-T \, \mathrm{grad} \, (g_k/T)$ and the resulting thermoelectric potential U_t has then the differential form

(31)
$$\frac{\partial U_t}{\partial T} = -\sum_k \left(\frac{t_k}{e_k}\right) \frac{q_k^* - h_k}{T}$$

where q_k^* is the transport heat and h_k the enthalpy. The difference between these values is just the activation energy for the partial conductivity $-k \, T \log \sigma_k$. Whereas conduction measurements at high and low frequency allows discrimination between the contributions of the ions and the Bjerrum defects, the thermoelectric potential gives some information on the prevalence of the defects of each sign, because the sum terms of the equation above have different signs. However, in addition to pure ice, we must consider ice doped with HF or with NH_3. In the first case, a low doping increases the ratio t_+/t_-, and a high doping increases the ratio t_L/t_D. In the second case the reverse is true, with the following assumption: an ammonium molecule is built in a regular site in the lattice, in which it brings almost without energy expenditure a D effect, and it can dissociate chemically, liberating hydroxyl ions. In both cases, the modified dissociation enthalpies are to be taken into account.

The thermoelectric potential has been measured by several investigators (Latham and Mason, 1961; Bryant and Fletcher, 1965; Takahashi, 1966;

Table 4 Thermoelectric Effect in Ice at −20°C (mV/°C)

Pure Ice	HF Doping Iᵃ	HF Doping IIᵇ	NH₃ Doping Iᵃ	NH₃ Doping IIᵇ	Electrodes, Crystallinity	Reference
-2.0 ± 0.2		-3 ± 0.3			Cu (contact) polycrystalline	Latham et al. (1961)
-3.5 ± 0.5	-1.6 ± 0.2	$+0.2 \pm 0.3$	$+2.0 \pm 0.3$	$+0.0 \pm 0.3$	Brass, Pd polycrystalline (contact)	Bryant et al. (1965)
-1.5 ± 0.5	-1 ± 0.5	$+0.5 \pm 0.2$			Pd (contact) monocrystalline	Takahashi (1966)
-2.3 ± 0.3^c					Brass (induction), monocrystalline	Brownscombe et al. (1966)
-1.1 ± 0.1	-1.1 ± 0.1	-0.3 ± 0.1	$+2.2 \pm 0.2$	$+0.4 \pm 0.2$	Pd (contact) monocrystalline	Bryant (1967)

[a] Low doping ($10^{-7} - 10^{-5}$ mole/liter, concentration of D and L unaffected).

[b] High doping ($>10^{-6}$ mole/liter, D or L majority).

[c] Electrochemical potential $dU_c/d \ln c \, (\mathrm{HF}) = -17 \pm 5$ mV, with low HF doping.

50

Bryant, 1967), but the agreement is only qualitative (Table 4). In pure ice, its value ranges around -2.3 ± 0.3 mV/°C (Brownscombe and Mason, 1966), increasing to -1.1 ± 0.1 mV/°C with low HF doping and to -0.3 ± 0.1 mV/°C with high HF doping (Bryant and Fletcher, 1967). In ammonium-doped ice, the sign is reversed, as can be expected from the model, $+2.2 \pm 0.2$ mV/°C in the low range and $+0.4 \pm 0.2$ mV/°C in the high range (Bryant, 1967). This behavior is qualitatively consistent with the hypothesis that the mobility is higher for positive ions than for negative ones, and that it is roughly the same for both Bjerrum defects. However, if energy values from the dielectric measurements are introduced, the magnitude of the steps between the different concentration ranges is too large by 50%. The same electronic model of the previously introduced metal–ice contact potential can be invoked to explain this discrepancy, but this time by appealing to the temperature dependence of the potential. The determination of these diffusive effects is difficult because only a potential difference is measurable, and all the elements placed in series in the measuring circuit contribute by some amount. Some improvement can be expected from vibrating-capacitor methods, which eliminate one of the contact potentials (Jaccard, 1969).

6 Molecular Diffusion

The first diffusion measurements in ice were performed in polycrystalline, isotopically enriched samples by Kuhn and Thürkauf (1958), who found an equal self-diffusivity for deuterium and for ^{18}O amounting to $(10 \pm 2) \times 10^{-11}$ cm^2/sec at -2°C. Apart from another experiment with ^{18}O in artificial monocrystals by Delibaltas et al. (1966), subsequent research has concentrated on the diffusivity of tritium in natural monocrystalline samples from the Mendenhall glacier, as in the work of Itakagi (1967) and Ramseier (1967), or in artificial monocrystals, either pure or doped with HF or NH$_4$F, as in the work of the Munich group (Blicks et al., 1966; Dengel et al., 1966). The results, summed up in Table 5, show a value at -10°C of about 2×10^{-11} cm^2/sec and an activation energy slightly above 14 kcal/mole. The impurities have no significant influence. The diffusion is anisotropic, the coefficient measured perpendicularly to the c-axis being larger by some 10%. Itakagi even reports an anisotropy of the activation energy, but this has not been confirmed by Ramseier, who used samples from the same source.*

The small disagreement among authors about the absolute value of the coefficient can be traced back to microstructural effects. Interfaces are

---

* Referring probably to an interview with Itakagi, Ramseier writes: "This author feels that the large scatter of data does not warrant the statistical analysis and the subsequent interpretation of the results."

Table 5   Molecular Diffusion in Ice (at $-10°C$)

Reference	Diffusing Species	Sample	Diffusion Constant (cm^2/sec)	Activation Energy (kcal/mole)	Temperature Range (°C)	Remarks
Kuhn et al. (1958)	^{2}H^{18}O	Polycrystalline artificial	$(10 \pm 2) \times 10^{-11}$			At $-2°C$
Itakagi (1967)	^{3}H	Monocrystalline natural	$(3 \pm 1) \times 10^{-11}$	14 ± 2	-10 to -30	$E_\perp/E_\parallel \simeq 1.2$
Blicks et al. (1966)	^{3}H	Monocrystalline artificial + HF	$(2.3 \pm 0.2) \times 10^{-11}$	14.5 ± 1	-2 to -30	$D_\perp/D_\parallel \simeq 1.1$
Dengel et al. (1966)	^{3}H	Monocrystalline artificial + NH$_4$F				
Delibaltas et al. (1966)	^{18}O	Monocrystalline artificial	$(2-5) \times 10^{-11}$	16 ± 3	-5 to -30	$D_\perp/D_\parallel \simeq \begin{cases} 1.07 \\ 1.12 \end{cases}$
Ramseier (1967)	^{3}H	Natural monocrystalline artificial	$(1.5 \pm 0.1) \times 10^{-11}$	14 ± 1	-2 to -40	
Kopp et al. (1965)	F	Monocrystalline artificial	$(2-20) \times 10^{-7}$	13 ± 2	-5 to -30	Nuclear magnetic relaxation time
Haltenorth et al. (1969)	F	Monocrystalline artificial	$(1.08 \pm 0.01) \times 10^{-7}$	4.61 ± 0.04	-5 to -90	Conducting melt water

present in any case in polycrystals, but they also occur in pure monocrystals, as shown by Truby (1955). He discovered a semimicroscopic substructure, the crystal being a mosaic of hexagonal prisms with dimensions of the order of 1μ and presenting minute angular deviations from each other. It is believed that part of the impurities could stay preferentially near the interfaces, since the blocks become smaller with increasing HF doping. These interfaces are expected to hinder somewhat the molecular transfer. On the other hand, it has been noticed by Itakagi (1967) that a mechanical surface treatment during the preparation of the samples produces near the surface a disordered layer extending to a depth of some 200μ. Since this is the distance over which the concentration is measured in most of the diffusion experiments, unless special care is taken, the structural perturbation is likely to be reflected by the results. Another effect, pointed out by Delibaltas et al. (1966), is the influence of the isotope constitution on the transfer probability: the rotational frequency is not the same in $^{1}H_{2}{}^{18}O$ as in $^{3}H^{1}H^{16}O$.

The experimental results obtained thus far give enough information to make up a picture of the microscopic diffusion mechanism. Since the diffusivity has the same magnitude for deuterium, tritium, and ^{18}O, Kuhn has pointed out that the only significant diffusion taking place may be the one of whole molecules, either by a vacancy or by an interstitial mechanism. According to the first measurements by Dengel and Riehl, the activation energy seemed to amount to 13.5 kcal/mole, as for the dielectric relaxation time. This led Haas (1962) to build up an ingenious model in which a Bjerrum defect is associated with an interstitial molecule, bonded to two regular molecules. But subsequent measurements strongly favored a different activation energy around 14 kcal/mole or higher, and the hypothesis of a mechanism common to Debye relaxation and to self-diffusion had to be revised, especially because the latter is independent of HF or NH_4F doping. As Gränicher (1958) has shown, such a diffusion mechanism, performed either by means of interstitials or of vacancies, is unable to change the protonic configuration: if a molecule jumps in an adjacent vacancy, causing the vacancy to advance by one step, it will have to make three bonds with neighbors for which the configuration is already settled. Therefore, the orientation in space of this molecule is fully determined and it is just the same as for the molecule that occupied the site before the vacancy occurred. The same can be said of a place exchange between a regular molecule and an interstitial, and of course of an interstitial moving in the lattice without exchange.

Gränicher (1958) assumes the formation energy to be about 12 kcal/mole for the vacancies (just the lattice energy of the molecule) and estimates it to be higher than 28 kcal/mole for Frenkel defects, strongly favoring the former species. This agrees with the lack of anisotropy for the activation energy: the maximum free diameter is 2.44 Å for diffusion along the c-axis and 1.84 Å perpendicular to it. Since the molecule has a diameter of 2.76 Å, the energy required to squeeze it would be different for the two cases. On the other hand, the observed anisotropy for the diffusivity itself has been invoked by Dengel et al. (1966) and by Ramseier (1967) to support the vacancy hypothesis; that is, the distance covered after a given number of steps is 12% longer perpendicular than parallel to the c-axis.

This relation is certainly correct for the maximum distance, but diffusion is related to the average squared distance, and here the probabilities must restore the equality. If not, the same argument could be applied to diamond ice, because the different stacking mode (cubic instead of hexagonal close packing) does not change the maximum distance ratio; but this leads to a contradiction, since cubic ice is necessarily isotropic. If the anisotropy is real, its cause should instead be sought in second-order effects produced by the different environment of the vacancy. In cubic ice, every molecule pair is bound in a centrosymmetrical way, but in hexagonal ice the pairs parallel to the c-axis are mirror symmetrical. This could have an influence on the vibrational frequencies of the vacancy neighbors, as well as on the electronic distribution involved in the effective charges.

Ramseier has shown that the atomic diffusion theory, together with the vacancy hypothesis and Zener's theory (1952) for activation entropy, accounts for the preexponential factor of the diffusivity. It has the following form for a vacancy:

(32) $$D_0 = f\nu a^2 \exp(S/R)$$

where f if a correlation coefficient equal to 0.5 for the diamond structure, ν is the vibration frequency, 4.7×10^{12}/sec (given by the Debye temperature determined from X-ray intensity measurements), and a is the jump distance. The entropy should then amount to (17 ± 1) cal/(mole)(deg). Zener's theory predicts

(33) $$S \simeq -\lambda \frac{Q}{T_m} \frac{d(\mu/\mu_0)}{d(T/T_m)}$$

where λ = factor near unity

Q = energy of activation

T_m = melting temperature

μ = Young's modulus at T (μ with respect to μ_0; T with respect to $T = 0$)

The last factor is about $\frac{1}{3}$, so the calculated entropy agrees with the value above.

If the enthalpy of formation of a Schottky defect is assumed to be around 12 kcal/mole, this leaves only 2 to 3 kcal/mole for the transfer energy, that is,

half of the hydrogen-bond energy (4–5 kcal/mole). This might be a consequence of the softness of the lattice, which allows a mechanical relaxation
around the vacancy to bring the neighboring molecules nearer to each other.
Besides this statical effect, local phonon modes may temporarily reduce the
intermolecular distance still further. A molecule then has the possibility of
jumping into the vacancy while making the new bonds before the old ones are
completely broken. This suggested mechanism has not yet been investigated
in detail.

Onsager and Runnels (1969) have examined the nuclear magnetic relaxation
time of the protons and compared it with the dielectric relaxation rate and the
diffusion time between two subsequent defect jumps. According to their
analysis, the magnetic relaxation is too fast by an order of magnitude to have
a common mechanism with the dielectric relaxation, but the comparison is
quite favorable with diffusion. This also precludes a common dielectric and
diffusive mechanism. From the correlation time and line shape, they infer
that the diffusive motion must have an rms displacement by a step significantly
larger than the intermolecular distance. This would imply an interstitial
mechanism, in contradiction to the vacancy mechanism assumed before.
However, the nuclear magnetic relaxation time depends significantly on HF
doping (see below), whereas diffusivity does not. The correlation between the
two is therefore questionable and probably of a more complex nature than
has been assumed until now. As Gränicher (1969) has shown, experiments on
the pressure dependence of the diffusivity will be the best means of indicating
unequivocally the nature of the involved defects, because the activation
volume is certainly positive for interstitial and negative for vacancies. Unfortunately, such information is not now available.

Using the HF dependence of the nuclear relaxation time, Kopp, Barnaal,
and Lowe (1965) have devised a nondestructive method for measuring the
diffusivity of fluorine in ice. They obtain a high value—around 10^{-6} cm^2/sec
at $-10°$C—with an activation energy of 13 kcal/mole, with the following
interpretation: the HF molecules occupy regular lattice sites and few of them
dissociate. The remaining F$^-$ ions are ejected to interstitiality, travel a long
distance, and eventually are trapped by a vacancy, thereby allowing the corresponding hydronium ion to follow or precede its partner. The activation
energy should account mainly for the acid dissociation (7 kcal/mole) and for
the ionic ejection (6 kcal/mole).

These measurements were repeated by Haltenorth and Klinger (1969) with
another method (successive ablations of the sample and determination of the
fluorine by the conductivity of the melt water). They obtain a lower value of
10^{-7} cm^2/sec at $-10°$C, with a much smaller activation energy of 4.6
kcal/mole, over the wider temperature range -5 to $-90°$C (instead of -4 to
$-30°$C in the previous measurements). Having taken special care to avoid

surface diffusion, they use this phenomenon to explain the discrepancy between their result and Kopp's, whose samples (8 mm diameter by 60 mm length) had a large surface-to-volume ratio. As in the case of pure ice, they also observe an anisotropy of 20%, the coefficient being larger perpendicular to the c-axis, and a significant dependence of the presence of small angle boundaries. Owing to the large value of the coefficient, which is 10^4 times higher than for self-diffusion, it seems improbable that fluorine migrates by a vacancy mechanism. Rather, an interstitial process should be invoked, but there is no explanation yet for the low activation energy.

7   Mass Transport by Plastic Deformation

The observation of glacier flow suggests that ice must be a deformable material identical in this respect to the soft metals. Experiments performed in the last century revealed that the plastic properties were strongly anisotropic: if a single crystal is subjected to a stress in an arbitrary direction, it glides along planes that are perpendicular to the c-axis (McConnel, 1891). Further investigators confirmed these early observations, but attempts to find a preferential glide direction failed. The glide planes are concentrated in regions that appear on the surface as slip lines 50 to 150 μ apart (Nakaya, 1956; Reading and Bartlett, 1968). Several experiments have been performed to study the glide dynamics by applying a tensile stress (Glen and Perutz, 1954), by shear (S. Steinemann, 1954; Rigsby, 1951), and by compression (Griggs and Coles, 1954).

As soon as a load is large enough to produce a plastic deformation, the motion, instead of coming to a stop, proceeds with time as a creep. If a constant stress τ is applied, the deformation being measured as a function of the time, two distinct ranges appear. In the first one, the angular deformation γ accelerates almost uniformly according to the law (Griggs and Coles, 1954):

$$(34) \qquad \gamma = 4a[(\tau - bT^2)t]^2$$

where $a = 0.62$, $b = 0.014$ if γ is expressed in percent, τ in kg/cm^2, T in °C, and t in hours.

In the second range, the acceleration decreases and the velocity depends on powers of the deformation and of the stress magnitude; that is, for the shear γ

$$(35) \qquad \dot{\gamma} = k(T)\gamma^m\tau^n$$

A third range, in which the deformation rate decreases with time, as it appears in metals, does not occur in ice for basal glide: there is no work hardening. In another type of experiment, the sample is deformed at a constant rate and the stress is measured as a function of the strain, which is then

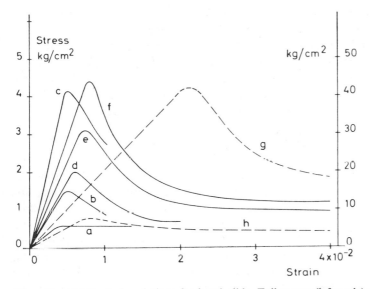

Figure 9. Stress–strain relations for basal glide. Full curves (left scale): *a–c*—temperatures of   15,   26, and −35°C, respectively, with a strain rate of 0.8 × 10⁻⁵/mins *d–f*—strain rates of 4, 10, 16 × 10⁻⁵/min, respectively, at −15°C. (After Higashi et al., 1964. Dotted curve (right scale): *g*—pure ice; *h*—ice doped with 3 ppm HF at −70°C with a strain rate of 1.6 × 10⁻⁵/min. After Jones, 1967.

proportional to the time. The stress increases linearly until it reaches a maximum, after which it decays rapidly and then more slowly. This behavior is also typical of the lack of work hardening (Figure 9).

Although the power law given above has been generally observed, there is some scatter in the value of the exponents dependent on the samples and the method. Near the melting point, n ranges between 1.5 and 4; it lies around 4 at low temperatures down to $-70°C$ (Glen and Jones, 1967), and m seems to be near unity. Experiments performed by Wakahama (1967) on well-annealed natural samples from the Mendenhall glacier show a value near 3, which agrees very well with the dislocation theory developed by this author. He assumes the existence of Frank–Read sources and finds a dislocation velocity proportional to the stress with a value of 3×10^{-6} m/sec for 1 kg/cm² at $-10°C$. The temperature-dependent constant obeys an Arrhenius law with an activation energy of 15 ± 2 kcal/mole (Readey and Kingery, 1964; Higashi et al., 1965). At $-10°C$, a resolved shear stress of 5 kg/cm² produces a shear rate of 3×10^{-5}/sec.

The apparent lack of preferential direction in the glide plane is so unusual that many investigators have been induced to look for it, but the search has been without success, except a faint tendency noticed by S. Steinemann

(1954): the creep velocity is slightly larger along the $\langle 11\bar{2}0 \rangle$ direction than along $\langle 10\bar{1}0 \rangle$. The situation has been cleared by Kamb (1961), who used a simple geometrical argument: with three preferential directions at 120° and a stress power law for the creep velocity, the anisotropy vanishes if (a) the exponent n is unity, in accordance with the superposition principle, (b) $n = 3$; otherwise it keeps small values. For $n = 2$ the angle between the applied shear stress and the glide direction is at most 2 degrees for a stress close to the bisector of two neighboring preferential directions. For $n = 4$, the deviation does not exceed 3 degrees, with the opposite sign. Since the true value of n lies around 3, the angular deviation is too small to be detected, and this explains the negative experimental results. From structural considerations, Kamb proposed the $\langle 11\bar{2}0 \rangle$ direction for the Burger vector. This has been verified by X-ray diffraction topography (Webb and Hayes, 1967; Fukuda and Higashi, 1968) in which a picture of the crystal is obtained from the X-rays diffracted by a particular plane system. Distorted lattice regions, such as those produced by dislocations, appear with a different intensity, making visible the dislocation bundles (provided their Burger vector is not parallel to the diffracting planes). Pictures made of the same region but using planes of different orientation then allow the complete determination of the dislocations, which are mainly of the screw type with the Burger vector along $\langle 11\bar{2}0 \rangle$.

Valuable information has also been supplied by the analysis of etch pits appearing on the surface after a thermal or a chemical treatment at the emergence point of the dislocations. Especially in the case of crystals strained with no shear stress component resolved along the basal plane, the shape of these etch pits and of etch channels has revealed the existence of other types of dislocations, probably with Burger vectors along $\langle 0001 \rangle$ and gliding in the planes $\{10\bar{1}0\}$ or $\{11\bar{2}0\}$ (Mugumura and Higashi, 1963; Levi et al., 1965). In tensile tests, the deformation is accompanied by the appearance of elongated hexagonal voids, which do not occur in compression tests. Mae (1968) has explained this phenomenon successfully by vacancies produced by dislocation climb and aggregating by diffusion. Higashi (1967) registered the stress–strain curve with the tensile axis exactly parallel to the basal plane (nonbasal glide). There is no drop in the stress after an initial linear increase, as for a shear in the basal plane, but the stress increases still more, indicating work hardening as in certain metals (Figure 10). This is attributed to the interaction of the dislocations in the different planes $\{10\bar{1}0\}$, $\{1\bar{1}00\}$, and $\{01\bar{1}0\}$. The yield stress is 20 times higher than in the basal case. This high value results from the pinning of the dislocations on the vacancies generated during the climb. The exponent n in the power law is higher too, near 7, but the activation energy is slightly smaller, around 12 kcal/mole. Since the ice is harder for the nonbasal glide, this mode does not appear in single-crystal experiments unless special

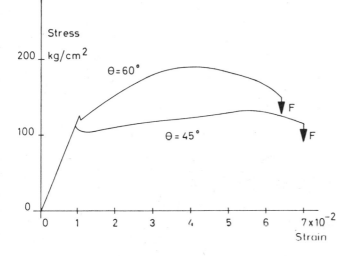

Figure 10. Stress–strain relations for nonbasal glide at $-20°C$ with a strain rate of 18×100^{-5}/min for different values of the angle θ between the tensile axis and the plane $\langle 10\bar{1}0 \rangle$ (F = fracture). After Muguruma *et al.*, 1966.

care is taken, but it is certainly significant in the deformation of polycrystalline aggregates. However, these are not considered here, since they pertain to glaciology rather than to the study of transport properties.

Although this section does not seem to have much in common with the mechanisms discussed previously, a recently discovered phenomenon bridges the gap: in constant-strain-rate experiments performed at $-70°C$, Jones (1967) observed that incorporation of hydrofluoric acid greatly increases the plasticity of ice. For instance, 3 ppm of HF reduce the maximum shear stress to one-fourth, and 40 ppm to one-tenth. This effect has been explained by Glen (1968), who pointed out that the displacement of a dislocation, consisting in breaking and making bonds, must be sensitive to the protonic configuration. A molecule involved in the process has to part from a nearest neighbor and then bind a third nearest neighbor, in the case of basal glide, with a Burger vector along $\langle 11\bar{2}0 \rangle$ (Figure 11). But if the configuration is disordered, there is a probability of 0.5 that the third nearest neighbor has an unfavorable orientation. Therefore, if no rearrangement mechanism is allowed, two of four steps will on an average create a D and an L defect. The energy required on a certain length is then translated by a critical stress, estimated to be of the order of several kilobars, whereas the observed values are below 1 bar.

Even if the defects created by this process are mobile, a significant reduction of the energy requires an annihilation of the defects quasi-simultaneous with

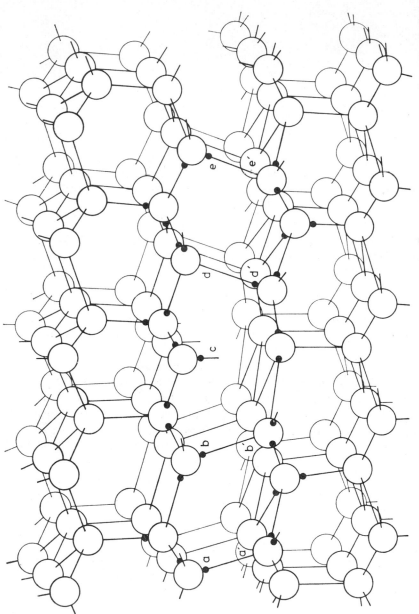

Figure 11. Edge dislocation in the basal plane, Burgers vector in the $\langle 11\bar{1}0 \rangle$ direction. The proton configuration is drawn only in the front layer near the dislocation. A shift to the right (c–d', then d–e') would create a DL pair, but not a shift to the left (c–b', then b–a').

the dislocation motion. In this case, the correlated defect motion would be much too fast for thermally activated jumps, and even a tunneling mechanism would not bring a reduction by a factor of 10^4. Consequently, a certain basic concentration of randomly migrating defects is necessary to modify the proton configuration. The dislocation lies parallel to $\langle 11\bar{2}0 \rangle$ directions, but with kinks where it jumps from such a line to a neighboring one. The kinks move on as long as they find the molecules suitably oriented; but before proceeding further, they have to wait until a passing defect turns a misoriented molecule. The average dislocation velocity is therefore strongly dependent on the defect concentration, which is very sensitive to a dopant such as HF at low temperature.

This explains the relative softness of fluoride-doped ice crystals, since at the temperatures of Jones's experiments thermal activation is still adequate to maintain a certain migration rate, but only a very low intrinsic dissociation for the defects. It is interesting to note that the observed activation energy for creep, which is the same as for dielectric relaxation, agrees with this picture of rate-limiting molecular reorientation. Experiments with other dopants have been recently reported (Jones and Glen, 1969). Ammonia does not soften the ice but produces a slight hardening, although it is believed to increase the D defect concentration. A tentative explanation introduces a trapping of the D's or of the OH^- ions on the dislocation lines. The binding energy can be electrostatic (if the dislocation core bears a net charge), but it can also be elastic (if the neighborhood of the dislocation accommodates a defect easily). As the average diffusivity of the trapped defects vanishes, they cannot contribute to the molecular reorientation near the moving kinks. Therefore, the ammonia restrains the movement of the dislocations, since it reduces the H_3O^+ and the L concentrations in the crystal.

It has been found also that ammonium fluoride has no effect on the plasticity. We know from electric measurements that NH_4F modifies the defect concentration (Levi et al., 1963; Brill and Camp, 1961; Zaromb and Brill, 1956); therefore its dissociation should shift the chemical equilibrium in the same direction as ammonia, since a small shift in the other direction would be easily observable as a softening. Glen has pointed out that, if his theory is correct, an ordered phase would be quite soft, since every molecule then has a favorable orientation. Although polarization measurements seem to indicate a kind of ferroelectric (or antiferroelectric) phase transition near $100°K$ for hexagonal ice (Dengel et al., 1964) and near $190°K$ for diamond ice (Cubiotti and Geracitano, 1967), with a trend to ordering of the proton configuration at low temperature, no structure determination or creep experiments connected with Glen's prediction have yet been reported.

REFERENCES

Auty, R. P., and R. H. Cole, *J. Chem. Phys.*, **20**, 1309 (1952).

Bernal, J. D., and R. H. Fowler, *J. Chem. Phys.*, **1**, 515 (1933).

Bertie, J. E., and E. Whalley, *J. Chem. Phys.*, **40**, 1637 (1964).

Bertie, J. E., and E. Whalley, *J. Chem. Phys.*, **46**, 1271 (1967).

Bjerrum, N., *Kgl. Danske Videnskab. Selskab.*, *Mat. Fys. Medd.*, **27**, 56 (1951).

Blicks, H., O. Dengel, and N. Riehl, *Phys. kondens. Materie*, **4**, 375 (1966).

Bradley, R. S., *Trans. Faraday Soc.*, **53**, 687 (1957).

Brill, R., H. Ender, and A. Feuersanger, *Z. Elektrochem.*, **61**, 1071 (1957).

Brill, R., and P. R. Camp, *SIPRE Res. Rep. No. 68* (1961).

Brownscombe, J. L., and B. J. Mason, *Phil. Mag.*, **14**, 1037 (1966).

Bryant, G. W., and N. H. Fletcher, *Phil. Mag.*, **12**, 165 (1965).

Bryant, G. W., *Phil. Mag.*, **16**, 495 (1967).

Bullemer, B., and N. Riehl, *Solid State Commun.*, **4**, 447 (1966a).

Bullemer, B., and N. Riehl, *Physics Letters*, **22**, 411 (1966b).

Bullemer, B., H. Engelhardt, and N. Riehl, in "Physics of Ice," *Proc. Intern. Symp. Munich 1968*, Eds. N. Riehl, B. Bullemer, and H. Engelhardt, Plenum Press, New York, 1969, p. 416.

Chan, R. K., D. W. Davidson, and E. Whalley, *J. Chem. Phys.*, **43**, 2376 (1965).

Cohan, N. V., M. Cotti, J. İribarne, and M. Weissmann, *Trans. Faraday Soc.*, **58**, 490 (1962).

Cohan, N. V., and M. Weissmann, *Nature*, **201**, 490 (1964).

Cubiotti, G., and R. Geracitano, *Physics Letters*, **24A**, 179 (1967).

Decroly, J. C., H. Gränicher, and C. Jaccard, *Helv. Phys. Acta*, **30**, 465 (1957).

Delibaltas, P., O. Dengel, D. Helmreich, N. Riehl, and H. Simon, *Phys. kondens. Materie*, **15**, 166 (1966).

Dengel, O., U. Eckener, H. Plitz, and N. Riehl, *Physics Letters*, **9**, 291 (1964).

Dengel, O., E. Jacobs, and N. Riehl, *Phys. kondens. Materie*, **1**, 58 (1966).

Dengel, O., N. Riehl, and A. Schleippmann, *Phys. kondens. Materie*, **5**, 83 (1966).

Dunitz, J. D., *Nature*, **197**, 860 (1963).

Durand, M., M. Deleplanque, and A. Kahane, *Solid State Commun.*, **5**, 759 (1967).

Eigen, M., L. De Mayer, and H. C. Spatz, *Ber. Bunsen. Ges.*, **68**, 19 (1964).

Eisenberg, D., and C. A. Coulson, *Nature*, **199**, 368 (1963).

Engelhardt, H., B. Bullemer, and N. Riehl, in "Physics of Ice," *Proc. Intern. Symp. Munich 1968*, Eds. N. Riehl, B. Bullemer, and H. Engelhardt, Plenum Press, New York, 1969, p. 430.

Fischer, S. F., and G. L. Hofacker, in "Physics of Ice," *Proc. Intern. Symp. Munich 1968*. Eds. N. Riehl, B. Bullemer, and H. Engelhardt, Plenum Press, New York, 1969, p. 369.

Fukuda, A., and A. Higashi, in "Physics of Ice," *Proc. Intern. Symp. Munich 1968*, Eds. N. Riehl, B. Bullemer, and H. Engelhardt, Plenum Press, New York, 1969, p. 239.

Glen, J. W., and M. F. Perutz, *J. Glaciol.*, **2**, 397 (1954).

Glen, J. W., *Advan. Phys.*, **7**, 254 (1958).

Glen, J. W., and S. J. Jones, in "Physics of Snow and Ice," *Proc. Intern. Conf. Sapporo 1966*, Ed. H. Oura, Institute of Low Temperature Sciences, Hokkaido University, 1967, p. 267.

Glen, J. W., *Phys. kondens. Materie*, **7**, 43 (1968).

Gosar, P., and M. Pintar, *Phys. Stat. Solidi*, **4**, 675 (1964).

Gosar, P., in "Physics of Ice," *Proc. Intern. Symp. Munich 1968*, Eds. N. Riehl, B. Bullemer, and H. Engelhardt, Plenum Press, New York, 1969, p. 401.

Gränicher, H., *Z. Kristallogr.*, **110**, 432 (1958).

Gränicher, H., in "Physics of Ice," *Proc. Intern. Symp. Munich 1968*, Eds. N. Riehl, B. Bullemer, and H. Engelhardt, Plenum Press, New York, 1969, p. 527.

Gränicher, H., in "Physics of Ice," *Proc. Intern. Symp. Munich 1968*, Eds. N. Riehl, B. Bullemer, and H. Engelhardt, Plenum Press, New York, 1969, p. 534.

Griggs, D. T., and N. E. Coles, *SIPRE Rept. 11* (1954).

Gross, G. W., *Science*, **138**, 520 (1962).

Gross, G. W., *Ann. N.Y. Acad. Sci.*, **125**, 380 (1965).

Haas, C., *Physics Letters*, **3**, 126 (1962).

Haltenorth, H., and J. Klinger, in "Physics of Ice," *Proc. Intern. Symp. Munich 1968*, Eds. N. Riehl, B. Bullemer, and H. Engelhardt, Plenum Press, New York, 1969, p. 579.

Higashi, A., S. Koinuma, and S. Mae, *Japan. J. Appl. Phys.*, **3**, 610 (1964); *Japan. J. Appl. Phys.*, **4**, 575 (1965).

Higashi, A., in "Physics of Snow and Ice," *Proc. Intern. Conf. Sapporo, 1966*, Ed. H. Oura, Institute of Low Temperature Sciences, Hokkaido University, 1967, p. 277.

Humbel, F., F. Jona, and P. Scherrer, *Helv. Phys. Acta*, **26**, 17 (1953).

Ida, M., N. Nakatani, K. Imai, and S. Kawada, *Sci. Rept. Kanazawa Univ.*, **11**, 13 (1966).

Itakagi, K., *J. Phys. Soc. Japan*, **22**, 427 (1967).

Itoh, R., *J. Phys. Soc. Japan*, **22**, 698 (1967).

Jaccard, C., *Helv. Phys. Acta*, **32**, 89 (1959).

Jaccard, C., *Phys. kondens. Materie*, **3**, 99 (1964).

Jaccard, C., in "Physics of Snow and Ice," *Proc. Intern. Conf. Sapporo, 1966*, Ed. H. Oura, Institute of Low Temperature Sciences, Hokkaido University, 1967, p. 173.

Jaccard, C., in "Physics of Ice," *Proc. Intern. Symp. Munich 1968*, Eds., N. Riehl, B. Bullemer, and H. Engelhardt, Plenum Press, New York, 1969, p. 348.

Johnstone, J. H. L., *Proc. Trans. Nova Scotian Inst. Sci.*, **13**, 126 (1912).

Jones, S. J., *Physics Letters*, **25A**, 366 (1967).

Jones, S. J., and J. W. Glen, *Phil. Mag.*, **19**, 13 (1969).

Kamb, W. B., *J. Glaciol.*, **3**, 1097 (1961).

Kim, D. Y., and V. H. Schmidt, *Can. J. Phys.*, **45**, 1507 (1967).

Kopp, M., D. Barnaal, and J. J. Lowe, *J. Chem. Phys.*, **43**, 2965 (1965).

Kopp, M., *Progr. Rept. No. 19*, Dept. Phys. Univ. Pittsburgh (1966).

Kuhn, W., and M. Thürkauf, *Helv. Chim. Acta*, **41**, 938 (1958).

Kydon, D. W., M. Pintar, and H. E. Petch, *J. Chem. Phys.*, **48**, 5348 (1968).

Lamb, L., and A. Turney, *Proc. Phys. Soc. (London)*, **62B**, 272 (1949).

Latham, J., and B. J. Mason, *Proc. Roy. Soc. (London)*, Ser. A, **260**, 523 (1961).

Levi, L., O. Milman, and E. Suraski, *Trans. Faraday Soc.*, **59**, 2064 (1963).

Levi, L., and D. Arias, *J. Chim. Phys.* 668 (1964).

Levi, L., E. M. de Achaval, and E. Suraski, *J. Glaciol.*, **5**, 691 (1965).

Levi, L., and L. Lubart, *Phys. kondens. Materie*, **7**, 368 (1968).

Mae, S., *Phil. Mag.*, **18**, 101 (1968).

McConnel, J. C., *Proc. Roy. Soc. (London)*, **49**, 323 (1891).

Mounier, S., and P. Sixou, in "Physics of Ice," *Proc. Intern. Symp. Munich 1968*, Eds. N. Riehl, B. Bullemer, and H. Engelhardt, Plenum Press, New York, 1969, p. 562.

Muguruma, J., and A. Higashi, *J. Phys. Soc. Japan*, **18**, 1261 (1963).

Nakaya, U., *SIPRE Res. Paper 13* (1956).

Onsager, L., and M. Dupuis, *Rendi. Scuola Intern. Fis. Enrico Fermi* (X. Corso) 234 (1960).

Onsager, L., and M. Dupuis, in *Electrolytes*, Pergamon Press, New York, 1962, p. 27.

Onsager, L., and L. K. Runnels, *Proc. Nat. Acad. Sci. U.S.*, **50**, 208 (1963).

Onsager, L., and L. K. Runnels, *J. Chem. Phys.*, **50**, 1089 (1969).

Pauling, L., *J. Amer. Chem. Soc.*, **57**, 2680 (1935).

Pauling, L., *Nature of the Chemical Bond*, Cornell University Press, Ithaca, N.Y., 1952, p. 301.

Peterson, S. W., and H. A. Levy, *Acta Cryst.*, **10**, 70 (1957).

Ramseier, R. O., *J. Appl. Phys.*, **38**, 2553 (1967).

Readey, D. W., and W. D. Kingery, *Acta Met.*, **12**, 171 (1964).

Readings, C. J., and J. T. Bartlett, *J. Glaciol.*, **7**, 479 (1968).

Rigsby, G. P., *J. Geol.*, **59**, 590 (1951).

Steinemann, A., *Helv. Phys. Acta*, **30**, 553 (1957).

Steinemann, S., *J. Glaciol.*, **2**, 404 (1954).

Takahashi, T., *J. Atmos. Sci.*, **23**, 74 (1966).

Truby, F. K., *Science*, **121**, 404 (1955).

Wakahama, G., in "Physics of Snow and Ice," *Proc. Intern. Conf. Sapporo, 1966*, Ed. H. Oura, Institute of Low Temperature Sciences, Hokkaido University, 1967, p. 291.

Webb, W. W., and C. E. Hayes, *Phil. Mag.*, **16**, 909 (1967).

Weissmann, M., and N. V. Cohan, *J. Chem. Phys.*, **43**, 124 (1965).

Whalley, E., in "Physics of Ice," *Proc. Intern. Symp. Munich 1968*, Eds. N. Riehl, B. Bullemer, and H. Engelhardt, Plenum Press, New York, 1969, p. 19.

Wollan, E. O., W. L. Davidson, and C. G. Shull, *Phys. Rev.*, **75**, 1348 (1949).

Workman, E. J., F. K. Truby, and W. Drost-Hansen, *Phys. Rev.*, **94**, 1073 (1954).

Young, I. G., and R. E. Salomon, *J. Chem. Phys.*, **48**, 1635 (1968).

Zaromb, S., and R. Brill, *J. Chem. Phys.*, **24**, 895 (1956).

Zener, C., in *Imperfections in Nearly Perfect Crystals*, Eds. W. Shockley, J. H. Holloman, R. Maurer, and F. Seitz, Wiley, New York, 1952, p. 306.

3 The Ice Interface

H. H. G. Jellinek,
Department of Chemistry,
Clarkson College of Technology,
Potsdam, New York

Water substance has very peculiar and interesting properties in both the liquid and solid state. The ice-solid, ice-ice$_{grain\ boundary}$, ice-H_2O_{liquid}, and ice-H_2O_{vapor} interfaces are no exceptions in this respect. They do not appear to be "normal" interfaces at all and have properties that are of great significance to surface science. However, it would not really be true to state that only ice has these apparently abnormal characteristics; as a matter of fact, all substances have such characteristics to a certain degree. It seems that a matter of magnitude in the case of ice makes the interface unique. Water, after all, is one of the most important substances, and since its freezing point is so conveniently located, these interfaces [(ice-$H_2O_{vapor\ or\ liquid}$ or ice-solid)] are very suitable for study. Each surface or interface is not just a geometrical area but has to be considered in some depth. Once this is done, most materials are seen to have some "abnormal" surface properties.

In solids, for instance, the interfacial layer often affects the substance to considerable depths; a transition layer is established in the interfacial region. In case of ice this transition layer seems to be of larger magnitude than for many other solids. This has important consequences. Faraday (1850)[1] discovered this "layer" on ice and with it the phenomenon of regelation or refreezing, to use the expression coined by Tyndall.[2] Faraday's views were superseded, at that time, by the theory of pressure melting put forth by Lord Kelvin[3] and his brother, J. Thomson.[4]

Since it would be impossible to discuss the vast amount of literature having some bearing on the ice–air or ice–solid interface in one chapter of a book, a selection has been made, and some aspects of the ice interface are discussed in more detail than others. Some are even completely bypassed. Thus the author's preference will be apparent in the choice of topics presented here.

65

1   The Ice–Water$_{Liquid}$ Interfacial Free Energy

The ice–H_2O_{liquid} interfacial tension is a fundamental parameter needed in many investigations, frequently, for example, in biological or biochemical studies. Numerous attempts have been made to calculate or estimate this interfacial tension quantitatively. A good summary of these attempts is given in a book by Dufour and Defay.[5] These authors present a table containing most of the data known until recently, and a version of it is given as Table 1. All values for this interface are of the same order of magnitude in spite of the diversity of methods used.

Volmer (1939)[6] suggested for the calculation of $\gamma_{i,l}$ a relationship as follows:

$$(1) \qquad \frac{\gamma_{i,l}}{L_l} = \frac{\gamma_{l,v}}{L_v}$$

where $\gamma_{i,l}$ and $\gamma_{l,v}$ are the interfacial free energies for ice–H_2O_{liquid} and H_2O_{liquid}–H_2O_{vapor}, respectively; L_l and L_v are the corresponding latent heats of fusion and vaporization. Actual values were calculated by Krastanov (1941).[7] McDonald (1953)[8] made corrections for variations in the latent heats with temperature. Volmer's relation gives only an estimate of the order of magnitude of the $\gamma_{i,l}$ values.

The interfacial tension of ice–H_2O_{liquid} can also be estimated by counting the number of hydrogen bonds that have to be ruptured to cut an ice crystal and by using Antonow's rule.[14]

$$(2) \qquad \gamma_{i,l} = \gamma_{i,g} - \gamma_{l,g}$$

This method was employed by McDonald (1953)[8] and by Briegleb (1949).[9] It is only approximate, for errors are introduced by the distortion near the ice surface and by the approximate nature of Antonow's rule. Thus these values will only be estimates of the order of magnitude of $\gamma_{i,l}$.

The lowering of the freezing point of water in narrow pores can also be used for the estimation of $\gamma_{i,l}$. This was done by Kubelka and Prokscha (1944).[11] Also the freezing of small water drops is suitable for this purpose [Mason, 1957;[10] also Jacobi (1955)[12]]. The experiments evaluated by Jacobi were recalculated by Dufour and Defay (1963);[5] see their book[5] for details of the differences in these calculations.

Roulleau (1955)[15] used his own experiments on the freezing of drops for calculating $\gamma_{i,l}$ and obtained 23.3 dynes/cm at $-33°C$. These calculations differ from those of Dufour and Defay in a number of ways.

Skapski[16] deduced 27 ergs/cm^2 for ice–H_2O_{liquid}.

Oura[17] calculated a value of $\gamma_{i,l} = 18.5$ dynes/cm, evaluating experiments by Smith-Johannsen (1948)[18] on the freezing of supercooled water. By a

Table 1 Ice–Water Interfacial Tension (dyne/cm) by Various Authors

Temperature (°C)	Krastanov[7,a]	McDonald[8,a]	McDonald[8] (uncorrected)	McDonald[8,b]	Briegleb[9]	Mason[10]	Kubelka and Proksch[11,d]	Jacobi's[12] Calculations[c,e]	Dufour and Defay's[5] Calculations[e]	Hobbs et al.[13] Experiments
0	10.0	10.0	49	21	23.5	—	—	23.1	23.8	Ice–water_liquid
−5	—	—	—	—	23.0	—	25.0	22	23.3	0°C:33
−10	10.2	9.6	45	19	21.5	—	—	21	22.8	−40°C:25
−15	—	—	—	—	—	—	—	—	22.3	
−20	10.4	9.1	43	16	18.5	—	—	19	21.8	Ice–water_vapor
−25	—	—	—	—	—	—	—	—	21.3	0°C:109
−30	10.6	8.5	40	14	16.5	—	—	17.1	20.75	Ice–ice_grain boundary
−35	—	—	—	—	15.5	—	—	16.1	20.24	0°C:65
−40	10.8	7.7	38	12	14.5	—	—	15.1	19.7	
−41	—	—	—	—	—	17.5	—	—	—	
−45	—	—	—	—	—	—	—	—	—	
−50	11.0	6.8	36	10	—	—	—	—	—	

From Dufour and Defay, 1963
a Calculated by Volmer's rule.
b Corrected for lattice distortion.
c Calculated from freezing of drops.
d Calculated from freezing temperature in pores.
e Jacobi's experiments.

similar procedure he arrived at a value of 98.5 dynes/cm for the ice–air interface on the basis of experiments by Schaefer (1949).[19] Usually a value of about 100 dynes/cm is accepted for the ice–air interface.

Very recently, Hobbs and Ketcham[13] carried out experiments in order to determine not only the interfacial free energy of ice–H_2O_{liquid}, but also of ice–H_2O_{vapor} and ice–ice$_{grain\ boundary}$.

Grain boundary grooves form on the surface of polycrystalline ice. If $\gamma_{s,v}$ is isotropic and $\gamma_{v,\ grain\ boundary}$ is independent of the orientation of the boundary, thermodynamic equilibrium prevails at the root of the groove. A relationship holds then as follows:

$$(3) \qquad \frac{\gamma_{s,v}}{\gamma_{v,\ grain\ boundary}} = \frac{1}{2\cos(\theta_{s,v}/2)}$$

where $\theta_{s,v}$ is the solid–vapor grain boundary groove angle. A similar equation can be deduced for the solid–liquid interface; however, $\theta_{s,l}$ must be larger than 0°C:

$$(4) \qquad \frac{\gamma_{s,l}}{\gamma_{v,\ grain\ boundary}} = \frac{1}{2\cos(\theta_{s,l}/2)}$$

If $\theta_{s,v}$ and $\theta_{s,l}$, respectively, can be determined experimentally, then the ratios $\gamma_{s,v}/\gamma_{v,\ grain\ boundary}$, $\gamma_{s,l}/\gamma_{v,\ grain\ boundary}$, and $\gamma_{s,v}/\gamma_{s,l}$ can be obtained.

Absolute values of these interfacial tensions are evaluated using Young's equation with the equilibrium contact angle ϕ formed at the intersection of the solid–vapor, solid–liquid, and liquid–vapor interfaces, respectively. If the liquid does not completely wet the solid surface, one has

$$(5) \qquad \gamma_{s,v} = \gamma_{s,l} + \gamma_{l,v}\cos\phi$$

Combination of (5) with (3) and (4) gives

$$(6) \qquad \gamma_{s,v} \geq \frac{\gamma_{l,v}\cos\phi\cos(\theta_{s,l}/2)}{\cos(\theta_{s,l}/2) - \cos(\theta_{s,v}/2)}$$

Hence $\gamma_{s,v}$ and by rearrangement also $\gamma_{s,l}$ and $\gamma_{g,\ grain\ boundary}$ can be obtained.

For complete wetting of the surface by the liquid, one has

$$(7) \qquad \gamma_{s,v} \geq \gamma_{s,l} + \gamma_{l,v}$$

Hence

$$(8) \qquad \gamma_{s,v} \geq \frac{\gamma_{l,v}\cos(\theta_{s,l}/2)}{\cos(\theta_{s,l}/2) - \cos(\theta_{s,v}/2)}$$

Thus one can only calculate a lower limit for $\gamma_{s,v}$ and the other parameters.

The experiments were carried out in a stainless steel box having a glass cover on top. The ice surface was viewed with an interference microscope.

Actually a formvar replica of the ice surface was made, and was evaluated in terms of grain boundary groove angles. The final values obtained at 0°C (Table 1) were as follows: $\gamma_{s,v} = (109 \pm 3)$ ergs/cm^2, $\gamma_{s,l} = (33 \pm 3)$ ergs/cm^2, and $\gamma_{v, \text{grain boundary}} = (65 \pm 3)$ ergs/cm^2.

2 The Interfacial Transition Layer on Ice

It is instructive to introduce this topic via some fundamental work on adhesion of ice performed by the author a number of years ago.[20] Work in this area has also been carried out by a number of other workers;[21] the various results obtained agree, on the whole, as far as type and regularities are concerned with the present author's findings. The actual numerical values are different, however, for these depend on the history of the ice samples and on complicated experimental technique, which is sensitive to slight variations. Many of these experiments were especially designed for finding indications of the transition layer on ice and for determining its properties.

Tensile and shear tests were carried out, for instance, with stainless steel substrates having ordinary smooth surfaces as obtained on a lathe and also with steel whose surfaces were polished and mirror polished. In addition, optically flat quartz disks were also used as substrates. Furthermore, plastic materials such as cast polystyrene and polymethylmethacrylate having smooth surfaces were investigated. Ice was sandwiched between one of these surfaces and a rough aluminum surface, which adhered much more strongly to ice than the smooth one. All tensile tests with stainless steel substrates gave cohesive breaks (bulk tensile strength of ice is ca. 15.8 kg/cm^2). These results were independent of the surface finish of the substrate. Decreasing the ice volume increases the tensile strength; each cross-sectional area yielded a separate curve. The highest tensile strength reached in these experiments was 70 kg/cm^2, but there is no reason why higher values cannot in principle be obtained, if ice samples of still smaller volume could be made. Berghausen et al.[21] found the same phenomenon. As a matter of fact, this is a general characteristic for all ordinary materials and is the result of imperfections. The probability of not finding an imperfection in a volume of material increases with decreasing volume. These imperfections are also the cause for the small experimental tensile strengths of materials, which are usually not larger than about 1 to 10% of the theoretical values. The polymer and quartz substrates gave cohesive breaks for tensile tests. All cohesive breaks were virtually independent of temperature.

The picture changes completely when shear tests are performed. The shear apparatus was arranged so that the applied force was acting exactly in the plane of the interface (ice–substrate). The type of break was in all instances "adhesive" (whether a monolayer is left on the substrate or ice, respectively, is not known). Even with stainless steel substrates, adhesive breaks were obtained until a temperature of approximately −13°C was reached, when the shear-strength versus temperature curve showed a sudden discontinuity and adhesive changed to cohesive breaks; Figure 1 shows this behavior. Hunsacker et al.[21] also found a similar break at approximately −12°C, using different types of joints. Raraty and Tabor[21] applying torque to their samples observed such discontinuity for ice annulus at −7°C. Thus this effect has been confirmed by several workers.

Adhesive breaks, which show a strong dependence on temperature, remain predominant for polymer substrates at temperatures even lower than −13°C. The effects of surface finish and rate of load application are of significance in this connection. At high rates of load application, strength increases linearly and quite steeply with time until a pointed maximum is reached, whereupon strength then decreases very rapidly. The "break" actually occurs at this maximum. The slope leading to the height and the maximum itself increase with increasing stress application, or for constant stress application with surface roughness of the substrate (Figure 2). However, for mirror-polished stainless steel or for optically flat quartz, stress–time plots change their characteristics if the rate of stress application is gradually decreased. The maxima become lower, the slopes become less steep, and the falling off of stress after the maximum becomes more gradual. Eventually, further decrease

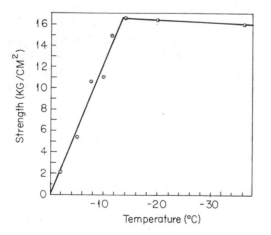

Figure 1. Strength as a function of temperature for snow-ice sandwiched between stainless steel disks, obtained by shear.[20] Cross-sectional area 1.54 cm², height of ice, 0.2 to 0.4 cm. Adhesive breaks down to −13°C; cohesive breaks below −13°C. Each point represents the average of at least 12 tests.

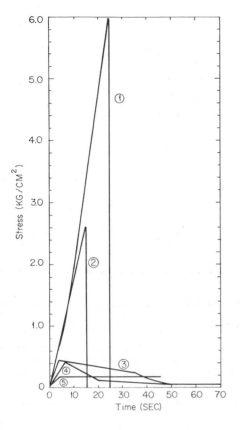

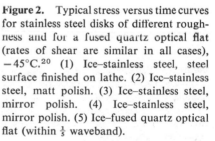

Figure 2. Typical stress versus time curves for stainless steel disks of different roughness and for a fused quartz optical flat (rates of shear are similar in all cases), $-45°C.$[20] (1) Ice–stainless steel, steel surface finished on lathe. (2) Ice–stainless steel, matt polish. (3) Ice–stainless steel, mirror polish. (4) Ice–stainless steel, mirror polish. (5) Ice–fused quartz optical flat (within $\frac{1}{5}$ waveband).

of the rates of stress application produces very low maxima and the stress remains constant and close to the maximum value for long periods of time. A very small force is sufficient to keep the ice sliding over the substrate surface. Even if the movement is stopped for some time, the rate is the same as before if the force is applied again; Figure 3 shows this effect.

It is, however, surprising that tensile strength tests with optical quartz plates as substrates give only cohesive breaks. Although at first this behavior of ice presented quite a puzzling feature, surveying the literature gave clues to its significance. The discussion presented here is condensed in many respects, and reference should be made to the original paper.[20]

As mentioned in the introduction, the story started with Michael Faraday (1850),[1] who observed more than one hundred years ago that near 0 °C snow sticks together (i.e., snowballs can be formed). He assumed a water layer on the surface of ice. As soon as such a layer touches that of another surface, the water layers freeze and the liquid disappears. The liquid layer was

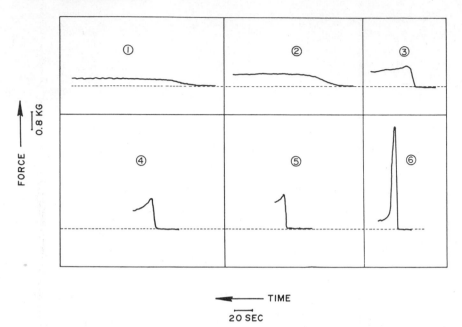

Figure 3. Typical recordings of force as a function of time at different rates of shear for snow-ice–fused quartz optical flat at $-4.5°C$.[20] (1) Rate 0.53×10^{-3} cm/sec; cross section 4.84 cm². (2) Rate 1.1×10^{-3} cm/sec; cross section 4.24 cm². (3) Rate 1.2×10^{-2} cm/sec; cross section 3.93 cm². (4) Rate 1.6×10^{-2} cm/sec; cross section 4.84 cm². (5) Rate 2.5×10^{-2} cm/sec; cross section 4.52 cm². (6) Rate 4.1×10^{-2} cm/sec; cross section 4.84 cm².

assumed to be stable at the ice–air interface. This phenomenon was termed by J. Tyndall[2] "regelation" (to become ice again). The concept was violently opposed by the Thomson brothers, J. Thomson[3] and W. Thomson[4] (later Lord Kelvin). The first had just elaborated pressure melting of ice on thermodynamic grounds (Clausius–Clapeyron). He maintained that pressure sufficient to melt some of the ice develops even when crystals touch each other only very slightly, because of the small area over which surfaces actually make contact; this leads eventually to the freezing together of the ice pieces. Faraday[1] performed a number of very ingenious experiments to disprove the pressure melting theory; also Helmholtz[23] took part in the controversy. However, at that time the different views could not be reconciled, for it was assumed that this liquid layer only exists at 0°C. The pressure-melting theory found its way into all the relevant textbooks and Faraday's views were forgotten.

The problem was not considered again for a century. Then Weyl[24] maintained, without performing any experiments and calculations, that at 0°C and even below this temperature, an ice surface does not have the expected random arrangement of water molecules (oxygen and hydrogen atoms alternately facing the air). He argued that a sample of ice acquires the minimum amount of free energy if all oxygen atoms face the air. This demands a reorganization of the surface molecules from a random to a nonrandom arrangement. Change in the uppermost layer of ice creates a disturbance in the underlying layers, forcing rearrangements that decay slowly with depth. Thus a transition layer ranging from bulk ice to liquid water at the very top is created on ice in this way. The disturbance may reach depths of about 100 Å into the ice crystal. Weyl advanced some known experimental facts in support of his views. The assumption of such a transition layer, even below 0°C, explains easily the phenomenon of regelation. It does not, of course, do away with the pressure-melting theory. Each case has to be considered to determine whether experimental conditions are such that pressure melting can occur or whether the transition layer is responsible for a particular experimental fact; in many cases both processes may be operative simultaneously.

The peculiar behavior of ice in tensile and shear tests, respectively, can now be satisfactorily accounted for. The following process is assumed to take place on applying a tensile force. Figure 4 shows ice adhering to a metal plate. A transition layer is present between substrate and bulk ice. This means that a

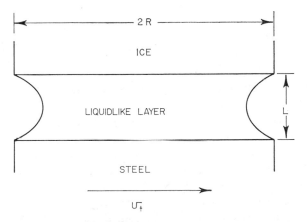

Figure 4. Liquidlike transition layer sandwiched between ice and stainless steel. Pressure difference Δp across curved surface is $\Delta p = 2\gamma t/L$, where γ_c is the surface tension of the transition layer of thickness L and v_t is the velocity of the stainless steel disk for shear experiments.[20]

transition layer is not only assumed to be present at ice–gas but also at ice–solid interfaces. Its thickness will be a function of temperature and of the nature of the substrate. This transition layer is indicated in Figure 4 and is assumed to wet the substrate completely. Thus there exists a pressure difference across the layer–air interface of the transition layer of a magnitude as follows:

$$(9) \qquad \Delta p = \gamma_t \left(\frac{1}{r_1} + \frac{1}{r_2} \right)$$

where Δp is the pressure difference across the air–transition layer interface, r_1 and r_2 are the radii of curvature of this interface, and γ_t its surface free energy.

One of the radii of curvature is quite large and can be neglected, hence (9) reduces to

$$(10) \qquad \Delta p = \frac{2\gamma_t}{d}$$

where d (L in Figure 4) is the diameter of the air–layer interface or the thickness of the transition layer. If $\gamma_t = 76.4$ dynes/cm and $d = 10^{-6}$ cm, then $\Delta p = 1.5 \times 10^8$ dynes/cm²; this value is larger than the bulk tensile strength of ice and the latter will break in cohesion long before the transition layer is ruptured.

The situation is quite different for shear experiments. Here the internal friction or the viscosity, respectively, of the transition layer has to be overcome. For a Newtonian liquid, one has

$$(11) \qquad \bar{S} = \frac{\eta}{d} v + a$$

where $\bar{S}$ is the average shear stress (adhesive strength), η the effective viscosity coefficient of the transition layer, d its thickness, and v the velocity of ice relative to the substrate, or vice versa, at time t; a is a constant indicating that the transition layer has a small yield point.

The shear experiments with ice very strongly suggest the existence of such a transition layer. It would be very difficult, indeed, to explain the "adhesion" results other than by the presence of such a layer; it decreases in thickness with decreasing temperature and eventually (at a low enough temperature) leads to cohesive instead of adhesive breaks. In addition, there is plenty of evidence from a variety of experiments and theoretical considerations for support of such a transition layer on ice.

As pointed out above, Weyl[24] was the first to take up the problem of the water and ice surfaces, respectively, after the controversy of the last century. He pointed out that electrical double layers are produced at the surfaces of crystals because of distortions of crystal lattices. In the case of sodium chloride, for instance, the anions are more polarizable than the cations. The surface-free energy is decreased by the displacement of anions with respect to cations. The electrical double layer is about 0.2μ thick in this case.

A similar process, involving dipoles, is supposed to take place in the ice crystal lattice. The free energy of ice is decreased by orienting all dipoles in such a way that all oxygen atoms are directed outward instead of having a random arrangement in the surface (see, however, Fletcher's revised paper[25] where the hydrogen atoms are assumed to face the air). A double layer is formed by this rearrangement. The consequence is that a disturbance is created in the ice surface layer extending about 50 to 100 Å into the ice, forming the transition layer. Liquid water also has a similar arrangement at its surface; of course, the disturbance in this case does not reach far into the liquid. An interesting survey entitled "The Depth of the Surface Zone of a Liquid," by Henniker,[26] is of relevance in this connection.

The first experimental evidence for the existence of a transition layer on ice was presented by Nakaya and Matsumoto in 1953.[27] They measured the force of two ice spheres "sticking together." No separation took place at the moment the spheres were pulled apart, but they rolled over each other before separation occurred. Such behavior is to be expected if the spheres have transition layers on their surfaces. More quantitative work was performed by Hosler, Jensen, and Goldshlak (1960),[28] who measured the force of separation as function of temperature and humidity. Figure 5 gives the arrangement of

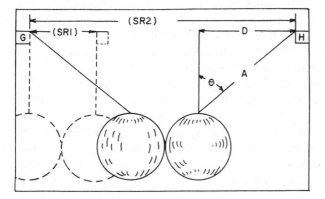

Figure 5. Initial and final positions of ice spheres.[28] Force of separation F = weight of sphere $\times \tan \theta$; the horizontal distance that one sphere has moved = $(SR2 - SR1)/2$ = D. L = length of suspending thread, hence $\tan \theta = D/L$.

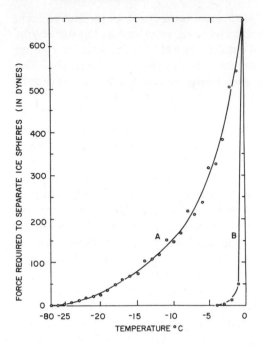

Figure 6. Separation force as a function of temperature:[28] *A*, water-vapor saturated atmosphere; *B*, dry atmosphere.

the spheres and Figure 6 shows the separating force as a function of temperature and humidity. In a dry atmosphere, the force vanishes at about $-4°C$, whereas in an atmosphere saturated with water vapor the separation force only disappears at about $-25°C$. These results indicate that the transition layer cannot fully develop in a dry atmosphere. That pressure melting is not involved here can be demonstrated by simple calculations.

A pressure P_{-5} of 600 atm is necessary to depress the melting point of ice by $-5°C$. If two spheres touch, the required contact area for this to happen is designated as A_{-5} cm^2; if the two spheres would touch each other under the same force at $-20°C$, the pressure would have to be P_{-20} or 1980 atm, and the relevant contact area has to be A_{-20} cm^2. Hence the forces of separation at these two temperatures are $F_{-5} = P_{-5}A_{-5}$ and $F_{-20} = P_{-20}A_{-20}$, respectively. Since $F_{-5} = F_{-20}$, one obtains

(12)
$$\frac{A_{-5}}{A_{-20}} = \frac{P_{-20}}{P_{-5}}$$

The contact area at $-5°C$ would have to be 3.2 times that at $-20°C$ for the same contact force in either case. The actual experimental ratio was 10.5. The actual contact areas were also computed, taking 17.4 kg/cm^2 as the tensile strength of bulk ice. The actual pressures on the contact areas were 3000 to 10,000 times smaller than those required for pressure melting to take place.

The first theoretical attempt to show the existence of the transition layer on water and ice was made by Fletcher.[29] The treatment was approximate only and was criticized by Watson-Tobin,[29] mainly on the grounds that dipole–dipole interaction between water molecules had been neglected. Recently Fletcher[25] revised his theory, taking account of the quadrupole moments of the water molecule and also of some new information on water structure. A number of quantum mechanical treatments of the water molecule result in a dipole moment of water that agrees with the experimental value (1.84×10^{-18} esu). Calculations can also be used to deduce the quadrupole moments of the water molecule ($Q_{xx} \simeq -6.5 \times 10^{-26}$; $Q_{yy} \simeq -5.3 \times 10^{-26}$; $Q_{zz} \simeq -5.6 \times 10^{-26}$ esu). The z-axis bisects the H–O–H angle in the plane of the molecule and the x-axis is at right angles to this plane. It is important that these moments be negative and of almost equal magnitude. The lowest energy configuration is achieved if the protons are directed outward. This is contrary to the assumption in Fletcher's first paper,[29] where oxygen was assumed to point outward. The energy differences between these two configurations (random and oriented) is of the order of -10^{-13} erg. However, in reality, this energy difference will be marked by adsorption of nonpolar molecules, such as air molecules, at the surface.

If it is nevertheless assumed that a fraction α_0 of the surface molecules have their dipoles directed out of the surface, then it can be calculated that, for n surface molecules per unit area ($n \simeq 1 \times 10^{15} \text{ cm}^{-2}$), the energy of the surface, due to this orientation, is larger than that of a random surface by

$$(13) \qquad\qquad \Delta U_1 = n(\alpha_0 - \tfrac{1}{2})\epsilon_1$$

where ϵ_1 is the energy difference for one molecule due to surface orientation ($\epsilon_1 \simeq -1 \times 10^{-14}$ erg); however, surface orientation can still take place. The orientation originates from long-range electrostatic interactions between water molecules caused by dipole and quadrupole moments. Just these moments were neglected in the first version of Fletcher's theory.[29]

It is clear that the free energy of the system in the new state must be lower than that in its former state if spontaneous orientation in a surface is to take place. There is a large amount of hydrogen bonding in liquid water and in ice. Hence this surface orientation must have an effect on the underlying layers owing to rupture and to formation of such bonds. An entropy loss will take place owing to the ordering of the surface structure, and this loss must be more than compensated for. The rearrangement of the dipoles in the surface will produce an electric field at right angles to it. The interaction produced by this orientation between molecular dipoles and quadrupoles must also be included in the free energy balance.

Rearrangement of the underlying surface layers (the first layer consists of oriented dipoles) will be facilitated if these lower layers are highly disordered, having a large number of defects, so that relaxation can take place. An estimate of the properties of the ice surface can be made by assuming as a first approximation a uniform surface zone of thickness h; this zone is in equilibrium with bulk ice. If this surface zone has a fraction α_0 of its molecules oriented, then the free energy resulting from this orientation can be calculated. In addition, the free energy of the bulk ice–(surface zone) interface has to be considered. Calculations indicate that at temperatures near the melting point, the equilibrium state of the surface has its lowest value, if it is covered with a liquid layer. The temperatures for the existence of such a transition layer ranges from about 0 to $-6°C$. Very near the melting point its thickness tends to infinity.

The existence of a surface potential on ice because of this dipole orientation was predicted by Fletcher.[25] The surface potential is 0 at low temperatures, but at higher temperatures a surface potential appears when the transition layer is formed. This potential will be of the order of 0.1 V; the exterior surface should be positive compared with that of the bulk ice. The transition layer should also have an effect on the surface conductivity of ice, the latter is expected to have a magnitude of about $10^{-8}/\Omega$. The calculations have actually been carried out assuming that the layer is uniform throughout; this is a simplified view and is certainly not true. Rather, there will be a gradual transition in properties from liquid water at the very surface to those of bulk ice at a certain distance below the surface.

The large surface conductivity predicted by Fletcher[25] is in fair agreement with experimental results by Jaccard.[30] The latter obtained $10^{-10}/\Omega$ at $-11°C$. Hence the layer persists approximately down to $-11°C$. The uncertainties in Fletcher's calculated values of the temperature range make the existence of the transition layer quite likely down to a temperature of $-11°C$ or even lower.

There is still more experimental evidence in favor of a transition layer on ice. Telford and Turner[31] carried out a detailed study of the migration of thin steel wires through ice in a range of temperatures from -0.5 to $-3.5°C$. Experiments of this kind had been performed repeatedly before, but not under such stringent conditions. The steel wire was 0.45 cm in diameter. The ice cubes had an edge 1 cm long and were polycrystalline; the load was constant at 2.1 kg, and the temperature was controlled to $\pm 0.02°C$. The movement of the wire was recorded; 1 cm of the chart corresponded to a migration of

4.0×10^{-3} cm. The migration of wires was measured in a number of ice cubes. A linear relation was obtained for the logarithmic plot of the migration-velocity versus temperature over a range of temperatures from -3.5 to $-0.7°C$. The velocities ranged from 10^{-7} to 10^{-6} cm/sec. An enormous increase in velocity occurred above $-0.7°C$: it became 200 times greater at about $-0.5°C$ than at below $-0.7°C$. This increase in velocity was attributed by the authors to pressure melting. Below about $-0.7°C$, however, pressure melting can no longer explain the experimental results. Since 4.5 atm is necessary to depress the melting point of ice to $-0.5°C$, the calculated pressure caused by the wire is simply not great enough. The authors explained this small rate of migration of the wire at $-0.7°C$ and below by assuming the existence of a transition layer between the wire surface and bulk ice. All previous workers in this field had completely missed this small movement of a wire through ice. If the transition layer is assumed to be Newtonian as a first approximation, then the slow motion through ice is given by the following expression:

(14)
$$v = \frac{F}{12\pi\eta} \left(\frac{d}{a}\right)^3 \text{ cm/sec}$$

where v is the velocity of the wire, F the acting load per unit length caused by the wire, η the viscosity of the transition layer, d its height, and a the radius of the wire.

The wire is assumed to move through the ice by squeezing the transition layer beneath it to its top, where eventually regelation takes place. Townsend and Vickery[32] and Nunn and Rowell[33] also performed experiments on migration of wires. Their results do not agree with the assumption of pressure melting made by Nye[34] in his theory of regelation.

A paper by Jellinek and Ibrahim[35] may be mentioned here. These authors sintered very small ice spheres (radius 0.5 μ), following surface area changes by the BET method over a range of temperatures. The experimental results are compatible with the assumption of a transition layer. Kingery[36] also performed sintering experiments that are in agreement with the presence of a transition layer. A paper by Mason, Bryant, and van der Heuvel[37] on growth habits and surface structure of ice crystals also leads to the assumption that a transition layer may exist on ice. Actually rough estimates can be obtained of the viscosity and thickness of the transition layer from the shear experiments carried out by Jellinek[20] and discussed above. If d varies from 10^{-5} to 10^{-6} cm at $-4.5°C$, estimated from the roughness of the substrate surface, a viscosity η is obtained for an interfacial layer adhering to a steel substrate of 70 to 700 poises and one adhering to a quartz substrate of 15 to 150 poises. For supercooled water at $-5°C$, η is 2.1×10^{-2} poises and for polycrystalline ice it is about 10^{14} poises.

3   Ice Growth and Diffusion

Growth of ice is also influenced by the transition layer; rapid surface diffusion takes place while growth occurs. Some preliminary work on diffusion of ions in an ice interface was carried out by Peck and Murriman.[38] The diffusion of radioactive sodium ions along an ice–aluminum interface was measured by these workers. Ice was frozen in a way that caused most of the dissolved air to be expelled. The substrate surface was thoroughly cleaned to remove all grease. Aluminum weighing cups were partly filled with distilled water, which was frozen at -5 or $-10°C$. The samples were allowed to stand for 24 hours, whereupon the cups were inverted and small holes were made in the centers of the bottom walls. Radioactive $^{22}Na^+$ ion solutions were injected through these holes in 1-μl droplets. The total activity was 0.05 μc in each case. Apparently diffusion took place mainly along the aluminum–ice interface. At the end of each diffusion experiment, the ice was sublimed, the aluminum bottom was cut into small sections, and their respective activities were measured. The solution of the diffusion equation is (Crank[39])

$$(15) \qquad C = \frac{M}{4\pi D_s t} \exp \left(\frac{-r^2}{4D_s t} \right)$$

where C is the activity at distance r from the point source (hole), M the total activity of the interface, D_s the surface diffusion coefficient in cm^2/sec, and t the diffusion time in seconds. Equation 15 can be expressed in logarithmic form:

$$(16) \qquad -\log \frac{C}{M} = \log (4\pi D_s t) + \frac{r^2}{4D_s t}$$

Hence from the slope of the plot of log C/M versus r^2, the value for D_s can be derived. The following average values for the diffusion coefficients were obtained: $-5°C \sim 1.7 \times 10^{-7} cm^2/sec$, $-10°C \sim 7.2 \times 10^{-8} cm^2/sec$. These results are preliminary. The energy of activation is 27 kcal/mole, a relatively large value. In comparison, the self-diffusion coefficient in water[55] is $1.4 \times 10^{-5} cm^2/sec$ at 5°C and the self-diffusion coefficient for tritons in single ice crystals[40] is about $10^{-4} cm^2/sec$ at $-7°C$; its energy of activation is 14 to 16 kcal/mole. Recent experiments on diffusion of ^{22}Na ions in polycrystalline ice by Jellinek and Chatterjee[41] are discussed below. The data for surface diffusion are not inconsistent with the assumption of a transition layer on ice. The high energy of activation probably results from the exponential decrease of δ, the thickness of the layer, with absolute temperature, which contributes to the overall energy of activation.

Recently detailed studies were carried out by Jellinek and co-workers on grain growth and diffusion in polycrystalline ice. Grain growth and grain size distributions were investigated as function of temperature and electrolyte concentration by Jellinek and Gouda,[42] and the influence of tensile stress on the growth rate was also studied.[47]

Growth of crystallites (grains) in polycrystalline ice was followed under a polarizing microscope. The ice was prepared under rigidly standardized conditions on microscope slides. The resulting ice layers were about 0.5 mm thick. Various locations of the ice surface were photographed at definite growth times, temperatures, electrolyte concentrations and tensile stresses. The magnified photographs were evaluated according to thin section analysis,[54] which gives the size of grains in terms of three-dimensional equivalent sphere diameters. From these data, volume percentage distribution curves can be constructed and volume average diameters can be obtained as a function of time and various other parameters.

It is interesting to note that the growth law obtained for ice is of the same type as that found for polycrystalline metals. The volume-average diameter $\bar{b}$ in the case of ice growth is given by

$$(17) \qquad\qquad \bar{b} = K_{exp}t^n$$

where K_{exp} and n are constants, characteristic of the ice. If $\bar{b}$ is plotted versus t in logarithmic terms, a straight line results whose slope is n, and $\log \bar{b} = \log K_{exp}$ for $t = 1$. Typical plots are shown in Figure 7 and some data are collected in Table 2.

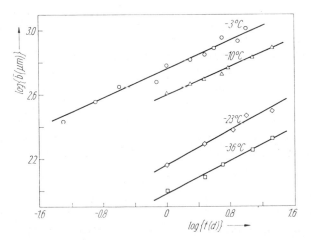

Figure 7. Pure ice:[42] number-average diameter $\bar{b}$ as function of time.

Table 2   Grain Growth Parameters Obtained from (17)

Temperature (°C)	Pure Ice		Doped Ice (1.7×10^{-2} M NaCl)	
	$10^2 K_{exp}$(cm/days^n)	n	$10^2 K_{exp}$(cm/days^n)	n
−3.0	4.85	0.31	5.69	0.23
−10.0	3.58	0.25	3.95	0.22
−23.0	1.95	0.35	1.46	0.27
−36.0	1.12	0.29	0.94	0.26
		Mean 0.30		Mean 0.25

From Jellinek and Gouda, *Phys. Stat. Sol.*, **31**, 413 (1969).

The data can be represented by Arrhenius equations as follows:

pure ice

$$(18) \qquad K_{exp} = 1.56 \times 10^3 \exp\left(-\frac{5600}{RT}\right) \text{cm/days}^n$$

ice made from 1.7×10^{-2} M NaCl solutions

$$(19) \qquad K_{exp} = 3.58 \times 10^4 \exp\left(-\frac{7200}{RT}\right) \text{cm/days}^n$$

Actually the growth rate is faster for initially pure ice than for doped ice above $-10°C$, and vice versa below this temperature.

The grain size volume percentage distribution curves start as normal curves; however, after some growth time, they develop positively skewed tails. The amount of time that passes before these curves start to deviate from normal ones depends on temperature (i.e., rate of growth).

The low energies of activation and small preexponential factors for grain growth indicate that diffusion is involved in the growth process. It is likely that, on a molecular scale, translational and rotational diffusions occur. Probably several water molecules constitute the diffusing unit. This was also indicated by sintering and diffusion experiments, discussed below.

The driving cause for grain growth is the free interfacial energy inherent in the ice–ice$_{grain boundary}$ interface; the growth rate can be expressed by the following relation:[43]

$$(20) \qquad \frac{d\bar{b}}{dt} = k\,\Delta F$$

where k is a growth rate constant and ΔF the free energy in the grain boundary–ice interfacial region. ΔF is assumed to be directly proportional to $2\gamma/\bar{b}$, where γ is the interfacial tension or free energy of the ice–ice$_{grain\ boundary}$ interface. If $\bar{b}$ is small at $t = 0$ and growth is fast, (20) leads to

$$(21) \qquad\qquad \bar{b} = kt^{1/2}$$

If $\bar{b}$ at $t = 0$ is not negligibly small, (21) becomes

$$(22) \qquad\qquad \bar{b}^2 = \bar{b}_0{}^2 + kt$$

which has actually been verified for "South Polar Firn," whose grains have grown for hundreds of years.[44]

In the case of short-term experiments with artificial ice, $\bar{b}_0$ is small. Moreover, n for such experiments is usually smaller than $\frac{1}{2}$; in the present case it ranges from 0.25 to 0.3.

The rate of grain growth can also be treated according to the theory of absolute reaction rates,[45] modified somewhat by Turnbull.[46] An equation can be deduced as follows:

$$(23) \qquad\qquad \frac{d\bar{b}}{dt} = \frac{e\bar{k}T\lambda\,\Delta F}{hRT}\, e^{\Delta S^*/R} e^{-E_a/RT}$$

and also

$$(24) \qquad\qquad \frac{d\bar{b}}{dt} = \frac{D_t}{\lambda}\ \text{cm/sec}$$

where $d\bar{b}/dt$ is the growth rate of the volume-average diameter for equivalent spheres, e the base of the natural logarithms, λ the distance between two successive equilibrium positions of the diffusing unit (usually taken as equal to the diameter of the diffusing unit), h and $\bar{k}$ Planck's and Boltzmann's constants, respectively; ΔF the molar free energy difference in the grain boundary region, R the ideal gas constant, T the absolute temperature, ΔS^* the entropy of activation, E_a the Arrhenius energy of activation, and D_t the diffusion coefficient at time t.

If only the very initial stage of grain growth is considered (far from equilibrium), then the reverse of the diffusion process is negligible and (24) can be written for $t = 1$ sec as

$$(25) \qquad \left(\frac{d\bar{b}}{dt}\right)_{t=1\ sec} = Kn(1\ \text{sec})^{n-1} = \frac{\bar{k}T\lambda}{h}\, e^{-\Delta F_f^*/RT}$$

$$= \frac{e\bar{k}T\lambda}{h}\, e^{\Delta S_f^*/RT} e^{-E_a/RT} = \frac{D_f}{\lambda}\ \text{cm/sec}$$

where ΔF_f^* and ΔS_f^* are the free energy and the entropy of activation, respectively, for the forward reaction; D_f is the diffusion coefficient for the forward growth direction only.

Strictly, (25) should also contain the driving pressure (stress), and this is discussed later. However, as long as $2\bar{k}T \gg$ (stress × molecular volume) of the diffusing unit, stress can be neglected in (25). The parameters in (25) have been evaluated.[47] The entropies in both cases (pure and doped ice) are negative; this suggests that the activated complex has a more ordered structure than that of the reactant. This "order" decreases in presence of sodium chloride. The free energy of activation is 10.6 to 9.8 kcal/mole for pure ice over a range of temperatures from -3 to $-36°C$, and 10.3 to 9.96 kcal/mole for the same range of temperatures for doped ice; the amounts are quite similar to the heat of vaporization of water (10.95 kcal/mole at $-20°C$). This indicates that hydrogen bonds are broken during the diffusion process. Evaluation[39,41] of λ indicates that the diffusing unit is not one water molecule but consists of several hundred molecules.

As mentioned above, sintering of ice spheres was studied by Jellinek and Ibrahim.[35] For the first time, the BET method was used for this purpose instead of observing two ice spheres growing together under a microscope. Ice powder consisting of spheres of an average diameter of $1\ \mu$ was prepared and its specific surface area was determined at liquid nitrogen temperature. The powder was then kept for a desired time interval at a temperature high enough for sintering to take place. After this interval the surface area was measured again at liquid nitrogen temperature and eventually the area was obtained as function of sintering time and temperature. Further evaluation yielded the number-average radius of the spheres as function of time. The actual cause for sintering of such small spheres (for larger spheres the mechanism changes completely) is the surface free energy, γ, of the ice or that of the transition layer on ice. The requisite equation is

$$(26) \qquad \frac{X^2}{R} = \frac{3\gamma t}{2\eta}$$

where X is the large radius of curvature of the neck joining two spheres, R the average radius of the spheres, η the bulk ice viscosity, and t the sintering time. This expression is similar to that for grain growth; see (21).

Three ranges of energies of activation were found for sintering: -8 to $-13°C$: 73.8 kcal/mole; -13 to $-28°C$: 7.2 kcal/mole; -28 to $-35°C$: 20 kcal/mole. The first energy of activation has an abnormally large preexponential factor, suggesting the presence of a transition layer. This, of course, is not present in the case of a bulk ice–grain boundary interface. The second temperature range, however, shows a similar magnitude for sintering as was found for grain growth, which was 5.6 kcal/mole (pure ice), 7.2 kcal/mole for sintering. The entropies of activation for sintering show a number of interesting features; the pressure difference caused by interfacial tension between the growing neck between two small spheres and air reaches

200 kg/cm^2. The diffusion unit calculated from sintering experiments amounts to about 30 water molecules. Here, again, the diffusing unit turns out to be larger than one water molecule.

It was noted above that grain growth was also studied when tensile (compressive) stress was applied to the ice specimens.[47] In this case, the stress has to be included in (17) and the growth equation becomes

$$(27) \qquad \log \bar{b} = (mS + \log k_0) + n \log t$$

where m is the slope of the log (K_{exp}) versus S (stress) plot and k_0 is the rate constant when $S = 0$.

The Arrhenius energy of activation and the preexponential factor increase with increasing tensile stress:[45] 0 kg/cm^2: $A = 6.15 \times 10^5$, $E = 8.7$ kcal/mole; 5.9 kg/cm^2: $A = 2.86 \times 10^6$, $E = 9.5$ kcal/mole; 12 kg/cm^2: $A = 4.8 \times 10^{-7}$, $E = 9.6$ kg/cm^2; temperature range -3.82 to $-21.0°$C. The value of A for 0 kg/cm^2 differs from that of the previous paper (Ref. 42).

The equation for the forward rate according to the theory of absolute reaction rates becomes for this case

$$(28) \qquad \left(\frac{d\bar{b}}{dt}\right)_{t=1 \text{ sec}} = Kn(1 \text{ sec})^{n-1} = \frac{2\lambda \bar{k} T}{\lambda_1 Sh} e^{-(\Delta F_{f,S=0}^{\ddagger} - cS)/RT}$$

$$= \frac{2e\lambda \bar{k} T}{\lambda_1 Sh} e^{+\Delta S_f^{\ddagger}/R} e^{-E_a/RT} = \frac{D_f}{\lambda} \text{ cm/sec}$$

where S is the tensile stress and c a constant of proportionality; actually $2c = \lambda \lambda_2 \lambda_3 N_A$ is the molar volume of the diffusing unit, and N_A Avogadro's number. Straight lines are obtained by plotting $(\Delta F_{f\ S=0}^{*} - cS)$ versus stress (Figure 8). The slopes are equal to half the molar volume of the diffusing unit. Here, again, units larger than one water molecule are obtained (78 and 53 molecules at -3.8 and $-16°$C, respectively). Applied tensile stress increases the rate of grain growth by lowering the free energy of activation for the diffusion process of water molecules through grain boundaries. The entropy of activation is again negative, but increases with increasing stress; this suggests that structural order of the activated complex decreases with increasing stress. (λ_{11} λ_2 and λ_3 are distances between neighboring molecules in three dimensions; ref. 45 p. 481).

One of the first attempts to measure diffusion in polycrystalline ice was made by Jellinek and co-workers.[41] Studies of diffusion in single ice crystals, however, have been made repeatedly.[40] The results of these workers lead to diffusion coefficients for single crystals of similar order of magnitude (2 to 3) $\times 10^{-11}$ cm^2/sec at $-10°$C and to similar energies of activation (ca. 14.5 kcal/mole for self-diffusion of ^{18}O, ^{2}H, ^{3}H, etc.). HF is an exception because

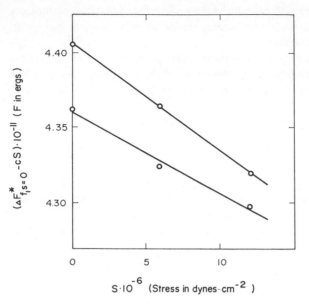

Figure 8. Free energy of activation versus stress at various temperatures (slope = $\frac{1}{2}$ molar volume of diffusing unit):[47] open circles: $-3.8°C$; solid circles: $-16.0°C$.

of its relatively great solubility in ice. A vacancy or an interstitial mechanism is generally assumed for the diffusion process. The formation energy for a vacancy is about 12 kcal/mole, leading to an estimate of the concentration of vacancies of about 10^{12} cm^{-3} at $-10°C$. Onsager and Runnels (1963)[48] considered the diffusion of interstitial water molecules as the mechanism of the process. The concentration of such molecules was estimated to be about 10^{10} to 10^{11} cm^{-3}. The formation energy for an interstitial molecule is about 14 to 15 kcal/mole. (Some further details can be found in Fletcher's book.[29]) Jellinek and Juznic[41] studied the diffusion of radioactive $^{137}Cs^+$ ions into polycrystalline ice, and Jellinek and Chatterjee[41] investigated that of radioactive $^{22}Na^+$. The diffusion coefficients were measured in initially pure polycrystalline ice and in ice doped with the respective chloride salt. Cylinders of polycrystalline ice were prepared and small amounts of radioactive aqueous solutions were frozen to the tops of these cylinders. After definite time intervals, the ice specimens were cut into thin disks and these were analyzed for their radioactivity. Diffusion coefficients were deduced from such measurements. Results indicate that similar to the case of polycrystalline metals, volume (lattice) diffusion and grain boundary diffusion take place. The logarithms of the radioactive counts/min, A, were plotted versus distance from the top of the cylinders first against the square of the distance, X^2, and second versus the first power of X. The volume (lattice) diffusion coefficient, D_L, can be derived from the first type of plot using a relation as follows:

$$(29) \qquad A = \frac{1}{S} \frac{M}{(\pi D_L t)^{\frac{1}{2}}} \exp -\frac{X^2}{4 D_L t}$$

where S is the cross-sectional area of the cylinder and M the total radio-activity at the top of the cylinder at time $t = 0$. Fisher[49] has given a derivation, yielding an equation for the evaluation of grain boundary diffusion, D_B:

$$(30) \qquad \frac{D_L}{D_B} = \frac{\delta(\pi D_L t)^{\frac{1}{2}}}{2} \left(\frac{d \ln A}{dX} \right)^2$$

where D_B is the grain boundary diffusion coefficient, and D_L the volume diffusion coefficient measured on the same polycrystal, and δ the grain boundary thickness. Actual D_L values obtained for single crystals at the same temperature[40] give more reasonable values for D_B.

The D_L values were derived from the initial slopes of the log A versus X^2 plots ($X \leq 0.3$ mm) and the products $D_B\delta$ are derived from the final slopes

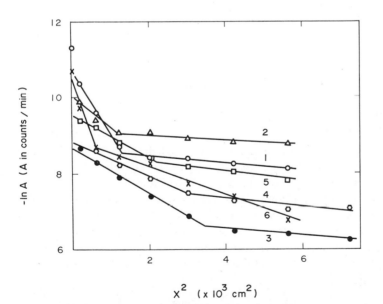

Figure 9. Typical log A (radioactive counts/min) versus X^2 plots:[41] (1) $-18°C$, salinity $s = 5 \times 10^{-3}$ mole/liter, $t = 7.6 \times 10^4$ sec; (2) $-24°C$, $s = 1 \times 10^{-3}$ mole/liter, $t = 1.8 \times 10^5$ sec; (3) $-12°C$, $s = 2 \times 10^{-3}$ mole/liter, $t = 7.5 \times 10^4$ sec; (4) $-6°C$, $s = 5 \times 10^{-3}$ mole/liter, $t = 7.5 \times 10^4$ sec; (5) $-6°C$, $s = 10^{-2}$ mole/liter, $t = 2.4 \times 10^4$ sec; (6) $-12°C$, $s = 0$ mole/liter, $t = 2.0 \times 10^5$ sec.

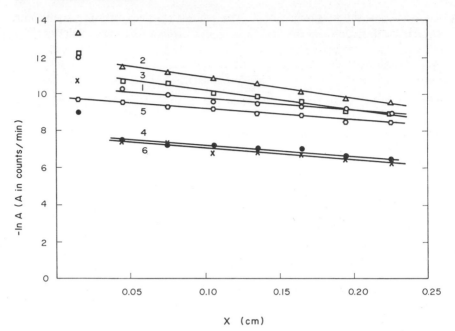

Figure 10. Typical log A versus X plots:[71] (1) $-24°C$, $s = 10^{-2}$ mole/liter, $t = 7.9 \times 10^4$ sec; (2) $-24°C$, $s = 10^{-3}$ mole/liter; $t = 1.3 \times 10^5$ sec; (3) $-18°C$, $s = 10^{-3}$ mole/liter, $t = 1.96 \times 10^5$ sec; (4) $-12°C$, $s = 10^{-3}$ mole/liter, $t = 3.2 \times 10^4$ sec; (5) $-6°C$, $s = 2 \times 10^{-3}$ mole/liter, $t = 1.4 \times 10^5$ sec; (6) $-12°C$, $s = 0$ mole/liter, $t = 7.9 \times 10^4$ sec.

of the log A versus X curves as function of temperature and salt concentration. In the latter case, $D_B\delta$ was evaluated with the help of D_L values obtained from the same polycrystalline ice and also by taking D_L values for diffusion of tritons in single crystals, obtained by Dengel and Riehl[40] (2×10^{-11} cm²/sec $\pm$ 10% at $-7°C$). Figures 9 and 10 give some typical plots for $^{22}Na^+$ ions. Arrhenius parameters were also obtained from the experimental data.

Both grain boundary and lattice diffusion can be treated by the theory of absolute reaction rates, which presupposes a vacancy mechanism. Eventually, the following equation is obtained (instead of $D_B\delta$, D_L and D_B, respectively, can also be evaluated):

$$(31) \qquad D_B\delta = \frac{\delta e \bar{k} T \lambda^2}{h} e^{\Delta S^*/R} e^{-\Delta H^*/RT} = \frac{\delta e \bar{k} T \lambda^2}{h} e^{-\Delta F^*/RT} \text{ cm}^2/\text{sec}$$

The symbols in (31) are the same as in (23), and $\Delta H^* = E_a - RT$ where E_a is the Arrhenius energy of activation.

Wang et al.[50] found that the energies of activation for viscous flow and dielectric relaxation in ice are similar to the energy of activation for self-diffusion in water and concluded that all three processes have essentially the same underlying mechanism. It is likely that the mechanism of diffusion in grain boundaries (coefficients $D_B\delta$) is similar in the case of ice to those for viscous flow and dielectric relaxation and probably also to that for electrical conductivity. The energies of activation and their standard deviations for grain boundary diffusion ($D_B\delta$) in initially pure ice are 15.9 ± 0.6 kcal/mole for $^{22}Na^+$ ions and 15.3 ± 3.2 kcal/mole for $^{137}Cs^+$ ions, respectively. From studies of the viscoelastic properties of polycrystalline ice, Jellinek and Brill[51] obtained 16.1 kcal/mole, assuming that slip takes place in grain boundaries. The grain boundary thickness is contained implicitly in the viscosity values.

The energy of activation for dielectric relaxation has to be compared with the energy of activation for volume diffusion, D_L, which is 8.8 ± 3.5 kcal/mole for Cs^+ ions and 9.8 ± 0.4 kcal/mole for Na^+ ions. The energy of activation for dielectric relaxation is 13.2 kcal/mole, which is not appreciably larger than the values for diffusion; however, this value was obtained for pure ice, whereas in the case of diffusion, the ice is always doped to a certain extent, even if it is initially pure. Kuroiwa[52] has very carefully determined the energy of activation for mechanical relaxation of pure polycrystalline ice (13.1 kcal/mole, $\tau_0 - 6.9 \times 10^{-16}$ sec) and found it equal to that for dielectric relaxation as determined by Auty and Cole.[55] As a matter of fact, data obtained for electrical and mechanical relaxation, respectively, fall on the same Arrhenius plots. The influence of various electrolytes was also studied. Mechanical and electrical relaxation are believed to be caused by proton migration through lattice defects. Doped ice has many more such defects than pure ice.

In order to obtain grain boundary diffusion coefficients without the grain boundary width δ, calculations have to be carried out utilizing the $NaCl-H_2O$ phase diagram and assuming that most of the salt is located in the grain boundary of the ice sample.[53] This seems to be true for ice of low salinity (up to about 0.1%). Hence a certain proportion of the water in a saline solution is needed to form the grain boundary if this solution is frozen, forming polycrystalline ice; all the salt accumulates in this boundary. Salinity of ice is defined here as grams of salt per 1000 g of saline melt (i.e., %₀). It may be assumed that one gram of saline solution has a salinity of s, and that the corresponding polycrystalline ice has a grain boundary of salinity p and a volume-average grain diameter of $\bar{b}$ cm, which can be obtained experimentally.[42,54]

Furthermore, X g are needed to produce the saline solution that constitutes the grain boundary according to the phase diagram. The grain boundary is assumed to be a liquid solution above and a solid one below the eutectic temperature. Hence $pX/1000 = s/1000$ g of salt are contained in the boundary. The grain boundary volume is $s/(p\rho_{gb,T})$ cm^3 if $\rho_{gb,T}$ is the boundary density in g/cm^3 at the absolute temperature of $T°K$. If $\rho_{i,T}$ is the density of the ice grains at the same temperature, then their total volume in the ice specimen is

$$\frac{1 - X}{\rho_{i,T}} = \frac{1 - s/p}{\rho_{i,T}}$$

Evaluation yields for ice containing grains considered as equivalent cubes of edge length $\bar{b}$ ($\delta \ll \bar{b}$), an expression as follows:

$$(32) \qquad \delta = \frac{s\bar{b}\rho_{i,T}}{3pp_{gb,T}} \text{ cm}$$

If grains are assumed to be equivalent spheres, δ equals twice the grain boundary thickness or if they are columns of quadratic cross section of edge length $\bar{b}$ ($\delta \ll \bar{b}$), then 3 in either case has to be replaced by 2 in (32). The grain boundary thickness is directly proportional to the salinity $\bar{s}$ and the volume-average grain diameter $\bar{b}$.

The volume-average grain diameter $\bar{b}$ for ice doped with 1‰ NaCl was measured by Jellinek and Gouda[42] for a range of temperatures from -3 to $-30°C$ and found to be an exponential function of the inverse absolute temperature. Hence the grain boundary width δ must also be an exponential function of the inverse absolute temperature. In the case under discussion ($s = 1‰$ or $1.7 \times 10^{-2} M$), the relation is as follows (spheres):

$$(33) \qquad \delta = \frac{s\rho_{i,T}t^n}{2pp_{b,T}} 3.56 \times 10^8 \exp\left(-\frac{7200}{RT}\right) \mu m$$

$\bar{b}$ is also expressed in μm in (33).

For example, the volume-average grain diameter after twelve days' growth at $-3°C$ was $\bar{b} = 1008 \ \mu m$ (after a half-day it was only 396 μm). The calculated double grain boundary thickness after 12 days was $\delta = 0.29 \ \mu m$ ($p = 64‰$ obtained from the phase diagram) and $s = 0.0586‰$ ($\rho_{i,T} = 0.9996$ g/cm^3, $\rho_{b,T} = 1.0494$ g/cm^3); after a half-day $\delta = 0.12 \ \mu m$. The corresponding values for a salinity $s = 0.586‰$ are 3.06 and 1.3 μm, respectively. It was found that the diffusion coefficients for volume and grain boundary diffusion for Cs$^+$ and Na$^+$ ions, respectively, are of the same order of magnitude. The values for Na$^+$ ions seem to be somewhat larger. The lattice diffusion constants for either case are smaller than those for single crystals, probably because there are many more defects per cubic centimeter in small grains than in single crystals. The corresponding energies of activation are approximately 9.4 kcal/mole for Na$^+$ ions and 8.8 kcal/mole for Cs$^+$ ions, whereas those for single crystals range from 13 to 16 kcal/mole. Grain boundary diffusion for either ion is faster than the corresponding lattice or volume diffusion. The energies of activation increase in either case with salinity.

The energies of activation for D_B values, in the case of Na^+ ions, are about half of those for $D_B\delta$. This is because log δ is linearly dependent on $1/T$ and δ contributes its share to the total energy of activation.

In fact, the D_B values for Na^+ ions evaluated with D_L values from single crystals[45] agree fairly well with diffusion coefficients for Na^+ ions in water at 0°C and infinite dilution: Na^+ in $H_2O \sim D = 0.64 \times 10^{-5}$ cm²/sec and NaCl in $H_2O \sim D = 0.78 \times 10^{-5}$ cm²/sec at 0°C. The grain boundary diffusion coefficients found in this work for Na^+ ions in ice at -6°C and 10^{-3} M NaCl is $D = 9.7 \times 10^{-5}$ cm²/sec.

It was indicated above that the mechanism of viscous flow (the grain boundary width δ is implicit in the viscosity coefficient) in polycrystalline ice may be similar to the diffusion process in grain boundaries; also the dielectric relaxation process[55] may be similar to the diffusion process in the grain boundary and to lattice diffusion, each making its own contribution to the energy of activation for dielectric relaxation. The energy of activation for the D_B values for Na^+ ions (initially pure ice) are actually of the same order of magnitude as those for viscous flow of water (5.5 kcal/mole).

It should also be pointed out that there is no discontinuity in the Arrhenius plot when temperatures below the eutectic temperature for NaCl are reached. This indicates that the grain boundary in polycrystalline ice does not have an ice structure, nor is it an aqueous saline solution; rather, it has properties lying between those of ice and water.

It is also of interest[61] that mechanical relaxation in polycrystalline ice[52] shows clearly the influence of grain boundaries. Energies of activation for this process of 60.0 kcal/mole were obtained by Kuroiwa[61] for pure ice and ranging to 30.0 kcal/mole for impure ice. The procedure adopted by this author for the evaluation of energies of activation is open to doubt. The conclusion is reached that there is a kind of quasi-viscous flow in the grain boundary, which is increased by impurities. It is likely that the mechanism of d-c electrical conductance in ice[56] is also similar to that of diffusion. The energy of activation amounting to 11 kcal/mole is near the energy of activation for volume diffusion for Cs^+ and Na^+ ions: 8.8 and 9.8 kcal/mole, respectively.

4 Surface Conductivity

Kopp[57] measured the electrical conductivity of snow and reached the conclusion that a transition layer should be postulated in order to account for the experimental results.

Experiments concerning the d-c surface conductivity of ice were carried out by Jaccard.[30] His measurements indicate a surface conductance for pure ice of about $10^{-10}/\Omega$ at $-11°C$. This value is equivalent to the conductance of bulk ice, 1 to 4 mm thick. The d-c surface conductance results are in accordance with the assumption of a transition layer on ice.

The surface conductivity of ice has also been measured by Bullemer and Riehl.[58] They found a fairly large Arrhenius energy of activation of 32.2 kcal/mole for temperatures above $-11°C$. Fletcher[25,29] tried to explain this large value by the assumption that a transition layer exists on ice at $-10°C$ and above. High concentrations of H_3O^+ ions and OH^- ions were calculated in this layer. The energy of activation for change in layer thickness with temperature must be added to the energy of activation for the conductivity; that is, it is included in the experimental value for the energy of activation.

Camp and co-workers[59] also performed experiments concerning the electrical conductance of ice. Workers in the past found large discrepancies in energies of activation. Three ranges of Arrhenius energies of activation can be distinguished: low values of about 7 kcal/mole, values of intermediate magnitude of about 13 kcal/mole, and high values of about 20 kcal/mole or more. The low range is thought to be owing to impurities. One of the other two groups of energy values is probably the result of surface conductance. The authors conclude, on the basis of their data, that the energy of activation (13.4 ± 0.2 kcal/mole) is caused by two processes: (a) surface conductance and (b) the exponential decrease in the transition layer thickness on the surface of ice contributing to the energy of activation. The assumption of the transition layer leads to an energy of activation for surface conductance which is theoretically reasonable. An approximate thickness of 10 Å at $-10°C$ is sufficient to obtain reasonable energies of activation for surface conductance. It is of interest to note that the thermoelectric effect in ice was found to be much greater than expected above $-7°C$. This was assumed to be due to a surface layer (transition layer) on ice.[60]

5 Freezing Potentials

The field of freezing potentials has grown quite extensively and is treated briefly here. There are a number of reviews available in the literature (e.g., Gross,[60] Drost-Hansen,[61] and Pruppacher[62]). Costa Riberro[30] observed in 1942 that the interface of ice with dilute electrolyte solution became electrically charged on freezing or melting. Workman and Reynolds,[64] who made the same discovery independently, investigated this phenomenon—often referred to as the Workman–Reynolds effect—more fully. Large potentials with a magnitude of about 30 V appear in the ice–solution interface on freezing of dilute aqueous solutions (ca. 10^{-4} mole/liter). It is assumed that ions are incorporated differentially into the ice while freezing takes place, causing an electric charge; the solution assumes the opposite charge. The ions have to overcome an appreciable energy barrier (30 V is equivalent to 70 kcal/mole). This large barrier cannot be cleared in one jump, but it has to be presumed that the barrier path is relatively long and consists of several steps. Single crystals of ice have charge transfers about 100 times as large as polycrystalline material. The c-axis also has an appreciable effect—the charge transfer is about 10 times larger in the c-direction than in the direction of the basal plane. The effect does not appear to be dependent on freezing rate. Three distinct aspects of the mechanism of the Workman–Reynolds effect concerning ions have to be considered: (a) differential adsorption at the phase boundary, (b) differential incorporation, and (c) differential diffusion away from the grain boundary.

An important question arises in this connection. Is the charge separation caused by "chemical" adsorption of ions at the ice–solution interface initially, or is it caused by polar orientation of water molecules? Workman and Reynolds prefer the latter view. Anions and cations are incorporated at different rates. Hence a liquid boundary layer of high ion concentration is formed rapidly at the interface. Since the diffusion coefficients for ions differ, those that diffuse slowly will accumulate at the interface. The charge separation is dependent on concentration. Truby,[65] who observed the advancing ice–solution interface with an electron microscope, found a hexagonal honeycomb microstructure whose cells were 0.5 to 20 μm in diameter. The center of each cell represents a dislocation that may result from strain. Here again the concept of the transition layer hovers in the background. Electron microscopy will most likely reveal only the lower layers near to bulk ice.

Although the ions F^-, Li^+, Na^+, and K^+ could, in principle, fit easily into the ice lattice, it appears that size does not play a significant role; electronegativity is the decisive factor. The electronegativity of halides is relatively large, that of alkali metals is small. The electronegativity of oxygen is large and that of the F^- ion still larger; hence the latter can replace oxygen in the ice lattice. The type of cation does not have a great influence on the freezing potential, and its electronegativity only changes little with its type. Drost-Hansen[61] has given quite an interesting qualitative picture of the state of affairs when ions are incorporated into ice. This picture (see Figure 11) should only be considered as a rough illustration of the processes involved, for in many respects these are not yet fully understood. The microstructure in the ice–solution interface modifies appreciably the mechanical properties of the ice near the interface, leading to a weakening of ice. This is also important in connection with ice adhesion.

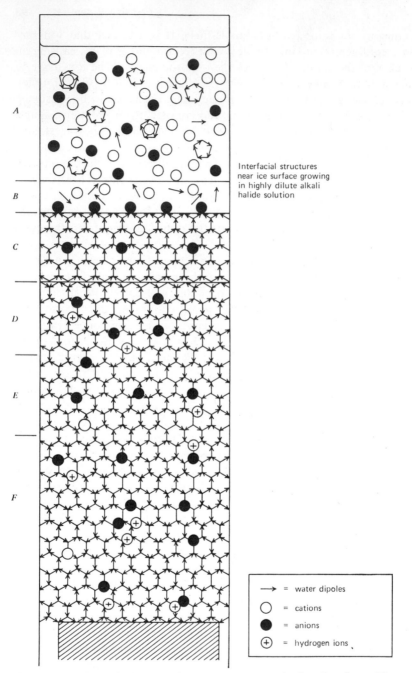

Interfacial structures near ice surface growing in highly dilute alkali halide solution

Legend:
→ = water dipoles
○ = cations
● = anions
⊕ = hydrogen ions

Figure 11. Schematic diagram of an advancing water–ice interface formed from a dilute solution of an alkali-halide.[61]

6   Ice Growth Habits on Solid Surfaces

Mason and co-workers[37] have studied in detail the growth habits of ice crystals and the underlying causes for the various habits obtained. They also refer to previous work in this very large area. The beautiful experiments by Nakaya[66] should be especially mentioned in this necessarily brief review.

It has been known for some time that molecules adsorbed on a surface are quite mobile.[67] The mobility of water molecules adsorbed on solid surfaces, including ice, is considered to be the underlying cause for the existence of different crystal habits under a variety of conditions.

Very characteristic crystal habits can always be obtained in well-defined temperature ranges: 0 to $-3°C$—thin hexagonal plates; -5 to $-8°C$—hollow prismatic columns; -8 to $-12°C$—hexagonal plates; -12 to $16°C$—dendritic, fernlike crystals; -16 to $-25°C$—hexagonal plates; -25 to $-50°C$—hollow, prismatic columns. Large variations in air supersaturation have no influence on crystal habits. However, supersaturation affects the anisotropy of the crystals and some of their secondary characteristics (e.g., thickness of plates, formation of solid or hollow prisms). Replacing air by helium, hydrogen, or carbon dioxide influences the rate of growth, but not the crystal habit. All experimental data obtained by these authors led them to conclude that migration of water molecules adsorbed on the surface of ice is determining the crystal habit. They also mention the possibility of a transition layer on ice, which would account for the rapid movement of water molecules on the surface. Kobayashi,[68] too, studied growth habits of ice crystals and came to conclusions similar to those of Mason et al.[37]

The type of migration of water molecules of importance here is actually confined to molecules within some characteristic distance from a growth step to which they can migrate; there they become incorporated into the ice lattice. The step height is of the order of 200 to 1000 Å; that is, the step consists of a few hundred molecular water layers.

Details of the process are not well understood. The main conclusion, however, is that the crystal habit is conditioned by variations in the mean displacement of adsorbed water molecules and by the relative rates of surface diffusion on basal and prism faces, respectively. The crystal habit is only dependent on temperature, but the growth rate is also a function of the degree of water supersaturation of the air and influences secondary characteristics of crystals. However, the observed temperature dependence of the diffusion length is not in agreement with theory, although the order of magnitude of this length is reasonable.

Hobbs and Scott[69] tried to resolve this difficulty by assuming that adsorption of water molecules is not as small as had been assumed in theory. Moreover, they believe that structural changes of the surface take place at temperatures above $-12°C$. These changes in effect amount to the assumption of a transition layer on ice. Actually, Kobayashi does not agree with Hobbs and Scott's conclusions.

7 Growth of Ice on Various Substrates

The growth of crystals on substrates of different compounds (epitaxy) is another rather large field; however, relatively little work has been done in the case of ice. Camp et al.[70] studied such growth processes. Solid surfaces were cooled to between 0 and $-5°C$ before starting the growth of ice. The experimental arrangement and technique consisted essentially of a polarizing microscope in conjunction with microphotography. Ice was grown on substrate surfaces arranged vertically so that ice was growing outward in the horizontal direction. Aluminum, lucite (polymethylmethacrylate), and glass surfaces were investigated. Ice layers are characterized with respect to ice morphology and growth rate, and appreciable differences were found for the morphology of ice growing on various solid surfaces. The dominant feature, which always seems to be observed first, is an ice layer rapidly spreading over the whole substrate surface before growth in the form of dendrites takes place. Eventually the ice layer thickens and incorporates the dendrites. It appears that the thermal properties of the substrate are not decisive for the growth process; rather, the nature and structure of the substrate surfaces (e.g., polar or nonpolar) influence the morphology and growth rate. The degree of supercooling of the substrate is also significant; however, it influences mainly the growth rate.

The growth rate, v, can be expressed by

$$(34) \qquad v = a(\Delta t)^n \text{ cm/sec}$$

where a and n are constants and t is the growth time.

Some of the substrates initiate several modes of growth, which are apparent in discontinuities of the $\log \Delta v$ versus $\log (\Delta t)$ plots. The mechanism of this epitaxial growth is not fully understood, but it is clear that the type of substrate has a profound influence on the shape and rate of growth of the ice crystal growing on it.

8 Adsorption of Vapors on Ice Surfaces

A number of interesting studies have been carried out by Adamson and co-workers[71] on adsorption of various vapors on ice surfaces. Adsorption gives information about the nature of the ice surface in relation to different absorbates for various temperature ranges.

First, adsorption isotherms of nitrogen on ice were determined at low temperatures using the BET method. All adsorption measurements were carried out at 77°K. The constants c derived from the BET equation can be compared with c-values obtained during similar experiments carried out during sintering of ice.[35]

The c-values derived from both studies are in fairly satisfactory agreement (Table 3). The adsorption isotherms suggest that the ice surface is quite inert

Table 3 Comparison of c-Values Derived from the BET-Equation for Adsorption and Sintering Experiments

Adamson et al.[71] adsorbed at 77°K			Jellinek and Ibrahim[35] Sintered at 77°K	
m^2/g	c	$\Delta Q_a{}^a$ cal/m	°C	$\bar{c}$
			-8	19.9
11.8	66	640	-12	14.9
4.5	39	560	-12	18.0
1.8	50	600	-20	12.0
9.6	52	602	-25	13.5
5.8	74	660	-25	14.2
9.0	31	525	-30	13.4
	—	—	-35.0	12.4
	Av. 52	Av. 597.6	-35.0	22.5
				Av. 15.6
			Av.^{a} $\bar{Q}_a$ = 500 cal/m	

^{a} Q_a is the heat adsorption in excess of the respective heat of vaporization of water.

in the vicinity of 77°K and also at appreciably higher temperatures. The behavior is similar to that occurring when polypropylene or polytetrafluroethylene is used as a substrate. There is no indication that the ice surface has any polarity at low temperatures. In effect ice has a low-energy surface; the latter term was introduced by Zisman[72] into surface science.

The explanation advanced by the authors for this behavior is on the following lines. Surface hydroxyl hydrogens form bent hydrogen bonds with nearby oxygen atoms because of the influence of the adsorbate. Thus the surface becomes completely nonpolar. (Ethane gives an anomalous adsorption isotherm owing to hydrate or clathrate formation.)

The work was extended[73] to n-alkanes in a second paper. These compounds were adsorbed on ice over a range of temperatures of -96.5 to

$-30°C$. Ethane and propane behave abnormally because of hydrate or clathrate formation, but n-pentane and n-hexane are quite normal. The latter two give type III isotherms. The heat of adsorption n-hexane is about equal to the heat of condensation of water for high surface coverages. However, in the submonolayer region, the molar heat of adsorption increases with decreasing coverage. The entropy of adsorption behaves similarly, particularly in the region from -33.3 to $-38.3°C$. Adsorption on liquid water actually shows similarities to that of n-pentane and n-hexane on ice. Hence this behavior of the heat of adsorption cannot be caused by surface heterogeneity. It is assumed by the authors that the ice surface suffers a restructuring of its surface at relatively high temperatures. A change in shape in the log P (pressure) versus $1/T$ plot is observed at $-35°C$ in the case of n-hexane. This indicates that the isosteric heat values change at this temperature for low coverages. Below $-35°C$, the isosteric heat remains approximately constant with extent of coverage.

According to these authors, at about $-35°C$ and above the surface of ice is restructured by the adsorbate, whereas below this temperature it is stable. This change in behavior at $-35°C$ is reversible. The actual transition temperature from a polar to a low-energy surface is a function of the nature of the adsorbate.

Thus the adsorptive properties of the ice surface lead naturally to the view of a surface transition layer on ice, which appears at temperatures near the melting point of ice and gives to its surface a polar nature owing to the special orientation of the H_2O molecules (i.e., protons pointing outward).

9   Contact Angles on Ice Surfaces

A very interesting study, which is closely related to the adsorption work discussed above, was performed by Adamson and co-workers[74] with respect to advancing contact angles of various liquids on ice. It was pointed out above that ice has a completely apolar, inert surface near liquid nitrogen temperatures, but under certain conditions at higher temperatures a restructuring of the surface takes place.

Contact angles reflect this behavior. Two types of substances can be distinguished, designated A and B. Liquids of class A do not affect the surface of ice; however, class B compounds interact with this surface and restructure it, increasing the interaction between the respective liquid and ice. Class A and B substances show relatively high and quite low contact angles, respectively. For instance, hexane vapor adsorbed on either H_2O or ice behaves as a class B substance; there is a tendency for compound formation in this case, and its contact angle on ice at $-5°C$ is almost 0. Pentane behaves similarly.

Jellinek[74] suggested carbon disulfide as an interesting case. It was investigated on a polycrystalline ice surface and was one of many liquids tested. Actually a Zisman plot[72] was obtained. The contact angle of CS_2 on ice is about 21° at −5°C.

For a water$_{\text{hydrogen bonding liquid}}$–hydrocarbon$_{\text{non-polar liquid}}$ interface, Good and Girifalco's[75] equation can be written, according to Fowkes,[76] as follows:

$$(38) \qquad \gamma_{\text{H}_2\text{O,Hydrocarbon}} = \gamma_{\text{H}_2\text{O}} + \gamma_{\text{Hydrocarbon}} - 2(\gamma_{\text{H}_2\text{O}}^w \gamma_{\text{Hydrocarbon}})^{1/2}$$

where w refers to dispersion forces.

In the case of ice–CS_2, there is no hydrogen bonding, hence only the dispersion forces resulting from water must be considered. At 20°C, $\gamma_{\text{H}_2\text{O}}^w = 25$ dynes/cm for water–CS_2. A critical surface tension $\gamma_c = 29$ dynes/cm (in the sense of Zisman[72]) was found for ice at −5°C, which suggests that the ice surface at −5°C is rather like that of water. This view agrees with the assumption of a transition layer on ice near its melting point.

The contact angles of carbon disulfide on ice were also determined on single ice crystals over a temperature range of −5 to −50°C. The results give a straight line when plotting $\cos \theta$ versus temperature having a slope of 0.35/°C. The important feature of these results is that there is no break in the curve indicating the onset or disappearance of a transition layer. This does not agree with the many experimental data that do indicate the presence of such a transition layer on ice. For these reasons, Adamson and co-workers[74] assumed compounds of type A (structure neutral) and of type B (structure changing), as mentioned above. Apparently carbon disulfide belongs to the A class and does not influence the surface structure from near 0°C to low temperatures. The contact angle actually "sees" only the uppermost layer, which in this case is the same at low and high temperatures.

10   Water Structure near Macromolecules

It is not the author's intention to give here a full account of "bound" water; this is done in another chapter. However, some studies concerning polymers are discussed, and these have some bearing on the polymer–ice or polymer–transition layer interface. Polyvinylpyrrolidone (PVP) is a water-soluble polymer of interest in this connection. Literature of some of the work on freezing of polymer solutions can be found in a book entitled *Cryobiology*.[77] More recently Luyet and Ramussen[78] and Luyet and Rapatz[79] carried out studies with aqueous polymer solutions.

A systematic study of freezing of PVP solutions was undertaken by Jellinek and Fok.[80] The freezing process was followed by calorimetry, DTA, and dilatometry. Measurements of heat of fusion indicated the amount of "bound" water in the solutions.

Grams of frozen water for each gram of PVP increase with increasing water activity. Actually the amount of "unfrozen" water or, better, "frozen" water (as will be apparent later) in grams plotted versus water activity gives a type III isotherm. The temperature coefficient for this bound water is almost 0, hence the amounts of bound water determined over a range of temperatures give this type III "isotherm" in the sense of the BET method. The "isotherm" is shown in Figure 12. The freezing point curve for PVP in solutions is presented in Figure 13. Heat of fusion values were determined for a whole range of PVP concentrations and over a temperature range from near 0 to $-45°C$. Actually the amount of "bound" water depends on the path of freezing. Thus if a 5% w/w PVP solution is frozen at a certain rate of cooling and is eventually kept at $-45°C$, more "bound" water is in the system for each gram of PVP than if a 46.4% w/w PVP solution is frozen at the same rate and is also kept at $-45°C$ (2.2 g and 0.62 g "bound" water for each gram of PVP, respectively). This behavior indicates that true equilibrium is not established. Apparently the "adsorption" process is not completely reversible. Thus the solution, having passed through a path of higher activities, binds finally a larger amount of water per gram of PVP than the initially more concentrated solution. This may also explain the discrepancies of Jellinek and Fok's[80]

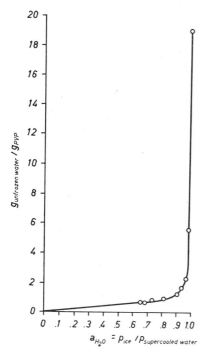

Figure 12. Grams of "unfrozen" (bound) water per grams of PVP versus a_{H_2O} (activity of water); type III isotherm.

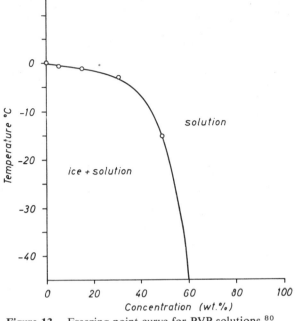

Figure 13. Freezing point curve for PVP solutions.[80]

values compared with those obtained by Luyet and Rasmussen,[78] who also carried out freezing experiments with PVP solutions.

The adsorption of water vapor on PVP shows some interesting features.[81] The BET method was used in this investigation. Very little information is available on sorption of water vapor below or near 0°C. The work under discussion is concerned with various polymers, including PVP. Figure 14 gives the sorption isotherms obtained for PVP at −2.0 and +10°C. The specific surface area of the sample was about 0.17 m²/g. Additional isotherms (specific surface area ca. 3.1 m²/g) for −20, +7, and −2°C are also shown. An isotherm obtained by Dole and Faller[82] for a coarser sample of PVP is included (+25°C). The interesting feature is that the temperature coefficient is 0, as was the case for the data obtained from freezing experiments with aqueous PVP solutions (Figure 12). This isotherm is also included in Figure 14. Plotting the isotherms according to the BET equation for type III isotherms gives straight lines whose slopes are almost the same. At quite low relative pressures, several hundred monolayers of water are adsorbed on PVP; at $p/p_0 = 0.7$, the polymer plus water becomes a sticky syrupy liquid. The results obtained by sorption are quite consistent with the freezing experiments. It is concluded that the bound water has an icelike structure near

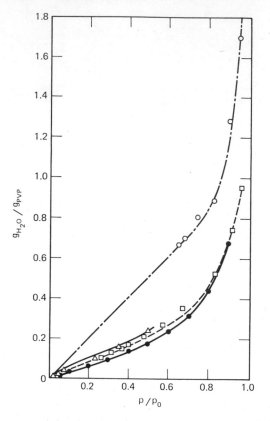

Figure 14. Sorption of water by PVP ($g_{H2O}/1\ g_{PVP}$):[81] open circles—freezing experiments, -45 to $-0.5°C$ (Jellinek and Fok[80]); solid circles—$25°C$ (Dole and Faller[82]); triangles—-2.0 and $+7.5°C$ (Jellinek and Nagajaran[81]); squares—-20 and $+7.0°C$ (Jellinek and Clerc[81]).

PVP molecules. This conclusion is supported by infrared measurements on PVP solutions by Klotz,[83] who found that the infrared band for water shifts for increasing PVP concentration to the infrared ice bands.

Freezing of water in cross-linked polymers[84] presents some very interesting features. The network in a gel prevents ice crystals from growing normally. This growth is inhibited by the size of the mesh of the gel network. A crystal can only continue to grow beyond such a mesh and envelop it by being cooled to a temperature appreciably lower than normal.

It can be shown by thermodynamic reasoning that a crystal prevented from its normal growth has a different vapor pressure from that of one that is not hindered at all in its development. The vapor pressure of a macroscopic amount of liquid water is given (Clausius–Clapeyron) by

$$(35) \qquad p_1 = p_0\left(1 + \frac{\Lambda}{RT_0^2}\,\delta T\right)$$

where p_1 and p_0 are the vapor pressures of the liquid at $T_0 + \delta T$ and T_0, respectively, and Λ is the molar heat of vaporization. A crystal that grows normally (e.g., an ice crystal of infinite dimensions) has a vapor pressure as follows:

$$(36) \qquad p_{cr,\infty} = p_0\left(1 + \frac{\Sigma}{RT_0^2}\,\delta T\right)$$

where Σ is the molar heat of sublimation.

For a crystal of microscopic dimensions (edge of cube $= a$) it can be shown that its vapor pressure is larger than that of an infinitely large crystal

$$(37) \qquad P_{cr,a} = p_0\left(1 + \frac{\Sigma}{RT_0^2}\,\delta T + \frac{4\gamma_{cr,1}}{aRT}\,\frac{M}{\rho_{cr}}\right)$$

where $\gamma_{cr,1}$ is the free energy of the crystal–liquid interface, M the molecular weight of the substance (H_2O), and ρ_{cr} the crystal density. Putting (35) equal to (37) gives the freezing point lowering δT_a (°C) as follows:

$$(38) \qquad \delta T_a = -\frac{4\gamma_{cr,1}}{a}\,\frac{T_0}{\Sigma - \Lambda}\,\frac{M}{\rho_{cr}}$$

This freezing point lowering increases directly proportional with the interfacial tension and is inversely proportional to the size of the crystal. It should not be confused with the ordinary freezing point lowering caused by the number of polymer molecules dissolved in a liquid. One can write δT_a in the following form:

$$(39) \qquad \delta T_a = -\frac{4\gamma_{cr,1}}{a\rho_{cr}}\,\frac{10^3}{RT_0}\,K_f$$

where

$$K_f = \frac{10^{-3}RT_0^2}{(\Sigma - \Lambda)}\,M$$

is the molar freezing point lowering.

This analysis can be pursued somewhat further. The freezing point lowering of a crystal (cube) because of inhibition of its growth by one chain molecule, which surrounds the crystal, can be calculated. It is given by

$$(40) \qquad \delta T_{a,\text{hindered}} = -\frac{20}{3}\,\frac{\gamma_{cr,1}}{a}\,\frac{T_0}{\Sigma - \Lambda}\,\frac{M}{\rho_{cr}}$$

This value for $\delta T_{a,\text{hindered}}$ has about twice the magnitude of that encountered before. It is again proportional to the interfacial tension and inversely proportional to the edge length, a. This result indicates that a crystal hindered in its growth by a chain molecule has to be cooled to about $T_0 - 2\delta T_a$ before it is in equilibrium with its surrounding melt and is able to continue growing. At this temperature it starts to engulf (envelop) the polymer chain.

During its growth the crystal exerts a calculable tensile stress on the chain molecule. This force amounts to $a \cdot \gamma_{cr,l}$. Tensile stresses of 5×10^{10} dynes/cm^2 are obtained for aqueous polymer solutions. This value is about 10 times larger than the tensile strength of the best steel. Hence chain rupture will occur. If rubber in benzene is considered, the tensile stress is hardly sufficient to rupture a primary bond in the backbone of the chain. This force is independent of the cross section of the molecule.

This brief survey covers only a very small part of the extensive and interesting investigations. For further details, reference should be made to the publications by Kuhn and co-workers.[84] It may be mentioned here that chain rupture of linear (i.e., not cross-linked) chain molecules by freezing proceeds by a mechanism quite different from that in networks; this has been studied by Jellinek and Fok.[85]

Since the topic of "bound" water is treated from a biological point of view in another chapter, no attempt is made here to discuss the underlying reasons for the structure of water near macromolecules (e.g., hydrophobic bonds).

Homogeneous and heterogeneous nucleation have been repeatedly dealt with in the literature (see, e.g., Refs. 29 and 5, and a symposium in *Z. Angew. Math. Phys.* **14**, 1963); these topics are not discussed here.

REFERENCES

1. M. Faraday, Lecture at Royal Institution, London, reported in *Athenaeum*, p. 640 (1850); also *Proc. Roy. Soc. (London)*, **10**, 440 (1860); *Phil. Mag.* (4th Ser.), **17**, 162 (1859); "Faraday's Diary," in *Researches in Chemistry and Physics*, Bell, London (1933), pp. 373–378.
2. J. Tyndall and T. H. Huxley, *Phil. Trans. Roy. Soc. (London)*, **147**, 327 (1857).
3. W. Thomson, *Proc. Roy. Soc. (Edinburgh)*, **2**, 267 (1850).
4. J. Thomson, *Proc. Roy. Soc. (Edinburgh)*, **10**, 152 (1860); also *Trans. Roy. Soc. (Edinburgh)*, **16**, 575 (1849).
5. L. Dufour and R. Defay, *Thermodynamics of Clouds*, International Geophysical Series, vol. 6, Academic Press, New York, 1963.
6. M. Volmer, *Kinetik der Phasenbildung*, vol. 4, *Die Chemische Reaktion*, Steinkopff, Dresden, 1939 (1941).
7. L. Krastanov, *Meteorol. Z.*, **57**, 357 (1940).
8. J. E. McDonald, *J. Meteorol.*, **10**, 416 (1953).
9. G. Briegleb, *Zwischenmolekulare Kraefte*, Braun, Karlsruhe, 1949.
10. B. J. Mason, *The Physics of Clouds*, Oxford University Press (Clarendon), London and New York, 1957.
11. P. Kubelka and R. Prokscha, *Kolloid-Z.*, **109**, 79 (1944).

12. W. Jacobi, *Z. Naturforsch.*, **10a**, 322 (1955).
13. W. M. Ketcham and P. V. Hobbs, *Phil. Mag.*, **19**, 1161 (1969).
14. G. Antonow, *J. Chim. Phys.*, **5**, 372 (1907).
15. M. Roulleau, *J. Sci. Meteorol.*, **10**, 12 (1958).
16. A. S. Skapski, *Proc. Eastern Snow Conf.*, **4**, 52 (1957).
17. H. Oura, *J. Phys. Soc. Japan*, **5**, 277 (1950).
18. R. Smith-Johannsen, G.E. Res. Lab., Project Cirrus, Final Report, pp. 79–81, 1948.
19. V. J. Schaefer, *Chem. Rev.*, **44**, 291 (1949).
20. H. H. G. Jellinek, *Proc. Phys. Soc.*, **71**, 797 (1958); *U.S. SIPRE, Res. Rept. 23* (1957); *J. Coll. Sci.*, **14**, 268 (1959); *U.S. SIPRE, Res. Rept. 62* (1960); *Can. J. Phys.*, **40**, 1294 (1962).
21. P. E. Berghausen et al., WADC Tech. Rept. 55-461 (1953), 55-44 (1955); W. D. Bascom, R. L. Cottington, and C. R. Singleterry, *J. Adhesion*, **1**, 246 (1969); T. F. Ford and O. D. Nichols, NRL Rept. 5662 (1961) and 5832 (1962); M. Landy and A. Freiberger, *J. Colloid Interface Sci.*, **25**, 231 (1967); also *Naval Eng. J.*, p. 63, Feb. 1968; L. E. Rarty and D. Tabor, *Proc. Roy. Soc. (London), Ser. A*, **245**, 184 (1958), also I. D. Bowden and D. Tabor, *Proc. 2nd Congr. Surface Activity, London*, **3**, 386 (1957); A. Sellerio, *Phys. Z.*, **34**, 180 (1933); D. L. Loughborough, *Z. Angew. Math. Phys.*, **3**, 460 (1952); J. C. Hunsacker et al., *Progr. Rept., Natl. Acad. Sci., MIT*, Rept. No. 1 (1940); P. V. Hobbs and B. J. Mason, *Phil. Mag.*, **9**, 187 (1964).
22. H. H. G. Jellinek, *J. Colloid Interface Sci.*, **25**, 192 (1967); also *U.S. CRREL, Spec. Rept.*, **70** (1964).
23. J. Tyndall, *Glaciers of the Alps . . .*, Tickner and Fields, Boston, 1861, Chapter IV.
24. W. A. Weyl, *J. Colloid Sci.*, **6**, 389 (1951).
25. N. H. Fletcher, *Phil. Mag.*, **17**, 1287 (1968).
26. J. C. Henniker, *Rev. Mod. Phys.*, **21**, 322 (1949).
27. U. Nakaya and A. Matsumoto, *U.S. SIPRE, Res. Rept. 4* (1953).
28. C. L. Hosler et al., *J. Meteorol.*, **14**, 415 (1957).
29. N. H. Fletcher, *Phil. Mag.*, **7**, 235 (1962); also *Chemical Physics of Ice*, Cambridge University Press, New York, 1970; see also R. J. Watson-Tobin, *Phil. Mag.*, **8**, 333 (1963) and N. H. Fletcher, *Phil. Mag.*, **8**, 1425 (1963).
30. C. Jaccard, *Proc. Intern. Conf. Cloud Phys., Tokyo and Sapporo, 1965*; "Physics of Snow and Ice," *Intern. Conf. Low Temperature Sciences, Sapporo, 1966*, **1**, 173; *Z. Angew. Math. Phys.*, **17**, 657 (1966).
31. J. W. Telford and J. S. Turner, *Phil. Mag.*, **8**, 527 (1963).
32. D. W. Townsend and R. P. Vickery, *Phil. Mag.*, **16**, 1275 (1967).
33. K. R. Nunn and D. M. Rowell, *Phil. Mag.*, **16**, 1281 (1967).
34. J. F. Nye, *Phil. Mag.*, **16**, 1249 (1967).
35. H. H. G. Jellinek and S. H. Ibrahim, *J. Colloid Interface Sci.*, **25**, 245 (1967).
36. W. D. Kingery, *J. Appl. Phys.*, **31**, 833 (1960).
37. B. J. Mason, G. W. Bryant, and A. P. van der Heuvel, *Phil. Mag.*, **8**, 505 (1963).
38. J. W. Peck and R. P. Murriman, *Tech. Note*, U.S. Army Terrestial Sci. Center, Hanover, N.H., June 1969.
39. J. Crank, *The Mathematics of Diffusion*, Oxford (Clarendon Press), London, 1956.
40. O. Dengel and N. Riehl, *Phy. Kondens. Materie*, **1**, 191 (1963).
41. H. H. G. Jellinek and A. Chatterjee, *Phys. Stat. Sol.*, (a) **4**, 173 (1971); H. H. G. Jellinek and K. Juznic, *Phys. Stat. Sol.*, 837 (1970).
42. H. H. G. Jellinek and V. Gouda, *Phys. Stat. Sol.*, **31**, 413 (1969).
43. J. G. Byrne, *Recovery, Recrystallization and Grain Growth*, Macmillan, Series in Materials Science, Macmillan, New York, 1965, pp. 98–102.

44. A. Gow, *J. Glaciol.*, **8**, 241 (1969).
45. S. Glasstone, K. J. Laidler, and H. Eyring, *The Theory of Rate Processes*, McGraw-Hill, New York, 1941.
46. D. Turnbull, *Trans. AIME*, **191**, 661 (1951).
47. H. H. G. Jellinek and K. Juznic, *Proc. 5th Intern. Congr. Rheol., Kyoto, 1968*, Tokyo University Press, **2**, 407 (1970).
48. L. Onsager and L. K. Runnels, *Proc. Natl. Acad. Sci. U.S.*, **50**, 208 (1963).
49. J. C. Fisher, *J. Appl. Phys.*, **22**, 74 (1951).
50. J. H. Wang, C. V. Robinson, and I. S. Edelman, *J. Amer. Chem. Soc.*, **75**, 466 (1953); J. H. Wang, *J. Phys. Chem.*, **69**, 4412 (1965).
51. H. H. G. Jellinek and R. Brill, *J. Appl. Phys.*, **27**, 1198 (1956).
52. D. Kuroiwa, *Contrib. Institute of Low Temperature Sciences, Hokkaido University*, No. 18, 1–62 (1964).
53. H. H. G. Jellinek and A. Chatterjee, *J. Glaciol.*, in press (1971).
54. H. Roethlisberger, *J. Geol.*, **63**, 579 (1953).
55. R. P. Auty and R. H. Cole, *J. Phys. Chem.*, **20**, 1309 (1952); see also D. Eisenberg and W. Kauzmann, *The Structure and Properties of Water*, Oxford University Press, New York and Oxford, 1969.
56. O. M. Eigen, L. D. Mayer, and H. Spatz, *Ber. Bunsenges.*, **68**, 19 (1964).
57. M. Kopp, *Z. Angew. Math. Phys.*, **13**, 431 (1962).
58. B. Bullemer and N. Riehl, *Solid State Commun.*, **4**, 447 (1966).
59. P. R. Camp, W. Kiszenik, and D. A. Arnold, *U.S. CRREL, Res. Rept. 198*, September 1967.
60. G. W. Gross, *Advan. Chem. Ser.*, **73** (27), 89 (1968).
61. W. Drost-Hansen, *J. Colloid Interface Sci.*, **25**, 131 (1967).
62. H. R. Pruppacher, E. H. Steinberger, and T. L. Wang, *J. Geophys. Res.*, **73**, 571 (1968).
63. Costa Riberro, Jr., Ph.D. Thesis, University of Brazil, Rio de Janeiro, 1945; *Anais Acad. Bras. Cienc.*, **19** (4), 6 (1947).
64. E. J. Workman and S. E. Reynolds, *Phys. Rev.*, **78**, 254 (1950); *Nature*, **169**, 1108 (1952).
65. F. K. Truby, *J. Appl. Phys.*, **26**, 1416 (1955); *Science*, **121**, 404 (1955).
66. U. Nakaya, *Snow Crystals (Natural and Artificial)*, Harvard University Press, Cambridge, 1954.
67. J. H. DeBoer, *The Dynamical Character of Adsorption*, 2nd ed., Oxford University Press (Clarendon), 1968.
68. T. Kobayashi, "Physics of Snow and Ice," in *Intern. Conf. Low Temperature Sciences, Sapporo, 1966*, **1**, 95 (1967).
69. P. V. Hobbs and W. D. Scott, *J. Colloid Interface Sci.*, **25**, 223 (1967).
70. P. R. Camp, *Ann. N.Y. Acad. Sci.*, **125**, 317 (1965).
71. A. W. Adamson, L. M. Dormant, and M. Orem, *J. Colloid Interface Sci.*, **25**, 206 (1967); A. W. Adamson and L. M. Dormant, *J. Amer. Chem. Soc.*, **88**, 2055 (1966).
72. W. A. Zisman, *Ind. Eng. Chem.*, **55**, 19 (1963).
73. M. W. Orem and A. W. Adamson, *J. Colloid Interface Sci.*, **31**, 278 (1969).
74. A. W. Adamson, F. Shirley, and K. T. Kunichika, *J. Colloid Interface Sci.*, in press (1970).
75. L. A. Girifalco and R. J. Good, *J. Phys. Chem.*, **61**, 904 (1957); see also R. J. Good, *Advan. Chem. Ser.*, No. 43, p. 74, Amer. Chem. Soc., Washington, D.C., 1964.
76. F. M. Fowkes, in *Hydrophobic Surfaces*, Ed., R. M. Fowkes, Academic Press, New York, 1969, p. 151.

77. *Cryobiology*, Ed., H. T. Meryman, Academic Press, New York, 1966.
78. B. Luyet and D. Rasmussen, *Biodynamica*, **10**, 137 (1967).
79. B. Luyet and G. Rapatz, *Biodynamica*, **10**, 149 (1968).
80. H. H. G. Jellinek and S. Y. Fok, *Kolloid-Z.* and *Z. Polym.*, **220**, 122 (1967).
81. H. H. G. Jellinek, M. D. Luh, and V. Nagarajan, *Kolloid-Z.* and *Z. Polym.*, **232**, 758 (1969).
82. M. Dole and I. L. Faller, *J. Amer. Chem. Soc.*, **72**, 414 (1950).
83. I. M. Klotz, *Federation Proc.*, **24**, 524 (1965); see also *Science*, **128**, 815 (1959).
84. W. Kuhn and H. Majer, *Z. Phys. Chem.*, **3**, 330 (1955); *Z. Phys. Chem.*, **30**, 289 (1961); W. Kuhn, *Helv. Chim. Acta*, **39**, 1071 (1956); W. Kuhn, R. Block, and P. Laeuger, *Kolloid-Z.* and *Z. Polym.*, **193**, 1 (1963); G. Kanig and H. Karge, *J. Colloid Interface Sci.*, **21**, 649 (1966).
85. H. H. G. Jellinek and S. Y. Fok, *Makrom. Chem.*, **104**, 18 (1967).

4 Nonaqueous Electrolyte Solutions

J. Padova, Soreq Nuclear Research Center, Yavne, Israel

1 Introduction

Aqueous solutions of electrolytes have traditionally received far more attention than the corresponding nonaqueous solutions, even though it has been realized for a long time that water cannot be regarded as the prototype of electrolytic solvents.[1] Nevertheless, during recent years, there has been an increasing interest in the behavior of electrolytes in nonaqueous solvents with a view to investigating ion–ion and ion–solvent interactions under varied conditions. Electrostatic theories have been tested in media of varying dielectric constants and theoretical implications of the present knowledge of nonaqueous electrolyte solutions have been considered in the current literature.[2] The study of the thermodynamics and transport properties of these solutions provides an opportunity for extending the range of properties from which the structure of electrolyte solutions may be inferred. This chapter is concerned mainly with thermodynamic behavior, although both reversible and irreversible phenomena are considered.

2 Classification of Solvents

Nonaqueous solvents, in a broad sense, may be defined as media other than water that will dissolve a reasonable number of compounds and permit the occurrence of chemical reactions.[3] They may be differentiated according to the characteristic features of their chemical bonding;[4] however, the present discussion is restricted to solvents that are liquid at, or near, room temperature.

Solvents may be divided into basic, acidic, amphiprotic, and aprotic categories. Brønsted[5,6] has focused attention on three properties of a solvent —the dielectric constant, the acidic strength, and the basic strength—as being of primary importance in affecting the manifestation of acidity or basicity in that solvent. Eight classes of solvents[7] were suggested to represent all gradations in these properties. This classification is presented in Table 1 where a

109

Table 1   Classification of Solvents According to Brønsted

Class	Dielectric Constant	Acidic Property	Basic Property	Example
Amphiprotic	+	+	+	Water
Protogenic	+	+	−	Hydrogen oxyanide
Protophilic	+	−	+	N-Methylpropionamide
Aprotic	+	−	−	Acetonitrile
Amphiprotic	−	+	+	Ethanol
Protogenic	−	+	−	Acetic acid
Protophilic	−	−	+	Pyridine
Aprotic	−	−	−	Benzene

plus or minus sign indicates the predominance or weakness, respectively, of the property considered. Solvents with dielectric constants greater than 30 are assigned a plus sign and those with less than 30 are given a minus sign. It has been pointed out[8] that the four classes of solvents—amphiprotic, protogenic, protophilic, and aprotic—are not clearly restrictive, and the proper classification of many solvents is in doubt.

Two different types of hydrogen-bonded solvents can be distinguished by considering the relative magnitude of the excess thermodynamic functions of mixing with water, ΔH_E and $T\Delta S_E$.[9] One type is nonaqueous mixtures, where deviations from Raoult's law are determined by $|\Delta H_E| > T|\Delta S_E|$ and the other has what may be termed a "typically aqueous" behavior in which the predominant factors appear to be nonpolar group–water interactions (hydrophobic hydration). The latter group was found to exhibit positive deviations from Raoult's law in spite of large negative heats of mixing for hydrocarbons, alcohols, amines, ketones, glycols, and ethers. Nitriles, dimethylsulfoxide (DMSO), amides, glycerol, and polyhydroxy compounds represent the former type.

This classification broadly agrees with Parker's differentiation[10–12] between protic and dipolar aprotic solvents. Protic solvents, such as fluoro-alcohols, methanol, formamide, ammonia, and hydrogen fluoride are strong hydrogen donors. Solvents with a dielectric constant larger than 15, which, although they may contain hydrogen atoms, cannot donate suitably labile hydrogen atoms to form hydrogen bonds, are classed as dipolar aprotic. This is an arbitrary choice, made because ion aggregation is so extensive at low dielectric constants that it becomes very difficult to observe the behavior of solvent separated pairs.[12] Common dipolar aprotic solvents are dimethyl-formamide (DMF), dimethylacetamide (DMA), dimethylsulfoxide (DMSO), hexamethylphosphoramide (HPA), tetrahydrothiophendioxide (Sulfolane

Table 2 Physical Properties of Protic Solvents at 25°C

Solvent	Molecular Weight	Melting Point (°C)	Boiling Point (°C)	Density (g/ml)	Viscosity (cp)	Dipole Moment (debye)	Refractive Index	Dielectric Constant	Trouton Constant
H_2O	18.02	0	100.0	0.9971	1.0	1.84	1.3325	78.3	26.0
H_2F_2	40.02	−89.37	19.51	1.002[a]	0.256[a]		1.90[c]	84.0[a]	24.7
H_2SO_4	98.02	10.37	300.00	1.8269	24.54			120.00	
$HCOOH$	44.1	8.4	100.80	1.214	1.637	1.19	1.3714	57.0	
$HCONH_2$	45.0	−2.45	193.10	1.129	3.30	3.68	1.4453	111.3	37.3
NH_3	17.1	−77.8	−33.50	0.681[b]	0.2543[b]	0.93	1.325[d]	23.0[b]	
N_2H_4	32.4	2.0	113.5	0.9955				51.7	
CH_3COOH	60.0	16.63	118.1	1.044	1.04	1.73	1.3698	6.2	
CH_3OH	32.0	−97.68	64.5	0.787	0.55	1.67	1.3288	32.6	25.0
C_2H_5OH	46.1	−114.5	78.3	0.785	1.08	1.70	1.3598	24.3	
C_3H_7OH	60.1	−126.2	97.2	0.804	2.3	1.66	1.3838	22.2	
C_4H_9OH	74.1	−89.3	117.5	0.810	3.9	1.68	1.3975	16.1	

[a] At 0°C.
[b] At −33°C.
[c] Gas.
[d] At 16.5°C.

Table 3 Physical Properties of Dipolar Aprotic Solvents at 25°C

Solvent	Molecular Weight	Melting Point (°C)	Boiling Point (°C)	Density (g/ml)	Viscosity (cp)	Dipole Moment (debye)	Dielectric Constant	Refractive Index	Trouton Constant
NMF	59.1	−3.8	111.2	0.99	1.65	3.86	185.5	1.4269	23.3
DMF	73.1	−61.0	153.0	0.9443	0.796	3.90	36.71		
NMA[a]	73.1	29.8	206.0	0.942	3.02	4.12	171.7		
DMA	87.0		165.0	0.937	0.92	3.79	37.8		22.2
DMSO	78.1	18.2	189.0	1.096	1.96	3.96	46.6	1.4773	29.6
HPA	179.4	7.2	23.00	1.02		5.54	30.5		
THO	120.1	27.5	283.00	1.2613	9.87	4.70	42.0	1.4816	
ACO	58.1	−95.4	56.2	0.785	0.30	2.84	20.7	1.3561	21.12
NMT	61.0	−28.5	101.0	1.130	0.612	3.52	35.9	1.3936	21.71
ACN	41.0	−45.7	81.6	0.7768	0.344	3.27	36.2	1.3416	22.09
EC	88.1		238.00	1.321		4.87	89.6		
PC	102.1	−49.2	241.7	1.2057	2.013	5.19	64.7	1.439	23.19
THF	72.1			0.880	0.46	1.71	7.4		
PNH	79.1		115.6	0.973	0.83	2.23	12.3	1.469	
N-Methyl-pyrrilidone	99.13	24.4	204.5	1.027	1.6	4.09	32.2		

[a] At 40°C.

THO), dimethylsulfone (DMO), acetone (ACO), nitromethane (NMT), acetonitrile (ACN), nitrobenzene, sulfur dioxide, ethylene carbonate (EC), propylene carbonate (PC), and N-methyl-2-pyrrolidone (NMP).

Physical constants for both classes of solvents are given in Tables 2 and 3. The sharp distinction between protic and dipolar aprotic solvents that originated from their respective influences on the rates of reactions[13] has particularly emphasized the role played by hydrogen bonding on ion–solvent interactions. Other factors, such as dipole–dipole interactions and dispersion forces that contribute toward solvent structure, must be taken into account as well. Four types of strong solute–solvent interactions were suggested:[10] electrostatic, π-complex forming, hydrogen bonding, and structure promoting.

We shall not discuss the general concept of acids and bases that has superseded Brønsted classification. Suffice it to say that in terms of the hard and soft acids and bases concept,[14] protic solvents are considered to be "hard," since they might display general hydrogen bonding with small anions, whereas dipolar aprotic solvents are "soft," since they exhibit a mutual polarizability interaction with large, polarizable anions.

3   Solubility as Experimental Evidence for Ion–Solvent Interaction

The dissolution of a salt in a nonaqueous solvent depends on the dielectric constant and polarity of the solvent and on the lattice energy. The process may be visualized as taking place by dissociation of the crystal lattice into gaseous ions which are then solvated into the solvent.

$$(1) \qquad\qquad C^+A^- \rightarrow C_g^+ + A_g^- \; \Delta G_{lattice}$$

$$(2) \qquad\qquad C_g^+ + A_g^- \rightarrow C_s^+ + A_s \, \Delta G_{solvation}$$

Since the free energy of solution ΔG_{sol} is thermodynamically given by

$$(3) \qquad\qquad \Delta G_{sol} = -\Delta G_{lattice} + \Delta G_{solvation}$$

the free energy of solvation of the ions must be considerable in order to overcome the lattice energy. This might possibly qualitatively explain why solubility in water is greater than in nonaqueous solvents. In practice, however, the standard free energy of solution is obtained from solubility measurements [cf. (19)], which afford one of the methods by which free energies of solvation and transfer are derived.[15]

Some representative solubility data in protic and dipolar aprotic solvents are shown in Tables 4 and 5.

The solubility trends in protic solvents are quite similar. In methanol,[15,16] ethanol,[17] formamide,[18,19] formic acid,[20] and ammonia,[21] solubilities decrease with increase in cation size and generally increase with increase in anion size

Table 4 Molal Solubilities of Electrolytes in Protic Solvents at 25°C

Salt	CH$_3$OH	(Ref.)	C$_2$H$_5$OH	(Ref.)	F	(Ref.)	HCOOH (Ref. 321)	NMF[a] (Ref. 322)	NH$_3$[b] (Ref. 323)
LiCl	4.95	(16)	6.6	(37)			5.55	5.05	
LiBr	3.95	(16)					7.35		
LiI	24.8	(37)	18.2	(37)			7.04		
NaF	0.16	(318)	0.04	(318)					
NaCl	0.24	(16)	0.02	(318)	1.58	(319)	0.85	0.544	2.20
NaBr	1.57	(16)	0.25	(318)	3.50	(319)	1.78	2.56	6.21
NaI	4.17	(16)	1.99	(318)	3.78	(319)	3.48	4.25	8.80
KF	0.39	(16)							
KCl	0.07	(16)	0.02	(318)	0.85	(319)	2.36	0.273	0.02
KBr	0.18	(16)			1.78	(319)	1.77	0.812	2.26
KI	1.04	(16)	0.11	(17)	4.17	(319)	1.90	2.57	11.09
RbCl							3.86		
RbBr							2.66		
RbI							1.97		
CsCl	0.22	(30)	0.08	(17)	1.38	(320)	5.5		
CsBr			0.023	(30)	2.05	(319)	2.82		
CsI			0.01	(17)	3.75	(319)	1.06		
NH$_4$Cl					7.2	(319)		0.906	
NH$_4$Br								2.20	
NH$_4$I									
LiNO$_3$	6.15	(16)							
LiClO$_4$	17.1	(36)	14.3	(36)	13.4	(36)			15.0
NaNO$_3$	0.31	(16)							
NaClO$_4$	4.16	(37)			4.27	(319)			
KNO$_3$									
KClO$_4$	0.0077	(15)	0.0009	(15)					1.04

[a] Molarity.
[b] At 0.11°C.

Table 5   Molal Solubilities of Electrolytes in Some Dipolar Aprotic Solvents at 25°C

Salt	DMF (Ref.)	DMSO (Ref.)	NMA[a] (Ref. 25)	DMA (Ref. 26)	ACN (Ref.)	PC (Ref. 27)
LiCl	2.67 (24)	2.132 (28)	0.413	2.03	0.79 (1)	2.43
LiBr	1.88 (24)	3.3 (22)		2.96	7.4 (1)	1.365
LiI		1.219 (28)				
NaCl		0.078 (28)	0.339	0.025		
NaBr	1.0 (24)			0.635		0.08
NaI	0.415 (24)	1.0 (22)	2.86	2.31	1.26 (1)	1.11
KCl	0.025 (24)	0.023 (28)	0.120	0.012		
KBr	0.069 (24)	0.50 (22)	0.429	0.049		
KI	2.5 (24)	2.5 (22)	1.556	0.093	0.10 (1)	0.223
RbCl		0.036 (22)				
RbBr					0.002 (1)	
RbI					0.062 (1)	
CsCl	0.003 (324)	0.037			0.005 (1)	
CsBr	0.026 (324)				0.03 (1)	
CsI						
NH$_4$Cl			0.896			
NH$_4$Br			1.745			
NH$_4$I			2.470			
NaNO$_3$	1.53 (23)	4.6	0.65			
KNO$_3$	0.20 (23)	1.0	0.247			
NH$_4$NO$_3$			3.231			
LiClO$_4$	7.05 (36)	2.65 (28)	5.0		1.53 (36)	
NaClO$_4$	3.52 (23)	1.8 (22)	0.413			
KClO$_4$	1.43 (23)	2.5 (22)	1.349			
NH$_4$ClO$_4$		10 (28)	3.54			
NaSCN	3.7 (24)	0.15 (22)	2.170			
KSCN	1.82 (35)	17 (22)	4.526			
NH$_4$SCN		4.5 (22)				

[a] Molarity at 40°C.

115

Solubilities are comparable to those in water, with halides following the same pattern of behavior, that is, iodide > bromide > chloride. The perchlorates, nitrates, and thiocyanates, however, are the most soluble salts. In non-hydrogen-bonding solvents, the solubility of the salts is initially dependent on the nature of the anion. Typically, only those salts containing univalent anions that are large or highly polarizable are reasonably soluble in such solvents.[1] Generally, salts are much less soluble in DMSO[22] than in water. For a given cation in DMF,[23,24] solubilities of the corresponding halides increase sharply when the size of the anion increases. In NMA[25] and DMA[26] the order of increasing solubilities with respect to the cation is $K^+ <$ $Na^+ <$ $Li^+ <$ NH_4^+, whereas for ammonium salts the order of increasing solubility is $C^- <$ $ClO_4^- <$ $Br <$ $I^- <$ $NO_3^- <$ SCN. Since DMA and DMF have very similar physical constants (dielectric constant and dipole moment), solubilities would be expected to be similar for a given salt in the two solvents. However, because such similarity does not exist, no dependence on the dielectric constant of the medium can be assumed. This is confirmed by the solubilities in EC, PC, and so on,[27,28] which are not a smooth function of the dielectric constant; yet solubilities of ionic salts in water and aliphatic alcohols definitely are a smooth function of the dielectric constant.[29-31]

It is suggested that solvents containing atoms with unshared electron pairs, such as nitrogen and oxygen, should make relatively good coordinating solvents for the cations, and it has indeed been confirmed that cations are strongly solvated in highly polar solvents with a negative charge localized on an oxygen atom, as in DMSO, pyridine N-oxide, 2-pyridones, DMF, DMA,[32] and so on. Cations are sometimes poorly solvated and give rise to low solubilities in acetonitrile[1] and nitromethane[33] because the negative charge of the solvent dipole is not localized on a favorable electron donor atom[34] or is protected by bulky groups.[10,35] Exceptions are usually explained in terms of very specific ion–solvent interactions, for example, the solubilities of lithium halide, and especially $LiClO_4$, in many solvents.[36]

Of the many theories that have been applied to predict changes in solubility with variation of solvent, the majority are dependent upon changes in the electrostatic properties of the solvent. Numerous relations between solubility of electrolytes and various functions of the dielectric constant of the solvent have appeared in the literature. The most commonly used equation is

$$(4) \qquad \log S = A_1 + \frac{A_2}{\varepsilon}$$

where S is the solubility, A_1 and A_2 are constants. It will be seen later that (4) is a simplified form of the expression proposed by Born for the calculation of solvation free energy. Walden[37] suggested the empirical relationship

$$(5) \qquad 3 \log \varepsilon = B + \log N$$

where N is the mole fraction, which is not obeyed for small electrolytes. Assuming the Debye–Hückel limiting law to hold in concentrated solutions, Ricci and Davis[38] derived a similar equation based on the observed constancy of the mean activity coefficient of the salt in its salinated solution in any solvent.

As far as the author is aware, the only attempt to develop a quantitative relation between the solubility of simple salts and the dielectric constant of the solvent was carried out by Bjerrum.[39] After the energy necessary to separate the ions of the solid salt was determined from the lattice energy, the pressure of the saturated ionic vapor in equilibrium with the solid was calculated from the Nernst heat theorem. Bjerrum assumed that a gaseous ion should be regarded as a rigid sphere of known radius immersed in a dielectric continuum. The partition coefficient of the ion without its charge was assumed to be that of the corresponding neutral molecule—compare the calculation of free energy taking into account electrostatic and noncoulombic contribution in the section, Extrapolation Methods—and the electric work given by Born's expression.

By this procedure, the following expression was obtained:

$$(6) \quad \log S = \frac{e^2}{4.6kT} \left[\left(\frac{1}{2r_1} + \frac{1}{2r_2} \right) \left(1 - \frac{1}{\varepsilon} \right) - \frac{1.57}{a} \right] + \log V_1 + \log V_2 + 8.0$$

where r_1 and r_2 are the radii of the ions in solution, V_1 and V_2 are the partition coefficients of the corresponding inert gases, a is the interionic distance in the crystal, ε the dielectric constant, and e and k are the electron charge and Boltzmann constant, respectively.

Unfortunately, this formula was found to give poor results in the solubility calculation,[40] but the correctness of Bjerrum's approach is apparent.[41] Any improvement should be based on free energy calculations, (33) and (34). It is apparent, however, that the dielectric constant is not the main determinant of solubility, but rather ion solvation, which depends on the solvent base strength of the hydrogen-bonding capability.

4 Determination of Thermodynamic Properties

The thermodynamics of transfer of a pair of gaseous ions into a solvent (solvation) has been widely used to describe the ion–solvent interactions that take place at infinite dilution. The thermodynamic solvation process is represented by the following equation:

$$(7) \quad C_g^+ + A_g^- \rightarrow C_{sol}^+ + A_{sol}^- \Delta Y_{sol}^\circ$$

where C^+A^- is the ion pair and ΔY_{sol}° represents the change in the thermodynamic function involved.

The various thermodynamic properties are now considered.

Solvation Enthalpies. Solvation enthalpies are obtained either through a Born–Haber cycle, as the difference between the enthalpy part of the lattice energy and the heats of solution at infinite dilution, or from a new method based on a mass spectrometric study of ion–solvent interaction in the gas phase.[42]

Heats of solvation obtained from heats of solution extrapolated to infinite dilution at 25°C are given in Tables 6 to 10.

Table 6   Heats of Solvation ΔH_S° in Methanol at 25°C (kcal/mole)

Salt	$-\Delta H_S^\circ$	Reference	Salt	$-\Delta H_S^\circ$	Reference
LiCl	214.0	325, 326	LiClO$_4$	188.0	325
LiBr	206.1	326	NaClO$_4$	161.1	325
LiI	199.0	326	Mg(ClO$_4$)$_2$	583.4	325
NaCl	187.9	326	Ca(ClO$_4$)$_2$	506.7	325
NaBr	180.8	326	Su(ClO$_4$)$_2$	475.8	325
NaI	173.7	325, 326	Ba(ClO$_4$)$_2$	443.3	325
KCl	167.80	326	Pb(ClO$_4$)$_2$	484.6	325
Kbr	167.7	326	LiNO$_3$	205.0	329
KI	152.9	326	NaNO$_3$	179.0	329
RbCl	161.6	326, 327	NH$_4$Br	157.3	329
RbBr	153.7	326	NH$_4$NO$_3$	155.5	329
RbI	146.6	326	AgNO$_3$	189.6	329
CsCl	152.9	325–327	N(CH$_3$)$_4$Cl	130.1	330, 331
CsBr	145.0	325	N(CH$_3$)$_4$Br	124.0	330, 331
CsI	137.8	325–327	N(CH$_3$)$_4$I	115.1	330, 331
ZnCl$_2$	686.6	328	N(C$_2$H$_5$)$_4$Cl	117.1	330, 331
CaCl$_2$	559.6	326	N(C$_2$H$_5$)$_4$Br	111.0	330, 331
SrCl$_2$	524.4	326	N(C$_2$H$_5$)$_4$I	102.1	330, 331
BrCl$_2$	504.8	326	N(C$_3$H$_7$)$_4$Br	115.0	330, 331

Table 7   Heats of Solvation $-\Delta H_S^\circ$ in Ethanol, Propanol, and Butanol at 25°C (kcal/mole)

Salt	Ethanol	Propanol	Butanol	Reference
LiClO$_4$	187.0	186.6	186.6	332
Pb(ClO$_4$)$_2$	478.1	476.4	475.5	332
Mg(ClO$_4$)$_2$	582.3	576.4	575.2	333
Ca(ClO$_4$)$_2$	500.3	496.7	490.1	333
Sr(ClO$_4$)$_2$	466.9	464.6	463 6	333
Ba(ClO$_4$)$_2$	435.4	433.1	432.6	333

Table 8 Heats of Solvation in Ethanol, Formic Acid, Dimethylsulfoxide, and Propylene Carbonate (kcal/mole)

Salt	Ethanol	Reference	Formic Acid	Reference	DMS	Reference	PC (Ref. 49)
LiCl	217.0	334			−214.3	337	−217.1
LiBr					−208.4	337	−198.9
LI					−201.2	337	−192.5
NaCI	−185.6	334	−186.0	327	−189.1	51	−181.1
NaBr	−178.0	334			−183.2	51	−176.1
NaI	−171.7	335			−176.0	51	−169.7
KCl	−166.7	336	−170.0	327	−170.7	51	−163.1
KBr	−159.4	336			−164.8	337	−158.6
KI	−153.1	336			−157.6	51	−152.2
CsCl			−164.0	327	−155.6	51	−151.2
CsBr			−157.0	327	−149.7	51	−146.0
CsI			−140.0	327	−142.6	51	−136.6
LiClO$_4$							−120.0
NaClO$_4$	−159.4	335					−161.4

To compare ionic solvation enthalpies, anionic and cationic contributions were separated making different extrathermodynamic assumptions.

In formamide solutions,[43] ionic contributions were determined by a modification of Verwey's method.[44] The solvation enthalpies of alkali halides for a given cation were found to be a linear function of the reciprocal crystal radii of the ions:

$$(8) \qquad \Delta H_{\mathrm{sol}}(C^+A^-) = A_{C^+} + \frac{B}{r_{A^-}}$$

Ahren's[45] ionic radii values were taken.

The constant A_{C^+} was identified with the experimental value of the cationic solvation enthalpy, assuming that a very large anion does not contribute to the solvation enthalpy: B was the same for all plots except that of cesium.

Latimer, Pitzer, and Slansky's modification[46] of Born's formula[47] was applied to the treatment of data on propylene carbonate solutions.[48,49] The results were expressed in terms of transfer properties, since the transfer enthalpies of cations could be reproduced very well. Assuming that the electrostatic part of the transfer enthalpy could be obtained from the Born equation and also that the structural contributions are nearly the same for Cs^+ and I^-, it was concluded[49] that

$$\Delta H_{\mathrm{tr}}(Cs^+) = \Delta H_{\mathrm{tr}}(I^-) = \tfrac{1}{2}\Delta H_{\mathrm{tr}}(Cs^+I^-) = -3.6 \text{ kcal/mole}$$

Table 9   Enthalpies of Solvation in Formamide, *N*-Methylformamide, *N*-Methylacetamide, *N*-*N*-Dimethylformamide, and *N*-Methylpropionamide (kcal/mole)

Salt	Formamide (Ref. 43)	*N*-Methyl-formamide	Refer-ence	*N*-Methyl-acetamide (Ref. 50)	*N*-*N*-Dimethyl-formamide	Refer-ence	*N*-Methyl-propionamide (Ref. 341)
LiF	−240.2						
LiCl	−210.6	−214.3	338		−215.7	339	−211.7
LiBr	−204.6			−198.7	−212.5	339	−206.1
LiI	−195.7	−198.5	50		−198.5	340	
NaF	−216.5						
NaCl	−187.9	−182.0	338				−186.3
NaBr	−181.0	−181.0	338		−184.0	339	−179.8
NaI	−171.9	−172.7	50	−171.6	−178.5	339	−171.3
KF	−196.7	−196.1	50	−195.4			−194.0
KCl	−168.1	−168.5	338	−167.6			
KBr	−161.2	−162.2	50	−161.1	−165.3	339	−161.3
KI	−152.1	−154.3	50	−153.3	−159.1	340	−152.8
RbF	−190.8						
RbCl	−162.1						
RbBr	−155.2						
RbI	−146.3	−148.2	50	− 47.0			−147.1
CaF	−180.5						
CsCl	−154.3	−154.3	338				
CsBr	−147.2						
CsI	−138.1	−139.6	50	−138.4			−138.9

Table 10   Single-Ion Enthalpies of Transfer from Water to Solvents (kcal/g-ion)

Ions	TMS	PC	DMF	DMSO	CH_3OH
Li^+	5.4	−0.05	1.5	−6.5	−3.7
Na^+	−4.05	−3.3	−5.9	−7.2	−4.8
K^+	−6.4	−6.0	−7.5	−8.45	−4.7
Rb^+	−6.8	−6.7	−7.2		−4.7
Cs^+	−6.3	−7.4	−6.8	−8.25	−3.8
$AsPh_1^+$	−2.85	−2.6	−2.15	−2.3	0.8
Cl^-	6.7	7.45	2.8	5.1	1.9
Br^-	3.35	4.35	−1.35	1.1	0.9
I^-	−1.6	0.1	−5.4	−2.6	−0.4
ClO_4^-	−4.7	−3.05	−7.9	−4.8	−1.1
BPh_4^-	−2.85	−2.6	−2.15	−2.3	0.8

From Ref. 55.

Similar assumptions for the transfer enthalpy of Cs^+ in formamide,[43] N-methylformamide,[50] and N-methylacetamide[50] yielded the values -2.9, -3.6, and 3.0 kcal/mole, respectively. Arnett and McKelvey[51] employed the assumption previously discussed by Grunwald and co-workers;[52] that is, that the tetraphenylarsonium cation and the tetraphenylboride anion have identical transfer enthalpies. The main difference between the two ions is the sign of their respective charge, which lies buried within similar large organic envelopes.

A similar pattern of anionic behavior in dimethylsulfoxide was observed[53] for ionic enthalpies of transfer as in propylene carbonate solutions. It was later found[54] that data on propylene carbonate solutions, as well as on sulfolane, dimethylformamide, and methanol,[55] could be interpreted in the same way.

Single ionic enthalpies of transfer from water to various organic solvents are listed in Table 10. It has already been pointed out[53] that the qualitative trend of the halide ions is the same in each case, whereas the varying trend of the alkali metal ions seems to reflect the differences in solvent basicity.

In considering ionic transfer with propylene carbonate (PC) as the reference solvent[54,55]—and PC is a nearly ideal solvent because of its low basicity—it should be mentioned that the coulombic interactions do not mainly govern the experimental transfer enthalpies and that anions and cations do not follow the same behavior pattern.

The order DMSO > DMF > TMS > PC is found in Table 10 for the solvating strength of these aprotic solvents toward Na^+, K^+, Rb^+; some specific interaction in DMSO toward the Li^+ accounts for its position. The fairly constant value of $AsPh_4^+$ for all aprotic solvents implies a small and nearly constant enthalpy of transfer for $AsPh_4BPh_4$.

The order of solvating strength of solvents toward anions is DMF > DMSO > TMS > PC; it is found that the enthalpy changes for small and large anions are strongly dependent on the nature of the aprotic solvent.

Solvation Entropies. The various methods of determining entropies of solvation have been reviewed by Conway and Bockris.[56] The entropies of the gaseous ions are calculated from the Sackur–Tetrode equation for monoatomic gases

$$(9) \qquad \bar{S}_g^\circ = R[\ln V + \tfrac{3}{2} \ln T + \tfrac{3}{2} \ln M] + C$$

where R is the gas constant, M the atomic weight, V is in cm^3/mole, T is the absolute temperature, and C is a constant. The entropies of the ions in solution can be obtained from EMF measurements, from solubility studies, or from a modified Born equation applied to free energies.[57] The entropy of solvation is then given by

$$(10) \qquad \Delta S^\circ = \bar{S}_2^\circ - \bar{S}_g^\circ$$

where $\bar{S}_2^{\circ}$ is the partial molal entropy of the electrolyte and $\bar{S}_g^{\circ}$ is the standard molal entropy of the gaseous electrolyte. Values for entropies of solvation obtained from data in the literature are given in Table 11.

Table 11   Entropies of Solvation $-\Delta S_S^{\circ}$ (cal/deg-mol–molal scale)

	Solvent				
Electrolyte	DMF (Ref. 62)	CH$_3$OH (Ref. 342)	C$_2$H$_5$OH (Ref. 342)	NMF (Ref. 125)	HCOOH (Ref. 343)
LiCl	77.3	86.2	89.8	63.7	
LiBr	73.7	80.2	83.1		
LiI	71.0	76.5	82.4		
NaCl	71.8	75.3	81.2	48.7	55.7
NaBr	68.2	69.9	74.4	47.7	
NaI	65.5	65.6	73.7		
KCl	68.3	58.5	73.6	45.4	60.1
KBr	64.7	54.5	66.8		
KI	62.0		66.1		
CsCl	62.1			42.7	88.8
CsBr	58.5				
CsI	55.8				111.5

The partial molal entropy of any particular electrolyte in the various solvents increases in the order

$$NH_3 < N\text{–}N\text{–}DMF \sim C_2H_5OH < CH_3OH < NMF < F < H_2O < D_2O$$

indicating that entropies are most positive for those solvents having the highest degree of internal order.

In the entropy correlations for liquid ammonia,[58] methanol,[59,60] and ethanol,[61] the total entropies of the solvated electrolytes were divided into ionic contributions by assigning, through a trial-and-error method, a value for S_2° of the hydrogen ion. The division was made in such a way that S_2° for both cations and anions in a given solvent, X, when plotted against S_2° for the same ions in water, fit on the same curve. Similar treatments were carried out by Criss et al.[62] for various solvents, and a linear relation was observed in each case.

$$(11) \qquad \bar{S}_2^{\circ}(X) = a + b\bar{S}_2^{\circ}(H_2O)$$

where a and b are constants characteristic of the solvent. These ionic entropies $S_2^{\circ}(X)$ are considered to be absolute entropies and are listed in Table 12. It is

Table 12 Standard Absolute Ionic Entropies $-\Delta S_S^\circ$ of Solvation (mole fraction scale) in Nonaqueous Solvents at 25°C (cal/deg-mol)

Ion	NH_3	F	NMF	DMF	CH_3OH	C_2H_5OH
Li^+	53.1	34.3	41.6	48.5	42.7	47.5
Na^+	50.0	33.2	31.8	43.2	39.5	44.6
K^+	45.7	25.9	28.5	39.6	33.2	38.5
Rb^+	41.0	24.7			30.0	
Cs^+	42.0	23.6	25.8	35.7	28.8	
Cl^-	40.4	25.6	28.1	38.6	32.4	37.8
Br^-	43.4	23.0	27.0	35.5	29.8	35.7
I^-	39.1	18.9		32.4	26.0	31.0

From Ref. 62.

further suggested that the absolute entropy of any given ion in a solvent is given by

$$(12) \qquad\qquad S_2^\upsilon = kS_{str} + C$$

where S_{str} is the part of the entropy of a solvent that arises because of its internal order and C is the partial molal entropy of the ion in a solvent with no internal structure. Equation 12 was tested by relating S_{str} to the deviation from the ideal boiling point of the solvent, ΔT_{bp}, and plotting S_2° versus ΔT_{bp}. The linear relation observed may be represented by

$$(13) \qquad\qquad \bar{S}_2^\circ = k' \Delta T_{bp} + C$$

Values of k' and C may be found in Table 13. The equation is useful for estimating ionic entropies for which no data are available. It was pointed out,

Table 13 Structural Constants of (13)

Ion	k' (cal/deg²-mol)	C (cal/deg-mole)
H^+	0.22 ± 0.04	-47.1
Li^+	0.19 ± 0.02	-39.7
Na^+	0.20 ± 0.02	-32.6
K^+	0.23 ± 0.02	-29.5
Cs^+	0.22 ± 0.02	-20.0
Cl^-	0.19 ± 0.02	-24.6
Br^-	0.23 ± 0.02	-24.5
I^-	0.25 ± 0.02	-20.7

however, that ΔT_{bp} is not an entirely satisfactory parameter, since it suggests that water is more ordered than D_2O, which is contrary to current views.[63] One interesting feature, nevertheless, is that k' is approximately the same for all ions, suggesting that the original solvent structure is completely disrupted in the vicinity of the ion. The negative values obtained for C simply imply that the ions would introduce some order in the solvent by orienting the solvent molecules. A similar behavior of ions in mixed solvents has been observed lately.[64]

Gibbs Free Energies. Free energies, G, may be obtained from the thermodynamic relation

$$(14) \qquad \Delta G = \Delta H - T\Delta S$$

but some methods give a direct determination of the free energy of solution. The EMF of a chemical cell is directly related to the change in the free energy involved in the solvation of the ion. In cells that are reversible with respect to both anions and cations, such as the amalgam cell

$$(15) \qquad M_yHg | MX | AgX - Ag$$

the standard free energy of solution of the electrolyte ΔG°_{sol} is obtained directly from the standard electrode potential of the cell E°:

$$(16) \qquad -\Delta G^\circ_{sol} = z\tilde{F}E^\circ$$

Free energies of transfer may be obtained by using electrochemical cells with transport. This may be realized by combining two amalgam cells in which the electrolyte is not in the same medium, for example, water and another solvent

$$(17) \qquad Ag - AgX | MX_{water} | M_yHg | MX_{solvent} | AgX - Ag$$

The net effect of the cell reaction is the transfer of MX from water to the other solvent. The standard potential of this double cell is given by the difference in the standard potential of the aqueous cell $_wE^\circ_{MX}$ and the nonaqueous cell $_sE^\circ_{MX}$. This is by definition the "medium effect"[65]

$$(18) \qquad \ln \gamma_0 = \frac{_wE^\circ_{MX} - {_sE^\circ_{MX}}}{RT}$$

By referring both standard states to the ion pair in the gaseous state, it may be shown that the medium effect expresses the difference between solvation and hydration energies.

In the solubility method, the standard free energy of solution for a solute is given by

$$(19) \qquad \Delta G^\circ_{sol} = -RT \ln K$$

where K is the equilibrium constant for the reaction

$$(20) \qquad MX \rightleftharpoons M^+ + X^-$$

In dilute solutions, the Debye–Hückel expression for the mean activity coefficients may be introduced, leading to:

$$(21) \qquad \Delta G_{sol}^\circ = -2.303[-\nu \log m - \log (\nu_+{}^{\nu+}\nu_-{}^{\nu-} + \nu A d_0^{1/2} m^{1/2})]$$

where ν_+ and ν_- are the numbers of positive and negative ions, respectively, A is the Debye–Hückel limiting slope, and d_0 is the density of the solvent. These methods were used by Izmailov[66] in the determinations and calculations of solvation free energies in various solvents; the data obtained are listed in Table 14, together with some recent values.

It is currently fully realized that there is no experiment whereby the free energy of transfer ΔG_{tr}° of a single ionic species can be determined.[12,67] Each of the methods that has been proposed rests on extrathermodynamic assumptions, which are examined next.

THE PLESKOV METHOD. An ideal reference ion should have the same free energy of solvation in every solvent. Pleskov[68] suggested that rubidium might serve as such a reference because of its low polarizability and relatively large radius. However, merely from electrostatic considerations it can be shown that noticeable free energies of transfer may be expected, hence Pleskov's assumption is not justified. Strehlow's approach,[69] based on the reduction–oxidation equilibrium between the ferrocene and the ferricinium ions, rested on the assumptions that only the ion is solvated (since the reduced form is neutral) and that the solvation free energy is the same,[70] irrespective of solvent. It was shown,[71] however, that the reduction–oxidation couple may be involved in some specific interaction with water and possibly with non-aqueous solvents.

MODIFICATION TO THE BORN'S EQUATION. Voet[72] suggested that a solvent characteristic term, R_+, be added to the crystal radii of cations and another term, R_-, to the crystal radii of the anions in order to obtain experimental values of the free energy of solvation ΔG from the Born equation, (26)

$$(22) \qquad -\Delta G = \frac{N_e{}^2}{2}\left(1 - \frac{1}{\varepsilon}\right)[(r_+ + R_+)^{-1} + (r_- + R_-)^{-1}]$$

where ε is the dielectric constant of the solvent, e is the charge of the ion, and r_+ and r_- are the crystal ionic radii of the cation and anion, respectively.

Latimer et al.[46] applied (22) to the alkali halides in water and methanol, and Strehlow[73] to other organic solvents. The ionic contributions to the free energy of solvation may be calculated from these equations using the appropriate values of R_+, R_-.

Table 14 Solvation Free Energies in Nonaqueous Solvents at 25°C (kcal/mole)

Electrolyte	NH$_3$ (Ref. 66)	N$_2$H$_4$ (Ref. 344)	CH$_3$OH (Ref. 66)	C$_2$H$_5$OH (Ref. 66)	C$_4$H$_9$OH (Ref. 78)	CH$_3$·COCH$_3$ (Ref. 344)	C$_5$H$_{11}$OH (Ref. 344)	CH$_3$CN (Ref. 344)	HCOOH (Ref. 66)	DMF (Ref. 57)	NMF (Ref. 125)
LiCl	188.9	187.0	187.7	184.8	186.0	185.0	186.0	178.5	194.3	193.5	203.2
LiBr	186.5	184.5	182.3	179.7	183.0	183.0	180.0	177.0	184.7	190.3	195.0
LiI	180.9	173.0	175.4	171.5	172.5	172.0	174.5	172.0	176.2	185.6	
NaCl	164.2	165.5	164.2	160.6	160.0	150.0	155.5	154.5	177.8	166.8	178.9
NaBr	161.8	163.0	159.9	156.8	157.0	148.0	149.5	153.0	168.2	163.6	170.7
NaI	156.2	152.5	152.2	148.1	146.5	137.0	144.0	148.0	158.7	158.9	
KCl		148.5			140.5	138.0	140.0	139.0	159.3	149.2	160.8
KBr		146.0			137.5	136.0	134.0	137.5	149.7	146.0	152.6
KI	136.8	135.5	135.2	131.5	127.0	125.0	128.5	132.5	141.2	141.3	
RbCl		143.5			131.5	132.0	135.5	184.5	156.3		
RbBr		141.0			128.5	130.0	129.5	133.0	146.7		
RbI	130.6	130.5		125.5[a]	118.0	119.0	124.0	128.0	138.2		
CsCl	130.9	135.0	131.4		128.5	127.0	127.5	125.5	143.6	135.8	139.7
CsBr	128.5	132.5	127.4		125.5	125.0	121.0	124.0	134.7	132.6	131.5
CsI	122.8	122.0	120.0	117.0[a]	115.0	113.0	115.5	119.0	126.2	127.9	
AgCl	199.7	202.0	178.9	178.6				179.0	197.1		
AgBr	196.3	199.5	173.6	173.6				177.5	188.7		
AgI	190.6	189.0	166.7	164.6				172.0	180.2		
CaCl$_2$	493.6	507.0						478.8	465.9		
ZnCl$_2$	666.0	647.5	624.6	617.5				600.0	644.9		
CdCl$_2$	677.0	590.5	560.4	558.4				540.0	567.0		

[a] Ref. 43.

Later investigators[71,74,76] fitted different values of R_+ and R_- in order to reproduce their experimental results; thus there are no single characteristic values of the increments for a given solvent. Therefore no theoretical significance can be attributed to (22); besides, the assumption of identity of the radii in the gaseous and solution states is not valid.

EXTRAPOLATION METHODS. These approaches rest on the assumption of vanishing free energy of solvation for infinitely large ions. Izmailov[77,78] has extrapolated data, either for pairs of cations to cations of infinite size or for combinations of isoelectronic alkali and halide ions to infinite principal quantum number. Similar values were obtained from both methods, but these long-range extrapolations involve some uncertainties, which are reflected by lack of internal consistency in the single ion medium effects determined.

Feakins et al.[79,80] tacitly assumed a rule of corresponding states between solvation and hydration (cf. Section 2) whereby a linear relationship

$$G_s^\circ(X^-) = G_w^\circ(X^-) + kr_c^{-1}$$

is assumed between ionic free energies of solvation, G_s°, and hydration G_w°, and the crystal ionic radius, r_c, thereby illustrating a particular case of the method of comparative measurement.[81]

Since this procedure does not allow for effects of structure making or breaking by the large cations, de Ligny and Alfenaar[82,83] have considered the free energy of transfer to consist of an electric part ΔG_{el}, expressed by the Born equation at low fields, and a neutral part $\Delta G_{neutral}$, corresponding to the contribution of an uncharged solute of the same size as the ion. They suggested extrapolating the difference $\Delta G_{tr} - \Delta G_{neutral}$ to $r^{-1} = 0$ to alleviate any discontinuity.

The results were claimed to be very accurate,[83] although very significant contributions of $\Delta G_{neutral}$ (calculated from the transfer of isoelectronic rare gases) were involved. These do not tally with Izmailov's estimate[78] that values of $\Delta G_{neutral}$ are only a small fraction of $\Delta G_{solvation}$. The linearity of the plots was not improved by this nonelectrolytic or structural effect correction; rather, a change of sign in the ΔG_{tr} of large cations was induced.

It should be noted[84] that in any event extrapolations in r^{-1} of the free energy quantities are invalid, in principle, since the leading term in the free energy of solvation is $Ze\bar{\mu}/r^2$, where $\bar{\mu}$ is the average dipole moment of the solvent. In addition, a Born charging energy will apply beyond the solvation shell. Hence no single, or single-power, term in reciprocal r can really represent free energies for ions in solutions. Compensation effects between dielectric saturation ionic radii and other factors may bring about an approximate representation in r^{-1}.

THE REFERENCE ELECTROLYTE METHOD. Popovych[85] proposed the choice of a reference electrolyte with a solvation energy that, in many common solvents, should be equally divisible between the anion and the cation. To meet these requirements, the reference electrolyte must consist of large symmetrical counterions, identical in size, surface charge density, and solvation properties, since it is assumed that the energy changes owing to solvent be equal for the two ions. Tetraphenylarsonium tetraphenylboride, tetraphenylphosphonium tetraphenylboride,[52] and triisoamyl-n-butylammonium tetraphenylboride [85,87,88] (TABPh$_4$) are composed of very large symmetrical ions with a central atom buried under insulating phenyl layers and are of approximately equal crystallographic size.

The electrolyte TABPh$_4$ was chosen,[71] for it had already been used in the evaluation of single-ion conductances in nonaqueous solvents.[89] Similar results were obtained with tetraphenylarsonium tetraphenylboride [86] as reference electrolyte.

Both reference electrolytes have a low solubility in many solvents, which makes it convenient to determine their free energies of solvation from solubilities without having to resort to uncertain activity corrections.

It should be pointed out that the criterion of equal free energies of solvation is based on the transport properties of the ions—say, equality of Stokes's radii—in various solvents, and this does not necessarily imply a direct correlation between transport and thermodynamic properties. Other factors such as dielectric relaxation, size, and secondary solvation effects should be taken into account. The problem of obtaining the ionic contribution would still persist, especially in dipolar aprotic solvents, which differentiate strongly between the relative solvation of cations and anions.

Table 15   Ionic Free Energies of Transfer from Water to Methanol at 25°C (kcal/mole)

Ion	Ref. 66	Ref. 73	Ref. 80	Ref. 83	Ref. 85	Ref. 86	Ref. 90
H$^+$	+3.1	0.07	−2.93	−1.974	1.85		−5.84
Li$^+$	+1.0		−3.8				−6.7
Na$^+$	+3.0	−0.55	−2.9	−1.830			−5.9
K$^+$	+2.0	−0.17	−2.0	−1.223	1.80	1.7	−4.9
Rb$^+$	2.8	−0.06			0.89		−5.3
Cs$^+$							−5.5
Ag$^+$					0.78	1.2	—
Cl$^-$	−3.0	5.46	8.57	7.768	2.08	1.9	11.20
Br$^-$	−1.0	4.81	9.05	7.299	−0.95		10.60
Z$^-$	0.2	4.03	7.02	6.222			9.70
SO$_4^{2-}$							15.18
Picrate$^-$						−1.2	
TAB$^+$					−4.30		
BPh$_4^-$					−4.30	−4.2	

It is instructive to look at the values quoted in the literature for the ionic free energies of transfer from water to methanol obtained by the various methods. They are compared with the results of Case and Parsons [90,91] for the real free energy of transfer ($\Delta\alpha^i$) in Table 15.

The disparity in results is only too well illustrated. There is little agreement on the magnitude of the free energy of transfer, and sometimes the sign differs, too. This is not too surprising, since the problem of evaluating individual ionic thermodynamic properties cannot be solved by thermodynamic means and each value obtained depends on the assumptions made.

Partial Molal Volumes.   The partial molal volume V_2° of an electrolyte at infinite dilution is thermodynamically defined [92] as the actual increase in volume that takes place when one gram-mole of electrolyte is added to an infinite amount of solvent. This is directly related to the effect of the ions on the solvent. If V_{in} is in the intrinsic or effective molal volume occupied by the ions in solution, the change in volume undergone by the solvent ΔV or electrostriction [93] is given by

$$(23) \qquad\qquad \Delta V = V_{in} - \overline{V}_2^\circ$$

The intrinsic volumes of electrolytes can be obtained experimentally from the concentration dependences of the apparent molal compressibility, ϕ_K, and the apparent molal volume, ϕ_V; the partial molal volume, $\overline{V}_2^\circ$, is identical with ϕ_V° at infinite dilution (cf. Chapter 15).

The paucity of data on compressibility of electrolytes in nonaqueous solvents does not permit the calculation of intrinsic volumes, although they may be obtained from the continuum theory as will be seen later.

The partial molal volumes of electrolyte at infinite dilution V_2° in various solvents are given in Table 16, where values for $\overline{V}_2^\circ$ in methanol have been obtained from the original density measurements recommended by Redlich and Meyer.[94]

Following Mukerjee's method [95] of dividing V_2° into its ionic components in water, Millero [96] studied the partial molal volumes of ions in methanol, water, and N-methylpropionamide (NMP),[97] using the Frank–Wen [98] model for ion–solvent interactions. The correspondence equation used by Criss et al.[62] for ionic solvation entropies was applied to the determination of ionic partial molal volumes in NMP and methanol, and reasonable agreement, within the experimental error, was found with the values obtained earlier through Mukerjee's procedure. No simple correlation to common physical properties such as dielectric constant could be made, and the empirical constants obtained are supposed to account for dielectric saturation effects as shown by electrostriction effects.

The tetraalkylammonium halides offer another method of assigning absolute $\overline{V}_2^\circ$ values to the halide ions in methanol.[99] Since these large cations

Table 16 Partial Molal Volumes of Electrolytes at Infinite Dilution in Various Solvents at 25°C (ml/mole)

Electrolyte	Methanol (Ref. 99)	Form-amide (Ref. 345)	NMA (Ref. 347)	NMP (Ref. 97)	Ethylene-diamine (Ref. 348)	HCOOH (Ref. 349)
LiCl	−9.69	19.60	20.5		13.5	
LiBr	−2.61					
NaCl	−6.00	21.10		30.7		15.5
NaBr	0.98	28.00		35.8		20.4
NaI	9.16	39.85				31.5
NaNO$_3$		33.55		39.6	35.5	
KCl	5.95	32.00		35.5		18.5
KBr	13.00	38.90		40.9		23.4
KI	21.13	50.75	45.9			34.5
KNO$_3$		44.10		44.7		
RbCl		35.90				28.3
RbBr		42.60				33.2
RbI		54.65				44.5
RbNO$_3$		48.25				
CsCl		42.30				33.4
CsBr		49.30				38.4
CsI		61.05				49.5
CsNO$_3$		54.65				
NH$_4$Cl	13.78	37.20				
NH$_4$Br	20.75	44.05				
NH$_4$I						
NH$_4$NO$_3$	29.83	49.24				
AgNO$_3$					20.0	
(CH$_3$)$_4$NCl	83.0					
(CH$_3$)$_4$NI						
(C$_2$H$_5$)$_4$NCl	140.7					
(C$_2$H$_5$)$_4$NBr	148.0					
(C$_2$H$_5$)$_4$NI		185.5[a]	175.4			
(C$_3$H$_7$)$_4$NBr	220.0					
(C$_3$H$_7$)$_4$NI		255.5[a]	247.5			
(C$_4$H$_9$)$_4$NBr	286.2					
(C$_4$H$_9$)$_4$NI		322.4[a]	322.9		325.0	
(C$_5$H$_{11}$)$_4$NI		394.0[a]	391.1			
(C$_6$H$_{13}$)$_4$NI			461.2			
(C$_7$H$_{15}$)$_4$NI			527.3			

[a] Ref. 346.

do not show any specific interaction with methanol,[100] a plot of $\overline{V}_2^\circ$ of these salts against the molecular weight of the cations should give, when extrapolated to 0 molecular weight, the absolute partial molal volume of the halide ion. Data published for tetraalkylammonium bromides in methanol yield a value of V_2° equal to 6 ml/mole for the bromide ion. This result is in very good agreement with the value obtained from density[100] and transport measurements and should provide a standard table of absolute ionic partial molal volumes.[101]

5   Theoretical Aspects of Ion–Solvent Interactions

Free Energy of Solvation.   If the free energy of solvation is defined as the change in energy undergone by an ion in passing from a point in a vacuum infinitely far from the solvent into the bulk of the solvent,[56] the ion has to pass through the vacuum–solvent interface. The real free energy of ionic solvation $\alpha_i{}^S$ is then defined as the total transfer energy, which can be split into two parts:[102] an electrostatic contribution owing to the passage of the ion through the surface potential at the vacuum–solvent interface and the free energy of the ion–solvent interaction or chemical solvation. Thus

$$(24) \qquad \alpha_i{}^S = \Delta G_i{}^S + Z_i \tilde{F} \chi(S)$$

where $\alpha_i{}^S$ and $\Delta G_i{}^S$ both refer to one mole of ions, Z_i is the valence of the ion, $\tilde{F}$ is Faraday's constant, and χ is the surface potential of the solvent.

The real free energy of solvation of the ion, i, may be determined from Volta potential difference measurements in galvanic cells. In the published measurements [90,91,103–106] of the real potentials of ions in nonaqueous solutions, the compensation potential [103] between an aqueous and a nonaqueous solution was determined in order to obtain the real free energy of transfer of ions from water to another solvent.

$$(25) \qquad \Delta^S \alpha_i{}^S = \alpha_i{}^S - \alpha_i{}^w$$

The absolute real free energies of solvation of the ions in nonaqueous solvents were obtained using Randles's values for aqueous solutions. The available data are given in Table 17. Some uncertainties in the values listed are the result of assumptions made in estimating the ionic activity coefficients in the various solvents.

Electrostatic Theories of Ion–Solvent Interactions.   Attempts to calculate the free energy of solvation in a given solvent have been only partially successful and were mostly directed towards the calculation of ionic contributions.

Table 17   Real Free Energies of Ions in Nonaqueous Solvents

Ion	Methanol (Ref. 90)	Methanol (Ref. 105)	Ethanol (Ref. 90)	Butanol (Ref. 90)	Form-amide (Ref. 90)	Formic Acid (Ref. 90)	Aceto-nitrile (Ref. 90)
H^+	266.3	265.1	265.9	264.5	264.0	253.8	257.8
Li^+	228.8	127.4	127.1	127.5		124.3	121.1
Na^+	104.1	103.0	102.0	101.5		102.3	98.8
K^+	86.1	87.1	86.7	82.0	85.0	83.8	83.0
Rb^+	80.8	81.1	80.7	73.0	77.9	80.8	78.2
Cs^+	76.5	71.1	72.7	70.0		71.8	70.3
Cl^-	59.5	60.5	58.8	58.5	66.8	68.2	55.7
Br^-	54.3	54.2	53.1	55.5		60.2	54.2
I^-	47.5	47.8	46.7	45.0		50.7	49.5
Tl^+	87.1		87.3		55.8		
Ag^+	120.8		120.7			121.8	124.5
Cu^+							154.0
Zn^{2+}	494.8		489.2		491.6	495.6	477.5
Cd^{2+}	442.8	434.3	439.7		438.0	434.6	427.6
Cu^{2+}	507.3		513.3		507.2		519.6
Pb^{2+}	372.5		369.3		368.3		351.6
Ca^{2+}						368.6	368.8

In Born's model,[47] the ion is represented by a rigid conducting sphere of radius, r_i, and charge, Ze, in a medium of dielectric constant, ε. The change in electrostatic free energy, experienced when the sphere is transferred from a vacuum into the solvent considered, yields the free energy of solvation ΔG of the ion in the given solvent as expressed by

$$(26) \qquad \Delta G = -\frac{(Z_i e)^2}{2r_i}\left(1 - \frac{1}{\varepsilon}\right)$$

This expression implies identical radii in the gas and solvent phases. The neglect of dielectric saturation caused in the medium by the intense field of the ions does not improve the correlation between experimental and calculated values. The use of empirical corrections, such as a modified set of radii or the addition of constant decrements in the ionic radii, is not considered here.[74]

Noyes's approach[107,108] has been generalized to apply to any solvent as follows. The ionic free energy of solvation ΔG_S° is defined by

$$(27) \qquad \Delta G_S^\circ = \Delta G_{con}^\circ + Z\,\Delta G_R^\circ$$

where ΔG_{con}° is the conventional free energy of solvation of the ion and ΔG_R° is a reference free energy of solvation. The value of ΔG_{con}° can be calculated unambiguously from thermodynamic data, if available. The solvation energy was considered to be strictly electrostatic, ΔG_{el}°, if the solvation process may be described as the sum of three steps: discharging the gaseous ion, solvating the then-neutral species, and recharging the solvated neutral species. This process implies that there is 0 free energy of solvation for neutral species, and therefore

$$(28) \qquad -\Delta G_{con}^{\circ} = Z\,\Delta G_{R}^{\circ} - \Delta G_{el}^{\circ}$$

Assuming that the actual electrostatic charging energy may be represented as a power series expansion in the reciprocal radius

$$(29) \qquad \Delta G_{el}^{\circ} = -\frac{Z^2 e^2}{2}\left(1 - \frac{1}{\varepsilon}\right) r^{-1} - Z^2 A r^{-2} - Z^2 B r^{-3} - Z^2 C r^{-4} - \cdots$$

with the Born equation holding in the limit for large ions, the missing constants in (29) may be calculated by the method of least squares for any solvent for which sufficient experimental data are available.

The empirical constants A, B, C are not assumed to have any specific physical significance; they reflect the deviations from the continuous dielectric medium model inherent in the Born equation. They may possibly be interpreted as arising from specific ion–solvent interactions (cf. Buckingham[109]) or simply from the fact that ions see solvent molecule entities.[110]

From the values of ΔG_{R}° obtained by this procedure, the absolute ionic free energies of solvation were calculated using (27). For ammonia, the best values of ΔG_{R}°, A, and B, in kcal/mole, were 80.009, -67.603, and 10.402, respectively. This procedure is an improvement over earlier methods but is open to the same criticisms.

A more rigorous approach[2,112] results from the use of the fundamental equation for the free energy of a dielectric continuum in an electrostatic field[111]

$$(30) \qquad \Delta G = \frac{1}{4\pi} \int \int \mathbf{E} \cdot d\mathbf{D}\, dv$$

where $\mathbf{E}$ and $\mathbf{D}$ are the field strength and electric displacement vectors, respectively, in a volume element, dv.

Equation 30 may be shown to reduce to Born's equation if the dielectric constant is assumed to be independent of field strength (i.e., that no dielectric saturation takes place), and if (30) is referred to a standard gaseous reference state in which the ion has the same radius as in the solvent considered. Assuming spherical symmetry around the ion, the element of volume will be given by $dv = 4\pi r^2\, dr$. Using the differential dielectric constant defined by

$$\varepsilon_d = \frac{d\mathbf{D}}{d\mathbf{E}}$$

and substituting in (30), we obtain the compact basic equation

$$(31) \qquad G = \frac{N}{2} \int_{r_e}^{\infty} \int_{0}^{E_r} \varepsilon_d \, d(E^2) r^2 \, dr$$

where E_r is the field strength at a distance r, and r_e is the intrinsic or effective radius of the bare ion in solution.[93] The double integral represents, the increase in the electrostatic part of the free energy of the dielectric, owing to the field of the ion. Since the localization of the free energy is a formal description, we may look upon it as the result of the interaction between the ion and the dielectric continuum. Referring it to the reference gaseous phase, a general expression for the free energy of solvation ΔG_{sol} was obtained

$$(32) \qquad \Delta G_{sol} = \Delta G - N \frac{Z^2 e^2}{2 r_g}$$

where r_g is the radius of the ion in the gaseous phase. The last term of the expression may be identified with the standard free energy of ionization of the ion,[113,114] thereby dispensing with the need for gaseous ionic radii. The procedure for the computation of the double integral has been described[93] and was applied to calculations of the free energy of ions in methanol solution[99] as a function of the intrinsic radius, r_e. The entropy and enthalpy of solvation are then obtained from standard thermodynamic relations

$$(33) \qquad \Delta S = \frac{N}{2} \int \int \left(\frac{\partial \varepsilon_d}{\partial T} \right)_{P,E} d(E^2) r^2 \, dr$$

$$-\Delta H = -\Delta G + T \Delta S$$

$$(34) \qquad \Delta H = -\frac{N}{2} \left[\int \int \varepsilon_d \, d(E^2) r^2 \, dr - T \int \int \left(\frac{\partial \varepsilon_d}{\partial T} \right)_{P,E} d(E^2) r^2 \, dr \right]$$

Thermodynamics of a Fluid in an Electrostatic Field. Many attempts have been made to treat the thermodynamics of a fluid in an electrostatic field rigorously. Only in 1955, as a result of Frank's clear thermodynamic analysis,[115] was a method found that permitted the field strength E of the electrostatic field to be treated as a variable of state of the fluid in as general a sense as the more usual variables of state, pressure P and temperature T. Other treatments were restricted either to constant volume[116–118] or to the specific example of a small condenser at a constant uniform field in a large vessel of fluid.[118–121] It was possible to extend the latter treatment to the case where the field changes continuously within the dielectric[2] and to calculate various thermodynamic properties of ions at infinite dilution.

From electromagnetic theory it is possible to show[2] that the isothermal, isobaric work done on a fluid is given by the expression

$$(35) \qquad W = \frac{1}{4\pi} \int \int \bar{E} \cdot d\bar{D} \, dv$$

In free space $D = \varepsilon_0 \bar{E}$, where ε_0 is called the permittivity of the free space. In a dielectric medium, such as a fluid, $\bar{D}$ is formally expressed in a similar manner, and ε is generally a symmetric tensor of the second rank or dyad. In the case of isotropic media, however, ε reduces to a scalar quantity and

$$(36) \qquad \bar{D} = \varepsilon \bar{E}$$

From (35) the elementary electrical work done on the fluid during any infinitesimal and reversible change is

$$(37) \qquad dW = \frac{1}{4\pi} \left[\int \bar{E} \cdot d\bar{D} \right] dv$$

The increment in internal energy, dW, is obtained by the application of the first law of thermodynamics to an infinitesimal reversible process involving changes in P, v, and E.

$$(38) \qquad dU = dQ - dW = dQ - P\,dv + \frac{1}{4\pi} \left(\int \bar{E} \cdot d\bar{D} \right) dv$$

where dU is the increment in internal energy of the fluid element, dW is the total work done by the sample, and dQ is the heat absorbed in the process.

By the second law of thermodynamics, $dQ = T\,dS$ and

$$(39) \qquad dU = T\,dS - P\,dv + \frac{1}{4\pi} \left(\int \bar{E} \cdot d\bar{D} \right) dv$$

We may define a characteristic function, H^*, analog to the enthalpy

$$H^* = U + Pv - \frac{v}{4\pi} \int \bar{E} \cdot d\bar{D}$$

where

$$(40) \qquad dH^* = T\,dS + v\,dP - \frac{v}{4\pi} \int \bar{E} \cdot d\bar{D}$$

and an analog Gibbs free energy, G^*, where

$$(41) \qquad dG^* = -S\,dT + v\,dP - \frac{v}{4\pi} E\,dD + \Sigma \mu_i\,dn_i$$

which has been generalized to a multicomponent fluid.

By introducing the differential dielectric constant, ε_d, the fundamental equation is obtained

$$(42) \qquad dG^* = -S\,dT + v\,dP - v\frac{E\varepsilon_d}{4\pi}\,dE + \Sigma \mu_i\,dn_i$$

from which differential thermodynamic relations will be obtained by cross differentiation. At this point it should be stressed that in the application of the thermodynamics of a fluid in the electrostatic field to the calculation of ionic thermodynamic properties, changes in the properties of the solvent are occurring, although they are assigned to the ions considered. The electrostatic role played by the ions as a source of field strength does not entail any change in their intrinsic properties on account of either a charging or a dissolution process, since they are supposed to be present in the solution initially in their final state.

The partial molal volume V_2° of an electrolyte at infinite dilution, already defined [92] as the actual increase in volume when one mole of electrolyte is added to an infinite amount of solvent, was shown to be caused by an increase in volume owing to the presence of the electrolyte, which at equilibrium occupies an effective or intrinsic volume V_{in}, and a decrease in volume owing to the contraction or electrostriction ΔV of the solvent under the influence of the electric field caused by the ion.

The change in volume undergone by the solvent due to the change in field strength dE is obtained from the fundamental equation

$$(43) \qquad v^{-1}\left(\frac{\partial v}{\partial E}\right)_{\mu,T} = \frac{E}{4\pi}\left(\frac{\partial \varepsilon_d}{\partial p}\right)_{E,T}$$

This is the isotomic electrostriction, since the change in volume must be carried out at constant chemical potential both inside and outside the field.[2] The total electrostriction produced by N ions is

$$(44) \qquad \Delta V = \frac{N}{2}\int_{r_e}^{\infty}\int_{0}^{E_r}\left(\frac{\partial \varepsilon_d}{\partial p}\right)_{E,T} d(E^2)r^2\, dr$$

Integration of the electrostatic volume change is performed up to r_e, the intrinsic radius of the ion, when by definition $2.52r_e^3 = V_{in}$ and by using Graham's relation [122]

$$(45) \qquad \varepsilon_d = n^2 + \frac{\varepsilon_0 - n^2}{1 + bE^2}$$

where n is the refractive index and b is a parameter independent of E. For methanol, $b = 2.2 \times 10^{-8}$. The limiting case when the dielectric constant, ε_0, is taken to be independent of the field strength may be shown to reduce to the expression derived from the Born equation

$$(46) \qquad \Delta V_{Born} = \frac{N}{2}\frac{Z^2 e^2}{\varepsilon_0^2}\left(\frac{\partial \varepsilon_0}{\partial p}\right)_T r^{-1}$$

which for methanol yields

$$(47) \qquad \Delta V_{Born} = 24.1\frac{Z^2}{r_e}$$

where ΔV is expressed in ml/mole and r_e in angstroms. The partial molal volume V_2° in methanol solutions was calculated from (44) for every r_e,[99] and the ionic contributions were evaluated taking $V_2^\circ = 6.0$ ml/mole for the Br^- ion.[100] It was found that cations are larger in methanol solutions than in aqueous solutions, whereas anions tend to be smaller. This holds even for large cations such as the tetraalkylammonium ions. The difference, however, grows smaller with increasing cation size.

The dependence of the differential constant ε_d of methanol and other alcohols on the distance r from the ion is similar to its behavior in water. The dielectric constant, above a certain initial value for r, is equal to the bulk dielectric constant and is independent of the field strength. If the dielectric constant is less than another critical value, for $r = r_0^\circ$, it is again field independent.

This picture of ions in solution extends the model suggested by Frank and Wen[98] in aqueous solutions to alcohol solutions as well. The charge of the ion is considered to be inside a small spherical cavity surrounded by three concentric regions. The innermost region is one of "immobilization" (its upper limit is the radius of dielectric saturation r_0°). The second is more randomly organized, and the third region corresponds to the bulk solvent. The radii of dielectric saturation, r_0°, for various alcohols are given in Table 18.

The previously mentioned calculation of the electrostatic part of the free energies of the dielectric due to the ions was also applied to the evaluation of the "medium effect."[123] The primary medium effect, usually denoted by $\log \gamma_0$ and given by Owen[124] as

$$(48) \qquad 2K \log \gamma_0 = {}^wE_N^\circ - {}^sE_N^\circ$$

where ${}^wE_N^\circ$ and ${}^sE_N^\circ$ stand for the standard potential of an ion pair in water and another solvent, respectively, represents the effect of transferring a pair of ions from one solvent at infinite dilution to another at infinite dilution

Table 18   Radii of Dielectric Saturation for Nonaqueous Solvents

Solvent	Radius, r_0° (Å)
Methanol	1.90
Ethanol	2.10
Propanol	2.35
Isopropanol	2.40
Butanol	3.20
Pentanol	3.60
Hexanol	3.80
Glycerol	1.97

where the only effects are the ion–solvent interactions. This is equivalent to the difference between the Gibbs free energies of solvation or the free energy of transfer. Indeed, the simplest expression for the primary medium effect was given by the Born equation, thereby implying that the standard cell potential is a linear function of the reciprocal of the dielectric constant.[125] This prediction, however, is not borne out by experiment and there is usually considerable departure from a straight line.

The application of the above treatment, which takes into consideration the change of the dielectric constant of the solvent within the ion field, leads to the following expression:

$$
\log \gamma^\circ = \frac{1}{2kT} \left(n^2 \int_{r_e}^{\infty} E_r^2 r^2 \, dr + \frac{\varepsilon_0 - n^2}{b} \int_{r_0}^{r_0^\circ} \ln \left(1 + b E_r^2\right) r^2 \, dr \right)_{\text{water}}
$$
$$
- \frac{1}{2kT} \left(n^2 \int_{r_e}^{\infty} E_r^2 r^2 \, dr + \frac{\varepsilon_0 - n^2}{b} \int_{r_0}^{r_0^\circ} \ln \left(1 + b E_r^2\right) r^2 \, dr \right)_{\text{solvent}}
$$
(49)

where n is the refractive index of the solvent, ε_0 is the static dielectric constant, b, r_0, and r_0° are constants characteristic of the solvent, and E_r is the field strength at a distance r from the ion. Calculated values of $\log \gamma^\circ$ for the transfer of HCl from water to methanol were found to be in reasonable agreement with experimental results.[123]

The calculation of the entropy of solvation should be carried out at a constant chemical potential of the solvent.[115] The change in entropy of the solvent when subjected to an electrostatic field of field strength E under these conditions is obtained from (42). Cross differentiation yields

$$
\left(\frac{\partial S}{\partial E} \right)_{\mu, T} = \frac{vE}{4\pi} \left(\frac{\partial \varepsilon_d}{\partial T} \right)_{P, E}
$$
(50)

By integrating around the ion using spherical symmetry, the electrostatic contribution to the entropy of solvation of N ions is found to be

$$
\Delta S = \frac{N}{2} \int \int \left(\frac{\partial \varepsilon_d}{\partial T} \right)_{P, E} d(E^2) r^2 \, dr
$$
(51)

This evaluation of the entropy, as well as the evaluation of the enthalpy, involves the still unknown temperature dependence of certain parameters. In addition, structural effects certainly contribute significantly to the entropy of solvation, so there is no way to appraise the above calculation for the time being.

It is believed that the most significant consequence of this treatment is that the picture of an ion in solution, as obtained from a continuum theory, corresponds very closely to that obtained by Frank and Wen[98] on molecular grounds; namely, that an ion is a charge contained in a cavity that is surrounded by three concentric regions. The importance of this correspondence between continuum and molecular models, whereby the dielectrically saturated layer matches the primary solvated layer and the other two regions correspond to the secondary solvated layer and the bulk solvent, respectively, should not be overlooked in structural considerations.

Structural Approaches. Eley and Pepper[126] attempted to apply an earlier treatment[127] to the solution of ions in methanol. The heat of solvation was evaluated by splitting the process into two stages.

1. A cavity the size of the ion was made in the solvent. The internal energy change was generally taken to be 8 kcal, and the reorientation energy of methanol molecules, due to the formation of the hole, was neglected.

2. The ion was removed from the gas phase and placed in the cavity, which involved a change in the energy of methanol molecules in the first coordination shell (tetrahedral configuration). The energy of interaction between the ion and the rest of the solvent was given by a modified Born equation

$$(52) \qquad E = \frac{NZ^2e^2}{2(r + 2r_s)} \left(1 - \frac{1}{\varepsilon} - \frac{T}{\varepsilon^2} \frac{\partial \varepsilon}{\partial T} \right)$$

where $2r_s$ is the diameter of a solvent molecule.

Fairly good agreement was obtained with experimental data. The entropies, however, could not be reproduced by this simple model, which stresses only the electrostatic ion–dipole energy.

It was suggested that the first shell of oriented dipoles may affect solvent molecules further away. A more detailed model[43] based on the Buckingham theory[109] involves a primary solvation layer composed of the nearest solvent molecules rigidly bound in either tetrahedral or octahedral configuration within a continuum having macroscopic properties of the pure solvent. Within the solvated layers, the energies of interaction between the ion and polarizable spheres with permanent multipole moments and between the spheres themselves are taken into account. The polarization of the solvent due to the solvated entity completes the description of the interaction energy. These detailed interaction energies computed for formamide solutions and the ionic enthalpies of solvation in formamide were calculated for both a tetrahedral and an octahedral model. Only the calculated enthalpies for the octahedral coordination were in fairly good agreement with the experimental values. The deviations probably result from neglect of the orientation effects outside the first solvated layer and the possibility of a continuous change in the number of solvated molecules. Hydrogen bonding of the anion and the libration of the molecules in the first shell were not taken into account.

The Solvation Approach to Ion–Solvent Interaction. The use of a specific model, the hard sphere model of a solvated ion, was suggested[128] for the correlation and interpretation of apparent molar properties of electrolytes in solution.

The model of a solvated ion consists of a spherical cavity containing the charge of the ion, with an outside spherical shell enclosing the solvation shell, n solvated molecules of solvent. The high electrostatic field predominating at the surface of the ion or the inner surface of the solvation shell causes dielectric saturation and makes this shell incompressible.[129] The molal volume V_s of the solvated electrolyte may be shown to be[130]

$$(53) \qquad V_s = \phi_v + n_s \overline{V}_0$$

where ϕ_v is the apparent molal volume of the solvent. At infinite dilution

$$V_s^\circ = \phi_v^\circ + n_s^\circ \overline{V}_0$$

Much confusion exists in the definition of solvation number.[101] Bockris[131] suggested the term "primary solvation number" for the number of molecules near an ion that have lost their translational degrees of freedom and move as an entity with the ion during its Brownian motion. Accordingly, we identify the solvation number, n_s°, defined by (54), as a primary solvation number. "Secondary solvation" is caused by electrostatic interaction beyond the first solvation shell and depends on the observed phenomena. If the molal volume of the solvated electrolyte is available from other sources, such as kinetic data, solvation numbers are then obtained from (54). Several methods have been suggested for the determination of primary solvation numbers.

DENSITY METHOD.   The application of density measurements to the calculation of primary solvation numbers is limited to the calculation of the apparent molal volume of the electrolyte. We have already seen from (54) that the molal volume of the solvated ions must also be known, and that it may be obtained from kinetic methods, which may or may not involve secondary solvation contributions.

Viscosity Method.   The viscosity of electrolytes in solution are usually interpreted in terms of the Jones–Dole equation[132]

$$(55) \qquad \frac{\eta}{\eta_0} = 1 + Ac^{1/2} + Bc$$

where η and η_0 are the viscosities of the solution and the solvent, respectively, A is a positive, computable constant[133,134] representing the contribution of the interionic forces, B is an empirical constant representing the contributions of cospheres of the ions,[135] and c is the molar concentration. It has been shown[136–138] that, at concentrations higher than 0.1 M, A may be neglected and B may be identified[136–141] with the product KV_s°, where K is a shape factor equal to 2.5 for spheres.[142] The solvation number, n_s^o, is then obtained from the relation

$$(56) \qquad n_s^o = \frac{0.4B - \phi_v^o}{V_0}$$

Results obtained from literature data are listed in Table 19.

Table 19   Solvation Numbers Obtained by the Viscosity Method

	Solvent	
Ion	Methanol	Formamide
Li^+		4.0
Na^+		4.0
K^+	6.0	2.0
Rb^+		1.5
Cs^+		1.5
NH_4^+		1.0
$(CH_3)_4N^+$	0	0
$(C_2H_5)_4N^+$	0	0
Cl^-	1.5	1.5
Br^-	1.0	1.0
I^-	1.0	0
SCN^-		0
NO_3^-		0

Mobility Method. It was suggested[143] that the conventional, primitive model[144] of Stokes should apply to nonaqueous solvents and yield the relation

$$(57) \qquad \lambda_i^o = \frac{|Z_i|\tilde{F}^2}{6\pi\eta_0 r_s}$$

where λ_i^o is the equivalent conductance of the ion at infinite dilution, and Z_i and r_s are the charge and radius of the ion, respectively. The assumption was generally made that the volume V_s^s of the solvation shell is

$$(58) \qquad V_s^s = \frac{4\pi}{3}(r_s^3 - r_c^3)$$

where r_c is the crystal radius of the ion. n_s would then be obtained from

$$(59) \qquad n_s = \frac{V_s^s}{V_0}$$

Assuming Stokes's relation to hold, the ionic solvated volume should be obtained, because of packing effects,[145] from

$$(60) \qquad V_s^\circ = 4.35r_s^3$$

when V_s° is expressed in mol/mole and r_s in angstroms. Then the solvation number may be obtained as before from (59). Limitations of Stokes's law make the mobility method inapplicable to ions of medium size. Various corrections, both empirical[134,146,147] and theoretical,[148-151] have been suggested in order to apply it to most of the ions. Corrected solvation numbers only are quoted from the literature data in Table 20.

Table 20   Solvation Numbers Obtained from the Mobility Method

				Solvent			
Ion	Meth-anol	Ethanol	DMSO	Acetone	Form-amide	DMF	N-Methyl-acetamide
H^+					3.5	3.0	
Li^+	5.0	5.0	4.3	2.9	5.4	3.2	5.1
Na^+	5.0	4.0	2.3	2.6	4.0	3.0	3.5
K^+	4.0	3.0	2.4	2.0	2.5	2.0	3.3
Rb^+	3.0	2.0			2.3		
Cs^+	3.0	2.0			1.9		2.6
$(C_4H_9)_4N^+$	0		0				
NH_4^+	2.0	2.0		1.0	2.0		2.7
Ba^{2+}					7.0		9.0
Sr^{2+}					6.0		8.6
Ca^{2+}					6.0		8.6
Mg^{2+}							10.3
Cl^-	1.5	2.0	0.6	1.0		0.5	2.1
Br^-	1.0	2.0	0.5	1.0	1.0	0.5	1.7
I^-	1.0	1.0	0.4		1.0	0.5	1.5
ClO_4^-	2.0	2.0	0			1.0	
NO_3^-	2.0	2.0	0				1.5
SCN^-			0				1.3

COMPRESSIBILITY METHOD.   Assuming the solvated ion to be incompressible, Passynski[129] pointed out the possibility of determining solvation numbers from the decrease in the compressibility of electrolyte solutions as compared to that of the solvent. If B_0 and β are the compressibility of the solvent and solution, respectively, the incompressible fraction of the solution[56] is $(1 - \beta/\beta_0)$. Since by definition[128]

$$(61) \qquad V_s = \frac{1000(\beta_0 - \beta)}{c\beta_0}$$

the solvation number is then

(62)
$$n_s = \frac{1000 d_0 (\beta_0 - \beta)}{c \beta_0 M_0} - \frac{\phi_v}{V_0}$$

$$n_s = \frac{\phi_K d_0}{M_0 \beta_0}$$

where d_0 and M_0 are the density and molal weight of the solvent, respectively, and ϕ_K is the apparent molal compressibility of the electrolyte. At infinite dilution this reduces to

(63)
$$n_s^\circ = -\frac{\phi_K^\circ d_0}{M_0 \beta_0}$$

The same relation may be obtained from ϕ_A[152]

$$\phi_K = \left(\frac{\partial \phi_v}{\partial p} \right)_T$$

Measurements were carried out in methanol,[153–157] ethanol,[153,156,157] butanol,[156,157] and formamide.[158] Results are given in Table 21.

NUCLEAR MAGNETIC RESONANCE METHOD. It was reported[159] that NMR techniques afford a direct method for the determination of the solvation numbers of certain paramagnetic or diamagnetic cations in nonaqueous solvents. Under certain circumstances it is possible to observe separate NMR signals from the bulk molecules and the solvation shell molecules, and the solvation number of the cation can be determined directly from the ratio of the areas under the absorption peaks. The most important conditions for the observation of separated NMR signals are that the electron spin relaxation time is short compared with the hyperfine interaction between the impaired electron (for paramagnetic ions) and the relevant nucleus, and that the rate of exchange of solvent and bulk molecules is relatively slow. Measurements were reported in dimethylsulfoxide,[160] dimethylformamide,[161–164] and methanol.[159,165–168] In other cases,[169–172] when proton magnetic resonance signals of the solvated and bulk solvent so nearly coincided that they could not be observed separately, the technique introduced by Jackson, Lemons, and Taube[173] was used. An addition of cupric ion to the solution selectively broadened the bulk CH_3 peak, thereby permitting observation of the solvation shell, CH_3. A primary solvation number of 6 was usually obtained.

THE ISOTOPE DILUTION TECHNIQUE. This method was applied[174] to the investigation of methanol exchange between solvated cations and the bulk solvent, thereby providing useful information about the composition of the ion–solvent complex. The conditions for the success of the method are that the solvated cation be a specific entity distinguished from the solvent and that

Table 21   Solvation Numbers Obtained from Compressibility Measurements at 25°C

	Solvent	
Electrolyte	Methanol	Ethanol
LiCl	4.2	2.7
LiBr	5.0	3.4
LiI	5.6	3.7
NaCl	4.7	
NaBr	5.6	2.9
NaI	6.2	3.2
KBr	5.2	
KI	6.0	
CsCl	3.0[a]	
NH_4Cl	4.9	
NH_4Br	5.0	2.5
NH_4I	5.5	3 3
$LiNO_3$	5.3	3.4
$NaNO_3$	5.9	
NH_4NO_3	5.2	2.9
$McCl_2$	12[b]	
$CaCl_2$	10[b]	
$ZnCl_2$	6[b]	

From Ref. 153.
[a] Ref. 154.
[b] Ref. 155.

the ion–solvent complex exchange not be complete by the time isotopic sampling is made. This may be considered equivalent to the definition of the ion–solvation complex given earlier, and it was indeed suggested[174,175] that the isotope dilution technique measures a "kinetic solvation number" comparable to the primary solvation number. For Co^{2+}, Ni^{2+}, Mg^{2+}, and Fe^{2+} a coordination number of 6 was obtained.

Activity Method.   Robinson and Stokes[65] have thermodynamically interpreted the deviations of experimental activity coefficients from the Debye–Hückel equation in terms of solvated ions, and Glueckauf[176] has applied statistical entropy calculations in using the same model. Some results are given in Tables 22 and 23. It must be pointed out, however, that in both methods the solvation number was assumed to be independent of concentration, which is not borne out either by experiment or theory.[128] Other factors are discussed later.

Table 22   Solvation Numbers Obtained by Robinson–Stokes Method

	Solvent		
Electrolyte	Formamide	Methanol	Ethanol
LiCl	5.1		
NaCl	4.2	3.7	
KCl	3.5		
NaBr		3.2	3.3
NaI	3.8	2.2	
KBr		2.7	2.5
RbCl	2.5		
CsCl	1.0		
KNO₃	0		
KI	4.4		

From Ref. 65.

Table 23   Solvation Numbers in Sulfuric Acid at 25°C

Ion	Reference[a]	Reference[b]
Li^+	2.6	2.3
Na^+	3.8	3.0
K^+	2.4	2.1
Ag^+	2.4	2.1
NH_4^+	1.2	1.2
Me_2COH^+	1.5	1.0
$MePhCOH^+$	3.8	1.4
Ph_2COH^+	7.2	1.3
$PhNH_3^+$	1.5	0.8
$Ph_2NH_2^+$	3.8	0.6
Ph_3NH^+	7.2	0.6
Ba^{2+}	11.5	6.5
H_3O^+	2.1	1.8

[a] Robinson–Stokes method, Ref. 65.
[b] Glueckauf's method, Ref. 176.

As far as the solvation numbers obtained by various methods can be compared, the agreement seems to be more than reasonable. As a whole, the results seem to confirm the model of a solvated ion that was seen to be consistent with various properties of the solution and to point to its usefulness as a working entity. However, this model is still open to the same criticisms that are directed against the "sphere in continuum" model.[177]

6   Spectral Aspects of Ion–Solvent Interactions

The results of studies of the ultraviolet and visible spectra of ions in various solvents have been interpreted in terms of charge transfer complexes. These have been considered in a very recent review[178] and are not discussed here. Nevertheless, infrared and Raman spectroscopy, as well as nuclear magnetic resonance investigations and dielectric constant measurements should shed some light on ion–solvent interactions, since the changes produced in the bulk solvent properties are attributed to the presence of the ions. The evidence adduced from Mossbauer effect spectra should contribute to the knowledge of the ion environment, and X-ray diffraction techniques should produce reliable ion–solvent molecular distances. This formidable arrangement of physical methods should provide us with a complete picture of ions in solution, but we are still far from this ultimate goal.

Vibration and Rotation Spectroscopy.   Infrared and Raman spectroscopy are concerned with vibrational and rotational transitions. However, since the selection rules are different, the information obtained from one kind of spectra, say infrared, sometimes complements that obtained from Raman spectroscopy, thereby providing valuable structural information on the ion–solvent interaction.

Infrared spectra indicate the characteristic state of individual bonds in the solvent molecules. Consequently, any study of the changes in position and intensity of the fundamental absorption bands caused by the addition of ionic solutes provides specific information concerning the structure and nature of ion–solvent interaction; furthermore, studying such changes permits the identification of the atoms that interact with the solvent. The effect of intrinsic properties of the ions (electron orbitals, radius, charges, etc.) on the nature of the interaction are determined by changes in position and intensity of the absorption bands. During the past few years, several far-infrared studies of electrolytes in various solvents have been reported. In most cases a new band, the frequency of which was dependent on the cation, was observed. In acetone solutions, perchlorates of silver,[179] lithium,[179,180] and magnesium,[180] the interaction of the acetone molecules with the cations led to a displacement of the absorption band of the group $C{=}C$ toward lower frequencies and a displacement of the group

toward higher frequencies. The shifts were found to be independent of the anion used,[180] pointing to an ion–solvent complex of the form

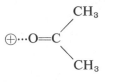

when two molecules of acetone are usually coordinated to the cation.[179] The appearance of a third band because of the splitting of the degenerate frequencies CH of the CH_3 groups was assigned to the formation of a hydrogen bond between the iodide ions.[180] These results have been confirmed by Raman spectra measurements of acetone solutions of sodium, lithium, and barium perchlorates,[181] where changes in the integral molal intensity of the lines show stronger interactions in the order $Na^+ < Li^+ < Ba^{2+}$. Similar results were obtained in methanol solutions[182–185] where infrared spectra[182] pointed to the formation of hydrogen bonds between alcohol molecules and the I^- anion. Raman spectra[184,185] have confirmed that anions have more of an effect on O–H bonds than cations. The ClO_4^- ion is most effective at breaking down the hydrogen bonding between methanol molecules, leading to large concentrations of monomers in these solutions. Acetonitrile solutions have been investigated rather thoroughly. All the available evidence indicates that coordination of acetonitrile to metal ions occurs through the lone pair of electrons of the nitrogen atom.

The observed blue shifts in the infrared spectra of the $C\equiv N$ and $C-C$ bands are attributed to the cations,[185,186] whereas the hydrogen bonding between the iodide anion and the CH_3 group has been confirmed by investigations of solutions of alkaline and tetrabutylammonium perchlorates and halides in deuteriacetonitrile,[187] as well as by Raman spectra[186,188–192] frequency shifts occurring on coordination of acetonitrile to cations that were found to be independent of the anion.

Silver nitrate solutions[186,190,193,194] were studied by both spectroscopic methods with a view to differentiating between various kinds of ion pairs and calculating solvation numbers.

In liquid ammonia solutions[195,196] it was shown that bulky ions like Bu_4N^+ or NO_3^- have little effect upon the infrared spectrum, thereby permitting the study of various cations or anions. Halide ions strongly perturb the stretching vibrations of ammonia by hydrogen bonding, and monoatomic ions cause an important increase in the symmetrical bonding frequency by interaction with the nitrogen electron pair $M^{2+}\cdots NH_3$. The shift increases with charge and decreases with ionic size.

It was shown that solvation of cations takes place as a result of their interaction with the oxygen of the amides[197] in formamide F, NMF, and DMF.

In the cases of F and NMF, the interaction takes place with the nitrogen in the amides as well. Here, again, there is hydrogen bonding between the iodide ion and the hydrogen atoms of the amino groups.

French and Wood[198] recently measured the far-infrared spectra of sodium tetraphenylborate in pyridine, dioxane, supendine, and tetrahydrofuran (THF) solutions. They reported a band at $175° cm^{-1}$ in all solvents and assigned it to the interionic vibration of an unsolvated ion pair. This is not in agreement with the reported specific solvation of the sodium ion by THF[199] and the substantial frequency of the C–O bond shift experienced in $Na^+[Co(CO)_4]$ solutions.[200] Further evidence from the infrared spectra of alkali metal ions in DMSO,[201,202] dialkylsulfoxide,[203] 2-pyrrolidone,[202,204] and pyridine[205,206] solutions shows that all cations are solvated. The study of the $S—O$[201,203] or $C═O$[204] frequencies indicated that the solvent dipoles are oriented with the oxygen atom in direct proximity to the metal ion, and that the far-infrared bands can be assigned to the vibrations of the cation in the solvent cage, in agreement with the model proposed by Maxey and Popov.[201,202,207]

The evidence provided by infrared and Raman spectra generally tends to confirm a strong interaction between cations and solvents, yielding solvated cations; it also points to the hydrogen-bond-forming properties of the halide ions, which seem to act specifically toward enhancing the structure of solvents.

Nuclear Magnetic Resonance Spectra. Further information about ion–solvent interaction has resulted from measurements of nuclear magnetic resonance in nonaqueous solutions of electrolytes. Nuclear magnetic resonance frequencies are determined by the electronic environment surrounding a nucleus owing to magnetic screening by the electron cloud.[208] An ion in solution causes an external perturbation in the electron density around the solvent molecule by electrostatic interactions, by affecting the solvent structure in some way or by forming bonds with the solvent molecules. This roughly corresponds to regions of primary solvation, secondary solvation, and bulk solvent. Changes produced by ions therefore change resonance frequency and cause a chemical shift.

$$(64) \qquad\qquad \delta = \frac{H - H_r}{H_r}$$

where H is the magnetic field strength at a given nucleus of the solution and any constant reference field. The chemical shift of a nucleus varies from molecule to molecule, and its distribution should yield important information.

There is, however, an important condition to be fulfilled if this effect is to be observable;[209] namely, that the exchange of solvent molecules between these various environments be slow or that $1/\tau \ll \Delta\delta\cdot\nu$, where τ is the mean residence time of the molecule in a given environment, $\Delta\delta$ is a representative difference in chemical shift between the different environments, and ν is the NMR frequency. In practice, this inequality does not hold, and only the average of the chemical shifts of all nuclei is observed. However, as explained in the section on the NMR method, in certain instances in nonaqueous solvents, such as methanol,[159] a separate resonance due to bonded solvent may be resolved and the cation solvation number determined from relative peak areas. An alternative method is based on molal shifts.[210] In most instances only a single averaged line is detected and the individual molar shifts are deduced quite arbitrarily.

The relative ion shifts have been explained in terms of structure-making and structure-breaking effects of the ions on the bulk solvent, including detailed arguments about the signs of the shifts.[211–213] Butler and Symons[214] suggested that the residual shift of the main solvent signal from that of the pure solvent is caused largely or even entirely by the rapid exchange between bulk solvent molecules and those bonded to the anions and represents a direct determination of the anion shift in the hydroxyl proton resonance spectrum of methanol at low temperature. Secondary effects[215] were shown to be small. Cation shifts[216] were first obtained from proton magnetic resonance (PMR) measurements of $Mg(ClO_4)_2$, $Zn(ClO_4)_2$, and $Al(ClO_4)_2$ in methanol, and a scale of ion shifts was established.[217] The large deshielding found for small cations with high charge density was expected. Anion shifts can be understood in terms of a hydrogen-bonding model in agreement with infrared spectra, as noted in the previous section, and hence strong bases give large downfield shifts. The model may become a better approximation at higher temperature.

In the PMR of electrolytes in liquid ammonia,[218] only a single proton resonance line was exhibited. The PMR shifts to lower fields in the order $K^+ < Na^+ < Li^+$ and $ClO_4^- < NO_3^- < Br^- < I^-$ for several series of salts runs parallel to the strength of the ammonia alkali bonds. The effect of decreasing cation size is explained by the relatively favorable solvation of small ions and the concomitant polarization of the lone pair of ammonia ions and deshielding of the protons. In DMF solutions, however, a downfield shift in the protons of coordinated DMF[219] could be induced by adding halide ions to solutions of perchlorate. An analysis of the formyl shifts points again to coordination through the oxygen atom of the amide group.

Using the proton shift of a DMSO molecule[220] as a parameter, the ion–solvent interaction of alkali salts was investigated by a chemical mole-ratio method with n-pentanol as inert solvent. The evidence obtained points to the existence of a bicoordinated cation in the cases of Li^+, NH_4^+, and Na^+. The nuclear magnetic resonance of solute nuclei offers further opportunity of investigation. There are nuclei of spin $\geq \frac{1}{2}$ for all alkali and halogens, which makes it possible to study the whole range of alkali halides in solution. However, the sensitivity of chemical shift to solvent decreases rapidly as ionic size increases.[210] The range of ^{7}Li chemical shifts of dilute solutions of lithium bromide and perchlorate in 11 organic solvents[221] was observed to be nearly comparable to the total range reported for various compounds and generally attributed to structural influences.

It is suggested that a specific carbonyl influence, resulting from coordination of a lithium ion at the oxygen atom of the carbonyl group and oriented so that the ion is in a region of similar negative shielding associated with the magnetic anisotropy of that group, might account for the grouping of carbonyl-containing solvents. Analogous explanations were advanced for the extremely high and low shieldings in the solvents acetonitrile and pyridine, respectively. A similar study of ^{23}Na chemical shifts[227] of solutions containing sodium iodide dissolved in 14 different oxygen or nitrogen-donor solvents has suggested a fair correlation between the magnitude of the paramagnetic term[223] of the general nuclear screening equation[224] and the Lewis basicity of the solvent.

The occurrence of the sodium resonance in acetonitrile at a markedly higher field strength than other nitrogen-containing solvents confirms infrared spectra and indicates that the main solvating influence of this solvent is via interactions of the π-electron of —C≡N with the solute rather than via the exposed sp lobe of the nitrogen atom.[222] Chemical shifts of ^{27}Al together with line widths in various organic solvents have determined a scale of affinities of bond strength as follows:

$$H_2O > C_2H_5OH, C_3H_7OH \gg C_2H_5NCS > C_6H_5CN > CH_3CN$$

and confirmed the existence of at least two species of $AlCl_3$ in acetonitrile.[225]

The applications of the electron-spin-resonance (ESR) technique to ion–solvent interaction, as well as general aspects of solvation and ion pairing,[227] have been reviewed recently[210,226] and are not considered here.

Mossbauer spectra have been claimed to yield information on the solvation of ferrous salt solutions.[228] Since recoilless resonance absorption can be produced only by crystals, the Mossbauer spectrum of the solutions can be recorded only after freezing them (assuming thereby that no changes occur in the immediate environment within the 10–12 sec it takes for rapid freezing). Little information may be obtained from the isomer shift data, which depend on the electron density at the Mossbauer nucleus; but changes taking place in the solvate layer were reflected in the value of quadrupole splitting, since it is determined by the symmetry of the charge distribution around the nucleus. The comparison of spectra in aqueous and nonaqueous solutions[229] confirmed the nephelauxetic effect of the chloride ion[230] in glycol, pyridine, and

DMF; methanol and formamide behaved like water. The affinities of methanol and formamide for ferrous ions in solution were supposed to be approximately the same, since mixed solvates[231] were observed in the Mossbauer spectra of methanol formamide solutions.

Dielectric Method.   Recent dielectric studies at microwave frequencies have provided information concerning the properties of molecules in the neighborhood of ions in solutions. When static dielectric constant decrements of salt solutions in methanol and ethanol were analyzed[232] in the manner used by Ritson and Hasted,[233] it became clear that the decrements were too large to be explained by the usual solvation concept, even when allowing for increased penetration of the ionic charge in material of low dielectric constant. However, as in water, calculations have shown that the ionic field does not appear to penetrate decisively beyond the first solvation number. Since solvation numbers not very different from those in water would not account for the observed decrements, it was suggested[232] that ion solvation causes sufficient bond breaking to cause a serious disturbance in the interaction between chains and to shorten the chain length of alcohols. Using a given similar discontinuous model where the dielectric constant varies with the solvation shell,[234] it was claimed[235] that Hasted and Roderick's results[232] in methanol solutions could be reproduced reasonably well. This actually takes into account the differentiation between field and volume effects,[236] which cause the dielectric constant depression.

Later measurements[237–241] have shown that two cases may be distinguished: solutions that show a dielectric relaxation analogous to that of the solvent, such as $LiClO_4$, $Mg(ClO_4)_2$, $LiCl$, and LiI in ethanol[241] and methanol,[238] and those for which an additional relaxation process must be assumed, such as $LiClO_4$ in ethylacetate.[239] The static permittivity, defined in terms of the relaxation spectrum, decreases greatly with increasing concentration in all cases. When the relaxation time of the solution is of the same order of magnitude as that of the solvent, it is usually assumed[242] to be owing to solvent molecules outside the solvation shell. These results were interpreted in the case of alcohols[238] in terms of solvent molecules whose rotation is hindered by the ion and a nonelectrostatic effect, leading to the hybridization of orbitals sp^3 and donor–acceptor bonds. The conclusion was drawn then that the three cations, Li^+, Na^+, and Mg^+, are "in contact" with the same number of solvent molecules and should similarly affect the permittivity. This was confirmed by comparing the various dielectric decrements due to these ions in the same solvent. It was further concluded that the solvent structure characterized by chainwise association is not disturbed in the presence of the salt and that solvation occurs with chains of the same kind.[238] A more general treatment was given for both polarization and charge transport relaxation without any restriction on concentration[237] which in the limiting case of infinite dilution reduces to the classical formula of dielectric theory. The static permittivity is still affected by the solvation-hindered rotation molecules, but an additional microscopic model is described in which the charged species are supplied by solvated ions and the polar species are represented by the nonsolvated solvent molecules. It was claimed[237] that good agreement is obtained between the calculated and experimental dielectric decrements.[243]

Diffraction Techniques. Diffraction formally offers an excellent method for identifying the degree of bonding in solutions by determination of the average environment through evaluation of the radial distribution function (RDF). The experimental results, however, are complicated to a considerable extent by the possible overlapping of ion–solvent, ion–ion, and solvent–solvent interactions. The RDF of the solution is usually referred to the RDF of the solvent, thereby increasing the uncertainty range.

Some solutions of ferric chloride in methanol,[244,245] of cobaltous chloride and bromide in methanol and ethanol,[244] and of potassium iodide in formamide[246] have been examined recently. The transition metal cations were found to be tetrahedrally coordinated and are usually associated in methanol and ethanol, except for $CoBr_2$, for which the average species if CoS_2Br_2 (S represents a solvent molecule). The cobalt halide contact distance was found to be the same in the crystalline state as in alcohol solutions within experimental error.[244] For ferric chloride the results are tentatively interpreted as pointing towards the existence of the dimer species Fe_2Cl_6.[245] In potassium iodide formamide solutions,[240] maxima in the distribution curves were assigned to potassium ion interactions with the carbonyl oxygen of the solvent. Solvation numbers were found to range from 4 in dilute solutions to 2 in the most concentrated regions.

The information from these diffraction studies is obtained at high concentrations, and it seems unlikely that models proposed to describe dilute solutions are still applicable. It seems that X-ray diffraction cannot be used as a direct method, for example, for the determination of ionic radii in solutions.

7   Transport Properties

Viscosity. In a very dilute solution, the interstitial solvent between the cospheres of the ions is unmodified and has the same properties as in the pure solvent.[135] Each species of ion would be expected to contribute toward a change in the viscosity, but electrostatic forces between opposite charge ions must be taken into account. The concentration dependence of electrolytic solutions has been interpreted in terms of the semiempirical Jones–Dole equation [132]

$$(65) \qquad \frac{\eta}{\eta_0} = 1 + Ac^{\frac{1}{2}} + Bc$$

where A is obtained from interionic theory [133,134] and B is an adjustable parameter, either positive or negative, that accounts for ion–solvent interaction.

The solvents in which negative B values have been found, such as water,[138] sulfuric acid,[247] glycerol, and ethylene glycol,[248] all have molecules capable of elaborating hydrogen bonds in three dimensions. The corresponding negative B values are found among monoatomic ions, for those of low surface charge density.

It is believed [249] that the balance between the electrostatic effect in orienting a first solvation shell around the ion and the tendency exhibited by the molecules to stay as a part of the three-dimensional structure, may lead to a structural collapse in that region, an effect tending to make B more negative. Such a disordered region may exist at the periphery of the region of dielectric saturation even for ions that are structure-makers and the B value of a particular ion may be seen as a resultant of positive and negative contributions.[137] In nonaqueous systems, the structure-breaking contribution is probably mostly negligible, as indicated by the large positive values found in NMF,[137] acetone,[250] N-methylpropionamide,[251] and so on.

Nevertheless, the analysis of viscosity data for potassium iodide–glycerol solutions and cesium iodide solutions in ethylene glycol and glycerol [248] has yielded highly negative B coefficients. This is attributed to considerable weakening of the hydrogen bonding in both ethylene glycol and glycerol. Cesium iodide may definitely be described as a structure breaker in both solvents, whereas the observed increase in dielectric relaxation time with KI concentration confirms an increase in the mobility of glycerol molecules or negative solvation.[252] The shape and size of electrolytes were correlated with the B coefficients through a requirement of Vand's equation.[253] Stokes's assumption [138,139] of a rigid solvated volume, independent of concentration, is only valid at high dilution. Assuming that the large tetraalkylammonium ions have spherical symmetry, a correlation between partial molal volumes and viscosity measurements was proposed through Einstein's equation.[142] For small ions, the viscosity data indicated larger radii than did the molal volumes, as expected from electrostriction considerations. For large ions, the agreement was more quantitative.

This difference in behavior is due to the solvation experienced by small ions. Evidence for additivity of ionic contributions has been conclusive [139] and various criteria were applied to the calculation of the ionic B coefficients. In sulfuric acid [247] the bisulfate anion HSO_4^-, similar in size and character to the solvent, was assumed to have a negligible effect on the solvent structure, and the experimental B was therefore assigned to the specific cation. In DMSO solutions,[141] the assumption of equal ionic B coefficients for $(i\text{-amyl})_3BuN^+$ and $B(Ph)_4^-$ ions was made on the basis of equal size in solution. It was suggested that the close correlation between ionic B coefficients and solvation properties for KCl in aqueous solutions confirmed that equal contribution of the KCl ions is a good approximation.[139]

There is no reason for not trying a similar division in other systems. In the first instance, as shown by Gurney,[135] the B coefficients for various salts both in aqueous and nonaqueous solutions change in a predictable manner with the molal entropy of solution. Second, ionic B coefficients in aqueous solutions that were computed by equating the contributions made by K^+ and Cl^- with the B coefficients of KCl at 25°C, showed a linear relation with partial molal ionic entropies. Then Asmus[218] and Nightingale[146] showed that a single linear relation may be used to correlate entropies of hydration with ionic B values for both monoatomic and polyatomic ions. Since it is usual to link ionic entropies with B coefficients,[137] the same correlation should be made in nonaqueous and mixed solvents.[254] As with the linear relationship shown to exist between ionic entropies in various solvents [cf. (11)], it may be assumed that in aqueous mixtures,[64] as well as in nonaqueous solvents, the equality of the ionic B coefficients for K^+ and Cl^- should be preserved.[254] Correlation with entropies of solvation were made accordingly, and ionic B coefficients could be evaluated.

It was argued [137] that the following factors are responsible for the variations of the B coefficients from solvent to solvent:

1. Structure breaking, which is more important in three-dimensional solvents than in other classes of solvents.

2. High molal volume and low dielectric constant, which yield high B values for similar solvents.

3. Reduced B values when the primary solvation of ions is sterically hindered in high molal volume solvents or if either ion of a binary electrolyte cannot be specifically solvated.

No satisfactory theoretical calculation of the B coefficient has been proposed,[255] and we must await the development of a quantitative theory in order to elucidate the various points raised. Values of B in nonaqueous solvents are found in Table 24.

Conductance. Most of our knowledge on nonaqueous solutions has come from conductivity measurements [256] on ionophores, completely dissociated

Table 24   Jones and Dole's B Coefficients in Nonaqueous Solvents at 25°C

Electrolyte	Methanol	DMSO	CH$_3$CN	CH$_3$NO$_2$	NMF
KCl	0.7635[a]				0.615[f]
KBr	0.7396[a]				0.584[f]
KI	0.6747[a]				
NaI					
NaCl					0.599[f]
NH$_4$Cl	0.661[a]				
NaSCN		0.62[d]			
NaClO$_4$		0.62[d]			
NaB(Ph)$_4$		1.14[d]			
(i-amyl)$_0$BuNPh$_4$		1.57[d]	1.47[e]		0.567[f]
Me$_4$NBr	0.42[b]				
Bu$_4$NBr	0.84[b]		0.93[b]	0.75[b]	
Et$_4$NBr	0.56[c]		0.69		
Pr$_4$NBr	0.70[c]				
Bu$_4$NBr	0.84[c]		0.87[b]		
Pr$_4$NI	0.66[c]		0.71[b]		
Bu$_4$NB(Ph)$_4$			1.35[b]		
Pr$_4$NB(Ph)$_4$			1.24[b]		
Bu$_4$NPi			1.13[b]		
Pr$_4$NPi			0.90		
Et$_4$PNi			0.85		
Me$_4$NPi			0.78		

[a] Ref. 132.
[b] Ref. 140.
[c] Ref. 290.
[d] Ref. 141.
[e] Ref. 350.
[f] Ref. 137.

in solutions, and ionogens consisting of neutral molecules that yield ions by reacting with suitable solvents.[257] The investigation of conductance as a function of concentration gives the conductance at infinite dilution, the dissociation constants of ionogens, the association constants of ionophores, ionic radii, and information about the structure of the solutions in the vicinity of the ion.[258] The analysis of the data has usually been carried out in terms of the Fuoss–Onsager conductance theory,[259] which is basically an application of the interionic attraction theory to the specific model of rigid charged spheres,[260] representing the ions in an electrostatic and hydrodynamic continuum, that is, the solvent.[177]

The conductance equation for unassociated electrolytes has two parameters —λ_0, the limiting conductance at infinite dilution, and $\mathring{a}$, the mean contact distance between the ions—and is able to reproduce the concentration dependence of the equivalent conductance up to a concentration determined by $\kappa \mathring{a} \leq 0.2$, where κ is the characteristic Debye–Hückel length. The equation is expressed as

$$(66) \qquad \Lambda = \Lambda_0 + Sc^{\frac{1}{2}} + EC \log c + (J - B\Lambda_0)C$$

where S and E are functions of ε_0, T and η_0 independent of $\mathring{a}$, J is an explicit function of $\mathring{a}$, and B is the Jones–Dole viscosity coefficient.

Discussion of the parameter $\mathring{a}$ is deferred until Walden's product[177] and the expression for the association constant K_A are considered. Fuoss's application of a more sophisticated approach to alkali halides in low-dielectric-constant media[177] included explicit retention of the Boltzmann factor.[261] However, the derivation of the Fuoss–Onsager equation[261] was queried in a discussion[262] of Pitts's equation,[263] and the numerical discrepancy between the two theories was pointed out.[264,265] The value obtained for the conductance at infinite dilution using the Fuoss–Onsager theory[259] differed from that obtained using Pitts's[263] theory by an amount that greatly exceeds the experimental error.

This observation was confirmed by Fuoss and Hsia,[266] who recalculated the relaxation field in the earlier theory. The resultant equation for Λ, which has a complicated concentration dependence with terms up to $c^{\frac{3}{2}}$, was successful with some experimental data.[266,267] There is some evidence[268] that the numerical difference between the predictions of the Fuoss–Hsia equation[269] and that of Pitts[263] has diminished to less than 0.04%[270] for the limiting conductances at infinite dilution, Λ_0. However, we shall consider only the values obtained through the Fuoss–Onsager extrapolation, since it has been most widely used.

In the state of infinite dilution to which A_0 refers, the motion of an ion is limited solely by its interactions with the surrounding solvent molecules, for there are no other ions within a finite distance. Under these conditions, the validity of Kohlrausch's law of the independent migration of ions is almost axiomatic.[65] Thus for a 1-1 electrolyte

$$(67) \qquad \Lambda_0 = \lambda_1^\circ + \lambda_2^\circ$$

where λ_1° is determined by measurements of transport numbers t_i, and $\lambda_i^\circ = t_i \Lambda_0$.

Reliable values of single-ion conductances are useful for the investigation of ion–solvent interactions. The division of electrolyte conductances into the limiting ionic conductances ideally requires transference numbers[271] that have only been determined accurately in a few anhydrous solvents such as nitromethane[272] and acetonitrile.[271] For several other solvents, provisional scales of single-ion conductances have been based on the assumption that the ionic contributions of certain reference electrolytes, such as tetrabutyl-ammonium tetraphenylboride,[273,274] triisoamylbutylammonium tetraphenyl-boride,[89] and tetraisoamylammonium tetraisoamylphenylboride,[271] have equal mobilities.

Limiting ionic conductances in nonaqueous solvents were obtained from transport numbers when available or through reference electrolytes. Their values are listed in Tables 25 and 26.

An inspection of the data shows that the conductances of alkali metals generally increase in the order

$$Li < Na < K < Rb < Cs$$

and those of the tetralkylammonium ions decrease in the order

$$Me_4N^+ > Et_4N^+ > Pr_4N^+ > Bu_4N^+$$

Table 25   Limiting Ionic Conductances in Protic Solvents at 25°C

Ion	MeOH (Ref. 290)	EtOH (Ref. 292)	PrOH (Ref. 292)	BuOH (Ref. 288)	HCOOH (Ref. 356)	HCONH (Ref. 357)	NMF at 30°C (Ref. 354)
Li$^+$	39.08^a	17.07^a		8.10^f	19.36	8.5	
Na$^+$	45.08^a	20.37^e	8.35		20.97	10.1	21.5
K$^+$	47.78^b	22.2^b	6.88^a		23.99	12.7	22.13
Rb$^+$	56.08^a					12.8	
Cs$^+$	61.33^a	26.46				13.5	24.39
Ag$^+$	50.07^c						
NH$_4$$^+$				6.68^f	27.01	15.6	
Me$_4$N$^+$	68.73	29.65	14.40	9.67	23.62	12.5	
Et$_4$N$^+$	60.05	29.27	15.05	10.40		10.0	26.20
Pr$_4$N$^+$	46.08	22.98	12.19	8.80			
Bu$_4$N$^+$	38.94	19.67	10.71	7.84		6.8	
Am$_4$N$^+$	34.8						
Cl$^-$	52.09	21.87	10.45	7.76	26.52	17.1	19.70
Br$^-$	56.43	23.88	12.22	8.23	28.30	17.2	21.56
I$^-$	62.62	27.0	13.81	9.52		16.6	22.76
SCN$^-$						17.2	
NO$_3$$^-$	61.13^d					17.4	
ClO$_4$$^-$	71.0^c		16.42	11.22	29.35		
Ph$_4$B$^-$	37.05				12.33		

^{a} Ref. 359.
^{b} Ref. 360.
^{c} Ref. 351.
^{d} Ref. 362.
^{e} Ref. 363.
^{f} Ref. 364.

Table 26   Limiting Ionic Conductances in Some Dipolar Aprotic Solvents at 25°C

Ion	DMF (Ref. 351)	DMSO (Ref. 352)	NMT (Ref. 365)	NMA at 35°C (Ref. 355)	N-Methyl-propion-amide at 30°C (Ref. 353)	ACN (Ref. 289)	TMS (Ref. 280)
Li^+	25.0	11.4[a]		5.65			4.33
Na^+	29.9	14.54		7.19	5.06	76.9	3.61
K^+	30.8	14.7		7.28	5.36	83.6	4.05
Rb^+		15.3[a]		8.33		85.6	4.16
Cs^+		16.1[a]				87.3	4.27
Ag^+	35.2					86.0[d]	
NH_4^+	38.7						4.97
Me_4N^+	38.9	18.6[b]	54.9	11.28		94.5[e]	
Et_4N^+	35.6	16.5	47.7	10.33		84.8[e]	3.98
Pr_4N^+	29.2	13.4	39.1	8.23		70.3[e]	
Bu_4N^+	26.2	11.0	34.1	7.11		61.4[e]	
Am_4N^+				6.68			
Cl^-	55.1	24.4[b]	62.7	10.60	6.24		9.30
Br^-	53.6	24.1[b]	62.9	11.72	7.06	100.7[f]	8.92
I^-	52.3	23.8[b]		13.42	8.34	102.1	7.22
SCN^-		29.50[c]					9.64
NO_3^-	57.3					106.4	
ClO_4^-	52.4	24.50				103.7	6.85
Ph_4B^-		11.05				58.1	

[a] Ref. 358.
[b] Ref. 361.
[c] Ref. 141.
[d] Ref. 366.
[e] Ref. 367.
[f] Ref. 368.

The former trend is similar to that shown by anions, where

$$ClO_4^- < I^- < Br^- < Cl^-$$

Inasmuch as the ionic conductance is inversely related to the size of the ion, it seems probable that the Li^+ and Na^+ are well solvated in most solvents and that they are less conducting than the alkylammonium salts. Evidently the conductance is determined by the interaction of the ion on the solvent molecules, rather than by just the size of the ion. Solvation of the anions seems to indicate that they are as conducting as cations of similar crystallographic radii in the hydroxylic solvents but much more so in the dipolar aprotic solvents.[34]

Stokes's Law and Walden's Rule. If the ions of an electrolytic solution are represented by spheres of radius R_i and the solvent is assumed to be a continuum described by its macroscopic viscosity, η_0,[275] then Stokes's hydrodynamics leads directly to the velocity

$$(68) \qquad\qquad v_0 = \frac{xz_ie}{6\pi\eta_0 R_i}$$

for an ion of charge z_ie moving in a field of field strength, x. In terms of the limiting conductance, λ_i°, this relation becomes

$$(69) \qquad\qquad \lambda_i^\circ = \frac{|z_ie|e\tilde{F}}{6\pi\eta_0 R_i}$$

for the conventional primitive model.[144]

Substituting numerical values for the constant yields

$$(70) \qquad\qquad \lambda_i^\circ = \frac{0.819z_i}{\eta_0 R_i}$$

where λ_i° is in mho/(cm)(M), η_0 is in poises, and R_i is expressed in angstroms. Stokes's law is believed to be valid only for a "cannon ball moving through treacle."[270] Hence, for bulky organic ions, ionic conductances should be inversely proportional to the ionic radius, R_i. If the radius is assumed to be the same in every solvent, the product

$$(71) \qquad\qquad \lambda_i^\circ\eta_0 = \frac{0.819|z_i|}{R_i}$$

is a constant. Equation 71 is known as Walden's rule.[277] The effective radii obtained from limiting ionic conductances have been used to estimate the number of solvent molecules associated with the moving ion, thereby assigning the whole of ion–solvent interaction to an effective solvation shell around the ion. The failure of the Stokes's radii to give the effective size of the solvated ions for small ions is often attributed to the inapplicability of Stokes's law to molecular motion.

Robinson and Stokes,[65] Nightingale,[146] and others[141,151,278-280] have suggested a method of correcting the radii. The tetraalkylammonium ions were assumed not solvated and by plotting the Stokes's radii against the crystal radii of these large ions, a calibration curve was obtained for each solvent. This approach, however, suffers from one serious flaw. The basic assumption on which this approach rests is that the Walden product is invariant with temperature.[281] Experimental results[282,283] indicate that this assumption is incorrect and that the resulting solvation numbers obtained from this type of correction to Stokes's law are meaningless.

The changes in the Stokes's radius, R_i, were attributed by Born[148] to ion–dipole reorientation that yields an apparent increase in the real radius, r, of the ion. The effect of dielectric saturation on R was later discussed by Schmick,[284] but no further improvements in this model were made until Fuoss noticed the dependence of the Walden product, $\Lambda_0\eta_0$, on the dielectric constant and considered the effect of the electrostatic forces on the hydrodynamics of the system.[275] He proposed that the dielectric relation in the solvent caused by ion motion leads to excess frictional resistance. In a heuristic argument he suggested the expression

$$(72) \qquad \lambda_{i,0}^{\circ} = \frac{\tilde{F}e|z_i|}{6\pi R_{\infty}(1 + A/\varepsilon R_{\infty}^2)}$$

from which the classical Stokes's radius, R_i, may be derived as

$$R_i = R_{\infty} + \frac{A}{\varepsilon}$$

where R_{∞} is the hydrodynamic radius of the ion in an hypothetical medium of dielectric constant where all electrostatic forces vanish and A is an empirical constant.

Then in 1961, Boyd[149] showed that the effect of dielectric relaxation on ionic motion can be treated theoretically and that the dielectric relaxation times for polar solvents lead to an effect of the correct magnitude in some aprotic solvent mixtures. The expression he obtained was

$$(73) \qquad \lambda_i^{\circ} = \frac{\tilde{F}e|z_i|}{6\pi\eta_0 r_i}\left[1 + \frac{2}{27}\frac{1}{\pi\eta_0}\frac{z_i^2 e^2 \tau}{r_i^4 \varepsilon_0}\right]$$

where τ, which was taken as unaffected by the hydrodynamic disturbance of the ionic motion, is the Debye relaxation time for the solvent dipoles. Finally, in a more rigorous derivation, Zwanzig[150] showed that the "ultimate that has been devised"[285] by an application of a continuum model for the solvent to ionic mobilities yielded the following expression for the Walden product

$$(74) \qquad \lambda_i^{\circ}\eta_0 = \frac{\tilde{F}e|z_i|}{6\pi r + B/r^3}$$

where B is a function of the solvent properties only and is given by

$$(75) \qquad B = \frac{2e^2}{3}\frac{\tau}{\eta_0}[(\varepsilon_0 - \varepsilon_{\infty})\varepsilon_0^2]$$

where ε_{∞} is the limiting high-frequency dielectric constant of the solvent.

It may be seen that Born's and Zwanzig's equations are very similar and both may be written in the form

$$(76) \qquad \lambda_i^{\circ} = \frac{AR_i^3}{C + R_i^4}$$

from which a maximum value of λ_i°, $\lambda_m^{\circ} = A(3^{3/4}/4)C^{-1/4}$ conductance units corresponding to the radius, $R_m = (3C)^{1/4}A^{\circ}$, may be obtained.

An excellent discussion of the above effect in aqueous solutions[144] indicates that neither slipping nor streaming through the solvent would improve the model, but the use of a local viscosity might produce an almost perfect fit to halide conductances.

In a revised calculation of the dielectric friction coefficient,[286] both hydrodynamic and electrostatic interactions were taken into account, but dielectric saturation was neglected. Improved results with either "sticking" or "slipping" boundary conditions provided somewhat more reasonable values for the maximum conductance.

For comparing results in different solvents, (74) can be used in the linear form suggested by Atkinson and Mori.[287] Rearranging terms and inserting numerical constants gives

$$(77) \qquad \frac{15.5}{\lambda_i^\circ \eta_0} = 18.8r + \frac{15.3 \times 10^{12}}{r^3} \frac{(\tau/\eta_0)(\varepsilon_0 - \varepsilon_\infty)}{\varepsilon_0^2}$$

To facilitate plotting we may write

$$(78) \qquad L^* = 18.8r + \frac{15.3 \times 10^{12}}{r^3} R^*$$

where r may be obtained both from the intercept and the slope of the line.

In order to test Zwanzig's theory,[150] (78) was applied to methanol,[285,289] ethanol,[285,287] acetonitrile,[285,287] butanol,[288] and pentanol[288] solutions, solvents for which accurate conductance[283,288–292] and transference data[284] are available. All the plots were found to be straight lines.[285,287] A quantitative test was obtained by comparing the value of r from the slope and the intercept of the straight line obtained from data for various alcohols.[285] These are given in Table 27.

It is quite obvious that the two radii are far from equal, indicating that the theory does not account quantitatively for mobility changes within the homologous series of alcohols. It was further suggested[285] that the relaxation effect is not the predominant factor affecting ionic mobilities and that these mobility differences could be explained qualitatively[289] if the microscopic properties of the solvent, dipole–moment, and free electron pairs were considered the predominant factors in the deviation from Stokes's law.

Association and Ion-Pairs. The revised Fuoss–Onsager equation for an associated electrolyte is given by

$$(79) \quad \Lambda = \Lambda_0 + S(c\gamma)^{1/2} + Ec\gamma \log{(c\gamma)} + (J - B\Lambda_0)c\gamma - K_A c\gamma y_\pm^2 \Lambda$$

where γ is the degree of dissociation and $y_\pm$ is the Debye–Hückel limiting mean molar activity coefficient. This equation involves three arbitrary parameters, Λ_0, J, and K_A, from each of which a distance of closest approach

Table 27   Ionic Radii Calculated from the
Intercept and Slope of a Plot of Zwanzig's
Equation in Alcohol Solutions at 25°C

Ion	$\overset{\circ}{r}$, Intercept (Å)	$\overset{\circ}{r}$, Slope (Å)
Li^+	3.0	6.1
Na^+	2.9	7.2
K^+	2.5	7.8
Cs^+	2.0	7.3
Me_4N^+	1.8	7.9
Et_4N^+	2.3	9.4
Pr_4N^+	3.2	15.1
Bu_4N^+	3.8	16.3
Cl^-	2.2	6.3
Br^-	2.1	6.7
I^-	1.9	7.1

$\overset{\circ}{a}$ may be calculated—a_Λ, a_J, and a_K, respectively. The latter value, however, rests on the approach taken to obtain a theoretical expression for the association constant of ion pairs, K_4. Ion-pair formation in particular has been considered at length,[293,294] and an essential requirement of all treatments is a knowledge of the extent of association (i.e., the associated species formed and the association constants). Attempts have been made to relate association constants to properties of the solvent and the solute, but all are limited in their application and can only be used under restricted conditions if the results are to have any meaning.[295]

BJERRUM'S APPROACH.   The ion-pair concept was introduced by Bjerrum[296] when he considered the probability of finding two ions of opposite charge at a distance, r, from each other. The resulting distribution function has a minimum at a characteristic distance, $q = e^2/2\varepsilon kT$, where k is the Boltzmann constant, and it would appear that an ion located at a distance $r < q$ from the reference ion of opposite charge tends to form an ion pair, whereas any other one outside a sphere of radius q would be considered as being free. Then the association constant was obtained from the calculated fraction of paired ions:

$$K_A = \frac{4\pi N}{1000} \left(\frac{e^2}{2\varepsilon kT}\right)^3 Q(b)$$

where $Q(b)$ is a function that has been computed by Bjerrum for $1 < b < 15$.
   This model has been corrected by Fuoss,[297] who allowed for the probability of finding one ion pair at a distance r, and it was still further improved[75] when the respective factor was introduced and the critical distance, q, could

have any desired value. The Bjerrum model has been criticized, especially for the unsatisfactory and artificial device imbedded in the critical distance, q.[294] Moreover, it yields different $\mathring{a}$ values for the same electrolyte in different media and it predicts a limiting dielectric constant beyond which no association occurs.[298] It has also been suggested that the parameter $\mathring{a}$ is not correct and depends on the dielectric constant, the temperature, and the nature of the solvent.[293] It may be said, however, that the concept of free ions and ion pairs is useful for dilute solutions of ionophores in media of low dielectric constant. It loses its usefulness in solvents of high dielectric constant or in very dilute solutions if only coulomb forces are considered.[294] Other bonding forces should be involved.

A THERMODYNAMIC APPROACH TO ION PAIRING. Denison and Ramsey[299] treated a system of hard spheres representing ions of radii r_1 and r_2 in a continuous medium of dielectric constant, ε, where only ions in contact were counted as ion pairs. From a thermodynamic consideration of a stepwise process for bringing about the change from an associated ion pair of a uni-univalent salt (in which the contact distance $a = r_1 + r_2$) to the two free ions at large interionic distances, the equation

$$(80) \qquad \ln K_A = \ln K_A^\circ + \frac{e^2}{a\varepsilon kT}$$

can be obtained[300] by assuming that the solvation free energy changes of the free ions and the ion pairs are approximately the same. K_A° is the association constant of the two uncharged ions.

It may be shown[298] that only for $\mathring{a} = 5.1$ Å would the Bjerrum and Denison–Ramsey approaches yield the same value for K_A. For smaller ions, the two theoretical values diverge widely. The Denison–Ramsey thermodynamic approach was elaborated further by Gilkerson.[301,302] However, the free volume formulation[301] of the ion-pair dissociation process to the hard sphere in a continuum approach to which a specific solvation is grafted[301,302] was not rigorous,[303] and the energy, E_S, should be replaced by the change in molar free energy of specific solvation, $\Delta G_S = \Delta H_S - T\Delta S_S$, leading to:

$$(81) \qquad \frac{1}{K_A} = \frac{3000}{4\pi N a^2} \exp\left(-\frac{\Delta H_s}{RT} + \frac{\Delta S_s}{RT} - \frac{e^2}{a\varepsilon kT}\right)$$

The experimental separation of the three terms in the exponential is almost impossible unless more assumptions are made.

ELECTROSTATIC THEORY. A different approach has been proposed by Fuoss.[304] His model consists of spherical cations of radius, a, and volume, v, and point-charge anions, both immersed in a continuum having a dielectric constant, ϵ. Only anions on the surface of or within the sphere of volume

$v = \frac{4}{3}\pi a^3$ of a cation are counted as ion pairs. By using the ion potential, as given by the Debye–Hückel theory and the Boltzmann method, Fuoss obtained the following expression:

$$(82) \qquad\qquad K_A = \frac{4\pi N a^3}{3000} \exp \frac{e^2}{\varepsilon a k T}$$

for the association constant of contact ion pairs. The expression obtained is similar to that proposed by Denison and Ramsey,[300] where the vaguely defined term K_A° is now expressed explicitly as an excluded volume. Other elaborations[87,305] are not considered here, since the shortcomings of (82) may well result from the neglect of dielectric saturation[306] and the improper use of the continuum theory.[307]

In order to consider the usefulness of the expressions proposed, let us go back to the definition of ion pairs as given by Griffiths and Symons.[308] Four classes of ion pairs were suggested.

1. Complexes—two or more ions held in contact by covalent bonds.
2. Contact ion pairs—ions in contact that do not present any covalent bonding.
3. Solvent-shared ion pairs—pairs of ions linked electrostatically by a single oriented solvent molecule.
4. Solvent-separated ion pairs—pairs of ions linked electrostatically but separated by more than one solvent molecule.

Class 1 may not be looked upon as an ion pair in the usual sense of the concept. The Denison–Ramsey and Fuoss treatments only consider class 2, whereas Bjerrum's treatment possibly includes classes 2 and 3. Class 4 is not taken into account in any of the theories. Conductance measurements, however, cannot distinguish among the various species and the association constant is really a statistical average of all the species present in the solution. Insofar as the molecular nature of solvents can be ignored, as in Fuoss's equation, the association constant of a salt, AB, in a series of solvents may be expected to be a smooth function of the dielectric constant, ε, of the solvents.[309] Hence a plot of log K_A versus $1/\varepsilon$ for any given salt should be a straight line, the slope of which is proportional to $(a_K^\circ)^{-1}$ and intercept equal to log K_A. Values of $\mathring{a}_K$ may be obtained[310] from either the slope or the intercept; however, they usually differ considerably.[288] The indication of association is shown by the values of $\mathring{a}_J$, which, compared with those of $\mathring{a}_K$, have led to various speculations about the class of ion pairing present in the solution.

In acetone solution there is a suggestion that free ions are heavily solvated, whereas the formation of either solvent-shared or solvent-separated ion pairs is promoted[311] by changes in temperature or pressure. Contact ion pairs are formed with LiCl; in LiI, the ion pair is obtained from solvated ions. LiBr is an intermediate case,[312] as shown by comparison of the Denison–Ramsey and Stokes radii, respectively. Similarly, the dissociation constants in THF were interpreted in terms of different ion pairing.[313]

For the alcohols, a fairly straight line is obtained for alkylammonium halides in normal alcohols.[288] It would seem that salts in hydroxylic solvents follow the simple exponential law predicted by electrostatics, despite the difference in $\mathring{a}_K$ values already mentioned. It could be that some factor, such as anionic solvation, is predominant in controlling the extent of pairing. This explanation, however, is made unlikely by the extremely high association of Bu_4NClO_4 compared to its association in an isodielectric solvent, acetone.[116] It was further shown that, by varying solvent composition at a fixed dielectric constant, K_A could be varied up and down by an order of magnitude.[314] This confirms that ϵ is not the predominant factor in association processes, even when the difference between ion–solvent and ion-pair–solvent interactions is taken into account.

It has been suggested that the simple electrostatic model with the solvent treated as a continuum successfully describes in broad outline the associative behavior of potassium iodide in a variety of solvents,[309] but that to understand the details it is necessary to take into account the molecular nature of the solvent. It is felt that the best alternative would be based on a multistep association process[315,316] involving various classes of ion pairs. The ultrasonic absorption of $MnSO_4$[287] has confirmed a three-step association process where only the first step can be adequately described using the classical continuum model. The second and third steps most probably involve the successive removal of solvent molecules from between the ions, giving a contact ion pair in the final state. The degree of dissociation of ion pairs in the hydroxylic solvents[288] was suggested to involve a two-step process,[317] as shown for acetone solutions.[317a] Substantial supporting evidence has come from ESR, EPR, and infrared spectra as reviewed by Swarcz.[294] Moreover, no physical significance could be assigned to the various $\mathring{a}_i$ parameters.

REFERENCES

1. J. F. Coetzee, in *Progress of Physical Organic Chemistry*, vol. 4, Eds., A. Streitureser, Jr., and R. W. Taft, Interscience, New York, 1965, p. 45.
2. J. Padova, *J. Phys. Chem.*, **72**, 692 (1968).
3. V. Gutman, *Coordination Chemistry in Nonaqueous Solutions*, Springer, New York, 1968.
4. C. C. Addison, *Internat. Conf. Nonaqueous Solvents Chemistry*, McMaster University, Hamilton, Canada, 1967.
5. J. N. Brønsted, *Rec. Trav. Chim.*, **42**, 718 (1923); *Chem. Rev.*, **5**, 231 (1928).
6. J. N. Brønsted, *Z. Angew. Chem.*, **43**, 229 (1930).
7. J. N. Brønsted, *Ber.*, **61**, 2409 (1928).

8. R. G. Bates, in *Solute–Solvent Interactions*, Eds., J. F. Coetzee and C. D. Ritchie, Dekker, New York, 1969, p. 45.

9. F. Franks, in *Hydrogen-Bonded Solvent Systems*, Eds., A. K. Covington and P. Jones, Taylor and Francis, London, 1968, p. 33.

10. A. J. Parker, *Quart. Rev.* (*London*), **16**, 163 (1962).

11. A. J. Parker, *Intern. Sci. Technol.*, **28**, August (1965).

12. A. J. Parker, *Chem. Rev.*, **69**, 1 (1969).

13. A. J. Parker, *J. Chem. Soc.*, (London), 1328 (1961).

14. R. G. Pearson, *J. Chem. Educ.*, **45**, 581, 1643 (1967).

15. C. L. De Ligny, D. Bax, M. Alfenaar, and G. L. Elferink, *Rec. Trav. Chim.*, **88**, 1183 (1969).

16. R. E. Harner, J. B. Sydnor, and E. S. Gilreath, *J. Chem. Eng. Data*, **8**, 411 (1963).

17. J. D. R. Thomas, *J. Inorg. Nucl. Chem.*, **24**, 1477 (1962).

18. F. J. Grigorovich, *Russ. J. Inorg. Chem.*, **8**, 507 (1963).

19. B. Becker, *J. Chem. Eng. Data*, **15**, 31 (1970).

20. K. Pavlopulos and H. Strehlow, *Z. Phys. Chem.*, **202**, 474 (1954).

21. G. Jander, in *Die Chemie in Wässerahnlichen Lösungsmitteln*, Springer, Berlin, 1949, p. 48.

22. J. N. Butler, *J. Electroanal. Chem.*, **14**, 89 (1967).

23. R. C. Paul and B. B. Sreenathan, *Ind. J. Chem.*, **4**, 382 (1966).

24. G. Pistoia, G. Pecci, and B. Scrosati, *Ric. Sci.*, **37**, 1168 (1968).

25. L. R. Dawson, J. E. Berger, J. W. Vaughn, and H. C. Eckstrom, *J. Phys. Chem.*, **67**, 281 (1963).

26. G. Pistoia and B. Scrosati, *Ric. Sci.*, **37**, 1173 (1968).

27. W. S. Harris, "Electrochemical Studies in Cyclic Esters," thesis, UCRL–8381, 1958.

28. Carlos A. Melendre, "Solubilities, Conductances, Viscosities, and Densities of Solutions of Selected Materials in Dimethyl Sulfoxide," M.Sc. thesis, UCRL–16330, 1965.

29. R. S. Biktimirov, *Russ. J. Phys. Chem.*, **37**, 1391 (1963).

30. N. A. Izmailov and V. S. Chernyi, *Russ. J. Phys. Chem.*, **34**, 58 (1960).

31. S. A. Voznesenski and R. S. Biktmirov, *Zh. Neorg. Khim.*, **2**, 942–945 (1957).

32. H. E. Zaugg, B. W. Harrom, and S. Borgwardt, *J. Amer. Chem. Soc.*, **82**, 2895 (1960).

33. H. Van Looy and L. P. Hammett, *J. Amer. Chem. Soc.*, **81**, 3872 (1959).

34. E. Price, in *The Chemistry of Nonaqueous Solvents*, vol. I, Ed., J. J. Lagowski, Academic Press, New York, 1966, Chapter 2.

35. J. Miller and A. J. Parker, *J. Amer. Chem. Soc.*, **83**, 117 (1961).

36. M. M. Markowitz, W. N. Hawley, D. A. Boryta, and R. F. Harris, *J. Chem. Eng. Data*, **6**, 325 (1961).

37. P. Walden, *Elektrochemie Nichtwässerige Lösungen*, Barth, Leipzig, 1924.

38. J. E. Ricci and T. W. Davis, *J. Amer. Chem. Soc.*, **62**, 407 (1940).

39. N. Bjerrum, in *Chemistry at the Centenary* (*1931*) *Meeting, Brit. Assn. Advan. Sci.*, p. 34.

40. H. N. Parton, in *Electrochemistry*, Pergamon Press, London, 1964, p. 455.

41. R. G. Bates, in *Solute–Solvent Interactions*, Eds., J. F. Coetzie and C. D. Ritchie, Dekker, New York, 1969, p. 45.

42. P. Kebarle, Arshadi, and Scarborough, *J. Chem. Phys.*, **49**, 817 (1968); P. Kebarle, *Advan. Chem. Ser.*, **72**, 24 (1968) and references therein.

43. G. Somsen, *Rec. Trav. Chim.*, **85**, 517 (1966).

44. E. J. F. Verwey, *Rec. Trav. Chem.*, **61**, 127 (1942).
45. L. H. Ahrens, *Geochim. Cosmochim. Acta*, **2**, 155 (1952).
46. W. M. Latimer, K. S. Pitzer, and C. M. Slansky, *J. Chem. Phys.*, **7**, 108 (1939).
47. M. Born, *Z. Physik [I]*, **45** (1920).
48. Y. C. Wu and H. L. Friedman, *J. Phys. Chem.*, **70**, 501 (1966).
49. Y. C. Wu and H. L. Friedman, *J. Phys. Chem.*, **70**, 2020 (1966).
50. L. Weeda and G. Somsen, *Rec. Trav. Chim.*, **86**, 263 (1967).
51. E. M. Arnett and D. R. McKelvey, *J. Amer. Chem. Soc.*, **88**, 2598 (1968).
52. E. Grunwald, G. Baughman, and G. Kohnstam, *J. Amer. Chem. Soc.*, **82**, 5801 (1960).
53. H. L. Friedman, *J. Phys. Chem.*, **71**, 1723 (1967).
54. C. V. Krishnan and H. L. Friedman, *J. Phys. Chem.*, **73**, 3934 (1969).
55. G. Choux and R. L. Benoit, *J. Amer. Chem. Soc.*, **91**, 6221 (1969).
56. B. E. Conway and J. O'M. Bockris, *Modern Aspects of Electrochemistry*, vol. I, Butterworths, London, 1954, Chapter 2.
57. C. M. Criss and E. Luksha, *J. Phys. Chem.*, **72**, 2966 (1968).
58. W. M. Latimer and W. L. Jolly, *J. Amer. Chem. Soc.*, **75**, 4147 (1953).
59. B. Jakuszewski and S. Taniewska-Osinska, *Lodz. Towarz. Nauk Acta Chim.*, **4**, 17 (1959).
60. B. Jakuszewski and S. Taniewska-Osinska, *Lodz. Towarz. Nauk Wydzial III, Acta Chim.*, **7**, 32 (1961).
61. B. Jakuszewski and S. Taniewska-Osinska, *Lodz. Towarz. Nauk Wydzial III, Acta Chim.*, **8**, 11 (1962).
62. C. M. Criss, R. P. Held, and E. Luksha, *J. Phys. Chem.*, **72**, 2970 (1968).
63. J. Greyson, *J. Phys. Chem.*, **66**, 2218 (1962).
64. F. Franks and D. S. Reid, *J. Phys. Chem.*, **73**, 3152 (1969).
65. R. A. Robinson and R. H. Stokes, *Electrolyte Solutions*, 2nd ed., Butterworths, London, 1959.
66. N. A. Izmailov, *Russ. J. Phys. Chem.*, **34**, 1142 (1960).
67. D. Feakins, in *Hydrogen-Bonded Solvent Systems*, Eds., A. K. Covington and P. Jones, Taylor and Francis, London, p. 243.
68. V. A. Pleskov, *Usp. Khim.*, **16**, 254 (1947).
69. H. H. Koep, H. Wendt, and H. Strehlow, *Z. Elektrochem.*, **64**, 483 (1960).
70. A. Lauer, *Electrochim. Acta*, **9**, 1617 (1964).
71. J. F. Coetzee and J. J. Campion, *J. Amer. Chem. Soc.*, **89**, 2513 (1967).
72. A. Voet, *Trans. Faraday Soc.*, **32**, 1301 (1936).
73. H. Strehlow, in *The Chemistry of Nonaqueous Solvents*, vol. I, Ed. J. J. Logowski, Academic Press, New York, 1966, Chapter 3.
74. J. F. Coetzee, J. M. Simon, and R. J. Bertozzi, *Anal. Chem.*, **41**, 776 (1969).
75. J. C. Poirier and J. H. DeLap, *J. Chem. Phys.*, **35**, 213 (1961).
76. G. R. Haugen and H. L. Friedman, *J. Phys. Chem.*, **72**, 4549 (1968).
77. N. A. Izmailov, *Dokl. Akad. Nauk, SSSR*, **149**, 884 (1963).
78. N. A. Izmailov, *Dokl. Akad. Nauk, SSSR*, **149**, 1364 (1963).
79. D. Feakins and P. Watson, *J. Chem. Soc.*, (London), 4734 (1963).
80. A. L. Andrews, H. P. Bennetts, D. Feakins, K. G. Lawrence, and Tomkins, *J. Chem. Soc.*, A, 1486 (1968).
81. N. E. Khomitov, *Russ. J. Phys. Chem.*, **39**, 336 (1965).
82. C. L. De Ligny and M. Alfenaar, *Rec. Trav. Chim.*, **84**, 81 (1965).
83. M. Alfenaar and C. L. De Ligny, *Rec. Trav. Chim.*, **86**, 929 (1967).
84. B. E. Conway, *Ann. Rev. Phys. Chem.*, **17**, 481 (1966).

85. O. Popovych, *Anal. Chem.*, **38**, 558 (1966).
86. R. Alexander and A. J. Parker, *J. Amer. Chem. Soc.*, **82**, 5549 (1967).
87. O. Popovych and A. J. Dill, *Anal. Chem.*, **41**, 456 (1969).
88. O. Popovych and R. M. Friedman, *J. Phys. Chem.*, **70**, 1671 (1966).
89. M. E. Coplan and R. M. Fuoss, *J. Phys. Chem.*, **68**, 1177 (1964).
90. B. Case and R. Parsons, *Trans. Faraday Soc.*, **63**, 1224 (1967).
91. B. Case, N. S. Hush, R. Parsons, and M. E. Peover, *J. Electroanal. Chem.*, **10**, 360 (1965).
92. F. H. MacDougall, *Thermodynamics and Chemistry*, 3rd ed., Wiley, New York, 1939.
93. J. Padova, *J. Chem. Phys.*, **39**, 1552 (1963).
94. O. Redlich and D. M. Meyer, *Chem. Rev.*, **64**, 221 (1964).
95. P. Mukerjee, *J. Phys. Chem.*, **65**, 740 (1961).
96. F. J. Millero, *J. Phys. Chem.*, **73**, 2417 (1969).
97. F. J. Millero, *J. Phys. Chem.*, **72**, 3209 (1968).
98. H. S. Frank and W. Y. Wen, *Discussions Faraday Soc.*, **24**, 133 (1957).
99. J. Padova, *J. Chem. Phys.*, to be published.
100. J. Padova and I. Abrahmer, *J. Phys. Chem.*, **71**, 2112 (1967).
101. J. E. Desnoyers and C. Jolicoeur, in *Modern Aspects of Electrochemistry*, vol. 5, Eds., J. O'M. Bockris and B. E. Conway, Butterworths, London, 1969, Chapter 1.
102. J. E. B. Randles, *Trans. Faraday Soc.*, **52**, 1573 (1956).
103. F. O. Koenig, *Compt. Rend., C.I.T.C.E.*, 3rd Reunion, Manfredi, Milan, 1952, p. 299.
104. N. A. Izmailov and Yu. F. Rybkin, *Dopovidi Akad. Nauk Ukr. RSR*, No. 1, 69 (1962).
105. N. A. Izmailov and Yu. F. Rybkin, *Dopovidi Akad. Nauk Ukr. RSR*, No. 1, 1071 (1962).
106. Yu. F. Rybkin, *Dopovidi Akad. Nauk Ukr. RSR*, No. 1, 1071 (1962) by ref. 41.
107. R. M. Noyes, *J. Amer. Chem. Soc.*, **84**, 513 (1962).
108. R. M. Noyes, *J. Amer. Chem. Soc.*, **86**, 971 (1964).
109. A. D. Buckingham, *Discussions Faraday Soc.*, **24**, 151 (1957).
110. J. A. Planbeck, *Can. J. Chem.*, **47**, 1401 (1969).
111. J. A. Stratton, *Electromagnetic Theory*, McGraw-Hill, New York, 1941, pp. 149–151.
112. J. Padova, *Electrochim. Acta*, **12**, 1227 (1967).
113. N. S. Hush, *Australian J. Sci. Res.*, **1**, 480 (1948).
114. W. A. Millen and D. W. Watts, *J. Amer. Chem. Soc.*, **89**, 6051 (1967).
115. H. S. Frank, *J. Chem. Phys.*, **23**, 2023 (1955).
116. M. Frohlich, *Theory of Dielectrics*, Clarendon, Oxford, 1949, pp. 9–13.
117. C. J. F. Böttcher, *Theory of Electric Polarisation*, Elsevier, Amsterdam, 1952, pp. 115–119.
118. E. A. Guggenheim, *Thermodynamics*, North Holland, Amsterdam, 1957.
119. J. K. Kirkwood and I. Oppenheim, *Chemical Thermodynamics*, McGraw-Hill, New York, 1961, pp. 233–237.
120. T. L. Hill, *J. Chem. Phys.*, **28**, 61 (1958).
121. K. S. Pitzer and L. Brewer, in *Thermodynamics*, Eds., Lewis and Randall, 2nd ed., McGraw-Hill, New York, 1961.
122. D. C. Grahame, *J. Chem. Phys.*, **21**, 1054 (1953).
123. J. Padova, *Israel J. Chem.*, **4**, 41 (1966).
124. B. B. Owen, *J. Amer. Chem. Soc.*, **54**, 1758 (1932).

125. E. Luksha and C. M. Criss, *J. Phys. Chem.*, **70**, 1496 (1966).

126. D. D. Eley and D. C. Pepper, *Trans. Faraday Soc.*, **37**, 581 (1941).

127. D. D. Eley and M. G. Evans, *Trans. Faraday Soc.*, **34**, 1093 (1938).

128. J. Padova, *J. Chem. Phys.*, **40**, 691 (1964).

129. A. Passinsky, *Acta Phys.-Chim. (URSS)*, **8**, 835 (1938).

130. J. Padova, *J. Chem. Phys.*, **39**, 2599 (1963).

131. J. O'M. Bockris, *Quart. Rev. (London)*, **3**, 173 (1949).

132. G. Jones and M. Dole, *J. Amer. Chem. Soc.*, **51**, 2050 (1929).

133. H. Falkenhagen and M. Dole, *Physik, Z.*, **30**, 611 (1929).

134. H. Falkenhagen and E. L. Vernon, *Physik, Z.*, **33**, 140 (1932).

135. R. W. Gurney, *Ionic Processes in Solution*, McGraw-Hill, New York, 1954.

136. J. Padova, *J. Chem. Phys.*, **38**, 2635 (1963).

137. D. Feakins and K. G. Lawrence, *J. Chem. Soc.*, A, 212 (1966).

138. R. H. Stokes, in *The Structure of Electrolyte Solutions*, Ed., Hamer, Wiley, New York, 1959.

139. R. H. Stokes and R. Mills, *Viscosity of Electrolytes and Related Properties*, Pergamon Press, New York, 1965.

140. D. F. T. Tuan and R. M. Fuoss, *J. Phys. Chem.*, **67**, 1343 (1967).

141. N. P. Yao and D. N. Bennion, UCLA Rept. No. 69-30, June 1969.

142. A. Einstein, *Ann. Phys.*, **19**, 289 (1906); **34**, 591 (1911).

143. H. Uhlich, *Trans. Faraday Soc.*, **23**, 388 (1927).

144. H. S. Frank, in *Chemical Physics of Ionic Solutions*, Eds., Conway and Barradas, Wiley, New York, 1966.

145. R. H. Stokes and R. A. Robinson, *Trans. Faraday Soc.*, **53**, 301 (1957).

146. E. R. Nightingale, *J. Phys. Chem.*, **63**, 1381 (1959).

147. R. A. Robinson and R. H. Stokes, *Electrolyte Solutions*, 2nd ed., Butterworths, London, 1959, p. 125.

148. M. Born, *Z. Physik*, **1**, 221 (1920).

149. R. H. Boyd, *J. Chem. Phys.*, **35**, 1281 (1961).

150. R. Zwanzig, *J. Chem. Phys.*, **38**, 1603, 1605 (1963).

151. E. J. Passeron, *J. Phys. Chem.*, **68**, 2728 (1964).

152. J. Padova, *Bull. Res. Council Israel*, **A10**, 63 (1961).

153. D. S. Allam and W. H. Lee, *J. Chem. Soc.*, A, 5 (1966).

154. A. S. Kaurova and G. P. Roshchina, *Akust. Zh.*, **12**, 118 (1966).

155. A. S. Kaurova and G. P. Roshchina, *Akust. Zh.*, **12**, 319 (1966).

156. G. P. Roshchina, A. S. Kaurova, and S. Sharapova, *Ukr. Fiz. Zh.*, **12**, 93 (1967).

157. G. P. Roshchina, A. S. Kaurova, and I. D. Kosheleva, *Zh. Strukt. Khim.*, **9**, 3 (1968).

158. I. G. Mikhailov, M. V. Rozina, and V. A. Shutilov, *Akust. Zh.*, **10**, 213 (1964).

159. J. H. Swinehart and H. Taube, *J. Chem. Phys.*, **37**, 1579 (1962).

160. S. Thomas and W. L. Reynolds, *J. Chem. Phys.*, **44**, 3148 (1966).

161. A. Fratiello and R. Schuster, *J. Phys. Chem.*, **71**, 1948 (1967).

162. N. A. Matwigoff, *Inorg. Chem.*, **5**, 788 (1966).

163. A. Fratiello, D. Miller, and R. Schuster, *Mol. Phys.*, **12**, 111 (1967).

164. N. A. Matwigoff and W. G. Movius, *J. Amer. Chem. Soc.*, **89**, 6077 (1967).

165. Z. Luz and S. Meiboom, *J. Chem. Phys.*, **40**, 1058 (1964).

166. Z. Luz and S. Meiboom, *J. Chem. Phys.*, **40**, 1066 (1968).

167. Z. Luz and S. Meiboom, *J. Chem. Phys.*, **40**, 2686 (1964).

168. S. A. Al-Baldawi and T. E. Gough, *Can. J. Chem.*, **47**, 1417 (1969).

169. S. Nakamura and S. Meiboom, *J. Amer. Chem. Soc.*, **89**, 1765 (1967).

170. H. H. Glaeser, H. W. Dodgen, and J. P. Hunt, *J. Amer. Chem. Soc.*, **89**, 3065 (1967).
171. A. M. Chmelnick and D. Fiat, *J. Chem. Phys.*, **49**, 2101 (1968).
172. T. D. Alger, *J. Amer. Chem. Soc.*, **91**, 2220 (1969).
173. A. Jackson, J. Lemons, and H. Taube, *J. Chem. Phys.*, **32**, 533 (1960).
174. J. H. Swinehart, T. E. Rogers, and H. Taube, *J. Chem. Phys.*, **38**, 398 (1963).
175. T. E. Rogers, J. H. Swinehart, and H. Taube, *J. Phys. Chem.*, **69**, 134 (1965).
176. E. Glueckauf, *Trans. Faraday Soc.*, **51**, 1235 (1955).
177. R. Fuoss, *Rev. Pure Appl. Chem.*, **18**, 125 (1968).
178. M. J. Blandamer and M. F. Fox, *Chem. Rev.*, **70**, 59 (1970).
179. A. D. E. Pullin and J. McC. Pollock, *Trans. Faraday Soc.*, **54**, 11 (1958).
180. I. S. Perelygin, *Opt. Spectr.*, **16**, 21 (1964).
181. S. Minc, Z. Kecki, and T. Gulik-krziwicki, *Spectrochim. Acta*, **19**, 353 (1963).
182. I. S. Perelygin, *Opt. Spectr. (USSR)*, **13**, 194 (1962).
183. S. Minc and S. Kurowski, *Spectrochim. Acta*, **19**, 339 (1963).
184. R. E. Hester and R. A. Plane, *Spectrochim. Acta*, **23A**, 2289 (1967).
185. Z. Kecki, *Spectrochim. Acta*, **18**, 1165 (1962); Z. Kecki and J. Witanowski, *Roczniki Chem.*, **38**, 691 (1964).
186. I. S. Perelygin, *Opt. Spectry. (USSR)*, **13**, 198 (1962).
187. J. P. Rocher and P. Van Huong, *J. Chim. Phys.*, **67**, 211 (1970).
188. L. C. Evans and G. Y. S. Lo, *Spectrochim. Acta*, **21**, 1033 (1965).
189. C. C. Addison, D. W. Amos, and D. Sutton, *J. Chem. Soc.*, A, 2285 (1968).
190. C. B. Baddiel, M. J. Tait, and G. J. Tait, *J. Phys. Chem.*, **69**, 3634 (1965).
191. C. D. Schmulbach, *J. Inorg. Nucl. Chem.*, **26**, 745 (1964).
192. K. F. Purcell and R. S. Drago, *J. Amer. Chem. Soc.*, **88**, 919 (1966).
193. G. J. Janz, K. Balasubrahmanyam, and B. G. Oliver, *J. Chem. Phys.*, **51**, 5723 (1969).
194. G. J. Janz, M. J. Tait, and J. Meier, *J. Phys. Chem.*, **71**, 963 (1967).
195. J. Corset, P. V. Huong, and J. Lascombe, *Spectrochim. Acta*, **24A**, 1385 (1968).
196. J. Corset, P. V. Huong, and J. Lascombe, *Spectrochim. Acta*, **24A**, 2045 (1968).
197. I. S. Perelygin, S. V. Izosimova, and Yu. M. Kessler, *Zh. Strukt. Khim.*, **9**, 390 (1968).
198. M. J. French and J. L. Wood, *J. Chem. Phys.*, **49**, 2358 (1968).
199. E. G. Höhn, J. A. Olander, and M. C. Day, *J. Phys. Chem.*, **73**, 3880 (1969).
200. W. F. Edgell, M. T. Yang, and N. Koizumi, *J. Amer. Chem. Soc.*, **87**, 2563 (1965).
201. B. W. Maxey and A. I. Popov, *J. Amer. Chem. Soc.*, **89**, 2230 (1967).
202. J. L. Wuepper and A. I. Popov, *J. Amer. Chem. Soc.*, **92**, 1493 (1970).
203. B. W. Maxey and A. I. Popov, *J. Amer. Chem. Soc.*, **91**, 20 (1969).
204. J. L. Wuepper and A. I. Popov, *J. Amer. Chem. Soc.*, **91**, 4352 (1969).
205. W. J. McKinney and A. I. Popov, *J. Phys. Chem.*, **74**, 535 (1970).
206. W. K. Thompson, *J. Chem. Soc.*, (London), 4028 (1964).
207. J. C. Evans and G. Y. S. Lo, *J. Phys. Chem.*, **69**, 3223 (1965).
208. J. F. Hinton and E. S. Amis, *Chem. Rev.*, **67**, 367 (1967).
209. H. G. Hertz, G. Stalidis, and H. Versmold, *J. Chim. Phys.*, Sp. No. 177 (1969).
210. J. Burgess and M. C. R. Symons, *Quart. Rev. (London)*, **22**, 768 (1968).
211. R. M. Hammaker and R. M. Clegg, *J. Mol. Spectry.*, **22**, 109 (1967).
212. B. S. Krumgal'z, K. P. Mishchenko, and B. I. Ionin, *Russ. J. Phys. Chem.*, **41**, 1045 (1967).
213. C. Franconi, C. Dejak, and F. Conti, in *Nuclear Magnetic Resonance in Chemistry*, Ed., B. Pesces, Academic Press, New York, 1965, p. 363.

214. R. N. Butler, E. A. Philpott, and M. C. R. Symons, *Chem. Commun.*, 371 (1968).
215. R. N. Butler and M. C. R. Symons, *Chem. Commun.*, 71 (1969).
216. R. N. Butler and M. C. R. Symons, *Trans. Faraday Soc.*, **65**, 945 (1969).
217. R. N. Butler and M. C. R. Symons, *Trans. Faraday Soc.*, **65**, 2559 (1969).
218. N. Asmus, *Z. Naturforsch.*, **4A**, 589 (1949).
219. W. G. Movius and N. A. Matwiyoff, *J. Phys. Chem.*, **78**, 3063 (1968).
220. B. W. Maxey and A. I. Popov, *J. Amer. Chem. Soc.*, **90**, 4470 (1968).
221. G. E. Maciel, J. K. Hancock, L. F. Lafferty, P. A. Mueller, and W. K. Musker, *Inorg. Chem.*, **5**, 554 (1966).
222. E. G. Bloor and R. G. Kidd, *Can. J. Chem.*, **46**, 3425 (1968).
223. W. G. Schneider and A. D. Buckingham, *Discussions Faraday Soc.*, **34**, 147 (1962).
224. A. Saika and C. P. Slichter, *J. Chem. Phys.*, **22**, 26 (1954).
225. H. Haraguchi and S. Fujiwara, *J. Phys. Chem.*, **73**, 3467 (1969).
226. M. C. R. Symons, *Amer. Rev. Phys. Chem.*, **20**, 219 (1969).
227. M. C. R. Symons, *J. Phys. Chem.*, **71**, 172 (1967).
228. A. Vèrtes, *Acta Chim. Acad. Sci. Hung.*, **63**, 9 (1970).
229. K. Burger, A. Vèrtes, and I. N. Czakó, *Acta Chim. Acad. Sci. Hung.*, **63**, 115 (1970).
230. K. Burger, A. Vèrtes, and E. Papp-Molnár, *Acta Chim. Acad. Sci. Hung.*, **57**, 257 (1968).
231. A. Vèrtes, K. Burger, and M. Suba, *Acta Chim. Acad. Sci. Hung.*, **63**, 123 (1970).
232. J. B. Hasted and G. W. Roderick, *J. Chem. Phys.*, **29**, 17 (1958).
233. D. M. Ritson and J. B. Hasted, *J. Chem. Phys.*, **16**, 17 (1948).
234. E. Glueckauf, *Trans. Faraday Soc.*, 1637 (1964).
235. P. S. K. Mohana Rao and D. P. Swarup, *J. Chem. Phys.*, **43**, 4530 (1965).
236. J. Padova, *Israel J. Chem.*, **2**, 279 (1964).
237. J. P. Badiali and J. C. Lestrade, *J. Chim. Phys.*, Spec. No. 107 (1969).
238. J. P. Badiali, H. Cachet, and J. C. Lestrade, *J. Chim. Phys.*, **46**, 1350 (1967).
239. J. P. Badiali, H. Cachet, F. Govaerts, and J. C. Lestrade, *Compt. Rend. Acad. Sci. Paris*, **265C**, 149 (1967).
240. H. Cachet, J. C. Lestrade, and J. Epelboin, *Compt. Rend. Acad. Sci. Paris*, **261**, 678 (1965).
241. H. Cachet, I. Epelboin, and J. C. Lestrade, *Electrochim. Acta*, **11**, 1759 (1966).
242. J. B. Hasted, *Progress in Dielectrics*, vol. 3, Wiley, New York, 1961, p. 103.
243. H. Cachet, thesis, Paris, 1968, as cited in ref. 237.
244. D. L. Wertz, Ph.D. thesis, University of Kansas, 1967.
245. D. L. Wertz and R. F. Kruh, *J. Chem. Phys.*, **50**, 4013 (1969).
246. R. J. DeSando and G. H. Brown, *J. Phys. Chem.*, **72**, 1088 (1968).
247. R. Gillespie, in *Chemical Physics of Ionic Solutions*, Eds., B. E. Conway and R. G. Barradas, Wiley, New York, 1966, p. 599.
248. K. Crickard and J. F. Skinner, *J. Phys. Chem.*, **73**, 2060 (1969).
249. H. S. Frank and M. W. Evans, *J. Chem. Phys.*, **13**, 507 (1945).
250. G. R. Hood and L. P. Hohlfelder, *J. Phys. Chem.*, **38**, 979 (1934).
251. T. B. Hoover, *J. Phys. Chem.*, **68**, 876 (1964).
252. E. A. Nikiforov, *Zh. Strukt. Khim.*, **10**, 137 (1969).
253. V. Vand, *J. Phys. Chem.*, **52**, 277 (1948).
254. J. Pavoda, unpublished results.
255. H. Falkenhagen and G. Kelb, in *Modern Aspects of Electrochemistry*, vol. 2, Ed., J. O'M. Bockris, Butterworths, London, 1959, p. 84.
256. C. W. Davies, *Ion Association*, Butterworths, London, 1962.

257. R. M. Fuoss, *J. Chem. Educ.*, **32**, 527 (1955).

258. J. Barthel, *Angew. Chem.*, **80**, 253 (1968).

259. R. M. Fuoss and F. Accascina, *Electrolytic Conductance*, Interscience, New York, 1959.

260. C. Evers and R. L. Kay, *Amer. Rev. Phys. Chem.*, **11**, 21 (1960).

261. R. M. Fuoss, in *Chemical Physics of Ionic Solutions*, Eds., B. E. Conway and R. G. Barradas, Wiley, New York, 1966, p. 462.

262. E. Pitts, R. E. Tabor, and J. Daly, *Trans. Faraday Soc.*, **65**, 849 (1969).

263. E. Pitts, *Proc. Roy. Soc. (London) Ser. A*, **217**, 43 (1953).

264. R. Fernandez-Prini and J. E. Prue, *Z. Physik. Chem. (Leipzig)*, **228**, 373 (1965).

265. R. Ferrandez-Prini, *Trans. Faraday Soc.*, **64**, 2146 (1968).

266. R. M. Fuoss and K. L. Hsia, *Proc. Natl. Acad. Sci. (U.S.)*, **59**, 1550 (1967).

267. K. L. Hsia and R. M. Fuoss, *J. Amer. Chem. Soc.*, **90**, 3055 (1968).

268. A. D. Pethybridge and J. E. Prue, *Am. Rep. (London)*, **65A**, 129 (1969).

269. R. M. Fuoss and K. L. Hsia, *Proc. Natl. Acad. Sci.*, **59**, 1550 (1967); *J. Amer. Chem. Soc.*, **90**, 3055 (1968).

270. R. Fernandez-Prini, *Trans. Faraday Soc.*, **65**, 3311 (1969).

271. C. M. Springer, J. F. Coetzee, and R. L. Kay, *J. Phys. Chem.*, **73**, 471 (1969).

272. R. L. Kay, S. C. Blum, and H. I. Schiff, *J. Phys. Chem.*, **67**, 1223 (1963).

273. R. M. Fuoss and E. Hirsch, *J. Amer. Chem. Soc.*, **82**, 1013 (1960).

274. H. Fowler and C. A. Kraus, *J. Amer. Chem. Soc.*, **62**, 2237 (1960).

275. R. M. Fuoss, *Proc. Natl. Acad. Sci. (U.S.)*, **45**, 807 (1959).

276. G. J. Hills, in *Chemical Physics of Ionic Solutions*, Eds., B. E. Conway and R. G. Barradas, Wiley, New York, 1966, p. 576.

277. P. Walden, H. Uhlich, and G. Busch, *Z. Physik. Chem.*, **123**, 429 (1926).

278. R. Gopal and M. M. Husain, *J. Ind. Chem. Soc.*, **40**, 981 (1963).

279. L. G. Longsworth, *J. Phys. Chem.*, **67**, 689 (1963).

280. M. Della Monica, U. Lamanna, and L. Senatore, *J. Phys. Chem.*, **72**, 2124 (1968).

281. R. L. Kay, in *Trace Inorganics in Water*, Amer. Chem. Soc., Washington, D.C., 1968, p. 1.

282. D. F. Evans and R. L. Kay, *J. Phys. Chem.*, **70**, 366 (1966).

283. R. L. Kay and D. F. Evans, *J. Phys. Chem.*, **70**, 2325 (1966).

284. S. Schmick, *Z. Physik*, **24**, 56 (1924).

285. R. L. Kay, G. P. Cunningham, and D. F. Evans, in *Hydrogen-Bonded Solvent Systems*, Eds., A. K. Covington and P. Jones, Taylor and Francis, London, 1968, p. 249.

286. R. Zwanzig, *J. Chem. Phys.*, **52**, 3625 (1970).

287. G. Atkinson and S. K. Koz, *J. Phys. Chem.*, **69**, 128 (1965).

288. D. F. Evans and P. Gardam, *J. Phys. Chem.*, **73**, 158 (1969).

289. R. L. Kay, B. J. Hales, and G. P. Cunningham, *J. Phys. Chem.*, **71**, 3925 (1967).

290. R. L. Kay, C. Zawoyski, and D. F. Evans, *J. Phys. Chem.*, **69**, 4208 (1965).

291. D. F. Evans and T. L. Broadwater, *J. Phys. Chem.*, **72**, 1037 (1968).

292. D. F. Evans and P. Gardam, *J. Phys. Chem.*, **72**, 3281 (1968).

293. C. A. Kraus, *J. Phys. Chem.*, **60**, 129 (1956).

294. M. Swarcz, *Carbanions, Living Polymers, and Electron Transfer Processes*, Interscience, New York, 1968.

295. L. D. Pettit and S. Bruckenstein, *J. Amer. Chem. Soc.*, **88**, 4783 (1966).

296. N. Bjerrum, *Kgl. Danske Videnskab. Selskab.*, **7**, No. 9 (1926).

297. R. M. Fuoss, *Trans. Faraday Soc.*, **30**, 967 (1934).

298. P. H. Flaherty and K. H. Stern, *J. Amer. Chem. Soc.*, **80**, 1034 (1958).

299. J. T. Denison and J. B. Ramsey, *J. Amer. Chem. Soc.*, **77**, 2615 (1955).

300. H. Y. Inami, H. K. Bodenseh, and J. B. Ramsey, *J. Amer. Chem. Soc.*, **83**, 4745 (1961).

301. W. R. Gilkerson, *J. Chem. Phys.*, **25**, 1199 (1956).

302. W. R. Gilkerson and R. E. Stamm, *J. Amer. Chem. Soc.*, **82**, 5295 (1960).

303. W. R. Gilkerson, *J. Phys. Chem.*, **74**, 746 (1970).

304. R. M. Fuoss, *J. Amer. Chem. Soc.*, **80**, 5059 (1958).

305. E. A. Guggenheim, *Trans. Faraday Soc.*, **56**, 1159 (1960).

306. J. Byberg, S. J. K. Jensen, and U. K. Kläning, *Trans. Faraday Soc.*, **65**, 3023 (1969).

307. J. Padova, unpublished results.

308. T. R. Griffiths and M. C. R. Symons, *Mol. Phys.*, **3**, 90 (1960).

309. J. E. Prue, in *Chemical Physics of Ionic Solutions*, Eds. B. E. Conway and R. G. Barradas, Wiley, New York, 1966, p. 163.

310. E. Schaschel and M. C. Day, *J. Amer. Chem. Soc.*, **90**, 503 (1968).

311. W. A. Adams and K. J. Laidler, *Can. J. Chem.*, **46**, 2005 (1968).

312. L. Savdoff, *J. Amer. Chem. Soc.*, **88**, 664 (1966).

313. J. Comyn, F. S. Dainton, and K. J. Ivin, *Electrochim. Acta*, **13**, 1851 (1968).

314. A. D'Aprano and R. M. Fuoss, *J. Amer. Chem. Soc.*, **91**, 211 (1969).

315. R. L. Kay and J. L. Harves, *J. Phys. Chem.*, **69**, 2787 (1965).

316. M. Eigen and K. Tamm, *Z. Elektrochem.*, **66**, 93, 107 (1962).

317. R. L. Kay, D. F. Evans, and G. P. Cunningham, *J. Phys. Chem.*, **73**, 3332 (1969).

317a. G. S. Darbari and S. Petrucci, *J. Phys. Chem.*, **74**, 268 (1970).

318. *Solubilities of Inorganic and Organic Compounds*, vol. 1, Eds., H. Stephen and T. Stephen, Pergamon Press, London, 1963.

319. R. Gopal and M. M. Husain, *J. Ind. Chem. Soc.*, **40**, 272 (1963).

320. J. W. Vaugh, in *The Chemistry of Nonaqueous Solvents*, vol. 2, Ed., J. J. Lagowski, 1967, p. 192.

321. H. Krauer, in *Chemistry in Nonaqueous Ionizing Solvents*, vol. 4, Eds., G. Jander, H. Spandau, and C. C. Addison, Interscience, New York, 1963, p. 225.

322. G. A. Strack, S. K. Swanda, and L. W. Bahe, *J. Chem. Eng. Data*, **9**, 416 (1964).

323. J. J. Lagowski and G. A. Moczygemba, in *The Chemistry of Nonaqueous Solvents*, vol. 2, Ed., J. J. Lagowski, Academic Press, New York, 1967, p. 320.

324. C. M. Criss, *U.S. At. Energy Comm.*, TID 22366 (1965).

325. S. I. Drakin and C. Yu-Min, *Russ. J. Phys. Chem.*, **38**, 1526 (1964).

326. B. Jakuszewski and S. Taniewska-Osinska, *Bull. Acad. Pol. Sci.*, **9**, 133 (1961).

327. G. P. Kotlyarova and E. F. Ivanova, *Russ. J. Phys. Chem.*, **38**, 221 (1964); **40**, 537 (1966).

328. M. Waycicka, *Roczniki Chem.*, **38**, 1207 (1964).

329. B. Jakuszewski, S. Taniewska-Osinska, and R. Logwinienko, *Bull. Acad. Pol. Sci.*, **9**, 127 (1961).

330. R. H. Boyd and P. S. Wang, 155th Meeting Amer. Chem. Soc., Abstr. No. 128, San Francisco, Calif., 1968.

331. R. H. Boyd, *J. Chem. Phys.*, **51**, 1470 (1969).

332. L. N. Erbanova, S. I. Drakin, and M. Kh. Karapet'yants, *Russ. J. Phys. Chem.*, **38**, 1450 (1964).

333. L. N. Erbanova, M. Kh. Karapet'yants, and S. I. Drakin, *Russ. J. Phys. Chem.*, **39**, 1467 (1965).

334. K. P. Mishchenko, *Acta Phys.-Chim. URSS*, **3**, 693 (1935).

335. K. P. Mishchenko and V. V. Sokolov, *Zh. Strukt. Khim.*, **5**, 819 (1964).

336. G. A. Krestov and V. I. Klopov, *Zh. Strukt. Khim.*, **5**, 829 (1964).

337. R. F. Rodewald, K. Mahendran, J. L. Bear, and R. Fuchs, *J. Amer. Chem. Soc.*, **90**, 6698 (1968).

338. R. P. Held and C. M. Criss, *J. Phys. Chem.*, **69**, 2611 (1965).

339. R. P. Held and C. M. Criss, *J. Phys. Chem.*, **71**, 2487 (1967).

340. L. Weeda and G. Somsen, *Rec. Trav. Chim.*, **86**, 893 (1967).

341. R. K. Wicker, Ph.D. thesis, University of Delaware, 1966.

342. G. A. Krestov, *Zh. Strukt. Khim.*, **3**, 516 (1962).

343. G. P. Kotlyarova and E. F. Ivanova, *Russ. J. Phys. Chem.*, **40**, 537 (1966).

344. N. A. Izmailov, *Dokl. Akad. Nauk SSSR*, **149**, 1103 (1963).

345. R. Gopal and R. K. Srivastava, *J. Phys. Chem.*, **66**, 2704 (1962); *J. Ind. Chem. Soc.*, **40**, 99 (1963).

346. R. Gopal and M. A. Siddiqui, *Z. Physik. Chem.*, *N.F.*, **67**, 122 (1969).

347. R. Gopal and M. A. Siddiqui, *J. Phys. Chem.*, **73**, 3390 (1969).

348. F. C. Schmidt, W. E. Hoffman, and W. B. Schaap, *Proc. Indiana Acad. Sci.*, **72**, 127 (1962).

349. V. N. Fesenko, E. F. Ivanova, and G. P. Kotlyarova, *Russ. J. Phys. Chem.*, **42**, 1416 (1968).

350. C. Treiner and R. M. Fuoss, *Z. Physik. Chem.*, **228**, 343 (1965).

351. J. E. Prue and P. J. Sherrington, *Trans. Faraday Soc.*, **57**, 1795 (1961).

352. D. Atlani, J. C. Justice, M. Quintin, and P. Dubois, *J. Chim. Phys.*, **66**, 180 (1969).

353. R. M. Gopal and O. N. Bhatnagar, *J. Phys. Chem.*, **70**, 4070 (1966).

354. R. M. Gopal and O. N. Bhatnagar, *J. Phys. Chem.*, **70**, 3007 (1966).

355. (a) R. D. Singh, P. P. Rastogi, and R. M. Gopal, *Can. J. Chem.*, **46**, 3525 (1968); (b) R. M. Gopal and O. M. Bhatnagar, *J. Phys. Chem.*, **69**, 2382 (1965).

356. T. C. Wehman and A. I. Popov, *J. Phys. Chem.*, **72**, 4031 (1968).

357. J. M. Notley and M. Spiro, *J. Phys. Chem.*, **70**, 1502 (1966).

358. J. S. Dunnet and R. P. H. Gasser, *Trans. Faraday Soc.*, **61**, 922 (1965); M. D. Archer and R. P. H. Gasser, *Trans. Faraday Soc.*, **62**, 3451 (1966); J. M. Crawford and R. P. H. Gasser, *Trans. Faraday Soc.*, **63**, 2758 (1967).

359. R. L. Kay, *J. Amer. Chem. Soc.*, **82**, 2099 (1960).

360. H. Brusset and T. Kikindai, *Bull. Soc. Chim. France*, 1150 (1962).

361. D. Arrington, Ph.D. thesis, University of Kansas, 1968.

362. G. C. Hammes and S. Petrucci, *J. Amer. Chem. Soc.*, **91**, 275 (1969).

363. M. Spivey and T. Shedlowsky, *J. Phys. Chem.*, **71**, 2165 (1967).

364. H. V. Venkatasetty and G. H. Brown, *J. Phys. Chem.*, **66**, 2075 (1962); **67**, 954 (1963).

365. (a) A. K. R. Unni, L. Elias, and H. I. Schiff, *J. Phys. Chem.*, **67**, 1216 (1963); (b) S. C. Blum and H. I. Schiff, *J. Phys. Chem.*, **67**, 1220 (1963); (c) R. L. Kay, S. C. Blum, and H. I. Schiff, *J. Phys. Chem.*, **67**, 1222 (1963).

366. H. L. Yeager and B. Kratohvil, *J. Phys. Chem.*, **73**, 1963 (1969).

367. D. F. Evans, C. Zaurogski, and R. L. Kay, *J. Phys. Chem.*, **71**, 3925 (1967).

368. S. Springer, Ph.D. thesis, University of Pittsburgh, 1968.

5   Fused Salts as Liquids

Elizabeth Rhodes, Department of
Chemical Engineering, University
College of Swansea, Wales

1   Introduction

The most significant characteristic of fused salts as compared with other liquids is that the constituent units are all electrically charged. The electrostatic Coulomb forces are long range, and although in nonelectrolytic solvents the properties of dilute solutions of electrolytes can be adequately represented by the Debye–Hückel theory,[1] fused salts represent a class of electrolytic solutions whose concentration is well outside the range of applicability of this theory.

The first studies of fused salts in the 1880s were measurements of their temperature of fusion;[2] systematic studies of such properties as viscosity, electrical conductivity, and thermodynamic functions commenced in the early 1900s for salts melting below about 1000°C.[3-5]

Data on salts melting at higher temperatures were obtained somewhat later when suitable furnaces and refractory container materials were available.[6] The real upsurge of interest in molten salts has been since 1950, when their potential use as heat-transfer fluids and homogeneous reaction media became apparent, although siliceous melts and certain oxide melts have always been of commercial importance in the extractive metallurgical industries and in glass manufacture.

Since 1955 there have appeared three review articles[7-9] on fused salts in the *Annual Review of Physical Chemistry*, and a number of books entirely devoted to fused salts[6,10-14] have been published, including a recent comprehensive compilation of data pertaining to molten salts.[15]

In 1961 a Faraday Society Discussion Meeting[16] on the structure and properties of ionic melts was held; this followed a much earlier meeting on electrode processes, which included a number of papers on the electrochemistry of fused salts.[17] Regular biennial Molten Salt Discussion Group

Meetings are held in America under the auspices of the Gordon Conferences and in Europe at the invitation of Euchem.

The preceding brief summary takes no account of major articles appearing in other scientific texts, for instance, "Melting and Crystal Structure," by A. R. Ubbelohde,[18] the chapter, "Electrodes in Fused Salt Systems," by R. W. Laity in *Reference Electrodes*, edited by Ives and Janz,[19] or "Statistical Mechanics of Ionic Systems," by M. Blander in *Advances in Chemical Physics*.[20] Thus it is apparent that it is impossible to cover the whole field of interest in fused salts. The author estimates that over the past ten years some 5000 scientific papers have been published on work in fused salts. In order to limit the field, four criteria have been chosen, namely:

1. The temperature of fusion should be less than 1500°C, that is, the molten salt is containable in Pyrex, silica, magnesium oxide, or precious metal vessels.

2. The very interesting but complex silicates and metallurgical slags are only briefly considered.

3. Reactions in molten salts and electrode process and electrode kinetic studies are only mentioned insofar as they help to elucidate melt structure.

4. The objective is to relate the fused salt liquid state to other liquid systems.

2   Typical Fused Salts

The range of fusion temperature is from about 400 to 3000°K (Table 1), therefore it is apparent that powerful molecular forces must modify the overall electrostatic bonding energies of the salts; the work of identifying, describing, and understanding these counteracting forces has been one of the main interests of molten salt research workers over the past two decades. The simplest inorganic salts are the alkali halides, which consist of assemblages of positive and negative, singly charged, spherical, ionic particles. Except for the lithium halides, the anions and cations are roughly of the same size, although the anions are somewhat larger and more readily polarizable. The melting points are in the range 800 to 1300°K and the volume changes on fusion are about 20%; the entropies of fusion are approximately 6 eu (i.e., 3 R)—the entropy increase required for positional randomization. Melting theories, notably as expounded by Ubbelohde,[18] seek to explain deviations in properties of melts from the alkali-halide norm in terms of additional modes of entropy uptake in the melt.

Table 1 Representative Melting and Boiling Points of Inorganic Salts

Salt	Melting Point (°K)	Boiling Point (°K)	$\dfrac{T_b}{T_f}$	Salt	Melting Point (°K)
NaF	1268	1977	1.56	NaNO$_3$	583
MgF$_2$	1536	2490	1.62	Ca(NO$_3$)$_2$	834
SbF$_3$	563	649	1.15	Na$_2$SiO$_3$	1362
TaF$_5$	380	602	1.59	CaSiO$_3$	1803
NaCl	1081	1738	1.60	NaBO$_2$	1239
MgCl$_2$	987	1691	1.71	NaNO$_2$	544
ZnCl$_2$	556	1005	1.80	K$_3$PO$_4$	1613
LiBr	820	1583	1.93	KSCN	450
HgBr$_2$	514	592	1.15	AgCN	623
CaBr$_2$	1003	1083	1.08	KCN	883
KI	958	1597	1.67	NaOH	591
MgO	3073	—	—	Ca(OH)$_2$	1108
PbO	1159	1745	1.50	K$_2$Cr$_2$O$_7$	671
V$_2$O$_5$	943	2325	3.45	KOOCH	441
K$_2$CO$_3$	1169	—	—	NaOOC·CH$_3$	602
CaCO$_3$	1613	—	—	KOOC·CH$_3$	577
Na$_2$SO$_4$	1157	—	—	NH$_4$CNS	361
BaSO$_4$	1623	1673	1.03	(n-Butyl)$_4$NBr	391
LiClO$_3$	391	543	1.39	(n-Butyl)$_4$NNO$_3$	394

From Ref. 15.

3 Structure of Ionic Melts

There have been only a limited number of X-ray diffraction studies on molten salts.[21-32] X-ray diffraction measurements on normal liquids are difficult to carry out with precision; experimental technique has to be very good in order to obtain reliable results at the high temperatures of molten salts, particularly since the effect of temperature broadening on diffraction peaks must be considered. The data obtained are in the form of the pair radial distribution function (RDF), which gives the probability that pairs of atoms (or ions) are to be found separated by a given distance. In pure fused alkali halides there are two species of ion present, cation and anion, and hence three RDFs are required to describe the cation-anion, cation-cation, and anion-anion distributions, respectively. So far, all three separate RDFs have not been resolved for any salt, although a modified RDF provides some structural information. Levy and Danford[21] give the appropriate modified radial distribution function, $D_{\alpha\beta}(r)$, in the following form:

$$(1) \qquad D_{\alpha\beta}(r) = 4\pi r \int_{-\infty}^{\infty} \mu \rho_{\alpha\beta}(\mu) T_{\alpha\beta}(\mu - r) \, d\mu$$

where

$$(2) \qquad T_{\alpha\beta}(r) = \frac{1}{\pi} \int_0^\infty f_\alpha f_\beta M(s) \cos{(sr)} \, ds$$

In these equations $T_{\alpha\beta}(r)$ is a known RDF; f_α and f_β are atomic scattering amplitudes for atoms α and β; $\rho_{\alpha\beta}$ is the volume density, and $\mu = 2$ for a binary salt; $M(s)$ is a modification function chosen so that $f_\alpha f_\beta M(s)$ is independent of s where

$$(3) \qquad s = \frac{4\pi}{\lambda} \sin\frac{\phi}{2}$$

with λ representing the wavelength and ϕ the angle of scattering.

A similar analysis was applied to neutron-scattering data;[28] the advantage of using neutron-scattering data is that the method is susceptible to isotope mass changes and hence there is the increased possibility of obtaining sufficient RDF data to resolve all three component RDF spectra.

Typical radial distribution functions for molten alkali halides are shown in Figure 1. Radial distribution functions for melts of polyatomic anions are more complex; Figure 2 shows the RDFs for Na_2CO_3 and Na_2SO_4. With polyatomic anions there are several alternative cation sites (Figure 2), and it is expected that the A sites should be more favored in the liquid, for they represent the closest anion-cation contact distances. However, the available data are not sufficiently precise to provide a definite answer to the question of siting of cations in liquids. The data, however, do conclusively prove that $CO_3{}^{2-}$, $SO_4{}^{2-}$, and $NO_3{}^{-}$ [29,30] exist as discrete entities in the melts.

On going from a high-temperature crystal to a melt of the alkali halides, the X-ray and neutron diffraction data show that there is a decrease in the cation-anion distance, that the next-nearest-neighbor distance (cation-cation or anion-anion) increases, and that, on the whole, except for the lithium salts, the coordination number decreases drastically from 12 to 8, indicating that very little long-range order remains in the melt (Table 2).

Generally the cation-anion distances in melts of alkali halides are very similar to those determined by microwave spectra for gaseous alkali halides.[33] For instance, r_{KCl} at one-quarter peak maximum for potassium chloride in melt is 2.69 Å, whereas in the gas phase r_{KCl} is 2.67 Å, which is reasonable experimental justification for assuming that in melts the preferred cation-anion distance is that of the monomeric gaseous ion pair. This assumption forms one of the keystones for the latest statistical mechanical theories of the thermodynamics of ionic melts.[20]

Apart from the sparse X-ray diffraction data, additional evidence for close-contact cation-anion pairs can be adduced from other spectroscopic data (see Section 8).

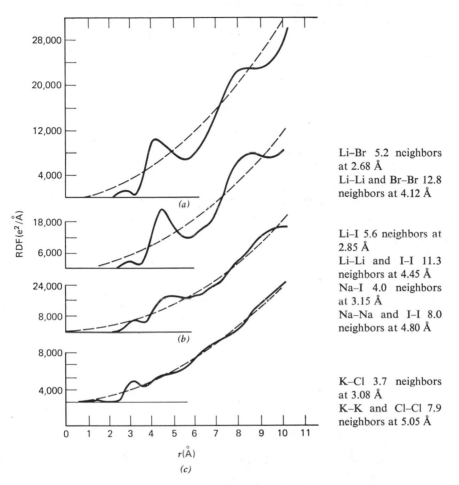

Li–Br 5.2 neighbors at 2.68 Å
Li–Li and Br–Br 12.8 neighbors at 4.12 Å

Li–I 5.6 neighbors at 2.85 Å
Li–Li and I–I 11.3 neighbors at 4.45 Å
Na–I 4.0 neighbors at 3.15 Å
Na–Na and I–I 8.0 neighbors at 4.80 Å

K–Cl 3.7 neighbors at 3.08 Å
K–K and Cl–Cl 7.9 neighbors at 5.05 Å

Figure 1. Radial distribution functions of alkali halides. From Ref. 28, The New York Academy of Sciences 1960; reprinted by permission.

Furukawa[30] suggests that the nearest-neighbor coordination numbers and distances may be related, for the alkali halides, with the experimental densities by the following expression:

$$(4) \qquad \frac{V_S}{V_L} = \left(\frac{r_s}{r_l}\right)^3 \frac{n_l}{n_s}$$

where V_S and V_L are the molar volumes of solid (at melting point) and liquid, and n_s, n_l, r_s, and r_l are the solid and liquid coordination numbers and distances. Lumsden[13] queries this correlation, noting that it is difficult to

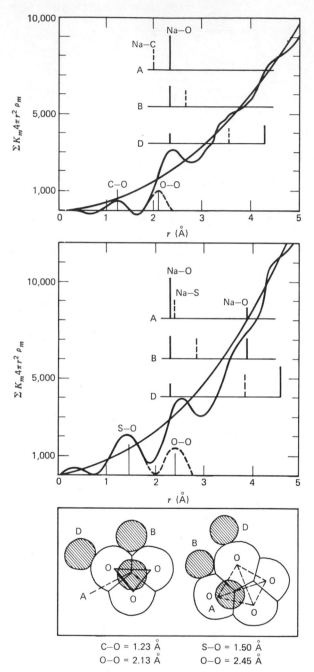

Figure 2. Radial distribution functions of molten alkali carbonates and sulfates. From Ref. 29. Reprinted by permission of the Faraday Society.

Table 2   First and Second Peak Distances and Coordination Numbers (CN) in Melts from Alkali Halide Radial Distribution Functions

| | Solid | | | | Liquid | | | | |
| | First Peak r_{+-} at Melting Point | | Second Peak r_{++} at Melting Point | | First Peak r_{+-} in Melt | | Second Peak r_{++} in Melt | | |
Salt	(Å)	CN	(Å)	CN	(Å)	CN	(Å)	CN	Method[a]
LiCl	2.66	6	3.76	12	2.47	4.0	3.85	12.0	X
					2.45	3.5	3.80	8.3	N
LiBr	2.85	6	4.03	12	2.68	5.2	4.12	12.8	X
LiI	3.12	6	4.41	12	2.85	5.6	4.45	11.3	X
NaI	3.35	6	4.74	12	3.15	4.0	4.80	8.9	X
KCl	3.26	6	4.61	12	3.10	3.7	—	12	X
					3.53	3.5	4.7	7.4	N
CsCl	3.57	6	5.05	12	—	4.6	4.87	7.1	X
	(3.70)	8	(4.27)	(6)					
CsBr	(3.72)	(6)	5.26	(12)	3.55	4.6	5.4	8.3	X
	3.86	8	4.46	6	3.55	4.7	5.2	7.9	N
CsI	3.94	(6)	5.57	(12)	3.85	4.5	5.5	7.2	X
	4.08	8	4.72	6					

Adapted from Ref. 21.

[a] X, X-ray diffraction; N, neutron diffraction.

distinguish between 4 and 6 coordination in the melt purely from measurements on densities because the calculated difference between NaCl (6) and ZnS (4) structures is only 3%. Lumsden is of the opinion that the 4-coordination ZnS lattice fits the concept of molten salts best, any distortion of the lattice resulting from the inclusion of interstitial ions rather than from the appearance of vacancies, which would be the natural happening on the 6-coordination NaCl model. This model requires less energy than the pure-hole model of fusion and is consistent with the idea of fluctuating cracks between clusters of ions.[27,34]

There is a little X-ray diffraction work on $AlCl_3$, InI_3, SnI_4, and CdI_2 by Wood and Ritter,[25] whose data show that the dimer Al_2Cl_6 accounts for the covalent character of the aluminum chloride melt. In the complex molten oxide system of B_2O_3–PbO–Al_2O_3, Zarzycki[32] claims that the RDFs show that molecular interaction only extends over 11 Å and concludes that association complexes are relatively small, with the lead atom at the center of the complex.

4   Theories of Fused Salts

Quasi-Lattice or Cell Theory. Various models have been proposed to describe the structure of molten salts and to enable calculations of thermodynamic and transport properties to be made. There has been extensive use of the *quasi-lattice theory*,[35] especially in the calculation of thermodynamic properties of mixtures of molten salts by Førland,[11] Blander,[12] and Lumsden,[13] who followed the original use of this theory by Temkin.[36] The cell theory or quasi-lattice theory[37,38] explains the melting process on the basis of the introduction of vacancies—in particular, Schottky vacancies or holes—into the crystal lattice. Stillinger[12,39] has modified the original theory and applied it to simple ionic liquids by taking into account polarization energy requirements. In the lattice theory of fused salts, the liquid is considered as split up into small volume elements or cells, each ion lying at the center of a cell and the cells of identical shape and size forming a regular lattice. Correlation functions, $g_{\alpha\beta}(r)$, for *pairs of cells* containing cations and anions are derived by applying the principles of statistical mechanics (see "Statistical-Mechanical Compilations of the Thermodynamic Properties of Molten Salts.")

"Hole" Theory of Liquids. Basically, this theory describes spontaneous density fluctuations in liquids as thermal motion causes the molecules or ions to move about.[40,41] The density fluctuations are assumed to be of molecular size, and these low-density regions or "holes" may drift about continuously in the liquid rather than moving by discrete jumps as in the solid. The movement of the holes is treated by the methods of fluid mechanics; the work required for movement being composed of the reversible work necessary to create a cavity of radius, r, and the kinetic energy for radial, (r), and translational motion, (t):

$$(5) \qquad f_h(r, P_r, P_t) = b \exp\left[-\frac{1}{kT}\left(W(r) + \frac{P_r^2}{2m_r} + \frac{P_t^2}{2m_t}\right)\right]$$

This is a Boltzmann distribution function, where

$$(6) \qquad W(r) = \frac{4\pi r^3}{3}(P - P_0) + 4\pi r^2 \sigma$$

and σ is the surface tension, P the atmospheric pressure, and P_0 vapor-pressure of the liquid.

Solving and normalizing (5) gives

$$(7) \qquad F_h(r) = \frac{16}{15/\pi}\left(\frac{kT}{4\pi\sigma}\right)^{-7/2} r^6 \exp\left(-\frac{W(r)}{kT}\right)$$

Hence the average hole volume, $\overline{V}$, can be computed from the hole radius. For the alkali halides this hole radius was found to be 2.1 Å. If on melting the volume increase is ΔV_m, then it is reasonable to assume that $\Delta V_m = N_h \overline{V}$. Let θ_f be fraction of overall liquid volume occupied by the dense ionic fluid, then

$$(8) \qquad\qquad \theta_f = \frac{V - \Delta V_m}{V}$$

The expansivity of holes is given by

$$(9) \qquad\qquad \alpha_h = \frac{3 N_h \overline{V}}{2kT}$$

and compressibility of the holes by

$$(10) \qquad\qquad \beta_h = 0.4713 \frac{N_h (\overline{V})^2}{kT}$$

Bockris and Richards[42] evaluated α_h and β_h using the hole theory, and the calculations compared reasonably with the ultrasonic compressibility data for many salts (generally to about 20%).

It should be noted that there is a random distribution of holes, and it is not necessary to postulate in the hole theory that the electrostatic forces on the ions cause behavior different from that of normal liquids. It is unlikely, however, that the use of the macroscopic surface tension, σ, is permissible in considerations of small spherical cavities in electrolytic liquids; hence any calculations are likely to be of dubious value.

Mott and Gurney's model of cybotactic or microcrystalline liquid structure has also been applied to melts, but it is difficult to see how it can explain the diffusion of ions in a melt.[43] The Bernal model of polyhedra in the liquid state would also imply the presence of holes in the melt; no calculations have been made using this model for fused salts.[44]

Liquid-Free-Volume Theory. A free-volume theory was first elaborated by Zernicke and Prins in 1927;[45] this additional (free) volume was thought to be associated with the increasing volume of the cell—that is, "cell free volume"—and could be considered as the extra volume requirement of the particular particle occupying that cell.[46,47] The increase in volume is associated with the atoms in a continuous way and is equally distributed among the atoms. Cohen and Turnbull[48] consider that the free volume should be associated with the liquid as a whole, that is, the concept of "liquid free volume, V_f" or the excess volume that can be distributed in the liquid with no energy change. Eventually this "liquid free volume" gives rise to voids, which are a prerequisite for any transport in the liquid. The free volume is randomly distributed throughout the liquid

(11) $$V_f = \bar{V} - V_0$$

where $\bar{V}$ is average volume per molecule and V_0 is volume per molecule when the molecule is in a hypothetical solid cell at the same temperature as the liquid.

Theory of Significant Structures. The significant structure theory of Eyring et al.[49] is based on the concept that a liquid consists of fragments of crystal structure in a fluid of a compressed gas of the same material. The thermodynamic partition function then consists of a hypothetical mixture of the solid and gaseous partition functions combined together by using two adjustable parameters, which are evaluated by means of a comparison liquid. Blomgren[50] considers the effect on the combined partition functions of the strong Coulomb forces in a molten salt. He assumes that: (a) the Coulomb forces only have an explicit effect on the potential energy part of the partition function, (b) the remainder of the partition function is the same as for a normal liquid, and (c) the potential energy consists of a separable Coulombic part that varies as $(V_s/V)^{\frac{1}{3}}$ (i.e., solidlike) and a non-Coulombic part that varies as V_S/V (i.e., gaslike).

The potential energy of the liquid, E_L, is given by

(12) $$E_L = E_1\left(\frac{V_s}{V}\right)^{\frac{1}{3}} + E_2\left(\frac{V_s}{V}\right)$$

It may be predicted that E_2, the gaslike potential energy, is about 2000 cal/mole for a salt such as potassium chloride, whose ion sizes and electronic configurations are similar to those of argon. If the coordination number on going from a solid to a liquid changes from 6 to 4, then $E_1 \sim \frac{2}{3}E_s$ (energy of sublimation). With these assumptions, the partition function, f, for a molten salt can be written in a form analogous to that of a molecular liquid

$$f = \left[\frac{1 + n(V - V_S)/V_S \exp\left[-a(E_1 + E_2)V_S/(V - V_S)RT\right]}{(1 - \exp - \theta/T)^3}\right]^{2NV_S/V}$$

(13) $$\times \left[\left\{\left(\frac{2\pi m_1 kT}{h^2}\right)^{\frac{3}{2}}\frac{eV}{N}\right\}\cdot\left\{\left(\frac{2\pi m_2 kT}{h^2}\right)^{\frac{3}{2}}\frac{eV}{N}\right\}\right]^{N(V - V_S)/V}$$

$$\times \exp N\left[E_1\left(\frac{V_S}{V}\right)^{\frac{1}{3}} + E_2\left(\frac{V_S}{V}\right)\right]$$

where n and a are adjustable parameters evaluated for the liquid at the melting point. In an evaluation of the molar volume, V, of molten potassium chloride, it was found using (13) that the deviation from experimental only increased to about 4% at temperatures 300°C above the melting point.

Statistical-Mechanical Computations of the Thermodynamic Properties of Molten Salts. The foregoing theories are, at best, attempts to evaluate approximations to the configurational integrals and partition functions of the statistical-mechanics formulations. Basically the statistical-mechanics theories depend on the theory of corresponding states, which has been applied to molten salts by Reiss, Mayer, and Katz[51] and by Reiss, Katz, and Kleppa.[52] Recent reviews by Blander[20,53] and by Luks and Davis[54] have continued the excellent article by Stillinger, "Equilibrium Theory of Pure Fused Salts."[12]

In dense ionic systems, dimensional-analysis methods lead to relative values of the partition functions. At the high temperatures of melts it is only necessary to consider the classical partition function, Q, that is

$$(14) \qquad A = -kT \ln Q$$

Now $Q = KZ$ where K is the kinetic energy integral and Z is the configurational integral. Using the theory of corresponding states it is possible to omit the kinetic energy integral; thus the configurational integral alone for a symmetric salt of N_c cations and N_a anions can be written

$$(15) \qquad Z = \frac{1}{(N!)^2} \int \cdots \int_v \exp\left(\frac{-U}{kT}\right)(d\tau_c)^N (d\tau_a)^N$$

where U is the potential energy of the particular ionic configuration in which the ion positions are specified by the respective volume elements, $d\tau$, in configurational space.

Blander[20] then evaluates the potential energy, U, by summing the pair-potential energies for cation-anion, cation-cation, and anion-anion interactions; U, U', and U'' are the pair potentials, respectively, such that c < c' and a < a' so that no pair potential is computed twice.

The total potential energy

$$(16) \qquad U = \sum_c \sum_a U + \sum_{c<c'} \sum U' + \sum_{a<a'} \sum U''$$

On a rigorous theory, the pair-potential energy for a cation-anion pair, γ, δ, is composed of several terms[12,54]

$$(17) \qquad U_{\gamma\delta}(r) = A_{\gamma\delta} e^{-B_{\gamma\delta}(r)} + \frac{z_\gamma z_\delta e^2}{r} - \left[\frac{(z_\gamma e)^2 \alpha_\delta + (z_\delta e)^2 \alpha_\gamma}{2r^4}\right] + \frac{C_{\gamma\delta}}{r^6} + \cdots$$

1. $A_{\gamma\delta} e^{-B_{\gamma\delta}(r)}$ is the short-range repulsion between electron clouds around ions as they are brought into contact; in ionic salts this value may be approximated by a rigid sphere. This interaction is additive.

2. $z_\gamma z_\delta e^2/r$, the Coulombic charge-charge interaction, is also additive.

3. $C_{\gamma\delta}/r^6$ is the dipole-dipole interaction. The coefficient, $C_{\gamma\delta}$, is given by the London formula for the dispersion energy[55]

$$(18) \qquad\qquad C_{\gamma\delta} = \frac{\frac{3}{2}\alpha_\gamma\alpha_\delta I_\gamma I_\delta}{I_\gamma + I_\delta}$$

where α's and I's denote the respective ionic polarizabilities and ionization potentials of the ions. The terms are additive.

4. $[(z_\gamma e^2)\alpha_\delta + (z_\delta e^2)\alpha_\gamma]/2r^4$ is the charge-induced dipole contribution, which is dependent on the ionic polarizabilities, α_γ and α_δ. It is *not additive* and depends on the local electric field (usually equated with the dielectric constant, D, of the fluid) set up by the charges and induced dipoles surrounding an ion.

Reiss, Mayer, and Katz[51] apply the corresponding-states theory[56] to a fused salt by assuming that the equilibrium properties of the salt can be related to a model ionic melt. The model ionic melt is defined as follows.

1. The long-range Coulomb forces create a locally ordered structure of positive ions surrounded by negative ions, and vice versa.

2. Only short-range interactions between ions of opposite sign are important. Interaction between ions of like charge is neglected owing to the locally ordered structure.

3. The pair potential between ions γ and δ of opposite sign is taken to be of the form $U_{\gamma\delta}(r) = \infty$ for $r \leq \lambda$, where $\lambda = r_+ + r_-$ (i.e., the sum of the ionic radii). For

$$(19) \qquad\qquad r > \lambda \qquad U_{\gamma\delta}(r) = \frac{z_\gamma z_\delta e^2}{Dr}$$

where D is a dielectric constant introduced to account for local electric field and the many body effects. With the pair potential in this form, each salt in any group of salts can be characterized by a single parameter, λ, which is conveniently taken to be the sum of the ionic radii, $r_+ + r_-$, and is often known as the hard-core cutoff parameter. Because of the change of dielectric constant, it is not possible to compare different groups of salts (e.g., alkali halides and alkaline earth oxides). By transferring to dimensionless quantities,[20,54] the following expressions for reduced pressure, Π, reduced temperature, τ, and reduced volume, θ, for a charge-symmetrical salt—that is, $|z_\gamma| = |z_\delta|$—can be obtained

$$(20) \qquad \Pi = \left(\frac{D\lambda^4}{z^2 e^2}\right) P \qquad \tau = \left(\frac{D\lambda kT}{z^2 e^2}\right) \qquad \theta = \left(\frac{V}{\lambda^3}\right)$$

Assuming that $D = 1$ and that the melting point, T_m, is a corresponding state, Reiss et al.[51] calculate that for the alkali halides $\lambda T_m = 3.0 \times 10^{-5}$

deg/cm. This relation was found to hold except for the lithium halides, which have λT_m values of $\sim 2.2 \times 10^{-5}$ deg/cm.

By a similar analysis, reduced expressions can be obtained for surface tension and vapor pressure.

Correlations for vapor pressure (except for lithium salts) are good, but calculated values for Σ_m (reduced surface tension at the melting point) show poor agreement with experimental values. For alkali halides, surface tension $\bar{\sigma}_m \lambda^3 \sim 3.3 \times 10^{-21}$ dyne/cm. An equation representing the reduced vapor pressure Π'' of the alkali halides has been evaluated and is shown in Figure 3. If association occurs in the vapor phase,

$$n\mathrm{AX} \xrightleftharpoons{K_n} \mathrm{A}_n\mathrm{X}_n$$

an expression for the association constant, K_n, for the formation of an n polymer from a monomer can be derived [20]

$$(21) \qquad \ln K_n(T) + 3(n-1) \ln g' = \ln K_{n0} \frac{T}{g'}$$

where g' is a scaling parameter, $g' = \lambda_0/\lambda$, evaluated from a comparison salt (e.g., NaI, where K_{20} is the dimer association constant for this salt).

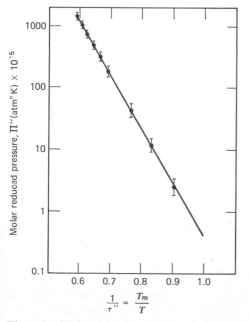

Figure 3. Reduced vapor pressure of alkali halides: $\log_{10} \Pi'' = 9.20/\tau'' - 0.894 \log_{10} \tau'$ $- 6.21$. From Ref. 20, by permission from *Molten Salt Chemistry*, ed. M. Blander, John Wiley and Sons, New York.

For asymmetric salts, $z_\gamma \neq z_\delta$, the reduced temperature and reduced pressure, respectively, are given by

$$(22) \qquad \tau = \frac{\lambda D k T}{e^2}$$

and

$$(23) \qquad \Pi = \frac{D\lambda^4 P}{e^2}$$

so that the reduced equation of state Π is now a function of the charges on the cation and anion, that is

$$(24) \qquad \Pi = \Pi(\tau_1 \theta_1 z_\gamma z_\delta)$$

Molten salts are very dense, and the structure of the liquid is determined primarily by the available packing possibilities or configurations of the anion-cation cores. The scaled particle theory of Stillinger[39] replaces the ion-pair correlation function, $g_{\alpha\beta}(r)$, by the number-average–pair-correlation function, $g_m(r)$, of a hypothetical nonelectrolytic fluid whose particle diameter, r, is the sum of the anion and cation radii of the salt

$$(25) \qquad g_m(r) = \left[\sum_{\alpha,\beta=1}^{\nu} \rho_\alpha \rho_\beta \right]^{-1} \sum_{\alpha,\beta=1}^{\nu} \rho_\alpha \rho_\beta g_{\alpha\beta}(r)$$

where $\rho = (\rho_\alpha + \rho_\beta)$; ρ, ρ_α, and ρ_β are the number densities of the pair particle and the respective ions, and ν is the number of salts in the system. The isothermal compressibility, β_T, is then given by the statistical-mechanics formula

$$(26) \qquad \rho k T \beta_T = 1 + \rho \int_0^\infty [g(r) - 1] 4\pi r^2 \, dr$$

In order to evaluate $g(r)$ it was assumed that the molecular interaction was given by the hard-sphere scaled-particle formula of Reiss, Frisch, and Lebowitz,[57] who obtained the following equation of state for fluids:

$$(27) \qquad P = \frac{6kT}{\pi\lambda^3} \left[\frac{y(1 + y + y^2)}{(1 - y)^3} \right]$$

where $y = (\pi\rho\lambda^3/6)$.

This theory involves a relation between the reversible work to create a spherical cavity in a fluid, the probability of a density fluctuation producing a cavity, and the distribution function, $g(r = \lambda)$. From (26) and (27)

$$(28) \qquad \beta_T = \frac{1}{\rho k T} \frac{(1 - y)^4}{(1 + 2y)^2}$$

Using similar formulas, Mayer[58] calculated values of the hard-core diameter, λ, for the alkali halides from the thermal-expansion relation

$$(29) \qquad \alpha = (1 - y^3)[T(1 + 2y)^2]$$

and the surface tension

$$(30) \qquad \sigma = \frac{kT}{4\pi\lambda^2}\left[\frac{12y}{1-y} + \frac{18y^2}{(1-y)^2}\right] - \frac{P\lambda}{2}$$

Values of λ_σ, λ_α, and λ_β calculated from experimental data were found to be nearer the gas-phase value than the ionic crystal values. For sodium chloride at 1073°K, these values were $\lambda_\sigma = 2.46$ Å, $\lambda_\alpha = 2.63$ Å, and $\lambda_\beta = 2.48$ Å. The sum of the Pauling crystal ionic radii is 2.76 Å, whereas λ in the gas phase is 2.36 Å.

Yosim and Owens[59] have further developed the rigid-sphere model of Reiss et al.[57] in order to calculate the entropy, enthalpy, entropy of fusion and compressibility of molten alkali halides. They assumed that the entropy of a molten salt could be simulated by compressing two moles of rigid spheres to the volume of the fused salt. The entropy change on further compression to the solid phase should then be equivalent to the entropy of fusion. The following state equation was used:

$$(31) \qquad P = \frac{6kTy(1 + y + y^2)}{\Pi a^3(1 - y)^3}$$

where $y = (\Pi a^3 \rho)/6$ with a the diameter of rigid sphere and ρ the number density; for an ionic salt of type MX, a is the sum of the radii of cation and anion, $\rho = 2N/V$, and for a uni-univalent salt, $y = \Pi a^3 N/3V$. The entropy of the liquid phase, S_L, is given by

$$(32) \qquad S_L = S° - 2R\left[\frac{\ln (V_g/V_L)}{1 - y} + \frac{3y(2 - y)}{2(1 - y^2)}\right]$$

where $S°$ is calculated from the Sackur–Tetrode equation and the entropy of vaporization of the fluid, S_{Vap}.

On solving for S_L, good agreement was obtained if the particle diameter, a, was taken as the mean ionic diameter obtained from X-ray measurements of the melt, but not if gas-phase or crystal values were used.

Corresponding-states melting curves have also been obtained for the alkali halides[60] at high pressures such that

$$(33) \qquad \tau = 3.22 + 7.28 \times 10^{-4}\Pi - 8.68 \times 10^{-8}\Pi^2$$

where $\Pi = D\lambda^4\rho/z^2$ and $\tau = D\lambda^4 T/z^2$, with λ taken as the sum of the ionic radii at 25°C and 1-atm pressure; lithium halides and alkali fluorides deviate markedly.

Similar equations have been evaluated to calculate the reduced volumes of alkali and silver nitrates, and again deviations are found for the lithium salts.[61] The failure of the corresponding-states laws for the lithium and fluoride salts is ascribed to interaction between anions due to the small size of the lithium cation, or to cation-cation interaction in the case of the fluoride salts.[51] Boyer, Fray, and Meadowcroft,[62] using experimental surface tensions and expansivities, have attempted to calculate phosphate chain lengths in pure molten phosphates of calcium, lithium, sodium, and zinc by applying the rigid-sphere model. Calculations show that the long chains are not stable entities and that the usual particles are $P_2O_7^{4-}$ or PO_4^{3-} ions.

Monte Carlo Computations for Fused Salts. Førland and Østvold[63] have recently developed a method to estimate the configurational thermodynamic properties of a fused salt using the Monte Carlo method. The method has been tested for a two-dimensional model of molten sodium chloride, and the calculations seem to give reasonable results when compared with experimental data.

The interaction potential used was of the usual *ionic* form but contained only two terms in order to cut computing time, that is

$$(34) \qquad U_{ij} = \frac{Z_i Z_j e^2}{r_{ij}} + \frac{\ell_{ij} e^2}{r_{ij} n}$$

where ℓ_{ij} is the Born–Landé repulsion coefficient and n is a number varying from 5 to 12, depending on the types of ions present. The Born–Landé coefficient for the sodium chloride crystal is $\ell_{ij} = 43.5$ Å. Scaling down to give appropriate values for a two-dimensional crystal, r_0 was found to be 2.65 Å for the equilibrium distance. The Monte Carlo cell chosen had an edge of 3.10 Å calculated from the density of fused sodium chloride at 1423°K scaled to correspond to a two-dimensional cell. Computations were then made of the energy as a function of the number of configurations for cells with 16, 32, and 64 ions in the cell. Convergence was found to be fair for the largest cell and there was reasonable correspondence with the X-ray data for fused sodium chloride given by Levy et al.[28]*

Tests of Theories of Fused Salts. As more accurate experimental data becomes available, check calculations on the various theoretical models of molten salts can be made. Vilcu and Misdolea[64] used the significant-structures theory to calculate C_V, C_p, β_T, and α_p at constant volume and at constant pressure. They found that the compressibility (β_T) values calculated on the significant-structure theory lay between the experimental values and those

---

* *Note added in proof.* More accurate computer simulations of molten salts have now been made by K. Singer and L. V. Woodcock (Ph.D. Dissertation, L. V. Woodcock, London, 1970) using the methods of molecular dynamics.

calculated by Yosim and Owens[59] using the corresponding-states theory. Experimental and theoretical values of C_V were in good agreement, C_V decreasing as the temperature increased; on the other hand, there were large discrepancies between experimental and theoretical values of C_p. There was also poor agreement for values of α_p. They concluded that significant-structures–partition-function calculations appear to forecast volume dependencies inaccurately but to give reasonable agreement when temperature-dependent functions are required.

Eyring and co-workers[65] have modified the significant-structures partition function used by Carlson et al.[49] to take *account of monomer and dimer gas-like molecules in the fluid*. For instance, if dimers are present

$$2MX \xrightarrow{K_2} (MX)_2$$
$$(n_1) \qquad (n_2)$$

where K_2 is the equilibrium constant. The absolute activities are related by the function $(n_1/f_1)^2 = (n_2/f_2)$, where n_1 and n_2 are the number of molecules of each sort present and f_1 and f_2 are the partition functions. In terms of the unit volume of gas-like molecules present, the partition functions may be written as $F_1 = f_1/(V - V_S)$, and $F_2 = f_2/(V - V_S)$ where V is the total volume and V_S is the volume of solid-like molecules, hence $K_2 = F_2/F_1^2$. The K_2 is also given by the Eyring absolute reaction rate equation (35).

$$(35) \qquad K_2 \sim \exp\left(-\frac{\Delta H_2}{RT}\right) \exp\frac{\Delta S_2}{R}$$

Now

$$(36) \qquad n_2 = \frac{1}{2}\left[\frac{N(V - V_S)}{V} - n_1\right]$$

$$n_1 = \left(\frac{V - V_S}{4K_2}\right)\left[-1 + \left(1 + \frac{8K_2N}{V}\right)^{\frac{1}{2}}\right]$$

$$(37) \qquad F_1 = \left(\frac{2\Pi mkT}{h^2}\right)^{\frac{3}{2}}\left(\frac{8\Pi^2 I_1 kT}{\sigma' h^2}\right)\left[1 - \exp\left(-\frac{h\nu}{kT}\right)\right]^{-1}$$

where I is moment of inertia, m is mass of molecule and σ' is symmetry number.

The liquid partition function, f_l, is then of the form

$$(38) \qquad f_l = \left\{\frac{\exp(E_s/2RT)(V/V_S)^{\frac{1}{3}}}{[1 - \exp(-\theta/T)]^3}\right.$$
$$\times \left[1 + n\left(\frac{V - V_S}{V_S}\right)\exp\left(\frac{-aE_s(V/V_S)^{\frac{1}{3}}}{2RT(V - V_S)/V_S}\right)\right]\right\}^{2NV_S/V}$$
$$\times \left\{\left(\frac{F_1 eV}{N}\right)^{N(V - V_S)/V}\left(1 + \frac{2n_2}{n_1}\right)^{n_1}\left[\frac{K_2N}{eV}\left(\frac{n_1}{n_2} + 2\right)\right]^{n_2}\right\}$$

where E_s = sublimation energy at m.p., $\theta = h\nu/k$ = Einstein temperature, a and n are adjustable parameters, $n = ZV_S/V_m$ where Z is the number of nearest neighbors in crystal, and V_m is volume of melt, $a \sim 0.005$.

The Helmholtz free energy is given by

$$(39) \qquad\qquad A = -RT \ln f_l$$

Calculations of C_p, C_V, α, β, η, and σ agreed with experimental values for the alkali halides using (38).

The significant-structures theory has also been applied to the molten mercuric halides,[66] and a partition function assuming the following equilibrium was constructed, incorporating the equilibrium constant, K:

$$2HgX_2 \rightleftharpoons HgX^+ + HgX_3^-$$

where $K = f_{HgX} f_{HgX_3}/f_{HgX_2}^2$.

The critical constants were calculated, and the results appear in Table 3.

Table 3   Critical Constants of Mercuric Halides

Salt	$T_c(^\circ K)$	$P_c(atm)$	$V_c(cm^3/mole)$
$HgCl_2$	1114	173.9	190.08
$HgBr_2$	1071	143.9	220.74
HgI_2	1149	125.2	272.12

Reprinted by permission of the American Chemical Society from *J. Phys. Chem.*, **72**, 4155 (1968).

Mercuric halides have low ionic conductivities, which accords with the use of the equilibrium constant, K, in the partition function.

There has lately been an attempt to use the cellular hole model to calculate the specific heat of the alkali halides.[67] Calculations were made of the energy of formation of holes using the idea of cooperative relaxation of the structure around the holes and McClellan's approximations[68] to calculate the free volume. It was found that the calculated specific heat required corrections because of the vibrational energy.

5   Thermodynamics of Fused Salt Mixtures

Introduction.   There is considerable interest in thermodynamic properties of mixtures of molten salts, particularly about the extent and reasons for their departure from ideality. Many measurements of the excess enthalpies, entropies, and volumes of mixing have been made, and there are excellent articles by Førland (including many details of experimental procedures),[11] Blander,[12]

Kleppa,[8] and Kirgintsev and Avvakumov,[69] and the book by Lumsden solely devoted to the thermodynamics of molten salt mixtures.[13]

Thė basic model that has been used to describe molten salt mixtures is the Temkin model.[36] In this model the postulate is that the different types of anions are randomly arranged on the anion sublattice, whereas the cations are randomly arranged on the corresponding interpenetrating cation sub-lattice. For uni-univalent salts, if a mixture contains n_A moles of A cations, n_B moles of B cations, n_x moles of X anions, and n_y moles of Y anions, and x_A, x_B, x_X, x_Y are the corresponding ionic fractions defined such that $x_A = n_A/(n_A + n_B)$ and $n_a = Nx_A$, and so on, then according to the Temkin model the integral Helmholtz energy of mixing, ΔA_M, is

(40) $$\Delta A_M = NkT(x_A \ln x_A + x_B \ln x_B + x_X \ln x_X + x_Y \ln x_Y)$$

A theory of random mixing leads to a similar equation. If the system is nonideal, the Helmholtz free energy of mixing

(41) $$\Delta A_M = NkT \sum x \ln x + \Delta A^E(x)$$

where

(42) $$\Delta A^E(x) = -kT \ln \left[\frac{1}{N!} \int \cdots \int e^{-\langle U \rangle / kT} \, dr_1 \cdots dr_N\right]$$

and $\langle U \rangle$ is the pair-potential energy.

Binary mixtures of molten salts may be classified into five main categories:

1. Charge-symmetrical systems containing a common ion (e.g., KCl–NaCl mixtures).

2. Charge-unsymmetrical systems with a common ion (e.g., KNO_3–$Ca(NO_3)_2$ mixtures).

3. Charge-symmetrical *reciprocal* molten salt mixtures containing two different cations and two different anions (e.g., NaCl–KBr mixtures).

4. Charge-unsymmetrical *reciprocal* systems (e.g., NaCl–K_2CO_3 mixtures).

5. True *ternary* systems having three different cations and a common anion, or vice versa [e.g., (Li, K, Na)NO_3].

Categories 3 and 4 are sometimes known as ternary systems, which can be a confusing nomenclature. An obvious difference between fused salt mixtures and molecular mixtures is that, even in the simplest so-called binary mixtures, there are in fact three distinct entities in the system. However, in the binary and ternary systems with common ions, the common ion can be regarded as the solvent with the other ions acting as solutes; and it is the differences in charge, size, and polarizability between these solutes that account for the excess mixing functions. One of the difficulties in molten salts is defining a frame of reference with regard to chemical potentials and ionic and equivalent fractions and activities.[69,70] This has often led to confusion, particularly when charge-unsymmetrical systems are considered. With that warning, the theories of *regular solutions* and the Duhem–Margules relationships have been applied successfully to mixtures of molten salts.[11,12] Blomgren[71] deals with regular solution theories for binary and reciprocal salt systems, whereas Ryabchikov[72] has extended the treatment of reciprocal salt systems to obtain an equation relating the derivatives of the integral excess thermodynamic quantities to the equivalent ionic fractions. Braunstein[73] derived equations for the composition dependence of activities in concentrated binary molten salt mixtures if association with ideal mixing of the associated and unassociated species is the cause of deviations from ideality. In dilute molten salt solutions with a common anion, the solvent obeys Raoult's law and the solute obeys Henry's law, as has been proved many times by cryoscopic and EMF activity coefficient measurements.[74–80]

During the last 10 years, reliable calorimetric heats-of-mixing data have become available and these can in part be supplemented and compared with entropy and enthalpy data obtained from EMF activity measurements or cryoscopic and phase-diagram data. For certain systems, vapor pressure data is obtainable, and calculations of activities of liquid mixtures can be made on the assumption that the vapor phase is ideal.

Charge-Symmetrical Mixtures with a Common Ion. The quasi-lattice theories of mixing proposed by Førland[81] and Blander[82] were based on the quasi-chemical theory of Guggenheim.[83] If salt 1 is AX and salt 2 is BX, then the potential energy of the ion triplets $A^+X^-A^+$, $B^+X^-B^+$, and $A^+X^-B^+$ are given by U_{11}, U_{22}, and U_{12}, respectively, where the potential energies are presumed to be independent of the local environment. The molar heat of mixing of the solution is

$$(43) \qquad H_M = N_1 N_2 \lambda \left(1 - N_1 N_2 \frac{2\lambda}{Z'RT} + \cdots \right)$$

where $\lambda = NZ'/2(2U_{12} - U_{11} - U_{22}) = \tfrac{1}{2}NZ'\,\Delta\epsilon'$, with N Avogadro's number and Z' the number of cation nearest neighbors of a given cation. On mixing cations in a mixture of three hard spherical ions, the Coulombic energy change, $\Delta\epsilon_C$, is given by

$$(44) \qquad \Delta\epsilon_C = -e^2\left(\frac{1}{d_1} + \frac{1}{d_2}\right)\left(\frac{d_1 - d_2}{d_1 + d_2}\right)^2$$

where e is electronic charge and d_1 and d_2 are distances between the center of the anion and the two respective cations.

On this basis it would be expected that the excess energy of mixing would be symmetrical; that is, λ should be independent of composition. A different approach has been used by Lumsden,[13,84] who considers that the energy of mixing ($\Delta\epsilon_p$) is due to the interaction of the anion with the polarization field. He deduces the following relationship:

$$(45) \qquad \Delta\epsilon_p = \frac{\alpha F^2}{2} = -\alpha e^2 \left(\frac{1}{d_1} + \frac{1}{d_2}\right)^4 \left(\frac{d_1 - d_2}{d_1 + d_2}\right)^2$$

where F is field intensity on an anion between two cations of different sizes, α is polarizability, and d_1 and d_2 are the respective cation-anion distances. Davis et al.[85,54] and Blander[20] have applied the conformal ionic solution theory[51,52] to charge-symmetrical mixtures of molten salts with the following result for the excess energy of mixing:

$$(46) \qquad \Delta A^E = x_1 x_2 \left(\frac{kT}{2}\right) \left(\frac{\lambda_1 - \lambda_2}{\lambda}\right)^2 N^3 [\epsilon - \omega + N(\omega - \alpha^2)]$$

where $(N^3 kT/2)[\epsilon - \omega + N(\omega - \alpha^2)]$ is a complex integral that is a negative function of temperature and pressure arising from the second derivatives of the hard-sphere repulsion forces, and λ_1 and λ_2 are the hard-core cutoff parameters equivalent to d_1 and d_2, the cation-anion distances.

Equation 46 is obviously very similar to those used by Kleppa et al.[8] to correlate calorimetric data on heats of mixing (e.g., alkali nitrates)

$$(47) \qquad \Delta H_M = -\frac{N_1 N_2}{40} \left(\frac{d_1 - d_2}{d_1 d_2}\right)^2$$

Blander[20] prefers to use the dimensionless size parameter, $g_i = d_0/d_i$, where d_0 and d_i are the hard-core parameters characterizing the salts. For a binary system AX + BY where g_1 and g_2 are the respective size parameters

$$(48) \qquad \Delta H_M = x_1 x_2 \Omega(T, P)(g_1 - g_2)^2$$

It was apparent that this simple theory (Figure 4) was only approximate when calorimetric data on mixtures with easily polarizable cations became available. Davis and Rice[85] proposed a model that included short-range dispersion interactions as a perturbation of the original pair potential. It was assumed that the pair-potential energy, $U_{\alpha\beta}(r)$, was the sum of the Coulombic polarization and van der Waals dispersion forces with the coefficient of the polarization term modified to account for the many-body effects. The dispersion coefficient was taken to be that of isolated interacting ions, and the short-range interaction between the ions of like sign was neglected because of local structure.

For like ions, where C = Coulombic, NC = non-Coulombic, and ξ is a coupling parameter

$$(49) \qquad U_L(R) = U_L^C(R) + \xi U_L^{NC}(R) \qquad R \geq 0$$

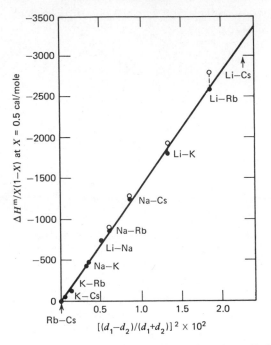

Figure 4. Excess heats of mixing of molten alkali halides. From O. J. Kleppa and L. S. Hersh, J. Chem. Phys., **34**, 351 (1961), reprinted by permission.

For unlike ions

$$(50) \qquad U_U(R) = U_U{}^C(R) + \xi U_U{}^{NC}(R) \qquad R > \lambda$$

Now the full solution for these equations contains a first-order term that is dependent on the non-Coulombic interactions but does not include the cation-anion hard-core diameter.[86,87] For melts $\Delta H_M \sim \Delta A^E$ and in symmetrical salt melts, if $\lambda_1 \sim \lambda_2$ and one cation is highly polarized (e.g., Ag^+), then

$$(51) \qquad \Delta H_M = x_1 x_2 U_0{}^{++}$$

If cation-cation and anion-anion dispersion effects are important (i.e., where both cations and anions are readily polarized), a term linear in $(\lambda_1 - \lambda_2)/\lambda_2$ should also be included

$$
(52) \qquad \Delta H_M = x_1 x_2 \left[U_0{}^{++} + (U_1{}^{+-} + U_1{}^{++} + x_1 U_1{}^{++} a)\left(\frac{\lambda_1 - \lambda_2}{\lambda_2}\right) \right.
$$
$$
\left. + \Omega\left(\frac{\lambda_1 - \lambda_2}{\lambda_2}\right)^2 \right]
$$

The term $x_1 U_1^{++} a$ causes asymmetry in the concentration dependence of ΔH_M; equations of this form[87] have been obtained for various mixtures of uni-univalent salts containing Ag^+ or Tl^+ ions (e.g., $AgNO_3$–alkali nitrates)

$$(53) \qquad \Delta H_M = x_1 x_2 \left[U_0^{++} + 12\left(\frac{\lambda_1 - \lambda_2}{\lambda_1 \lambda_2}\right) - 340\left(\frac{\lambda_1 - \lambda_2}{\lambda_1 \lambda_2}\right)^2 \right]$$

There are large differences among the coefficients of the $(\lambda_1 - \lambda_2)/\lambda_1 \lambda_2$ terms for other systems, and it is suggested that these are due to covalent effects in the silver and thallium mixtures. Nakamura[88] has investigated the excess energy of mixing of molten mixtures of silver and thallium chlorides with the alkali chlorides. He calculates that

$$(54) \qquad \Delta E_{mix}^E = E_c^E + E_d^E + E_p^E = (k_c + k_d + k_p) x_1 x_2$$

where the subscripts c, d, and p refer to Coulombic interaction, van der Waals dispersion, and polarization, respectively. He assumes that the cation-anion interactions are not modified by their environment. Summing the $\sum k$ terms, he finds that their signs but not their values agree with the data of Hersh and Kleppa.[87] Nakamura ascribes the change from endothermic to exothermic heats of mixing to van der Waals interaction.

Murgulescu et al.[89] have determined a number of heats of mixing for uni-univalent salts by calorimetric methods obtaining similar values to those of Kleppa et al.[87,90] Marchidan and Telea[91] have continued these investigations, and their results for the common cation systems are interesting in that, for the large polarizable Cs^+ cation in CsBr–CsI mixtures, there was a positive heat of mixing, whereas for the NaI–NaCl system there was a negative heat of mixing. Sternberg and Herdlicka[92] evaluate the excess heats of mixing for a large number of alkali-halide systems and compare them with experimentally determined values. If

$$(55) \qquad \Delta H_{mix}^E = B x_1 x_2$$

where B is the interaction parameter, then in any rigorous calculation

$$(56) \qquad B = B_{Coulombic} + B_{repulsive} + B_{London} + B_{polarization}$$

To evaluate (55) they use a parameter, d, denoting change in distance on mixing where

$$(57) \qquad 0 < d < \left[\frac{\Delta R}{2} - \frac{R_{AB} - R_{AC}}{2}\right]$$

and

$$R_{AB}^* = R_{AB} - d \qquad \text{and} \qquad R_{AC}^* = R_{AC} + d$$
$$\text{(mixture)} \qquad\qquad\qquad \text{(mixture)}$$

The values of the various interaction parameters were calculated using NaBr–KBr as the basis of a comparison system. On summing the parts of the total interaction, parameter B, they found that the summation with all four interaction terms gave the best fit to the data but that the original Lumsden interaction parameter, $B = B_{\text{Coulombic}} + B_{\text{London}}$ also fitted the experimental data quite well.

Charge Unsymmetrical Mixtures with a Common Ion. Davis[93] has derived a theoretical model for these mixtures using the conformal-ionic-solution (CIS) theory. To a first approximation it is permissible to consider the Coulombic interactions only, for the multiple charge on one of the ions means that the lower order Coulombic term dominates ΔA^E

$$
(58) \quad \Delta A^E = -kT \ln \left\{ \frac{Z_{12}\lambda_2}{[Z_1\lambda_2]^{x_1}[Z_2\lambda_2]^{x_2}} \right\} + x_1 N^2 kT \left(\frac{\lambda_1 - \lambda_2}{\lambda_1 \lambda_2} \right)
$$
$$
\times \{[1 + x_2(z - 1)]\alpha_{12}\lambda_2 - \alpha_1\lambda_2\}
$$

where Z_1, Z_2, Z_{12} are the configurational partition functions and the ionic charges are $z_1 = z_A = 1$ and $z_2 = z$; λ_1 and α_1, and so on, are the hard-core diameters and polarizabilities, respectively. Kleppa and McCarty[94] found that the heats of mixing of $MgCl_2$ with the alkali chlorides and $AgCl$ were negative except for $AgCl$, where there was a slight positive deviation. The systematic deviations obey a fairly simple linear equation dependent on the Coulombic interaction only.

Considerable use has been made of phase-diagram studies by Førland,[12] Kirgintsev and Avvakumov,[69] and Doucet and Vallet.[95] Doucet and Vallet conclude that, for alkali nitrate alkaline earth nitrate systems, the energy of interaction, W, is mainly Coulombic and that the excess chemical potential

$$
(59) \quad \mu_{M^{+2}} = W N_{M^{+2}} = -\frac{Nze^2}{2} \left(\frac{N_2^2}{\rho} \right)
$$

where

$$
(60) \quad \frac{1}{\rho} = \frac{1}{2d_1} + \frac{2}{d_2} - \frac{4}{d_1 + d_2}
$$

A plot of W, which was negative and varied from -2240 cal/mole for mixtures of potassium and barium nitrates to -4730 cal/mole for mixtures of potassium and calcium nitrates was a linear function of $1/\rho$. A similar function was obtained for sodium–alkaline-earth-nitrate mixtures, but the negative deviations in heats of mixing were smaller than for potassium mixtures. These values may be compared with the extensive calorimetric data of Kleppa and co-workers,[96,97] who obtained values of -1100 cal/mole and -3000 cal/mole, respectively, for the $K–Ba–NO_3$ and $K–Ca–NO_3$ mixtures.

Many of the studies on charge-unsymmetrical systems have been made by the electromotive force cell method first used for molten salts by Hildebrand

and Salstrom[98] and by various Russian workers.[10,12] Recent workers have used the cell[99]

$$\text{graphite Cl}_2 \mid \text{NaCl} \parallel \text{Na}^+ \parallel \text{NaCl, MCl}_2 \mid \text{Cl}_2 \text{ graphite}$$
$$\quad\ \ \text{gas} \qquad \text{liq.} \quad \text{glass} \qquad \text{liq.} \qquad\qquad \text{gas}$$

where the two halves of the cell are separated by a glass containing Na^+ ions. This method has also been successfully applied to potassium halide mixtures and the values of ΔH_M agreed with Kleppa and McCarty's result.[94] The partial molar entropies of mixing for potassium chloride were ideal for BaCl_2–KCl mixtures and showed negative deviations for SrCl_2–KCl mixtures with larger negative deviations for CaCl_2–KCl mixtures. If the alkali chlorides were mixed with MgCl_2,[100] the partial molar entropy curves showed an inflection at $\text{X}_{\text{MgCl}_2} = 0.33$ mole fraction for all systems except sodium chloride. It is suggested that these S-shaped curves are indicative of high local ordering corresponding to the formation of MgCl_4^{2-} complexes.

Nonsymmetrical salt systems can be very complex owing to the formation of associated species. Braunstein[101] evaluates the approximate Margules equations for use in conditions where associated species can give rise to deviations from ideal and notes that, for instance, the formation of MgCl_4^{2-} complexes in KCl–MgCl_2 melts cannot account quantitatively for the observed deviations from ideality. Complex systems such as silicates, borates, and cryolites have received much attention because of their technological importance. The results are not easy to interpret and are often difficult to obtain. Østvold and Kleppa[102] found marked dependence of the enthalpies on composition for the system PbO–silica at $\text{X}_{\text{SiO}_2} = 0.35$ and typical S-shaped entropy curves consistent with the appearance of the SiO_4^{4-} orthosilicate ion were obtained. Similar S-shaped entropy curves are also reported by Holm and Kleppa[103] in mixtures of BeF_2 with alkali fluorides, and these are attributed to the appearance of BeF_4^{2-} ions and the breakdown of the network structure of BeF_2.

Common-Cation–Mixed-Anion Systems.   Departures from ideality are less marked in common-cation–mixed-anion systems than in common-anion mixtures. For instance, the system Na_2SO_4–Na_2CO_3[104] forms an ideal mixture with no excess heat of mixing. By a series of exchange reactions between hydrogen halide and halides of lithium, sodium, potassium, rubidium, cesium, calcium, and magnesium, the equilibrium constants for the reactions

$$\text{M}^+\text{Cl}^- + \text{HBr}_g \leftrightarrows \text{M}^+\text{Br} + \text{HCl}_g$$

can be evaluated with respect to the composition of the melt, and the excess free energy of mixing in the liquid phase can be calculated[105]

(61) $\qquad -\Delta G_M^\circ = RT \ln K_M = \Delta \bar{G}_{\text{MBr}} - \Delta \bar{G}_{\text{MCl}} + RT \ln P_{\text{HCl}} - RT \ln P_{\text{HBr}}$

For a regular solution and constant cation composition

$$(62) \qquad RT \ln \left[\frac{P_{HCl} \cdot x_{Br^-}}{P_{HBr} \cdot x_{Cl^-}} \right] = b(x_{Br^-} - x_{Cl^-}) + RT \ln K_M$$

For the alkali halides, small positive deviations from ideality were found of the order 100 to 500 calories, increasing as the cation radius increased. On the other hand, the magnesium and calcium mixtures showed small negative deviations from ideal. Negative deviations are expected on the CIS theory if the anion sizes are appreciably different. However, the small size of the cations compared with the anions may mean that there will be significant van der Waals interaction between nearly contacting anions leading to positive heats of mixing (Figure 5 shows the variation of heats of mixing for common cation or anion systems).

TERNARY SYSTEMS WITH A COMMON ANION AND THREE DIFFERENT CATIONS. Ternary systems are of importance as low-melting-point eutectics. Although determinations of enthalpies or heats of fusion of the mixtures have been made (e.g., mixtures of alkali carbonates[106]), there have been few determinations of the heats of mixing. Guion[107] has analyzed ternary mixtures of alkali

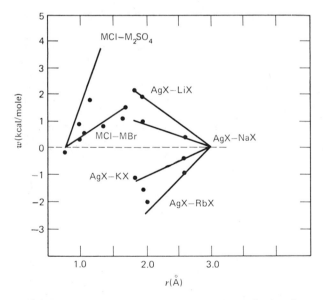

Figure 5. Heats of mixing: exchange energy, w, is given by $w = 31 \left[(r_1 - r_2)/(r_1 + r_2) \right]$ $\times (1 - r/r_0)$, where r is the radius of a common ion, r_1 and r_2 are radii of other ions; $r_0 = 3.06$ for a common anion and 0.78 for a common cation. X = Cl, Br, NO$_3$; M = Li, Mg, Na, Tl, Ag, Pb, K, Cs. From Ref. 69, reproduced by permission from *Russian Chemical Reviews*, **34**, 58 (1966).

and silver nitrates. He finds that the partial molar free energy of solution of silver nitrate in a mixture of alkali nitrates is an additive function of the mixing terms from the binary systems

$$\Delta G_{23}{}^E = N_2 N_3 f(T, P) \left(\frac{\lambda_2 - \lambda_3}{\lambda_2 \lambda_3} \right)^2 \tag{63}$$

Smirnov and Khaimenov[108] found that ΔH_M for the formation of zirconium and titanium complexes in alkali-halide melts was a function of the average total radii of the alkali ions. A systematic investigation of ternary alkali and silver nitrate systems has been carried out by Meschel and Kleppa.[109] The ternary enthalpy of mixing is a sum of two terms

$$\Delta H_{\text{tern}}^M = \frac{\Delta H_{\text{obs}}}{(n_a + n_b + n_c)} + \left(\frac{n_b + n_c}{n_a + n_b + n_c} \right) \Delta H_{\text{BC}}{}^M \tag{64}$$

The first term is the observed heat of mixing, and the second term is calculated from the enthalpy of mixing equation of the appropriate binary. The ternary excess enthalpy term, $\Delta H_{\text{tern}}^E / X_A X_B X_C$, was calculated from the experimental data for mixtures of silver nitrate with binary nitrate systems, and a linear dependence on the size parameter, δ_{BC}, was found

$$\delta_{\text{BC}} = \frac{d_{\text{BNO}_3} - d_{\text{CNO}_3}}{d_{\text{BNO}_3} d_{\text{CNO}_3}} \qquad 10^2/\text{Å} \tag{65}$$

Values of $\Delta H_{\text{tern}}^E / X_A X_B X_C$ varied from about -50 cal/mole for Na–K systems to 550 cal/mole for Li Cs systems.

Reciprocal Molten Salt Systems. The metathetical mixing reaction, $NaCl + KBr = NaBr + KCl$, can be considered a conventional ternary thermodynamic system, since a mixture of any composition can be made by combining suitable amounts of any three of the following salts: NaCl, NaBr, KBr, KCl. Blander and Yosim[110] used CIS theory, which in this case gives almost the same answer as quasi-lattice theory, apart from the $\Delta E/RT$ term, which is only important in dilute solutions. The following equation was obtained for the excess energy of mixing of ions A^+, B^+, X^-, Y^-:

$$\Delta A_M{}^E = X_A X_X \Delta A^\circ + X_X \Delta A_{12}{}^E + X_Y \Delta A_{34}{}^E + X_A \Delta A_{13}{}^E$$
$$+ X_B \Delta A_{24}{}^E + X_A X_B X_X X_Y P [g_1 + g_4 - g_2 - g_3]^2 \tag{66}$$

where

$$P = \frac{1}{2} \left[\left(\frac{N^2}{Z'} \right) M - \left(\frac{A}{Z'} \right)^2 \right] \tag{67}$$

A and M are derivatives of configurational integrals, and X_A, $A_{12}{}^E$, g_1, and so on are the respective ionic fractions, binary interaction energies and size parameters. The P term is equivalent to nonrandom mixing and can be evaluated by comparison. $\Delta A°$ is the standard energy of the reaction. The Flood–Førland–Grjotheim[111] quasi-lattice theory is usually limited to the first term of

$$(68) \qquad \Delta A_M{}^E = A_A A_X (Z' \Delta E_1) + X_A X_B X_X X_Y \left(\frac{Z' \Delta E_1{}^2}{2RT} \right) \cdots$$

The final term is equivalent to the P term of (66). Equation 68 is generally used in the form

$$(69) \qquad RT \ln \gamma_{AX} = \pm (1 - N_A)(1 - N_X) \Delta \mu°$$

where $\Delta \mu°$ is the change of chemical potentials for ions A and X in (68). Equation 69 has been modified by Førland,[81] who introduced terms corresponding to the binary additive terms of (66). Blander and Topol[112] compare the calculated values from CIS theory with those from EMF and phase-diagram data. Nonrandom mixing may result from Coulombic forces as in the LiF–KCl system or non-Coulombic forces as in the NaNO$_3$–AgCl system (Figure 6).

A number of ternary reciprocal systems have been studied by Kleppa and colleagues,[113,114] who found a linear correlation between the deviations from the ideal quasi-lattice mixing theory and the size difference of the ions. There are many phase-diagram studies reported in the literature, and a correlation derived from the CIS theory[115] should prove useful in predicting immiscibility gaps

$$(70) \qquad T_C = \frac{\Delta G°}{5.5R} + \left(\frac{\lambda_{12} + \lambda_{24} + \lambda_{13} + \lambda_{34}}{11R} \right)$$

where T_C is the consolute temperature and λ_{12}, and so on, are the hard-core parameters and $\Delta G°$ is the standard free energy of reaction for formation of the salt pair. If the calculated T_C exceeds the liquidus temperature for the stable salt pair, there is a miscibility gap.

Sternberg and Bejan[116] used a formation cell to study heats and entropies of mixing of AgI–KCl and calculations were made using the Førland equations,[81] modified by entropy corrections to account for deviations from random mixing. Plambeck[117] has measured the free energies of solvation in water, ammonia, and fused LiCl–KCl eutectic and reports that the Born equation for solvation is obeyed in all three solvents. In this situation it appears that the fused salt behaves in a normal ligand manner. Førland,[11] Blander,[12] and Lumsden[13] give comprehensive accounts of the many reciprocal molten salt mixtures investigated and their reports should be referred to for further information.

Vapor-Pressure Measurements of Mixtures of Fused Salts. A discussion of the vapor phase of ionic salts is outside the scope of this chapter; however, knowledge of the structure of the vapor phase is important if the correct

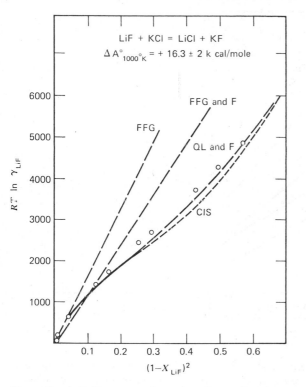

Figure 6. Activities on reciprocal molten salt mixtures: experimental (circles) and theoretical values of $RT \ln \gamma_{LiF}$. FFG—Flood–Førland–Grjotheim model; F—Førland corrections; QL—quasi-lattice model; CIS—conformal ionic solutions model.

deductions are to be made about the composition of mixed melts in equilibrium with the vapors. Salt vapors are discussed by Bauer and Porter,[12] Bloom,[14] and Yarym-Agaev.[118] Certain salts, such as $AlCl_3$, are almost completely dimerized in the vapor phase, whereas others, such as Hg_2Cl_2, dissociate. The simple alkali halides generally vaporize as monomers but also form dimers in the vapor phase and sometimes trimers. Generally speaking, lithium halide dimers are more stable than sodium halide dimers; for a common cation, the fluoride dimers are more stable than the chlorides, and so on. Figure 7 shows the dimerization energy of alkali-halide molecules.

Transpiration vapor-pressure measurements above mixed melts of alkali halides and halides of group IIB metals indicate the formation of complexes such as PbClBr in the vapor phase which can be identified by mass spectrometry.[119,120] Hagemark et al.[121] have measured the density of vapors of $KCl–PbCl_2$ and $RbCl–PbCl_2$ using a gold isotensiscope and observe large

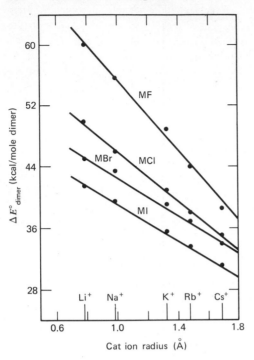

Figure 7. Dimerization energy of alkali halide molecules in vapor phase. From Bauer and Porter, in Ref. 12, by permission from *Molten Salt Chemistry*, ed. M. Blander, John Wiley and Sons, New York.

deviations from ideal gas behavior consistent with the formation of associated species in the vapor. From the vapor-pressure measurements they have calculated the association constants, K, for the formation of the mixed species as log $K = 3.45$ for $KPbCl_3$ and 3.66 for $RbPbCl_3$. Work on mixed alkali-halide vapors has shown that the simple conformal ionic solution formula

$$(71) \qquad RT - \ln \frac{K}{4} = M\left(\frac{1}{d_{AX}} - \frac{1}{d_{BX}}\right)^2$$

for the formation of the dimer $A_2X_2 + B_2X_2 \leftrightarrows 2ABX_2$ does not hold, and modifications to the interaction parameter are required to account for the van der Waals cation-cation and anion-anion interactions.[122] Figure 8 shows association constants of mixed alkali-halide vapors.

Transpiration vapor-pressure measurements on $PbCl_2$–$CsCl$ and $CdCl_2$–$CsCl$ systems[123] are interpreted as being due to formation of complex ions such as $CsCdCl_3$ in the *melt* as well as in the vapor. A typical S-shaped partial

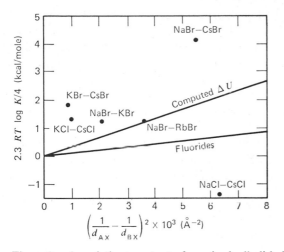

Figure 8. Association constants for mixed alkali-halide vapors: computed ΔU represents the Coulombic energy change for the reaction $A_2X_2 + B_2X_2 = 2ABX_2$ where $\Delta U = -330 \, (1/d_{AX} - 1/d_{BX})^2$. From Ref. 122. Reprinted by permission of the American Chemical Society from *J. Phys. Chem.*, **72**, 4620 (1968).

entropy curve was calculated from the vapor pressure data using the Gibbs–Duhem equation. Adams and Quan[124] used the transpiration vapor pressure method to investigate $Rb_2O-B_2O_3$ equilibria in the liquid phase. The activity of $RbBO_2$ in the melt showed a discontinuity at 17 mole % Rb_2O; at this composition the integral enthalpy of mixing was -5.8 kcal. The S-shaped entropy curves obtained corresponded to the formation of BO_4^- tetrahedra in the melt.

In calculations of activity coefficients, and enthalpies and entropies of mixing in melts from vapor-pressure data, corrections to the apparent vapor pressures due to the formation of complexes in the vapor phase are often of little importance. Correction is essential in the system $MgCl_2 + KCl$, however, since about 50% of the vapor consists of the $KMgCl_3$ dimer.[125] Formation of such dimers can cause loss of relatively nonvolatile components from molten salt mixtures.

Volume Changes on Mixing Molten Salts. Many of the earliest measurements of the densities of molten salts were made using the Archimedean principle.[5] This method is still used, but the technique and resulting measurements are more precise than in the 1900s. Pyknometry is also extensively used especially in studies of changes of volume on mixing. The first volumes of mixing systematically studied were for the halides of cadmium, lead, and silver; no excess volume change was found except for the $PbCl_2 + CdCl_2$ mixture where ΔV^E was -0.84% at 66% $PbCl_2$.[126] Mixtures of silver salts with alkali halides show very slight positive excess volumes of about $+0.3$ cm³/mole for all except the (Li–Ag)Br mixtures, where ΔV^E is -0.3 cm³/mole.[12] Powers, Katz, and Kleppa[127] report that the excess volumes of mixing for sodium nitrate with nitrates of lithium, potassium, rubidium, and cesium are all positive, the maximum deviation ΔV^E being related to the mole fractions, x, and interionic distances by the familiar Tobolsky parameter; however, comparable temperatures are not specified. The ΔV^E values range from about 0.2 to 1.4 cm³/mole. James and Liu[128] reported smaller deviations from ideal for alkali nitrate melts but they measured the temperature dependence of ΔV^E and find that it can change sign as the temperature is increased. Typical excess volume data are shown in Table 4. A similar effect was found

Table 4   Excess Volumes of Mixing

Salt System	ΔV^E_{max} (%)	Temperature (°C)	V^E_{max} (%)	Temperature (°C)
$LiNO_3$–$CsNO_3$	-0.3	280	$+0.5$	450
$NaNO_3$–$RbNO_3$	-0.9	250	$+0.4$	450
$NaNO_3$–KNO_3	$+0.15$	250	$+0.3$	350
Li_2SO_4–K_2SO_4 (Eutectic)	$V^E = +0.71\%$ at 750°C			

for $LiClO_3$–$LiNO_3$ mixtures.[129] Markov and Prisyazhny[130] say that the change in molar volume on mixing reciprocal pairs of salts AX and BY is given by an equilibrium constant dependent on the volumes

$$(72) \qquad K_p = \left(\frac{V_M - \frac{1}{2}(V_{AX} + V_{BY})}{\frac{1}{2}(V_{AY} + V_{BX}) - V_M} \right)^2$$

The behavior of mixtures of molten carbonates is similar to that of molten nitrates with small positive deviations in molar volumes of the order of 1 to 2%.[131] Spindler and Sauerwald[132] found that ΔV^E for binary mixtures of alkali halides was temperature dependent. McAuley, Rhodes, and Ubbelohde[133] measured the equivalent volumes of mixtures of molten alkali and alkaline earth nitrates and found ideal volumes of mixing. Extrapolation of the data to 100 eq % of the alkaline earth nitrates enabled them to calculate the equivalent volumes of these compounds, and from the data they postulated that the major contribution to the molar volume is the cubic close packing of the large pseudo-spherical nitrate anion. The small alkaline earth cations are then able to fit into the interstitial tetrahedral holes left in the nitrate anion lattice. When the available volume is too small for free rotation of the nitrate ion, as at lower temperatures, a system such as KNO_3–$Ca(NO_3)_2$

cannot rearrange itself and crystallize out and, hence, can readily supercool to a glass. The equivalent volume occupied by freely rotating close-packed nitrate spheres is 42.6 cm³/mole, and the extrapolated equivalent volumes of the alkaline earth nitrates were found to be Mg, 41.3 cm³; Ca, 41.3 cm³; Sr, 44.1 cm³; Ba, 44.9 cm³.

The system $AgNO_3$–$RbNO_3$ is interesting in that ΔV^E is positive and asymmetric ($+0.4$ cm³/mole); the heat of mixing is negative, and the partial molar entropy of $AgNO_3$ is also negative.[134] It seems that in mixtures containing Ag^+ ions the entropy terms are most important and that second-order CIS theory is needed to explain the data. Congruent compound formation in the solid is not shown on density isotherms of melts, and molar volumes of reciprocal salt systems are additive to within fairly reasonable limits.[135–137] However, if there is marked compound and dimerization formation in the vapor phase, molar volume composition isotherms show definite deviations from ideality; lead chloride–alkali-halide systems, for example, show positive deviations from the ideal, with a maximum deviation at 0.33 mole fraction of $PbCl_2$.[138,139] The peaks in Figure 9 (of the order of

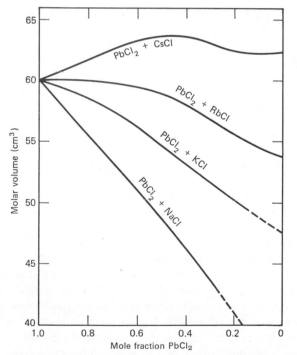

Figure 9. Excess volumes of mixing for lead chloride with alkali-halide melts at 720°C. Ref. 138, *Australian Journal of Chemistry*, reprinted by permission.

$+1.2$ cm^3/mole) are interpreted as due to the presence of complex ions such as $PbCl_3^-$ or $PbCl_4^{2-}$ in the melt. In the system KBr–TlBr,[140] negative deviations ($\sim 10\%$) in the excess volume curves are found at mole fractions corresponding to $2KBr \cdot TlBr$. These negative deviations decrease with increasing temperature and eventually positive deviations appear. It is more likely, therefore, that the negative deviations are due to local crystal field ordering in the melt and to van der Waals interaction rather than to specific compound formation. Molten NaCl–AlCl$_3$ and KCl–BiCl$_3$[141] systems both show very large negative deviations from additivity. $\Delta V^E \sim 20\%$ for the NaCl–AlCl$_3$ system, which is unlikely to be entirely explained by compound formation and is possibly due to the covalent character of AlCl$_3$, which exists as Al_2Cl_6 in the melt. The molar volumes of silicates and borates are not additive either, deviations usually corresponding to the formation of SiO_4^{4-} and BO_4^{3-} tetrahedra in the melts.[142,143]

Interaction of ionic salts (up to $0.1\ M$) with high-temperature water has been investigated, and it is shown that above 100°C the apparent partial molar volume of water changes abruptly.[144] Braunstein et al.[145] have investigated concentrated solutions of LiNO$_3$–KNO$_3$–H$_2$O and KNO$_3$–Ca(NO$_3$)$_2 \cdot$ 4H$_2$O above 100°C and found that the specific volumes are additive and can be described by a quadratic equation depending on the mole ratio of water present. The extrapolated partial molar volume of water in the melts was $\overline{V}_{H_2O} = 16.3$ cm compared with $\overline{V}_{H_2O} = 19.00$ cm^3 for pure water at 119°C. The partial molar volumes of KNO$_3$ and LiNO$_3$ were in approximate agreement with extrapolated pure molten salt values, however. This behavior of water is ascribed to breakdown of the hydrogen bonding in the compound by the high electrostatic fields of the concentrated ionic salt solutions.

6   The Fusion Process and Physical Properties of Molten Salts

Fusion.   A frame of reference is required to understand the properties of ionic melts, and the reference state usually considered is the crystalline solid. On going from solid to liquid changes in molar volume, entropy, enthalpy, and electrical conductivity are most marked and these properties provide a useful means of categorizing the various salts. Table 5 is compiled from data given by Janz,[15,146] Klemm,[12] and Ubbelohde.[18] Analysis of fusion parameters is not straightforward, since some account has to be taken of transformations in the solid phase before the solid melts; premelting phenomena have been extensively discussed by Ubbelohde.[18] The volume change on fusion is equivalent to the extra volume required for freedom of movement, and this results in loss of long-range order of the solid, although a local residual quasi-lattice remains in the melt with cations surrounded by anions, and vice versa. The properties of a fused salt are thus those that would be expected of a liquid, that is, mass and charge transport and fluidity.[147] There is a vast difference between a simple ionic melt such as sodium chloride and melts of an inorganic compound such as zinc chloride, which is almost completely polymerized in the liquid state.

From Tables 1 and 5 it can be seen that fusion temperatures range from 400 to 3000°K. The temperature of fusion $T_f = H_f/S_f$; hence if S_f has its normal value of 3 R = 6 eu (for positional randomization) H_f should be of the order of 6 kcal for a uni-univalent alkali halide, and this holds approximately. Lithium nitrate has the unusually high value of 11.6 eu; Ubbelohde[148] ascribes this high entropy to other modes of entropy uptake on melting. S_f to a first approximation becomes $S_f = S_{positional} + S_{orientational} + S_{vibrational} + S_{configurational} + S_{association} + \cdots$. Some or all of these modes of entropy uptake may be present in different salts, and much experimental effort has been directed to studying which of these effects are important in the various melts. Determinations of heats and entropies of fusion have usually been by drop calorimetry or cryoscopy.[6,75-77,149,150] Volume changes of fusion are

Table 5a   Properties of Molten Salts at their Melting Points

Salt	$T_f(°K)$	$\dfrac{\Delta V_f}{V_S}$ (%)	H_f (kcal/mole)	S_f (eu)	Molar Volume (cm³)	$\alpha \times 10^4$
LiCl	883	26.2	4.8	5.4	28.3	2.87
NaCl	1081	25.0	6.7	6.2	37.5	3.49
NaBr	1020	22.4	6.2	6.1	44.1	3.50
KF	1129	17.2	6.8	6.0	30.6	3.41
KI	958	15.9	5.7	6.0	68.0	3.89
CsCl	918	10.0	4.8	5.3	60.2	3.84
AgCl	728	8.9	2.9	4.0	26.5	1.76
CaCl$_2$	1045	0.9	6.8	6.5	45.0	2.07
BaCl$_2$	1233	3.5	3.9	3.2	53.2	2.20
AlCl$_3$	465	83.0	8.5	18.4	—	—
LaCl$_3$	1131	—	13.0	11.5	61.7	2.50
HgCl$_2$	549	18.8	4.2	7.5	46.7	6.80
LiNO$_3$	527	21.4	6.1	11.7	38.7	—
KNO$_3$	607	3.3	2.8	4.6	56.1	3.99
Na$_2$CO$_3$	1127	—	6.7	5.9	43.5	2.33
Na$_2$SO$_4$	1155	—	5.7	4.9	—	—
KSCN	450	5.4	3.1	7.0	61	4.86
AgNO$_3$	485	0.7	3.0	6.2	38.5	2.61
ZnCl$_2$	556	11.64	2.3	4.1	51.0	2.14
PbCl$_2$	771	—	4.4	5.7	58.0	4.0

Table 5b   Properties of Molten Salts at their Melting Points

Salt	$\beta_T \times 10^{12}$ (cm^2/dyne)	C_p [cal/(deg)(mole)]	σ (dynes/cm)	η (cp)	Λ Ω^{-1}cm^2 eq^{-1}
LiCl	19	15.0	137	1.9	183
NaCl	28.7	16.0	116	1.5	150
NaBr	31	—	100	1.4	148
KF	—	16.0	141	—	124
KI	49.9	—	79	2.4	104
CsCl	39	18.0	92	1.7	86
AgCl	—	16.0	178	2.3	118
CaCl$_2$	14	23.6	151	4.5	56
BaCl$_2$	15	26.3	165	5.5	66
AlCl$_3$	—	31.2	—	—	15
LaCl$_3$	—	37.7	—	—	29
HgCl$_2$	47	25.0	—	1.9	0.03
LiNO$_3$	18	26.6	116	5.4	28
KNO$_3$	21	29.5	110	2.9	36.5
Na$_2$CO$_3$	—	34	218	—	154
Na$_2$SO$_4$	—	—	192	—	150
KSCN	—	—	101	13.2	8.6
AgNO$_3$	—	30.6	148	4.2	30.4
ZnCl$_2$	49	24.1	—	$\sim$200.0	0.02
PbCl$_2$	—	23.6	137	4.3	40

determined directly by the capillary method of Sauerwald and Schinke[151] or by measuring the density of solid and liquid below and above the melting point using a pyknometer and densitometer.[152]

Compressibility. The effect of pressure on the melting transformation often provides useful information on how ions pack in the liquid phase. Pioneering work in this respect was done by Bridgman[153] and much recent work is by Pistorius and Rapoport.[154-156] Information on the change of melting point with pressure can be used to obtain either V_f or S_f using the Clausius–Clapeyron equation. Generally the experimental values of V_f are greater than those derived from pressure data.[157] The Simon equation is often used to correlate data

$$(73) \qquad P - P_0 = A\left[\left(\frac{T}{T_0}\right)^C - 1\right]$$

where T_0 is the triple-point melting temperature, P is the pressure, and P_0 is usually 0; A and C are constants.

In determinations of the melting curves of halides of sodium, potassium, and rubidium, it was found that at high pressures the order of melting was reversed: fluorides melted lowest and iodides highest. Furthermore, dT/dP

for fluorides is about half the value of that for the iodides (e.g., 0.0232 deg/atm for KF and 0.051 deg/atm for KI). Rapoport[156] found that the melting curves for $CsNO_3$, $RbNO_3$ III, KNO_3 VI, and $NaNO_3$ rose rapidly with pressure, indicating the ready compressibility of the liquid nitrates. At high pressure the volume changes on fusion for NaCl and KCl decrease to about 4% (from 20%[157]) and the melting temperatures rise to 4000°K, indicating the relative incompressibility of these hard-sphere liquids.[158]

There have been several recent attempts to obtain a melting-point relationship of greater applicability than the Simon equation. An equation using the Gruneison constant, γ, in a version of the Lindemann melting equation has the great advantage that no arbitrary parameter appears

$$(74) \qquad T_m = T_m^\circ \left[1 + 2(\gamma - \tfrac{1}{3})\left(\frac{\Delta V}{V}\right) \right]$$

In (74), however, the correct values of γ and $\Delta V/V$ to use are not well established.[159]

A few of the high-pressure melting curves for polyatomic anions (e.g., KNO_3, KNO_2, $NaClO_3$) have maxima. Rapoport considers that the contraction in volume is due to the blocking of free rotation of the nitrate ion and could account for increased supercooling with pressure.[160] Compressibility measurements[161] up to 9 kbar on alkali nitrates agreed with ultrasonic measurements of Bockris and Richards.[42]

Determinations of the thermal expansions, α (readily obtained from density measurements), adiabatic and thermal compressibility coefficients, β_S and β_T, are of special value in testing theories of molten salts.[162] β_S and β_T are available for most alkali halides and nitrates and for group II halides. For alkali salts, β_S ranges from about 15×10^{12} to 35×10^{12} cm²/dynes at the melting point, increasing by about 2 to 5×10^{12} cm²/dynes for every 100°C. The velocity of sound in molten salts is about 2000 meters/sec, that is, comparable with that of normal liquids.

There have been few adiabatic compressibility determinations on mixtures of fused salts, but Sternberg and Vasilescu[163] report that in mixtures of $PbCl_2$ and alkali chlorides there are marked negative deviations from additivity in the order Li < Na < K < Rb < Cs. Heric[164] calculates β_T/V (where V is the molar volume) for alkali salt melts at 200°C intervals and finds that at each temperature β_T/V is sensibly constant for the halides. This is not true for the alkali nitrates, however; in this case the value of β_T/V decreases with increasing cation radius. Heric deduces, therefore, that in fused salt mixtures with nitrates there should be an asymmetric interaction term dependent on the molar volumes.

Gas Solubility. The solubilities of inert gases in molten salts are very low; because of its small size, helium is the most easily dissolved. Gas solubility is proportional to pressure in accordance with Henry's law. In molten fluorides the solubility of inert gases increases with temperature; the heat of solution is endothermic and is the work required to form cavities large enough to accommodate the gas atoms in the melt. In LiF–NaF–KF (46.5–11.5–42.0 mole %) eutectic, the solubility of helium as given by Henry's law constant, Kp, is 11.3×10^{-8} mole/(cm^3)(atm) at 600°K and 23.0×10^{-8} mole/(cm^3) (atm) at 800°K, whereas in the same eutectic argon has a solubility of only 0.90×10^{-8} mole/(cm^3)(atm) at 600°K and 3.4×10^{-8} mole/(cm^3)(atm) at 800°K.[165]

In molten sodium nitrate, which has a relatively low surface tension and hence a low energy requirement for the formation of holes, argon and nitrogen dissolve with liberation of heat. The heat of solution for argon was -1.7 kcal/mole and for nitrogen was -2.7 kcal/mole; Kp was about 19×10^{-7} mole/(cm^3)(atm) for both gases at 642°K, although nitrogen was slightly more soluble. Exothermic heats of solution are thought to be due to solvation of the gas atom in the melt, the magnitude of this effect depending on the polarizabilities of the dissolved gas molecules.[166]

Water dissolves to a limited extent in alkali chlorides[167] and alkali nitrates[168] up to about 5×10^{-3} mole/mole, but it can be removed reversibly by back-flushing with hydrogen chloride or nitrogen gas. The solubility of water in lithium chloride melts decreases with increasing temperature and the heats of solution are exothermic. Solubilities of carbon dioxide and oxygen in molten carbonates have been investigated because of their use in fuel cells. Henry's law constant for the solution of oxygen in $(Li–K–Na)_2CO_3$ eutectic is about 8×10^{-6} mole/(cm^3)(atm) and for carbon dioxide is 9.0×10^{-5} mole/(cm^3) (atm) at corresponding temperatures.[169] A gas such as hydrogen fluoride is comparatively soluble in molten fluorides, the solubility increasing as the percentage of alkali fluoride is increased. The value of Kp varies from 1 to 8×10^{-5} mole/(cm^3)(atm), and the solubility of hydrogen fluoride decreases with increasing temperature, indicating that chemical reaction has occurred.[170]

Extensive chemical reaction may of course occur when certain organic gases are dissolved in fused salts. Although the subject is outside the scope of this chapter, the work on organic chemical reactions in fused salts by Sundermeyer[171] should be consulted in this respect.

Surface Tension.   Values of surface tensions for melts are of some interest, for they enable calculations to be made of V_h, the hole volume as defined by Fürth,[41] where

$$(75) \qquad\qquad V_h = 0.68N \left(\frac{kT}{\sigma}\right)^{3/2}$$

σ is the surface tension of the liquid. The few values listed in Table 6 show how small the nominal hole sizes are for nitrates as compared with

Table 6   Hole Sizes in Alkali Chlorides and Nitrates

	$T = 850°C$			$T = 430°C$		
Salt	LiCl	KCl	CsCl	LiNO$_3$	KNO$_3$	CsNO$_3$
V_h (10^{23} cm^3)	3.17	4.61	6.25	1.88	1.95	2.35

From Ref. 15.

alkali halides. The hole size increases very little with cation size for the nitrates.

Values of the surface tension, σ_m, at the melting point are about the same for all the alkali nitrates (110 dynes/cm); for alkali halides, however, they change markedly, with cation size from 138 dynes/cm for lithium chloride to 91 dynes/cm for cesium chloride. The surface tension decreases in the order $F \gg Cl > Br > I$ for the same cation. Surface tension data uphold the view that with big anions the physical characteristics of the melt are determined by the packing requirements of the anion. Reiss and Mayer[172] have calculated the surface tensions of melts from the theories of corresponding states and conformal ionic solutions using the hard-core diameters and densities of the salts; they found good agreement with experiment except for the halides of mercury and bismuth.

Most determinations of surface tensions in molten salts are by the maximum bubble pressure method. There is a comprehensive review of surface tension data by Sokolova and Voskresenskaya, and many surface tension composition isotherms for various mixtures are illustrated.[173] The simpler systems are additive, but complex curves showing negative deviations in additivity and/or points of inflection are found for such systems as KCl–PbCl$_2$.[174] Reciprocal salt systems show definite evidence of compound formation. At 60 mole % potassium bromide, deviations in KBr–CdCl$_2$ systems corresponding to the ionic complexes CdCl$_4^{2-}$ and CdBr$_4^{2-}$ were marked, and the depth of the deviations increased with increasing temperature.[175] Sokolova and Voskresenskaya[176] determined the surface tension composition isotherms for BaCl$_2$–NaCl and K$_2$SO$_4$–NaBr and found that Reshetnikov's equation was applicable to the curves,[177] where x is mole fraction, σ_1 and σ_2 are surface tensions, and k is a constant

$$(76) \qquad \sigma = \sigma_1 + \frac{(\sigma_2 - \sigma_1)x}{x + k(1 - x)}$$

As Figure 10 shows, for any given cation there is an approximately linear dependence of surface tension on the generalized moment of the anion $(\mu = z/r)$.[178]

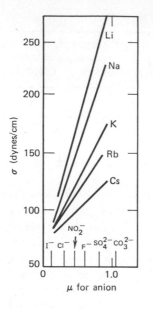

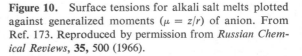

Figure 10. Surface tensions for alkali salt melts plotted against generalized moments ($\mu = z/r$) of anion. From Ref. 173. Reproduced by permission from *Russian Chemical Reviews*, **35**, 500 (1966).

Calculations of the dependence of the surface tension of molten metaphosphates on the generalized moments of the anion were interpreted as suggesting that values of the surface tension of the metaphosphates were independent of the length of the $(PO_3^-)_n$ chains.[179]

Bertozzi et al.[180] have investigated binary systems of the alkali nitrates, alkali halides, and alkaline-earth–alkali-metal halides, and find behavior similar to that observed by Sokolova.

Values of the surface tension for melts at their melting point range from about 27 dynes/cm for $GaCl_3$ to 300 dynes/cm for Na_2SiO_3, and 690 dynes/cm for Al_2O_3. Most uni-univalent salts, however, have surface tensions of about 100 dynes/cm. The temperature dependence of surface tension $d\sigma/dT$ is usually positive and varies from 0.01 to 0.1 dynes/(cm)(deg). Metaphosphates and silicates, however, have a negative temperature dependence, probably because of the breaking down of the network structure as the temperature is increased.

7   Transport Properties of Fused Salts

Introduction.   Theories of transport are mostly dependent on some idea of extra space (i.e., the "hole" theory of Fürth), vacant cells in the quasi-lattice theory, or liquid-free-volume theory. Many conductivity and viscosity results are correlated using the Arrhenius equation:

$$(77) \qquad \chi = A \exp\left(\pm\frac{E\chi}{RT}\right)$$

where χ is transport property.[181] Calculations of A, the preexponential term, have been made using the Eyring absolute reaction rate theory.[182] For instance the equivalent conductivity, Λ, is given by

$$(78) \qquad \Lambda = 5.18 \times 10^{18} z_i (D + 2) l_i^2 \exp\left(\frac{\Delta S_\Lambda^*}{R}\right) \exp\left(\frac{-\Delta H_\Lambda^*}{RT}\right)$$

where z_i is the charge on an ion, D is the dielectric constant of molten salt (usually taken as 3 or 4), l_i is the half-migration distance of the conducting ion, and ΔS_Λ^* and ΔH_Λ^* are entropy and energy of activation for conduction, respectively.

Rice[183] has applied statistical theories and pair-potential correlations to the kinetic theory of transport in fused salts. Using the theories of hydrodynamics and irreversible thermodynamics, Sundheim[11] has derived formulas relating the various transport parameters to the phenomenological coefficient, L_{ik} (mobility), and Klemm[184,12] and Laity[185] have elaborated the idea of friction coefficients, using irreversible thermodynamics formulations. The friction coefficient, r_{ik}, is the inverse of the phenomenological coefficient or mobility, L_{ik}. Using this formulation, the conventional ionic transport parameters can be defined in terms of L_{ik} or r_{ik}.

If the ith species is defined as the cation $(+)$ and the kth species as the anion $(-)$, then there are three possible phenomenological coefficients or mobilities, namely

$$(79) \qquad L'_{++} = \frac{L_{++}}{C_+} \qquad L'_{--} = \frac{L_{--}}{C_-} \qquad \text{and} \qquad L'_{+-} = \frac{z_+ L_{+-}}{C_-} = \frac{z_- L_{+-}}{C_+}$$

where C_+ and C_- are equivalent concentrations of cation and anion. The terms with like sign are equivalent to diffusional mobilities, whereas the unlike sign term is really an interionic friction.

In pure binary salts an equation for the equivalent conductance, where $\mathscr{F}$ is Faraday's number, is given by

$$(80) \qquad r_{+-} = \left(\frac{z_+ + z_-}{\Lambda}\right)\mathscr{F}^2$$

and self-diffusion coefficients, D_{++} (or D_{--}), by

$$(81) \qquad r_{++} = \frac{1}{z_-}\left[\left(\frac{z_+ + z_-}{D_{++}}\right)RT - z_+ r_{+-}\right]$$

Friction coefficients are really proportionality constants and at high temperatures, where random thermal motion becomes very much more important, they should be independent of the applied forces. Angell[186] has adapted

Cohen and Turnbull's liquid-free-volume theory[48] to viscosity and conductance measurements in mixtures of fused salts.

Two equations have been employed to check the adequacy of conventional theories of transport in melts as compared with dilute aqueous solutions. The Stokes–Einstein equation, (82a), relating self-diffusion coefficients, D_i, with viscous flow holds reasonably well for melts, and any deviations that occur are probably due to deviations from Stokes's law.

$$(82a) \qquad\qquad D_i = \frac{kT}{6\pi\eta_i}$$

Mass diffusion and viscous flow processes do not involve interionic friction. From an analysis of singlet and doublet distribution functions, Rice and Allnatt[187] show that an ionic liquid behaves similarly to a molecular fluid with respect to mass transport. Using the theory of corresponding states, they find that molten potassium chloride and argon have similar self-diffusion coefficients at corresponding temperatures. On the other hand, the Nernst–Einstein equation, (82b), relating the diffusion coefficient and equivalent conductivity does not hold

$$(82b) \qquad\qquad D_i = \frac{RT}{z^2\mathscr{F}^2}\Lambda_i$$

Angell and Tomlinson[188] quote values for the quotient $(\mathscr{F}^2\sum z_i^2 D_i/RT\Lambda_i)$, which on the Nernst–Einstein theory should equal 1, and in all cases—even for alkali halides—the quotient is greater than 1. Bockris et al.[189] have used the hole theory to deduce the work of hole formation, ΔH_h, and calculate that ΔH_h should equal $3.74RT$. The observed energy for self-diffusion $\Delta H_D^* = \Delta H_h + \Delta H_j$, where ΔH_j (energy of jump) should be very much less than ΔH_h. The derivation has now been proved to be faulty, but measurements of the diffusion coefficients at high pressure show that the activation energy for diffusion at constant volume is independent of temperature and approximately 20% of the activation energy at constant pressure. The experimental "activation volume" is approximately equal to the most probable volumes of the holes, ΔV_h, calculated by Fürth's formula, (75), from surface tension data.[190] The "free-volume" theory states that

$$(83) \qquad\qquad D = AT^{-\frac{1}{2}}\exp\left[\frac{-k}{(T-T_0)}\right]$$

which, on differentiating, becomes

$$(84) \qquad\qquad \frac{d\ln D}{d(1/T)} = \frac{E_D}{R} = -\tfrac{1}{2}T - k\left[\frac{T}{T-T_0}\right]$$

where T_0 is the temperature below which the liquid contains no free volume, E_D is the activation energy for diffusion, and k and A are constants.

Angell considers that at T_0 the metastable liquid should obey the Nernst–Einstein theory, departures from which should increase as $T - T_0$ increases.[186] If the configurational entropy, S_c, is 0 at T_0 and translational movement is by cooperative rearrangement of particles whose probability of movement is given by

$$(85) \qquad \omega(T) = \exp\left(\frac{-C}{TS_c}\right)$$

then the inclusion of the configurational entropy term in the Arrhenius equation should modify the diffusion parameters so that the Nernst–Einstein equation is obeyed.[191,192]

Self-Diffusion. One of the difficulties in diffusion measurements in molten salts is that of fixing a suitable frame of reference: the fixed-volume frame is equivalent to the solvent fixed frame for solutions, whereas the porous plug fixed frame is an external reference frame and, since the reciprocal relations between the phenomenological coefficients, L_{ik}, do not hold, comparative measurements cannot be made. Usually the rate of diffusion into a vertical capillary, filled with melt, and immersed in the active salt melt of a radio-active tracer ion is measured. Self-diffusion coefficients are comparable with those of liquids at normal temperatures and are of the order of 1 to 10 $\times$ 10^{-5} cm^2/sec for most cations and anions.[193]

Interdiffusion coefficients in binary salt melts are of the same order of magnitude as in pure salts and are measured by techniques including chrono-potentiometry and voltametry.[12,194] Activation energies for diffusion are about 11 kcal in molten carbonates[195] and about 5 kcal for alkali nitrates and halides; there are no obvious trends in activation energies. Diffusion techniques have also been applied to systems such as cryolite in order to determine the presence of complex ions. The diffusion coefficients are similar for sodium and fluoride ions: $D_{Na^+} = 9.50 \times 10^{-5}$ cm^2/sec; $D_{F^-} = 5.88 \times 10^{-5}$ cm^2/sec; whereas $D_{Al_2F_6^{3-}} = 1.8 \times 10^{-5}$ cm^2/sec, which is markedly less. Mobilities are: $U_{Na^+} = 96 \times 10^{-5}$ cm^2/(V)(sec), whereas $U_{F^-} = 3.6 \times 10^{-5}$ cm^2/(V)(sec) only, showing that it must form part of the complex ion $Al_2F_6^{3-}$.[196]

Ionic Mobilities and Transport Numbers. In discussing mobilities, it is important to be certain that one is considering either the internal mobility, U_{+-}, defined by

$$(85) \qquad \Lambda = \mathscr{F}U_{+-} = \frac{\kappa}{z_+ C_+}$$

or the external mobility, U_{iw}, defined in terms of the external transport number, t_{iw}, as

$$(86) \qquad t_{iw} = \frac{z_i \mathscr{F} C_i U_{iw}}{\kappa}$$

The internal mobilities in molten salts are of the order 1×10^{-3} cm^2/(V)(sec) and can be calculated from conductance data of pure salts.

Cation-cation or anion-anion *relative* mobilities (internal) are measured using such electrolysis and electromigration techniques as observing the position of a migrating boundary under impressed current. Boundary conditions and liquid junction potentials have to be carefully defined in order that diffusion potentials and hence mobilities may be accurately calculated. External transport numbers are usually obtained from electrophoresis experiments on asbestos strips or on packed column diffusion cells, using the Hittorf method and determining the amount of ions transported from one electrode compartment to the other during electrolysis.

Transport numbers in pure salts have meaning only when an external frame of reference is used and can be computed from Hittorf experiments. In mixtures of salts, the solvent provides the frame of reference.[12,197] Transport measurements on pure fused salts are to a certain extent dependent on the type of porous plug used—porcelain or porous glass, for example.[198] Duke and Victor[199] used a U-shaped cell with porous membrane and platinum electrolysis electrodes to determine the transport numbers of the alkali metal cations in pure fused nitrates and nitrites. Potassium, rubidium, caesium, and thallium all had the same transport number of about 0.59, but sodium had a larger transport number of 0.68 in NaNO$_3$ and 0.75 in NaNO$_2$ melts. In AgNO$_3$ and AgCl the transport numbers were given by the following temperature-dependent relationships, where $\theta = t$ (°C):

$$\text{in AgNO}_3 \ t_{\text{Ag}^+} = 0.798 - 5.6 \times 10^{-4}(\theta - 200)$$

$$\text{in AgCl } t_{\text{Ag}^+} = 0.682 - 4.8 \times 10^{-4}(\theta - 500)$$

A Hittorf cell with porous plug was used and the weight change during electrolysis was measured. Note that the relative transport numbers decrease with increasing temperature.[200] However, measurements of the transport number of nitrate ion in NaNO$_3$ showed that $t_{\text{NO}_3^-}$ was independent of temperature.[201] In binary systems of alkali silicate melts, the transport number of the smaller cation is less than that of the larger cation; this is apparently because the coordination number of the Na$^+$ ion in the melt is only 6 (i.e., tighter binding), whereas that of the Rb$^+$ ion or the Cs$^+$ ion is 12.[202] In mixtures of alkali nitrates, however, the transference number of the larger cation is less than that of the small cation: $t_{\text{Na}^+} \gg t_{\text{K}^+} > t_{\text{Rb}^+} > t_{\text{Cs}^+}$. Conductance is by both cation and anion, and determinations of relative mobilities showed a marked concentration dependence for the NO$_3^-$ ion in (Rb–Cs)NO$_3$ melts; but in (Na–Rb)NO$_3$ melts, the concentration dependence of NO$_3^-$ was negligible. The mobilities of both Na$^+$ and Rb$^+$ ions were greatest in sodium-rich melts.[203]

Electromigration experiments in molten and solid binary sulfate mixtures seem to indicate that in $(Li-K)_2SO_4$ mixtures the cation with the highest mobility is the one present in highest concentration.[204] Experiments on mixtures of alkaline earth halides indicate that the mobilities of the individual cations decrease linearly as the concentration of the largest cation is increased.[205] Murgulescu and Topor[206] have measured the mobilities of a number of cations in various binary alkali melts using a horizontal porous diaphragm in a Hittorf cell. They conclude that the mobility depends on the relative radii of the cations, with Na^+ being a positive exception, and on the free volume available to the main cations.

Comparisons of diffusion coefficients, D, and ionic mobilities, U,[207,208] for Li^+ ions in various alkali nitrate systems indicate that, although the relative mobilities depend on composition in the ratio $(D/U)_{Li} > (D/U)_{Na} > (D/U)_K$, the value of $(D/U)_{Li}$ is not constant for different melts and the activation energy for diffusion of Li^+ ions is greater than that for Na^+ or K^+ ions. This is ascribed to ion association of the small lithium cation. Forcheri[209] correlates the observed mobilities of binary mixtures with differences in the molar volume of the components and suggests that an appropriate parameter to use in describing excess transport properties would be the volume fractions, f_a and f_b, in (88). $U_{a,1}$, $U_{b,0}$ are mobilities of ions A and B in pure melt of A; similarly $U_{a,0}$ and $U_{b,1}$ are the respective mobilities in pure melt B, V_a, V_b, and V are the molar volumes, and ΔU_a is excess mobility at 0.5 mole fraction

$$(88) \qquad \frac{\Delta \Lambda}{\mathscr{F} f_a x_b} = U_{a,1} - U_{b,0} + (U_{b,1} - U_{a,0})\frac{V_b}{V_a} + \Delta U_a \frac{x_a V_b}{0.25\ V}$$

Electrical Conductivity.   Methods employed in conductance measurements depend on the temperature and corrosiveness of the fused salt. Fused salt conductivities are about 1 mho/cm, hence cell designs have to restrict the area of the path and increase the path length. At moderate temperatures, Pyrex, silica, sapphire, and magnesium oxide capillaries are used; thus the cell constant is of the order 200 to 800 cm^{-1}. Accurate a-c conductance bridges or impedance comparators are required in order that small changes in resistance ($<0.1\ \Omega$) may be measured. At high temperatures, a second type of cell is used. A crucible of platinum acts as one electrode, and a carefully positioned second electrode dips into it; all resistances are normally of the order of 0.1 Ω, and hence precision resistance measurements using a modified a-c Kelvin bridge are needed. Critical accounts of apparatus and compilations of conductance results are given in Refs. 6, 12, 15, 193, and 210 to 212. In a-c conductance measurements, particularly with low cell constants, there is a danger of frequency dispersion effects; these are mitigated at frequencies in the range 10 to 100 kHz/sec and for cell constants greater than 100 cm^{-1}.[213,214] Direct-current conductance measurements are rarely used in molten salts because of polarization effects.[215] Conductance measurements at high pressures have recently been carried out on fused alkali-metal nitrates and these are of special importance in any understanding of conductance mechanisms.[216,217]

Data on pure fused salts are either presented as specific conductances, κ, at particular temperatures—or, if the temperature dependence is known, in the form

$$(89) \qquad \kappa = a + bT + cT^2 \qquad \Omega^{-1}\,\text{cm}^{-1}$$

or as the Arrhenius equation

$$(90) \qquad \kappa = A \exp\left(\frac{-E_\kappa}{RT}\right)$$

where T is usually in °C [(89)] and in °K [(90)].

If molar volume and expansivity data are also available, the equivalent conductance, Λ, is a better parameter

$$(91) \qquad \Lambda = A \exp\left(\frac{-E_\Lambda}{RT}\right) \qquad \Omega^{-1}\,\text{cm}^2\,\text{eq}^{-1}$$

The temperature dependence of equivalent conductance is approximately linear, with slight curvature over 100-degree temperature ranges for the alkali halides; however, this is not nearly so good an approximation for salts with lower melting points, such as the nitrates.[218,219,220] For compounds such as HgI_2, $InCl_3$, $GaCl_3$, $BiCl_3$, and $SnCl_2$, the curvature of the specific conductivity reciprocal temperature plot increases until a maximum is reached (usually some 200–400°C above the melting point).[10,221] Values of E_Λ are about 10 kcal in the case of trivalent salts; alkali halides and nitrates range between about 3 and 5 kcal, and polymeric melts have activation energies of the order 15 to 20 kcal.

Values of equivalent conductances can be compared at the melting point, T_m, or $1.1T_m$, which are approximately corresponding temperatures. For purely ionic salts there is a correlation between the temperature, cation radius, and the absolute equivalent conductance. Without invoking any mechanism of transport such as the "hole" theory, which is implicit in any calculation of activation energies on the Arrhenius theory, one can simply postulate that, as in a gas, the conductance of the system will be proportional to the velocity at which the ions move. Since under a-c conditions of measurement there is no net transfer of charge due to the applied voltage, the conductance is purely proportional to the *thermal velocity* of movement of ions, which is proportional to $\sqrt{T_m}$ °K. The correlation is also dependent on cation size, because the resistance to movement is proportional to the diameter of the cation. For alkali metals

$$(92) \qquad \Lambda_m r_c{}^+ = -230 + 11.1\sqrt{T_m} \pm 7\ \Omega^{-1}\ cm^3\ eq^{-1} \times 10^8$$

For alkaline earth metals

$$(93) \qquad \Lambda_m r_c{}^{2+} = -490 + 16.6\sqrt{T_m} \pm 8\ \Omega^{-1}\ cm^3\ eq^{-1} \times 10^8$$

As in other correlations, salts of silver, thallium, lead, and cadmium deviate considerably from the equation, as do fluorides.[222] Reduced equivalent conductivities of various salts are shown in Figure 11.

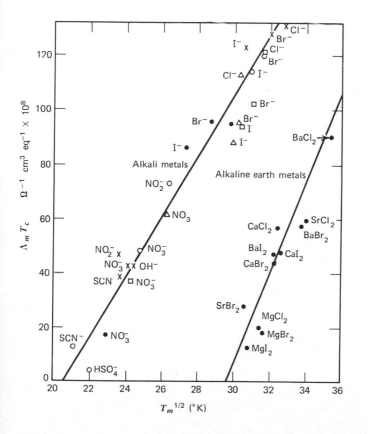

Figure 11. Reduced equivalent conductivities of alkali metal and alkaline earth metal salts at their freezing points. Symbols for salts and radii for corresponding positive ions (Å): solid circles—lithium, 0.60; crosses—sodium, 1.01; open circles—potassium, 1.33; squares—rubidium, 1.48; triangles—cesium, 1.69.

Determinations of conductance parameters at constant volume have been made using either a piston cell or a liquid barrier cell in order to prevent the absorption of high pressure gas by the melt.[216]

Table 7   Constant-Volume Conductance Parameters of Various Salts

Parameters	LiNO$_3$	NaNO$_3$	KNO$_3$	RbNO$_3$	CsNO$_3$
ΔV_Λ^* (ml/mole)	0.5	3.8	7.0	7.8	8.3
$\alpha_T \Delta V_\Lambda / \beta RT$ (cal/mole)	110	1100	1930	2220	2640
E_V (cal/mole)	3290	2110	1680	1640	1040
E_P (cal/mole)	3400	3210	3610	3860	3660

From Ref. 216. Reprinted by permission of the Faraday Society.

Activation volumes, ΔV_Λ^*, and activation energies for conductance at constant volume, E_V, were evaluated (Table 7). Melt conductance in lithium nitrate is almost exclusively energy dependent, whereas the available volume seems to be the important factor in cesium nitrate melt

$$(94) \qquad \Lambda_V = A \exp - \left[\frac{E_V}{RT} + \frac{\alpha_T \Delta V_\Lambda^*}{\beta RT} \right]$$

Many binary systems have been investigated and a comprehensive compilation is given by Janz.[15] The main features of the equivalent conductance–composition isotherms are summarized in an article by Klemm.[12] Ideal conductance isotherms are additive and linear (e.g., NaNO$_3$ + NaNO$_2$ systems). However, most systems show deviations from additivity, the majority being negative deviations; systems that show deviations in thermodynamic properties also show deviations in transport properties (Figure 12).

Deviations that are not too large (e.g., 5%) can usually be expressed[223] as differences from ideal mixing, that is for ideal mixing

$$(95) \qquad \Lambda_m = x_1{}^2 \Lambda_1 + x_2{}^2 \Lambda_2 + 2x_1 x_2 \Lambda_1$$

Then, in accordance with Ref. 224, the deviation in conductivity is given by

$$(96) \qquad \Delta \Lambda = \frac{\Lambda_{\text{exp}} - \Lambda_m}{\Lambda_m}$$

In common alkali-cation-mixed halide systems, deviations increase as differences in size and polarizability of the anions, r_{a_1}, r_{a_2}, increase and the size of the common cation, r_c, decreases[225]

$$(97) \qquad \Delta \Lambda_{0.5} = f(T) \left(\frac{r_{a_1} - r_{a_2}}{r_c} \right) x_1 x_2$$

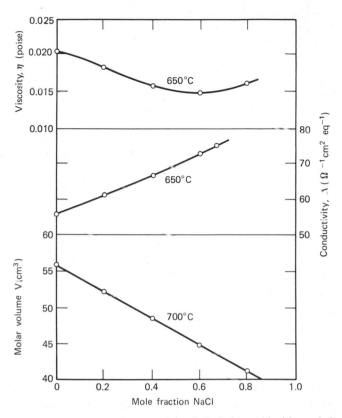

Figure 12. Transport properties of Cadmium chloride and Sodium chloride melts. From Ref. 139. Reprinted by permission of the Faraday Society.

There is also some asymmetry in these systems. Very large negative deviations are found where it is thought that complex ions exist in the melt. Usually the phase diagrams show that solid compounds are formed, although this is not the case for LiCl–KCl systems where $\Delta\Lambda = 26\%$. Definite minima in conductance isotherms—$(PbCl_2 + NaCl)$, $(AlF_3 + NaF)$—are usually considered attributable to compound formation.[226]

Some of the binary systems, on reducing the temperature, show marked decreases in conductivity and a tendency to supercool and form glasses when the temperature is reduced. In addition to the well-known silicate, phosphate, and borate glasses,[227] certain mixtures of salts of polyatomic ions will readily supercool. One of the mixtures that has been extensively studied is KNO_3–$Ca(NO_3)_2$ melt.[228–231,186,191,219] Non-Arrhenius behavior of these melts appears in Figure 13.

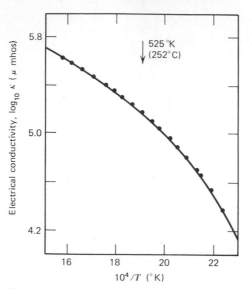

Figure 13. Non-Arrhenius behaviour of potassium nitrate and calcium nitrate melts. From E. Rhodes, W. E. Smith, and A. R. Ubbelohde, *Proc. Roy. Soc. (London), Ser.* A, **285**, 262 (1965), reproduced by permission of the Royal Society.

Zinc chloride–alkali-halide systems also form glasses on supercooling,[232] as do some sulfate–chloride systems.[233] These systems have been interpreted by Angell in the light of the "free-liquid-volume" theory. The temperature dependence of conductance is given by

$$(98) \qquad \Lambda = AT^{-\frac{1}{2}} \exp\left[\frac{-k}{(T - T_0)}\right]$$

where T_0 is the temperature at which free volume, V_f, appears

$$(99) \qquad V_f = \alpha \overline{V}_m(T - T_0)$$

where α is the thermal expansion and $\overline{V}_m$ is mean molecular volume. Cohen and Turnbull[48] give k as $(\gamma V^*/\alpha \overline{V}_m)$ where V^* is the critical void volume and γ is a correction function for overlap of free volume. The constants k and T_0 in (98) are usually calculated empirically by successive approximations from the experimental data. An approximate linear correlation between T_0 and the mean cationic strength, $\sum n_i z_i/r_i$, has been found; usually the preexponential term $T^{-\frac{1}{2}}$ is of little importance except for lithium salts. An extensive number of glass-forming molten salt systems have been studied and it appears that T_0 is now a well-established thermodynamic parameter characterizing the salt system.

Viscosity. Many of the remarks about mechanisms of self-diffusion and conductance can be applied to shear viscosity. Methods of determination of viscosities of the order 0.5 to 5 cp are similar to those used for aqueous systems. Capillary viscometers of either the Ostwald or Ubbelohde type (sometimes with a form of electrical timing device) are frequently used. Falling-body devices, damped oscillations of a suspended bob, and torsional rotating cylinders for melts of very high viscosity are other methods that have been used.[6,193] Viscosity measurements are usually expressed in the form of Arrhenius equations or quadratic temperature-dependence equations. Compilations of data may be found elsewhere.[12,15,193]

Viscosity data follow the same trends as those exhibited by conductance measurements and are usually treated in the same manner, using the Arrhenius equation and liquid-free-volume theory. The Batchinski function, $\mathscr{B}$, in fact applies very well to molten salts at low temperatures

$$(100) \qquad \eta = \frac{\mathscr{B}}{V - b}$$

where b is the incompressible volume and V is the specific volume of the melt.

A correlation dependent on the absolute temperature does not apply to viscosity. There have only been one or two determinations of bulk viscosity in molten salts using ultrasonic waves,[234] but the thermal expansion coefficient, α, is related to the sheer viscosity, η, and bulk viscosity, θ, by

$$(101) \qquad \alpha = \frac{2\pi^2 f}{C^3 \rho} \left(\frac{4\eta}{3} + \theta \right)$$

where f is the frequency of ultrasonic relaxation, C is velocity of sound waves, and ρ is melt density. Bulk viscosities decrease with increasing temperatures, as demonstrated in Table 8.

Table 8 Viscosities of Salt Melts

	Salts				
	$LiNO_3$	$NaNO_3$	$AgNO_3$	KNO_3	$CdCl_2$
Temperature (°C)	300	350	250	400	600
θ (cp)	7.95	8.2	19.9	21.6	3.8
η (cp)	3.9	2.4	3.6	2.1	2.3

In order to classify various types of melts, correlations of the activation energies for viscous flow and electrical conductance (Table 9) are frequently made. The three classes that follow are due to Janz.[193]

class I	$E_\eta > E_\Lambda$	alkali and alkaline earth halides, most completely ionized salts
class II	$E_\eta \sim E_\Lambda$	low-melting salts (e.g., alkali, Tl, and Ag nitrates); quaternary ammonium salts
class III	$E_\eta < E_\Lambda$	mercuric halides (i.e., nonionizing solvents similar to H_2O)

Table 9   Comparative Activation Energies for Viscous Flow and Electrical Conductance

	Salts						
	NaCl	KCl	Na_2CO_3	K_2CO_3	$NaNO_3$	$TlNO_3$	$HgCl_2$
E_Λ (kcal)	2.99	3.41	4.18	4.65	3.21	3.34	6.2
E_η (kcal)	9.30	6.58	25.7	29.1	3.88	3.66	3.4
E_η/E_Λ	3.11	1.93	6.1	6.3	1.21	1.02	0.45

Murgulescu and Zuca[235] formulate a theory based on the Eyring reaction rate theory in which they introduce the concept of ion pairs as the unit of flow. Comparison of the ion-pair theory with experiment when the radius of the ion pair is introduced into the pair-correlation equation for viscous flow (computed by the Enskog method) shows good agreement for alkali chlorides and bromides and confirms the idea that mass movement in a melt is sensibly an electrically neutral phenomenon.[236] Copeland and Christie[237] have investigated the effect on fluidity of fused sodium nitrate when helium, argon, and nitrogen gases at high pressures are used to compress the melt. As the pressure, P, of helium was increased, the fluidity, ϕ_T, decreased; with argon, as P increased ϕ_T increased at first, but the increase in fluidity decreased as the temperature was raised. With nitrogen, an inversion temperature at 328°C occurred, whereupon the increase in fluidity with increasing pressure changed to a decrease in fluidity with increasing pressure and increasing temperature. The interpretation seems to be that argon and N_2 (below 328°C) *dilute* the melt, making it less viscous, whereas the small helium atom fills the holes in the melt, decreasing the activation volume for viscous flow. At 400°C the activation volume for viscous flow, ΔV_η^*, was calculated to be 8.8 cm³/mole and $E_{p(\eta)} = 3.93$ kcal/mole, whereas $E_{V(\eta)} = 1.50$ kcal/mole at 1 atm (i.e., a very similar effect to that found for conductance and additional evidence that the "liquid-free-volume" theory is of special importance in describing transport processes in molten salts).

The viscosity composition isotherms of binary mixtures of fused salts show similar departures from additivity as in conductivity measurements (see

Figure 12). The extent of the deviations appears to be dependent on differences in the ionic radii in common ion systems. The majority of viscosity isotherms show negative deviations from linearity, but the positions of the minima do not necessarily correspond to those of the conductance isotherms; the system $CdCl_2$ + $CdBr_2$ is the exception showing positive deviations.[230]

Departures from linearity of viscosity reciprocal temperature plots are found for binary and ternary mixtures of salts. When the temperatures are very low, viscosities can increase by several orders of magnitude, particularly if the melts supercool. Again the KNO_3–$Ca(NO_3)_2$ systems previously cited are typical of binary systems. Ternary nitrate melts also show marked curvature of the reciprocal temperature viscosity plots and have very low liquidus temperatures, for example, 111°C for $LiNO_3$–KNO_3–$Cd(NO_3)_2$ eutectic.[239] At low temperatures and small volumes it is difficult for the nonspherical ions to rotate and move freely, hence the systems tend to lock in such a way that more energy is required to crystallize out than to supercool. The Angell free volume theory and the idea of restricted volume for movement seem to be adequate explanations for non-Arrhenius behavior without having to invoke the idea of clusters in other than the limited sense of ion-pair groupings.[240]

Thermal Conductivity. Although thermal conductivity data, K, are of great importance in heat-transfer uses of molten salts, few measurements have been made because of the experimental difficulties. Turnbull[241] used a transient method employing a thin platinum wire heated by a constant current. The rate of heating depends on the thermal conductivity of the liquid within which the wire is immersed. Electrolysis and polarization effects cause difficulties in using this method. Bloom et al.[242] used the steady-state method, but also with electrical heating, heat being transferred across the thin melt annulus between two concentric vertical silver cylinders.

$$(102) \qquad K = \frac{Q \ln (r_2/r_1)}{2\pi l(t_1 - t_2)}$$

where Q is quantity of heat transferred per second, r_1, t_1 and r_2, t_2 are radii and temperature of inner and outside cylinders, respectively, and l is the length of the molten salt annulus.

Mixtures of salts were also studied and negative deviations from additivity of the thermal conductivities are explained in terms of packing on a quasi-lattice model. Data are also available for mixed fluoride melts, which are important in atomic reactor technology.[243] Thermal conductivities are of the order of 1.3×10^{-3} cal/(cm)(sec)(deg) for alkali nitrates and 1×10^{-3} cal/(cm)(sec)(deg) for the LiF–NaF–KF eutectic, that is, a factor of 2 to 5 greater than for organic liquids.

8   Spectroscopy of Fused Salts

There have been two main applications of the techniques of spectroscopy in fused salts: (a) in studies of the properties and interactions of pure fused salts and (b) in examining spectra of dilute solutions of chromophoric entities in fused salt solvents. Data are reported for pure melts or solids as absorbance or optical density, $A = \log_e (I_0/I)$, or absorption constant, $k = (2.303/b) \times \log_{10} [(I_0/I)]$, where I_0 is the intensity of incident light, I is the intensity of the transmitted light, and b is the path length. In solution spectra the molar extinction coefficient, ε, is often used where $\varepsilon = A/Mb$, and M is the number of gram ions of chromophore (e.g., transition metal ion) per liter of melt.

Electronic Absorption Spectra—Ultraviolet and Visible.   The problems in obtaining electronic absorption spectra involve shielding the spectrophotometer from the heat generated by the high-temperature furnace, which must be small enough to fit into the spectrometer and yet have a large enough constant temperature volume to ensure reasonable temperature control over the total area and volume exposed to the light beam. Various cell designs are illustrated and discussed by Gruen,[11,244] Smith,[12] Janz,[15] and Rhodes and Ubbelohde.[245,246]

At temperatures below 400°C it is possible to use the normal absorbance spectrophotometer assembly. At higher temperatures (1000°C), where the thermally emitted radiation from thermostat and fused salt is $\sim 20\%$ of the light signal, it is essential to use a configuration such as the following:[244]

Light Source—Chopper—Sample—Monochromator—
Detector—Amplifier—Recorder

Transparent windows and cell container materials that have been used include Pyrex, fused silica with enhanced ultraviolet transmission, single-crystal sapphire, and magnesium oxide. Cells for solution experiments must be vacuum tight. So-called windowless cells have also been used, in which the salt is suspended vertically in a ring of platinum.[247]

SOLUTION SPECTRA.   Molten salts form very useful solvents for observing the spectra of transition metals, lanthanides, and actinides. Melts such as the LiCl–KCl or $LiNO_3$–$NaNO_3$–KNO_3 eutectics are transparent in the near-infrared region between 10,000 and 4000 cm^{-1}, where many low-temperature solvents (especially water) have marked absorption bands. The observed bands and changes in their positions and intensities with solvent and temperature are described in terms of crystal-field or ligand-field theory.[248,249]

It is not possible to discuss this theory here. Suffice it to say that the effects observed are due to the splitting of the ground-state terms of the free ions by the surrounding electric field set up by the ligands. The amount of splitting is denoted by the crystal-field splitting parameters Dq or Ds. For instance, the ground states of the free ions of the $3d$ transition metal ions are ^2D, ^5D, ^3F, and so on; in an octahedral (6 coordination) ligand field, a D state is split into two levels (T_2 and E) separated by an energy 10 Dq, whereas in a tetrahedral field (4 coordination) the splitting is only $(-\frac{4}{9})$ Dq of octahedral. The actual value of Dq depends on the electric field of the solvent; for example, for Co^{II}, Ni^{II}, and Cr^{III}, the absorption maxima shift to progressively lower energies on changing from an aqueous solution to a molten Al_2Cl_6 solvent. For Co^{II}, Dq (energy of absorption maximum for the transition $[T_1(F) \rightarrow T_1(P)]$) decreases in the order as follows:

H_2O 510 nm > NO_3^- melt 560 nm > F^- melt 580 nm > Cl^- melt 634 nm

where λ = wavelength (nm), and ν = frequency (cm^{-1}) $\equiv$ energy (eV). Since $\lambda \propto 1/\nu$ if the wavelength increases, the frequency and the energy of splitting decreases.

Because of the above effects, shifts in spectra of dilute solutions of transition metals[244] are of great help in elucidating the structure of the fused salt solvent, and they are of spectroscopic and practical interest (e.g., spectra of uranium) as well. One of the first investigations of transition metal spectra, concerning the formation of manganates in sodium hydroxide melts, was reported by Lux and Niedermaier in 1956.[250]

Temperature variations of transition metal spectra have been fairly extensively studied, for instance, $NiCl_2$ dissolved in LiCl–KCl eutectic. Boston and Smith[251] noted systematic changes in the spectrum with changing temperature and attributed these to the changes in ratio of two nickel species in the melt, namely, hexachloro-Ni^{II} ions and tetrachloro-Ni^{II} ions (Figure 14).

Gruen and McBeth, however, considered that the spectral changes were due to tetragonal distortion of the $NiCl_4^-$ ion in the strong crystal field of the LiCl–KCl solvent.[252] The question of assigning correct coordination geometries to ions in molten salts raises a certain amount of controversy, since many of the arguments are based on the theory of isosbestic points in spectra —that is, crossing of spectra at two or more different temperatures (see Figure 14 and Ref. 253). In later work, Smith et al.[254] considered that the coordination geometry of the Ni^{2+} ion is strongly influenced by the charge potential of the surrounding shell of cations. They suggest that at low temperature the tetrahedral species having a cation shell rich in K^+ ions is in equilibrium with the octahedral species having an Li^+-rich cation shell; at high temperatures the two-entity behavior disappears. Similar effects are observed in $MgCl_2$–KCl melts,[255] and in $ZnCl_2$–KCl melts,[256] where at temperatures below 300°C octahedral Ni^{II} predominates, whereas above 300°C a distorted tetrahedral $NiCl_4^-$ complex exists.

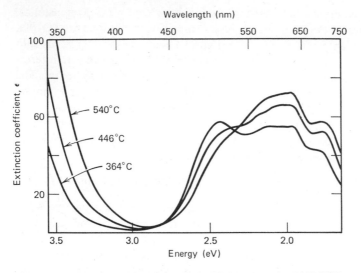

Figure 14. Temperature variation of Ni(II) spectra in LiCl–KCl eutectic melt. From Ref. 251. Reprinted by permission of the American Chemical Society from *J. Phys. Chem.*, **62**, 409 (1958).

It is fairly certain, however, that the octahedral complex is the most stable entity at low temperatures and low available volumes. Structures of nickel compounds in the low-melting (125°C) systems dimethyl sulfone and Li–Na–KNO_3 eutectic confirm the hypothesis that an octahedral complex is the stable low-temperature form. In the nitrate melt, NO_3^- behaves as a bidentate ligand, forming the $Ni(NO_3)_3^-$ complex.[257] Similar effects are found for other transition metal ions in fluoride,[258] chloride,[259] and sulfate[260] melts. Johnson[261] proposes a spectrochemical and nephelauxetic series in molten salts. The spectrochemical series corresponds to that found in an aqueous solution; the ligand field splitting Δq (or Dq) is least for $Br^- < Cl^- < F^- < SO_4^{2-} < NO_3^- < CNS^- \ll CN^-$. The nephelauxetic factor, β, is 0.9 in melts compared to 0.75 in aqueous and solid solutions. Examples of spectra of metal ions other than the transition metals are Tl^+, Pb^{2+}, and Bi^{3+} ions studied in LiCl–KCl eutectic by Smith et al.,[262] and uranium and its oxidation states in various solvents.[263] Most simple metallic ions, however, do not absorb light in the visible and ultraviolet regions.

CHARGE-TRANSFER SPECTRA. Nonmetallic ions in fused salts exhibit very intense charge-transfer spectra because of the transfer of p electrons into an expanded orbital confined by surrounding cations. The absorption maxima for the halides lie in the region 40,000 to 80,000 cm^{-1} and molar extinction coefficients are of the order 10^4 to 10^5.[264–266] Because of the high energies and intensities involved, there have been virtually no determinations of fundamental charge-transfer spectra of fused salts. Shifts of the band edge with temperature have been reported by Retschinsky,[267] Mollwo,[268] and Sundheim and Greenberg,[269] who noted a pronounced shift to the red (~ 1000 cm^{-1}) with each 100°C rise in temperature in both solid and fused halide, nitrate, and sulfate salts, and on melting. On melting there is apparently a shift to the blue of the absorption peak maximum of the charge-transfer band at 50,000 cm^{-1} of the alkali nitrates,[270] whereas the peak maximum for lithium iodide moves to the red.[245]

Most of the spectra of nonmetallic ions lie at energies too high to be readily studied in the fused state; however, the lowest-lying nitrate ion transformation at approximately 3000 Å ($n-\pi^*$ transition) has been extensively studied by Smith and Boston[271] and Cleaver, Rhodes, and Ubbelohde,[216] and the observed changes in the spectrum correlated with the ion interactions in the melt and phase transformations. Changes in the position of the maximum of the absorption spectrum, $h\nu_{max}$, with cationic environment are given by

$$(103) \qquad h\nu_{max} = 0.34 \sum_i \left(\frac{z}{r_0}\right)_i \bar{X}_i^+ + 3.81$$

where $(z/r_0)_i$ is the cationic potential and $\bar{X}_i$ is the equivalent fraction of the cation.[271] Temperature changes are quite small and can be correlated by an expression $dE_2/dT \propto (1/V)\, dV/dT$; the shifts in $E_{2,max}$ with temperature are dependent on the thermal expansion of the melt and changes in the cation-anion distance. Blue shifts that are found on melting are interpreted as meaning closer cation-anion pairings in the melt such as statistical averaged "ion pairs." (Figure 15).

Small deviations from additivity were found for $E_{2,max}$ on mixing various alkali nitrates when one of the cations was the lithium ion.[272] Doucet et al.[273] report rather different values for the constants of the Smith and Boston equation [(103)] for alkali–alkaline-earth nitrate mixtures, especially at the eutectic composition. Rhodes, Smith, and Ubbelohde[220] noted that, on supercooling KNO_3–$Ca(NO_3)_2$ systems, there was no abrupt change in $E_{2,max}$ at the nominal freezing point, but rather a gradual curvature corresponding to increased constriction in the glassy state.

Vibrational Spectra—Infrared and Raman. Infrared and Raman spectroscopy are means of examining the vibrational spectra and measuring the molecular characteristics and interactions of polyatomic ions. Raman spectra of melts were first measured systematically by Bues in 1955 for alkali nitrate, phosphate, and arsenate glasses and melts.[274,275] Details of Raman and infrared experimental methods and apparatus are given in Refs. 12, 6, and 15.

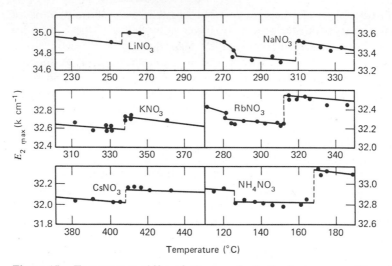

Figure 15. Temperature shifts of $E_{2,\text{max}}$. of alkali nitrates on melting. From Ref. 246, reproduced by permission of the Royal Society.

Since it is difficult to obtain suitable windows for use in the infrared region, Raman and infrared measurements often use reflectance techniques. Light sources are conventional, but there is increasing use of lasers instead of Toronto arcs in Raman spectroscopy.

Depending on the symmetry of the polyatomic ion and on selection rules, certain frequencies may, theoretically, be active, in either infrared *or* Raman spectra, but not in both. The number of possible frequencies increases with lack of symmetry until there will be no degeneracy in the least symmetrical systems and all the vibrational frequencies will be active in either the infrared or Raman spectra. In condensed systems, the selection rules do not strictly apply, and degenerate frequencies may be split by the crystal field, causing the appearance of relatively weak forbidden vibrational bands. Assignments of bands are often ambiguous, and structural information derived from vibrational data may be somewhat inconclusive.

In pure ionic melts investigations of ionic-type compounds have mainly been concerned with following effects on the nitrate ion of various cation environments. In covalent-type melts (e.g., halides of groups II and III), the mercuric salts are the most interesting.[276] In early work Hexter[277] studied the infrared spectrum of solid and molten sodium nitrate and found that only the nitrogen–oxygen-stretching frequency, ν_3, persisted in the melt. From Raman spectra data, Wait et al.[278] suggest that the D_{3h} symmetric symmetry is retained in most alkali nitrate melts, as opposed to the C_{2v} symmetry suggested by selection rules. There is loss of degeneracy of the ν_3 vibration because of lack of freedom to rotate in lithium and silver nitrate melts. In other nitrate melts, however, the symmetrical stretching frequency, ν_1, shows a definite correlation with the polarizing power of the cation.

A "contact ion-pair" model is advanced for the nearest-neighbor pairwise association in which an M^+ ion interacts with oxygen atoms at unidentate sites in the ion pair; $M^+–O$ frequencies in the region 100 to 250 cm^{-1} are predicted on this model. James and Leong[279] found that the Raman ν_1 frequency of sodium nitrate was sensitive to temperature variation, showing a marked jump at the melting point from 1064 to 1058 cm^{-1}. Infrared and Raman spectra data on mixtures of alkali metal nitrates with alkaline earth nitrates (Hester et al.[280]) also suggest interionic association; for instance, the appearance of a new stretching frequency at 332 cm^{-1} in infrared spectra is attributed to the Ca–ONO$_2$ stretch in KNO$_3$–Ca(NO$_3$)$_2$ melts. On the other hand, Devlin et al.,[281] from total reflection infrared spectra, suggest that all the observed vibrational spectrum features can be accounted for by a perturbed orthorhombic lattice structure in the melt. It is probably still too early to say which possible interpretation is correct. More data in the far-infrared region might help in this matter, for this is the region in which M–O stretching frequencies should be found. Vallier[282] reports nonadditivity of Raman frequencies in binary melts of NaNO$_3$–Ca(NO$_3$)$_2$ and KNO$_3$–Ca(NO$_3$)$_2$.

Other polyatomic melts that have been investigated are hydroxides (infrared),[283] sulfates (Raman),[284] thiocyanates (infrared),[285] lithium perchlorate (infrared and Raman).[286] Tetrahedral SO$_4^{2-}$ ions were found to be present in the melts, and shifts in the Raman (r_1a_1) SO$_4^{2-}$ frequency were dependent on the cationic potential, (z/r^+). In thiocyanate melts the relative bond strengths and force constants of the M–S–C≡N and S=C=N–M bonds in complexes were evaluated. In lithium perchlorate the infrared spectrum of the melt would suggest a type of distorted cubic packing with free rotation of the ClO$_4^-$ ion and the presence of polyatomic aggregates instead of ion pairs.

The covalent IIB halides are interesting because the observed bands depend on the type of species present in the melt, such as MX^+, MX_3^-, MX_4^{2-}. In gallium chloride melts, for instance, the species GaCl$_4^-$ was confirmed by Raman spectroscopy.[276] In mixed mercury halide melts there is a marked decrease in the ν_1 stretching frequency on going from the vapor phase to melt, indicating that appreciable ionic character still exists in these predominately covalent melts.[287]

In mixtures of molten salts experimental data are mainly concerned with studying complex formation in, for instance, zinc halide–metal halide systems. Results are interpreted in terms of (ZnCl$_2$)$_n$ polymers and ZnCl$_4^{2-}$ ions, depending on the amount of alkali halide present.[288,289] Tin, cadmium, and aluminum halide systems with alkali halides have also been investigated.[276]

Nuclear-Magnetic-Resonance and Electron-Spin-Resonance Spectra. There have been few studies of molten salts by NMR spectroscopy, since an NMR active entity is required as a probe. So far most of the data refer to ^{205}Tl, which is NMR active,[290,291] although there is one report of NMR investigations on $Ca(NO_3)_2$ tetrahydrate melts where the proton ^1H is used as the probe.[292]

Hafner and Nachtrieb[291] describe a small high-temperature furnace that fits in a modified NMR spectrometer head without heating the magnet and resonance coil. Relatively large chemical shifts, δ, for the resonance frequency of the ^{205}Tl nucleus are found for the different solid and molten thallium salts. The nitrate appears to be most ionic of the thallium salts, and the ^{205}Tl resonance frequency was found to increase in the order $NO_3^- < Cl^- < Br^- < I^-$. The NMR shifts in the pure melts are associated with increasing covalent character of the salts. In mixtures of thallium halides and alkali halides, the direction of the chemical shift was dependent on the cation radius. The covalent interaction between Tl^+ and the halide ions was decreased by small alkali metal ions and increased by large cations for a given mole fraction, shifts being proportional to Δr where $\Delta r = r(Tl^+X^- - M^+X^-)$. Linear temperature dependence of the chemical shift was found and, with due allowance for change in coordination number on melting, the cation-anion shrinkage on melting, TlCl, was calculated as being 0.18 Å from the observed jump in resonance frequency.

Electron-spin-resonance fused salt data are almost nonexistent, although Swanson[293] has reported ESR studies on $MnCl_2$ in LiCl–KCl eutectic and Kukk has done some preliminary work on transition-metal halides in fused halide solvents.[294]

Refractometry. Zarzycki and Naudin[295,296] have combined molar refractivity measurements on fused salts with magneto-optical measurements of Verdet's constant (rotation effect of a magnetic field on the plane of polarized light). Measurements of the refractive index are made using a reflecting metal prism fitted with silica windows as the melt container and refractivity cell. From the refractive index measurements and magneto-optical measurements at long wavelengths, it is possible to calculate the fundamental absorption frequencies in the ultraviolet region, although a number of assumptions have to be made, particularly with regard to the optically active electrons. Calculations for the first and second bands have been made for alkali salt melts: $h\nu_1 \sim 7$ to 10.6 eV and $h\nu_2 \sim 43.0$ to 51 eV (NaBr to KF); whereas for alkali nitrates, $h\nu_1$ was about 12.0 eV and $h\nu_2 \sim 34$ eV. On melting, jumps to the red for $h\nu_1$ were calculated and these are greatest for KF (9.8%) and CsI (8.2%); the average value is about 5%.

Molar refractivities (R_D) were found to be approximately additive for mixtures of molten salts, but slight positive deviations were found for $PbCl_2$–KCl and $CdCl_2$–KCl mixtures. In the melt, $dR_D/dT \sim 0.002$ cm³/deg, which is greater than for the solid.[297] For alkali halides Vesnin and Batsanov[298] report an increase of approximately 5% in R_D on melting, and this increase is roughly proportional to the heat of fusion.

9 Conclusion

Two vast areas of interest in a study of fused salts have been omitted except in passing reference to their use in evaluating thermodynamic parameters of melts. Cryoscopy and solid-liquid phase equilibria can be used to calculate activities and heats of fusion of melts. Activities of ions in fused salts can also be derived from electromotive force data on suitable reversible electrochemical cells. With regard to binary and ternary phase equilibria studies, there are an immense number of phase diagrams reported in the literature and a particularly good summary is given by Ricci in *Molten Salt Chemistry*.[12] Earlier Russian work is included in the Consultants Bureau publication of Russian research in fused salts.[212] Sinistri, Franzosini, and Rolla[299] have recently published an atlas of some 200 molten salt systems showing the occurrence of miscibility gaps in these systems. Janz[15] tabulates references to binary, ternary, and quarternary molten salt systems and also to metal–metal-salt mixtures, and Lumsden[13] quotes much cryoscopic, phase equilibria, and EMF activity data in his book, *Thermodynamics of Salt Mixtures*, whilst Voskresenskaya[300] has edited a comprehensive compilation of solid-liquid equilibria data in systems of anhydrous inorganic salts.

Much work with molten salts deals with electrochemical properties of the systems.[301,302] Electrolysis cells and EMF cells using fused salts all have practical importance as fuel cells or in metal extraction and refining operations.[10,12,19] Electromotive force series in fused salts have been constructed, and orders slightly different from the aqueous electrochemical series are obtained: notably, the alkaline earth metals are more electropositive than the alkali metals in fused salts. Electrode potentials are of the order of -2.7 V for alkaline earth metals in fused chlorides.[15] Thermal decompositions and reactions in molten salts have not been discussed, either, and it has been taken for granted that the measurements described refer to thermally stable systems. Except in the discussion of transport properties, an equilibrium situation has been assumed.

Overall, the objective has been to present fused salts as physical liquids with liquid state properties similar to any other liquid except that fused salts are composed of *charged particles* and the temperature range is much higher. With care most liquid-state experimental procedures can be adapted to the study of fused salts. In theoretical considerations comparisons have been made with the solid state rather than the gas state; this is justified by the observation that molten salts are very dense liquids with, in many cases, a long stable temperature range of some $1.6T_m$ for the liquid state, which is considerably greater than for most organic liquids. Structural, physical, thermodynamic, and transport properties have been described, and theoretical concepts with regard to the nature of fused salts have been explained in some detail.

REFERENCES

1. P. Debye and E. Hückel, *Z. Physik*, **24**, 185, 305 (1923).
2. F. D. Rossini, *Natl. Bur. Std. (U.S.)*, Circ. No. 500 (1952).
3. P. Drossbach, *Electrochemistry of Molten Salts*, Springer, Berlin, 1938.
4. W. Biltz and W. Klemm, *Z. Anorg. Allgem. Chem.*, **152**, 267 (1926).
5. F. M. Jaeger, *Z. Anorg. Allgem. Chem.*, **101**, 1 (1917).
6. J. O'M. Bockris, J. L. White, and J. D. Mackenzie, *Physico-Chemical Measurements at High Temperatures*, Butterworths, London, 1959.
7. G. E. Blomgren and E. R. Van Artsdalen, *Ann. Rev. Phys. Chem.*, **11**, 273 (1960).
8. O. J. Kleppa, *Ann. Rev. Phys. Chem.*, **16**, 187 (1965).
9. S. J. Yosim and H. Reiss, *Ann. Rev. Phys. Chem.*, **19**, 59 (1968).
10. Y. K. Delimarskii and B. F. Markov, *Electrochemistry of Fused Salts*, Transl. A. Peiperl, Sigma Press, Washington, D.C., 1961.
11. B. R. Sundheim, *Fused Salts*, McGraw-Hill, New York, 1964.
12. M. Blander, *Molten Salt Chemistry*, Wiley, New York, 1964.
13. J. Lumsden, *Thermodynamics of Molten Salt Mixtures*, Academic Press, London, 1966.
14. H. Bloom, *Chemistry of Molten Salts*, Benjamin, New York, 1967.
15. G. J. Janz, *Molten Salts Handbook*, Academic Press, New York, 1967.
16. "Structure and Properties of Ionic Melts," *Discussions Faraday Soc.*, **32** (1961).
17. "Electrode Processes," *Discussions Faraday Soc.*, **1** (1947).
18. A. R. Ubbelohde, *Melting and Crystal Structure*, Clarendon, Oxford, 1965.
19. D. J. G. Ives and G. J. Janz, *Reference Electrodes, Theory and Practice*, Academic Press, New York, 1961.
20. M. Blander, *Advan. Chem. Phys.*, **11**, 83 (1967).
21. H. A. Levy and M. D. Danford, *Diffraction Studies of the Structure of Molten Salts*, Chapter 2, Ref. 12 (1964).
22. K. Lark-Horowitz and E. P. Miller, *Phys. Rev.*, **51**, 61 (1937).
23. M. Kunitomi, *J. Chem. Soc. Japan*, **64**, 74 (1943); *J. Chem. Soc. Japan*, **65**, 74 (1944); *J. Chem. Soc. Japan*, **70**, 344 (1949).
24. R. L. Harris, R. E. Wood, and H. L. Ritter, *J. Amer. Chem. Soc.*, **73**, 3151 (1951).
25. R. E. Wood and H. L. Ritter, *J. Amer. Chem. Soc.*, **74**, 1760, 1763 (1952); *J. Amer. Chem. Soc.*, **75**, 471 (1953).
26. V. I. Danilov and S. I. Krasnitskii, *Dokl. Akad. Nauk SSSR*, **101**, 661 (1955).
27. G. Zarzycki, *J. Phys. Radium*, **19A**, 13 (1958).
28. H. A. Levy, P. A. Agron, M. A. Bredig, and M. D. Danford, *Ann. N.Y. Acad. Sci.*, **79**, Art. 11, 762 (1960).
29. G. Zarzycki, *Discussions Faraday Soc.*, **32**, 38 (1961).

30. K. Furukawa, *Discussions Faraday Soc.*, **32**, 53 (1961).
31. R. Hodge and J. Trotter, *Can. J. Chem.*, **43**, 2692 (1965).
32. G. Zarzycki and F. Naudin, *Compt. Rend.*, **266C**, 1005 (1968).
33. H. A. Barrett and M. Mandel, *Phys. Rev.*, **109**, 1572 (1958).
34. J. A. Barker, *Lattice Theories of the Liquid State*, Pergamon Press, Oxford, 1963.
35. J. E. Lennard-Jones and A. F. Devonshire, *Proc. Roy. Soc. (London), Ser. A*, **169**, 317 (1939); *Proc. Roy. Soc. (London), Ser. A*, **170**, 464 (1940).
36. M. Temkin, *Acta Physicochem. URSS*, **20**, 411 (1945).
37. J. Frenkel, *Kinetic Theory of Liquids*, Dover, New York, 1955.
38. J. G. Kirkwood, *J. Chem. Phys.*, **3**, 300 (1935); *J. Chem. Phys.*, **18**, 380 (1950).
39. Γ. H. Stillinger, *J. Chem. Phys.*, **35**, 1581 (1961); *Phys. Rev.*, **126**, 1239 (1962).
40. W. Altar, *J. Chem. Phys.*, **5**, 577 (1937).
41. R. Fürth, *Proc. Cambridge Phil. Soc.*, **37**, 252 (1941).
42. J. O'M. Bockris and N. E. Richards, *Proc. Roy. Soc. (London), Ser. A*, **241**, 44 (1957).
43. N. F. Mott and R. W. Gurney, *Rept. Progr. Phys.*, **5**, 46 (1938).
44. J. D. Bernal, *Nature*, **185**, 68 (1960).
45. F. Zernicke and J. A. Prins, *Z. Physik*, **41**, 184 (1927).
46. H. Eyring and J. O. Hirschfelder, *J. Phys. Chem.*, **41**, 249 (1937).
47. J. Kirkwood, *J. Chem. Phys.*, **7**, 919 (1939).
48. M. H. Cohen and D. Turnbull, *J. Chem. Phys.*, **31**, 1164 (1959).
49. H. Eyring, T. Ree, and N. Hirai, *Proc. Natl. Acad. Sci. (U.S.)*, **44**, 683 (1958); C. M. Carlson, H. Eyring, and T. Ree, *Proc. Natl. Acad. Sci. (U.S.)*, **46**, 333 (1960).
50. G. E. Blomgren, *Ann. N.Y. Acad. Sci.*, **79**, Art. 11, 781 (1960).
51. H. Reiss, S. W. Mayer, and J. Katz, *J. Chem. Phys.*, **35**, 820 (1961).
52. H. Reiss, J. Katz, and O. J. Kleppa, *J. Chem. Phys.*, **36**, 144 (1962).
53. M. Blander, *J. Chem. Phys.*, **41**, 170 (1964).
54. K. D. Luks and H. T. Davis, *Ind. Eng. Chem. Fundamentals*, **6**, 194 (1967).
55. F. London, *Z. Physik. Chem.*, **B11**, 222 (1930).
56. K. S. Pitzer, *J. Chem. Phys.*, **7**, 583 (1939).
57. H. Reiss, H. L. Frisch, and J. L. Lebowitz, *J. Chem. Phys.*, **31**, 369 (1959).
58. S. W. Mayer, *J. Chem. Phys.*, **40**, 2429 (1964).
59. S. J. Yosim and B. B. Owens, *J. Chem. Phys.*, **41**, 2032 (1964).
60. B. B. Owens, *J. Chem. Phys.*, **44**, 3144 (1966).
61. B. B. Owens, *J. Chem. Phys.*, **44**, 3918 (1966).
62. A. J. G. Boyer, D. J. Fray, and T. R. Meadowcroft, *Phys. Chem. Glasses*, **8**, 96 (1967).
63. T. Førland and T. Østvold, *Acta Chem. Scand.*, **22**, 2415 (1968).
64. R. Vilcu and C. Misdolea, *J. Chem. Phys.*, **46**, 906 (1967).
65. Wei-chen Lu, T. Ree, V. G. Gerrard, and H. Eyring, *J. Chem. Phys.*, **49**, 977 (1968).
66. Mu Shik Jhon, G. Clemena, and E. R. Van Artsdalen, *J. Phys. Chem.*, **72**, 4155 (1968).
67. I. G. Murgulescu and G. Vasu, *Rev. Chim. Rep. Populaire Roumaine*, **11**, 681 (1966).
68. A. C. McClellan, *J. Chem. Phys.*, **40**, 567 (1964).
69. A. N. Kirgintsev and E. G. Avvakumov, *Russ. Chem. Rev.*, **34**, 58 (1965).
70. R. Haase, *J. Phys. Chem.*, **73**, 1160 (1969).
71. G. E. Blomgren, *J. Phys. Chem.*, **66**, 1500 (1962).
72. I. D. Ryabchikov, *Russ. J. Phys. Chem.*, **41**, 168 (1967).

73. J. Braunstein, *J. Chem. Phys.*, **49**, 3508 (1968).
74. E. R. Van Artsdalen, *J. Phys. Chem.*, **60**, 172 (1956).
75. E. Kordes, W. Bergmann, and W. Vogel, *Z. Elektrochem.*, **55**, 600 (1951).
76. A. S. Dworkin and M. A. Bredig, *J. Phys. Chem.*, **64**, 269 (1960); *J. Phys. Chem.*, **67**, 697 (1963).
77. C. Sinistri and P. Franzosini, *Ric. Sci.*, **A3**, 419 (1963).
78. M. Blander, F. F. Blankenship, and R. F. Newton, *J. Phys. Chem.*, **63**, 1259 (1959).
79. S. N. Flengas and T. R. Ingraham, *Can. J. Chem.*, **35**, 1139, 1254 (1957); *Can. J. Chem.*, **36**, 780, 1103, 1662 (1958).
80. I. G. Murgulescu and S. Sternberg, *Discussions Faraday Soc.*, **32**, 107 (1961).
81. T. Førland, "Properties of Some Mixtures of Fused Salts," *Norges Tek. Vitenskapsakad. Ser.*, No. 2, **4** (1957).
82. M. Blander, *J. Chem. Phys.*, **34**, 697 (1961).
83. E. A. Guggenheim, *Mixtures*, Oxford University Press, London, 1952.
84. J. Lumsden, *Discussions Faraday Soc.*, **32**, 138 (1961).
85. H. T. Davis and S. A. Rice, *J. Chem. Phys.*, **41**, 14 (1964).
86. M. Blander, *J. Chem. Phys.*, **36**, 1092 (1962); *J. Chem. Phys.*, **37**, 172 (1962).
87. (a) L. S. Hersh and O. J. Kleppa, *J. Chem. Phys.*, **42**, 1309 (1965); (b) L. S. Hersh, O. J. Kleppa, and A. Navrotsky, *J. Chem. Phys.*, **42**, 1522 (1965).
88. Y. Nakamura, *Compt. Rend.*, **262C**, 1459 (1966); *Compt. Rend.*, **266C**, 241 (1968).
89. I. G. Murgulescu, D. I. Marchidan, and C. Telea, *Rev. Chim. Rep. Populaire Roumaine*, **11**, 1027 (1966).
90. S. V. Meschel and O. J. Kleppa, *J. Chem. Phys.*, **43**, 4160 (1965).
91. D. I. Marchidan and C. Telea, *Rev. Chim. Rep. Populaire Roumaine*, **13**, 291, 1141 (1968).
92. S. Sternberg and C. Herdlicka, *Rev. Chim. Rep. Populaire Roumaine*, **13**, 13 (1968).
93. H. T. Davis, *J. Chem. Phys.*, **41**, 2761 (1964).
94. O. J. Kleppa and F. G. McCarty, *J. Phys. Chem.*, **70**, 1249 (1966).
95. Y. Doucet and C. Vallet, *Compt. Rend.*, **261**, 2884 (1965); *J. Chim. Phys.*, **63**, 988 (1966).
96. O. J. Kleppa, *Discussions Faraday Soc.*, **32**, 99 (1961); *J. Phys. Chem.*, **66**, 1668 (1962).
97. O. J. Kleppa and S. V. Meschel, *J. Phys. Chem.*, **67**, 902 (1963).
98. J. H. Hildebrand and E. J. Salstrom, *J. Amer. Chem. Soc.*, **54**, 4257 (1932).
99. (a) T. Østvold, *Acta Chem. Scand.*, **20**, 2187, 2320 (1966); (b) G. D. Robbins, T. Førland, and T. Østvold, *Acta Chem. Scand.*, **22**, 3002 (1968).
100. T. Østvold, *Acta Chem. Scand.*, **23**, 688 (1969).
101. J. Braunstein, *J. Chem. Phys.*, **49**, 3508 (1968).
102. T. Østvold and O. J. Kleppa, *Inorg. Chem.*, **8**, 78 (1969).
103. J. L. Holm and O. J. Kleppa, *Inorg. Chem.*, **8**, 207 (1969).
104. H. Flood, T. Førland, and K. Motzfeldt, *Acta Chem. Scand.*, **6**, 257 (1962).
105. H. Flood, T. Førland, and J. M. Toguri, *Acta Chem. Scand.*, **16**, 2429 (1962); *Acta Chem. Scand.*, **17**, 1502 (1963).
106. M. Rolin and J. M. Recapet, *Bull. Soc. Chim. France*, 2504, 2511 (1964).
107. J. Guion, *J. Chim. Phys.*, **64**, 1635 (1967).
108. M. V. Smirnov and A. P. Khaimenov, *Teoria Eksp. Khim.*, **4**, 68 (1968).
109. S. V. Meschel and O. J. Kleppa, *J. Chem. Phys.*, **48**, 5146 (1968).
110. M. Blander and S. J. Yosim, *J. Chem. Phys.*, **39**, 2610 (1963).
111. H. Flood, T. Førland, and K. Grjotheim, *Z. Anorg. Allgem. Chem.*, **276**, 289 (1954).

112. M. Blander and L. E. Topol, *Electrochim. Acta*, **10**, 1161 (1965).

113. S. V. Meschel and O. J. Kleppa, *J. Chem. Phys.*, **46**, 1853 (1967).

114. S. V. Meschel, J. M. Toguri, and O. J. Kleppa, *J. Chem. Phys.*, **45**, 3075 (1966).

115. M. Blander and L. E. Topol, *Inorg. Chem.*, **5**, 1641 (1966).

116. S. Sternberg and L. Bejan, *Rev. Chim. Rep. Populaire Roumaine*, **14**, 303 (1969).

117. J. A. Plambeck, *Can. J. Chem.*, **47**, 1401 (1969).

118. N. L. Yarym Agaev, *Zh. Fiz. Khim.*, **40**, 513 (1966).

119. I. G. Murgulescu and L. Topor, *Rev. Chim. Rep. Populaire Roumaine*, **11**, 1353 (1966); *Rev. Chim. Rep. Populaire Roumaine*, **12**, 1077 (1967).

120. H. Bloom and J. W. Hastie, *Australian J. Chem.*, **21**, 583 (1968).

121. K. Hagemark, D. Hengstenburg, and M. Blander, *J. Phys. Chem.*, **71**, 1819 (1967).

122. J. Guion, D. Hengstenburg, and M. Blander, *J. Phys. Chem.*, **72**, 4620 (1968).

123. H. Bloom and J. W. Hastie, *J. Phys. Chem.*, **72**, 2361 (1968).

124. C. E. Adams and J. T. Quan, *J. Phys. Chem.*, **70**, 331 (1966).

125. E. E. Shrier and H. M. Clark, *J. Phys. Chem.*, **67**, 1259 (1963).

126. N. K. Boardman, F. H. Dorman, and E. Heymann, *J. Phys. Chem.*, **19**, 1591 (1966).

127. B. F. Powers, J. L. Katz, and O. J. Kleppa, *J. Phys. Chem.*, **66**, 103 (1966).

128. D. W. James and C. H. Liu, *J. Chem. Eng. Data*, **8**, 469 (1963).

129. A. N. Campbell and M. K. Nagarajan, *Can. J. Chem.*, **42**, 1137 (1964).

130. B. F. Markov and V. D. Prisyazhnyi, *Ukr. Khim. Zh.*, **29**, 47 (1963).

131. A. T. Ward and G. J. Janz, *Electrochim. Acta*, **10**, 849 (1965).

132. H. Spindler and F. Sauerwald, *Z. Anorg. Chem.*, **335**, 267 (1965).

133. W. J. McAuley, E. Rhodes, and A. R. Ubbelohde, *Proc. Roy. Soc. (London), Ser. A*, **289**, 151 (1966).

134. J.-M. Simon and S. Brillout, *Compt. Rend.*, **264C**, 241 (1967).

135. P. I. Protsenko and R. P. Shisholina, *Zh. Fiz. Khim.*, **41**, 71 (1967).

136. B. F. Markov, V. D. Prisyazhnyi, and G. P. Prikhodko, *Ukr. Khim. Zh.*, **34**, 126 (1968).

137. V. D. Prisyazhnyi and E. V. Zvagol'skaya, *Ukr. Khim. Zh.*, **34**, 773 (1968).

138. H. Bloom, W. D. Boyd, J. L. Laver, and J. Wong, *Australian J. Chem.*, **19**, 1591 (1966).

139. B. S. Harrap and E. Heymann, *Trans. Faraday Soc.*, **51**, 259, 268 (1955).

140. E. R. Buckle, P. E. Tsaoussoglou, and A. R. Ubbelohde, *Trans. Faraday Soc.*, **60**, 484 (1964).

141. C. R. Boston, *J. Chem. Eng. Data*, **11**, 262 (1966).

142. J. W. Tomlinson, M. S. R. Heynes, and J. O'M. Bockris, *Trans. Faraday Soc.*, **54**, 1822 (1958).

143. E. F. Riebling, *J. Amer. Ceram. Soc.*, **50**, 46 (1967).

144. A. J. Ellis, *J. Chem. Soc., Ser. A*, 1579 (1966); *J. Chem. Soc., Ser. A*, 1138 (1968).

145. J. Braunstein, L. Orr, and W. MacDonald, *J. Chem. Eng. Data*, **12**, 415 (1967).

146. G. J. Janz, C. Solomons, and H. J. Gardner, *Chem. Rev.*, **58**, 461 (1958).

147. J. P. Frame, E. Rhodes, and A. R. Ubbelohde, *Trans. Faraday Soc.*, **55**, 2039 (1959).

148. A. R. Ubbelohde, *Quart. Rev. (London)*, **4**, 356 (1950); *Quart. Rev. (London)*, **11**, 246 (1957).

149. O. J. Kleppa and F. G. McCarty, *J. Chem. Eng. Data*, **8**, 331 (1963).

150. A. S. Dworkin and M. A. Bredig, *J. Phys. Chem.*, **64**, 269 (1960).

151. F. Sauerwald and H. Schinke, *Z. Anorg. Chem.*, **304**, 25 (1960).

152. F. J. Hazlewood, E. Rhodes, and A. R. Ubbelohde, *Trans. Faraday Soc.*, **62**, 3101 (1966).

153. P. Bridgman, *Proc. Amer. Acad. Sci.*, **51**, 55 (1915).
154. C. W. F. T. Pistorius, *J. Chem. Phys.*, **43**, 2895 (1965); *J. Chem. Phys.*, **44**, 4532 (1966); *J. Chem. Phys.*, **45**, 3513 (1966); *J. Chem. Phys.*, **46**, 2167 (1967); *J. Chem. Phys.*, **47**, 4870 (1967).
155. E. Rapoport and C. W. F. T. Pistorius, *J. Chem. Phys.*, **44**, 1514 (1966).
156. E. Rapoport, *J. Phys. Chem. Solids*, **27**, 1349 (1966).
157. H. Schinke and F. Sauerwald, *Z. Anorg. Chem.*, **287**, 313 (1956).
158. S. B. Kormer, M. V. Sinitsyn, G. A. Kirillov, and V. D. Urbin, *Sov. Phys. JETP (Engl. Transl.)*, **21**, 689 (1965).
159. S. N. Vaidya and E. S. R. Gopal, *Phys. Rev. Letters*, **17**, 635 (1966).
160. E. Rapoport and G. C. Kennedy, *J. Phys. Chem. Solids*, **26**, 1995 (1965).
161. B. B. Owens, *J. Chem. Phys.*, **44**, 3918 (1966).
162. H. Bloom and J. O'M. Bockris, in *Modern Aspects of Electrochemistry*, Ed., Bockris, London, 1959, Chapter 3.
163. S. Sternberg and V. Vasilescu, *Rev. Chim. Rep. Populaire Roumaine*, **13**, 265 (1968).
164. E. L. Heric, *J. Phys. Chem.*, **69**, 2785 (1965).
165. G. W. Watson, R. B. Evans, W. R. Grimes, and N V. Smith, *J. Chem. Eng. Data*, **7**, 285 (1962).
166. J. L. Copeland and L. Seibles, *J. Phys. Chem.*, **70**, 1811 (1966); *J. Phys. Chem.*, **72**, 603 (1968).
167. W. J. Burkhard and J. D. Corbett, *J. Amer. Chem. Soc.*, **79**, 6361 (1957).
168. J. P. Frame, E. Rhodes, and A. R. Ubbelohde, *Trans. Faraday Soc.*, **57**, 1075 (1961).
169. M. Schenke, G. H. J. Broers, and J. A. A. Ketelaar, *J. Electrochem. Soc.*, **113**, 404 (1966).
170. J. H. Schaeffer, W. R. Grimes, and G. M. Watson, *J. Phys. Chem.*, **63**, 1999 (1959).
171. W. Sundermeyer, *Angew. Chem. (Engl. ed.)*, **4**, 222 (1965).
172. H. Reiss and S. W. Mayer, *J. Chem. Phys.*, **34**, 2001 (1961).
173. I. D. Sokolova and N. K. Voskresenskaya, *Russ. Chem. Rev.*, **35**, 500 (1966).
174. N. K. Bordman, A. R. Palmer, and E. Heymann, *Trans. Faraday Soc.*, **51**, 277 (1957).
175. R. Ellis, J. Smith, and E. Baker, *J. Phys. Chem.*, **62**, 766 (1958).
176. I. D. Sokolova and N. K. Voskresenskaya, *Zh. Fiz. Khim.*, **36**, 955 (1962).
177. M. A. Reshetnikov, *Izv. Sektora Fiz.-Khim. Analiza, Inst. Obshch. Neorg. Khim., Akad. Nauk SSSR*, **19**, 174 (1949).
178. I. D. Sokolova, *Zh. Neorg. Khim.*, **11**, 933 (1966).
179. I. D. Sokolova, E. L. Krivovyazov, and N. K. Voskresenskaya, *Zh. Neorg. Khim.*, **8**, 2625 (1963).
180. G. Bertozzi et al., *J. Phys. Chem.*, **68**, 2908 (1964); *J. Phys. Chem.*, **69**, 2606 (1965); *J. Phys. Chem.*, **70**, 1838 (1966).
181. H. Bloom and E. Heymann, *Proc. Roy. Soc. (London), Ser. A*, **188**, 392 (1947).
182. J. O'M. Bockris, J. A. Kitchener, S. Ignatowicz, and J. Tomlinson, *Trans. Faraday Soc.*, **48**, 76 (1952).
183. S. A. Rice, *Trans. Faraday Soc.*, **58**, 499 (1962).
184. A. Klemm, *Z. Naturforsch.*, **7A**, 417 (1957).
185. R. W. Laity, *Discussions Faraday Soc.*, **32**, 172 (1961).
186. C. A. Angell, *J. Phys. Chem.*, **68**, 218, 1917 (1964); *J. Phys. Chem.*, **69**, 399 (1965); *J. Phys. Chem.*, **70**, 2793 (1966).
187. S. A. Rice and A. R. Allnatt, *J. Chem. Phys.*, **34**, 2144 (1961).
188. C. A. Angell and J. W. Tomlinson, *Trans. Faraday Soc.*, **61**, 2312 (1965).

189. J. O'M. Bockris and G. W. Hooper, *Discussions Faraday Soc.*, **32**, 218 (1961).
190. M. K. Nagarajan and J. O'M. Bockris, *J. Phys. Chem.*, **70**, 1854 (1966).
191. C. A. Angell, *J. Chem. Phys.*, **46**, 4673 (1967).
192. G. Adams and J. H. Gibbs, *J. Chem. Phys.*, **43**, 139 (1965).
193. G. J. Janz and R. D. Reeves, *Advan. Electrochem. Electrochem. Eng.*, **5**, 137 (1967).
194. C. E. Thalmayer, S. Bruckenstein, and D. M. Gruen, *J. Inorg. Nucl. Chem.*, **26**, 347 (1964).
195. P. L. Spedding and R. Mills, *J. Electrochem. Soc.*, **112**, 594 (1965).
196. F. Lantelme and M. Chemla, *Compt. Rend.*, **267C**, 281 (1968).
197. R. W. Laity and C. A. Sjöblom, *J. Phys. Chem.*, **71**, 4157 (1967).
198. G. Harrington and B. R. Sundheim, *J. Phys. Chem.*, **62**, 1454 (1958).
199. F. R. Duke and G. Victor, *J. Electrochem. Soc.*, **110**, 91 (1963).
200. P. Digby and H. H. Kellog, *J. Electrochem. Soc.*, **111**, 1181 (1964).
201. R. J. Labrie and V. A. Lamb, *J. Electrochem. Soc.*, **110**, 810 (1963).
202. V. I. Malkin and V. V. Pokidyshev, *Dokl. Akad. Nauk SSSR*, **127**, 1253 (1959).
203. V. P. Shvedov and I. A. Ivanov, *Elektrokhim.*, **3**, 95 (1967).
204. V. Ljubimov and A. Lunden, *Z. Naturforsch.*, **21A**, 1592 (1966).
205. F. Menes, *J. Chim. Phys.*, **63**, 983 (1966).
206. I. G. Murgulescu and D. Topor, *Rev. Chim. Rep. Populaire Roumaine*, **12**, 1279 (1967); *Rev. Chim. Rep. Populaire Roumaine*, **13**, 979 (1968).
207. M. Chemla, *Rev. Chim. Minérale*, **3**, 993 (1966).
208. J. A. A. Ketelaar and C. T. Kwak, *Trans. Faraday Soc.*, **65**, 139 (1969).
209. S. Forcheri, *Euchem. Conf. Molten Salts*, Italy, 1968.
210. I. S. Yaffe and E. R. Van Artsdalen, *J. Phys. Chem.*, **60**, 1125 (1956).
211. R. Winand, *Electrochim. Acta*, **3**, 106 (1960).
212. *Electrochemistry of Molten and Solid Electrolytes*, vols. 1–4, transl. Russ. Inst. Electrochem. J., Consultants Bureau.
213. E. R. Buckle and P. Tsaoussoglou, *J. Chem. Soc.*, **132**, 667 (1964).
214. G. D. Robbins and J. Braunstein, *Symp. Molten Salts, Atlantic City, N.J.*, Amer. Chem. Soc., 1968.
215. L. A. King and F. R. Duke, *J. Electrochem. Soc.*, **111**, 712 (1964).
216. A. F. M. Barton, B. Cleaver, and G. J. Hills, *Trans. Faraday Soc.*, **64**, 208 (1968).
217. C. A. Angell, L. J. Pollard, and W. Strauss, *J. Chem. Phys.*, **50**, 2694 (1969).
218. Y. Doucet and M. Bizouard, *Compt. Rend.*, **248**, 1328 (1959).
219. B. Cleaver, E. Rhodes, and A. R. Ubbelohde, *Discussions Faraday Soc.*, **32**, 1 (1961).
220. E. Rhodes, W. E. Smith, and A. R. Ubbelohde, *Proc. Roy. Soc.* (*London*), *Ser. A.* **285**, 263 (1965).
221. L. F. Grantham and S. J. Yosim, *J. Phys. Chem.*, **67**, 2506 (1963); *J. Chem. Phys.*, **45**, 1192 (1966).
222. E. Rhodes, *Euchem. Conf. Molten Salts, Italy*, 1968.
223. B. F. Markov and L. A. Shumina, *Zh. Fiz. Khim.*, **31**, 1767 (1957).
224. I. G. Murgulescu and S. Zuca, *Rev. Chim. Rep. Populaire Roumaine*, **4**, 227 (1959).
225. S. Zuca and M. Olteanu, *Rev. Chim. Rep. Populaire Roumaine*, **13**, 1567 (1968).
226. H. Bloom and E. Heymann, *Proc. Roy. Soc.* (*London*), *Ser. A*, **188**, 392 (1947).
227. R. E. Tickle, *Phys. Chem. Glasses*, **8**, 101, 113 (1967).
228. A. Dietzel and H. J. Poegel, *Atti 3rd Cong. Int. Vetro, Venice*, 219 (1953).
229. S. Urnes, *Glastech. Ber.*, **31**, 337 (1958).
230. O. Borgen, K. Grjotheim, and S. Urnes, *Glastech. Ber.*, **33**, 52 (1960).

231. E. Rhodes, W. E. Smith, and A. R. Ubbelohde, *Trans. Faraday Soc.*, **63**, 1943 (1967).
232. J. D. MacKenzie and W. K. Murphy, *J. Chem. Phys.*, **33**, 366 (1960).
233. C. A. Angell, *J. Amer. Ceram. Soc.*, **48**, 540 (1965).
234. R. W. Higgs and T. R. Litovitz, *J. Acoust. Soc. Amer.*, **32**, 1108 (1960).
235. I. G. Murgulescu and S. Zuca, *Rev. Chim. Rep. Populaire Roumaine*, **10**, 123 (1965).
236. G. Vasu, *Rev. Chim. Rep. Populaire Roumaine*, **14**, 167 (1969).
237. J. L. Copeland and J. R. Christie, *J. Phys. Chem.*, **73**, 1205 (1969).
238. B. S. Harrap and E. Heymann, *Chem. Rev.*, **48**, 45 (1951).
239. P. I. Protsenko and O. N. Razumovskaya, *Zh. Fiz. Khim.*, **38**, 2680 (1964).
240. A. R. Ubbelohde, *Advan. Chem. Phys.*, **6**, 459 (1963).
241. A. G. Turnbull, *Australian J. Appl. Sci.*, **12**, 30, 324 (1961).
242. H. Bloom, A. Doroszkowski, and S. B. Tricklebank, *Australian J. Appl. Sci.*, **18**, 1171 (1965).
243. W. R. Gambill, *J. Chem. Eng.*, **66**, 129 (1959).
244. D. M. Gruen, *Quart. Rev. (London)*, **19**, 349 (1965).
245. E. Rhodes and A. R. Ubbelohde, *Proc. Roy. Soc. (London)*, Ser. A, **251**, 156 (1959).
246. B. Cleaver, E. Rhodes, and A. R. Ubbelohde, *Proc. Roy. Soc. (London)*, Ser. A, **276**, 437, 453 (1963).
247. J. P. Young, *Analyt. Chem.*, **36**, 390 (1964).
248. C. J. Ballhausen, *Introduction to Ligand Field Theory*, McGraw-Hill, New York, 1962.
249. C. K. Jørgensen, *Absorption Spectra and Chemical Bonding in Complexes*, Pergamon Press, London, 1961.
250. H. Lux and T. Niedermaier, *Z. Anorg. Chem.*, **285**, 246 (1956).
251. C. R. Boston and G. P. Smith, *J. Phys. Chem.*, **62**, 409 (1958).
252. D. M. Gruen and R. L. McBeth, *J. Phys. Chem.*, **63**, 393 (1959).
253. J. R. Morrey, *J. Phys. Chem.*, **66**, 2169 (1962).
254. J. Brynestad, C. R. Boston, and G. P. Smith, *J. Chem. Phys.*, **47**, 3179 (1967); *J. Chem. Phys.*, **47**, 3193 (1967).
255. J. Brynestad and G. P. Smith, *J. Chem. Phys.*, **47**, 3190 (1967).
256. C. A. Angell and D. M. Gruen, *J. Phys. Chem.*, **70**, 1601 (1966).
257. C. H. Liu, J. Hasson, and G. P. Smith, *Inorg. Chem.*, **7**, 2244 (1968).
258. J. P. Young and G. P. Smith, *J. Chem. Phys.*, **40**, 913 (1964).
259. D. M. Gruen and H. A. Øye, *Inorg. Chem.*, **3**, 836 (1964); *Inorg. Chem.*, **4**, 1173 (1965).
260. K. E. Johnson, R. Palmer, and T. S. Piper, *Spectrochim. Acta*, **21**, 1697 (1965).
261. K. E. Johnson, *Electrochim. Acta*, **11**, 129 (1966).
262. G. P. Smith, D. W. James, and C. R. Boston, *J. Chem. Phys.*, **42**, 2249 (1965).
263. J. R. Morrey, *Inorg. Chem.*, **2**, 163 (1963).
264. R. Hilsch and R. Pohl, *Z. Physik*, **59**, 812 (1930).
265. H. Fesefeldt, *Z. Physik*, **64**, 626 (1930).
266. W. Martienssen, *J. Phys. Chem. Solids*, **2**, 257 (1957).
267. T. Retschinsky, *Ann. Phys.*, **27**, 100 (1908).
268. E. Mollwo, *Z. Physik*, **124**, 118 (1947).
269. B. R. Sundheim and J. Greenberg, *J. Chem. Phys.*, **29**, 1029 (1958).
270. E. Rhodes and A. R. Ubbelohde, unpublished data.
271. G. P. Smith and C. R. Boston, *J. Chem. Phys.*, **34**, 1396 (1961).
272. C. R. Boston, D. W. James, and G. P. Smith, *J. Phys. Chem.*, **72**, 293 (1968).
273. Y. Doucet, R. Bailleux, and N. Camoin, *Compt. Rend.*, **267**, 1342 (1968).

274. W. Bues, *Z. Anorg. Chem.*, **279**, 104 (1955); *Z. Physik. Chem.* (*Frankfurt*), **10**, 1 (1957).
275. W. Bues, K. Bühler, and P. Kuhnle, *Z. Anorg. Chem.*, **325**, 8 (1968).
276. S. C. Wait and G. J. Janz, *Quart. Rev.* (*London*), **17**, 225 (1963).
277. R. M. Hexter, *Spectrochim. Acta*, **10**, 291 (1958).
278. S. C. Wait, A. T. Ward, and G. J. Janz, *J. Chem. Phys.*, **45**, 133 (1966).
279. D. W. James and W. H. Leong, *Chem. Commun.*, 1415 (1968).
280. R. E. Hester and K. Krishnan, *J. Chem. Phys.*, **46**, 3045 (1967); *J. Chem. Soc.*, A, 1955 (1968).
281. K. Williamson, P. Li, and J. P. Devlin, *J. Chem. Phys.*, **48**, 3891 (1968).
282. J. Vallier and J. Ricodeau, *Compt. Rend.*, **267B**, 890 (1968).
283. J. K. Wilmshurst, *J. Chem. Phys.*, **35**, 1800 (1961).
284. G. E. Walrafen, *J. Chem. Phys.*, **43**, 479 (1965).
285. R. E. Hester and K. Krishnan, *J. Chem. Phys.*, **48**, 825 (1968).
286. W. H. Leong and D. W. James, *Australian J. Chem.*, **22**, 499 (1969).
287. J. H. R. Clarke and C. Solomons, *J. Chem. Phys.*, **48**, 528 (1968).
288. R. B. Ellis, *J. Electrochem. Soc.*, **11**, 485 (1966).
289. J. R. Moyer, J. C. Evans, and G. Y.-S. Lo, *J. Electrochem. Soc.*, **113**, 158 (1966).
290. T. J. Rowland and J. P. Bromberg, *J. Chem. Phys.*, **29**, 626 (1958).
291. S. Hafner and N. H. Nachtrieb, *J. Chem. Phys.*, **40**, 2891 (1964); *J. Chem. Phys.*, **42**, 631 (1965); *Z. Naturforsch.*, **20a**, 321 (1965).
292. C. T. Moynihan and A. Fratiello, *J. Amer. Chem. Soc.*, **89**, 5546 (1967).
293. T. B. Swanson, *J. Phys. Chem.*, **72**, 4701 (1968).
294. M. Kukk, *Dissertation Abstr.*, **25**, 1602 (1964).
295. J. Zarzycki and F. Naudin, *Compt. Rend.*, **256**, 1282, 5344 (1963).
296. J. Zarzycki and F. Naudin, *Compt. Rend.*, **257**, 3163 (1963); *Compt. Rend.*, **258**, 1488 (1964).
297. H. Bloom and B. M. Peryer, *Australian J. Chem.*, **18**, 777 (1965).
298. Yu. I. Vesnin and S. S. Batsanov, *Zh. Strukt. Khim.*, **6**, 522 (1965).
299. C. Sinistri, P. Franzosini, and M. Rolla, *Atlas of Miscibility Gaps in Molten Salt Systems*, National Research Council, Italy, 1968.
300. N. K. Voskresenskaya, *Handbook of Solid Liquid Equilibria in Systems of Anhydrous Inorganic Salts*, trans. from Russian by J. Schmorak, Keter Press, Jerusalem, 1970.
301. G. Mamantov, Ed., *Molten Salts*, Marcel Dekker, New York, 1969.
302. D. Inman, A. D. Graves and R. S. Sethi, "*Electrochemistry of Molten Salts*" in Specialist Periodical Report *Electrochemistry Vol.*, *1* G. J. Hills, Ed., Chemical Society, London, 1970.

6 Seawater

P. Kilho Park,
Oceanography Section,
National Science Foundation,
Washington, D.C.

1 Introduction

Many great men have used the vast oceans to describe their philosophical views. For instance, Mohandas Gandhi said, "You must not lose faith in humanity. Humanity is the ocean. If a drop of the ocean water gets dirty, the ocean does not become dirty." However, man is now capable of irreversibly altering his environment. As evidenced by recent human activities, the perturbation of the ocean does not appear negligible any longer. The ever-growing human population, demanding an ever-increasing water supply, has to realize that the magnitude of the earth's hydrosphere is limited to about 1.4×10^{21} liters. Intelligent and vigorous study of natural water chemistry, which includes ocean chemistry, by elite physical chemists and other best brains of mankind is mandatory if we are to have an acceptable future. Few pioneers have appeared in this field. Some, including the late Lars Gunner Sillén of Sweden, have already gone. In this chapter, I wish to present briefly the extent of our knowledge and our ignorance about the chemical properties of seawater, with a hope that this writing may help to focus the critical eyes of our colleagues on an important scientific problem of this century.

2 History of Seawater

In the past, the oceanic condition has been stable enough to let life exist on our planet for probably 2.7 billion years. Thermally and chemically, the ocean has proven to be an effective buffer solution: it resisted rapid climatic changes by absorbing and releasing heat. Any sudden input of new chemical species by submarine volcanism, river runoffs, and other phenomena was diffused swiftly into the large volume of the ocean. Carbon-14 dating of deep-ocean water showed that the physical turnover time of the ocean is in the order of 1000 years (Broecker et al., 1961; Broecker, 1963), which, geologically speaking, is a very short time. Thus the ocean was a haven for life for a very prolonged period.

The chemical medium called seawater is a product of the various chemical processes that occur in nature. Cosmochemically, our curiosity goes far back to the genesis of hydrogen and oxygen, and their subsequent transportation into the earth's gravitational field. Our present ocean is the product of the numerous encounters among various chemical species, living and nonliving, throughout geological history. It therefore contains the keys to understanding the past history of the earth. To illustrate this point, note that recent explorations of the ocean floor have revealed that the ocean floor is younger than the ocean itself. Below the ancient seawater, the sediments and the underlying rocks are constantly renewed by ocean-floor spreading. Up to the present, we have not found any marine sediment or rocks on the seafloor that are older than 200 million years, although the ocean has probably existed for the past 3 billion years.

Chemical evolution of the ocean is difficult to ascertain. Sampling of ancient water from the past geological periods is not possible. To understand the evolution of the ocean, we have been relying on whatever indirect evidence we can gather (Brancazio and Cameron, 1964; Conway, 1942, 1943; Holland, 1971; Rubey, 1951). For example, the intimate dependency of living organisms on the chemical and physical conditions of their environments is used as a tool. If we know the mode of evolutionary adaptations of the organism under study, we can estimate ancient environmental conditions such as temperature and salinity by studying fossil organisms. Another example is the study of marine calcareous deposits among ancient sediments. This study indicates that, at the time of deposition, the ocean was probably supersaturated with calcium carbonate, which in turn indicates that some carbon dioxide existed in the ancient atmosphere.

The origin of water in the ocean has been studied by Rubey (1951) and others. There are two hypotheses now, and both may explain how the present oceanic volume is realized. The hypotheses are: (a) early releasing of water from the crustal rocks and (b) continual releasing of water throughout geological time.

The first hypothesis contends that the earth could have undergone some drastic thermal metamorphosis in its early geological time, and water vapor was released from the interior to form the ocean. Hydrated minerals can be forced to release water by heating them. The sources of this heating may have been gravitational energy or the energy released by short-lived radioactive nuclides existing in the early years of the earth's history.

The second hypothesis explains that water was released from the interior of the earth in a continual process over geological time as a result of volcanic and tectonic actions. During the crystallization of magmas, volatiles such as H_2O and CO_2 accumulate in the remaining melt and are largely expelled, via volcanic eruptions, lava flows, and hot springs, throughout geological time.

The analysis of the earth, including the ocean, as a chemical system has as its ultimate aim the development of a set of differential equations linking all the environmental variables and their time derivatives (Holland, 1971). The time integration of such a set of equations would define the excursion of the environmental variables (Weyl, 1966). This modeling requires sufficient fundamental geochemical data to develop the equations. In reality, we are just beginning and still in the stage of establishing crude geochemical reservoir modeling.

3   Chemical Composition of Seawater

Seawater composition is not too far from an aqueous solution of 0.5-M NaCl and 0.05-M $MgSO_4$. It contains other substances in lesser quantities. Major chemical species, based on the chemical model of seawater by Garrels and Thompson (1962), are 55-M H_2O followed by 0.55-M Cl^-, and others

Table 1   Major Chemical Species in Seawater at 34.3 g/kg Salinity, pH 8.1, at 25°C, and 1 Atm

Substances	Molality	Activity Coefficient	Activity
H_2O	55	0.98	54
Cl^-	0.55	0.64	0.35
Na^+	0.47	0.76	0.36
Mg^{2+}	0.047	0.36	0.017
SO_4^{2-}	0.015	0.12	0.0016
K^+	0.0099	0.64	0.0063
Ca^{2+}	0.0095	0.28	0.0027
$MgSO_4^0$	0.0060	1.13	0.0068
$NaSO_4^-$	0.0058	0.68	0.0039
HCO_3^-	0.0016	0.68	0.0011
$CaSO_4^0$	0.00084	1.13	0.00095
$MgHCO_3^+$	0.00050	0.68	0.00034
$MgCO_3$	0.00017	1.13	0.00022
KSO_4^-	0.00012	0.68	0.00008
$NaHCO_3^0$	0.0001	1.13	0.00011
$CaHCO_3^+$	0.0001	0.68	0.00007
$NaCO_3^-$	0.00005	0.68	0.00003
CO_3^{2-}	0.00002	0.20	0.000004
$CaCO_3^0$	0.00002	1.13	0.00002

Based on Garrels and Thompson, 1962.

Table 2 Secondary Constituents[a] in Seawater; Secondary Solutes ($1\ \mu M \sim 1\ \mathrm{m}M$)

Secondary Solutes	Concentration (mM)	Species
Br	0.8	Br^-
N_2	0.5	N_2 (gas)
H_3BO_3	0.4	H_3BO_3, $H_4BO_4^-$, organocomplexes
O_2	0.0–0.3	O_2 (gas)
CO_3^{2-}	0.2	$MgCO_3^0$, $NaCO_3^-$, CO_3^{2-}, $CaCO_3^0$
SiO_2	0.0–0.15	$Si(OH)_4$
Sr	0.1	Sr^{2+}, $SrSO_4^0$
F	0.07	F^-
Nitrate	0.00–0.04	NO_3^-, (NO_2^-, NH_4^+)
Al	0.04	$Al(OH)_3$
Li	0.024	Li^+
Ar	0.01	Ar (gas)
CO_2	0.01	CO_2 (gas)
Phosphate	0.00–0.003	HPO_4^{2-}, PO_4^{3-}, $H_2PO_4^-$
Rb	0.0014	Rb^+
OH^-	0.001	OH^-

[a] Note that the major dissolved gases in seawater (N_2, O_2, Ar, CO_2) belong to this group.

Table 3 Microcomponents in Seawater

Elements	Concentration (μM)	Probable Species
I	0.5	IO_3^-, I^-
Fe	0.2	$Fe(OH)_3$
Ba	0.2	Ba^{2+}
In	<0.2	
Zn	0.15	Zn^{2+}
Mo	0.1	MoO_4^{2-}
Organic C	0.1	Various kinds of organic matter
Cu	0.05	Cu^{2+}, $Cu(OH)^+$
V	0.04	$VO_2(OH)_3^{2-}$
Mn	0.04	$Mn(OH)_4$
As	0.04	$HAsO_4^{2-}$
Ni	0.03	Ni^{2+}
Ti	0.02	
H^+	0.01	$(H_3O)^+$
U	0.01	
Sn	0.007	
Ne	0.005	Ne (gas)
Se	0.005	SeO_4^{2-}

Table 3 (cont.)

Elements	Concentration (μM)	Probable Species
Kr	0.004	Kr (gas)
Sb	0.004	
Cs	0.004	Cs^+
Y	0.003	
Co	0.002	Co^{2+}
He	0.001	He (gas)
Sc	0.001	
Cr	0.001	
Cd	0.001	
Ge	0.0008	$Ge(OH)_4$
Xe	0.0008	Xe (gas)
W	0.0005	WO_4^{2-}
Ga	0.0004	
Ag	0.0004	$AgCl_3^{2-}$
Th	0.0002	
Nb	0.0001	
Hg	0.0001	$HgCl_4^{2-}$
Pb	0.0001	Pb^{2+}, $PbOH^+$, $PbCl^+$
Bi	0.0001	
La	0.00009	La^{3+}
Be	0.00007	
Nd	0.00006	
Tl	<0.00005	
Ce	0.00004	
Pr	0.00002	
Gd	0.00002	
Dy	0.00002	
Au	<0.00002	$AuCl_2^-$
Sm	0.00001	
Er	0.00001	
Yb	0.00001	
Ho	0.000005	
Eu	0.000003	
Tm	0.000003	
Lu	0.000003	
Pa	10^{-8}	
Ra	4×10^{-10}	
Rn	3×10^{-15}	

(Table 1). Horne (1970) proposes an imaginative classification of the constituents of seawater, dividing the constituents by the categories of solutes, colloids, and heterogeneities, and by the fundamental difference in the nature of the water envelopes surrounding the substances.

The variability of concentrations of minor constituents of seawater is appreciable where these substances are involved in the biogeochemical processes occurring in the ocean. Their approximate concentrations and probable speciations are shown in Tables 2 and 3.

Since the ocean is well mixed, the ratios of major constituents in seawater are similar (Table 4). The constancy of composition of seawater among major constituents helped oceanographers to express the total salt content of seawater by chlorinity, which is an approximate measure of the chloride concentration in seawater. Chlorinity is measured by the silver nitrate titration. The conversion of chlorinity into salinity is carried out by the following empirical equation

(1) $$\text{salinity} = 1.80655 \times \text{chlorinity g/kg}$$

Beginning in 1960, oceanographers adopted the use of electrolytic conductance measurement to estimate the salinity of seawater (Park and Burt, 1965).

Table 4   Major Solute–Chlorinity Ratios in Seawater

Substance	Mole/Liter	$\dfrac{\text{Solute}}{\text{Chlorinity}}$ (g/g)	Species
H_2O	(54.9)	—	$(H_2O)_n$ $n = 1, 2, 4, 8, \cdots$
Cl^a	0.548	1.00	Cl^-
Na^a	0.470	0.5555 ± 0.0007	Na^+ (99%); $NaSO_4^-$ (1%)
Mg^a	0.0536	0.06692 ± 0.00004	Mg^{2+} (87%); $MgSO_4^0$ (11%); $MgHCO_3^+$ (1%)
SO_4^a	0.0282	0.139 ± 0.001	SO_4^{2-} (54%); $MgSO_4^0$ (22%); $NaSO_4^-$ (21%); $CaSO_4^0$ (3%); KSO_4^- (0.5%)
Ca^a	0.0102	0.02126 ± 0.00004	Ca^{2+} (84%); $CaSO_4^0$, $CaHCO_3^+$, $CaCO_3^0$
K^a	0.0100	0.0206 ± 0.0002	K^+ (99%); KSO_4^- (1%)
HCO_3^a	0.0023	pH dependent	HCO_3^- (69%); $MgHCO_3^+$ (19%); $NaHCO_3^0$ (4%); $CaHCO_3^+$ (4%)
Br	0.0008	0.0035	Br^-
H_3BO_3	0.0004	0.00023 ± 0.00001	$H_3BO_3^0$; $H_4BO_4^-$ (pH dependent)
Sr	0.0001	0.00040 ± 0.00002	Sr^{2+}, $SrSO_4^0$

From Culkin, 1965, and Culkin and Cox, 1966; chemical species data from Garrels and Thompson, 1962, at pH 8.1.
[a] Primary solutes.

4   The Oceanic Environment

Three common major variables used by physical chemists to study the ocean chemistry are temperature, pressure, and ionic strength (salinity). The temperature variation is considerable for near-surface seawater; it varies with place and time and has a typical range of -2 to $30°C$. Annual variations of up to $18°C$ are reported from the northwestern parts of the Atlantic and Pacific. Vertical distribution of temperature shows that the deep ocean possesses a relatively constant temperature of between 0 to $3°C$ below a depth of 3000 m. The average temperature of the entire ocean seems to be about $5°C$.

Hydrostatic pressure increases with increasing depth at an approximate rate of 1 atm/10 m. Since some deep trenches are close to 10,000 m deep, the oceanic pressure range is between 1 atm at the sea surface to 1000 atm at the deepest. When we consider the average oceanic depth of 4000 m, an average pressure of the entire ocean becomes about 200 atm.

A crude way to convert the salinity of seawater to ionic strength, μ, is by

$$(2) \qquad\qquad \mu = 0.02 \times \text{salinity}$$

The refinement of this relationship is under way with the increase of our knowledge about the chemical speciation of seawater. Since an average salinity of seawater is 35 g/kg, an average ionic strength of seawater is about 0.7.   In the ocean, salinity is a function of space and time, a result of different physical processes in the ocean. Salinity increases when seawater evaporates or freezes, and it decreases with precipitation, river runoffs, and sea-ice melting. In general, the surface-salinity range is within 32 to 37 g/kg, a comparable maximal ionic strength change of 0.1. In the deep ocean salinity is essentially constant at near 34.7 g/kg.

The fourth oceanic variable, which is becoming very important, is the population of marine organisms (Redfield, 1959). Since their material is drawn mostly from the water, certain generalizations can be made regarding the interchanges of chemical substances between seawater and the population it supports. Because marine life may move through the water independent of advection and mixing, it provides a mechanism for producing diversity of seawater chemical composition.

The basic generalization starts from the average elementary composition of marine life. Fleming (1940) gave this as $C:N:P = 106:16:1$ at. In addition, it is estimated that 138 molecules of oxygen are required for the complete oxidation of carbon and nitrogen associated with each atom of phosphorus. Since the organic matter is formed from seawater, these ratios provide a stoichiometric relation between the chemical compositions of marine life and seawater. A stoichiometric biochemical oxidation model in seawater, prepared by Richards (1965), is:

$$
(3) \quad (CH_2O)_{106}(NH_3)_{16}H_3PO_4 + 138O_2
$$
$$
= 106CO_2 + 122H_2O + 16HNO_3 + H_3PO_4
$$

where $(CH_2)_{106}(NH_3)_{16}H_3PO_4$ is a hypothetical organic molecule of marine life. Equation 3 is useful for estimating the potential fertility of seawater at a given time and place (Park, 1967) and for understanding the mechanism of biochemical control of pH in seawater (Park, 1968a, b).

5 Properties of Seawater

Two recent review articles essentially summarize the state of our knowledge on the chemical properties of seawater. Pytkowicz and Kester (1971) exhaustively reviewed what we know on matters including seawater composition, structure, thermodynamic properties, chemical equilibria, gas solubilities, rates of reactions, colligative properties, surface properties,

Table 5   Comparison of Pure Water and Seawater Properties

Property	Seawater (35 g/kg salinity)	Aqueous NaCl Solution (0.5 M)	Aqueous NaCl Solution (0.6 M)	Pure Water
Density (g/cm³ at 25°C)	1.02412	1.01752	1.02172	1.0029
Equivalent conductivity at 25°C [cm²/(Ω)(eq)]	—	93.62	91.58	—
Specific conductivity at 25°C (Ω⁻¹ cm⁻¹)	0.0532	0.0468	0.0458	—
Viscosity at 25°C (mp)	9.02	9.32	9.41	8.90
Vapor pressure (mm Hg at 20°C)	17.4	17.27	17.18	17.34
Isothermal compressibility at 0°C (unit vol/atm)	46.4×10^{-6}	46.6×10^{-6}	45.9×10^{-6}	50.3×10^{-6}
Temperature of maximum density (°C)	−3.52	—	—	+3.98
Freezing point (°C)	−1.91	−1.72	−2.04	0.00
Surface tension at 25°C (dynes/cm)	72.74	72.79	72.95	71.97
Velocity of sound at 0°C (m/sec)	1450	—	—	1407
Specific heat at 17.5°C [J/(g)(°C)]	3.898	4.019	3.998	4.182

From R. A. Horne, 1969.

electrolytic conductance, and dielectric constant. Horne (1970) prepared his review from a systems analysis approach, starting with the ocean as a major feature of the earth, enhancing the role of seawater in biogenesis as well as the origin of the ocean. In addition, extensive tables of seawater properties are listed in Horne's book, *Marine Chemistry* (1969). A brief review of the chemical properties of seawater is presented here.

Thermodynamic Properties. Reliable thermodynamic properties of sea-water for *in situ* conditions, low temperature and high pressure, are lacking. This is partially because the ionic strength of seawater, 0.7, is too concentrated to be applicable to the elaborate theories that have been developed for very dilute solutions. Furthermore, previous work has been carried out mainly under 1 atm, which applies only to the sea surface, not to the great depths of the ocean. In addition, the complexity of seawater composition (Table 1) forced previous workers to treat seawater as a simple two-component system—water (solvent) and sea salt (solute) (Fofonoff, 1962). For rough approximation, representing seawater by 0.5-M aqueous sodium chloride solution appears permissible in certain cases, as shown in Table 5. Montgomery (1957) compared the transport phenomena between pure water and 35 g/kg salinity seawater (Table 6).

Density of seawater is estimated as a function of salinity, pressure, and temperature on the basis of empirical equations. Wilson and Bradley (1966)

Table 6   Transport Phenomena in Water at 1 Atm

Name, Symbol, Units	Pure Water		Seawater (Salinity 35 g/kg)	
	0°C	20°C	0°C	20°C
Dynamic viscosity, η [g/(cm)(sec)] = poise	0.01787	0.01002	0.01877	0.01075
Thermal conductivity, κ [W/(cm)(°C)]	0.00566	0.00599	0.00563	0.00596
Kinematic viscosity, $\nu = \eta/\rho$ (cm²/sec)	0.01787	0.01004	0.01826	0.01049
Thermal diffusivity, $\kappa = \kappa/c_{p\rho}$ (cm²/sec)	0.00134	0.00143	0.00139	0.00149
Diffusivity, D (cm²/sec)				
NaCl	0.0000074	0.0000141	0.0000068	0.0000129
N₂	0.0000106	0.0000169		
O₂		0.000021		
Prandtl number, $N_P = \nu/\kappa$	13.3	7.0	13.1	7.0

From R. B. Montgomery, 1957.

obtained very careful measurements of the compressibility (Figure 1) and coefficients of thermal expansion (Figures 2 and 3) and prepared the tables for specific volume of seawater as a function of salinity, pressure, and temperature.

An equation of state for seawater has been proposed by Li (1967), based on the Tait–Gibson equation.

Among the colligative properties of seawater, direct measurements have been made for freezing-point lowering and vapor-pressure lowering. Both osmotic pressure and boiling point elevations have been calculated from these two measured properties (Fabuss and Korosi, 1966; Stoughton and Lietzke, 1967). The salinity dependency of these properties is shown in Table 7. One of the interesting oceanic phenomena that deal with the colligative properties of seawater is the osmoregulation of marine life. The cell walls of the organisms consist mostly of semipermeable membranes. Whenever salinity changes occur, these organisms must regulate the internal pressure quickly if they are to survive. Biogeography of the ocean is regulated, in part, by the organisms' osmoregulatory ability.

In the ocean, in addition to the macroscopic water mass movements, the transport processes include conduction of heat, diffusion of dissolved salts and gases, and rates of exchange of water with the vapor and solid phases (Fofonoff, 1962). The investigations of the various movements of solvent and solutes have been valuable in increasing our understanding of the structure

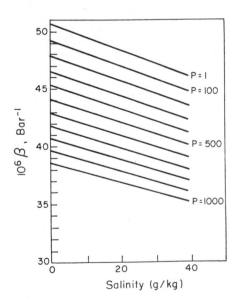

Figure 1. The compressibilty of seawater at 0°C. From Wilson and Bradley (1966).

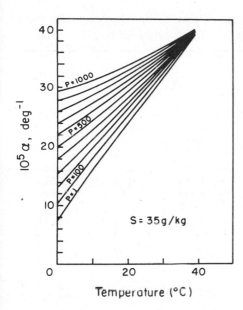

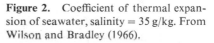

Figure 2. Coefficient of thermal expansion of seawater, salinity = 35 g/kg. From Wilson and Bradley (1966).

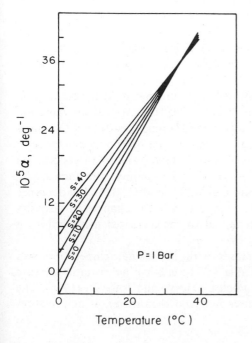

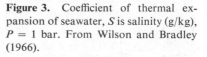

Figure 3. Coefficient of thermal expansion of seawater, S is salinity (g/kg), $P = 1$ bar. From Wilson and Bradley (1966).

255

Table 7   Osmotic Pressure, P, Freezing Point, v, Boiling-Point Elevation, Δt_s, and Vapor-Pressure Lowering, b, as a Function of Salinity

Salinity (g/kg)	P_0 (atm)	P_{20} (atm)	v (°C)	Δt_s (°C)	b (mm Hg)
4	2.59	2.78	−0.214	0.06	1.69
8	5.16	5.54	−0.427	0.12	3.39
12	7.73	8.30	−0.640	0.19	5.08
16	10.34	11.10	−0.856	0.25	6.82
20	12.97	13.92	−1.074	0.31	8.55
24	15.63	16.78	−1.294	0.38	10.30
28	18.31	16.65	−1.516	0.44	12.13
32	21.02	22.56	−1.740	0.51	13.97
36	23.76	25.50	−1.967	0.57	15.79
40	26.53	28.48	−2.196	0.64	17.73

From G. Dietrich, 1963.

of seawater. Horne (1969, pp. 88–125) summarized what we know about the seawater transport properties, covering dielectric relaxation, diffusion, viscosity, and electrolytic conductivity.

Oceanographers use electrolytic conductivity to estimate the salinity and density of seawater. Since the conductivity of a solution is a solute property, it provides information about the electrolyte species and about its movement. Conversely, viscosity is a property that is useful to study the solvent structure. Although seawater is not a very dilute solution, Walden's rule is applicable in first approximation (Horne and Courant, 1964). Recent review articles on the electrolytic conductivity of seawater include those of Cox (1965) and Park and Burt (1965, 1966).

The specific conductance of seawater as a function of temperature and salinity was determined by Thomas et al. (1934) (Figure 4). New measurements of the conductance are under way at the National Institute of Oceanography in England. The effect of hydrostatic pressure on the conductance of seawater was measured by Horne and Frysinger (1963) and by Bradshaw and Schleicher (1965). For typical seawater, 35 g/kg salinity, the increase in conductance by 1000 bars of pressure is 12% at 0°C. This conductance increase is greater than what one can expect from the changes in viscosity and volume, and it is partially attributed to the increased dissociation of magnesium sulfate.

The viscosity of seawater decreases with increasing temperature and increases with increasing salinity (Table 8). The effect of hydrostatic pressure on the viscosity of seawater was studied by Horne and Johnson (1966). The relative viscosity, $N_P/N_{1\,atm}$, decreases first and then increases at pressures higher than 1000 kg/cm^2 (Figure 5).

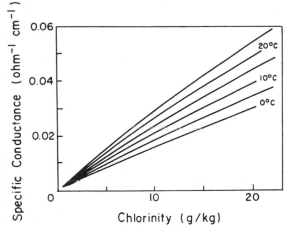

Figure 4.   Specific conductance of seawater. From Thomas et al. (1934).

Structure of Water in Seawater.   Most of our knowledge on the structure of water in seawater has been reviewed by Horne (1969, 1970). Since the structure of pure water is a highly controversial subject, a satisfactory structural model of water in seawater cannot be given at present. Horne's (1970) personal preference has been that of "flickering cluster" theory proposed by Frank and Wen (1957) for the water structure, with an average cluster containing about 40 water molecules (Némethy and Scheraga, 1962).

Table 8   Relative Viscosity of Seawater at 1 Atm[a]

Temperature (°C)	Salinity (g/kg)				
	5	10	20	30	40
0	1.009	1.017	1.032	1.056	1.054
5	0.855	0.863	0.877	0.891	0.905
10	0.783	0.745	0.785	0.772	0.785
15	0.643	0.649	0.662	0.675	0.688
20	0.568	0.574	0.586	0.599	0.611
25	0.504	0.510	0.521	0.533	0.545
30	0.454	0.460	0.470	0.481	0.491

From N. E. Dorsey, 1940.

[a] η/η_0, where η_0 is the viscosity of pure water at 0°C (1.787 cp).

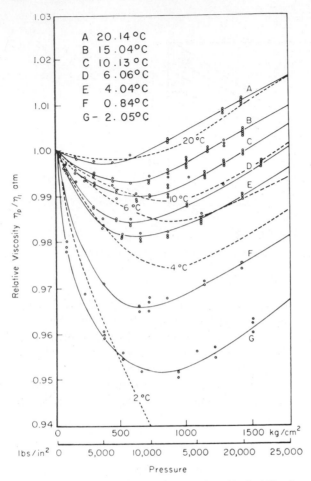

Figure 5. Relative viscosity of pure water (dashed lines) and 35-g/kg salinity seawater (solid curves). Temperatures (°C): curve *A*, 20. 14; curve *B*, 15.04; curve *C*, 10.13; curve *D*, 6.06; curve *E*, 4.04; curve *F* 0.84; curve *G*, 2.05. From Horne and Johnson (1966).

When a solute is added, a hydration atmosphere is formed near the solute. Horne (1970) suggests that at least two different types of hydration atmosphere exist. They are hydrophobic hydration for nonpolar solute and Coulombic hydration for polar or ionic solutes. The complexity of the Coulombic hydration sphere is illustrated by Horne (1970), using Na^+ as an example (Figure 6). Near the Na^+ ion, four primary hydrated water molecules exist (point **A**, Figure 6), which in turn are supported by Frank–Wen cluster (point **B**). Broken water structure surrounds the cluster (point **C**,

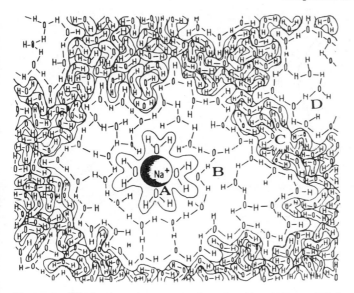

Figure 6. Water structure near a sodium ion. From Horne (1970).

Figure 6). A neighboring ion forms a structured water pattern at some distance (point **D**, Figure 6). At 20°C, the number of water molecules structured by the presence of the Na^+ ion is about 40. The nature of the hydration spheres of anions is little understood.

The effect of temperature on the Coulombic hydration atmospheres of Na^+ and Cl^- in seawater appear to be mild (Horne, 1970).

The probable effect of hydrostatic pressure on the structure of seawater is summarized by Horne (1969). When a pressure of about 1000 atm is reached, all the structured envelopes are dissolved, so that viscosity shows a minimum at that pressure. Beyond that, viscosity of seawater increases as unassociated liquids do. At a pressure of 5000 kg/cm^2, perhaps even the primary hydration sphere may be destroyed (Figure 7).

6 Recent Advances in Marine Chemistry

Two major contributions in recent years to the understanding of the chemical equilibrium processes occurring in the ocean were by Sillén (1961) and by Garrels and Thompson (1962). Sillén's approach was unique, for he considered the effects of both atmosphere and lithosphere (sediments) in his model. He started with one liter of water and gradually added atmosphere (3 liters), sediments (0.6 kg), and solutes. Although some objections to his equilibrium model still exist, Sillén (1961) demonstrated that there is a

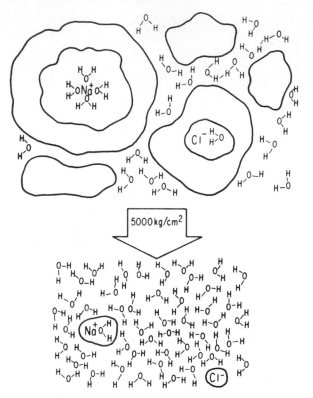

Figure 7. The effect of hydrostatic pressure on the structure of seawater. From Horne (1969).

Table 9 Species Distribution of Some Major Seawater Constituents

Ion	Molality	Free Ion (%)	With Sulfate (%)	With Biocarbonate (%)	With Carbonate (%)
Ca^{2+}	0.0104	91	8	1	0.2
Mg^{2+}	0.0540	87	11	1	0.3
Na^+	0.4752	99	1.2	0.01	—
K^+	0.0100	99	1	—	—

		With Calcium (%)	With Magnesium (%)	With Sodium (%)	With Potassium (%)
SO_4^{2-}	0.0284	54	3	21.5	21	0.5
HCO_3^-	0.00238	69	4	19	8	—
CO_3^{2-}	0.000269	9	7	67	17	—

From R. M. Garrels and M. E. Thompson, 1962.

considerable resemblance between his equilibrium model and the real system. His work has stimulated others to refine and refute his modeling by obtaining new knowledge on various equilibria in liquid and solid solutions. For instance, the reactivity of silicates with seawater has been shown to be considerable and fast (Garrels, 1965; Mackenzie and Garrels, 1965).

Major chemical species that are present in seawater, by a consideration of ion-pair formation, was estimated by Garrels and Thompson (1962) (Table 9). Their work has been continually updated by others (Thompson and Ross, 1966; Kester and Pytkowicz, 1969; Lafon, 1969).

7 pH Value of Seawater

What keeps the pH value of the present ocean at near 8 has been a topic of discussion in recent years. Sillén (1961), Garrels (1965), and Mackenzie and Garrels (1965) stressed that the reactions between silicate minerals and sea-water are of great importance in maintaining the seawater pH near 8. Before Sillén's proposal, the carbon dioxide system of the ocean was considered to control its pH value (Harvey, 1960).

The silicate theory contends that enough geological time, probably 3 billion years, has elapsed for seawater to interact with silicate minerals. Recent studies in oceanic tectonics indicate that the ocean floor is vigorously spreading in such a way that no existing marine sediments are older than 200 million years. Thus exposure of seawater to silicate minerals is facilitated not only by river runoffs but also by ocean-floor spreading. Sillén (1961) used the following as a schematic example of the pH-dependent silicate mineral equilibria

$$
(4) \quad 3Al_2Si_2O_5(OH)_4 + 4SiO_2 + 2K^+ + 2Ca^{2+} + 9H_2O \\
= 2KCaAl_3Si_5O_{16}(H_2O)_6 + 6H^+
$$

$$
(5) \qquad \log K = 6 \log [H^+] - 2 \log [K^+] - 2 \log [Ca^{2+}]
$$

When we know enough about the rates, routes, and reservoirs of the silicate mineral interactions, we may more accurately assess the silicate-buffer theory.

Later Pytkowicz (1967) pointed out that within the mixing time of seawater, in the order of 1000 years, the carbon dioxide buffer system, rather than the silicate system, is the primary buffering agent in the ocean.

Apart from the long-range, large-scale pH governing system, fine pH distribution in the present ocean, horizontally and vertically, can only be explained by the carbon dioxide system relationship. By simple mathematical analysis, I have shown that the apparent oxygen utilization by marine organisms is the major factor controlling the vertical distribution of pH in the northeastern Pacific Ocean, followed by a lesser contribution by carbonate mineral dissolution in the deep ocean (Park, 1968a, b).

8   Concluding Remarks

In this chapter, my attempt has been to give the reader a brief résumé of what we know about the important environmental solution called seawater. Since the ocean is a major entity in our environment, its study is no longer a mere scientific curiosity. The sooner we know more about the important aspects of marine chemistry, the more we scientists can contribute to social welfare. The sea has never been stoical; it is a part of the interwoven total environment that enhances atmosphere, lithosphere, hydrosphere, biosphere, and anthroposphere. We should increase, vigorously, our research efforts in marine chemistry.

Acknowledgments

I am grateful for the assistance and encouragement of R. A. Horne, A. C. Neumann, and M. M. Howell during the preparation of this chapter.

REFERENCES

Brancazio, P. J., and A. G. W. Cameron, Eds., *The Origin and Evolution of Atmospheres and Oceans*, Wiley, New York, 1964.

Broecker, W. S., R. D. Gerard, M. Ewing, and B. C. Heezen, "Geochemistry and Physics of Ocean Circulation," in *Oceanography*, Ed., M. Sears, Amer. Assoc. Advan. Sci., Publ. No. 67, Washington, D.C., 1961, pp. 301–322.

Broecker, W. S., "Radioisotopes and Large-Scale Oceanic Mixing," in *The Sea*, Ed., M. N. Hill, vol. 2, Wiley-Interscience, New York, 1963, pp. 88–108.

Bradshaw, A., and K. E. Schliecher, "The Effect of Pressure on the Electrical Conductance of Seawater," *Deep-Sea Res.*, **12**, 151–162 (1965).

Conway, E. J., "Mean Geochemical Data in Relation to Oceanic Evolution," *Proc. Roy. Irish Acad.*, **48B**, 8, 119–159 (1942).

Conway, E. J., "The Chemical Evolution of the Ocean," *Proc. Roy. Irish Acad.*, **48B**, 9, 161–212 (1943).

Cox, R. A., "The Physical Properties of Sea Water," in *Chemical Oceanography*, Eds., J. P. Riley and G. Skirrow, vol. 1, Academic Press, London, 1965, pp. 73–120.

Culkin, F., "The Major Constituents of Sea Water," in *Chemical Oceanography*, Eds., J. P. Riley and G. Skirrow, vol. 1, Academic Press, London, 1965, pp. 121–161.

Culkin, F., and R. A. Cox, "Sodium, Potassium, Magnesium, Calcium and Strontium in Seawater," *Deep-Sea Res.*, **13**, 789–804 (1966).

Dietrich, G., *General Oceanography*, Wiley, New York, 1963.

Dorsey, N. E., *Properties of Ordinary Water-Substance*, Reinhold, New York, 1940.

Fabuss, B. M., and A. Korosi, "Boiling Point Elevation of Sea Water and Its Concentrates," *J. Chem. Eng. Data*, 11, 606–609 (1966).

Fleming, R. H., "The Composition of Plankton and Units for Reporting Populations and Production," *Proc. 6th Pacific Sci. Cong.*, 3, 535–540 (1940).

Fofonoff, N. P., "Physical Properties of Sea-water," in *The Sea*, Ed., M. N. Hill, vol. 1, Wiley, New York, 1962, pp. 3–30.

Frank, H. S., and W.-Y. Wen, "Structural Aspects of Ion–Solvent Interaction in Aqueous Solutions," *Discussions Faraday Soc.*, 24, 133–140 (1957).

Garrels, R. M., "Silica: Role in Buffering Natural Waters," *Science*, 148, 69 (1965).

Garrels, R. M., and M. E. Thompson, "A Chemical Model for Sea Water," *Amer. J. Sci.*, 260, 57–66 (1962).

Harvey, H. W., *The Chemistry and Fertility of Sea Waters*, Cambridge University Press, London, 1960.

Holland, H. D. *The Chemistry of the Atmosphere and Oceans, Past and Present*, in preparation (1971).

Horne, R. A., *Marine Chemistry*, Wiley, New York, 1969.

Horne, R. A., "Sea Water," in *Advan. Hydrosci.*, 6, 107–140, Academic Press, New York, 1970.

Horne, R. A., and R. A. Courant, "Application of Walden's Rule to the Electrical Conduction of Sea Water," *J. Geophys. Res.*, 69, 1971–1977 (1964).

Horne, R. A., and G. R. Frysinger, "The Effect of Pressure on the Electrical Conductivity of Sea Water," *J. Geophys. Res.*, 68, 1967–1973 (1963).

Horne, R. A., and D. S. Johnson, "The Viscosity of Compressed Seawater," *J. Geophys. Res.*, 71, 5275–5277 (1966).

Kester, D. R., and R. M. Pytkowicz, "Sodium, Magnesium, and Calcium Sulfate Ion-Pairs in Seawater at 25°C," *Limnol. Oceanogr.*, 14, 686–692 (1969).

Lafon, G. M., "Some Quantitative Aspects of the Chemical Evolution of the Oceans," Ph.D. dissertation, Northwestern University, 137 pp., 1969.

Li, Y. H., "Equation of the State of Water and Sea Water," *J. Geophys. Res.*, 72, 2665–2678 (1967).

Mackenzie, F. T., and R. M. Garrels, "Silicates: Reactivity with Sea Water," *Science*, 150, 57–58 (1965).

Montgomery, R. B., "Oceanographic Data," in *American Institute of Physics Handbook* Sec. 2, "Mechanics," 115–124, McGraw-Hill, New York, 1957.

Nemethy, G., and H. A. Scheraga, "A Model for the Thermodynamic Properties of Liquid Water," *J. Chem. Phys.*, 36, 3382–3400 (1962).

Park, K., "Nutrient Regeneration and Preformed Nutrients off Oregon," *Limnol. Oceanogr.*, 12, 353–357 (1967).

Park, K., "Seawater Hydrogen-Ion Concentration: Vertical Distribution," *Science*, 162, 357–358 (1968a).

Park, K., "The Processes Contributing to the Vertical Distribution of Apparent pH in the Northeastern Pacific Ocean," *J. Oceanol. Soc. Korea*, 3, 1–7 (1968b).

Park, K., and W. V. Burt, "Electrolytic Conductance of Sea Water and the Salinometer," *J. Oceanogr. Soc. Japan*, 21, 69–80, 124–132 (1965).

Park, K., and W. V. Burt, "Electrolytic Conductance of Sea Water and the Salinometer: An Addendum to the Review," *J. Oceanogr. Soc. Japan*, 22, 25–28 (1966).

Pytkowicz, R. M., "Carbonate Cycle and the Buffer Mechanism of Recent Oceans," *Geochim. Cosmochim. Acta*, 31, 63–73 (1967).

Pytkowicz, R. M., and D. R. Kester, "The Physical Chemistry of Sea Water," in *Oceanogr. Mar. Biol. Ann. Rev.*, Ed., H. Barnes, George Allen and Unwin, London, in press, 1971.

Redfield, A. C., "Biological Consideration—The Fourth Phase," in *Physical and Chemical Properties of Sea Water*, 101–112, Natl. Acad. Sci.–Natl. Res. Council, Publ. 600, Washington, D.C., 1959.

Richards, F. A., "Anoxic Basins and Fjords," in *Chemical Oceanography*, Eds., J. P. Riley and G. Skirrow, vol. 1, 611–645, Academic Press, London, 1965.

Rubey, W. W., "Geologic History of Sea Water," *Bull. Geol. Soc. Amer.*, **62**, 1111–1147 (1951).

Sillén, G., "The Physical Chemistry of Sea Water," in *Oceanography*, Ed., M. Sears, Ed.), 549–581, Amer. Assoc. Advan. Sci., Publ. No. 67, Washington, D.C., (1961).

Stoughton, R. W., and M. H. Lietzke, "Thermodynamic Properties of Sea Salt Solutions," *J. Chem. Eng. Data*, **12**, 101–104 (1967).

Thomas, B. D., T. G. Thompson, and C. L. Utterback, "The Electrical Conductivity of Sea Water," *J. Conseil. Perm. Intern. Exploration Mer*, **9**, 28–35 (1934).

Thompson, M. E., and J. Ross, "Calcium in Sea Water by Electrode Measurement," *Science*, **154**, 1643–1644 (1966).

Weyl, P. K., "Environmental Stability of the Earth's Surface; Chemical Conditions," *Geochim. Cosmochim. Acta*, **30**, 663–679 (1966).

Wilson, W., and D. Bradley, "Specific Volume, Thermal Expansion and Isothermal Compressibility of Sea-Water," U.S. Naval Ord. Lab. Rept. No. NOL TR-66-103. AD-635-120 (1966).

7 Biofluids

D. A. T. Dick,
Department of Anatomy,
University of Dundee,
Dundee, Scotland

1 Introduction

Human blood plasma contains 141 mM/liter of Na^+ and 4.1 mM/liter of K^+ ions. The range of these values in normal individuals is small (136–147 mM/liter and 3.4–4.8 mM/liter for Na^+ and K^+, respectively). The concentrations of most other plasma constituents also lie within narrow limits (Altman, 1961). Plasma is thus not merely an arbitrary mixture but a well-defined fluid of constant composition. This characteristic of constancy is typical of biofluids in general and is indeed what enables us to speak meaningfully of the individual fluids and sometimes to identify them precisely, away from their normal site in a living organism.

Two questions immediately arise from this specific composition of human plasma: (a) why does the human body select these particular concentrations of ions in preference to others? and (b) how does the body maintain these concentrations within such narrow limits in spite of large variations in dietary intake and environmental conditions? As we shall see, answers to the first question are still speculative; the answer to the second question is that the body possesses intricate mechanisms for preserving constancy or, as it is called, *homeostasis*. For example, when we are too warm, we sweat and evaporation cools us; when we are too cold, we shiver and this muscular activity warms us. So far as animal body fluids are concerned, the main homeostatic mechanism determining their composition lies in the kidney.

The homeostatic abilities of organisms vary very widely—for many marine organisms such mechanisms are unnecessary, since the environment is virtually unchanging. The range of properties that are regulated also varies widely. Thus although many lower organisms keep constant the ionic composition, pH, and osmotic pressure of their internal fluids, only birds and

265

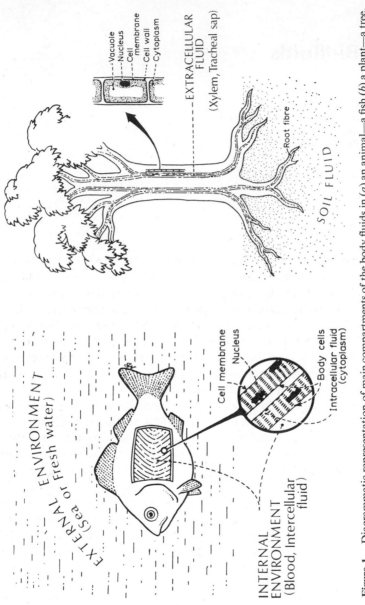

Figure 1. Diagrammatic representation of main compartments of the body fluids in (*a*) an animal—a fish (*b*) a plant—a tree.

mammals are able to maintain constant temperature against environmental fluctuations. Homeostatic mechanisms give rise to what was called by Claude Bernard the *milieu interne*, that is, the extracellular fluid in which the cells of the animal body actually live, whose properties are maintained constant by homeostatic mechanisms for the benefit of the individual cells. Small changes in the extracellular fluid cause malfunction or even death of the body cells.

The biofluids may be divided into two categories (Figures 1*a* and *b*). The first comprises the extracellular fluids, which are isolated from the external environment by skin or other cellular covering but do not form part of the cells themselves; these extracellular fluids are easy to sample because they are quite readily accessible, and thus their physical and chemical properties have been comprehensively analyzed. Extracellular fluids include both the blood and the tissue fluid that lies between the cells. The second category comprises the intracellular fluids, which lie within the phospholipid cell membrane or plasmalemma which bounds every living cell; it is very difficult to sample these without disturbing the functioning of the cell, so it is not surprising that even such elementary properties as intracellular osmotic pressure and pH are still the subject of controversy. Nevertheless, since the intracellular fluids are the actual site of most vital processes, they are even more important than the extracellular fluids; indeed the latter may be regarded as existing only to preserve the former.

In addition to constancy in relation to time, biofluids exhibit a remarkable similarity between different members of the same group of organisms. Although organisms are classified on morphological grounds, it frequently happens that the body fluids of members of similar morphological groups are themselves similar. For example, the plasma of most amphibia has roughly the same sodium chloride concentration, 100 meq/liter, while that of birds and mammals is higher but again uniform at about 150 meq/liter. It can only be supposed that the chemical composition of animal body fluids and their morphological characteristics alike trace the lines of evolutionary development; that is to say, in the course of evolution, groups of animals that are closely related have not merely developed similar anatomical characters but they have developed kidneys of similar functional characteristics, which are responsible for producing similar body fluids.

This chapter first describes the specific properties of biofluids that are of most physiological importance, especially in higher animals; second, the composition of various biofluids in animals and plants and of artificial fluids that are used as substitutes are briefly surveyed. Although this survey is not intended to be complete, it should give an impression of the range and variety of biofluids. For detailed information the reader is referred especially to *Blood and Other Fluids*, edited by P. L. Altman (1961), *Comparative*

Animal Physiology by C. L. Prosser and F. A. Brown (1961), and *Plant Physiology*, vol. II, edited by F. C. Steward (1959).

2   Physiologically Important Properties of Biofluids

Composition of Ionic Solutes.   The main ionic solutes are the alkali cations Na^+ and K^+, with Ca^{2+} and Mg^{2+} in smaller amounts; the common anions are Cl^-, HCO_3^-, small amounts of SO_4^{2-} and HPO_4^{2-} (especially in plants), and proteins whose isoelectric points mostly lie to the acid side of the body pH. Ionic solutes have two most outstanding functions: (a) they are normally, although not always, almost entirely responsible for the osmotic pressure of biofluids (this function is discussed later) and (b) the two ions Na^+ and K^+ are mainly responsible for the striking difference in the composition of extra-cellular and intracellular fluids in animals and of soil and sap fluids in plants.

The typical composition of extracellular and intracellular fluids in animals is shown in Figure 2; the predominance of Na^+ in extracellular fluid is charac-teristic of animals except for fresh-water protozoa, whose cells are directly exposed to solutions of low ionic concentration. In plants, K^+ is generally more concentrated than Na^+ in both extracellular and intracellular fluids; hence the name potash, since potassium was originally found in plant ash residues.

In aquatic animals, the main sites of interchange between the external environment and the extracellular fluid are the gut, the gills, and the kidney; in terrestrial animals they are the gut, the kidney, the lungs, and the sweat glands of the skin (Figure 3). Active transport of cations may occur at these sites to produce overall concentration differences (i.e., osmotic pressure differences, discussed separately) but little change is produced at this stage in the relative concentrations of cations; Na^+ remains the principal cation in the internal environment (Figure 2). On the other hand, at the boundary between the extracellular and the intracellular fluid—that is, at the cell membrane—marked selection occurs, K^+ being preferred to Na^+. The mechanism of this selection is referred to as the *sodium pump*. There are two types of sodium pump, depending on whether the intracellular anions can or cannot readily diffuse across the cell membrane. If the predominant anion is diffusible (e.g., Cl^-), then the pump performs a forced exchange of intra-cellular Na^+ for extracellular K^+ and there is little difference of electrical potential across the membrane (e.g., the erythrocyte). If the bulk of intra-cellular anions are nondiffusible, as in muscle, then the sodium pump simply expels Na^+, and K^+ is taken up passively by the high negative electrical potential consequently produced within the cell. In either case, an intra-

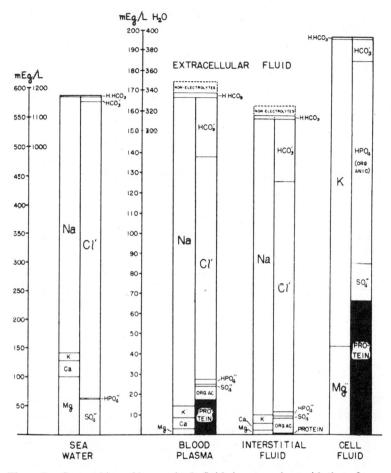

Figure 2. Composition of human body fluids in comparison with that of sea water. Note that the concentration scale used for sea water is different from that for the body fluids. Blood and interstitial or extracellular cellular fluid are high in Na$^+$, while intracellular fluid is high in K$^+$ (Gamble, 1947; by Courtesy of Harvard University Press).

cellular fluid low in Na$^+$ but high in K$^+$ is produced without any difference in osmotic pressure from the extracellular fluid.

Land animals have no gills and their gut has little regulatory function, since almost all cations taken by mouth are absorbed. Thus regulation of the extracellular fluid depends largely on the kidney. Various controlling influences are exerted on the kidney, mainly by two endocrine glands, the adrenal cortex, and the posterior lobe of the pituitary gland, so that excretion, especially of sodium chloride and water, is exactly matched with intake and the concentration of the internal environment is kept constant.

The reason for the difference in the composition of extracellular and intracellular fluid in animals is not entirely clear. On a functional level, a number of enzyme systems require K^+ for activation, especially those concerned with transphosphorylating reactions (Ussing, 1960), and this may be the purpose of the high intracellular K^+ level. The cell K^+ has also been shown to be important in protein synthesis in bacteria (Lubin and Ennis, 1964). On an evolutionary level, it has been supposed that the potassium in intracellular fluid represents the concentration present in the primeval ocean when life originated (Conway, 1942, 1943). As potassium was reabsorbed from the ocean in organic sediments while sodium continued to accumulate, the protoplasmic enzymes surrounded themselves by a phospholipid cell membrane, thus maintaining within it the original high potassium level at which they were most active. At a later stage, when sodium had accumulated in the oceans, cells developed two adaptive mechanisms: (a) an ion pump in the cell membrane to maintain the cell interior high in potassium and low in sodium, and (b) the internal environment or extracellular fluid, created by surrounding the fluid immediately around the cells by a skin or other covering. The internal environment served two purposes: first, to maintain the fluid surrounding the cells at a sodium level corresponding to primitive seawater and lower than that of the present ocean, and second, to retain an aqueous environment around the cells when organisms finally left the sea and began to live on land.

In higher plants, the main site of contact between external and internal environments is the root; as in animals, however, the cell membrane separates extracellular and intracellular fluids. As pointed out earlier, the position is different from that in animals; between the external and extracellular compartments (i.e., in the root), major cation selection may or may not occur, although in most plants sodium tends to be excluded in favor of K^+. Although a large uptake of water and ions occurs in the root, there is frequently no marked increase of concentration of ions, the fluid of the xylem being sometimes actually lower in total ion concentration than the soil fluid (Table 6). In contrast, marked increase of ion concentration occurs at the cell membrane and the raised osmotic pressure produced within the cell is counteracted by the presence of a tough cellulose cell wall outside the cell membrane.

There is still uncertainty about the mechanism of ion uptake in the root. The classical explanation was that of Lundegardh (1954), who discovered that anion uptake was coupled with the respiration rate and interpreted this to mean that anion uptake was directly linked with electron transfer. However, Epstein (1965) has demonstrated that ion uptake is mediated by a carrier, since saturation and competition effects occur. Epstein (1966) produced evidence of a dual carrier mechanism for cations; the first carrier, highly potassium-selective, operating at concentrations below 1 mM, and the second operating at higher concentrations with roughly equal affinity for sodium and potassium. The first carrier is responsible for selective potassium uptake in the root, and the second is probably essentially a mechanism for osmotic adjustment to high soil salinity. Laties (1969) has further suggested that the first carrier lies at the plasmalemma and the second at the tonoplast membrane (or vacuolar membrane) of the cortical cells of the root hairs from which absorbed fluid diffuses passively to the xylem. Steward and Mott (1970) have also pointed out the role of growth in ion accumulation by plants.

Ion uptake into plant cells appears to be governed by two different mechanisms. Uptake of K^+ uses adenosine triphosphate (ATP) as its energy source and does not depend directly on photosynthesis (Robertson, 1960; MacRobbie, 1965; Rains, 1968). On the other hand, in photosynthetic tissues Cl^- uptake appears to be directly linked to electron transfer reactions associated with the second light reaction of photosynthesis (MacRobbie, 1965).

As in animals, the reason for the preference for K^+ in the cell fluid of plants is not entirely clear, although the activation of various cellular enzymes by K^+ has been put forward as at least a teleological explanation (Evans and Sorger, 1966). It is clear, however, that despite many substantial differences among the biochemical mechanisms of plants and animals, they are fundamentally similar; their common preference for K^+ as the predominant intracellular cation is one of the most important pieces of evidence in support of this.

Composition of Other Solutes. Although other solutes are present in relatively small amounts (by molar concentration) as compared with small ions, they are nevertheless of great importance, since they often consist of substances in transit from one part of the organism to another. For example, there are food substances, amino acids and glucose, hormones or chemical messengers, such as noradrenalin and thyroxin, or waste substances on their way to be excreted, such as urea, bilirubin, and creatinine. Most of these substances exert no physical influence on the biofluids, but the plasma proteins in higher animals form an exception. The proteins are heterogeneous in nature; a first classification is into smaller albumin and larger globulin molecules. They have various functions; some act as immune bodies to resist infection and others participate in blood coagulation. Irrespective of their chemical or functional significance, the plasma proteins exert an important physical effect—their osmotic pressure. The protein or colloid osmotic pressure is small compared with that due to electrolytes. But across certain

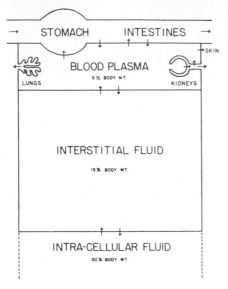

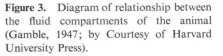

Figure 3. Diagram of relationship between the fluid compartments of the animal (Gamble, 1947; by Courtesy of Harvard University Press).

membranes, such as the capillary membrane, to which the electrolytes are permeable, the impermeable proteins exert the only effective osmotic pressure; thus they control water flow between the blood and tissue compartments of the extracellular fluid (see the section on osmotic pressure and Figure 3).

In some animal biofluids there are colored proteins whose function is to carry oxygen. Primitively, these proteins, normally red iron compounds (e.g., hemoglobin) or green or blue copper compounds (e.g., hemocyanin), are merely in solution in the coelomic fluid or in the blood, as in many invertebrates such as worms or crabs; but in higher animals the oxygen-carrying proteins are concentrated in small cells floating in the blood, the red cells or erythrocytes. This development, which enables the animal to increase the hemoglobin concentration without unduly increasing the blood viscosity, is discussed further in the section on viscosity.

Because they are insoluble in aqueous fluids, neutral fats are often transported in animals in a colloidal suspension, particularly in the lymph which drains from the intestinal tract, and carries most of the absorbed fat.

pH.   In animals the normal pH of the blood and tissue fluids is around 7; human plasma normally measures 7.4. Complex mechanisms involving bicarbonate and protein buffers and selective excretion by the kidney maintain the constancy of this value. If it is disturbed even by a tenth of a pH unit, severe reactions including difficult breathing, muscular spasms, and unconsciousness result.

In contrast to the blood, the fluids of the digestive tract of vertebrates show dramatic changes of pH. The lowest pH occurs in the gastric juice, which can reach pH 1.5. The highest pH occurs in the secretion of the pancreas, pH 8.6 to 8.8, which enters the duodenum, the part of the intestine just below the stomach. These changes of pH appear essentially to be related to the optimal conditions of activity of the digestive enzymes. Pepsin in the stomach acts best at a pH of about 2.0, whereas trypsin of the pancreas has maximum activity at pH 8.0 to 9.0. The high H^+ ion concentration in gastric juice appears to be derived from H_2CO_3, formed in the gastric mucosal cell by the action of carbonic anhydrase on carbon dioxide and water. The H^+ ions are actively transported into the stomach across the membrane of the mucosal cell (Rehm, 1967) and are accompanied by Cl^-, forming hydrogen chloride. The high pH in pancreatic secretion is due to the presence of sodium bicarbonate, but the mechanism of production of this is not known.

The pH of the intracellular fluid of animals has been the subject of dispute. Measurements based on the distribution of weak acids and bases such as HCO_3^- and DMO (dimethyloxazolidinedione) have given consistent results in the region of 7.0, and the bulk of measurements by glass microelectrodes have also given values in this region (Waddell and Bates, 1969). However, Carter et al. (1967) have recently obtained in rat muscle, by means of double-barreled microelectrodes, an intracellular pH of 5.99. It must be noted that, since the cell interior is 70 to 90 mV negative to the exterior, an internal pH of around 6.0 implies passive equilibrium of H^+ ions across the cell membrane; an internal pH of 7.0 would imply that there must be a mechanism of active expulsion of H^+ ions from the cell.

Osmotic Pressure. As pointed out in the section, "Composition of Ionic Solutes," the osmotic pressure is mainly controlled by the alkali cations Na^+ and K^+ and their counterions, especially Cl^-. In aquatic animals, the osmotic pressure of the extracellular fluid may be equal to, lower than, or greater than that of the external environment. In most marine invertebrates the extracellular fluid is similar in composition to seawater and has the same osmotic pressure. On the other hand, most vertebrates have the power of maintaining the extracellular fluid at a relatively constant osmotic pressure independently of the environment. Thus fresh-water fish have a higher osmotic pressure, or are hyperosmotic to their environment, whereas marine fish are mostly hypoosmotic to their seawater environment. The main regulating organs are the gills; in fresh-water fish these actively absorb ions from the very low external concentration, and in marine fish they excrete excess sodium chloride that has been taken in by mouth. In fish which can live in either fresh or seawater, such as salmon and trout, the gills appear to be able to change their function from absorption to excretion of sodium chloride as required, and the blood concentration remains almost constant. In fresh-water fish, excess water in the blood is also eliminated as dilute urine by the kidney.

Another means of adjusting the osmotic pressure of the extracellular fluid is used by cartilaginous fish, such as sharks and rays. Although the sodium chloride content is kept below that of seawater as in fresh-water fish by means of the gills, the osmotic difference is made up by retention of the waste product, urea, at a far higher concentration than in most animals, so that the total blood osmotic pressure is similar to that of seawater.

In animals no osmotic problem arises with the intracellular fluid since, owing to the high water permeability of the cell membrane, this is in osmotic equilibrium with the extracellular fluid (Appelboom et al., 1958; Maffly and Leaf, 1959). If a change occurs in the osmotic pressure of the extracellular fluid, the cells swell or shrink accordingly, so that the osmotic pressure of the intracellular fluid is adjusted to be equal to the external pressure. This simple rule is adhered to in all cases. It must be noted, however, that the changes of cell volume are not in simple proportion to the changes of osmotic pressure.

The equation conventionally used to relate cell volume and osmotic pressure at equilibrium is

$$(1) \qquad \Pi(V - b) = \phi R T n_2$$

where Π is the osmotic pressure, V the cell volume, b the nonsolvent volume, ϕ the average osmotic coefficient of the intracellular solute, and R and T the gas constant and absolute temperature, respectively.

It may be shown that the ratio of fractional cell volume change to fractional osmotic pressure change is given by

$$(2) \qquad \frac{\Pi}{V}\frac{dV}{d\Pi} = -\left[1 - \frac{b}{V}\right]\left[1 - \frac{\Pi}{\phi}\frac{d\phi}{d\Pi}\right]$$

The first term on the right-hand side expresses the effect of the nonsolvent volume, b; since b averages approximately 20 to 30% of cell volume (Dick, 1966), this term has a value of 0.7 to 0.8. The second term is important because in concentrated protein solutions (muscle and red blood cells contain 25 and 33% protein, respectively), $\Pi \, d\phi/\phi \, d\Pi$ is large and positive. Although $\Pi \, d\phi/\phi \, d\Pi$ in (2) refers to the average osmotic coefficient of all the cell solutes, when the proteins are sufficiently concentrated, the value of the second term of this equation lies in the range 0.85 to 0.95 (Dick, 1966). Overall, therefore, cells change their volume by only 60 to 70% of the change in external osmotic pressure; this effect helps to limit cell swelling and shrinkage.

In plant cells and bacteria, the intracellular is always higher than the extra-cellular osmotic pressure. As pointed out previously, this is because of active transport of ions into cells from the external fluid. The excess of internal osmotic pressure is counteracted by the inward hydrostatic pressure produced by a strong cellulose cell wall. The cell volume thus remains constant unless, as occasionally happens in dry conditions, the extracellular osmotic pressure rises above the intracellular; in this case—a condition known as plasmolysis—the cell shrinks within the cell wall, leaving a space underneath it. Only in this plasmolytic state is the osmotic pressure of the plant cell fluid equal to that of the external solution.

Ionic Strength. Since the predominant salts of both extracellular and intracellular fluids, sodium chloride and potassium chloride, are both uni-univalent, the ionic strength, $\frac{1}{2} \sum C_i Z_i^2$, is, to a first approximation, numerically equal to the sum of the molar concentrations of both salts. And since the osmotic pressure, which is uniform in both fluids (in animals at any rate), is mainly caused by sodium chloride and potassium chloride, it follows that the ionic strength is essentially uniform in the extracellular and intracellular fluids of animals. In plants, however, it is distinctly higher within the cells than in the extracellular fluid of the xylem.

The ionic strength determines the solubility of proteins in biological fluids although, since the ionic strength is so constant, it is not usually a significant factor in physiological changes. Adair (1967) has, however, given evidence that the osmotic coefficient of proteins in concentrated solution is affected by the ionic strength, so that it rises with rise of concentration more steeply than usual when the ionic strength is high. This effect may influence the osmotic behavior of cells in hypertonic solutions (Dick, 1969), since cell shrinkage will be reduced if $d\phi/d\Pi$ is increased as predicted by (2).

Viscosity. The viscosity of biological fluids is, of course, fundamentally governed by the solvent, water; that is, where solutes do not interfere substantially, the viscosity is approximately 1 cp at 20°C. There are only a few fluids in living organisms that move sufficiently fast for a viscosity of this order to introduce significant resistance to flow in relation to that imposed by resistive membranes of capillaries or cells. Two of these are the blood and the synovial fluid of joints.

The viscosity of blood depends not merely on its fluid part, the plasma, but on the fluid-filled red cells that occupy approximately half of the blood volume. Thus although the viscosity of plasma at 37°C is 1.1 to 1.6 cp, the viscosity of whole blood is approximately twice as much (2.2–3.6 cp) (Altman, 1961). The viscosity of plasma depends mainly on the content of plasma proteins. The viscosity of blood is complex owing to the presence of the red cells; it depends on the following factors.

1. The concentration of red cells in the blood—important effects are produced both in excess or in deficiency of red cells; in the former the increased blood viscosity increases the resistance against which the heart

works, in the latter the reduced viscosity and resistance to flow increases the heart output. In both cases the heart works at a disadvantage.

2. The diameter of the blood vessel through which flow occurs—in small vessels, such as capillaries, arterioles, and venules, whose diameter is only a few times that of the red blood cells, the effective resistance to flow is less than expected from the viscosity measured in wider vessels (Fahraeus–Lindquist effect). This has been explained on the basis that flow in these circumstances must be measured as the integral of a finite number of fluid laminae and not an infinite number, as assumed in the normal integration used in obtaining Poiseuille's law (the so-called sigma effect; Dix and Scott-Blair, 1940).

3. The velocity of blood flow—at all normal rates of flow, red cells tend to accumulate in the center of the blood vessel, and the peripheral blood is relatively depleted of cells. This phenomenon reduces the effective viscosity of the blood; but since the maximal effect occurs at a flow rate less than that in any part of the cardiovascular system, the result is merely to produce a low apparent viscosity coefficient, but not to produce anomalous flow, nonlinear with the pressure gradient (Haynes and Burton, 1959).

4. The temperature—as with all other fluids, the viscosity of blood varies inversely with the temperature. Although this effect is not normally important in homeothermic animals, it becomes important in extremities exposed to cold and probably plays a part in reducing the circulatory rate in frostbite.

The viscosity of joint synovial fluid varies from joint to joint in a way that is not understood. Thus in cattle, Davies (1944) found that the average viscosity in the tibio-tarsal and hip joints of cattle was around 5 cp, whereas that in the knee was around 39 cp. Sundblad (1953) found the average viscosity of fluid in the human knee joint was 46 cp. The viscosity of the synovial fluid appeared to depend primarily on the quantity of mucopolysaccharide present, although the degree of complexing of the macromolecules was probably also important. The synovial fluid was thixotropic, since the viscosity decreased with increase in the shear rate, presumably because of orientation of the linear macromolecules with increasing shear rate (Davies, 1966). A peak in the viscosity at the beginning of movement is presumably attributable to a gel-like state in the synovial fluid in the resting joint.

3   Extracellular Fluids in Animals

Comparison of Various Body Fluids.   The composition and properties of some body fluids are shown in Table 1. The values given are for man except those for endolymph and perilymph. The following comments are intended to indicate the significance of some of the differences among them.

Table 1 Composition of Human Body Fluids

	Sodium (mM/liter H_2O)	Potassium (mM/liter H_2O)	Calcium (mM/liter H_2O)	Magnesium (mM/liter H_2O)	Chlorine (mM/liter H_2O)	Carbonate (mM/liter H_2O)	Protein (g/100 ml)	Carbohydrate (g/100 ml)	pH	Other Values	Reference
Blood plasma	149	5.1	2.7	2.0	110	27	7.2	0.08	7.4	—	Altman (1961)
Tissue fluid	141	4.4	2.0	1.0	126	28	0.9	0.09	7.5	—	Altman (1961)
Lymph	145	3.5	2.3	—	103	—	4.9	0.14	—	—	Folk et al. (1948)
Cerebrospinal fluid	147	2.9	1.1	1.1	113	23	0.03	0.007	7.3	—	Altman (1961)
Endolymph[a]	26	142	1.5	0.5	107	—	0.02	—	7.4	—	Davson (1967)
Perilymph[a]	148	5	1.5	1.0	122	—	0.06	—	7.9	—	Altman (1961)
Saliva	10	21	1.5	0.3	16	6.4	0.4	—	6.8	I up to 27 μM/liter	Altman (1961)
Gastric juice	60	10	1.8	—	78–159	0–21	0.3	—	1.5–8.0		Altman (1961)
Pancreatic secretion	141	4.6	1.4	0.2	77	70	0.19–0.34	0.013	8.7		Altman (1961)
Bile											
Liver	148	7.3	1.7	0.8	93	40	0.3	0.06	7.5	2.3–3.3 } Dry matter	Altman (1961)
Gallbladder	—	—	3.0	—	23	10	0.4	0.24	6.0	18 } (g/100 ml)	Altman (1961)
Sweat	33	4.5	13	0.2	33	—	—	—	4.0–6.8		Robinson and Robinson (1954)
Milk	7.4	13	8.5	0.2	10	—	1.1	7.1			Bell et al. (1969)
Urine	260	39	3.8	4.2	169	—	0	0	4.8–7.8	39 mM HPO_4^{2-} 18 mM SO_4^{2-}	Altman (1961) Bell et al. (1969)

[a] Guinea pig.

277

The basic extracellular fluid is the blood plasma. From this, tissue fluid, which lies in the spaces between the body cells, is formed by filtration across the capillary wall. The motive force is the blood pressure. Since the capillary wall has pores in it of 60 Å diameter (Pappenheimer, 1953), probably lying between the cells of the endothelium, the filtrate contains cations in almost unchanged concentration (reduced slightly by the requirements of a Donnan equilibrium); the main change is a great reduction in the quantity of protein. The so-called colloid osmotic pressure of the protein that is retained in the plasma is of great importance, since when the hydrostatic pressure of the blood falls toward the venous end of the capillary, the colloid osmotic pressure causes withdrawal of fluid from tissue fluid back into the plasma. This circulation of fluid between blood plasma and tissue fluid is of great importance in the nutrition and oxygenation of the tissues and the removal of waste products.

Lymph is formed from tissue fluid by a process of absorption into blind-ended lymphatic capillaries, which drain eventually into the venous system. The mechanism of formation is not fully understood, but the residual hydrostatic pressure in the tissues plays a part. An important part in the absorptive process is played by pinocytosis, an active process by which the cells of the lymphatic capillary engulf droplets of tissue fluid and transfer them to the lymph. Again, little change of salt composition takes place between tissue fluid and lymph, but there is an increase in protein content. The lymphatic cells have the important function of removing protein from the tissue fluid. This is very important, since if the protein content of the tissue fluid is not kept low, the colloid osmotic pressure of the plasma is unable to withdraw tissue fluid back into the blood and an accumulation of fluid in the tissues, known as edema, results. Any blockage or failure of lymph flow thus causes edema.

The cerebrospinal fluid that bathes the brain is basically similar to tissue fluid and has many of the appearances of an ultrafiltrate of plasma. However, the K^+ and Ca^{2+} levels are too low, and the Na^+, Mg^{2+}, and Cl^- levels too high for a simple filtrate, and it has been concluded that some process of active secretion in the choroid plexuses of the brain is responsible for the formation of cerebrospinal fluid (Davson, 1967). The perilymph and endolymph of the internal ear are very remarkable. The outer perilymph resembles the cerebrospinal fluid, but the inner endolymph, which lies in the semicircular ducts, the saccule, the utricle, and the duct of the cochlea, has very high K^+ and low Na^+ levels and resembles closely the intracellular fluid (see Section 4). The endolymph appears to be formed by a process of active secretion in the stria vascularis of the duct of the cochlea (Kuipers and Bonting, 1969).

The next group of fluids consists of secretions into the alimentary tract.

Saliva has a low sodium chloride level, but the I⁻ concentration is normally 30 to 40 times that of plasma (Brown-Grant, 1961). Gastric juice and pancreatic secretion are at the opposite extremes of pH reached in the mammalian body, 1.5 and 8.7, respectively. These striking pH levels are attained for the purpose of activating the important proteolytic enzymes, pepsin in the stomach and trypsin in the pancreas. The bile is distinguished mainly by the remarkable degree of concentration it undergoes in the gall bladder, whereby the dry matter content rises from 3 to 18%. The concentrating mechanism involves absorption by the gall bladder epithelium of sodium chloride solution of the same concentration as plasma (Tormey and Diamond, 1967). The solids left behind consist of bile salts, lecithin, and bilirubin, and these substances have a powerful emulsifying action on fats in the food.

Sweat and milk are related in being the products of skin glands. Both have a considerably lower sodium chloride content than plasma, and an active process involving osmotic work is therefore involved in their production. The high Ca^{2+} content is important, since it can give rise to Ca^{2+} deficiency in the body when sweat or milk production is excessive. The important constituents of milk for nutritive purposes are protein, carbohydrate, fat (4%), and Ca^{2+}.

The urine is more variable in composition than the other body fluids, and the figures represent merely an average normal situation; great variations can occur normally with differences in diet and fluid intake, since the urine is an overspill designed to maintain the constancy of the other body fluids.

Variations in Body Fluids between Different Animals. A summary of the electrolyte composition of the body fluids of various animals is given in Table 2. The first striking point to notice is the remarkable uniformity of the electrolyte composition of the body fluids of mammals and birds. This uniformity is not confined merely to sodium and potassium but also extends to calcium, magnesium, and chlorine as well. This uniformity is mainly due to the homeostatic mechanism of the kidney.

The bony fishes also maintain a remarkably uniform sodium chloride concentration in the body fluids, even though some of them live in the sea and others in fresh water. In this case, the regulating mechanism is mainly in the gills. The body fluids of bony fishes are thus maintained much below the osmotic pressure of seawater. In cartilaginous fishes, the gills also maintain the sodium chloride concentration at approximately half that of seawater; but in this case the total osmotic pressure of the blood is the same as that of seawater, the difference being made up by a high concentration of urea in the blood.

Some insects, such a cockroaches, maintain ion levels in their body fluids not unlike those of mammals. Others, however, such as the silk moth, have a much lower osmotic pressure and also have unusually high concentrations

Table 2   Composition of Body Fluids of Various Animals

	Sodium	Potassium	Calcium (mM/liter)	Magnesium	Chlorine	Sulfate	Protein (g/100 ml)	Reference
Mammals								
Cattle	145	5.9	2.5	0.8	—	—	8.3	
Cat	157	5.2	—	1.1	112	—	—	
Dog	152	5.2	2.7	0.9	112	1.0	6.7	Altman
Dolphin	153	4.3	2.3	1.1	110	—	7.8	(1961)
Horse	154	5.7	3.1	0.9	108	—	6.8	
Seal	142	4.5	—	—	—	—	—	
Sheep	156	5.4	2.9	0.9	116	—	5.7	
Birds								
Chicken	154	5.6	5.0	1.2	117	—	3.6	
Bony fishes								
Eel (marine)	212	2.0	3.9	2.4	188	5.7	8.0	
Mackerel (marine)	188	9.8	—	—	167	—	3.5	
Trout (fresh water)	149	5.1	—	—	141	—	—	
Cartilaginous fishes								
Skate (marine)	254	8	6	2.5	255	—	—	
Insects								Prosser and
Cockroach	161	7.9	4.0	5.6	144	—	—	Brown (1961)
Silk moth	14	46	24	81	—	—	6.0	
Crustaceans								
Crab	468	12.1	17.5	23.6	524	—	4.0	
Lobster	472	10	15.6	6.8	470	—	—	
Crayfish (fresh water)	146	3.9	8.1	4.3	139	—	—	
Echinoderms								
Starfish	444	9.6	9.9	50	522	34	—	
Annelids								
Earthworm	105	8.9	—	—	43	—	—	
Seawater	470	10.0	10	54	548	28	—	

of potassium, calcium, and magnesium in their body fluids and relatively little sodium.

Crustaceans show great adaptation to the environment. Marine forms such as the crab and lobster have sodium chloride concentrations similar to that of seawater, but the fresh-water crayfish maintain blood sodium chloride levels higher than the environment and similar to those of fresh-water bony fish. This regulation is maintained partly by the gills, but mainly by the kidneys, which excrete a very dilute urine as a result of a powerful ion-absorbing mechanism, which withdraws ions from the plasma filtrate that is the initial source of the urine.

Starfish, like most marine invertebrates, maintain a sodium chloride concentration similar to that of seawater.

We shall see that there is no correlation between evolutionary development and the ability to maintain an ionic composition different from that of the environment. Although it is true that the most developed forms are able to maintain the most stable ionic concentrations, from the fish downward there is great variation among forms that are more or less in equilibrium with their environment and those that maintain substantial differences of concentration between their body fluids and the environment.

4   Intracellular Fluids in Animals

Composition of Cytoplasm of Various Mammalian Cells.   As mentioned in Section 1, the intracellular fluid is much less accessible than the extracellular. Data on its composition are therefore much more sparse and also inexact. Some of the available data are shown in Table 3.

The difference between the sodium and potassium levels of extracellular and intracellular fluid has already been noted. On comparison between different cells it will be seen that, although potassium levels are fairly uniform with the exception of leucocytes, sodium levels vary considerably and are high especially in brain, leucocytes, and ascites tumor cells. These variations may be real, but it must also be remembered that pathological leakage of sodium into the cell can very readily occur, so that these high sodium levels may well not represent the situation in the living healthy cell. The level of calcium in the cell is lower than that in the plasma, but the level of magnesium is higher. So far as calcium is concerned, the intracellular as well as the extracellular level is regulated by the parathyroid hormone and by vitamin D (Bianchi, 1968). In muscle, calcium is involved in excitation and contraction. During excitation, there is an increase in calcium in the muscle fiber, both by influx from outside and from a release of bound calcium from an internal store within the muscle. During relaxation, calcium is removed from the

Table 3 Composition of Intracellular Fluids of Mammalian Cells

	Sodium	Potassium	Calcium (mM/liter H_2O)	Magnesium	Chlorine	Carbonate	References
Mammalian plasma (average)	150	5	2.5	1.0	110	27	—
Muscle (rat)	16	152	1.9	16	5	1.2	Conway and Hingerty (1946)
Erythrocyte (human)	12–20	150	0.7	2.8	73.5	27	Davson (1964); Altman (1961)
Leucocyte (rabbit)	68	105	—	—	—	—	Hempling (1954)
Ascites tumor cells (mice)	50	134	—	—	64	—	Hempling (1958)
Liver (dog)	2.8	161	0.8	16	—	—	Harris (1960)
Brain (dog)	67	170	—	—	34	—	From van Harreveld (1966); calculated on basis of 20% extracellular space

cytoplasm by absorption into the sarcoplasmic reticulum within the muscle cell (Bianchi, 1968).

Chloride levels within cells are extremely variable. These are determined chiefly by the quantity of protein and organic ions present in the cell. A contrast is provided by muscle and erythrocyte. In muscle, the quantity of organic anion in the cell is very large, so that the chlorine content of the muscle fiber is very low. On the other hand, in the erythrocyte, although there is of course a large amount of protein hemoglobin, this protein is near its isoelectric point; thus it is not very strongly ionized, and the chloride level in the cell is only a little lower than in the external medium. In brain and ascites tumor cells, the high chloride level appears to be associated with the high sodium level that is also present as already mentioned. This may be an artifact due to pathological leakage of sodium chloride into the cell.

Intracellular Fluids of Protozoa. In the case of unicellular organisms, the intracellular fluid is in direct contact with the environment through the cell membrane. This has the same function as in higher animals in keeping intracellular potassium high and sodium low. Data on ion levels in fresh-water amoebae are given in Table 4.

Table 4   Intracellular Fluids of Amoebae

	Sodium	Potassium (mM/liter)	Chlorine	Reference
External medium	0.05–0.2	0.35–0.43	0.35–1.0	
Chaos chaos	0.5	33	—	Chapman-Andresen and Dick (1962)
Chaos chaos	0.35	28	17	Bruce and Marshall (1965)

Sodium, potassium, and chloride all appear to be accumulated from the external medium, but potassium and chlorine concentrations are much higher that that of sodium. Since the interior of the amoeba is electrically 80 mV negative to the exterior, potassium is roughly in electrochemical equilibrium with the external fluid. However, sodium is much lower and chloride much higher than would be required for electrochemical equilibrium. There thus appears to be some mechanism of active accumulation of chloride and the expulsion of sodium. The sodium expulsion may take place in the contractile vacuole (Chapman-Andresen and Dick, 1962; Riddick, 1968). Another important function of the contractile vacuole is to expel water that enters the cell from a hypotonic environment, as in the fresh-water amoebae (Schmidt-Nielsen and Schrauger, 1963).

Fluids in Intracellular Organelles.   If the intracellular fluid of the cytoplasm as a whole is relatively inaccessible, then that of the organelles is even more so. Analyses have been made of fluid from the giant nucleus of amphibian oöcytes (Table 5), but the results are highly variable. Very high figures for Na^+ and K^+, such as those obtained by Naora et al. (1962), suggest that some degree of ionic binding may take place in the nucleus under certain conditions; the results may, however, be artifacts of the technique employed.

Table 5   Ionic Composition of Cell Organelles

	Sodium	Potassium	Calcium (mM/liter H_2O)	Magnesium	Reference
Nucleus (amphibian	33	129	—	—	Riemann, Muir, and MacGregor (1969)
oöcyte)	281	258	—	—	Naora et al. (1962)
Mitochondria (rat liver)	2.2	45	1.9	15	Lehninger (1964)[a]

[a] Assuming 1.9 ml H_2O/g mitochondrial dry matter or 2.9 ml H_2O/g mitochondrial protein.

The ionic concentrations in mitochondria are relatively low (Table 5), but it must be remembered that the figures given refer to mitochondria isolated in 0.25 M sucrose solution after homogenization and centrifugation of the cell contents; therefore, some degree of electrolyte leakage has almost certainly occurred. Nevertheless, the magnesium content of mitochondria is strikingly high. This may be associated with the activation of many enzymes, especially those of adenosine triphosphate by magnesium.

Intracellular Fluids during Cell Freezing.   It has been pointed out by Meryman (1956) and Mazur (1963a) that when electrolyte solutions surrounding cells are lowered to $-10°C$ and the rate of temperature decrease does not exceed 1°C/min, then ice formation takes place only in the extracellular and not in the intracellular fluid. Since the former becomes more concentrated by separation of ice, the cell shrinks osmotically and, in the case of yeast (Mazur, 1961, 1963b), 90% of the original water content can be lost. Only if cooling is too rapid, so that the cell does not have time to shrink osmotically and thus becomes greatly supercooled in relation to the external solution, does intracellular freezing occur. Since water loss possibly takes place through minute water-filled channels in the cell membrane (Solomon, 1968), the question of why ice nucleation does not travel through the pores arises. Mazur (1963a) has suggested that, since the channels are probably only

8 to 12 Å in diameter, the radius of curvature required would preclude the existence of such ice nuclei above − 10°C. The cytoplasm itself does not appear to provide nuclei for the initiation of freezing unless the degree of supercooling exceeds 10°C.

State of Water in Intracellular Fluids.   Hinke (1970) has recently reported the presence in barnacle muscle cells of a water fraction, 32 to 42% of fiber water, which neither took part in osmotic movements nor acted as solvent for intracellular solutes. In smooth muscle of guinea pig, Elford (1970) found no nonsolvent water at 37°C, but at − 7°C (0.5°C above the freezing point of the bathing medium), 8.4% of the muscle water did not act as solvent for dimethyl sulfoxide. Elford's nonsolvent water constitutes 0.36 g/g of muscle protein and is in line with estimates of protein hydration obtained *in vitro*, but Hinke's estimate is much more than this and is not easily explained.

In view of the increased structure of water in the vicinity of a surface (Bernal, 1965), it has been proposed that intracellular structures cause some alteration in the average structure of the intracellular water (Bernal, 1965; Ling, 1965; Hechter, 1965). Recent nuclear-magnetic-resonance studies (Hazlewood et al., 1969; Cope, 1969) have indeed suggested that as much as 70% of the intracellular water in muscle is restricted in some way. This figure is more in line with the results of Hinke (1970). Such complexed water in the neighborhood of protein molecules may have the effects both of stabilizing the protein (Némethy and Scheraga, 1962) and of masking active groups in the protein (Klotz, 1960). Pauling (1961) has even suggested that the anesthetic action of a heterogeneous group of nonreactive and hydrophobic substances, such as chloroform, nitrous oxide, ethylene, argon, and xenon, is due to the increased structure of the surrounding water having a masking effect on neighboring proteins (see also Catchpool, 1965).

5   Fluids in Plants

Relation between External and Extracellular Fluids.   The ionic compositions of soil fluid, tracheal sap (xylem fluid), and the cell fluid of some typical higher plants are shown in Table 6. However, it must be remembered that plant fluids are less accurately regulated than the fluids of most animals, so far greater variations occur within physiological limits in response to environmental and seasonal changes. Table 6 shows typical average values. Although the formation of xylem fluid in the root is not necessarily associated with an increase of ionic concentration, nevertheless the root is a powerful organ of ion and water uptake from the soil, possible mechanisms of which were discussed in the section "Composition of Ionic Solutes."

Table 6   Fluids of Higher Plants

	Potassium	Calcium	Magnesium	Sulfate (mM/litre)	Phosphate	Reference
Soil fluid	3.6	0.5	0.3	0.4	0.01	⎫ Lundegardh
Tracheal sap (pear)	1.5	2.1	1.0	0.3	0.3	⎬ (1966)
Cell fluid (potato)	69.7	13.0	42.1	20.1	10.0	Steward (1959)

Various plants differ more in relation to their uptake of sodium from the soil than in perhaps any other feature of their fluid metabolism (Figure 4). Variations of sodium uptake are responsible for the ability of plants to survive in media of vastly different concentrations of sodium.

Relation between Extracellular and Intracellular Fluids.   For convenience, the mechanisms of ion transfer into the plant cell have been mainly studied in algae, which possess giant cells that are directly exposed to the external

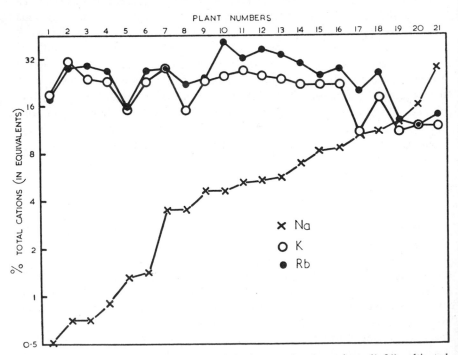

Figure 4.   Concentration of Na, K, and Rb in the sap of various plants (1–21) cultivated in a solution containing equivalent amounts of these ions (From Collander, 1941; by kind permission of the American Society of Plant Physiologists).

medium. The situation is complicated, however, by the presence of a cell wall outside the cell membrane of plasmalemma and by the existence of a large vacuole within the cytoplasm which indeed reduces the actual living cell to a thin layer of protoplasm lining the inside of the cell wall (Figure 1*b*). There is a significant difference between algae living in fresh water and in brackish water. Thus Hoagland (1944) showed that although in *Nitella*, a fresh-water species, all ions are accumulated within the plant cell, in the salt-water species *Valonia*, K^+ is accumulated and Na^+, Ca^{2+}, and Mg^{2+} are excluded from the cell (Figure 5). MacRobbie and Dainty (1958) measured in the brackish water alga, *Nitellopsis*, the concentrations of sodium, potassium, and chloride in external medium, protoplasm, and vacuolar sap, and also the fluxes across the plasmalemma (cell membrane) and the tonoplast membrane (which separates the vacuole from the protoplasm) (Figure 6). The Na^+ and K^+ concentrations in the protoplasm and vacuolar sap are equal and higher than those in the external medium. They are retained by nondiffusible anions in the protoplasm and by actively retained Cl^- in the vacuole (see below). In *Nitella* the vacuole is 120 mV negative to the exterior, but it is still necessary to postulate active transport of potassium since the potassium concentration is higher than required for electrochemical equilibrium [MacRobbie (1970)]. Since the sodium concentration is less than that required for electrochemical equilibrium, active expulsion of Na^+ is

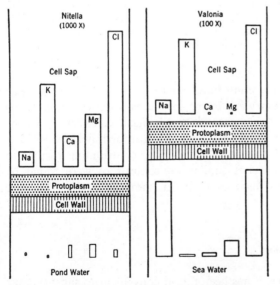

Figure 5. Concentrations of ions in Nitella (fresh water plant) and Valonia (salt water plant) in comparison with those in their environments. In Nitella all ions are accumulated in the plant but in Valonia Na, Ca, and Mg are lower than that in the external medium (Hoagland, 1944; by permission of Chronica Botanica Co.).

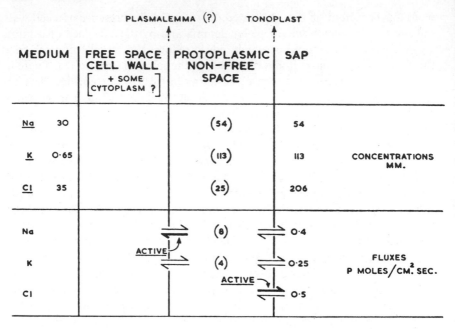

Figure 6.   Diagram of ion concentrations and fluxes in fluid compartments of the brackish water alga Nitellopsis. Figures in brackets show protoplasmic concentrations which are uncertain owing to uncertainty as to the volume of the protoplasm. Arrows indicate probable active and passive fluxes across the plasmalemma and tonoplast membranes (MacRobbie and Dainty, 1958; by courtesy of Rockefeller University Press).

postulated, and this probably contributes to the electrical potential. The chloride concentration is somewhat lower in the protoplasm than in the external medium but is probably near electrochemical equilibrium. In the vacuolar sap, however, chloride is very high, which probably indicates that it is pumped in across the tonoplast membrane, contributing further to the vacuolar potential. Somewhat similar concentrations were found by MacRobbie (1962, 1964) in *Nitella*, and similar pumping mechanisms were proposed.

6   Artificial Biological Fluids

For experimental purposes, it has been found convenient to substitute for natural animal body fluids artificial mixtures whose ionic composition is designed to be similar to that of the natural biological fluids they replace (Table 7). The first such fluid was that designed by Ringer for amphibians.

Table 7 Artificial Biological Fluids (All Concentrations mM Except Glucose)

Name	Animal	NaCl	KCl	CaCl₂	MgCl₂	MgSO₄	NaH₂PO₄	Na₂HPO₄	KH₂PO₄	NaHCO₃	Glucose (g/liter)
Ringer–Locke	Mammals, birds	154	5.6	2.2						2.4	1.0
Tyrode	Mammals, birds	137	2.7	1.8	0.5					11.9	1.0
Gey	Mammals, birds	137	4.0	2.5	1.0			0.84	0.18	3.0	2.0
Ringer	Amphibians	111	1.9	1.1						2.4	
Holtfreter	Fresh-water fish	60	0.7	0.9						2.4	
Locke	Insects	154	5.6	2.3			0.36			2.4	2.5
Seawater	Marine invertebrates	466	10.2	11.0	11.6	28.6				2.5	

From Cameron, 1950.

This solution was later modified by Locke for mammals by raising the sodium and potassium concentrations in it. Ringer's and Locke's solutions were buffered by sodium bicarbonate. Later Tyrode and Gey introduced other solutions for mammals, and these solutions contained phosphate buffer in addition to bicarbonate. In other solutions shown in Table 7, the electrolyte concentration has been adjusted to be suitable for different animals, as may be seen by comparing the concentrations in the artificial fluids with those of the body fluids of the animals concerned as shown in Table 2.

The solutions mentioned above replace only the electrolytes of body fluids, with the addition of glucose as an energy source. Such fluids are remarkably successful in maintaining the vitality of cells and tissues for periods of up to

Table 8   Composition of Morgan, Morton, and Parker's Medium No. 199[a]

	Mg/1000 ml		Mg/1000 ml
l-Arginine	70.0	Thiamin	0.010
l-Histidine	20.0	Riboflavin	0.010
l-Lysine	70.0	Pyridoxine	0.025
l-Tyrosine	40.0	Pyridoxal	0.025
dl-Tryptophane	20.0	Niacin	0.025
dl-Phenylalanine	50.0	Niacinamide	0.025
l-Cystine	20.0	Pantothenate	0.01
dl-Methionine	30.0	Biotin	0.01
dl-Serine	50.0	Folic acid	0.01
dl-Threonine	60.0	Choline	0.50
dl-Leucine	120.0	Inositol	0.05
dl-Isoleucine	40.0	*p*-Aminobenzoic acid	0.05
dl-Valine	50.0	Vitamin A	0.10
dl-Glutamic acid	150.0	Calciferol (vitamin D)	0.10
dl-Aspartic acid	60.0	Menadione (vitamin K)	0.01
dl-Alanine	50.0	α-Tocopherol phosphate (vitamin E)	0.01
l-Proline	40.0	Ascorbic acid	0.05
l-Hydroxyproline	10.0	Glutathione	0.05
Glycine	50.0	Cholesterol	0.2
Cysteine	0.1	Tween 80 (oleic acid)	20.0
Adenine	10.0	Sodium acetate	50.0
Guanine	0.3	*l*-Glutamine	100.0
Xanthine	0.3	Adenosine triphosphate	10.0
Hypoxanthine	0.3	Adenylic acid	0.2
Thymine	0.3	Ferric nitrate	0.1
Uracil	0.3	Ribose	0.5
		Desoxyribose	0.5

From Parker, 1950.
[a] This medium also contained a modified Tyrode's solution.

24 hr or even more in some cases. However, for the long-term maintenance and growth of biological tissues, not only the electrolytes but the organic constituents of body fluids must be provided. These organic materials are often supplied in the form of the actual blood plasma or body fluid of the animal concerned and also as animal tissue extracts, usually mixed into an artificial saline. However, for certain experimental purposes it has been desirable to provide a medium of accurately defined chemical composition. For such a medium to maintain successfully the growth of living tissues over long periods, a highly complex array of amino acids, vitamins, and other tissue constituents must be provided. The composition of a typical solution is shown in Table 8. The constituents shown are intended to be dissolved in Tyrode's solution, which provides the inorganic constituents.

7　Conclusion

The ability of simple inorganic solutions to maintain the vitality of living tissues and cells for long periods is perhaps the most striking evidence of the predominant importance of the electrolytes in biofluids. Provided the ionic balance, the pH, and the osmotic pressure are correct, and a small amount of glucose is added as an energy supply, all the other constituents of the body fluids can be dispensed with for a substantial time without detriment to the living cell. The critical importance of electrolytes is an indication that life originated in an aqueous electrolyte solution and that its complex processes still require an environment roughly similar to that of its origin.

Acknowledgments

I wish to thank Dr. H. A. P. Ingram and Dr. Janet Sprent for helpful suggestions and criticisms of the sections on plant fluids. For Figures 1a and b I am indebted to Miss M. Benstead.

REFERENCES

Adair, G. S., unpublished data quoted by D. A. T. Dick in *Physical Bases of Circulatory Transport*, Eds., E. B. Reeve and A. C. Guyton, Saunders, Philadelphia, 1967, p. 220.

Altman, P. L., *Blood and Other Body Fluids*, Federation of American Societies for Experimental Biology, Washington, D.C., 1961.

Appelboom, J. W. T., W. A. Brodsky, W. S. Tuttle, and I. Diamond, *J. Gen. Physiol.*, **41**, 1153 (1958).

Bell, G. H., J. N. Davidson, and H. Scarborough, *Textbook of Physiology and Biochemistry*, 7th ed., Livingstone, Edinburgh, 1968.

Bernal, J. D., *Symp. Soc. Exptl. Biol.*, **19**, 17 (1965).

Bianchi, C. P., *Cell Calcium*, Butterworths, London, 1968.

Brown-Grant, K., *Physiol. Rev.*, **41**, 189 (1961).

Bruce, D. L., and J. M. Marshall, *J. Gen. Physiol.*, **49**, 151 (1965).

Cameron, G., *Tissue Culture Technique*, Academic Press, New York, 1950.

Carter, N. W., F. C. Rector, D. S. Campion, and D. W. Seldin, *J. Clin. Invest.*, **46**, 920 (1967).

Catchpool, J. F., *Ann. N.Y. Acad. Sci.*, **125**, 595 (1965).

Chapman-Andresen, C., and D. A. T. Dick, *Compt. Rend. Trav. Lab. Carlsberg*, **32**, 445 (1962).

Collander, R., *Plant Physiol.*, **16**, 691 (1941).

Conway, E. J., *Proc. Roy. Irish Acad.*, **48B**, 119 (1942).

Conway, E. J., *Proc. Roy. Irish Acad.*, **48B**, 161 (1943).

Conway, E. J., and D. Hingerty, *Biochem. J.*, **40**, 561 (1946).

Cope, F. W., *Biophys. J.*, **9**, 303 (1969).

Davies, D. V., *J. Anat. London*, **78**, 68 (1944).

Davies, D. V., *Federation Proc.*, **25**, 1069 (1966).

Davson, H., *Textbook of General Physiology*, 3rd ed., Churchill, London, 1964.

Davson, H., *The Physiology of Cerebrospinal Fluid*, Churchill, London, 1967.

Dick, D. A. T., *Cell Water*, Butterworths, Washington, D.C., 1966.

Dick, D. A. T., *J. Gen. Physiol.*, **53**, 836 (1969).

Dix, F. J., and G. W. Scott-Blair, *J. Appl. Physics*, **11**, 574 (1940).

Elford, B. C., *Nature*, **227**, 282 (1970).

Epstein, E., "Mineral Metabolism," in *Plant Biochemistry*, Eds., J. Bonner and J. E. Varner, Academic Press, New York, 1965, pp. 438–466.

Epstein, E., *Nature*, **212**, 1324 (1966).

Evans, H. J., and G. J. Sorger, *Ann. Rev. Plant Physiol.*, **17**, 47 (1966).

Folk, B. P., K. L. Zierler, and J. L. Lilienthal, *Amer. J. Physiol.*, **153**, 381 (1948).

Gamble, J. L., *Chemical Anatomy, Physiology and Pathology of Extracellular Fluid*, Harvard University Press, Cambridge, 1947.

Harris, E. J., *Transport and Accumulation in Biological Systems*, 2nd ed., Butterworths, London, 1960.

Haynes, R. H., and A. C. Burton, *Amer. J. Physiol.*, **197**, 943 (1959).

Hazlewood, C. F., B. L. Nichols, and N. F. Chamberlain, *Nature*, **222**, 747 (1969).

Hechter, O., *Ann. N.Y. Acad. Sci.*, **125**, 625 (1965).

Hempling, H. G., *J. Cell. Comp. Physiol.*, **44**, 87 (1954).

Hempling, H. G., *J. Gen. Physiol.*, **41**, 565 (1958).

Hinke, J. A. M., *J. Gen. Physiol.*, **56**, 521 (1970).

Hoagland, D. R. (1944), quoted by P. J. Kramer in *Plant and Soil Water Relationships*, McGraw-Hill, New York, 1949.

Klotz, I. M., *Circulation*, **21**, 828 (1960).

Kuipers, W., and S. L. Bonting, *Biochim. Biophys. Acta*, **173**, 477 (1969).

Laties, G. G., *Ann. Rev. Plant Physiol.*, **20**, 89 (1969).

Lehninger, A. L., *The Mitochondrion*, Benjamin, New York, 1964.

Ling, G. N., *Ann. N.Y. Acad. Sci.*, **125**, 401 (1965).

Lubin, M., and H. L. Ennis, *Biochim. Biophys. Acta*, **81**, 614 (1964).

Lundegardh, H., *Symp. Soc. Exptl. Biol.*, **8**, 262 (1954).

Lundegardh, H., *Plant Physiology*, Oliver and Boyd, Edinburgh, 1966.

MacRobbie, E. A. C., *J. Gen. Physiol.*, **45**, 861 (1962).

MacRobbie, E. A. C., *J. Gen. Physiol.*, **47**, 859 (1964).

MacRobbie, E. A. C., *Biochim. Biophys. Acta*, **94**, 64 (1965).

MacRobbie, E. A. C., *Quart. Rev. Biophys.*, **3**, 251 (1970).

MacRobbie, E. A. C., and J. Dainty, *J. Gen. Physiol.*, **42**, 335 (1958).

Maffly, R. H., and A. Leaf, *J. Gen. Physiol.*, **42**, 1257 (1959).

Mazur, P., *J. Bacteriol.*, **82**, 662 (1961).

Mazur, P., *J. Gen. Physiol.*, **47**, 347 (1963a).

Mazur, P., *Biophys. J.*, **3**, 323 (1963b).

Meryman, H. T., *Science*, **124**, 515 (1956).

Naora, H., M. Naora, M. Ezawa, V. G. Allfrey, and A. E. Mirsky, *Proc. Natl. Acad. Sci. U.S.*, **48**, 853 (1962).

Némethy, G., and H. A. Scheraga, *J. Phys. Chem.*, **66**, 1773 (1962).

Pappenheimer, J. R., *Physiol. Rev.*, **33**, 387 (1953).

Parker, R. C., *Methods of Tissue Culture*, Hoeber, New York, 1950.

Pauling, L., *Science*, **134**, 15 (1961).

Prosser, C. L., and F. A. Brown, Eds., *Comparative Animal Physiology*, 2nd ed., Saunders, Philadelphia, 1961.

Rains, D. W., *Plant Physiol.*, **43**, 394 (1968).

Rehm, W., *Federation Proc.*, **26**, 1303 (1967).

Riddick, D. H., *Amer. J. Physiol.*, **215**, 736 (1968).

Riemann, W., C. Muir, and H. C. MacGregor, *J. Cell. Sci.*, **4**, 299 (1969).

Robertson, R. N., *Biol. Rev.*, **35**, 231 (1960).

Robinson, S., and A. H. Robinson, *Physiol. Rev.*, **34**, 202 (1954).

Schmidt-Nielsen, B., and C. R. Schrauger, *Science*, **139**, 606 (1963).

Solomon, A. K., *J. Gen. Physiol. Suppl.*, **51**, 335S (1968).

Steward, F. C., Ed., *Plant Physiology*, vol. II, Academic Press, New York, 1959.

Steward, F. C., and R. L. Mott, *Intern. Rev. Cytol.*, **28**, 275 (1970).

Sundblad, L., *Acta Soc. Med. Upsalien.*, **58**, 113 (1953).

Tormey, J. M., and J. M. Diamond, *J. Gen. Physiol.*, **50**, 2031 (1967).

Ussing, H. H., "The Alkali Metal Ions in Biology," in *Handbuch der Experimentellen Pharmakologie*, Eds., O. Eichler and A. Farah, Heidelberg, Springer, 1960.

van Harreveld, A., *Brain Tissue Electrolytes*, Butterworths, Washington, D.C., 1966.

Waddell, W. J., and R. G. Bates, *Physiol. Rev.*, **49**, 285 (1969).

8 Aspects of the Statistical–Mechanical Theory of Water

A. Ben-Naim and F. H. Stillinger, Jr.,
Bell Telephone Laboratories, Incorporated,
Murray Hill, New Jersey

Abstract

The prospects are examined for construction of a fundamental and systematic theory of liquid water, utilizing the techniques of classical statistical mechanics for rigid asymmetric rotors. For that purpose, a tentative molecular pair potential is proposed which exhibits the known tendency toward tetrahedral coordination and which fits the measured water-vapor second virial coefficient reasonably well. Several potential curves are displayed for the more important classes of pair configurations. By means of indirect calculations, we have established a local cooperative tendency for orientational correlation of neighboring water molecules in arrangements suitable for hydrogen bonding. Finally, we stress the relevance and importance of Monte Carlo calculations (with electronic computers) designed literally to provide submicroscopic pictures of the random hydrogen-bond networks in liquid water and aqueous solutions.

1 Introduction

The study of liquid water could superficially be considered as a single branch of the entire field of liquid-state research. However, it is obvious that this one substance occupies a place of special prominence, not only because of its unique physical characteristics but also because it seems to be the only fluid medium capable of supporting biochemical processes. No doubt these peculiar properties arise from the same molecular feature that has thus far prevented development of a serious first-principles theory of liquid water, namely, the noncentral forces operative between the molecules.

The earliest attempts to understand the behavior of water apparently stemmed from Röntgen's[1] suggestion that the liquid contained "ice molecules." Chadwell[2] has reviewed a number of these phenomenological treatments of water in terms of association complexes. Since little was known

295

about molecular structure and intermolecular forces during the early period, however, these treatments were necessarily very limited in scope.

In 1933, Bernal and Fowler[3] provided a major conceptual advance by pointing out the propensity for water molecules to bond to one another with locally tetrahedral geometry. In varying degrees, this structural feature has seemed to affect, if not to dominate, all subsequent attempts to explain the properties of liquid water (and aqueous solutions) in a statistical-mechanical context.[4] Even so, mere emphasis on tetrahedral coordination amounts to far less than complete mechanical description of the nature of water-molecule interactions.

The modern trend in formal liquid-state theory seeks to establish a clear quantitative connection between carefully specified intermolecular potentials (as the starting point) and various molecular distribution functions and thermodynamic properties implied by those potentials.[5] Satisfying success has been achieved in this approach for simple substances such as argon, not only because the relevant central pair potentials are rather well established but also because reliable integral equation methods are available for computation of the requisite radial distribution functions.

In this chapter we attempt to lay the groundwork for a corresponding formal statistical-mechanical theory of liquid water. For that purpose we presume (at least initially) that the total potential energy is composed of a pairwise-additive sum of pair potentials. Even in the case of argon this is not rigorously true, but it is reasonable to regard three-body forces, and so on, as mild perturbations on the pairwise-additivity model that may be accounted for at the end of the primary calculation. The next section is devoted to certain immediate implications of the pair-potential assumption, and we record there the corresponding exact formal expressions for the mean energy, the pressure, the compressibility, and some further quantities requiring at most knowledge of the water pair-distribution function.

Section 3 exhibits what we consider to be analytically one of the simplest water-molecule pair potentials that still retains certain essential features of the actual situation. It represents a modification and extension of Bjerrum's four-point-charge electrostatic model of the water molecule,[6] and inherently favors tetrahedral coordination.

The second and third virial coefficients for water vapor are examined in Section 4. By demanding that the theoretical expression for the first of these (with our suggested potential function) agree with experiment, certain free parameters in the potential are determined.

Section 5 is devoted to preliminary theoretical investigation of the pair-correlation function for liquid water. Whereas for substances such as argon with central molecular forces, this quantity (at fixed temperature and pressure) depends only on scalar distance, the full pair-correlation function for water is vastly more complicated. In order to fix the relative configuration and orientation of two water molecules, a minimum of *six* variables must be specified. With this formidable feature in mind, we assess the practical utility of the Percus–Yevick integral equation for the water pair-correlation function. In addition, we examine in Section 5 a semiempirically determined pair-correlation function in order to estimate the cooperative character of orientational ordering in liquid water.

The final discussion, Section 6, attempts to predict the most useful course of future research directed to the construction of a full fundamental statistical mechanical theory of liquid water.

2   Pair-Potential Assumption

The free water molecule is a nonlinear triatomic species exhibiting C_{2v} symmetry. The average bond angle is about 105°, only slightly less than the geometrically ideal angle

$$\text{(1)} \qquad\qquad \theta_T = 109°28'$$

between lines connecting the center of a regular tetrahedron to its vertices. The oxygen–hydrogen bond lengths in the isolated molecule are 0.96 Å, but in condensed phases such as ice and liquid water these lengthen perhaps to 1.00 Å on the average.[7]

In order to describe the position and orientation of a water molecule in space, six variables are required. We shall take these to be, first, the vector position $\mathbf{r} = (x, y, z)$ of the oxygen nucleus, and second, the set of Euler angles ϕ, θ, ψ required to fix the orientation of the molecule, regarded as a rigid body. Figure 1 and Figures 2a to c establish the particular Euler angle convention that we have employed.

Our primary intention is to describe liquid water by the techniques of classical statistical mechanics. The central quantity in that discipline is the canonical partition function, Q_N. Since we regard the individual water molecules as acting toward one another as rigid asymmetric rotors, the partition function has the following form:

$$\text{(2)} \qquad Q_N = \frac{1}{N!} \left[\frac{(2\pi kT)^3 m^{3/2} (I_1 I_2 I_3)^{1/2} Q_{\text{vib}}}{h^6} \right]^N$$
$$\times \int d\mathbf{x}_1 \cdots \int d\mathbf{x}_N \exp\left[-\beta V_N(\mathbf{x}_1 \cdots \mathbf{x}_N) \right], \qquad \beta = (kT)^{-1}$$

In this expression, m is the molecular mass, the I's are the three moments of inertia, Q_{vib} is the partition function for the vibrational degrees of freedom

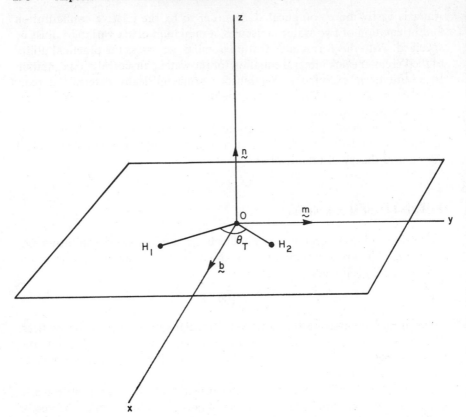

Figure 1.   Coordinate axes for the rigid nonlinear water molecule. Cartesian axes x, y, z are fixed; orthogonal unit vectors **b**, **m**, **n** rotate with the molecule. **b** is the molecule's symmetry axis, **m** is in the molecular plane, and $\mathbf{n} = \mathbf{b} \times \mathbf{m}$.

of an isolated water molecule, and V_N is the total potential of interaction between the N molecules in the system.

Vector $\mathbf{x}_j$ in (2) stands for the six coordinates specifying position and orientation of molecule j, and the $\mathbf{x}$ integrations in more detail must be carried out as follows:

$$(3) \qquad \int d\mathbf{x}_j \equiv \int_V d\mathbf{r}_j \int_0^{2\pi} d\phi_j \int_0^{\pi} \sin \theta_j \, d\theta_j \int_0^{2\pi} d\psi_j$$

where V is the vessel volume containing the N water molecules.

Molecular distribution functions $\rho^{(n)}(\mathbf{x}_1 \cdots \mathbf{x}_n)$ give the probabilities that a set of differential volume-and-orientation elements $d\mathbf{r}_j \, d\phi_j \, d\theta_j \, d\psi_j \, (j = 1 \cdots n)$

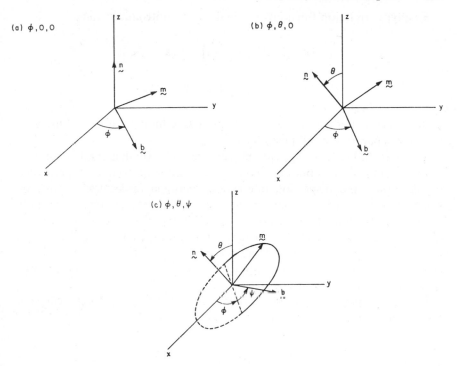

Figure 2. Euler angle convention for the water molecule. All three angles are 0 in the configuration shown in Figure 1. (*a*) φ increases from 0 by rotation about the initial **n** axis. (*b*) Rotation about the *new* **b** axis defines θ. (*c*) The last Euler angle, ψ, describes rotation about the *new* **n** direction. To generate all orientations uniquely, the limits $0 \le \varphi, \psi \le 2\pi$, and $0 \le \theta \le \pi$ must be imposed.

are simultaneously occupied by any of the water molecules. The precise definitions of the $\rho^{(n)}$ involve multiple integrals of the full configuration space canonical density

$$(4) \qquad \rho^{(n)}(\mathbf{x}_1 \cdots \mathbf{x}_n) = \frac{N! \int d\mathbf{x}_{n+1} \cdots \int d\mathbf{x}_N \exp\left[-\beta V_N(\mathbf{x}_1 \cdots \mathbf{x}_N)\right]}{(N-n)! \int d\mathbf{x}_1 \cdots \int d\mathbf{x}_N \exp\left[-\beta V_N(\mathbf{x}_1 \cdots \mathbf{x}_N)\right]}$$

Although this set of functions specifies the full orientation dependence for sets of n molecules, it suffices for some purposes merely to know the orientationally averaged densities; for that reason we also define the "contracted" molecular distribution functions

$$(5) \qquad \bar{\rho}^{(n)}(\mathbf{r}_1 \cdots \mathbf{r}_n) = \int \sin \theta_1 \, d\phi_1 \, d\theta_1 \, d\psi_1 \cdots \int \sin \theta_n \, d\phi_n \, d\theta_n \, d\psi_n \, \rho^{(n)}(\mathbf{x}_1 \cdots \mathbf{x}_n)$$

Finally, correlation functions $g^{(n)}$ and $\bar{g}^{(n)}$ are introduced thus

(6)
$$\rho^{(n)}(\mathbf{x}_1\cdots\mathbf{x}_n) = \left(\frac{N}{8\pi^2 V}\right)^n g^{(n)}(\mathbf{x}_1\cdots\mathbf{x}_n)$$

$$\bar{\rho}^{(n)}(\mathbf{r}_1\cdots\mathbf{r}_n) = \left(\frac{N}{V}\right)^n \bar{g}^{(n)}(\mathbf{r}_1\cdots\mathbf{r}_n)$$

with the properties (in the infinitely large system limit) of approaching unity for wide separation of all positions $\mathbf{r}_1\cdots\mathbf{r}_n$.

The potential energy, V_N, viewed from the most fundamental standpoint, is an enormously complicated function. Of course it comprises permanent dipole–dipole and dispersion interactions between molecules at moderate range, as well as hydrogen bonding at close range. But also it contains subtle many-body potentials, one aspect of which is the dielectric modification of dipole-dipole forces. In view of the separation of Q_{vib} (defined for *isolated* molecules) in (2) we must furthermore be prepared to admit that V_N will contain contributions from coupling of vibrations between neighboring molecules in the liquid phase.

In spite of these complications, there ought still to be a "best" choice of pair potential $v(\mathbf{x}_i, \mathbf{x}_j)$ such that the assumption

(7)
$$V_N(\mathbf{x}_1\cdots\mathbf{x}_N) \simeq \sum_{i<j=1}^{N} v(\mathbf{x}_i, \mathbf{x}_j)$$

retains most essential features of the liquid-water problem. A reasonable sort of criterion for choice of v would be minimization of the squared deviation between Boltzmann factors for V_N and its pairwise-additive approximation by (7)

(8)
$$\int d\mathbf{x}_1 \cdots \int d\mathbf{x}_N \left\{ \exp\left[-\frac{\beta}{2} V_N(1\cdots N)\right] \right.$$
$$\left. - \exp\left[-\frac{\beta}{2} \sum_{i<j=1}^{N} v(i,j)\right] \right\}^2 = \text{minimum}$$

By setting the first variation with respect to v of this last expression equal to 0, we obtain the condition

(9)
$$\int d\mathbf{x}_3 \cdots \int d\mathbf{x}_N \exp\left\{ -\beta\left[\tfrac{1}{2} V_N(1\cdots N) + \tfrac{1}{2}\sum_{i<j=1}^{N} v(i,j)\right] \right\}$$
$$= \int d\mathbf{x}_3 \cdots \int d\mathbf{x}_N \exp\left[-\beta \sum_{i<j=1}^{N} v(i,j)\right]$$

which must be obeyed for *all* $\mathbf{x}_1$ and $\mathbf{x}_2$. If we first take $\mathbf{r}_1$ and $\mathbf{r}_2$ to be far apart in (9), we see that the canonical partition functions (and hence the Helmholtz free energies) for the two potential functions

$$(10) \qquad \tfrac{1}{2}V_N + \tfrac{1}{2} \sum_{i<j=1}^{N} v(i,j)$$

and

$$(11) \qquad \sum_{i<j=1}^{N} v(i,j)$$

must be identical. Furthermore, by taking $\mathbf{r}_1$ and $\mathbf{r}_2$ close to one another in (9), we also conclude that interactions (10) and (11) will produce identically the same pair distribution function $\rho^{(2)}(\mathbf{x}_1, \mathbf{x}_2)$.

Although we cannot conclude that the optimal pairwise-additivity approximation (7) causes *no* change in free energy or in the various $\rho^{(n)}$ for water, the invariances just mentioned in passing from (11) to (10), or "halfway" from pairwise additivity to the true V_N, indicate that (7) is generally an excellent approximation.

Variational principle (8) is unfortunately not suited for direct construction of a liquid-phase v. Furthermore, the pair potentials it requires might exhibit small temperature and density dependence. We take the point of view in the following paragraphs that a fixed $v(\mathbf{x}_i, \mathbf{x}_j)$ can be determined for liquid water by alternative means and that residual temperature and density dependence is negligible if we restrict attention to the behavior of liquid water at (or near) room temperature and 1 atm pressure.

The primary practical advantage of pairwise-additive potentials is that most of the usual thermodynamic properties can be expressed in terms of just v and $\rho^{(2)}$. The most straightforward of these is the mean energy per molecule

$$(12) \qquad \begin{aligned} \frac{E}{N} &= \frac{E^{(0)}}{N} + \frac{1}{2N} \int d\mathbf{x}_1 \int d\mathbf{x}_2 v(\mathbf{x}_1, \mathbf{x}_2) \rho^{(2)}(\mathbf{x}_1, \mathbf{x}_2) \\ &= \frac{E^{(0)}}{N} + \frac{N}{16\pi^2 V} \int d\mathbf{x}_2 v(\mathbf{x}_1, \mathbf{x}_2) g^{(2)}(\mathbf{x}_1, \mathbf{x}_2) \end{aligned}$$

where $E^{(0)}/N$ is the mean molecular energy at the ambient temperature for the infinitely dilute vapor. The integral term in (12) merely counts molecular pairs in all possible pair configurations and accumulates the corresponding potential energy contributions.

The expression for the pressure, p, in virial form may be derived by a trivial generalization of Green's volume-scaling procedure for spherically symmetric molecules.[8] One obtains

$$(13) \qquad \frac{pV}{NkT} = 1 - \frac{1}{6NkT} \int d\mathbf{x}_1 \int d\mathbf{x}_2 [\mathbf{r}_{12} \cdot \nabla_{\mathbf{r}_{12}} v(\mathbf{x}_1, \mathbf{x}_2)] \rho^{(2)}(\mathbf{x}_1, \mathbf{x}_2)$$

$$\mathbf{r}_{12} = \mathbf{r}_2 - \mathbf{r}_1$$

The isothermal compressibility

(14)
$$\kappa_T = -\frac{1}{V}\left(\frac{\partial V}{\partial p}\right)_T$$

has the unique advantage of being exactly expressible in terms of the pair-distribution function, quite irrespective of the pairwise-additivity assumption (7). Furthermore, only the angle-integrated quantity $\bar{\rho}^{(2)}$ is required. Provided that we first understand the infinite-system-size limit to have been taken for $\bar{\rho}^{(2)}$, the general compressibility relation is

(15)
$$kT\kappa_T = \frac{V}{N} + \int d\mathbf{r}_{12}[\bar{g}^{(2)}(\mathbf{r}_{12}) - 1]$$

The chemical potential, μ, may in principle be obtained via the Gibbs–Duhem relation

$$dp = \rho\, d\mu$$

at constant temperature, by integrating the pressure with respect to density $\rho = N/V$ from the ideal gas limit. Alternatively, the potential decoupling procedure[9] for a single molecule, 1, say, may be employed. In this latter approach, the partially coupled molecule 1 is presumed to interact with its neighbors with potential $v(\mathbf{x}_1, \mathbf{x}_j; \xi)$, where

(16)
$$v(\mathbf{x}_1, \mathbf{x}_j; \xi = 0) \equiv 0$$

represents full decoupling of 1, and

$$v(\mathbf{x}_1, \mathbf{x}_j; \xi = 1) = v(\mathbf{x}_1, \mathbf{x}_j)$$

is the actual "physical" pair potential, fully coupled. By computing the reversible work required to "switch on" $v(1, j; \xi)$, that is, to increase ξ from 0 to 1, one finds

(17)
$$\mu = \mu^{(0)} + kT\ln\left(\frac{N}{V}\right) + \frac{1}{N}\int_0^1 d\xi \int d\mathbf{x}_1 \int d\mathbf{x}_2 \frac{\partial v(\mathbf{x}_1, \mathbf{x}_2; \xi)}{\partial \xi} \rho^{(2)}(\mathbf{x}_1, \mathbf{x}_2; \xi)$$

The first two terms in the right member of (17) are the ideal gas contributions, and the integral term accounts for interactions. Note that $\rho^{(2)}$ must be suitably defined for a pair of particles, one of which displays the partial coupling feature.

The dielectric properties of polar fluids such as liquid water are intimately related to the orientational correlations between neighboring molecules. Kirkwood's theory of polar dielectrics[10] leads to the following expression for the static dielectric constant, ε_0:

(18)
$$\frac{(\varepsilon_0 - 1)(2\varepsilon_0 + 1)}{3\varepsilon_0} = \frac{4\pi N}{V}\left(\alpha + \frac{\mu_d^2 g_K}{3kT}\right)$$

where α is the molecular polarizability (assumed to be isotropic), and μ_d is the permanent dipole moment. The specific form for orientational correlation, g_K, for water is again a pair-correlation function integral

$$(19) \qquad g_K = 1 + \frac{N}{8\pi^2 V} \int dx_2 (\mathbf{b}_1 \cdot \mathbf{b}_2) g^{(2)}(\mathbf{x}_1, \mathbf{x}_2)$$

and gives the average cosine of the angle between permanent dipole moment directions, $\mathbf{b}_1$ and $\mathbf{b}_2$, for neighboring molecules.

The definition of the hydrogen bond is somewhat arbitrary. Surely the various experimental techniques that are employed in its study are not precisely equivalent and need not quite agree on the concentration of hydrogen bonds in a given material sample. From our present point of view, we shall suppose that the existence of a "hydrogen bond" between two molecules of water means simply that their coordinates $\mathbf{x}_i$ and $\mathbf{x}_j$ lie between certain specified limits. This is equivalent to defining a characteristic bond function, $B(\mathbf{x}_i, \mathbf{x}_j)$, such that

$$(20) \qquad B(\mathbf{x}_i, \mathbf{x}_j) = 1$$

if i and j are so placed in space to form a hydrogen bond, and

$$(21) \qquad B(\mathbf{x}_i, \mathbf{x}_j) = 0$$

if not. Clearly B should be invariant to all but the relative positions and orientations of i and j. The average number, n_{HB}, of hydrogen bonds per molecule in water then is yet another example of a pair-distribution function quadrature

$$(22) \qquad n_{\text{HB}} = \frac{1}{2N} \int dx_1 \int dx_2 B(\mathbf{x}_1, \mathbf{x}_2) \rho^{(2)}(\mathbf{x}_1, \mathbf{x}_2)$$

Section 5 utilizes this general hydrogen-bond density expression with a specific set of B functions.

Finally, we note that the leading quantum-mechanical corrections to classical partition function (2), of order h^2, can also be reduced to pair-distribution function integrals. For the "asymmetric top" water molecule, the requisite expressions are quite complicated, and we refer the reader to a paper by Friedmann,[11] rather than reproducing the result here. Nevertheless it is worth pointing out that these quantum corrections are the key to understanding the small differences in equilibrium behavior of H_2O, D_2O, and T_2O.

3   Approximate Pair Potential

Our aim in this section is to exhibit a relatively compact analytical expression as an approximation to the "best" liquid-water pair potential. In doing so we are fully aware that our proposed form will eventually be supplanted by more accurate approximations. Nevertheless it seems to us important to develop, even on the present rather crude basis, an intuitive grasp for the way water molecules in various orientations exert forces on one another. Since our proposed potential is forced to fit certain key experimental data, we feel that its predictions will ultimately prove not to be in serious quantitative error.

To the best of our knowledge, the only other type of pair potential that has seriously been considered in description of the fluid states of water is the Stockmayer potential.[12] This potential consists of a sum of a Lennard-Jones potential and the potential of interaction between permanent point dipoles. Doubtless this potential accurately portrays the interaction between pairs of water molecules at large distance in the dilute vapor. Indeed Rowlinson[13,14] has used the Stockmayer potential to calculate the second and third virial coefficients for water vapor.

Still, there is good reason to question the aptness of the Stockmayer potential for understanding condensed phases. It has, for example, been proven by Onsager[15] that the minimum energy for a set of point dipoles is attained in the hexagonal close-packed crystal, not the tetrahedrally coordinated ice lattice. (The Lennard-Jones potentials would merely add extra relative stability to the former.) Also, in the wide variety of hydrate crystals loosely termed "clathrates," the water molecules stoutly maintain the local tetrahedral coordination observed in ordinary ice, even though the larger geometric aspects of the water networks change considerably.[16]

We believe (consistent with Bernal and Fowler[3]) that the marked propensity for water molecules to hydrogen bond into networks with local tetrahedral coordination is the single most important observation bearing on selection of a suitable approximate pair potential. Therefore we have chosen for detailed consideration a model potential that manifestly favors tetrahedral coordination. Like the Stockmayer potential, it combines a spherically symmetric Lennard-Jones interaction with a noncentral electrostatic contribution. Instead of relying on point dipoles, though, our angle-dependent part is based on Bjerrum's four-point-charge model of the water molecule.[6]

As Figure 3 shows, these four charges are placed at the vertices of a regular tetrahedron whose center is presumed coincident with the oxygen nucleus. The distance from this center to each of the four charges has been chosen to be 1.00 Å. Two of the charges, with magnitude $+\eta e$, may be identified as the water-molecule protons partly shielded by the electron cloud. The remaining two charges, $-\eta e$, represent crudely the unshared pairs of valence-shell electrons in the molecule. Bjerrum has pointed out that the choice 0.17 for η will reproduce the dipole moment known for the free-water molecule, but owing to polarization effects in the liquid, we have elected to regard η as an adjustable parameter.

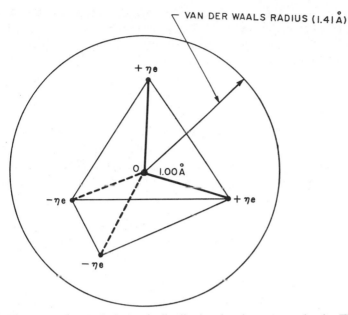

VAN DER WAALS RADIUS (1.41 Å)

Figure 3. Tetrahedral charge distribution for the water molecule. The oxygen nucleus, 0, is at the center of the regular tetrahedron with circumradius 1.00 Å. The positive charge $+\eta e$, are shielded protons, and the negative charges simulate unshared electron pairs.

In order to avoid having simultaneously to determine a large number of adjustable parameters such as η, we have presumed that the Lennard-Jones 12,6 part is the same as for neon, which is isoelectronic with water

$$(23) \qquad v_{\mathrm{LJ}}(r_{12}) = 4\varepsilon\left[\left(\frac{\sigma}{r_{12}}\right)^{12} - \left(\frac{\sigma}{r_{12}}\right)^{6}\right];$$

for neon[17]

$$(24) \quad \varepsilon = 5.01 \times 10^{-15}\,\mathrm{erg} = 7.21 \times 10^{-2}\,\mathrm{kcal/mole}, \qquad \sigma = 2.82\,\text{Å}$$

Specifically, v_{LJ} will refer to oxygen nuclei as the force centers. We see that the four Bjerrum charges $\pm\eta e$ are well buried inside the van der Waals radius (1.41 Å) of the molecule.

The electrostatic interaction between two tetrahedral charge distributions like the one shown in Figure 3 will consist of 16 separate charge-pair terms. It may be written thus

$$(25) \qquad v_{\mathrm{el}}(\mathbf{x}_1, \mathbf{x}_2) = (\eta e)^2 \sum_{\alpha_1,\alpha_2=1}^{4} \frac{(-1)^{\alpha_1+\alpha_2}}{d_{\alpha_1\alpha_2}(\mathbf{x}_1, \mathbf{x}_2)}$$

where α_1 and α_2, respectively, run over the four charges of molecules 1 and 2 such that even and odd values correspond to positive and negative charges.

The quantity $d_{\alpha_1\alpha_2}(\mathbf{x}_1, \mathbf{x}_2)$ is the scalar distance between the charges α_1 and α_2, and it obviously depends on the full set of molecular variables $\mathbf{x}_1$ and $\mathbf{x}_2$.

Except for one modification, the combination of (23) and (25) constitutes our water pair potential. The modification is required by the unphysical divergences that occur when two molecules move together in such a way that

$$(26) \qquad\qquad d_{\alpha_1\alpha_2}(\mathbf{x}_1, \mathbf{x}_2) = 0$$

Although this is not a serious matter when α_1 and α_2 have the same parity, it is catastrophic when they do not. For this reason, we multiply v_{el} by a "switching function," S, that is unity at large r_{12}, but vanishes when r_{12} is small enough that condition (26) might occur. Our complete water pair potential therefore has the form

$$(27) \qquad\qquad v(\mathbf{x}_1, \mathbf{x}_2) = v_{\text{LJ}}(r_{12}) + S(r_{12})v_{\text{el}}(\mathbf{x}_1, \mathbf{x}_2)$$

The specific form utilized for the switching function consists of three separate parts

$$(28) \qquad
\begin{aligned}
S(r_{12}) &= 0 & \text{for} \quad & 0 \le r_{12} \le R_1 \\[4pt]
&= \frac{(r - R_1)^2(3R_2 - R_1 - 2r)}{(R_2 - R_1)^3} & \text{for} \quad & R_1 \le r \le R_2 \\[4pt]
&= 1 & \text{for} \quad & R_2 \le r_{12} \le \infty
\end{aligned}$$

The cubic polynomial in interval $[R_1, R_2]$ renders $S(r_{12})$ a nondecreasing, continuous function with continuous first derivative. We must have $R_1 \ge 2.00$ Å to avoid the charge overlap catastrophe.

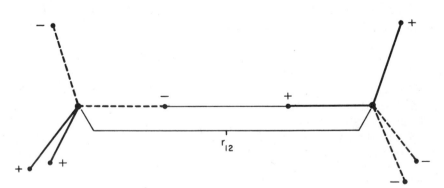

Figure 4.   The symmetrical eclipsed (SE) approach of two water molecules. Looking down the "bond" axis, the two triads of charges not along this axis would seem to fall upon (or eclipse) one another.

Pair potential $v(\mathbf{x}_1, \mathbf{x}_2)$ contains three adjustable parameters, η, R_1, and R_2. These quantities were determined by requiring first, for fixed η, that the potential have a minimum in one of the nearest-neighbor pair configurations occurring in the ice lattice. The specific ice lattice configuration used is the "symmetrical eclipsed" configuration illustrated in Figure 4. We required that the minimum occur at $r_{12} = 2.76$ Å, the observed neighbor distance in ice. The value of $v(\mathbf{x}_1, \mathbf{x}_2)$ at that minimum was also fixed at several different trial values $v_{\min}$ so that the conditions

(29)
$$v(\mathbf{x}_1, \mathbf{x}_2)\big|_{2 \cdot 76\,\text{Å}} = v_{\min}$$
$$\frac{\partial v(\mathbf{x}_1, \mathbf{x}_2)}{\partial r_{12}}\bigg|_{2 \cdot 76\,\text{Å}} = 0$$

uniquely determined R_1 and R_2. The strategy was to choose $v_{\min}$ to provide a good fit to the water-vapor second virial coefficient, $B(T)$, computed by the method of Section 4.

It was not possible to reproduce the experimental $B(T)$ when η was preset at the Bjerrum value 0.17. Instead, it proved necessary to increase η to 0.19 to permit adequate fit. (This increase in η beyond the value of 0.17 may be ascribed to a polarization effect operative at small distances.) One then obtains

(30)
$$\eta = 0.19$$
$$R_1 = 2.0379\,\text{Å}$$
$$R_2 = 3.1877\,\text{Å}$$

to complete the specification of $v(\mathbf{x}_1, \mathbf{x}_2)$. With this set of parameters, $v_{\min}$ for the symmetrical eclipsed, (SE), configuration is -6.50 kcal/mole, which is within the range of values quoted for hydrogen-bond energies.[18] Figure 5 presents the full r_{12} dependence of v in the SE configuration, along with the separate components v_{LJ}, S, and v_{el}.

Besides the SE configuration, there are three other nearest-neighbor configurations that occur in the ice lattice. They are illustrated in Figure 6. The nonsymmetrical eclipsed, NSE, configuration is obtained from SE by a 120° rotation about the oxygen–oxygen axis, and like SE its charges are in line with (i.e., eclipse) one another when viewed along this axis. The symmetrical staggered, SS, and nonsymmetrical staggered, NSS, cases have charges midway between one another when viewed along the oxygen–oxygen axis; SS is obtained from SE by a 180° rotation, and NSS from SE by a 60° rotation.

Figure 7 displays together the four potential curves, for varying r_{12}, for each of the ice configurations SE, NSE, SS, and NSS. The minima for the latter three numerically turn out to occur at the same distance, 2.76 Å, that

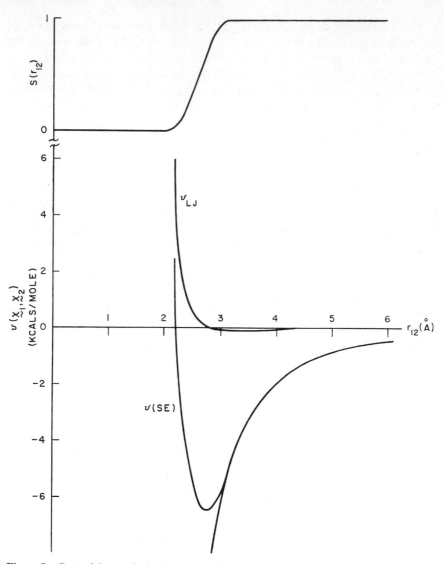

Figure 5. Potential curve for hydrogen bonding in the SE configuration; note three separate component functions v_{LJ}, S, and v_{el} related to $v(\mathbf{x}_1, \mathbf{x}_2)$ by (26).

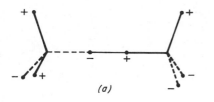

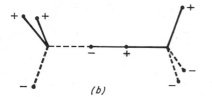

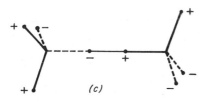

Figure 6. Ice-lattice neighbor configurations in addition to SE: (*a*) NSE; (*b*) SS; (*c*) NSS.

was forced upon the first one. However, the corresponding energies at those minima differ somewhat

$$
\begin{array}{lll}
& \text{SE} & -6.50 \text{ kcal/mole} \\
& \text{NSS} & -6.13 \text{ kcal/mole} \\
(31) & \text{NSE} & -5.58 \text{ kcal/mole} \\
& \text{SS} & -5.34 \text{ kcal/mole}
\end{array}
$$

The numerical values of the curvatures $\partial^2 v / \partial r_{12}^2$ for each of the four potential curves, evaluated at the minima, are

$$
\begin{array}{ll}
& v''(\text{SE}) = 22.53 \text{ kcal/(mole)(\AA}^2) \\
& v''(\text{NSS}) = 28.55 \text{ kcal/(mole)(\AA}^2) \\
(32) & v''(\text{NSE}) = 26.58 \text{ kcal/(mole)(\AA}^2) \\
& v''(\text{SS}) = 25.78 \text{ kcal/(mole)(\AA}^2)
\end{array}
$$

In the ice lattice, the staggered arrangements for nearest-neighbor pairs occur exactly three times as often as the eclipsed arrangements. Subject to

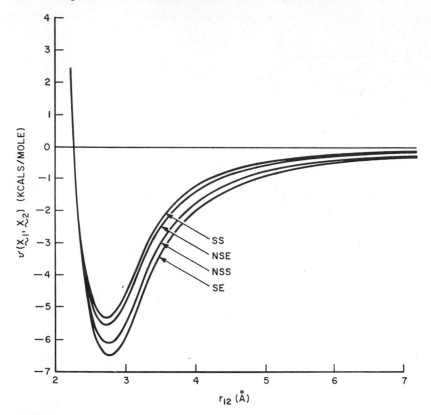

Figure 7. Potential curves for hydrogen bonding in the four ice lattice configurations.

this a priori restriction, Pauling's view of the residual entropy of ice[19] implies equal weights to the arrangements. We therefore calculate the "average curvature" to be

$$(33) \qquad \begin{aligned} (v'')_{av} &= \tfrac{1}{8}[v''(SE) + v''(NSE)] + \tfrac{3}{8}[v''(SS) + v''(NSS)] \\ &= 26.51 \text{ kcal/(mole)(Å}^2) \end{aligned}$$

This average curvature may also be estimated from the measured isothermal compressibility of ice, found by Jona and Scherer[20] to be

$$(34) \qquad \kappa_T = 1.11 \times 10^{-11} \text{ cm}^2/\text{dyne}$$

at $-16°C$. Assuming all nearest-neighbor distances to contract equally under compression, this κ_T is equivalent to

$$(35) \qquad (v'')_{av} = 25.25 \text{ kcal/(mole)(Å}^2)$$

Actually, the neighbors with less potential curvature should move together more rapidly than the "stiffer" neighbors, hence tending to weight the former more. This might explain part of the discrepancy between (33) and (35).

On the basis of this rough comparison, we conclude that our water potential is not grossly in error for description of the condensed phases.

Several other potential curves have been computed for different classes of water molecule pair configurations. Figure 8 displays the results for a pair of "two-bonded" configurations, TB_1 and TB_2; for both of these, two pairs of opposite charges simultaneously approach one another. Nevertheless, the distances involved are such that the repulsive part of v_{LJ} comes into play before the charge attractions get very large, so the net potential at the two-bonded minima is considerably higher than the single-bond results in (31). It therefore seems quite unlikely that two-bonded configurations play any significant role in liquid water.

Several structural models for liquid water, such as those proposed by Pauling,[21] Frank and Quist,[22] and Samoilov,[23] postulate the existence of

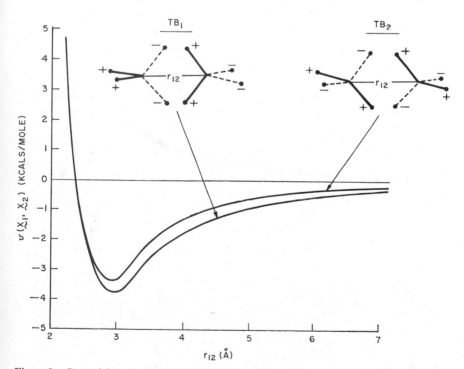

Figure 8.   Potential curves for "two-bonded" configurations.

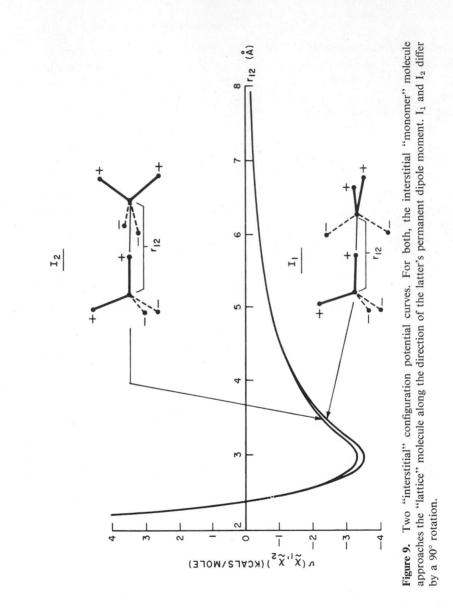

Figure 9. Two "interstitial" configuration potential curves. For both, the interstitial "monomer" molecule approaches the "lattice" molecule along the direction of the latter's permanent dipole moment. I_1 and I_2 differ by a 90° rotation.

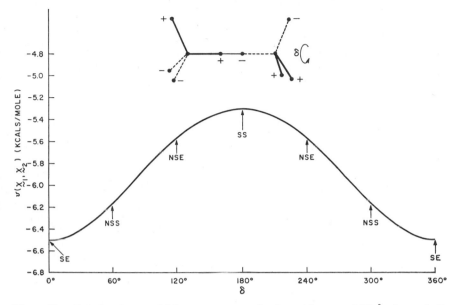

Figure 10. Rotational potential for two water molecules, with r_{12} = 2.76 Å. As angle δ increases from 0, the SE configuration deforms first to NSS (δ = 60°), then to NSE (120°), and then SS (δ = 180°). The curve is symmetric at about δ = 180°.

"monomeric interstitial" water molecules, trapped in cavities formed by hydrogen-bonded lattices. It is therefore pertinent to calculate potential curves for one molecule (the monomer) approaching another (a member of the lattice) along directions expected from the interstitial picture. Figure 9 presents two such curves. Again we see that the energy is substantially higher at the minima than for the four ice-lattice configurations, SE, NSE, SS, NSS. Nevertheless, such interstitial molecules can be somewhat stabilized by relatively free rotation.

Figure 10 exhibits a rotational potential energy curve. The two molecules start in the SE configuration, with r_{12} = 2.76 Å, and are rotated about the oxygen–oxygen axis. This process produces in turn the following sequence of configurations during a full 360° rotation: SE, NSS, NSE, SS, NSE, NSS, SE.

Finally, Figure 11 provides the predicted potential energy curve for a particular hydrogen-bond bending mode. This bend (also performed at fixed r_{12} = 2.76 Å) carries SE continuously to SS through a substantial potential barrier. From Figure 9, we see that the top of the barrier corresponds to interstitial configuration, I_1.

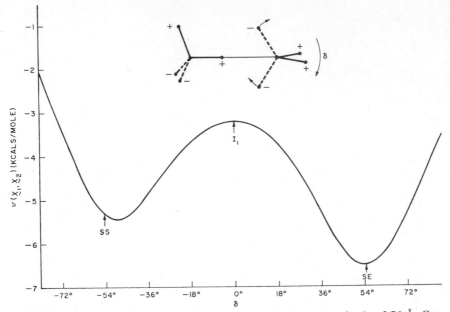

Figure 11. Potential curve for hydrogen-bond "bending," with r_{12} fixed at 2.76 Å. Configurations SS and SE occur at one-half the ideal tetrahedral angle, or 54.7°.

A survey of the various potential curves indicates that the absolute minimum potential energy for two water molecules is predicted by approximation (27) to occur at very nearly the geometrically ideal SE configuration, with $r_{12} \simeq 2.76$ Å.

4   Virial Coefficients

In the low-density vapor phase, the pressure equation of state (13) may be expanded into a density series

$$(36) \qquad \frac{pV}{NkT} = 1 + B(T)\left(\frac{N}{V}\right) + C(T)\left(\frac{N}{V}\right)^2 + D(T)\left(\frac{N}{V}\right)^3 + \cdots$$

The second, third, fourth,..., virial coefficients (B, C, D,..., respectively) convey the extent of imperfection of the gas, and they are intimately related to the interactions among clusters of ascending numbers of molecules.

The specific statistical-mechanical expressions for the virial coefficients in (36) may be obtained from (13) by insertion of an appropriate density series for $\rho^{(2)}(\mathbf{x}_1, \mathbf{x}_2)$, followed by partial integrations. However, a more direct route to the same results is provided by the standard Mayer cluster theory.[24] The second virial coefficient has the form

$$(37) \qquad B(T) = -\frac{1}{16\pi^2} \int d\mathbf{x}_2 \, \{\exp\left[-\beta v(\mathbf{x}_1, \mathbf{x}_2)\right] - 1\}$$

Knowledge of the total temperature dependence of $B(T)$ obviously does not in itself permit complete determination of $v(\mathbf{x}_1, \mathbf{x}_2)$. However, this knowledge can be useful in fixing free parameters in an assumed analytical approximation. On account of the exponential character of the integrand in (37), the value of $B(T)$ measured at low temperature is especially helpful in fitting $v(\mathbf{x}_1, \mathbf{x}_2)$ near its absolute minimum.

Reference to (3) shows that $B(T)$ for water is a sixfold integral. It is therefore much more difficult to evaluate with comparable accuracy than the single integration required for argon, the archetypal structureless spherical molecule. Still, methods have been developed for numerical integration of expressions such as (37); examples are the Haselgrove[25] and Conroy[26] techniques.

The few configurations for which the water pair potential was evaluated in the previous section were sufficiently special that convenient expressions could be written out, for computational purposes, for the charge-charge distance $d_{\alpha_1 \alpha_2}$ in v_{el}. But regardless of the specific integration scheme to be used for $B(T)$, it is necessary to evaluate $v(\mathbf{x}_1, \mathbf{x}_2)$ for very many configurations, the majority of which are irregular. For that reason we were compelled to develop a computer-coded system for finding v, given an arbitrary $\mathbf{x}_1$ and $\mathbf{x}_2$. This procedure utilizes transformation matrices $\mathbf{E}(\phi_i, \theta_i, \psi_i)$, depending on Euler angles for molecule i, which convert any vector $\mathbf{t}$ in the laboratory coordinate system, to the same vector $(\mathbf{t})^{(i)}$ resolved into components parallel to the right-handed coordinate system ($\mathbf{b}, \mathbf{m}, \mathbf{n}$ in Figure 1) attached to molecule i

$$(38) \qquad (\mathbf{t})^{(i)} = \mathbf{E}(\phi_i, \theta_i, \psi_i) \cdot \mathbf{t}$$

Since we have

$$(39) \qquad \begin{aligned} d_{\alpha_1 \alpha_2}^2 &= [\mathbf{r}_{12} + \mathbf{t}_{\alpha_2} - \mathbf{t}_{\alpha_1}]^2 \\ &= r_{12}^2 + t_{\alpha_2}^2 + t_{\alpha_1}^2 + 2\mathbf{r}_{12} \cdot \mathbf{t}_{\alpha_2} - 2\mathbf{r}_{12} \cdot \mathbf{t}_{\alpha_1} - 2\mathbf{t}_{\alpha_1} \cdot \mathbf{t}_{\alpha_2} \end{aligned}$$

where $\mathbf{t}_{\alpha_1}$ and $\mathbf{t}_{\alpha_2}$ are the displacements of charges α_1 and α_2 relative to their respective oxygen nuclei, it follows that

$$(40) \qquad \begin{aligned} d_{\alpha_1 \alpha_2}^2 &= r_{12}^2 + t_{\alpha_2}^2 + t_{\alpha_1}^2 + 2\mathbf{r}_{12} \cdot [\mathbf{E}^{-1}(2) \cdot (\mathbf{t}_{\alpha_2})^{(2)} - \mathbf{E}^{-1}(1) \cdot (\mathbf{t}_{\alpha_1})^{(1)}] \\ &\quad - 2[\mathbf{E}^{-1}(1) \cdot (\mathbf{t}_{\alpha_1})^{(1)}] \cdot [\mathbf{E}^{-1}(2) \cdot (\mathbf{t}_{\alpha_2})^{(2)}] \end{aligned}$$

The charge-position vectors, $(\mathbf{t}_{\alpha_i})^{(i)}$, can all easily be expressed in terms of the unit vectors $\mathbf{b}$, $\mathbf{m}$, and $\mathbf{n}$, so that the matrices $\mathbf{E}^{-1}$ in (40) account for arbitrary molecular rotations.

The Haselgrove method was actually the one used in our $B(T)$ calculations because it permits use of varying numbers of points, thus allowing estimation of integration error. For $0 \leq r_{12} \leq 2.00$ Å, the integrand shown in (37) is essentially constant at -1, so the contribution from this range to $B(T)$ is trivial to take into account. The actual numerical integration therefore was restricted to the range

(41) $$2.00 \text{ Å} \leq r_{12} \leq 15.00 \text{ Å}$$

since for larger separations the contributions are negligible. Typical computations at a given temperature involve 12,000 distinct pair configurations.

Table 1 shows the values computed for $B(T)$. We estimate the error to be about 5%. For comparison the table also includes Rowlinson's results for the Stockmayer potential,[13] as well as Kell, McLaurin, and Whalley's recent measurements.[27] As explained in Section 3, it was necessary to try different values of the charge magnitudes $\pm \eta e$. The numbers quoted in the table refer to $\eta = 0.19$.

Table 1   Second Virial Coefficient for Water (cm³/mole)

Temperature (°C)	$B(T)$[a]	Stockmayer Potential (Ref. 13)	Experiment (Ref. 27)
100	-466	-450	-450[b]
200	-190	-205	-197
300	-107	-122	-112
400	-65	-80	-72

[a] Computed from (37), using pair potential (27).
[b] Extrapolated from 150°C.

When η was preset as low as 0.17 (the Bjerrum value), and $v_{\min}$ varied, it was not possible to fit the experimental $B(T)$. The error was most significant at low T, and reflected too weak an attraction. On the other hand, when η was increased substantially beyond 0.19, it became impossible to find R_1 and R_2 to satisfy (29). The value chosen for η therefore seemed to represent a satisfactory intermediate value.

It would be unwarranted at this stage to spend a considerable effort to improve agreement between the theoretical and experimental $B(T)$'s. Surely this eventually could be done by a combination of the following: (a) shifts in positions of the charges from the regular tetrahedron vertices, (b) change in shape of the switching function, and (c) variation of ε and σ in v_{LJ} from the neon values.

But even if numerical error in integration of $B(T)$ were to be made negligibly small, one would still be confronted by quantum corrections and by the fact that the best pair potential for the liquid state, our primary object of interest here, very likely deviates somewhat from the true pair potential for isolated molecules. We believe, however, that our approximate fit to the experimental $B(T)$ serves to force upon $v(\mathbf{x}_1, \mathbf{x}_2)$ in (27) nearly the correct energy for hydrogen-bond formation. Since the fits to the measured $B(T)$ obtained with both our potential and the Stockmayer potential are about equally good, we see how insensitive $B(T)$ is to angular variations, and this serves to stress the importance of seeking other information to determine those angular variations.

The analog of $B(T)$ expression (37) for the third virial coefficient is a twelvefold integral

$$C(T) = -\frac{1}{3(8\pi^2)^2}$$

$$\times \int d\mathbf{x}_2 \int d\mathbf{x}_3 \{\exp\left[-\beta V_3(\mathbf{x}_1, \mathbf{x}_2, \mathbf{x}_3)\right]$$

(42)
$$-\exp\left[-\beta(v(\mathbf{x}_1, \mathbf{x}_2) + v(\mathbf{x}_2, \mathbf{x}_3))\right]$$
$$-\exp\left[-\beta(v(\mathbf{x}_1, \mathbf{x}_2) + v(\mathbf{x}_1, \mathbf{x}_3))\right]$$
$$-\exp\left[-\beta(v(\mathbf{x}_1, \mathbf{x}_3) + v(\mathbf{x}_2, \mathbf{x}_3))\right]$$
$$+\exp\left[-\beta v(\mathbf{x}_1, \mathbf{x}_2)\right] + \exp\left[-\beta v(\mathbf{x}_1, \mathbf{x}_3)\right]$$
$$+\exp\left[-\beta v(\mathbf{x}_2, \mathbf{x}_3)\right] - 1\}$$

In the event that the three-molecule potential energy, V_3, is composed just of pair potential contributions, the $C(T)$ expression simplifies considerably

(43)
$$C(T) = -\frac{1}{3(8\pi^2)^2} \int d\mathbf{x}_2 \int d\mathbf{x}_3 \, f(\mathbf{x}_1, \mathbf{x}_2) f(\mathbf{x}_1, \mathbf{x}_3) f(\mathbf{x}_2, \mathbf{x}_3)$$

where

(44)
$$f(\mathbf{x}_i, \mathbf{x}_j) = \exp\left[-\beta v(\mathbf{x}_i, \mathbf{x}_j)\right] - 1$$

The third virial coefficient is very sensitive to intermolecular potentials because it involves cancellation between large positive and large negative contributions coming from different regions of triplet configuration space.

It is much more difficult to carry out the integrations demanded for $C(T)$, even numerically, than for $B(T)$. Even so, we thought it worthwhile to seek a rough evaluation of $C(T)$ for comparison with experiment. Conroy's

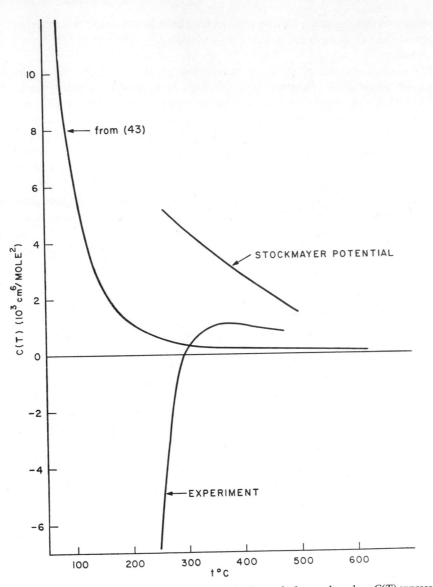

Figure 12. Third virial coefficient for water: the theoretical curve based on $C(T)$ expression (43) uses pair potential (27).

numerical integration procedure[26] was employed using potential (27) in (43). This procedure involved consideration of 9644 separate triplet configurations. The results (probably only accurate within a factor of 2) are presented in Figure 12, along with Rowlinson's Stockmayer potential $C(T)$,[14] as well as Kell, McLaurin, and Whalley's experimental values.[27]

It is clear that neither the Stockmayer potential nor our present form in (27) comes anywhere near to predicting an experimentally acceptable $C(T)$. The chief source of error is probably the fact that V_3 in general $C(T)$ expression (42) is not precisely a sum of v's for the separate pairs, but contains as well a true three-body potential. This inherent three-body part should consist mainly of a polarization effect; that is, one molecule interacts with the dipole moment induced in a second one by the third. Even though the extra three-body energy may be small, its effect can be very large, for it acts as a multiplier for

$$(45) \qquad \exp\{-\beta[v(\mathbf{x}_1, \mathbf{x}_2) + v(\mathbf{x}_1, \mathbf{x}_3) + v(\mathbf{x}_2, \mathbf{x}_3)]\}$$

to form the first integrand term in (42). If the three molecules are arranged to form two SE hydrogen bonds, with energy -13 kcal/mole, (45) at room temperature is about 2×10^9, and one of the succeeding integrand terms in (42) will have comparable magnitude but opposite sign. With such huge numbers, $C(T)$ ends up being exceedingly sensitive to inclusion of very small three-body potentials.

Although a refinement of our water pair potential·might improve agreement in Figure 12 somewhat, it should be stressed that prediction of a $C(T)$ in close agreement with experiment is not a prerequisite to formulation of an adequate liquid-state theory.

5   Pair-Correlation Function

We turn now to survey some theoretical aspects of the water pair-correlation function. This quantity may be expressed as

$$(46) \qquad g^{(2)}(\mathbf{x}_1, \mathbf{x}_2) = \exp\{-\beta[v(\mathbf{x}_1, \mathbf{x}_2) + w(\mathbf{x}_1, \mathbf{x}_2)]\}$$

The function w is both temperature and density dependent, and it comprises the average influence of the water medium surrounding fixed molecules 1 and 2.

In the large-r_{12} limit, of course, both the direct interaction, v, as well as the indirect correlation quantity, w, will vanish, to give unity for $g^{(2)}$. However, we can also specify the precise way in which $v + w$ goes to 0, because at large distances only the molecular dipole moments interact. Neglecting molecular polarizability, we see that molecule 1 will be surrounded by a polarization field at large distance equal to

$$(47) \qquad \mathbf{P}_1 = \frac{\varepsilon_0 - 1}{4\pi} \mathbf{E}_1$$

where $\mathbf{E}_1$ is the electric field due to molecule 1 and its orientationally correlated near neighbors. Classical electrostatics subsequently gives the following expression for $\mathbf{E}_1$[10]

$$(48) \qquad \mathbf{E}_1(\mathbf{r}_{12}) = -\frac{3\mu_d g_K}{2\varepsilon_0 + 1} \nabla_2 \frac{\mathbf{b}_1 \cdot \mathbf{r}_{12}}{r_{12}{}^3}$$

If this is substituted into (47) and the result identified as a deviation from isotropy of the distribution of directions for vector $\mathbf{b}_2$ in molecule 2, we must have

$$(49) \qquad \beta[v(\mathbf{x}_1, \mathbf{x}_2) + w(\mathbf{x}_1, \mathbf{x}_2)] \sim \frac{9 g_K(\varepsilon_0 - 1)}{4\pi\rho(2\varepsilon_0 + 1)} \mathbf{b}_1 \cdot \mathbf{T}_{12} \cdot \mathbf{b}_2$$

where $\mathbf{T}_{12}$ is the dipole-dipole tensor

$$(50) \qquad \mathbf{T}_{12} = \frac{1}{r_{12}{}^3}\left(1 - \frac{3\mathbf{r}_{12}\mathbf{r}_{12}}{r_{12}{}^2}\right)$$

In the low-density limit applicable to water vapor, w vanishes, so that the pair-correlation function, $g^{(2)}$, reduces to the Boltzmann factor for direct interaction, v

$$(51) \qquad \lim_{\rho \to 0} g^{(2)}(\mathbf{x}_1, \mathbf{x}_2) = \exp[-\beta v(\mathbf{x}_1, \mathbf{x}_2)]$$

The simpler correlation function $\bar{g}^{(2)}(r_{12})$, also in the zero-density limit, is equal to an angular average of (51)

$$(52) \qquad \lim_{\rho \to 0} \bar{g}^{(2)}(r_{12}) = \frac{1}{(8\pi^2)^2} \int \sin\theta_1 \, d\phi_1 \, d\theta_1 \, d\psi_1 \int \sin\theta_2 \, d\phi_2 \, d\theta_2 \, d\psi_2$$
$$\times \exp[-\beta v(\mathbf{x}_1, \mathbf{x}_2)]$$

This quantity gives the relative density of oxygen nuclei (regardless of molecular orientations) at distance r_{12} from the oxygen nucleus of a fixed molecule in the dilute vapor.

With the specific interaction (27), the integral in (52) has been evaluated at 4°C, and the result is plotted in Figure 13. The very high peak at about the ice-lattice spacing (2.76 Å) reflects the very strong attraction due to hydrogen bonding when the molecules are suitably oriented. Although the function has the same qualitative features as the pair Boltzmann factor for, say, a pair of argon atoms, the peak height is very much larger than for argon at the corresponding temperature. (The maximum of $\exp[\beta v_{LJ}(r)]$ for argon at

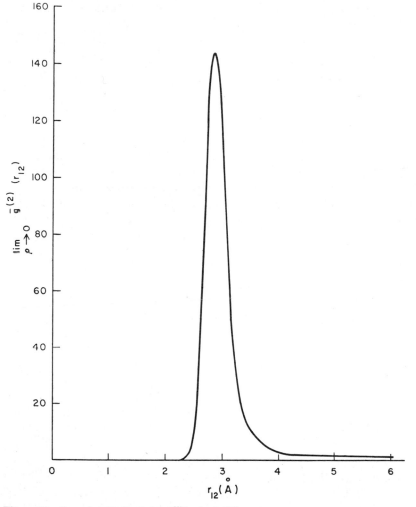

Figure 13. Low-density limit for $\bar{g}^{(2)}(r_{12})$ at 4°C.

83.8°K, its triple-point temperature, is only about 4.2.) Even in this orientationally averaged sense, we see that liquid water must be regarded as an
extremely strongly interacting many-body system.

One of the more important trends in liquid-state theory in recent years
has been the development of integral equation techniques for prediction of
pair-correlation functions.[5] So far, this approach has been applied numerically
only to fluids composed of spherical structureless particles, but it is important
to establish the extent of its possible utility for water. By most measures, the

Percus–Yevick integral equation[28] is the most reliable of those available; for water it is

$$
\begin{aligned}
\text{(53)}\quad \exp\left[\beta v(\mathbf{x}_1, \mathbf{x}_2)\right]g^{(2)}(\mathbf{x}_1, \mathbf{x}_2) = 1 + \frac{N}{8\pi^2 V}\int d\mathbf{x}_3[g^{(2)}(\mathbf{x}_1, \mathbf{x}_3) - 1] \\
\times \left\{1 - \exp\left[\beta v(\mathbf{x}_3, \mathbf{x}_2)\right]\right\}g^{(2)}(\mathbf{x}_3, \mathbf{x}_2)
\end{aligned}
$$

The speed and memory capacity of modern computers has reached a point where iterative solution of the Percus–Yevick equation for simple fluids such as argon is a relatively simple task.[29] However, it has already been pointed out that the water $g^{(2)}$ is a function of no less than six variables, rather than just radial distance as for simple fluids. It should be mentioned parenthetically, that one of the two molecules may be regarded as fixed in position and orientation. The six variables may then be taken to be the three polar coordinates of oxygen nucleus 2 relative to 1, and the three Euler angles for molecule 2. The corresponding numerical task for water, therefore, is orders of magnitude more difficult. To convey the structural information implicit in $g^{(2)}(\mathbf{x}_1, \mathbf{x}_2)$, a minimum of about 10 discrete values for each of the six variables should be considered, thus requiring a table of one million entries!

Also, our experience with trial integrals has shown that about four minutes is required to carry out the integral in (53), for each pair-configuration $\mathbf{x}_1, \mathbf{x}_2$, with even modest numerical accuracy. Therefore each iteration of (53) in seeking a numerical $g^{(2)}$ for water would require more than one thousand hours of computing time! Similar estimates apply to the other integral equations currently in vogue, so the difficulty is not to be regarded as a special intractability of the Percus–Yevick equation.

In spite of the present virtual impossibility of solving the Percus–Yevick integral equation (or any of the others) for liquid water, this general approach can still provide some indirect insight. For example, we can observe the effect on local structure of the hydrogen-bonding part of $v(\mathbf{x}_1, \mathbf{x}_2)$ by first integrating the Percus–Yevick equation with just the first term in the potential expression (27), and then comparing the result with the *experimental* $\bar{g}^{(2)}(r_{12})$. Figure 14 shows the two curves together, and both refer to 4°C, and the 1-atm molecular density for real water. The theoretical curve represents highly compressed, supercritical neon gas (the pressure would be about 5000 atm), since with only central interactions operative the molecular rotations are free and irrelevant. The experimental curve was determined by Narten, Danford, and Levy;[30] it probably contains slight artifacts (from Fourier inversion of the scattering data) such as the bump at 3.7 Å.

Not only does the hydrogen-bond part (Sv_{el}) of the water pair potential cause an enormous pressure reduction from 5000 to 1 atm, but we also see that the first peak of the pair-correlation function undergoes very marked narrowing. As a result, the number of nearest neighbors, defined by the area under

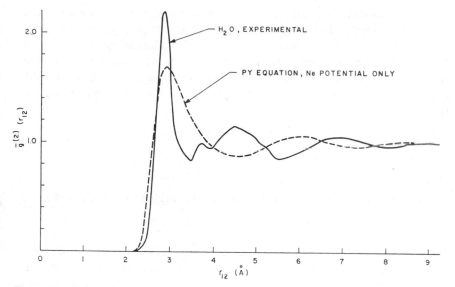

Figure 14. Comparison of the experimental $\bar{g}^{(2)}$ for water at 4°C, with the radial distribution function calculated from the Percus–Yevick equation (same temperature and density) using only the Lennard-Jones part of water potential (27).

the first peak out to the subsequent minimum, reduces from around eight for neon to roughly four for water.

Figure 14 also shows that the first-peak narrowing is accompanied by inward movement of the subsequent $\bar{g}^{(2)}(r_{12})$ peaks, so that the oscillations about the first peak for neon and for water are "out of phase." In the neon case, the second and third peaks are very nearly at two and three times the first-peak distance, respectively. Although the second peak for water is quite broad, its maximum lies very close to the position expected for perfect tetrahedral coordination, namely, $2 \sin(54°44') \simeq 1.633$ times the first-peak distance. The third water peak is too diffuse to identify uniquely and likely represents contributions from a wide variety of local structures that are possible with predominantly tetrahedral linkage.

The function $w(\mathbf{x}_1, \mathbf{x}_2)$ in (46), the general expression for the water pair-correlation function, may in principle be expanded in some complete ortho-normal set of functions, F_α, of just the Euler angle variables

$$(54) \qquad w(\mathbf{x}_1, \mathbf{x}_2) = w_0(r_{12}) + \sum_{\alpha=1}^{\infty} w_\alpha(r_{12}) F_\alpha(\phi_1 \cdots \psi_2)$$

where the coefficient functions, w_α, depend only on r_{12}. Equation 49 indicates the behavior of this expansion at large r_{12}, but greater structural significance attaches to the case of small r_{12}, when the molecules are first, second, or third neighbors, roughly. For such close pairs, the direct pair potential $v(\mathbf{x}_1, \mathbf{x}_2)$ exerts very strong forces and torques, and it is unclear what relative importance the angle-dependent ($\alpha \geq 1$) parts of $w(\mathbf{x}_1, \mathbf{x}_2)$ would have. In order to get some information on this point, we shall tentatively disregard the α sum in (54), and observe the consequences. We therefore assume for the moment that

(55) $$g^{(2)}(\mathbf{x}_1, \mathbf{x}_2) \simeq y(r_{12}) \exp\left[-\beta v(\mathbf{x}_1, \mathbf{x}_2)\right]$$

where

(56) $$y(r_{12}) = \exp\left[-\beta w_0(r_{12})\right]$$

If (55) is averaged over Euler angles $\phi_1 \cdots \psi_2$ at fixed r_{12}, the result is

(57) $$\bar{g}^{(2)}(r_{12}) = y(r_{12}) \lim_{\rho \to 0} \bar{g}^{(2)}(r_{12})$$

The left-hand member is the measured radial correlation function (shown in Figure 14), and the zero-density limit appearing in the right-hand member has already been computed and displayed in Figure 13. We can therefore combine these two pieces of information to produce a semiempirically determined $y(r_{12})$, which is shown in Figure 15. The most important feature of $y(r_{12})$ is its small value ($\simeq 0.015$) at the nearest-neighbor distance; this value prevents more than about four neighbors from fitting around any one molecule.

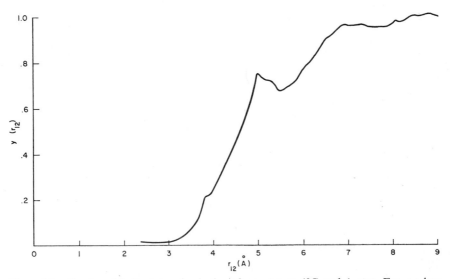

Figure 15. Semiempirically determined $y(r_{12})$ for water at 4°C and 1 atm. For r_{12} less than 2.5 Å, relation (57) for computing $y(r_{12})$ becomes effectively indeterminate.

Now that $y(r_{12})$ has been determined, it may be used in (55) along with our pair potential, (27), to yield an approximate pair-correlation function. The latter then may be utilized in (19) to calculate an approximate dielectric g_K. The result for water at 4°C is

$$(58) \qquad\qquad g_K = 1.94$$

On the other hand, Harris, Haycock, and Alder[31] have shown that g_K must be about 2.6 to be consistent with the measured dielectric constant. We are therefore forced to conclude that approximation (55) fails to give a proper account of orientational correlation between neighboring water molecules; in other words, it produces too little angular correlation.

This conclusion may be confirmed by calculating the number of hydrogen bonds per molecule n_{HB}, from (22), again employing the approximate water $g^{(2)}$ in (55). By "hydrogen bond" between two molecules, we shall mean that the pair potential for them shall be less than some preassigned upper limit, u. In other words, the characteristic bond function $B(\mathbf{x}_1, \mathbf{x}_2)$ may be written

$$(59) \qquad\qquad B(\mathbf{x}_1, \mathbf{x}_2) = U[u - v(\mathbf{x}_1, \mathbf{x}_2)]$$

where

$$(60) \qquad \begin{aligned} U(s) &= 0 \qquad s < 0 \\ &= 1 \qquad s \geq 0 \end{aligned}$$

is the unit step function. Table 2 presents the values calculated in this way for n_{HB} in 4°C water at 1-atm pressure.

Table 2   Number of Hydrogen Bonds per Water Molecule, n_{HB}, as the Cutoff Energy, u, is varied[a]

u (kcal/mole)	n_{HB}
-4.75	7.0×10^{-1}
-5.00	5.0×10^{-1}
-5.25	3.4×10^{-1}
-5.50	2.1×10^{-1}
-5.75	1.0×10^{-1}
-6.00	1.8×10^{-2}
-6.25	3.0×10^{-3}
-6.45	4.1×10^{-5}

[a] Pair-correlation function provided by (55); 4°C and 1 atm.

As u moves downward toward -6.50 kcal/mole, the number of hydrogen bonds per molecule declines rapidly to 0, because so little configuration space corresponds to formation of a bond. By referring to (31), we see that for low-temperature ice, n_{HB} should equal 2 when u exceeds -5.3 kcal/mole.

Since the energy of sublimation of ice at its melting point is 11.65 kcal/mole, and the heat of melting is 1.44 kcal/mole,[32] the molar energy of 4°C water compared to separated molecules will be close to 10.2 kcal/mole. Since, furthermore, the coordination number upon melting remains about 4, there is no way of accounting for this energy unless more than three neighbors of any molecule, on the average, participate in hydrogen bonds of no less than 4-kcal/mole energy each. However, the $u = -5$ kcal/mole entry in Table 2 is only 0.5, instead of roughly 1.5, as it should be. Once again we see that approximation (55) produces too little angular correlation between neighbors.

Evidently the local networks of hydrogen bonds that form in liquid water provide extra orientational correlation between nearest neighbors. It is easy to imagine that two such neighbors are embedded in a framework of other water molecules and that this framework acts like a machinist's jig to align the molecules in the correct orientation for hydrogen-bond formation. The implication is that a proper pair-correlation-function theory would specify the $w_\alpha(r_{12})$ in (54) to encompass this "jig effect," and the resultant $g^{(2)}(\mathbf{x}_1, \mathbf{x}_2)$ would yield more realistic g_K and n_{HB} predictions.

Finally we mention that if (53), the Percus–Yevick equation, is integrated over Euler angles for molecules 1 and 2 (while r_{12} is held fixed), and then trial form (55) inserted for $g^{(2)}$, the result constitutes an integral equation for $y(r_{12})$. We have spent considerable effort (though with some simplification of the angular integrations) attempting to solve that integral equation, to compare the result with the "experimental" y in Figure 15. That effort was entirely unsuccessful; no stable solution was found. In retrospect, it seems likely that this failure is symptomatic of the inability of the trial $g^{(2)}$ form in (55) to represent the cooperative aspects of hydrogen-bond network formation in liquid water.

6   Conclusions

Having thus surveyed the applicability of modern techniques in statistical mechanics to liquid water, we now attempt to identify the most likely course of significant progress in the near future.

The specific potential function, $v(\mathbf{x}_1, \mathbf{x}_2)$ in (27), was introduced largely for illustrative purposes. Although we believe it exhibits the major features of the correct water potential, some quantitative revisions will surely be warranted as further information comes to light. Direct quantum-mechanical calculations of the water pair potential could be of especial value in this regard; but because even a full hydrogen bond represents but a small fraction of the total energy involved in such calculations, this type of investigation must be implemented with great care and precision.

Morokuma and Pedersen[33] have recently carried out a quantum-mechanical computation of the interaction energy of two water molecules in a few selected configurations. They employed molecular orbital theory with a Gaussian basis set of modest proportions. Since their results predict a minimum value of $v(\mathbf{x}_1, \mathbf{x}_2)$ below -12.6 kcal/mole (roughly twice the accepted energy of the hydrogen bond, and certainly much larger than allowed by the measured second virial coefficient), it is certain that more extensive calculations are required. Still, the Morokuma–Pedersen work sets a valuable precedent that should be followed up in the near future.

One aspect of the quantum-mechanical calculations of particular interest would be the tracing out of the angular variation of $v(\mathbf{x}_1, \mathbf{x}_2)$ that is shown in Figure 11 for our own approximate potential. The potential barrier shown in Figure 11 between the SE and SS configurations is clearly a result of idealizing the lone pairs of unshared electrons by point charges, whereas in reality they are rather smeared out spatially. Morokuma and Pedersen[33] calculate that, rather than a barrier, a very shallow minimum should lie between SE and SS; so in fact the most stable configuration for two water molecules would not lie near SE, but near the "interstitial" configuration, I_1, instead. At present it is impossible to tell if this particular aspect of the Morokuma–Pedersen work is an artifact. More accurate molecular orbital calculations could help resolve the issue, although additionally it should be asked if electron correlation might tend to localize the unshared pairs along the characteristic tetrahedral directions. Future quantum-mechanical investigations should not overlook the nuclear distortion of the water molecules as they move relative to one another. This effect conceivably could vary with angle in a way that produces significant effects on the shape of potential curves such as the one in Figure 11. If it should turn out ultimately that the barrier shown in Figure 11 is either too high or altogether absent, the potential (27) could accordingly be modified by use of *three* negative charges, two in the same position as indicated in Figure 3 and one along their bisector (direction $-\mathbf{b}$). Of course the molecule must remain electrically neutral, so with two shielded protons having charges $+\eta e$, the total charge of the three negatives would be $-2\eta e$, as before.

Under the plausible assumption that our quantitative knowledge of $v(\mathbf{x}_1, \mathbf{x}_2)$ will continue to improve, it is important to prognosticate the significant uses of this function. In view of the pessimistic estimates for the direct solution of $g^{(2)}$ integral equations in the near future, an attractive alternative is the Monte Carlo simulation of the thermal behavior of a sample of the molecular fluid.[34,35] In this procedure, an electronic computer moves a set of several hundred molecules about inside a "vessel," according to Markov chain transition probabilities that are selected to realize a canonical distribution for the temperature of interest (these transition probabilities perforce depend on potential, v). With the presumption of ergodicity, the random motion of these molecules eventually allows any thermodynamic or structural feature of interest to be calculated as an average over the Markovian sequence of system states.

In principle, the full water pair-correlation function $g^{(2)}(\mathbf{x}_1, \mathbf{x}_2)$ could be evaluated by the Monte Carlo technique; but as mentioned earlier, the large number of configurational variables involved makes this impractical. Instead, $\bar{g}^{(2)}(r_{12})$ could be obtained for comparison with experiment, plus a more detailed analysis of the angular correlation of just nearest neighbors to establish the extent of hydrogen-bond bending in the liquid. The importance of hydrogen-bond bending in liquid water has been stressed by Pople.[36]

Since the Monte Carlo method actually produces "typical" liquid-water molecular arrangements, the strongest benefit to be derived from this method would be the pictures that could be made of a small portion of the liquid. Output from an electronic computer can nowadays routinely be used to produce stereoscopic images, and the student interested in water would have the opportunity literally to see how water molecules arrange themselves. The computer could be programmed not only to make the molecules clearly visible but also to indicate where the hydrogen bonds have formed (with suitable choice for B). A result of this submicroscopic view would be a deeper appreciation of the statistics of random hydrogen-bond networks according to probability of formation of polygons of different sizes (squares, pentagons, hexagons, etc.), and according to concentration of various types of faults in the random network (interstitial molecules, free ends, Bjerrum faults, etc.).

The Monte Carlo technique furthermore could be adapted to study of aqueous solutions. A fixed "impurity" can be placed inside the vessel which interacts suitably with the water molecules. For instance, if methane were the solute of interest, a central potential of the Lennard-Jones type would not be inappropriate. One could then study the change in extent and geometrical character of hydrogen bonding around this solute molecule and proceed to assess the current ideas about hydrophobic bonding.[37,38]

Although the Monte Carlo simulation of real water will very likely play an important role in future developments, it certainly must not be considered as an utterly definitive and complete source of knowledge. It is, after all, only a refined (and highly magnified!) sort of experiment on water, and for the most part will only tell us "what," not "why." The Monte Carlo results will eventually require explanations based on analytical theory, in the same way that the integral equation formalism for $g^{(2)}$ nowadays affords explanations for simple fluids (like argon).

There are two reasonable possibilities for "analytical" advances.

1. It is quite conceivable that trial $g^{(2)}$ form (55) could be generalized to represent better the cooperative nature of hydrogen bonding. One possibility would be

$$(61) \qquad g^{(2)}(\mathbf{x}_1, \mathbf{x}_2) \simeq y(r_{12}) \exp \left[-\beta z(r_{12}) v(\mathbf{x}_1, \mathbf{x}_2) \right]$$

involving now *two* dimensionless functions just of scalar distance r_{12}. The Percus–Yevick equation (or an alternative integral equation) could then be transformed into a pair of coupled integral equations for the functions y and z, and, one hopes, solved numerically. It is worth noting that, unlike (55), more general expression (61) is consistent with the large-r_{12} limiting pair-correlation behavior determined by (49). In addition, the requirement inferred in the previous section that neighboring particles require more orientational correlation than (55) provides can be accommodated by z exceeding unity at those distances.

2. Bernal's[39,40] admirable intuitive ideas about the coordination geometry of simple liquids deserve an incisive adaptation and application to liquid water. Unlike the simple liquids of spherical molecules, water has the advantage (at least at low temperature) of having definite coordination number 4. The object of a Bernal-type analysis therefore would primarily consist in description of the various types of polyhedra that occur surrounding voids and the statistics of fitting together these polyhedra to form a space-filling network.

Of course it is always risky to predict the future. There is a large chance that our projection for theoretical liquid-water research will prove somewhat misdirected. Nevertheless, significant progress will not come easily in this complicated field, so there is wisdom in attempting to plan effort intelligently. We hope that the present survey will aid scholars of the subject in that important endeavor.

Acknowledgments

The authors are indebted to Dr. J. Rasaiah for suggestions concerning the most effective numerical procedure for evaluating multidimensional integrals encountered in this investigation. We are also grateful to Mr. R. L. Kornegay and Mrs. Z. Wasserman for assistance in the computational aspects of the work.

REFERENCES

1. W. C. Röntgen, *Ann. Phys. Chim.* (*Wied.*), **45**, 91 (1892).
2. H. M. Chadwell, *Chem. Rev.*, **4**, 375 (1927).
3. J. D. Bernal and R. H. Fowler, *J. Chem. Phys.*, **1**, 515 (1933).
4. A convenient survey is provided by J. L. Kavanau, *Water and Solute-Water Interactions*, Holden-Day, San Francisco, 1964.
5. S. A. Rice and P. Gray, *The Statistical Mechanics of Simple Liquids*, Interscience, New York, 1965, Chapter 2.
6. N. Bjerrum, *Kgl. Danske Videnskab. Selskab, Mat.-Fys. Medd.*, **27**, 1 (1951).
7. L. Pauling, *The Nature of the Chemical Bond*, 3rd ed., Cornell University Press, Ithaca, N.Y., 1960, p. 466.
8. H. S. Green, *Proc. Roy. Soc.* (*London*), *Ser. A*, **189**, 103 (1947).
9. F. H. Stillinger, Jr., *Phys. Fluids*, **3**, 725 (1960).
10. J. G. Kirkwood, *J. Chem. Phys.*, **7**, 911 (1939).
11. H. Friedmann, *Physica*, **30**, 921 (1964).
12. W. H. Stockmayer, *J. Chem. Phys.*, **9**, 398 (1941).
13. J. S. Rowlinson, *Trans. Faraday Soc.*, **45**, 974 (1949).
14. J. S. Rowlinson, *J. Chem. Phys.*, **19**, 827 (1951).
15. L. Onsager, *J. Amer. Chem. Soc.*, **43**, 189 (1939).
16. G. A. Jeffrey and R. K. McMullan, *Progr. Inorg. Chem.*, **8**, 43 (1967).
17. J. Corner, *Trans. Faraday Soc.*, **44**, 914 (1948).
18. Ref. 7, Chapter 12.
19. Ref. 7, pp. 467–468.
20. F. Jona and P. Scherrer, *Helv. Phys. Acta*, **25**, 35 (1952).
21. Ref. 7, p. 473.
22. H. S. Frank and A. S. Quist, *J. Chem. Phys.*, **34**, 604 (1961).
23. O. Y. Samoilov, *Structure of Aqueous Electrolyte Solutions*, transl. D. J. G. Ives, Consultants Bureau, New York, 1965.
24. J. E. Mayer and M. G. Mayer, *Statistical Mechanics*, Wiley, 1940, Chapter 16.
25. C. B. Haselgrove, *Math. Comput.*, **15**, 323 (1961).
26. H. Conroy, *J. Chem. Phys.*, **47**, 5307 (1967).
27. G. S. Kell, G. E. McLaurin, and E. Whalley, *J. Chem. Phys.*, **48**, 3805 (1968).
28. J. K. Percus and G. J. Yevick, *Phys. Rev.*, **110**, 1 (1958).
29. A. A. Khan, *Phys. Rev.*, **134**, A367 (1964).
30. A. H. Narten, M. D. Danford, and H. A. Levy, *Discussions Faraday Soc.*, **43**, 97 (1967); see also, A. H. Narten, M. D. Danford, and H. A. Levy, Oak Ridge Natl. Lab. Rept. ORNL-3997, "X-Ray Diffraction Data on Liquid Water in the Temperature Range 4°C–200°C," September 1966.
31. F. E. Harris, E. W. Haycock, and B. J. Alder, *J. Chem. Phys.*, **21**, 1943 (1953).
32. G. Némethy and H. A. Scheraga, *J. Chem. Phys.*, **41**, 680 (1964).
33. K. Morokuma and L. Pedersen, *J. Chem. Phys.*, **48**, 3275 (1968).
34. N. Metropolis, A. W. Rosenbluth, M. N. Rosenbluth, A. H. Teller, and E. Teller, *J. Chem. Phys.*, **21**, 1087 (1953).
35. W. W. Wood, *J. Chem. Phys.*, **48**, 415 (1968).
36. J. A. Pople, *Proc. Roy. Soc.* (*London*), *Ser. A*, **205**, 163 (1951).
37. W. Kauzmann, *Advan. Protein Chem.*, **14**, 1 (1959).
38. G. Némethy, *Angew. Chem.*, **6**, 195 (1967).
39. J. D. Bernal, *Proc. Roy. Inst. G. Brit.*, **37**, 355 (1959).
40. J. D. Bernal, *Sci. Amer.*, **203** (August, 1960), p. 124.

9 Continuum Theories of Liquid Water*

G. S. Kell, *Division of Chemistry, Montreal Road Laboratories,*
National Research Council of Canada,
Ottawa, Ontario, Canada K1A 0R9

Abstract

A continuum theory of liquid water sees the structure as showing single distributions of oxygen–oxygen distances, oxygen–oxygen–oxygen angles, and energies; it is to be distinguished from mixture theories, which see water as a solution of distinguishable species. This chapter is a discursive survey of the evidence (a) for possible configurations for water molecules, (b) for the structure of the liquid that is derived from such configurations in the case of materials such as vitreous silica, and (c) from the properties of fluid water itself. The tetrahedral model is the ball-and-stick model most closely approximating the probable structure of liquid water. A classical treatment of the directional properties of the water molecule is illustrated by consideration of the quadrupole moment; the intermolecular potential is poorly known, even for water vapor. The intramolecular and intermolecular vibration spectra are related to those of ice, and their interpretation should start from ice, which is better understood. It is considered that the concept of broken hydrogen bond is inherently vague, even in the context of mixture theories. The potential energy of the liquid, which can be determined by combining statistical and thermodynamic data, promises to be a useful correlating parameter. The structure of aqueous solutions appears as a nine-dimensional problem, too difficult for rigorous treatment at the present time.

1 General Considerations

Background of Theories of Water. There is no agreement about the basic physical phenomena that produce the properties of liquid water. What is the

---

* N.R.C.C. No. 11982.

simplest model that will be waterlike? Some favor mixture models. Others favor continuum models; these appear incompatible with mixture models, however. This chapter outlines the evidence that supports continuum theories. Comparison with Chapter 10 will show the extent of the supposed incompatibility.

Continuum theories attempt to discover the structure of liquid water by asking:

1. What are the nature and magnitude of the forces between water molecules where the intermolecular geometry is known?
2. From the relation between energy and structure in such cases, what structure would be likely for liquid water?
3. Conversely, what can we deduce of forces and structure from the properties of the liquid itself?

Since there is no comprehensive treatment of the structure of liquid water from the continuum point of view, this chapter is a simple introduction to continuum theories, and the reader is directed to more specialized papers, which are not summarized here. The continuum approach is shown in relation to the whole range of liquid properties, and depth is sought only in a few places. In several cases purely formal developments are given which clarify understanding without leading to calculation. The distinction between mixture and continuum theories has been emphasized.

MIXTURE AND CONTINUUM THEORIES.   The molecular motions of water involve several time scales, as was emphasized by Eisenberg and Kauzmann.[1] The OH-stretching vibration near 3500 cm^{-1} takes 10^{-14} sec; the dielectric relaxation time is 10^{-11} sec. A process taking 3×10^{-12} sec is slow compared with the OH vibration but fast compared to the dielectric relaxation. The distinction between mixture and continuum theories may depend on the time scale. Macroscopic experiments see water as homogeneous and isotropic. Is there a shorter time scale on which water is composed of distinguishable species?

A *mixture model* is understood to describe liquid water as an equilibrium mixture (or solution) of species that are distinguishable in an instantaneous picture. Many mixture theories have been proposed. Two classes may be distinguished. In theories of class A, at least one component is thought of in definite structural terms, as an ice, for example, and is considered to occur in regions of at least some tens of molecules. Class A theories include those of Frank and Quist[2] or Kamb.[3] In class B theories, the species are considered to differ simply in the number of broken and unbroken hydrogen bonds, and extended structures may not be considered. The number of broken bonds found by various authors was collected by Falk and Ford,[4] and the corresponding thermodynamic parameters were investigated by Davis and Jarzynski.[5] In mixture theories, each water molecule might participate in from zero to four hydrogen bonds, so species with different numbers of hydrogen bonds are sometimes distinguished. A notation for such species has been given by Stevenson.[6] Any mixture theory that admits an infinite number of components, so that they have overlapping properties, becomes a continuum theory. Mixture theories normally consider only two or three components, but a few consider more because of the large number of phases of ice.

A *continuum theory* describes water as having essentially complete hydrogen bonding, at least at low temperatures, but as having a distribution of angles, distances, and bond energies. Convincing theoretical work on this model has been performed by Pople.[7] Whereas a mixture model implies different proportions of the species at different temperatures and pressures, a continuum model considers the average bond energy to change with temperature and pressure because of changes in the distributions of bond lengths and of distortions of the angles. I shall call the continuum model considered here the *tetrahedral network* model, for such a model is the static object that best represents my view of liquid water; a more accurate description would require both probability and dynamics.

Mixture and continuum theories may be illustrated by the potential energy. Figure 1a illustrates a mixture theory, showing the potential energy as a function of volume. In two-state theories of water, the structured or hydrogen-bonded component has greater volume (and lower entropy) than the non-structured one, and the barrier between the two wells is high enough that molecules remain on one side for perhaps the 3×10^{-12} sec mentioned above. Each well has its own energy levels. In continuum theory (Figure 1b) volume and entropy are not suitable parameters, but the number of molecules with a given potential energy shows a single peak; the curve need not—and in view of the complexity of the situation, will not—be symmetrical, and it may have shoulders; but it still reflects a single distribution. Each potential energy can be produced by several local geometries, but these geometries are themselves part of a distribution of geometries. For convenience, the distribution may be approximated by a number of discrete values (or as the sum of constituent curves) and calculations made by the methods used in mixture theories; but it must be realized that the choice of discrete values is largely arbitrary and that their use does not mean that water really has discrete components rather than a distribution.

Various measures may be used to describe the average structure of liquid water, including the pair distribution function.

In an obvious extension to molecules with angular dependence of the

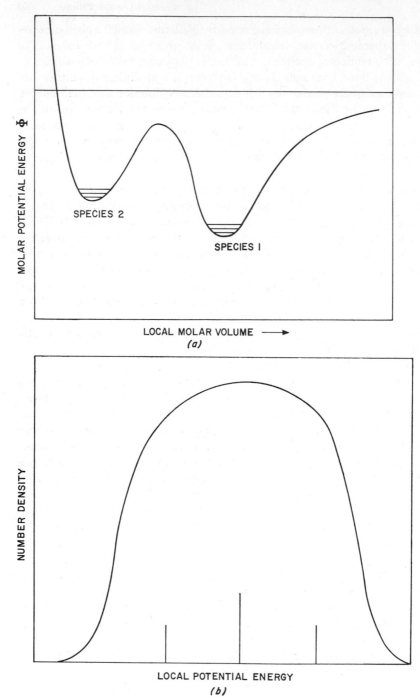

SPECIES 2

SPECIES I

LOCAL MOLAR VOLUME ⟶

(a)

LOCAL POTENTIAL ENERGY

(b)

334

molecular distribution functions for spherically symmetrical molecules, described, for example, by Egelstaff,[8] mixture and continuum theories may also be illustrated in terms of molecular distribution functions. Figure 2 shows the position of a second water molecule relative to a fixed one. The position of the oxygen is given by the three components of vector $\mathbf{R}$; the direction of the dipole axis requires two angles (only the number, not the coordinate system, is important here); and a further angle is needed to specify the rotation of the molecule about the dipole axis. Thus six components, one distance and five angles, are needed to specify the position and orientation of the second molecule.

The mean density of centers of mass of water molecules in ordinary three-dimensional space is $^3\rho$, and the density of molecules with a given configuration is $^6\rho$. The dipole may point in any direction, that is, over 4π steradians. The angle about the dipole axis is 2π, but because water has a plane of symmetry, only half of this is included. Hence $^3\rho = 4\pi^2\,^6\rho$. The probability of finding a water molecule in the position and configuration given by the six-dimensional vector, τ, is $^6n^{(1)}(\tau)\,d\tau$, and for water $^6n^{(1)}(\tau) = {}^6\rho$.

Define $^6n^{(2)}(\tau_1, \tau_2)\,d\tau_1\,d\tau_2$ as proportional to the probability of finding a molecule in the element $d\tau_2$ at τ_2, if there is a molecule in the element $d\tau_1$ at τ_1, irrespective of the coordinates of other molecules. The spherical analog is $^3n^{(2)}(\mathbf{R}_1, \mathbf{R}_2)\,d\mathbf{R}_1\,d\mathbf{R}_2$. In the limit of large separations $n^{(2)}(1, 2) \rightarrow n^{(1)}(1)n^{(1)}(2)$. The *pair-distribution function* is defined as

$$g(\tau) = \frac{1}{{}^6\rho^2}\,{}^6n^{(2)}(\tau)$$

which goes to unity for large separations, where $\tau = \tau_2 - \tau_1$. The spherical analog is $g(R) = n^{(2)}(R)/\rho^2$. For liquid water, $g(\tau)$ gives the probable position and orientation of first neighbors rather closely; more distant molecules are less precisely located, and $g(\tau)$ approaches unity at about 8 Å according to the radial distribution function from X-ray diffraction. Over long times or over all molecules in the liquid there is a single $g(\tau)$. Mixture models imply that on a local scale there are three distribution functions. That is, if $^6\rho_1$ is

---

Figure 1. Potential energy of water. (*a*) In two-state mixture theory there are two wells, and the well holding the denser species is deep enough to hold a significant population of molecules. (*b*) In continuum theory the molecules have a distribution of energies (with the broadness of an infrared band). The distribution may be approximated by the discrete energies shown, but the discrete values do not give a true picture of water.

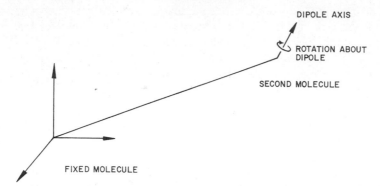

Figure 2. Six coordinates are required to describe the position and orientation of a second water molecule relative to a first one. Three coordinates are required to give the length and direction to the center of the second molecule, two to specify the orientation of the dipole axis, and another to specify the orientation of the molecule about the dipole axis.

the density of species 1 and $^6\rho_2$ the corresponding density of species 2, we can consider the three pair-distribution functions

$$g_{12}(\tau) = \frac{1}{^6\rho_1\,^6\rho_2}\; {}^6n_{12}^{(2)}(\tau)$$

which give the relation of molecules of species 1 to those of 2, and the corresponding $g_{11}(\tau)$ and $g_{22}(\tau)$. No extensive $g(\tau)$ is known for liquid water, but the concept provides a clear description of a mixture theory.

Evolution of theories of liquid water. With the development of notions of chemical equilibrium in the nineteenth century, it came to be realized that the maximum density of water could be explained by an equilibrium between a bulky icelike component and a component like normal liquids. Röntgen[9] sought to extend the idea to explain the decrease of compressibility from 0 to 50°C, the increase of thermal expansion with increasing pressure in the same temperature range, and the decrease of viscosity with increasing pressure. He argued qualitatively that the concentration of dissolved ice molecules must be less at high temperatures and high pressures, which explains the phenomena.

Theories of the composition of water made little progress for the next 40 years, as can be seen by comparing the report of a 1910 meeting on the constitution of water[10] with a 1927 review.[11] In one theory, water was considered a solution of hydrol (H_2O), dihydrol ($H_2O)_2$, trihydrol ($H_2O)_3$, and so on, the species being considered to have different properties and to be in equilibrium (see also, Dorsey[12]).

The equations for equilibrium do not depend on the identifications made of the species. A two-component theory relating bulk water, W (no subscript is used for bulk properties) to species W_1 present in mole fraction x_1 and species W_2 in mole fraction x_2 can express the stoichiometry:

$$W = x_1 W_1 + x_2 W_2 \qquad x_1 + x_2 = 1$$

and the Gibbs energy, G

$$G = x_1 G_1 + x_2 G_2$$

(Thermodynamic properties are on a per-mole basis and marks indicating this are omitted.)

There is equilibrium between the two species

$$W_1 \rightleftarrows W_2$$

and hence

$$G \equiv G_1 - G_2$$

Assuming an ideal solution, and as the species are distinguishable [13]

$$G_i = G_i^\circ + RT \ln x_i$$

where G_i° is the Gibbs energy of the pure component. Hence

$$\Delta G_{21}^\circ - G_2^\circ \qquad G_1^\circ - RT \ln \frac{x_1}{x_2}$$

(1)

$$= RT \ln \frac{x_1}{1 - x_1}$$

and

(2)
$$x_1 = \frac{1}{1 + \exp\left(-\Delta G_{21}^\circ / RT\right)}$$

Equation 2 shows that the mole fractions depend only on ΔG_{21}° and temperature and that ΔG_{21}° depends only on G_1° and G_2°, the properties of the pure components. Thus only two of G_1°, G_2°, and x_1 are independent. For example, G_1° and G_2° might be predicted from the properties of other phases of water, or G_1° might be predicted and x_1 measured by some spectroscopic method.

The other thermodynamic properties are related to G by the standard relationships. The entropy, S, the enthalpy, H, the specific heat at constant pressure, C_p, and the volume, V, may be obtained by differentiation of G

$$\left(\frac{\partial G}{\partial T}\right)_p = -S$$

$$\left(\frac{\partial H}{\partial T}\right)_p = \left(\frac{\partial (G + TS)}{\partial T}\right)_p = C_p$$

$$\left(\frac{\partial G}{\partial p}\right)_T = V$$

Similar relations, with appropriate subscripts, apply to the assumed species. Equation 2 shows that x_1 changes with temperature even if ΔG° remains unchanged. Since the thermodynamic functions are related to partition functions, these equations can also be given in the language of statistical mechanics.[14] These equations seem basic to all mixture models. There are no corresponding general equations for transport properties, which are treated, when they are considered at all, on an ad hoc basis.

The modern period of structural theories was introduced by Bernal and Fowler,[15] who proposed three chief molecular arrangements for water: water *I*, tridymite-like, or ice-like, present to a certain extent at temperatures below 4°C; *water II*, quartzlike, predominating at ordinary temperatures; and *water III*, a close-packed, ammonia-like liquid, predominating above 200°C. The radial distribution function was known from X-ray diffraction, and their theory incorporated two important features: that the nearest-neighbor coordination number is 4 and that the density maximum is produced by variations in the distance to second neighbors. Bernal and Fowler were explicit that there is not a mixture of volumes of different structures, but that the liquid is homogeneous at all temperatures.

Pople[7] established the modern continuum theory of water, providing an algebraic theory in which the majority of hydrogen bonds between neighboring molecules are regarded as distorted or bent; and it is the energy of this distortion that is to be determined, rather than the bonds to be classified as broken or unbroken by some arbitrary criterion. The building of tetrahedral networks with balls and sticks leads to a similar view of liquid water, and Monte Carlo calculations appear equivalent.

METHODS OF LIQUID-STATE PHYSICS. It is much simpler to calculate the properties of a fluid with a given intermolecular potential than it is to obtain the potential from experimental properties. Hence it is usual first to assume a potential, then to compute a property as well as possible in view of the mathematical difficulties, third, to compare the result with experiment, and finally to try again, improving either the model or the mathematics. It should be our aim to calculate the properties of all phases of water from the potential. Theory is difficult for water, with its strong directional forces, and the most fruitful method appears to be not to approach the liquid directly, but to make more rigorous calculations on the vapor or solid and to extrapolate as well as possible to the liquid.

The properties of liquid water might be calculated by several methods. We may distinguish from more empirical methods four fundamental approaches that start with the intermolecular potential. Not much has yet been published on the application of these four to water.

1. Molecular dynamics and Monte Carlo methods require strong computational facilities and apply to the whole range of fluid densities. Both methods attempt a rigorous calculation of fluid properties from an assumed

intermolecular potential. Both equilibrium and nonequilibrium systems may be studied by molecular dynamics, in which the trajectories of individual molecules are followed over a period of time. Monte Carlo methods are more economical but limited to equilibrium systems. If either method found water to be a mixture, it would prove mixture theories, but at present continuum theories seem supported. Barker and Watts[16] reported Monte Carlo calculations for liquid water, and Beshinske and Lietzke[17] have studied steam.

2. Integral equations involving molecular correlation functions such as $g(R)$ have an important place in theories of simple fluids such as argon. The underlying theory is exact, but mathematical difficulties lead to a number of approximate equations,[18] which are expected to fail at liquid densities.[19] The method has less promise for water.

3. Perturbation theories,[20,21] which derive the properties of a real fluid from the properties of hard spheres taken as known from another source such as (1) or (2), seem capable, in principle, of dealing with water.

4. Lattice theories, now felt to be more solidlike than liquidlike, have had a success with mixtures that has supported their use for pure liquids. Such a theory has been applied to liquid water by Weissmann and Blum.[22]

Less fundamental than those four, but capable of giving insight into the instantaneous structure of liquids (although not the dynamic properties) is model building such as that of Ordway[23] or Bernal.[24] This gives a picture of the environment in which the OH vibrations take place in liquid water but can give no indication of the diffusive motions that make it a liquid. Pople's method[7] may be classed here as algebraic model building.

There are also theories specific to water that see it as a mixture of more-ordered, bulky and less-ordered, less-bulky components. Mixture theories, if they do not postulate solely icelike or gaslike species, will involve species whose properties must be calculated by methods (1) to (4).

Structure of Liquid Water: Qualitative Considerations. Authors of continuum interpretations of the structure of liquid water start from the properties of the ices or hydrates. We consider here, with as little mathematics as possible, the evidence to be given great weight in deducing the structure of liquid water. This evidence leads to a picture of liquid water as a tetrahedral network with a local lifetime of about 10^{-12} sec. On the other hand, there are a number of kinds of data that seem unsuitable starting points because they lead to ambiguous conclusions. These include the infrared spectrum, the rate of chemical reactions, and the equilibrium and transport properties of electrolyte solutions. Rather, understanding of the structural aspects of these phenomena must be built on knowledge of the structure of water. When the approximate structure of water is agreed, such data will be useful for refining it.

A new form of liquid water, called "anomalous" water, condensing in capillary tubes, was reported by N. N. Fedyakin,[25] and its investigation pursued by B. V. Deryagin.[26] Its properties include high density, low vapor pressure, and low freezing point. Key aspects of this discovery appear to have been confirmed,[27] and the English name "polywater" has been proposed. Seemingly, this new form of water, although stoichiometrically H_2O, has a different electronic structure from ordinary water and an intermolecular potential that is different from all the phases considered in this chapter. Hence its properties do not bear on those of the ordinary liquid. It is not clear what relation this anomalous water has to the peculiar thermal expansion of water in quartz capillaries.[28]

EVIDENCE FOR THE STRUCTURE OF WATER.   The forces between water molecules seem the same in all the solid phases and in the vapor. It is thus legitimate to take information from nonliquid systems to see what geometrical relations are possible for water molecules and to deduce as well as possible what the energies of the various configurations are.

Extensive data bearing on the structure of liquid water are available from the following areas:

1. The ices.
2. Solid hydrates and hydroxides.
3. Model-building experiments.
4. Vitreous ice.
5. Vitreous silica.
6. Water vapor.
7. Liquid water itself.

The well-characterized crystalline phases of ice include ordinary hexagonal ice, Ih—the oxygen atoms form the hexagonal lattice—the corresponding cubic phase, ice Ic, and high-pressure ices. The properties of the ices have been summarized by Kamb.[3] In ices other than VII and VIII the oxygen–oxygen distances lie in the range 2.75 to 2.87 Å. In ices VII and VIII the distances are 2.86 Å at 25 kbar and 2.95 Å at atmospheric pressure; the structure of the oxygens in these two phases corresponds to two interpenetrating ice Ic lattices. In all the ices each water molecule is hydrogen bonded to four neighbors. In the cubic phases (Ic and VII) the angles are exactly tetrahedral. In other phases several oxygen–oxygen–oxygen angles are found: in ice II the range is 80 to 128°; in III, 87 to 141°; in V, 84 to 135°; and in VI, 76 to 128° (Ref. 1, p. 85). Ices like II, VIII, and IX have ordered hydrogens. All the forms known to exist in equilibrium with liquid water show disorder in the hydrogens. Features common to all the ices to be considered properties of liquid water are:

1. Each water molecule is hydrogen bonded to four others. The liquid contains more defects, including interstitials, than ices, but work should start from an essentially completely hydrogen-bonded system.

2. The oxygen–oxygen distance, known for liquid water from X-ray diffraction, varies rather little, but angles vary considerably.

Water can be treated as icelike only to a degree. Liquid water can be undercooled below the equilibrium line, with respect not only to ice Ih but also to ice III and the higher ices, and therefore it cannot contain aggregates large enough to be identified as nuclei by the liquid. In another area, water is a good solvent in contrast to the ices, which dissolve few substances. This suggests that the local structure of water is more variable than that of ice because of the lack of periodicity.

A great deal is known about the interactions of water molecules in the solid hydrates (or of OH groups in the solid hydroxides).[29] The information deduced is similar to that from the ices. The spectrum of the fundamental intramolecular oxygen–hydrogen vibrations in hydrates was used as a key to interpret liquid water by Wall and Hornig[30] and by Schiffer and Hornig.[31]

Model building gives a static picture from which liquidlike properties are absent and corresponds to the determination of the radial distribution function, $g(\tau)$. Model building is only easy at high densities where diffusion is slow, but examination of a model can suggest possible diffusive motions. For spherical molecules, Bernal[32] studied the random packing of plasticine balls, and Bernal and Mason[33] and Scott[34] considered that of steel spheres. The effect of vacancies has also been studied.[35]

The construction of a random tetrahedral network without defects (intended as a model for either fused quartz or liquid water) was described by Ordway.[23] In water the hydrogens do not occur midway between the two oxygens, as Ordway's model implies, but nearer to one of them. The stoichiometry, which requires that there be two hydrogens near each oxygen, can be assured by the use of tetrahedral elements with two vertices labeled 1 and two labeled 2; a 1 vertex of one element must then be joined to a 2 of another element. It appears that a model such as Ordway's can easily have the vertices labeled to satisfy the stoichiometry.

From the ices and hydrates we expect that the distance between neighboring oxygens will be about 2.8 Å and that there will be a wide variation of O–O–O angles. The radial distribution functions of Narten and Levy[36] give the distance of the first maximum in liquid water (i.e., the most probable distance between neighboring oxygens) as $R/\text{Å} = 2.82 + 6.0 \times 10^{-4} t$ from 0 to 200°C, with t the Celsius temperature. The range of variation of such distances is given by the halfwidth of the first peak, which is more than 0.25 Å. The variation of angles is less well understood, but the range found in ices and hydrates appears great enough to accommodate the liquid; no results from ball-and-stick models relate the density to the distribution of distances and angles. Since the internal vibrations of water molecules are faster than the diffusional and relaxational motions of the liquid, model building, in which liquidlike motions are ignored, elucidates the environment in which these internal vibrations take place.

Liquid water can be undercooled below the ice Ih—liquid freezing point, but it freezes to ice Ih before the glass-transition point is reached. Accordingly, evidence for the transition from glass to liquid is indirect. By extrapolating the glass-transition temperature of water–glycerol solutions to pure water, Yannas[37] estimated the glass-transition temperature as $127 \pm 4°$K; Miller[38] estimated $162 \pm 1°$K by extrapolating the viscosity from higher temperatures. Similarly, Dass and Varshneya[39] found that self-diffusion and spin-lattice relaxation times point to a glass-transition temperature of 150 to 155°K. All these numbers involve long extrapolations.

Amorphous ice can be prepared by condensing water vapor on a surface cooled below about 100°K. It is not certain that the material obtained is truly amorphous, for the observed broadening of the infrared and X-ray spectra would also be obtained if the material were microcrystalline. Ghormley[40] has expressed doubt that this amorphous ice is the vitreous form expected if the liquid were undercooled sufficiently. The behavior of amorphous ice on heating depends on the sample and rate of heating. At 135 to 150°K it transforms irreversibly to ice Ic, which in turn transforms irreversibly to ice Ih above 200°K. No glass-liquid transition occurs below the temperature of conversion to ice Ic.[40,41]

Although the glass-liquid transition in water has not been observed—because both liquid and glass transform spontaneously to crystalline ice in this region—the properties of the glass may, nevertheless, be estimated. The Gibbs energy of bulk water may be written as the sum of vibrational and potential parts

$$(3) \qquad G = G_{\text{vib}} + G_{\text{pot}}$$

As before, other properties may be obtained by differentiation. In particular, the specific heat at constant volume, C_V, and the compressibility, κ, separate into sums

$$C_V = C_{V,\text{vib}} + C_{V,\text{pot}}$$

$$\kappa = \frac{-1}{V}\left(\frac{\partial V}{\partial p}\right) = \frac{-1}{V}\left(\frac{\partial V_{\text{vib}}}{\partial p}\right) - \frac{1}{V}\left(\frac{\partial V_{\text{pot}}}{\partial p}\right) = \kappa_{\text{vib}} + \kappa_{\text{pot}}$$

With change of conditions, the change of G_{vib} is the change of G with the molecular configuration of the glass held constant; it includes the changing occupancy of vibrational levels with change of temperature. The G_{pot} is produced by the relaxation of the whole network to a new state of internal equilibrium. (Davis and Jarzynski[5] used "glass" instead of "vibrational" and "relaxational" instead of "potential." Eisenberg and Kauzmann[1] used "configurational" instead of "potential.") The internal equilibration of the water network requires only about 10^{-12} sec, which is too fast for conventional velocity-of-sound measurements to determine the elastic properties of an unrelaxing network. Slie, Donfor, and Litovitz[42] determined the velocity of sound in water–glycerol solutions and extrapolated the results to pure water. At 0°C, where the isentropic compressibility, $\kappa_S = 51.3 \times 10^{-6}$/bar, they obtained $\kappa_{\mathrm{vib},S} = 17.8 \times 10^{-6}$/bar, and hence $\kappa_{\mathrm{pot},S} = 33.5 \times 10^{-6}$/bar.

The idea of relaxation of the network is also useful in understanding the specific heat. For water at 0°C, $C_v/R = 9$. Each molecule has 3 of its 9 degrees of freedom involved in internal vibrations that contribute little to the specific heat. This leaves 6 degrees of freedom to be taken up in translational vibrations and rotational vibrations in the liquid. The most that these may contribute to the specific heat is R/mole, so that $V_{\mathrm{vib},V}/R \leq 6$. This leaves $C_{\mathrm{pot},V}/R \geq 3$.

The considerations embodied in (3), given here in the context of a continuum theory, are independent of any model of liquid water and were used by Davis and Litovitz[43] in connection with their two-state theory.

Vitreous silica is a tetrahedral network with four oxygens about each silicon. In this case all vertices of the tetrahedron are equivalent. By analysis of the vibrations of constructed models, Bell, Bird, and Dean[44] estimated the vibrational spectrum of vitreous SiO_2, GeO_2, and BeF_2, taking the geometries from the models and the force constants from other sources. A calculation of the configurational entropy of vitreous silica by Bell and Dean[45] led to about 1 cal mole^{-1} °K^{-1}. A similar value should apply to a vitreous network in water. The entropy of transformation of tetrahedra with unlabeled vertices from a crystal to a random network is as reported by Bell and Dean. The placing of hydrogens in the solid has entropy $R \ln \frac{3}{2}$;[46] the corresponding term for the glass will be rather similar. For water the entropy of the glass-liquid transition would be wanted in addition.

The thermal expansion of vitreous silica is low. This results from near cancellation of an increase in silicon–oxygen distance with increasing temperature, which tends to increase the volume, and a decreased silicon–silicon distance because of increased bending of the Si–O–Si and O–Si–O angles, which tends to decrease the volume.[47] The thermal expansion of liquid water at low temperatures may be explained similarly.

Evans and King[48] constructed a random tetrahedral model for vitreous silica in which the Si–O–Si angle is obtained by linking the tetrahedra with bent springs having an initial angle of 163°. For a hydrogen bond, such as the OH–O of water, an angle of 180° is likely preferred.

Most of the remaining sections treat the determination of structural data from the properties of vapor and liquid.

NEIGHBORS.   The ices contain only a limited number of environments, and it is clear which oxygens are hydrogen bonded. It is, accordingly, possible to sketch the topology of a small region of the lattice to show the hydrogen bonds.[49] In Figure 3a, 0 is the central molecule, and 1, 2, 3, and 4 are its first neighbors; 5, 6, and 7 are in turn first neighbors of 1, and hence second neighbors of 0; 8, 9, and 10 are also second neighbors of 0. Eleven is a third neighbor of 0 by either the path 1, 7, 11 or 2, 8, 11. Although 8 is a second neighbor, it is also a fourth neighbor by the path 1, 7, 11, 8. (In ice Ih there are 12 second neighbors, but in ice VI, which has four-member rings, the number is different; in lattices with nonequivalent atoms the number of neighbors of a given degree may depend on the molecule chosen as origin.)

A similar figure is possible for a small region of a tetrahedral network. In such a network there are rings of various sizes. Figure 3b shows a five-member ring: 0, 1, 5, 6, and 2. Four-member rings involve more strain and are probably rare. The real liquid will have defects present, and 3 represents a molecule with only three normal hydrogen bonds; the fourth one is interacting with both molecules 11 and 12. We expect this fourth one to shift to one or the other in less than the dielectric relaxation time of 10^{-11} sec, but this may leave strain elsewhere; such lability accounts for the low viscosity of the liquid. Ten is a fifth neighbor of 0, topologically speaking, but, instead of being hydrogen bonded to 0 to make a six-member ring, it is in an interstitial position relative to 0. Thirteen and 14 represent part of a hydrogen-bonded network that is, locally at least, interstitial relative to the first network. (The representation of interstitial molecules in Figure 3b is purely schematic; their apparent neighbors in the figure are not their neighbors in three dimensions.) Zero, 3, 15, 16, 17, and 4 represent a six-member ring approximately like those found in ice Ih.

Each molecule has open regions beside it. Interstitial molecules in these spaces are first interstitial neighbors. Although the tetrahedral coordination of water molecules makes the concept of topological second neighbor perfectly definite, cavities are irregular and the definition of a second interstitial neighbor is not clear. In ice Ic each cavity is surrounded by four molecules; in ice VII a second cubic lattice fills this cavity network and each molecule has four topological first neighbors and four interstitial first neighbors at the same distance.

About 4.4 first neighbors are estimated from X-ray diffraction.[36] If there are four topological first neighbors, as we assume throughout, there must be about 0.4 interstitial first neighbors. If all molecules are essentially equivalent, these interstitials must be tetrahedrally coordinated in turn. At high pressures

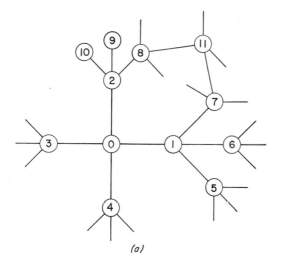

(a)

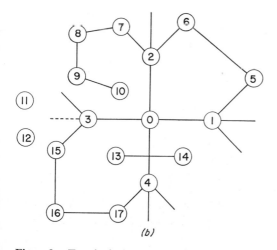

(b)

Figure 3. Topological relations in ice Ih and in liquid water. Circles represent the oxygens of the molecules, and hydrogens must still be added in such a way that the stoichiometry is preserved. (a) Ice Ih, showing the tetrahedral coordination and a six-membered ring. (b) Possible features (see text) of the local structure of liquid water. The representation of interstitial molecules is purely schematic; their apparent neighbors are not possible neighbors in three dimensions.

near the freezing point of ice VII, we expect most interstitial positions to be filled, with a complicated network interpenetrating itself. Estimation of the elastic energies of interstitial structures would permit a calculation of their probability. Preliminary calculations have been made by Kamb.[50]

2   Relations Involving Few Molecules

Intermolecular Geometry.   Each molecule is considered to have embedded in it a coordinate system that is convected with the molecule. The Cartesian coordinate system used here for a water molecule in the gas phase places the Z, or dipole, axis in the plane of the molecule bisecting the H–O–H angle with the positive direction from the oxygen toward the hydrogens. The Y axis is in the plane of the molecule, is perpendicular to the Z axis, and passes through the oxygen atom. The X axis is perpendicular to the other two. In this coordinate system the hydrogens are at $(0, \pm 0.757 \text{ Å}, 0.586 \text{ Å})$ corresponding to an oxygen–hydrogen length of 0.957 Å and an H–O–H angle of 104.5°. For $^1H_2{}^{16}O$ ($^1H = 1.008$ and $^{16}O = 15.995$) the center of mass is at $(0, 0, 0.066 \text{ Å})$. In static problems, and in the description of tetrahedral structures, it is usual to take the oxygen as origin; in dynamical problems, the center of mass is taken as origin. The intramolecular geometry in condensed phases can be found for the ordered ices and hydrates by neutron scattering. As the hydrogen bond forms, the OH distance increases slightly. It may be taken as 1.00 Å in condensed phases, and the H–O–H angle appears to lie between the gas-phase value of 104.5° and the tetrahedral angle of $\cos^{-1}(-\frac{1}{3})$ = 109.5°.[29] The tetrahedral unit directions are $(0, \pm 0.817 \text{ Å}, 0.577 \text{ Å})$ and $(\pm 0.817, 0, -0.577)$ in Cartesian coordinates.

The configuration of two dipoles (or other charge distributions symmetrical about an axis) is usually expressed in terms of the coordinate system illustrated in Figure 4. The distance R lies along the line between centers. $\phi_1(1)$ and $\phi_1(2)$ are the colatitudes and ϕ_3 is the azimuthal angle about the oxygen–oxygen axis between the dipoles.

If symmetry about the dipole axes were not assumed, the orientation of the molecules about their dipole axes could be given by $\phi_2(1)$ and $\phi_2(2)$, the azimuthal angles about those axes, and the whole configuration described by the six coordinates R, $\phi_1(1)$, $\phi_2(1)$, ϕ_3, $\phi_1(2)$, and $\phi_2(2)$. These correspond to the six coordinates mentioned in connection with the distribution function, $g(\tau)$.

NONPOLAR MOLECULES.   Water molecules show strong directional properties in all solid phases, and this must also be true in the fluid state. However, theory usually starts with molecules of spherical symmetry and then adds the directional properties.

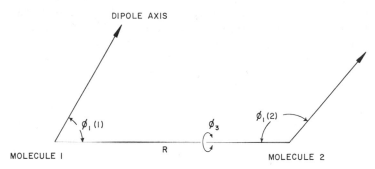

Figure 4. Coordinate system for axially symmetrical systems such as dipole-dipole interactions. Four parameters describe the separation and configuration.

The Lennard-Jones $(m - n)$ potential[51] is commonly used for the spherical part; it is of the form

$$\Phi_{LJ}(m, n; R) = \frac{m}{m - n} \left(\frac{m}{n}\right)^{m/(m-n)} \varepsilon \left[\left(\frac{\sigma}{R}\right)^m - \left(\frac{\sigma}{R}\right)^n\right]$$

For this potential $\Phi = 0$ at $R = \sigma$, and there is a minimum $\Phi = -\varepsilon$ at $R = \sigma' = \sigma(m/n)^{1/(m-n)}$ where the force is 0; at greater separations the force is attractive and at smaller repulsive.

Commonly m is given the value 12, and there are physical reasons for giving n the value 6; this gives the $(12 - 6)$ potential

$$(4) \qquad \Phi_{LJ}(r) = \frac{4\varepsilon\sigma^{12}}{r^{12}} - \frac{4\varepsilon\sigma^6}{r^6}$$

with its minimum at 1.12σ. The first term is the repulsive potential caused by overlap, and the second is the attraction arising from dispersion forces.

In nonpolar gases both the repulsive and attractive terms have quantum-mechanical origins. There is no adequate theory for the repulsive forces. The attractive, dispersion, or van der Waals force was first calculated by London in 1930.[52] Because dispersion forces give n a value of 6, the second term of (4) could be set equal to the dispersion force

$$4\varepsilon\sigma^6 = \tfrac{3}{4}U\alpha^2$$

where α is the static polarizability and U is a characteristic energy such as the ionization energy.[53]

POLAR MOLECULES. Polar molecules are distinguished from nonpolar ones by a permanent asymmetric distribution of electric charge that produces electrostatic forces; the leading polar forces fall off more slowly with distance than do the dispersion forces. For convenience, we first treat charge distributions as rigid and introduce polarization effects later. The potential energy of two molecules is considered the sum of a Lennard-Jones potential energy and an electrostatic one

$$\Phi = \Phi_{LJ} + \Phi_{elect}$$

The charge distribution of a molecule can be described in terms of multipole moments. Several nonequivalent definitions have been used. Following Kielich[54] we define the multipole moments in terms of r_i the vector from the center of the molecule to charge e_i.

$$M^{(0)} = q = \sum_i e_i$$

$$M^{(1)}_\alpha = \mu_\alpha = \sum_i e_i r_\alpha$$

$$M^{(2)}_{\alpha\beta} = \Theta_{\alpha\beta} = \tfrac{1}{2} \sum_i e_i (3r_\alpha r_\beta - r^2 \delta_{\alpha\beta})$$

$$M^{(3)}_{\alpha\beta\gamma} = \Omega_{\alpha\beta\gamma} = \tfrac{1}{2} \sum_i e_i [5r_\alpha r_\beta r_\gamma - r^2(r_\alpha \delta_{\beta\gamma} + r_\beta \delta_{\alpha\gamma} + r_\gamma \delta_{\alpha\beta})]$$

where $M^{(n)}$ is the 2^n-pole moment, $\sum_i$ indicates summation (or integration) over all the charges of the molecule, $r^2 = \mathbf{r} \cdot \mathbf{r}$, there is summation over repeated indices, and $\delta_{\alpha\beta} = 1$ if $\alpha = \beta$ and $\delta_{\alpha\beta} = 0$ if $\alpha \neq \beta$. We consider only uncharged molecules for which $q = 0$.

The first non-zero moment is independent of the origin. For water this is the dipole moment which has components $\mu_x = \mu_y = 0$, $\mu_z = 1.84$ D (where 1 debye = 1×10^{-18} esu cm). The quadrupole moment has only two independent terms, for the definition shows that

(5) $$\Theta_{xx} + \Theta_{yy} + \Theta_{zz} = 0$$

The first non-zero moment can be obtained from experiment. In water, quadrupole and higher moments are best obtained from wave functions.[55] If a point is given by $\boldsymbol{\lambda}^O$ with oxygen as origin and by $\boldsymbol{\lambda}^M$ with the center of mass as origin

(6) $$\boldsymbol{\lambda}^M = \boldsymbol{\lambda}^O + \delta\boldsymbol{\lambda}$$

and $\delta\lambda_x = \delta\lambda_y = 0$, $\delta\lambda_z = -0.066$ Å. By substitution of (6) into the defining equation, and since q is 0, this transformation produces the following change of quadrupole moments:

(7)
$$\Theta_{xx}{}^M = \Theta_{xx}{}^O - \mu_z \, \delta\lambda_z$$
$$\Theta_{yy}{}^M = \Theta_{yy}{}^O - \mu_z \, \delta\lambda_z$$
$$\Theta_{zz}{}^M = \Theta_{zz}{}^O + 2\mu_z \, \delta\lambda_z$$

Equation 5 is still satisfied. Surveys of the values of multipole moments have been given by Krishnaji and Prakash[56] and Stogryn and Stogryn.[57]

Table 1   Quadrupole Moments of Water[a]; μ, esu cm; Θ, esu cm^2

	Oxygen at Origin[a]			Center of Mass at Origin[b]			
$10^{18}\,\mu_z$	$10^{26}\,\Theta_{xx}{}^O$	$10^{26}\,\Theta_{yy}{}^O$	$10^{26}\,\Theta_{zz}{}^O$	$10^{26}\,\Theta_{xx}{}^M$	$10^{26}\,\Theta_{yy}{}^M$	$10^{26}\,\Theta_{zz}{}^M$	References[c]
Calculated from Wavefunctions							
1.52	−1.149	0.775	0.394	−1.048	0.855	0.193	1
1.72	−1.304	0.940	0.364	−1.190	1.054	0.137	1
1.66	−1.211	0.853	0.358	−1.101	0.963	0.139	1
1.42	−1.055	0.660	0.395	−0.961	0.754	0.207	1
1.67	−1.228	0.868	0.360	−1.117	0.987	0.139	1
1.53	−1.143	0.729	0.414	−1.042	0.830	0.212	1
(1,84)	−0.907	0.694	0.213	−0.785	0.815	−0.030	2
Obtained from Point-Charge Models							
(1.84)	−1.61	1.35	0.26	−1.49	1.47	0.02	3
2.10	−1.49	1.49	0.0	−1.35	1.63	−0.28	4
(1.87)	−1.26	1.26	0.0	−1.13	1.38	−0.25	4, 5
Recommended Values							
1,84	−1.24	0.88	0.36	−1.12	1.00	0.12	

[a] Most values of Θ^O were given in Ref. 56.
[b] Most values of Θ^M were given in Ref. 57.
[c] Numbers refer to references at end of table.

1. R. M. Glaeser and C. A. Coulson, *Trans. Faraday Soc.*, **61**, 389 (1965).
2. A. B. F. Duncan and J. A. Pople, *Trans. Faraday Soc.*, **49**, 217 (1953).
3. J. S. Rowlinson, *Trans. Faraday Soc.*, **47**, 120 (1951).
4. A. Ben-Naim and F. H. Stillinger, Chapter 8, this volume.
5. N. Bjerrum, *Kgl. Danske Videnskab. Selskab., Mat.-Fys. Medd.*, **27**, 1 (1951).

Some calculations of quadrupole moments of water are given in Table 1; for octopole moments, see Stogryn and Stogryn.[57] The values in Table 1 apply to an unperturbed gas-phase molecule. For self-consistency, the transformations of (7) must be made using the dipole moment from the same source.

Here it is appropriate to note that Glaeser and Coulson[55] define their quadrupole moment $Q_{\alpha\beta}$ as

$$Q_{\alpha\beta} = \sum_i e_i r_\alpha r_\beta$$

and with this definition the quadrupole components found from the four-charge model of Rowlinson[58] are quite out of line with the values from wavefunctions. However, with the definition of Θ given here, in which Θ measures the departure from a spherical charge distribution and is the natural quantity to use to describe the interaction of two polar molecules, Rowlinson's model has the right signs and a numerical error no greater than 0.4×10^{-26}. Campbell, Gelernter, Heinen, and Moorti[59] have pointed out that point charges can give a satisfactory account of the polar interactions in ice without having the correct $Q_{\alpha\beta}$.

Attention is sometimes drawn (e.g., by Ives and Lemon[60]) to the extreme position of the bulk properties of water among the isoelectronic "hydrides" CH_4, NH_3, H_2O, HF, and neon. For CH_4 and neon, $\Theta_{xx} = \Theta_{yy} = \Theta_{zz} = 0$. Molecules with a cylindrical charge distribution about the Z axis, such as NH_3 and HF, satisfy the relation $\Theta_{xx} = \Theta_{yy} = -\frac{1}{2}\Theta_{zz}$. In water, Θ_{xx} and Θ_{yy} have opposite signs, so the forces between a molecule and its neighbor change markedly with rotation about the Z axis; this produces the restricted rotation in three dimensions which in turn produces the "anomalous" properties of bulk water.

Several computer calculations[16,61] have been made using point-charge models. Multipole moments adequately describe the charge distribution, but inclusion of the higher moments presents computational problems and the use of point-charge models simplifies the computations. Point-charge models are easy to visualize and are therefore useful in molecular dynamics or Monte Carlo calculations. One is offered that satisfies the dipole moment and the recommended quadrupole moments of Table 1. All charges lie in the tetrahedral directions from the oxygen.

Charges of $+0.665 \times 10^{-10}$ esu ($0.138 \, e$) are at 1.097 Å in the directions near the hydrogens, and of -0.665×10^{-10} esu at 1.302 Å in the directions toward the lone pairs.

The striking feature, because $|\Theta_{xx}{}^o|$ is greater than $|\Theta_{yy}{}^o|$, is that the charges of the lone pairs are further away from the oxygen. At an oxygen–oxygen distance of 2.8 Å, there is only 0.4 Å between the nearest charges.

The electrostatic interaction energy of two water molecules at separation **R**—it is necessary that the two distributions do not overlap—is obtained by considering the potential, ϕ, at the origin of molecule 2 and the multipole moments of 2. The potential is obtained using the multipole moments of 1 and an expansion in the distance **r** from the origin of molecule 1 to the charges used in the evaluation of its multipole moments.[62] The electrostatic potential energy, Φ_{elect}, is

$$\Phi_{elect} = q\phi + \mu_\alpha \left(\frac{\partial\phi}{\partial r_\alpha}\right) + \tfrac{1}{3}\Theta_{\alpha\beta}\left(\frac{\partial^2\phi}{\partial r_\alpha \, \partial r_\beta}\right) + \cdots$$

where there is summation over repeated indices and where the potential and its derivatives are evaluated at the origin of molecule 2. The potential produced is

$$\phi = \frac{q}{R} + \frac{\mu_\alpha r_\alpha}{R^2} + \frac{\Theta_{\alpha\beta}}{3R^3}(3r_\alpha r_\beta - r^2\,\delta_{\alpha\beta}) + \cdots$$

and we are interested in the case where $q = 0$.

The preceding treatment has considered rigid charge distributions. In fact, all molecules are polarizable and field gradients induce moments. The change of dipole moment produced by a neighboring dipole is given by the treatment of Barker[63] and Buckingham and Pople.[64] If μ° is the dipole moment of an unperturbed molecule, with polarizability tensor α, then when a field, ϕ, is present the dipole moment becomes

$$\mu(1) = \mu^\circ(1) - \alpha \cdot \nabla\phi(1)$$

where $\phi(1)$ is the electric potential produced at molecule 1 by molecule 2. For water, there is little indication of variation of α with direction,[65] so we replace it by the scalar α, whose value is about 1.4×10^{-24} cm^3.

In the case of two molecules, the resultant dipole is obtained in closed form and has components

$$\mu_x(1) = \frac{\mu_x^\circ(1) - \alpha R^{-3}\mu_x^\circ(2)}{1 - \alpha^2 R^{-6}}$$

$$\mu_y(1) = \frac{\mu_y^\circ(1) - \alpha R^{-3}\mu_y^\circ(2)}{1 - \alpha^2 R^{-6}}$$

$$\mu_z(1) = \frac{\mu_z^\circ(1) + 2\alpha R^{-3}\mu_z^\circ(2)}{1 - 4\alpha^2 R^{-6}}$$

TWO-BODY INTERACTIONS.   The simplest "realistic" potential for a dipolar molecule such as water is that of Stockmayer,[66] with a rigid dipole at the center of each molecule and an $(m - 6)$ Lennard-Jones potential to represent the spherical part. Polarization and higher moments are ignored. The Stockmayer potential is

$$\text{(8)} \qquad \begin{aligned} \Phi_S(R, \mu) = \ &\Phi_{LJ}(R) + \mu^2 R^{-3}(2\cos\phi_1(1)\cos\phi_1(2) \\ &+ \sin\phi_1(1)\sin\phi_1(2)\cos\phi_3) \end{aligned}$$

where the coordinate system is that of Figure 4. The equation of state and viscosity of the vapor have been interpreted with this potential. For example, the equation of state Z may be expanded as a power series in the molar density, ρ

$$Z = \frac{p}{\rho RT} = 1 + B^{(2)}\rho + B^{(3)}\rho^2 + \cdots$$

The nth virial coefficient is $B^{(n)}$. According to Hirschfelder et al. (Ref. 53, p. 151), experimental values of $B^{(2)}$ are related to the potential energy by

$$B^{(2)}(T) = \frac{N_A}{4} \int_0^\infty \int_0^{2\pi} \int_0^\pi \int_0^\pi [1 - \exp(-\Phi/kT)]R^2$$
$$\sin \phi_1(1) \sin \phi_1(2) \, d\phi_1(1) \, d\phi_1(2) \, d\phi_3 \, dR$$

where N_A is Avogadro's number. However, Φ itself is not found by this expression; a particular functional form is assumed, and the free parameters are adjusted to give the best fit. Using the potential of (8), the experimental value of μ is used, and ε and σ of the Lennard-Jones part of the potential are adjusted.

Some values of the potential parameters for water are shown in Table 2. Here $t^* = \mu^2/2\sqrt{2}\varepsilon\sigma^3$ is the reduced dipole function. With polar molecules, experimental and analytical problems combine to produce an uncertainty of about 0.5% in $B^{(2)}$. The variation in the parameters obtained in this table illustrates the uncertainty about details of the intermolecular potential.

Table 2 Intermolecular Potential Parameters for Steam[a]

Potential	m	ε/k (°K)	t^*	σ (Å)	Experimental	Data[b]	References[b]
Rigid dipole	24	—	—	2.76	$B^{(2)}$	1	2
Rigid dipole	12	383	1.2	2.68	$B^{(2)}$	1	3
Rigid dipole	12	375	1.3	2.65	$B^{(2)}$	1	4
Polarizable dipole	12	357	1.3	2.69	$B^{(2)}$	1	4
Rigid dipole	12	506	0.9	2.71	Viscosity		5
Rigid dipole	12	380	1.2	2.65	$B^{(2)}$	1	6
Rigid dipole	18	330	1.30	2.71	$B^{(2)}$	1	6
Polarizable dipole	12	—	—	2.59	$B^{(2)}$	1	7
Polarizable dipole	24	—	—	2.68	$B^{(2)}$	1	7
Rigid dipole	12	331	1.4	2.52	$B^{(2)}$	8	8
Rigid quadrupole	12	277	1.2	2.98	$B^{(2)}$		9

[a] Except for the values from authors 5, 8, and 9, all entries are based on the same values of the second virial coefficient, $B^{(2)}$.
[b] Numbers refer to references at end of table.

1. F. G. Keyes, L. B. Smith, and H. T. Gerry, *Proc. Amer. Acad. Arts Sci.*, **70**, 319 (1936); S. C. Collins and F. G. Keyes, *Proc. Amer. Acad. Arts Sci.*, **72**, 283 (1938).
2. W. H. Stockmayer, *J. Chem. Phys.*, **9**, 398 (1941).
3. J. S. Rowlinson, *Trans. Faraday Soc.*, **45**, 974 (1949).
4. A. D. Buckingham and J. A. Pople, *Trans. Faraday Soc.*, **51**, 1173 (1955).
5. L. Monchick and E. A. Mason, *J. Chem. Phys.*, **35**, 1676 (1961).
6. S. C. Saxena and K. M. Joshi, *Phys. Fluids*, **5**, 1217 (1962).
7. S. Carrà and L. Zanderighi, *Nuovo Cimento*, **B49**, 133 (1967).
8. G. S. Kell, G. E. McLaurin, and E. Whalley, *J. Chem. Phys.*, **48**, 3805 (1968).
9. A. L. Tsykalo, *Russ. J. Phys. Chem.*, **42**, 261 (1968).

Klein and Hanley[67] have shown that a wide temperature range and good precision are needed to discriminate between representations of spherical potentials, and similar restrictions apply with polar molecules (Max Klein, private communication). The (12 − 6) Stockmayer potential gives an adequate representation of $B^{(2)}$ of water, except over the widest temperature range. Water was treated as composed of polarizable dipoles by Carrà and Zanderighi.[68] A formal treatment for polarization and higher multipoles was given by Buckingham and Pople,[64] but calculations involving the quadrupole moment were not made because values were not available. Such calculations do not appear to have been made since.

Since there can be exchange of energy between the internal vibrations and the kinetic energy of two colliding water molecules, treating these as rigid bodies for discussion of the viscosity and other transport properties involves mild approximations. The viscosity, η, is given by

$$\eta = \frac{5}{16}\left(\frac{mkT}{\pi}\right)^{1/2}\frac{1}{\sigma^2\langle\Omega^{(2,2)*}\rangle}$$

where $\langle\Omega^{(2,2)*}\rangle$ is a dimensionless collision integral averaged over all orientations; Monchick and Mason[69] tabulated this collision integral for the Stockmayer potential. The potential parameters they obtained for water are given in Table 2. Their estimate of the depth of the potential well is significantly greater than those from $B^{(2)}$ and corresponds to values of $B^{(2)}$ more negative than the experimental ones. Knowledge of the intermolecular potential of two water molecules is not fully satisfactory. With the rather primitive potential functions used, this is not surprising.

INTERACTIONS INVOLVING SEVERAL MOLECULES. The third virial coefficient, $B^{(3)}$, gives information about three-body interactions. Rowlinson[58] calculated $B^{(3)}$ for the Stockmayer rigid dipole model. Values of $B^{(3)}$ calculated with the Stockmayer parameters for $B^{(2)}$ agreed neither with the available experimental values of $B^{(3)}$ nor with more recent measurements.[70] Rather, the observed $B^{(3)}$ is somewhat like that for nonpolar gases. This is reasonable qualitatively, for the Stockmayer potential is unable to reflect the tetrahedral orientation that must be preferred when three or more molecules interact.

Since the relative position and orientation of two water molecules is a function of six variables, the relative positions and orientations of three are a function of 12. The complexities of the dynamics of such a system suggest that a treatment of less rigor but wider scope may be particularly useful for water. One such theory is the polymer theory of vapor-phase imperfections, which has been developed by many authors.

The interpretation of $B^{(2)}$ in terms of dimers is well developed, and treatments of dimers in water vapor have been given by O'Connell and Prausnitz[71] and Bolander, Kassner, and Zung.[72] Polymer theory has been used for dimers, trimers, and tetramers in water and methanol vapor by Kell and McLaurin.[73] The polymerization approach shows the relation between different virial coefficients and indicates the variety of interactions, ranging from the addition of a monomer to another monomer to form a dimer, through the addition of a monomer to a small drop, to the condensation of a monomer in the bulk liquid.

TETRAHEDRAL DEFORMATIONS.   In the cubic ices Ic and VII the oxygens in the hydrogen-bonded lattices are tetrahedral. In liquid water both oxygen–oxygen distances and oxygen–oxygen–oxygen angles are irregular. Consider the deformation possible through change of oxygen–oxygen–oxygen angle in a tetrahedral network of fixed oxygen–oxygen distance. Only resistance to bending keeps such a network from collapsing, and even when repulsive forces are present between nonhydrogen-bonded neighbors, large deformations are needed before these repulsions become important. Rather, there must be resistance to the bending of angles because the real liquid does not collapse. The preferred angle is close to tetrahedral, and we take it as exactly that.

Figure 5 shows a central molecule and four neighbors each at distance R. Denoting the O_1–O_0–O_2 angle by ϕ_{12}, and the distance from O_1 to O_2 by R_{12}, the distance R_{ij} is given by

$$R_{ij} = R\sqrt{2 - 2 \cos \phi_{ij}}$$

There are six distances between the four molecules. The mean distance is greatest when each angle is the tetrahedral angle; in that case $R_{ij}/R = 1.663$. The effect of deformation on the mean R_{ij} may be illustrated by the symmetrical case in which O(2), O(3), and O(4) lie in a plane with O(0), at $120°$ from each other, with the line O(1)–O(0) at right angles to the plane. There are then three angles of $90°$ and three of $120°$, well within the range occurring in the ices; the average distortion from tetrahedral is $15°$. For $120°$, $R_{ij}/R = 1.732$, whereas for $90°$, $R_{ij}/R = 1.414$, with the mean 1.573. Such

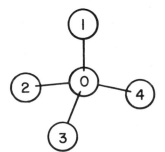

Figure 5.   Four-coordinated group of water molecules. The oxygens of each molecule are shown, and the distances to the central oxygen are constant. As the angles at the central molecule change from tetrahedral, the mean distance between the outer oxygens decreases.

a small decrease in the mean distance between second neighbors can account for the observed negative thermal expansion below 4°C which reflects a balance between the increase of oxygen–oxygen distance with temperature, which would decrease density, and the increasing deformation of oxygen–oxygen–oxygen angles, which would increase density. Similarly, although ices II or III are 20% denser than Ih, in them the oxygen–oxygen distances are greater than in Ih, and the increase in density is produced by distortion of the oxygen–oxygen–oxygen angles.

An estimate of the energy of this degree of deformation can be made. Kamb[3] described the deformation of the various ices by

$$U = U_1 + K_\theta \langle \theta^2 \rangle$$

where $K_\theta \simeq 1$ cal mole^{-1} deg^{-2}, although K_θ from the shearing constant, C_{44}, of ice Ih is ~ 5 cal mole^{-1} deg^{-2}. For the distortion just described

$$U - U_1 = 1[3(120 - 109)^2 + 3(90 - 109)^2]$$
$$= 1.5 \text{ kcal mole}^{-1}$$

The change of internal energy on melting of ice Ih is 1.4 kcal mole^{-1}. Thus the observed change in internal energy can be provided for by a distribution of O–O–O angles in the range occurring in the ices.

POPLE'S THEORY. In 1951, Pople offered a theory[7] in which liquid water was not described in terms of crystalline nor quasi-crystalline structure, nor were many hydrogen bonds described as broken. Rather, hydrogen bonds were considered as distorted. The theory was in quantitative (though not perfect) agreement with the radial distribution function, with the volume changes associated with the melting of ice and with the dielectric constant of water. Pople's paper has not received the study it deserves. Much of this chapter may be regarded as an introduction to that paper, which, after two decades, remains the most thorough treatment of water available.

The only deformations allowed in ice are those that are consistent with the preservation of long-range order. Deformations in the liquid are less restricted, and Pople treats the four bonds about an oxygen as bending independently. The preferred directions are tetrahedral. For any hydrogen-bonded oxygen–oxygen line, the OH will be at angle θ_1 and the lone pair at θ_2. The energy, E, of the deformation is given by the symmetrical relation

(9) $$E = F(\theta_1) + F(\theta_2)$$

and F can be approximated by

$$F(\theta) = F_0 - g \cos \theta$$

where θ is the angle by which either the hydrogen or the lone pair depart from linearity. Since

$$\cos\theta = 1 - \frac{\theta^2}{2!} + \frac{\theta^4}{4!} - \cdots$$

(where θ is in radians), $F(\theta)$ can also be expressed as

$$F(\theta) = \frac{g}{2!}\,\theta^2 - \cdots$$

taking F_0 equal to g. (Although in this section we follow Pople and talk of θ, in other sections we employ θ^2.) Pople assigned to g a value corresponding to a librational frequency of 230 cm^{-1}; the observed value is near 600 cm^{-1}. A change of g in this direction would give a narrower distribution of angles in liquid water. The average cosine of the angle of deformation is

$$\overline{\cos(\theta)} = \coth\frac{g}{kT} - \frac{kT}{g}$$

Pople used $g/kT = 10$ at 0°C, but a change of frequency to 600 cm^{-1} raises g/kT to 26. This gives an average distortion of 16°, rather than the 26° he obtained.

The determination of the distribution function was made by reference to Figure 6. A coordinate system was established, and by use of (9) the probability of different configurations could be found. Even powers of the moments were calculated, and Gaussian distribution functions were used to

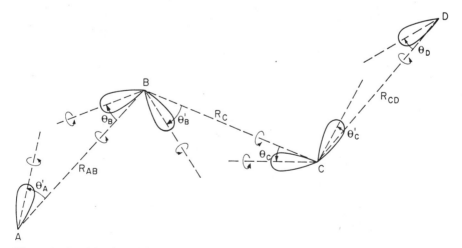

Figure 6.   Pople's view of four molecules connected by bent hydrogen bonds.

approximate the various shells of the distribution function. Pople assumed that the number of second neighbors was 11 and the number of third neighbors 22.

Because of the dipole moment induced by the field of neighboring molecules, the dipole moment in a condensed phase is not the vacuum value of 1.84 D, but is higher. Coulson and Eisenberg[49] showed that the total dipole moment in ice Ih is about 2.60 D, and that it is necessary to consider up to the octopole moment of first neighbors, to the quadrupole moment of second neighbors, but only the dipole moment of third neighbors. Eisenberg and Kauzmann (Ref. 1, p. 191) applied the ratio of the total moment to that from the first neighbors to Pople's results to obtain a value of $\varepsilon_0'(T)$ close to experiment.

3 Inferences from Bulk Properties

Diffusive Motions

FAR-INFRARED SPECTRUM. There is only one dielectric relaxation time for water, about 10^{-11} sec, and therefore the lifetime of species in any mixture theory must be somewhat less than this. The time taken for a vibration of ω wavenumbers is $1/\omega c$, where c is the velocity of light. A relaxation time, τ_0, of 10^{-11} sec corresponds to frequency $1/2\pi\tau_0 = 1.7 \times 10^{10}$ Hz or 0.6 cm^{-1}. Thus the spectrum between the lowest internal vibrational band near 1500 cm^{-1} and the far infrared could be interpreted according to a mixture theory, and this has been done, for example, by Walrafen.[74]

The infrared and Raman spectra of condensed phases have been reviewed by Wilkinson[75] and effect of proton disorder in phases of water has been discussed by Whalley.[76] Dean[77] has emphasized that the vibrations of disordered systems are very complicated and can be described simply only in special cases or in particular regions of the spectrum. With liquid water, at frequencies lower than about 0.5 cm^{-1}, the dipole of water molecules can follow an applied electric field and the dielectric constant is high; at higher frequencies it cannot follow and the dielectric constant is low.

The spectrum from 10 to 1000 cm^{-1} shows two important bands, ν_R, the librational or rotational vibrational band from 400 to 900 cm^{-1}, with maximum near 600 cm^{-1}, produced by molecules twisting in the cage of their neighbors, and the ν_T, or translational vibrational band near 160 cm^{-1} produced by the molecules vibrating in the cage of their neighbors. The broadness of these bands reflects the wide variety of local environments present in liquid water. It is the occasional movement of a molecule from one potential well to another that produces translational or rotational diffusion.

In the ν_R band in ice, and presumably in liquid water, the peak in the density of states is about 200 wavenumbers less than the infrared maximum.[78] Accordingly, in the liquid the Raman band is believed by some to show less dependence of optical activity on frequency and to show a distribution closer to the density of states. The ν_R vibrations of ice, liquid water, and the hydrates have been reviewed by Brun.[79]

The spectrum of the liquid in the ν_T range resembles that of vitreous ice, described by Bertie and Whalley,[80] which is in turn broader than the spectrum of ice I. Clearly, understanding of the liquid spectrum in this range must follow that of ice. Whalley[81] has indicated that the broadening in the liquid is produced by reduction of short-range order, and by the occurrence of rotational and translational diffusion; he has emphasized that, in the present state of knowledge, structural conclusions deduced from the ν_T spectrum are pure speculation.

DIELECTRIC RELAXATION.   Below about one wavenumber, the spectrum is usually described in terms of frequencies, less often of wavelengths. The frequency corresponding to 1 cm^{-1} is 3×10^{10} Hz. Most dielectric measurements in liquid water have been made below 10^8 Hz. Nuclear magnetic resonance studies of liquid water[82] provide information in the same frequency range. The dielectric properties of liquid water and of the ices under pressure have been described by Whalley.[83] Here we consider only the relaxation time.

The dielectric properties of a material may be described by the complex permittivity, ε^*

$$\varepsilon^* = \varepsilon' - i\varepsilon''$$

where ε' is the (real) dielectric constant, and ε'' is the imaginary part of the permittivity. The optical properties are characterized by the complex refractive index

$$n^* = n' - in''$$

which is related to the complex permittivity by

$$\begin{aligned} n^{*2} &= \varepsilon^* \\ &= n'^2 - 2in'n'' - n''^2 \\ &= \varepsilon' - i\varepsilon'' \end{aligned}$$

and hence

$$\varepsilon' = n'^2 - n''^2$$
$$\varepsilon'' = 2n'n''$$

The parameters ε', ε'', n', and n'' are functions of temperature, density, and frequency. The complex refractive index, n'', is related to the absorption of optical energy. The infrared absorption is given by[75]

$$(10) \qquad\qquad I = I_0 \exp\left(-4\pi n'' \omega x\right)$$

where I is the transmitted energy, I_0 the incident energy, and x is distance into the sample. Thus in the dielectric region the language appears different from that in the optical region above one wavenumber, but the information obtained is similar.

If ε_0' is the limiting low-frequency dielectric constant, ε_∞' the high-frequency limit, and if the relaxation is produced by a single exponential decay

$$(11) \qquad\qquad \varepsilon^* = \varepsilon_\infty' + \frac{\varepsilon_0' - \varepsilon_\infty'}{1 + i\omega_a \tau_0}$$

where τ_0 is the relaxation time and ω_a is the angular frequency. For H_2O the dielectric relaxation occurs with a narrow range of distribution times that narrows with increasing temperature,[84,85] and the same appears to be true for D_2O.[86]

The presence of a single relaxation time, τ_0, at each temperature shows that, if a mixture model is used, the two species must interconvert in less than τ_0. In a continuum model there is a distribution of angles and distances, and a single τ_0 means that a typical configuration must be able to change to another typical configuration—although some configurations may be inaccessible—in less than τ_0. Table 3 shows values of τ_0 obtained by different observers; agreement is excellent.

Table 3 Dielectric Relaxation Time of Liquid Water, $10^{11} \tau_0$ (sec)

Tempera-ture (°C)	H_2O			D_2O
	From Ref. 84	From Ref. 85	From Ref. 87	From Ref. 86
0	1.87	—	—	—
1	—	1.57	—	—
5	—	—	—	1.856 ± 0.017
10	1.36	1.20	1.35	1.599 ± 0.011
20	1.01	0.92	0.99	1.189 ± 0.008
30	0.75	0.72	0.735	0.923 ± 0.008
40	0.59	0.57	0.595	0.726 ± 0.005
50	0.47	0.48	0.49	0.592 ± 0.004
60	—	0.40	—	0.495 ± 0.004

The behavior at higher frequencies is in dispute. Chekalin and Shakhparonov[87] have claimed that at 7×10^{10} Hz the absorption is greater than predicted from (11), indicating another relaxation process. Certainly, at some frequency the range of environments that produced the broadness in ν_T will cause (11) to fail.

Important information about the equilibrium structure of water can be obtained from the dielectric constant by the Onsager–Kirkwood equation. A correlation function, g, is defined by

$$g = \frac{(\varepsilon_0' - \varepsilon_\infty')(2\varepsilon_0' + \varepsilon_\infty')kT}{4\pi N_A \mu^2 \varepsilon_0'} \left(\frac{3}{\varepsilon_\infty' + 2}\right)^2 = 1 + \sum \cos \gamma$$

where N_A is Avogadro's number, μ is the dipole moment *in vacuo*, and $\cos \gamma$ is the cosine of the angle between the central dipole and a neighboring dipole.

Because of stoichiometric requirements, water molecules can only be positioned in a tetrahedral lattice (or network) in a few ways. Hence there are only a few possible angles between dipoles.[88] If the dipoles were oriented as randomly as possible, g would be 2.3. If dipoles tend to align in the fields of their neighbors, g will be higher. For ice Ih, $g = 3.3$,[83] and for liquid water at 0°C $g = 2.6$. This high value of g shows that the tetrahedral structures of ice are well preserved in liquid water. With increasing pressure or increasing temperature g falls, which is to be interpreted in terms of increased distortion of O–O–O angles.

NEUTRON SCATTERING.   Since there are no selection rules for neutron scattering, observations of the inelastic scattering of cold neutrons offer, in principle, a direct way to determine the density of states in the liquid. The interpretation of the results is most natural with a continuum theory, but an interpretation in terms of a mixture model has been given, for example, by Szkatula and Fulinski.[89] The promise of neutron scattering as a source of detailed data about the motion of atoms and molecules in liquid water has not yet been fulfilled. Rather, the details of the scattering process and the elimination of multiple scattering are still the center of attention, the agreement among different laboratories is often poor, and the identification of processes producing features in the spectrum is uncertain.

The gross features of the spectrum from neutron scattering agree with the infrared and Raman spectra. (Energies in neutron scattering are usually reported in electron volts; we continue to use wavenumbers, 1 meV = 8.07 cm^{-1}; the useful range of neutron scattering is usually 10–1000 cm^{-1}.) In addition, neutron scattering gives the amplitude of vibrational motion. It is agreed that in the neutron spectrum of liquid H_2O a peak near 518 cm^{-1},[90] or 490 ± 15 cm^{-1},[91] is to be identified with ν_R as seen in the infrared or Raman spectra. Features at lower frequencies (e.g., 170 ± 10, 64 ± 10,

40 ± 3 cm^{-1}) as reported by Hughes et al.[91] are less certain in their interpretation but are probably part of ν_T. The rms amplitude of proton vibration in H_2O is given by Harling[90] as 0.23 Å.

Intramolecular Vibration Spectrum

NORMAL VIBRATIONS.   Much of the controversy between mixture and continuum theories has centered on the interpretation of the infrared and Raman spectra, especially in the 3400-cm^{-1} (3-μ) band and its overtones. Earlier parts of this chapter have considered molecules as rigid bodies, and the internal motions were ignored. However, the intramolecular spectrum derived from these internal vibrations is sensitive to changes in the local environment, but interpretation of the observations is not always unanimous, for the absorption bands of liquid water are particularly broad.

In the vapor, the water molecule belongs to the point group C_{2v}. This molecule is nonlinear with three atoms and it has three normal vibrations. The three normal vibrations, shown in Figure 7, are: symmetric stretching, ν_1; bending, ν_2; and antisymmetric stretching, ν_3. In the vapor the wavenumbers of these vibrations are well established and are shown in Table 4. From the table it is seen that ν_1, ν_3, and twice ν_2 are near 3300 cm^{-1}.

Although infrared properties may be described in terms of real and imaginary refractive indices (see, e.g., Ref. 92), it is more usual to consider only the absorption, which, for liquid H_2O and D_2O, occurs to some extent across this whole range, notwithstanding the existence of regions of comparative transparency. In D_2O with a few tenths mole % of H_2O, equilibrium produces a solution of HOD (leaving a negligible amount of H_2O), and the OH

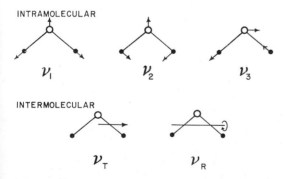

Figure 7.   Vibrations of water molecules. The intramolecular vibrations take place without translation or rotation of the molecule. The ν_1 is symmetric stretching, ν_2 is bending, and ν_3 is antisymmetric stretching. The intermolecular mode, ν_T, involves motion of the center of mass of the molecule, and ν_R involves rotation about an axis through the center of mass.

Table 4   Vibration Frequencies of Water (wavenumbers)

	ν_1	ν_3	ν_2	ν_R	ν_T
Vapor					
H_2O	3657[1]	3756[1]	1595[1]	—	—
D_2O	2671[1]	2788[1]	1178[1]	—	—
HOD	2727[1]	3707[1]	1402[1]	—	—
Liquid[b]					
H_2O	3440[1]		1200[2]	685[3]	193[3]
D_2O	2515[1]		1000[2]	505[3]	187[3]
HOD	2492[4]	3392[4]			
(dilute)					
Ice Ih[b]					
H_2O	3300[1]		1600[1]	840[1]	230[5]
D_2O	2425[1]			640[1]	220[5]
HOD	2440[4]	3300[4]			
(dilute)					

[a] Superior numbers refer to references at end of table.
[b] Values are for band maxima, and secondary features are omitted.

1. E. Whalley, *Develop. Appl. Spectr.*, **6**, 277 (1968).
2. M. Falk and T. A. Ford, *Can. J. Chem.*, **44**, 1699 (1966).
3. D. A. Draegert, N. W. B. Stone, B. Carnutte, and D. Williams, *J. Opt. Soc. Amer.*, **56**, 64 (1966).
4. T. A. Ford and M. Falk, *Can. J. Chem.*, **46**, 3579 (1968).
5. J. E. Bertie and E. Whalley, *J. Chem. Phys.*, **46**, 1271 (1967).

vibrations are largely uncoupled from the collective motions of the D_2O, which happen to absorb little near 3400 cm^{-1} where the OH ν_3 band is centered. Similarly, the uncoupled ν_1 vibrations of OD can be seen in dilute solution of HOD in H_2O near 2500 cm^{-1}. The bands are still several hundred wavenumbers wide, but are simpler than those of pure H_2O or D_2O. The band maxima change smoothly with temperature, moving to higher frequencies (i.e., becoming more gaslike). The same technique of isotope dilution is effective with the ices and hydrates, and because these have narrower bands, their interpretation is less obscure. The infrared spectrum of ice Ih has been discussed by Whalley,[81] who finds that the coupling of ν_1 and ν_3 is so large that description of the band in terms of ν_1 and ν_3 is probably not very useful, and furthermore that there is no evidence that $2\nu_2$ is important in the ice spectrum.

If the description of the infrared spectra of the ices in terms of ν_1 and ν_3

is not very useful, that of the liquid should be even less so. This is implied by the use of R and θ^2; in their use no specification of the position of the second hydrogen is made. The correctness of this simplification has been considered by Schiffer and Hornig.[31] These authors distinguish three kinds of interaction: the first two correspond to the coefficients $(\partial\omega/\partial R)_{\theta^2}$ and $(\partial\omega/\partial\theta^2)_R$ considered below; the third is the perturbation in an OH–O group by a neighboring interstitial molecule that is expected to change the energies of the hydrogen bond and the vibration frequency without a corresponding change of OH–O geometry.

It was early recognized[93] that the infrared vibrational spectrum could give evidence of association (i.e., hydrogen bonding), and that it was preferable to work with the fundamentals ν_1 and ν_3 rather than with overtones; that choice, however, is strongly disputed by Luck and Ditter.[94] In any event, the studies that give strongest support to the continuum theory have been made in the fundamental region. It seems fair to say that proponents of mixture theories have been most struck by the systematic changes of the spectra with temperature, which appear to indicate changes of composition, whereas proponents of continuum theories have been most struck by the need to explain the broadness of the bands and are agreed that this property implies a broad distribution of interaction configurations.

MIXTURE MODELS.   The changes of shape of infrared and Raman bands with temperature (or pressure) have led a number of workers to conclude that different regions of the bands can give the concentrations of different species in chemical equilibrium, and that this can provide a basis for the calculations of the thermodynamic properties of water. Thus Buijs and Choppin[95] interpreted the infrared spectrum near 10,000 cm^{-1} as showing three species which they identified as having 0, 1, and 2 hydrogen bonds. Worley and Klotz[96] interpreted the HOD spectrum near 7000 cm^{-1} as showing two species, which they identified as free and hydrogen bonded. Extensive work has been done by Walrafen with HOD solutions,[97] resolving the infrared and Raman bands into the sum of two Gaussian components, one increasing and one decreasing with temperature. The libration band ν_R[98] was likewise decomposed into Gaussian bands having the same temperature dependence. Satisfactory thermodynamic properties could be calculated from these intensities.[99] In short, this work offers a strong *prima facie* case for interpreting liquid water in terms of a mixture model. However, the following objections weaken that case:

1. The spectrum has not been put into relation with the rather similar spectrum of ice.

2. The decomposition of the spectral band into components is non-unique.[100]

3. The presence of isosbestic (Greek *iso-*, equal, *sbest-*, extinction) points has been taken as proving the existence of components having opposite variations of intensity with temperature.[97] However, ordinary ice Ih, in which all bonds are nearly equivalent, shows isosbestic points, and they also occur in other solids where the structure is known to have only one component.[101]

4. The calculations are not closely linked to any structures. It is true that Walrafen offered a structural model, his C_{2v} model, which agrees well with the Raman depolarization.[74] There is no suggestion that *only* the C_{2v} model can explain the observations. Indeed, Schiffer[102] believes that similarities between the spectra of hydrates and liquid water permit explanation of the liquid spectrum in terms of local distortions; this explanation is quite different from that of the C_{2v} model.

5. The intensity, I_0, of incident radiation decreases with distance, x, according to [cf. (10)]

$$I = I_0 \exp(-\mathscr{E}x)$$

where the absorptivity, $\mathscr{E}$, contains the information about water and is a function of temperature, liquid density, ρ, and frequency, ω; $\mathscr{E}$ is not itself the density, $\mathscr{D}$, of vibrational states—it is $\mathscr{D}$ that is related to thermodynamic properties—but is related to it by a factor $\mathscr{F}_{IR}$.

$$\mathscr{E}(T, \rho, \omega) = \mathscr{F}_{IR}(T, \rho, \omega)\mathscr{D}(T, \rho, \omega)$$

A similar relation introduces $\mathscr{F}_R$ to relate Raman scattering to the density of states. Although $\mathscr{F}_{IR}$ and $\mathscr{F}_R$ change across a band, infrared and Raman spectra have usually been interpreted assuming them to be constant. However, with the Raman spectrum of uncoupled ν_3 of HOD in D_2O, $\mathscr{F}_R$ appears to increase with increasing frequency (J. Schiffer, private communication). This accounts for the maximum of the Raman band occurring at a higher frequency than the infrared maximum and may account for the shoulder on the high-frequency side of the band. If a full description of the molecular geometry were available, the dependence of $\mathscr{F}$ and $\mathscr{D}$ on configuration τ could be considered to give

$$\mathscr{E}(T, \rho, \omega) = \int \mathscr{F}(T, \rho, \omega, \tau)\mathscr{D}(T, \rho, \omega, \tau)\, d\tau$$

This equation remains purely formal. Somewhat similar considerations are expressed in different language by Wyss and Falk.[103]

These empirical equations have been introduced to display the relation between $\mathscr{E}$ and $\mathscr{D}$. A more detailed analysis of the absorption process would require the use of wave vectors; see the treatment of the optical effects of the disorder in ice given by Whalley.[76]

FREQUENCY AND INTERMOLECULAR GEOMETRY. A two-component mixture involves at least three kinds of interaction. The structural questions to be answered are: what are the configurations and distances between neighboring molecules of species 1? What are they for neighboring molecules of species 2? What are they between a molecule of species 1 and a neighboring molecule of species 2? On the other hand, a continuum theory considers a range of interactions of one class also dependent on distance and configuration.

What is the change of frequency of a particular vibration, ω, or of the band maximum ω_{max}, with change of oxygen–oxygen distance, R, and of the departure of a hydrogen bond oxygen–hydrogen–oxygen from straightness, θ^2, which can be estimated from the oxygen–oxygen–oxygen angles? Write the expansion

$$\Delta\omega(OH) = \left(\frac{\partial\omega}{\partial R}\right)_{\theta^2} \Delta R + \left(\frac{\partial\omega}{\partial\theta^2}\right)_R \theta^2$$

The value of $(\partial\omega/\partial\theta^2)_R$ is not known; its value must be small, and will be taken as zero, and the problem is to determine

$$\Delta\omega(OH) = \left(\frac{\partial\omega}{dR}\right) \Delta R$$

A number of authors[30,104–106] have investigated the relation between $\omega(OH)$ and R. According to Wall and Hornig,[30] who sought a correlation that could be applied to liquid water, the slope changes rapidly in the region of 2.8 Å of interest in liquid water. They found that for R greater than 2.9 Å, $\Delta\omega(OH)/\Delta R \approx 200$ cm^{-1} Å^{-1}, whereas for R less than 2.75 Å, $\Delta\omega(OH)/\Delta R \approx 6000$ cm^{-1} Å^{-1}. From the Raman and infrared spectra of the ices Kamb[3] found

$$\frac{\Delta\omega_{max}(OH)}{\Delta R} \approx 750 \text{ cm}^{-1}\text{ Å}^{-1}$$

An estimate of $\Delta\omega_{max}(OH)/\Delta R$ may also be obtained from the relation

$$\frac{\Delta\omega}{\Delta R} = \left(\frac{\Delta\omega}{\Delta T}\right)\left(\frac{\Delta T}{\Delta R}\right)$$

Falk and Ford[4] found that for OH in liquid D_2O

$$\left(\frac{\Delta\omega_{max}}{\Delta T}\right)_{1 \text{ atm}} = 0.68 \text{ cm}^{-1}\text{ °K}^{-1}$$

and from Narten and Levy[36]

$$\left(\frac{\Delta R}{\Delta T}\right)_{saturation} = 0.0006_0 \text{ Å °K}^{-1}$$

based on their X-ray results.[107] Saturation pressure may be considered nearly constant over this range, so these equations may be combined to give

$$\left(\frac{\Delta\omega_{max}(OH)}{\Delta R}\right)_{1\ atm} = 1100\ cm^{-1}\ Å^{-1}$$

It seems clear that $\partial\omega_{max}(OH)/\partial R$ must be about 1000 cm^{-1} Å^{-1}. The half-width of the uncoupled OH vibration in D$_2$O is about 250 cm^{-1}.[108] Hence the halfwidth of the distribution of oxygen–oxygen distances is computed to be 0.25 Å, which agrees with the radial distribution function and with the amplitude of thermal motion from neutron diffraction. Thus the temperature dependence of the first peak in the molecular distribution function and the halfwidth of the maximum agree with the temperature dependence of the Raman and infrared bands and their halfwidth.

VIBRATIONS IN HYDRATES. In ice II, the uncoupled vibration of OD in H$_2$O has a halfwidth of 5 cm^{-1} at 100°K. In ice Ih at the same temperature it is 18 cm^{-1}, increasing to 40 cm^{-1} at 0°C.[108] In the liquid at 0°C it has a halfwidth of 149 cm^{-1}. The extreme narrowness in ice II is possible because the hydrogens are ordered in that ice. Similarly, narrow bands are observed in the spectra of some isotopically substituted hydrate crystals. Since the hydrogen bonding in a hydrate depends on the degree of coordination with the cation and on the interplay of the structure of the water molecule, the sizes of the ions, and other packing considerations, the hydrates show a range of hydrogen-bond strengths and perturbations.

The uncoupled $\omega(OH)$ in NaCl·2D$_2$O is near the frequency of the band maximum of $\omega(OH)$ in liquid D$_2$O, and this has been the starting point for a theory of the vibrations in liquid water by Schiffer and Hornig.[31,109]

In gypsum, CaSO$_4$·2H$_2$O, a hydrate in which the fundamentals of H$_2$O, D$_2$O, and HOD can all be observed by infrared, the interpretation of the spectra is agreed;[110,111] all water molecules are equivalent and are bonded through the hydrogens to oxygens in such a way that one oxygen–oxygen distance is 0.02 Å greater than the other. Kling and Schiffer[112] believe that the water molecule in gypsum is much like the average molecule in liquid water.

In a few hydrates bifurcated hydrogen bonds[29] are present in which one OH group bonds to two oxygen at an OH–O distance greater than normal. Bifurcated bonds are found in MgSO$_4$·4H$_2$O and violuric acid monohydrate. These bifurcated hydrates may be taken as approximate models of a step in the diffusion of liquid water, and they indicate that a hydrogen bond pointing midway between one oxygen and another must still be rather stable, despite the distortion of the bonding.

Other Inferences

THE CONCEPT OF BROKEN HYDROGEN BOND.   A prominent feature of mixture theories of liquid water has been the calculation of the fraction of broken hydrogen bonds. Falk and Ford[4] tabulated 23 estimates of the percentage of broken hydrogen bonds in water, with values ranging from 0.1 to 70%; and other values, such as those of O'Neil and Adami,[113] have been offered since. Without further specification, "broken hydrogen bond" is devoid of meaning. The further specification required may be seen from (2). The abundance of a species depends on G_i° for each species present. One species is commonly taken as related to ice I; variation in the choice of the other produces variation in the percentage of broken hydrogen bonds. Little edification is likely to be achieved by the use of "broken hydrogen bond," even within the context of mixture theories.

References to broken hydrogen bonds are so pervasive that authors whose algebra is in accord with continuum ideas speak of broken hydrogen bonds to satisfy, one supposes, those who feel that a theory of liquid water is incomplete unless it mentions them.

The total bonding energy of water molecules consists of two parts: directional forces, which give the hydrogen bond energy proper, and nondirectional forces. There is no unambiguous way to separate the two, so they are generally considered together, and the sum is also called hydrogen bond energy. Thus with the ν_1 vibration of HOD, the molecule may be described as "free" if the frequency is near 2720 cm^{-1} as in the vapor. Extrapolation of the data of Franck and Roth[114] for a density of 1 g cm^{-3} to infinite temperature gives a value of 2640 to 2670 cm^{-1}, which corresponds to the directional forces being averaged over all directions. At 0°C the band maximum is at 2488 cm^{-1}. Band maxima between 2488 and 2650 cm^{-1} reflect weakening of the mean hydrogen bond. Description of this weakening in terms of the breaking of bonds seems arbitrary.

THERMAL ANOMALIES.   In the early 1960s there seemed to be evidence[115] of thermal anomalies or kinks in the properties of water. This was questioned by Falk and Kell,[116] who pointed out that a number of reported kinks had not been found on reinvestigation, and that, in any event, the size of the reported kinks was not much larger than experimental error.

Since then one anomaly has been reported in the surface and interfacial tension of water[117] and one in the electrical conductivity of acid solutions,[118] where the kink found in the energy of activation seems to me to be imposed on the data rather than to follow from them. The temperature dependence of the iodide ultraviolet absorption spectrum in water has been reported to show[119,120] and not to show[121] an anomaly.

Most investigations of supposed anomalies have failed to find them and have shown that observations made at closely spaced temperature intervals scatter about the curve interpolated between measurements made at greater intervals. Properties investigated include the temperature dependence of the dielectric constant,[122] surface tension,[123] viscosity,[124] velocity of sound,[125] proton spin-lattice relaxation time,[126] the apparent molal volume of dissolved salts,[127] and some hydrolysis reactions.[128]

Authors who accept the evidence for the existence of kinks hold that kinks support mixture theories and that their existence is a key factor in discriminating among theories of liquid water. It has never been made clear how a mixture theory can accept them any more than a continuum theory. A solution theory cannot produce kinks in the Gibbs energy, G, of the bulk phase unless there are kinks in the Gibbs energies, G_1, G_2, of the components. A mechanical mixture theory cannot accommodate phenomena such as the maximum density, which the solution theory is invoked to explain.

MEAN POTENTIAL ENERGY AND SPECIFIC HEAT.   The potential energy of two water molecules has been described in terms of a Lennard-Jones potential plus terms arising from multipolar and induced multipolar forces. In the case of N molecules having positions and orientations given by $\tau^N \equiv (\tau_1, \tau_2, \ldots, \tau_N)$ the potential energy may be approximated by a sum of two-molecule potentials.

$$\Phi_N(\tau^N) = \sum_{i<j} \Phi(|\tau_i - \tau_j|)$$

The probability P of configuration τ^N is proportional to the Boltzmann term

$$P(\tau^N) \propto \exp\left[-\frac{\Phi_N(\tau^N)}{kT}\right]$$

and the mean potential energy $\langle\Phi\rangle$ is given by

(12)
$$\langle\Phi\rangle = \frac{\sum \Phi_N(\tau^N)P(\tau^N)}{\sum P(\tau^N)}$$

where the sums are over all configurations in the liquid. τ^N involves $6N$ coordinates, and although $\langle\Phi\rangle$ can be obtained by Monte Carlo calculations, it is simpler to obtain an equivalent quantity by a combination of thermodynamic and spectroscopic data.

The internal energy, U, is the sum of kinetic and potential parts

(13)
$$U = U_{\text{kin}} + U_{\text{pot}}$$

and $\langle\Phi\rangle$ as defined by (12) is to be identified with U_{pot} as defined by (13). A value for U can be obtained from thermodynamic measurements, U_{kin} can be calculated by statistical-mechanical methods from spectroscopic data, as shown by Ford and Falk,[108] and hence U_{pot} can be found.

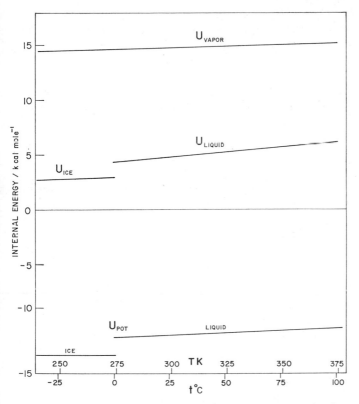

Figure 8. Internal and internal potential energies of water, showing U for vapor, liquid, and solid, and U_{pot} of liquid and solid.

In Figure 8, the zero for energy corresponds to molecules at large separation, with the translational, rotational, and vibrational energy appropriate to the temperature subtracted out. Inclusion of these terms gives U_{vapor} the internal energy for the gas at zero pressure

$$U_{vapor} = 3RT + U_{vib}$$

where

(14)
$$U_{vib} = \tfrac{1}{2}N_A hc \sum \omega + RT \sum \frac{\xi}{e^\xi - 1}$$

and where N_A is Avogadro's number, h Planck's constant, c the velocity of light, ω the frequency of vibration in wavenumbers, $\xi = hc\omega/kT$ where k is Boltzmann's constant, and summation is over the three internal vibrations of the molecule. In these equations the $3RT$ term is translational and rotational, the first of (14) is the zero-point energy and the second is the remaining vibrational energy. Once a single value of U_{vapor} has been obtained by this equation, other values may be obtained from the steam tables, which tabulate H (relative to their own zero), P, and V, using the relation

$$U = H - PV$$

At higher temperatures, where the saturation pressure is not small, $U_{sat. vapor}$ would lie below U_{vapor}, but the difference cannot be seen on the scale of Figure 8. The difference $U_{liquid} - U_{vapor}$ is the change of internal energy on condensation to the liquid. In Figure 8, U for the saturated liquid cannot be distinguished, in this temperature range, from U for a constant density of 18 cm³ mole⁻¹.

The vibrational energy of condensed water is also given by U_{vib}, using frequencies appropriate to the condensed phase and summing over the three internal vibrations and six intermolecular vibrations of the water molecule. Translational and rotational motions proper are almost absent from liquid water. Since the total internal energy is the sum of vibrational and potential parts, the potential energy U_{pot} of a condensed phase is given by (13) and U_{pot} is shown in Figure 8.

The average vibrational energy of a harmonic oscillator consists of two equals parts, kinetic and potential. Ford and Falk were interested in the rotational and translational contributions to U_{vapor} as well as the vibrations in the liquid, so they refer to all contributions as kinetic, their use including the potential part of the vibrational energy for both zero-point and higher vibrations.

The specific heat is readily obtained

$$C_V = \left(\frac{\partial U}{\partial T}\right)_V = \left(\frac{\partial U_{vib}}{\partial T}\right)_V + \left(\frac{\partial U_{pot}}{\partial T}\right)_V$$

where $(\partial U_{vib}/\partial T)_V$ is the $C_{V,glass}$ of the section, "Structure of Liquid Water: Qualitative Considerations" and $(\partial U_{pot}/\partial T)_V$ is the $C_{V,pot}$. Ford and Falk obtained an estimate of U_{pot} [and hence of $(\partial U_{pot}/\partial T)_V$] from spectral shifts. They showed that U_{pot} was correlated with $\Delta\omega = \omega - \omega_0$, the shift of frequency upward from the gas-phase value, by a function of the form

(15) $$U_{pot} = a[\exp(b\,\Delta\omega) - 1]$$

where $\Delta\omega$ was obtained by the use of isotopic dilution in the liquid and ice. For H_2O the values obtained from the shift of ν_3 gave $a = -15.089$ kcal mole⁻¹, $b = -5.00 \times 10^{-3}$ cm; from the ν_1 shift, the values for D_2O are $a = 15.080$ kcal mole⁻¹, $b = -7.19 \times 10^{-3}$ cm. The spectra entering into this correlation were not obtained at constant density but, at least approximately,

$$\frac{dU_{pot}}{dT} = \frac{dU_{pot}}{d\omega} \cdot \frac{d\omega}{dT}$$

From Ford and Falk, $d\omega/dT = 0.64$ cm^{-1}/$°$K, and $dU_{pot}/d\omega = 12.1$ cal mole^{-1} cm at $0°$C. Hence, $dU_{pot}/dT = 7.7$ cal mole^{-1} $°$K^{-1}. As $dU_{vib}/dT = 9.6$ cal mole^{-1} $°$K^{-1}, the sum 17.3 cal mole^{-1} $°$K^{-1} is the estimate of the specific heat, C, obtained by the Ford–Falk approach. The experimental value is 18.2 cal mole^{-1} $°$K^{-1} at $0°$C, in close agreement considering the approximations made in the calculation.

If only information from the liquid had gone into the determination of dU_{pot}/dT, the specific heat of the liquid would have been obtained exactly. The degree of agreement shows that the functional form of (15) is reasonable and that U_{pot} from ice can be combined with U_{pot} from the liquid in determining the coefficients of (15). The inclusion of ice data appears most responsible for the difference. For example, the spectral quantities do not refer to conditions of constant density. Furthermore, the Einstein oscillators used in (14) are only approximate, and (15) is only a correlation and not an exact relation. Further work seems desirable to determine the dependence of U_{pot} on pressure, to see if (15) can be made more exact.

In (3) the Gibbs energy, G, of liquid water was intuitively divided up into G_{vib} and G_{pot}. Since

$$G = G_{vib} + G_{pot}$$
$$= U + PV - TS$$
$$= U_{vib} + U_{pot} + P(V_{vib} + V_{pot}) - T(S_{vib} - S_{pot})$$

the division of U into U_{vib} and U_{pot} contains equivalent information.

4   Conclusions

Value of Tetrahedral Network Model. It is agreed that the ice lattice breaks down, in some sense, on melting. The typical mixture model explains this by saying that some hydrogen bonds break and some remain. We have preferred the alternate explanation that the melting involves a loss of long-range order made possible by distortion of the oxygen–oxygen–oxygen angles, and to a lesser extent by distortion of oxygen–oxygen distances, while preserving tetrahedral coordination at the typical oxygen atom. A static picture of such a structure is provided by the random tetrahedral network model. This model appears advantageous as a starting point for work on liquid water for at least two reasons:

1. The tetrahedral network can be put into intelligible relation with the properties of water in other phases, with the structure of other liquids, and

with the observed properties of liquid water itself. On the other hand, mixture models tend to explain water as something *sui generis*, not readily integrated with other knowledge.

2. The tetrahedral network is flexible enough to permit its refinement in the light of improved knowledge. Ice has come to be a part of solid-state physics (see, e.g., Ref. 129). It is time that the study of liquid water became a part of liquid-state physics, and the material covered in this chapter indicates that this is possible. On the other hand, the mixture theory of liquid water has existed since 1892, able to describe everything, but unable to produce new insights.

Narten and Levy[36] have emphasized that models of liquid water must agree with the X-ray diffraction pattern and that most proposed models leave much unspecified that must be known to establish the expected pattern. The present ideas are certainly an inadequate specification in their sense. However, these ideas are in general agreement not only with the diffraction pattern but also with other evidence, such as the infrared spectrum and the thermodynamic properties.

The mathematics of this chapter has been kept as simple as possible, and tools such as the wave vector and the Fourier transform, although they are necessary in the modern physics of the liquid and solid states, have been avoided entirely. One can be sure, however, that the mainstream of future work on the structure of water will be highly mathematical.

Aqueous Solutions.   The section on the structure of liquid water showed that the pair distribution function, $g(\tau)$, for liquid water is six-dimensional. The corresponding function that is the minimum to specify water structure around a spherical ion is nine-dimensional. Water structure is not specified unless the probable position of one water molecule is given relative to another water molecule; and in a solution, the position of both must be known relative to a solute particle. Consider one water molecule fixed. Specification of a second requires six dimensions as before. Specification of the center of a spherical solute particle requires a further three, for a total of nine.

It seems clear that, at a time when the competing existence of continuum and mixture theories shows the six-dimensional case to be poorly understood, no progress can be expected in the nine-dimensional one. Certainly, as Holtzer and Emerson[130] have shown, "structure making" and "structure breaking" are ambiguous terms as commonly used. Until the structure of water itself is better understood, it seems best to use only concepts that have a clear meaning, such as entropy, rather than to try to reduce a complex problem to a few catch phrases.

REFERENCES

1. D. Eisenberg and W. Kauzmann, *The Structure and Properties of Water*, Oxford, 1969, p. 150.
2. H. S. Frank and A. S. Quist, *J. Chem. Phys.*, **34**, 604 (1961).
3. B. Kamb, in *Structural Chemistry and Molecular Biology*, Eds., A. Rich and N. Davidson, Freeman, San Francisco, 1968, p. 507.
4. M. Falk and T. A. Ford, *Can. J. Chem.*, **44**, 1699 (1966).
5. C. M. Davis and J. Jarzynski, *Advan. Mol. Relaxation Proc.*, **1**, 155 (1967–1968).
6. D. P. Stevenson, in *Structural Chemistry and Molecular Biology*, Eds., A. Rich and N. Davidson, Freeman, San Francisco, 1968, p. 490.
7. J. A. Pople, *Proc. Roy. Soc. (London)*, *Ser. A.*, **205**, 163 (1951).
8. P. A. Egelstaff, *An Introduction to the Liquid State*, Academic Press, London, 1967, p. 12.
9. W. C. Röntgen, *Ann. Phys. Chem. (Wied)*, **45**, 91 (1892).
10. *Trans. Faraday Soc.*, **6**, 71–123 (1910).
11. H. M. Chadwell, *Chem. Rev.*, **4**, 375 (1927).
12. N. E. Dorsey, *Properties of Ordinary Water-Substance*, Reinhold, New York, 1940, pp. 161–175.
13. R. M. Noyes, *J. Chem. Phys.*, **34**, 1983 (1961).
14. T. L. Hill, *Statistical Mechanics*, McGraw-Hill, New York, 1956, p. 59.
15. J. D. Bernal and R. H. Fowler, *J. Chem. Phys.*, **1**, 515 (1933).
16. J. A. Barker and R. O. Watts, *Chem. Phys. Letters*, **3**, 144 (1969).
17. R. J. Beshinske and M. H. Lietzke, *J. Chem. Phys.*, **51**, 2278 (1969).
18. G. H. A. Cole, *Rept. Progr. Phys.*, **31**, (2) 419 (1968).
19. L. Verlet and D. Levesque, *Physica*, **36**, 254 (1967).
20. J. A. Barker and D. Henderson, *J. Chem. Phys.*, **47**, 2856, 4714 (1967).
21. E. Praestgaard and S. Toxvaerd, *J. Chem. Phys.*, **51**, 1895 (1969).
22. M. Weissmann and L. Blum, *Trans. Faraday Soc.*, **64**, 2606 (1968).
23. F. Ordway, *Science*, **143**, 800 (1964).
24. J. D. Bernal, *Proc. Roy. Soc. (London)*, *Ser. A.*, **280**, 299 (1964).
25. N. N. Fedyakin, *Kolloid Zh.*, **24**, 497 (1962).
26. B. V. Deryagin and N. V. Churayev, *Priroda*, **1968** (4), 16.
27. E. R. Lippincott, R. B. Stromberg, W. H. Grant, and G. L. Cessac, *Science*, **164**, 1482 (1969).
28. J. A. Schufle and M. Venugopalan, *J. Geophys. Res.*, **72**, 3271 (1967).
29. W. C. Hamilton and J. A. Ibers, *Hydrogen Bonding in Solids*, Benjamin, New York, 1968, p. 188.
30. T. T. Wall and D. F. Hornig, *J. Chem. Phys.*, **43**, 2079 (1965).
31. J. Schiffer and D. F. Hornig, *J. Chem. Phys.*, **49**, 4150 (1968).
32. J. D. Bernal, *Nature*, **183**, 141 (1959).
33. J. D. Bernal and J. Mason, *Nature*, **188**, 910 (1960).
34. G. D. Scott, *Nature*, **188**, 908 (1960).
35. J. D. Bernal and S. V. King, in *Physics of Simple Liquids*, Eds., H. N. V. Temperley, J. S. Rowlinson, and G. S. Rushbrooke, North-Holland, Amsterdam, 1968, p. 231.
36. A. H. Narten and H. A. Levy, *Science*, **165**, 447 (1969).
37. I. Yannas, *Science*, **160**, 298 (1968).
38. A. A. Miller, *Science*, **163**, 1325 (1968).
39. N. Dass and N. C. Varshneya, *J. Phys. Soc. Japan*, **26**, 873 (1969).
40. J. A. Ghormley, *J. Chem. Phys.*, **48**, 503 (1968).

41. M. Sugisaki, H. Suga, and S. Seki, *Bull. Chem. Soc. Japan*, **41**, 2591 (1968).
42. W. M. Slie, A. R. Donfor, and T. A. Litovitz, *J. Chem. Phys.*, **44**, 3712 (1966).
43. C. M. Davis and T. A. Litovitz, *J. Chem. Phys.*, **42**, 2563 (1965).
44. R. J. Bell, N. F. Bird, and P. Dean, *J. Phys.*, **C1**, 299 (1968).
45. R. J. Bell and P. Dean, *Phys. Chem. Glasses*, **9**, 125 (1968).
46. L. Pauling, *J. Amer. Chem. Soc.*, **57**, 2680 (1935).
47. P. H. Gaskell, *Trans. Faraday Soc.*, **62**, 1493, 1505 (1966).
48. D. L. Evans and S. V. King, *Nature*, **212**, 1353 (1966).
49. C. A. Coulson and D. Eisenberg, *Proc. Roy. Soc.* (*London*), *Ser. A*, **291**, 445, 454 (1966).
50. B. Kamb, *J. Chem. Phys.*, **43**, 3917 (1965).
51. J. E. Lennard-Jones, *Proc. Roy. Soc.* (*London*), *Ser. A*, **106**, 463 (1924).
52. F. London, *Z. Physik. Chem.*, **B11**, 222 (1930).
53. J. O. Hirschfelder, C. F. Curtiss, and R. B. Bird, *Molecular Theory of Gases and Liquids*, Wiley, New York, 1954, p. 960.
54. S. Kielich, *Physica*, **31**, 444 (1965).
55. R. M. Glaeser and C. A. Coulson, *Trans. Faraday Soc.*, **61**, 389 (1965).
56. Krishnaji and V. Prakash, *Rev. Mod. Phys.*, **38**, 690 (1966).
57. D. E. Stogryn and A. P. Stogryn, *Mol. Phys.*, **11**, 371 (1966).
58. J. S. Rowlinson, *J. Chem. Phys.*, **19**, 827 (1951).
59. E. S. Campbell, G. Gelernter, H. Heinen, and V. R. G. Moorti, *J. Chem. Phys.*, **46**, 2690 (1967).
60. D. J. G. Ives and T. H. Lemon, *RIC Rev.*, **1**, 62 (1968).
61. A. Ben-Naim and F. H. Stillinger, Chapter 8 of this volume.
62. A. D. Buckingham, *Quart. Rev.* (*London*), **13**, 183 (1959).
63. J. A. Barker, *Proc. Roy. Soc.* (*London*), *Ser. A*, **219**, 367 (1953).
64. A. D. Buckingham and J. A. Pople, *Trans. Faraday Soc.*, **51**, 1173 (1955).
65. J. F. Harrison, *J. Chem. Phys.*, **49**, 3321 (1968).
66. W. H. Stockmayer, *J. Chem. Phys.*, **9**, 398 (1941).
67. M. Klein and H. J. M. Hanley, *Trans. Faraday Soc.*, **64**, 2927 (1968).
68. S. Carrà and L. Zanderighi, *Nuovo Cimento*, **B49**, 133 (1967).
69. L. Monchick and E. A. Mason, *J. Chem. Phys.*, **35**, 1676 (1961).
70. G. S. Kell, G. E. McLaurin, and E. Whalley, *J. Chem. Phys.*, **48**, 3805 (1968).
71. J. P. O'Connell and J. M. Prausnitz, *Ind. Eng. Chem.*, *Fundamentals*, **8**, 453, 460 (1969).
72. R. W. Bolander, J. L. Kassner, and J. T. Zung, *J. Chem. Phys.*, **50**, 4402 (1969).
73. G. S. Kell and G. E. McLaurin, *J. Chem. Phys.*, **51**, 4345 (1969).
74. G. E. Walrafen, *J. Chem. Phys.*, **40**, 3249 (1964).
75. G. R. Wilkinson, in *Molecular Dynamics and Structure of Solids*, NBS Spec. Publ. 301, 1969, p. 77.
76. E. Whalley and J. E. Bertie, *J. Chem. Phys.*, **46**, 1264 (1967); E. Whalley and H. J. Labbé, *J. Chem. Phys.*, **51**, 3120 (1969).
77. P. Dean, *J. Phys. Soc. Japan Suppl.*, **26**, 20 (1969).
78. A. J. Leadbetter, *Proc. Roy. Soc.* (*London*), *Ser. A*, **287**, 403 (1965).
79. G. Brun, *Rev. Chim. Minérale*, **5**, 899 (1968).
80. J. E. Bertie and E. Whalley, *J. Chem. Phys.*, **46**, 1271 (1967).
81. E. Whalley, *Develop. Appl. Spect.*, **6**, 277 (1966).
82. J. A. Glasel, *Develop. Appl. Spect.*, **6**, 241 (1966).
83. E. Whalley, *Advan. High Pressure Res.*, **1**, 143 (1966).
84. J. A. Saxton, *Proc. Roy. Soc.* (*London*), *Ser. A*, **213**, 473 (1952).

85. E. H. Grant and R. Shack, *Brit. J. Appl. Phys.*, **18**, 1807 (1967).
86. E. H. Grant and R. Shack, *Trans. Faraday Soc.*, **65**, 1519 (1969).
87. N. V. Chekalin and M. I. Shakhparonov, *Zh. Strukt. Khim.*, **9**, 896 (1968); *J. Struct. Chem.*, **9**, 789 (1968).
88. H. Fröhlich, *Theory of Dielectrics*, 2nd ed., Oxford, 1958, p. 137.
89. A. Szkatula and A. Fulinski, *Physica*, **36**, 35 (1967).
90. O. K. Harling, *J. Chem. Phys.*, **50**, 5279 (1969).
91. D. J. Hughes, H. Palevsky, W. Kley, and E. Tunkelo, *Phys. Rev.*, **119**, 872 (1960).
92. M. R. Querry, B. Carnutte, and D. Williams, *J. Opt. Soc. Amer.*, **59**, 1299 (1969).
93. J. Errera, R. Gaspart, and H. Sack, *J. Chem. Phys.*, **8**, 63 (1940).
94. W. A. P. Luck and W. Ditter, *J. Mol. Struct.*, **1**, 261 (1967–1968).
95. K. Buijs and G. R. Choppin, *J. Chem. Phys.*, **39**, 2035 (1963).
96. J. Worley and I. M. Klotz, *J. Chem. Phys.*, **45**, 2868 (1966).
97. G. E. Walrafen, *J. Chem. Phys.*, **48**, 244 (1968).
98. G. E. Walrafen, *J. Chem. Phys.*, **47**, 114 (1967).
99. G. E. Walrafen, in *Hydrogen-Bonded Solvent Systems*, Eds., A. K. Covington and P. Jones, Taylor and Francis, London, 1968, p. 9.
100. J. W. Perram, in *Hydrogen-Bonded Solvent Systems*, Eds., A. K. Covington and P. Jones, Taylor and Francis, London, 1968, p. 135.
101. M. Falk and H. R. Wyss, *J. Chem. Phys.*, **51**, 5727 (1969).
102. J. Schiffer, *J. Chem. Phys.*, **50**, 566 (1969).
103. H. R. Wyss and M. Falk, *Can. J. Chem.*, **48**, 607 (1970).
104. R. C. Lord and R. E. Merrifield, *J. Chem. Phys.*, **21**, 166 (1953).
105. G. C. Pimental and A. L. McClellan, *The Hydrogen Bond*, Freeman, San Francisco, 1960, p. 88.
106. H. Ratajczak and W. J. Orville-Thomas, *J. Mol. Struct.*, **1**, 449 (1967–1968).
107. A. H. Narten, M. D. Danford, and H. A. Levy, *Discussions Faraday Soc.*, **43**, 97 (1967).
108. T. A. Ford and M. Falk, *Can. J. Chem.*, **46**, 3579 (1968).
109. J. Schiffer, Ph.D. thesis, Princeton University, 1963.
110. V. Seidl, O. Knop, and M. Falk, *Can. J. Chem.*, **47**, 1361 (1969).
111. R. Kling and J. Schiffer, *Chem. Phys. Letters*, **3**, 64 (1969).
112. R. Kling and J. Schiffer, *J. Chem. Phys.*, **54**, 5331 (1971).
113. J. R. O'Neil and L. H. Adami, *J. Phys. Chem.*, **73**, 1553 (1969).
114. E. U. Franck and K. Roth, *Discussions Faraday Soc.*, **43**, 108 (1967).
115. W. Drost-Hansen, *Ann. N.Y. Acad. Sci.*, **125**, 471 (1965).
116. M. Falk and G. S. Kell, *Science*, **154**, 1013 (1966).
117. R. Cini, G. Loglio, and A. Ficalbi, *Nature*, **223**, 1148 (1969).
118. V. V. Morariu and R. V. Bucur, *J. Mol. Struct.*, **2**, 349 (1968).
119. M. J. Blandamer, M. F. Fox, and M. C. R. Symons, *Nature*, **214**, 163 (1967).
120. M. J. Blandamer, T. A. Claxton, and M. F. Fox, *Chem. Phys. Letters*, **1**, 203 (1967).
121. S. D. Hamann and N. K. King, *Nature*, **215**, 1263 (1967).
122. E. W. Rusche and W. B. Good, *J. Chem. Phys.*, **45**, 4667 (1966).
123. G. J. Gittens, *J. Colloid Interface Sci.*, **30**, 406 (1969).
124. L. Korson, W. Drost-Hansen, and F. J. Millero, *J. Phys. Chem.*, **73**, 34 (1969).
125. W. Senghaphan, G. O. Zimmerman, and C. E. Chase, *J. Chem. Phys.*, **51**, 2543 (1969).
126. L. O. Bowen, *Phys. Letters*, **26A**, 150 (1968).

127. F. J. Millero and W. Drost-Hansen, *J. Chem. Eng. Data*, **13**, 330 (1968).
128. R. J. Miller, C. Pinkham, A. R. Overman, and S. W. Dumford, *Biochim. Biophys. Acta*, **167**, 607 (1968).
129. N. Riehl, B. Bullemer, and H. Engelhardt, Eds., "Physics of Ice," *Proc. 3rd Intern. Symp. 1968*, Plenum Press, New York, 1969.
130. A. Holtzer and M. F. Emerson, *J. Phys. Chem.*, **73**, 26 (1969).

10   Mixture Models of Water*

C. M. Davis† and J. Jarzynski†,
Department of Physics, The American
University, Washington, D.C.

Abstract

Various mixture models of water are discussed in terms of recent experimental evidence (structural and dielectric relaxation times, X-ray and neutron diffraction, and Raman spectra). The evidence presented strongly indicates the existence in liquid water of distinguishable species of water molecules whose percentages change with temperature. The time that a water molecule spends in a given state is on the order of 10^{-12} sec. In particular, a two-state approach is shown to be highly successful in accounting for the properties of water over a wide range of temperatures. These two states appear to be separated by enthalpy, entropy, and volume differences of approximately 2600 cal/mole, 9.0 eu, and 7.0 cm³/mole, respectively.

1   Introduction

At present there exists no universal agreement with regard to the structure of water. Indeed, the unsettled nature of structural theories is best illustrated by the large number and variety of models that have been and still are being proposed and by the large number of articles discussing these models. (In a recent bibliography[1] covering the years 1957–1968, more than 120 articles dealing with the structure of pure water are listed.) The various models can be lumped into two main classes: continuum models and mixture models. The first class is discussed in the previous chapter; this chapter deals with mixture models.

The concept of water as a mixture of two or more species in chemical

* The work reported here was supported in part by a grant from the Office of Saline Water, U.S. Department of Interior (Grant No. 14-30-2518).
† Present address: Physical Acoustics Branch, Naval Research Laboratory, Washington, D.C. 20390.

equilibrium has been recognized for many years. The first published statement to this effect was that of Röntgen[2] in 1892. Within the past several years, however, a number of facts have appeared which drastically restrict the nature of the species that may exist in liquid water. The most important data have been provided by investigation of structural relaxation properties and Raman spectra (especially stimulated Raman spectra). This chapter places particular emphasis on these investigations, but no attempt is made to present an exhaustive review of the mixture models that have been proposed. Mixture models are discussed in terms of the molecular species of water involved, the method of obtaining thermodynamic parameters, and the partition functions appropriate to the model. An attempt is made to point out the assumptions involved (both explicit and implicit). Finally, the predictions of the various models are compared with experiment. A critical review of theories prior to 1961 is included in the work of Némethy and Scheraga.[3] A particularly simple group of structural models of water is based on a two-stage approach. Models of this type (proposed before 1967), which lend themselves to thermodynamic calculations, have been discussed in some detail by Davis and Jarzynski.[4] Finally, a recently published book by Eisenberg and Kauzmann[5] deals with the various structural models of water in some detail.

It has become popular to emphasize the "anomalous" aspect of the properties of water. This vogue has tended to obscure the fact that these "anomalies" are for the most part merely slight perturbations of the normal properties of associated liquids. The only truly anomalous property is the thermal expansion, and even that is generally misinterpreted. Thus, since water is a liquid, its structure must exhibit the gross features common to all liquids, which account for the fluid properties. A brief discussion of the structure of liquids follows.

Statistical theories of the liquid state lag considerably behind those of either crystalline solids or dilute gases. This is because in the case of solids or gases theoretical problems may be simplified to a considerable extent. In dilute gases the kinetic energy (KE) is much greater than the potential energy (PE). Thus to a first approximation PE may be neglected and the ideal gas theory may be derived. The influence of PE may be introduced in the form of a perturbation expansion leading to the virial equation. In crystalline solids, on the other hand, $PE \gg KE$. The crystallographic structure is determined by PE and the influence of KE is introduced in the form of small perturbations (or vibrations) about the equilibrium position. In liquids $KE \approx PE$ and neither of the above simplifications is applicable. Recently developed statistical-mechanical approximations[6] allow the equilibrium and transport properties of liquefied inert gases to be calculated. These approximations do not apply to directionally bonded liquids such as water, and it is therefore necessary to employ empirical models.

Liquids may be grouped into two rather broad classes, depending on the intermolecular potentials involved and the type of radial distribution curves (RDC) exhibited. The first class consists of liquids such as liquefied inert gases and molten metals that interact by means of spherically symmetric, nonsaturating potentials. The corresponding RDCs are similar to those obtained by Bernal,[7] on the basis of random close packing of spheres. The number of neighbors in the first coordination shell varies from 9 to 11.[8] The second class of liquids is composed of those which interact by means of highly directional, saturating covalent potentials. Simple inorganic glasses, water, and other associated liquids belong to this class. These substances are usually characterized by three or four nearest neighbors in the crystalline state. This tendency persists in the liquid state, where the first coordination shell possesses approximately the same number of nearest neighbors as in the crystalline state. Thus the short-range order in liquids of both classes appears to be determined primarily by the nature of the intermolecular potentials.

The molecules of a liquid are thus assumed to occupy positions relative to their nearest neighbors reminiscent of the positions they occupy in the solid. The structure, however, is assumed to lack long-range order; in addition, the short-range order that does exist is imperfect. At every pressure and temperature, moreover, a degree of order that describes its geometric state is associated with the liquid. Thus in the liquid, a molecule may be thought of as surrounded by nearest neighbors in the same relative positions and interacting through the same intermolecular potential as in the solid. One important difference, however, is that the relatively large KE allows the individual molecules to change position. It is the ease with which the molecules in a liquid change position that accounts for its ability to flow.

2 Evidence for Tetrahedral Coordination in Water

In view of the preceding discussion it is not surprising that many mixture models of water assume that one of the species is icelike. Thus it will be helpful to consider the structure of ice.[9] The oxygen layers above and below are mirror images (Figure 1). A hydrogen atom lies along the line of centers. Oxygen atoms are tetrahedrally coordinated to three others in the same layer and one other in an adjacent layer (Figure 2). The tetrahedral structure is body-centered cubic, with the reference molecule surrounded by four other molecules situated at alternate corners of the cube, as shown in Figure 3 where the hydrogen atoms are shaded. The reference molecule (designated C) is connected to neighbors of two different types: (a) through its hydrogen atom as in case CA, or (b) through its lone electron pair[10] as in case CB.

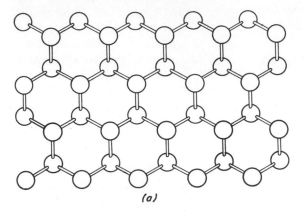

(a)

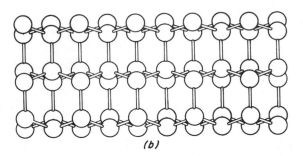

(b)

Figure 1. Positions of oxygen atoms in the ice lattice. Between each pair of adjacent oxygen atoms there is one hydrogen atom (not shown) situated one-third of the way along the line of centers. Hydrogen bond directions are indicated by the sticks. (a) Top view; (b) Side view.

The most direct information about the microscopic structure of a liquid is given by the RDC. A number of investigators have obtained RDCs for water by means of X-ray diffraction.[11-13] The most complete results are those of Narten and Levy (Figure 4), who found that the number of nearest neighbors is approximately 4.4. There is no unique technique for determining the number of nearest neighbors in a liquid. Recently, however, Pings[14] has presented a discussion of various criteria that have been used.

Unfortunately, the RDC of a liquid presents a one-dimensional picture, and although the number of molecules in a spherical shell of thickness Δr and radius r centered on the reference molecule is given, the angular orientation of these molecules within the spherical shell is not. However, the numbers of nearest neighbors in water is consistent with the assumption that the

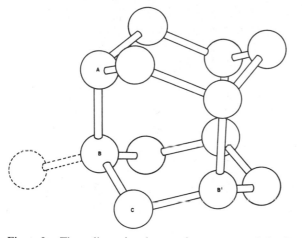

Figure 2.   Three-dimensional array of oxygen atoms in ice. Each oxygen atom is surrounded by four others: three in the same plane and one in an adjacent plane.

nearest-neighbor coordination in water is similar to that in ice. The coordination number is slightly greater than 4, which may indicate that some of the vacant cube corners (see Figure 3) are occupied by molecules not hydrogen bonded to the molecule at the center.

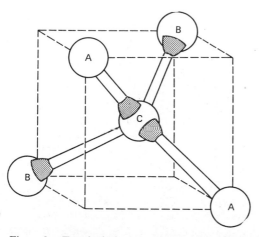

Figure 3.   Tetrahedral arrangement of nearest neighbors in ice. Hydrogen atoms are shaded. The relation of the tetrahedral structure to a body-centered-cubic structure is indicated. The nearest neighbors of the reference molecule (C) are of two types: those connected through one of the hydrogen atoms belonging to the reference molecule (A), and those connected through one of the lone electron pairs of the reference molecule (B).

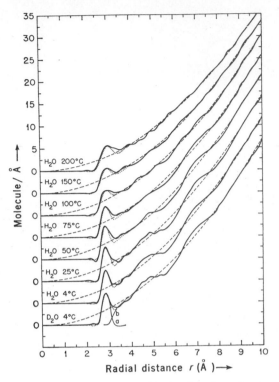

Figure 4. X-ray radial distribution functions of water: solid curves, observed data; dashed curves, $4\pi r^2 \times$ density; dotted curves, Gaussian components; area corresponding to 4.4 nearest neighbors is superimposed on the first major peak of each curve. [Reproduced by permission from A. Narten and H. Levy, *Science*, **165**, 450 (1969).]

Walrafen[15] has considered the intermolecular vibrations associated with a five-molecule hydrogen-bonded system similar to that in Figure 3. These vibrations are divided into two bending, four stretching, and three librational motions of the hydrogen bond. The first two types of motions correspond to restricted translations, whereas the latter type corresponds to restricted rotations. Evidence from inelastic neutron scattering,[16] infrared absorption,[17] and hyper-Raman and stimulated Raman scattering[18] was compared with the frequencies, selection rules, and depolarization ratios appropriate to point groups of various symmetries. The point group of lowest symmetry consistent with intermolecular spectroscopic data is the C_{2v}. Many point groups of higher symmetry were also examined and found to be inconsistent with the data. It should be emphasized that Walrafen[19,20] has been quite explicit in stating that *only* the C_{2v} point group can explain the observed selection rules and depolarization ratios associated with the *inter*molecular vibration.

In Chapter 9 Kell has apparently confused the arguments referring to *inter*molecular vibrations with those referring to *intra*molecular vibrations. Schiffer's[21] discussion referred to by Kell involves *intra*molecular vibrations. Finally, Kell has objected that Walrafen's interpretation of the water spectrum has not been properly related to the spectrum of ice. However, although the arrangement of nearest neighbors in a liquid is governed primarily by the *inter*molecular potential, the lattice is irregular and random. The supercooling of water clearly argues against the presence of an extended ice structure in water. Therefore, it does not seem necessary to use the spectrum of ice as a strict guide to an understanding of the spectrum of water.

Walrafen[19] has also pointed out, however, that a C_s model is required to account for the polarization and intensities observed for the *intra*molecular vibrations. In water the *inter*molecular hydrogen-bond-stretching vibrations observed in Raman and infrared spectra occur at approximately 170 cm^{-1}, whereas the corresponding *intra*molecular vibrations occur near 3400 cm^{-1}. Thus approximately 20 *intra*molecular vibrations occur during a single period of the corresponding *inter*molecular vibration. The C_{2v} and C_s models therefore refer not only to different structural features but also to different time scales, and the C_{2v} structure is not sensitive to the short-lived C_s perturbations.

When taken together, the number of nearest neighbors indicated by the RDCs and the symmetry considerations of Walrafen are a strong argument in favor of tetrahedrally coordinated water molecules, especially at lower temperatures. The various species of water molecules that may arise when hydrogen bonds are broken are considered in the next section.

3 Classification of Mixture Models

Considered from the standpoint of the reference molecule, a total of nine molecular species may be obtained by the successive breaking of hydrogen bonds. These species are distinguished both in terms of the number of molecules hydrogen bonded to the reference molecule and in terms of the bonds involved. A complete listing of the possible combinations is given in Figure 5. (Comparison should be made with Figure 3.)

Many of the mixture models that have been proposed[22-26] assume the existence of two states, one consisting of tetrahedrally coordinated icelike structure (state CAABB in Figure 5) and the other consisting of a close-packed array of monomers (state C in Figure 5). All the partially hydrogen-bonded species shown in Figure 5 are neglected. Furthermore, in each case, no partition function was given.

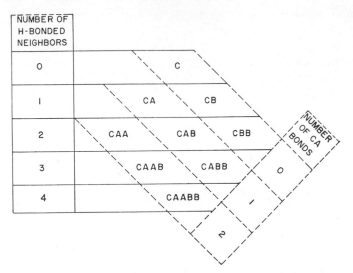

Figure 5. Species of broken-down ice structure obtainable from the tetrahedrally arranged group of molecules shown in Figure 3 (letters refer to the molecule in the indicated figure). Species with the same number of hydrogen-bonded neighbors are grouped horizontally; those with the same number of CA bonds are grouped along an incline.

Davis and Litovitz[27] (DL)* demonstrated that the close-packed structure could also be partially hydrogen-bonded. Their model consisted of a mixture of open-packed (species CAABB) and close-packed (species CAA, CAB, and CBB) structures. All other species were neglected. The absence of molecules that are three-, one- and non-hydrogen-bonded is an unrealistic aspect of this model. Specific volume data and RDC data were employed to calculate the two state parameters. On the basis of this model, various relaxational parameters (the relaxational portion of various thermodynamic parameters and the structure relaxation time) and the mole fraction of hydrogen bonds at each temperature were calculated. A nonequilibrium thermodynamics approach was used. Davis and Jarzynski[4] have since shown that the partition function corresponding to the thermodynamic expressions obtained is

$$Y = \sum_{N_1=0}^{N_0} \frac{N_0!}{N_1! \, N_2!} \left[\exp\left(-\frac{F_1}{RT}\right) \right]^{N_1} \left[\exp\left(-\frac{F_2}{RT}\right) \right]^{N_2}$$

(1)
$$\times \exp\left[\frac{-p(N_1 V_1 + N_2 V_2)}{RT} \right]$$

* Initials are used to refer to models mentioned often in the text and with which several names are associated. For the convenience of the reader, these initials and the names to which they correspond are collected in Appendix 1.

where subscripts 1 and 2 refer to the open- and close-packed structures, respectively, and N_0 is the total number of molecules present; F is the molar Helmholtz free energy and V is the molar volume. At atmospheric pressure the last factor in (1) may be neglected. Notice that in this partition function no provision is made for surface effects; thus the physical model corresponding to this partition function must be one in which molecules move from one structure to the other without passing through a surface. Such a partition function and the corresponding thermodynamic expressions (derived below) are not appropriate for a system consisting of a mixture of distinct clusters.

The values of N_1 and N_2 corresponding to the maximum term in the series for Y may be obtained from

$$(2) \qquad \ln \frac{N_1}{N_2} = \left(\frac{F_2 + pV_2}{RT}\right) - \left(\frac{F_1 + pV_1}{RT}\right)$$

or

$$(3) \qquad \frac{N_1}{N_2} = \exp\left(\frac{-\Delta G}{RT}\right)$$

where $\Delta G - (F_1 + pV_1) - (F_2 + pV_2) = \Delta F + p\,\Delta V = \Delta H - T\,\Delta S$ is the change in the Gibbs free energy between the open- and close-packed structures; H is the enthalpy, and ΔF, ΔH, and ΔS are the changes in the respective parameters between the open- and close-packed structures. The equilibrium values of the numbers of molecules in the open- and close-packed structures are N_1 and N_2, respectively, at temperature, T, and pressure, p. The equilibrium distribution may also be expressed in terms of the mole fractions of the open- and close-packed structures, x_1 and x_2 as

$$(4) \qquad \frac{x_1}{x_2} = \frac{x_1}{1 - x_1} = \exp\left(-\frac{\Delta G}{RT}\right) = \exp\left(\frac{-\Delta H + T\,\Delta S}{RT}\right)$$

where $N_1 = N_0 x_1$, and $N_2 = N_0 x_2$.

The molar volumes, V_1 and V_2 (corresponding to the open- and close-packed structures, respectively), are related by

$$(5) \qquad V = x_1 V_1 + x_2 V_2 = V_2 + x_1\,\Delta V$$

The existence of two states results in a relaxational (or structural) contribution to the specific heat, $C_{p,0} = (\partial H/\partial T)_p$, the isothermal compressibility, $\kappa_{T,0} = -(1/V)(\partial V/\partial P)_T$, and the thermal expansion coefficient, $\beta_0 = (1/V) \times (\partial V/\partial T)_p$. The subscript 0 indicates the parameter is the static or 0-frequency value. The contribution due to structural rearrangement is characterized by a structural relaxational time, τ_p. (The subscript p indicates that the measurements were carried out at constant pressure.) A discussion of structural relaxation in liquids has been given by Litovitz and Davis,[28] and in water by Davis and Jarzynski.[4] By using high-frequency ultrasonic waves, measurements may be made over a range of frequencies whose periods extend from times long compared to τ_p, to times short compared to τ_p. When measurements are made over time intervals long compared to τ_p, the structure has sufficient time to rearrange and the measured parameter contains the relaxational contribution (indicated by the subscript r). On the other hand, when measurements are made over time intervals small compared to τ_p, the structure does not have time to rearrange and the relaxational contribution is absent. This high-frequency contribution is indicated by the subscript ∞. The static parameters are simply the sum of the high-frequency and relaxational contributions.

The thermal expansion coefficient, β_0, is given by

$$(6) \qquad \beta_0 = \frac{1}{V}\left(\frac{\partial V}{\partial T}\right)_p = \frac{1}{V}\left[\left(\frac{\partial V_2}{\partial T}\right)_p + x_1\left(\frac{\partial \Delta V}{\partial T}\right)_p\right] + \frac{\Delta V}{V}\left(\frac{\partial x_1}{\partial T}\right)_p$$

The last term on the right results from structural rearrangement and is known as the relaxational thermal expansion coefficient, β_r

$$(7) \qquad \beta_r = \frac{\Delta V}{V}\left(\frac{\partial x_1}{\partial T}\right)_p$$

Similarly, the expression for $\kappa_{T,r}$ is found to be

$$(8) \qquad \kappa_{T,r} = -\frac{\Delta V}{V}\left(\frac{\partial x_1}{\partial P}\right)_T$$

The partial derivatives in (7) and (8) are obtained from (4). The final expressions are

$$(9) \qquad \beta_r = \frac{\Delta V \Delta H}{VRT^2}x_1(1-x_1)$$

and

$$(10) \qquad \kappa_{T,r} = \frac{(\Delta V)^2}{VRT}x_1(1-x_1)$$

Expressing the total enthalpy in the form of (5), the following expression for $C_{p,r}$ is obtained:

$$(11) \qquad C_{p,r} = \frac{(\Delta H)^2}{RT^2}x_1(1-x_1)$$

Applying the relation $C_v = C_p - TV\beta^2/\kappa_T$ twice in the right-hand side of $C_{v,r} = C_{v,0} - C_{v,\infty}$ yields the expression

$$(11') \qquad C_{v,r} = C_{p,r} + TV\left[\frac{\beta_\infty^2}{\kappa_{T,\infty}} - \frac{\beta_0^2}{\kappa_{T,0}}\right]$$

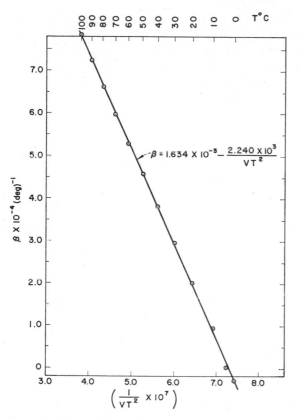

Figure 6. The variation of the thermal expansion coefficient of water versus $1/VT^2$. The points are experimental results. [From L. Smith and F. Keyes, *Proc. Am. Acad. Arts Sci.*, 69, 285 (1934).]

Equations 9 to 11 agree with the results obtained by DL[27] from nonequilibrium thermodynamics. Using the X-ray data of Morgan and Warren,[11] DL evaluated ΔV and x_1 at 1.5°C. These are 7.24 cm³/mole and 0.593 respectively. The value of $\kappa_{T,r}$ obtained from (10) is 30.9×10^{-12} cm²/dyne. This compares favorably with the value of 33.5×10^{-12} cm²/dyne estimated by Slie et al.[29] In order to evaluate β_r and $C_{p,r}$, it is necessary to know ΔH. Davis and Litovitz plotted β_0 versus $1/VT^2$ (Figure 6) and observed that it fit an expression of the form

(12)
$$\beta_0 = A + \frac{C}{VT^2}$$

It is desirable at this point to emphasize that the curve in Figure 6 displays actual experimental data. Similar curves plotted at higher pressures exhibit the same characteristics. Furthermore, although β_0 does not continue to follow a straight line at higher temperatures (and slightly elevated pressures to prevent boiling), it does show a large temperature dependence to well beyond 200°C. Indeed, the truly anomalous aspect of the β_0 is not the negative value below 4°C but the large temperature dependence. The model of DL accounts for this curve, but regardless of whether their model is correct, any model of the structure of water must also account for this curve over the entire temperature range.

The values of A and C in (12), obtained by a least-squares fit, are 1.634×10^{-3}/deg and -2240 cm³/deg, respectively. Following Litovitz and Davis,[28] β_0 may be written in the form

$$(13) \qquad\qquad \beta_0 = \beta_\infty + \beta_r$$

where β_∞ is the instantaneous or glasslike component. Comparing (12) with (13) and (9), they conclude that

$$(14) \qquad\qquad A = \beta_\infty$$

$$(15) \qquad\qquad C = \frac{\Delta V \Delta H x_1 (1 - x_1)}{R}$$

and

$$(16) \qquad\qquad \beta_r = \frac{C}{VT^2}$$

Using (15) and the value of ΔV and x_1 calculated from X-ray data, the value of ΔH is found to be -2547 cal/mole. Finally, using (4), they found ΔS to be -9.57 eu. Assuming that ΔH and ΔS remain constant with temperature, the values of x_1 and x_2 were calculated at each temperature from (4). These values of ΔH and x_1 were then used in (15) to calculate ΔV versus temperature. Finally, using (10) and (11), $\kappa_{T,r}$ and $C_{p,r}$ were calculated at each temperature.

Equations 4 to 11 also apply to the two-state structural models described in Refs. 22 to 26, and we shall now discuss the methods used by these authors to estimate ΔV, ΔH, and ΔS. Hall,[22] in 1948, accounted for the excess sound absorption in water versus temperature by means of such a two-state model. He assumed V_2 to be 10.4 cm³/mole, which is the volume for closest packing of spheres whose diameters are the same as the position of the first peak in the X-ray radial distribution curve. He chose V_1 to lie between the molar volumes of water and ice, 18.0 and 19.6 cm³, respectively. The average value obtained for ΔV in this manner is 8.46 cm³/mole. Trial values of $\kappa_{T,\infty}$ (16×10^{-12} and 18×10^{-12} cm²/dyne) were chosen for the purpose of calculation. Finally, using an equation corresponding to (4) and (10) combined, he estimated ΔG to be -500 cal/mole. Hall defined ΔG as $G_2 - G_1$, here it is defined as $G_1 - G_2$. Thus Hall's value of ΔG was positive. For consistency, however, we are following our own notation. Furthermore, Hall referred to ΔF instead of ΔG, but at atmospheric pressure $\Delta F \approx \Delta G$.

Holding ΔG constant, good agreement was obtained with measured values of α/f^2 versus temperature, where α is the ultrasonic absorption coefficient and f is the frequency in Hertz. Hall did not determine ΔH and ΔS and thus was unable to calculate β_r and $C_{p,r}$. The assumption that ΔG remains constant requires x_1 to also remain relatively constant over the entire temperature range [see (4)]. Furthermore, since both β_r and $C_{p,r}$ are proportional to $\partial x_1/\partial T$, a nearly constant value of x_1 leads to values of β_r and $C_{p,r}$ which are small at all temperatures. Litovitz and Carnevale[26] (LC) showed that a two-state model leading to (4) accounts for the excess sound absorption as a function of pressure. However, they found it necessary to use a temperature dependent, ΔF, whose value at $0°C$ was equal to that used by Hall but opposite in sign.

Smith and Lawson[25] (SL) measure the velocity of sound in water as a function of T and P. They discussed their results in terms of a two-state model characterized by (4), (5), and (9) to (11). Estimates of ΔV, ΔH, and ΔS were obtained on the assumption that at $0°C$ $x_1 = 0.5$, $\kappa_{T,m} = 12 \times 10^{-12}$ cm^2/dyne, and $C_{p,i} = 8.5$ cal/(C°) (mole). The results given in their paper are not consistent with the assumptions made. Values given below for SL are those calculated by the present authors.

In 1954 Grjotheim and Krogh-Moe[23] (GK) suggested that the first peak in the radial distribution curve is in reality two unresolved peaks corresponding to different types of nearest neighbors. They assumed that above $\approx 170°C$ water behaved "normally" and extrapolated the portion of the volume-temperature curve between 170 and 260°C to 0°C. From this curve they obtained V_2. The value of V_1 was obtained from the extrapolated volume of ice. In view of the ability of water to supercool, it does not seem likely that the open-packed structure is that of ice extrapolated to higher temperatures. In order to determine ΔH and ΔS, they considered several cases corresponding to whether the four bonds of water molecules break singularly, in pairs, or all at once. In each case they used an equation similar to (4), which they combined with (5) to calculate ΔH and ΔS. The best agreement with the heat of fusion and $C_{p,r}$ was obtained for $\Delta H = -2600$ cal/mole and $\Delta S = -8.9$ cal/(mole) (deg). Although they had all of the parameters necessary to calculate the values of β_r and $\kappa_{T,r}$, they did not do so.

Wada[24] assumed that V_1 and V_2 were linear functions of temperature. Equation 5 was written

$$(5') \qquad V = x_1 V_1°(1 + \beta_1 T) + (1 - x_1)V_2°(1 + \beta_2 T)$$

where $V_1°$ and $V_2°$ are volumes of the open- and close-packed structures at 0°C. Values of $V_1°$ and β_1 were chosen to agree with those of ice. The comment made previously with regard to the presence of ice in water also applies here.

The values of V_2° and β_2 were chosen to be consistent with the heat of sublimation of ice and the heats of fusion and vaporization of water. The RDC was discussed in terms of these structures. Unfortunately, the data of van Panthaleon van Eck et al.[30] were used. These data do not agree with those of Morgan and Warren. Brady and Romanow,[12] as well as Narten and Levy,[13] have repeated the measurements and have obtained reasonable agreement with Morgan and Warren.[11]

Frank and Quist[31] (FQ) proposed a two-state model of water based on a structure of water suggested by Pauling.[32] According to Pauling, liquid water has a structure similar to that of the gas hydrates. Such structures are known as clathrates. Thus liquid water can be regarded as a "water hydrate," consisting of some such framework as is found in the gas hydrates, with the enclosed sites or voids occupied by unbonded water molecules. Thus, according to this model, the two states are framework water of species CAABB and the interstitial water of species C. The calculations of FQ contain several unique features. Account is taken of the interaction between molecules in the different states, and the combinatorial factor, g, is determined using a method of counting the various possible microstates different from that used in (1).

Frank and Quist considered a system of N molecules, of which the fraction x_1 composes the framework, and the fraction $x_2 = (1 - x_1)$ the nonframework portion. The framework is assumed to provide a number of interstitial sites, equal to Nx_1/ν where ν is the number of framework waters per site. Among these sites, $N(1 - x_1)$ monomeric molecules are distributed. The resulting partition function is

$$
(17) \quad Y = \sum_{N_2=0}^{N_2=N_1/\nu} g \left[\exp\left(-\frac{F_1}{RT}\right) \right]^{N_1 - N_2\nu} \left[\exp\left(-\frac{F_1'}{RT}\right) \right]^{N_2\nu}
$$
$$
\times \left[\exp\left(-\frac{F_2}{RT}\right) \right]^{N_2} \exp\left[-\frac{pN_1V_1}{RT} \right]
$$

where F_1 is the Helmholtz free energy per mole of the molecules in the framework structures that surround an empty site, F_1' is the free energy of the molecules in the framework structure that surround a site containing a monomeric molecule, and F_2 is the free energy of the monomeric molecules. Only the volume, V_1, of the framework structure is included in (17), since the interstitial water occupies no volume. The limit $N_2 = N_1/\nu$ in the series for Y results from the physical property of the model that it must break down if more of the framework "melts" than corresponds to filling all of the sites in what remains.

In determining the combinatorial factor, g, it is assumed that the probability that a given framework site will be occupied is the same whether any other site is occupied. The factor, g, taken to be the number of ways in which the monomers can be distributed among sites, is

(18)
$$g = \frac{(Nx_1/\nu)!}{[N(1 - x_1)]! \, [Nx_1/\nu - N(1 - x_1)]!}$$

Thus monomers are treated as a lattice gas occupying the sites provided by the framework structure, and the method of counting microstates is the same as would be used, for example, in considering monolayer absorption from a gas onto a linear polymer chain. Equation 18 implies the existence of Nx_1/ν static sites. It is questionable whether this approach is valid for liquid water, since it is difficult to reconcile the concept of a framework structure sufficiently static to provide sites fixed in space with the concept of "flickering clusters" of varying extent, continually forming and breaking up.

The equilibrium value of x_1 is obtained from the maximum term in the series for Y. The relation obtained by this procedure is

(19)
$$\frac{\Delta G - \alpha}{RT} = \frac{1}{\nu} \ln \left[\frac{x_1/\nu}{x_1/\nu - (1 - x_1)} \right] + \ln \left[\frac{1 - x_1}{x_1/\nu - (1 - x_1)} \right]$$

where ΔG is the difference in the Gibbs free energy between the empty framework structure and the interstitial water, and $\alpha = \nu(F_1' - F_1)$. In (19) x_1 acts as an ordering parameter. It is interesting to note that the effect of the interaction between the interstitial water and the framework water is simply to replace the true ΔG by an effective value $(\Delta G - \alpha)$.

Equation 19 is used to obtain expressions for β_r, $\kappa_{T,r}$, and $C_{p,r}$. The most significant difference between (19) and (4) is that the former leads to values of x_1 considerably less sensitive to changes in temperature and pressure. This results in smaller values for the relaxational parameters. Frank and Quist assume for the framework structure that of chlorine hydrate with a volume $V_1 = 22.0 \text{ cm}^3/\text{mole}$ at 4°C and $\nu = 3.83$. Setting $x_1 = V/V_1$, where V is the molar volume of water, yields $x_1 = 0.818$ at 4°C. In addition, $\Delta H - \alpha$ is chosen as -22.0 cal/mole and the temperature dependence of V_1 (V_1 and ΔV are equal in this model) is chosen to be $V_1 = 22.0[1 + 33 \times 10^{-5}(T - 4) + 33 \times 10^{-7}(T - 4)^2]$, where T is the temperature in °C. Both of the above are chosen to yield agreement with the experimental volume-temperature curve of water. Finally, ΔS is obtained from (19) at 4°C.

A five-state model for the structure of water has been proposed by Némethy and Scheraga[3] (NS). They assume, following Frank and Wen,[33] that the cooperative character of hydrogen bonding results in formation and dissolution of clusters of hydrogen-bonded molecules. These clusters are short lived, forming and melting as a consequence of local energy fluctuations. The clusters are considered to be imbedded in, and in equilibrium with, monomeric "unbonded water." In the latter, the hydrogen bonds are broken but each molecule is participating in strong dipole-dipole and London interactions with neighboring molecules. The molecules in the interior of the clusters are tetracoordinated.

Némethy and Scheraga stress that the condition of a high degree of hydrogen bonding is fulfilled by a large number of tetrabonded structures involving interconnected networks of five- and six-membered rings, and so the model does not postulate a particular semicrystalline structure. The tridymitelike arrangement of ordinary ice is considered merely one of several possible, nearly equivalent structures. The model has the advantage that rather irregular arrangements of the molecules in the clusters are not excluded, since the clusters do not extend beyond a few molecular diameters. Thus there is no need to confine the possible structures to long-range repeating arrangements, an idealization made by a number of mixture models of water.

In this model there are two main structures—namely, the clusters and the unbonded water—but the molecules themselves are divided into five species of varying energy and internal freedom, depending on the number of hydrogen bonds in which they are participating. These species correspond to the five horizontal rows in Figure 5. No distinction is made between hydrogen bonds of types CA and CB. The interior of the clusters contains molecules with all four hydrogen bonds unbroken, but molecules with three, two, and one hydrogen bonds can be found on the surface of clusters, and the molecules occupying the space between the clusters have all four bonds broken. An energy level is assigned to each species, depending on the number of its hydrogen bonds. The energy levels are assumed to be equally spaced: $E_4 = 0$, $E_3 = 0.5 E_H$, $E_2 = 1.0 E_H$, $E_1 = 1.5 E_H$ and $E_u = 2.0 E_H$. Level 4 corresponds to state CAABB in Figure 5 and $2E_H$ is the energy required to bring a molecule from state CAABB to state C.

In order to set up a partition function for the above model, it is necessary to establish expressions for the fractions of molecules in each of five energy levels as a function of the average cluster size. The fractions of tetra-, tri-, and dibonded molecules, respectively, were determined as functions of n_0, the average number of molecules in the cluster, by first counting and averaging over models of small compact clusters with tridymite or similar structure over the range $n_0 = 12$ to 150. The actual clusters used in the model are then formed by adding singly bonded molecules to the above clusters. Occupancy by the singly bonded molecules converts some of the molecules counted initially as having two or three bonds into tri- or tetrabonded species, respectively. The final fractions of the various species determined by

Némethy and Scheraga (NS) are given in terms of the number of tetra-, tri-, and dibonded species, the number of surface sites available for singly bonded molecules, y_t, and the number of such sites actually occupied by singly bonded molecules, y_1. In deriving the expressions for the various species, it is assumed that occupation of any one of the available sites, y_t, by a singly bonded molecule is equally probable. It should be pointed out that the resulting average number of molecules of each species is dependent on the particular means used to construct the clusters. (Were the bond lengths all equal? Were the angles all tetrahedral, etc.?) Némethy and Scheraga did not give details about the explicit method used nor about the number of clusters constructed. Since there are obviously many ways of proceeding and it is quite unlikely that all possibilities were considered, the results obtained can not be unique. Furthermore, a finite number of clusters were considered and energy minimization procedures were not employed; therefore, the averages obtained are not averages in the statistical mechanical sense.

The numbers of all species existing in clusters are thus described in terms of the two parameters, n_0 and y_1. The description of the model is completed by introducing a third parameter, x_u, for the mole fraction of the unbonded molecules in the entire system. In this model n_0, y_1, and x_u play the role of ordering parameters. The values taken by these three parameters must satisfy the conditions of minimum free energy for the system. The actual mole fractions in the liquid become x_u for the unbonded species and $x_i = y_i(1 - x_u)$, with $i = 1, 2, 3, 4$, for the hydrogen-bonded species, where y_i is the fraction of molecules having i hydrogen bonds. Némethy and Scheraga assumed that all clusters are of uniform size, n_0. The use of a single n_0 and of a single set of x_i's is justified by ascribing to them the nature of averages over the possible range of structures occurring in the liquid. The partition function corresponding to the model is

$$(20) \qquad Z = \sum_{\{n_0, y_1, x_u\}} g(n_0, y_1, x_u) \prod_{i=1}^{4} \left[f_i \exp\left(-\frac{E_i}{RT} \right) \right]^{N_0 x_i}$$

where g is a combinatorial factor

$$(21) \qquad g(n_0, y_1, x_u) = \frac{N_0!}{N_4! \, N_3! \, N_2! \, N_1! \, N_u!}$$

and $N_i = N_0 x_i$ is the number of molecules of each species ($i = 4, 3, 2, 1, u$), u referring throughout to the unbonded species. Avogadro's number is represented by N_0, x_i are the mole fractions of the various species in the liquid, and E_i are the minimum potential energies of the various species.

The factors f_i are partition functions of the individual molecules in the various states of bonding and are determined as follows: for the molecules involved in hydrogen bonding, the three translational and three rotational degrees of freedom are represented by six vibrations. This corresponds to a self-consistent field treatment of the motion of the molecules, with each molecule moving in the effective field of all its neighbors. The unbonded molecules are assigned two vibrational modes, corresponding to vibrations changing the orientation of the dipole axis with respect to the dipole field of the neighboring molecules, and one rotational mode, corresponding to free one-dimensional rotation around the dipole axis. The unbonded species is also assumed to have translational freedom within a cage defined by its nearest neighbors as in the free-volume theories of normal liquids. Thus the molecular partition functions for the various species are

(22) $$f_i = f_{\text{vib}} \qquad \text{for} \qquad i = 1, 2, 3, 4$$

(23) $$f_u = f_{\text{tr}} f_{\text{rot}} f_{\text{vib}}$$

where

(24) $$f_{\text{vib}} = \prod_{j=1}^{s} \left[1 - \exp\left(-\frac{h\nu_{ij}}{kT}\right) \right]^{-1}$$

with $s = 6$ for $i = 1, 2, 3, 4$, and $s = 2$ for $i = u$.

(25) $$f_{\text{tr}} = \left(\frac{2\pi m k T}{h^2}\right)^{3/2} V_f$$

and

(26) $$f_{\text{rot}} = \prod \left(\frac{2\pi I k T}{h^2}\right)^{1/2}$$

The subscripts tr, rot, and vib refer to translation, rotation, and vibration, respectively, m is the mass of the molecule, and I is its moment of inertia around the dipole axis. The frequencies ν_{ij} used in the partition functions were assigned on the basis of infrared and Raman frequencies reported for water. Several other plausible reasons for the particular choice of the ν_{ij} made are given in the original paper.[3] Némethy and Scheraga admit that the assignment made cannot be considered definite, however they suggest that it is a physically reasonable one. Actually for the ν_{ij}'s chosen, the corresponding Debye temperatures are well below room temperature. Thus the value of f_{vib} obtained by NS in water corresponds very nearly to the classical value, and the particular assignment of ν_{ij}'s is unimportant. The quantity V_f represents a "free volume" for the translation of the unbonded molecule. According to the cell theory of liquids[34] it is given by

(27) $$V_f = \int \exp\left(\frac{-W}{kT}\right) d\tau$$

where $d\tau$ represents a volume element and W is the interaction potential in $d\tau$ due to the neighboring molecules.

The energy, E_H, and the free volume, V_f, depend on the interactions between the molecules in the unbonded liquid. Since there is no precise information about the magnitude of these interactions, it is not possible to derive values of these parameters from theory, thus E_H and V_f were treated as adjustable parameters that could be varied within the limits of physically reasonable values, to give the best fit to experimental data at 40°C.

The best values were found to be $V_f = 4.4 \times 10^{-25}$ cm³/molecule and $E_H = 1.32$ kcal/mole. Both values were taken to be constant over the temperature range 0 to 100°C.

To determine the equilibrium values of n_0, y_1, and x_u (and thus the average size of the clusters), it is necessary to evaluate the partition function Z. This is done by equating Z to the maximum term in the sum of the right-hand side of (20). In contrast to (4) and (19), the solution in this case cannot be obtained in a closed algebraic form, and a stepwise approximation procedure was used to find the maximum of the partition function with respect to n_0 and y_1.

The average value of n_0 was found to decrease with an increase in temperature from 90 at 0°C to 21 at 100°C, whereas y_1 and the mole fraction of unbonded water increase at the same time, as would be expected. The cluster sizes obtained are of a physically reasonable magnitude for the Frank–Wen[33] model of "flickering clusters," being of the order of a few molecular diameters and hence large enough for the argument of cooperative bonding to hold. Adjacent clusters are assumed to be separated on the average by one to two layers of unbonded water at 0°C.

Various thermodynamic functions may be calculated from Z, using standard statistical mechanical relations between thermodynamic properties and partition functions.[34] The Helmholtz free energy, F, the internal energy, U, the entropy, S, and the specific heat at constant volume, C_v, were calculated. The results for the internal energy and the entropy are within 3% of the experimental data, the deviation increasing at high temperatures. The agreement is less good for the specific heat, whose calculated and experimental values agree only at 40°C. Here the deviations are within 20% except at high temperatures.

It should be noted that for water at 1 atm, $pV = 0.44$ cal, hence $F \simeq G$, the Gibbs free energy, and $U \simeq H$, the enthalpy.

Finally, the model was used to calculate $\kappa_{T,r}$. Rigorously, each of the different species should be assigned a different molar volume, but this would introduce several additional adjustable parameters into the model. To avoid this, NS introduce a two-state approximation by assuming that the observed molar volume of water can be represented at any temperature in terms of two volume contributions: the mole volume of the clusters, V_{cl}, and for the unbonded liquid, V_L.

$$(28) \qquad V = (x_4 + x_3 + x_2)V_{cl} + (x_1 + x_u)V_L$$

Equation 28 may be partly justified as follows. The molar volume contributions of the molecules having three, two, or one hydrogen bonds is expected to lie between V_{cl} and V_L, corresponding to the increase in the coordination number and the change from hydrogen-bond distance to the distance corresponding to the van der Waals radii. Assigning V_{cl} to the tri- and di-hydrogen-bonded species and V_L to the monohydrogen-bonded molecules results in partial compensation of the errors. The molar volume of the clusters was assumed to be the same as that for ice extrapolated to temperatures above 0°C. Thus

$$(29) \qquad V_{cl} = 19.657(1 + 1.55 \times 10^{-4} \, T)$$

with T given in °C. Since the open-packed structure was assumed to be different from that of ice, the justification of this assumption is questionable.

For the unbonded liquid the thermal expansion equation was assumed to be of the form

$$(30) \qquad V_L = V_{L,0}(1 + aT + bT^2)$$

in analogy with other liquids. The coefficients $V_{L,0}$, a, and b, which are three additional adjustable constants, were calculated by fitting (28) to the experimental values at 0, 4, and 25°C. The values obtained were $V_{L,0} = 16.0726$ cm³/mole, $a = 1.419 \times 10^{-3}$, and $b = 8.895 \times 10^{-7}$. Using these values in (28) the V–T curve was calculated over the range 0 to 100°C. The agreement with the experimental curve is within 0.5% up to 70°C, and the calculated curve has its minimum at 4°C.

The two-state approximation is retained for calculations of the compressibility. From (28) it follows that the isothermal compressibility is

$$(31) \qquad \kappa_{T,0} = -\frac{1}{V} \left[x_L \left(\frac{\partial V_L}{\partial p} \right)_T + (1 - x_L) \left(\frac{\partial V_{cl}}{\partial p} \right)_T + (V_L - V_{cl}) \left(\frac{\partial x_L}{\partial p} \right)_T \right]$$

The relaxational part of the compressibility is given by the last term of (31). Némethy and Scheraga do not calculate β_r and $C_{p,r}$, although their paper contains sufficient data to do so. These calculations were made in Ref. 27. Also in their calculations of $\kappa_{T,r}$, NS held $V_{cl}-V_L$ constant versus T. This is not correct; therefore, in the following discussions $V_{cl}-V_L$ is determined from (29) and (30) and used in the calculation of both $\kappa_{T,r}$ and β_r.

Eucken[35] proposed a theory in which water is assumed to be an equilibrium mixture of octomers, quadrimers, dimers, and monomers. The structure of the octomers is shown in Figure 7. Because of the space at its center, Eucken assumed the volume per molecule of the octomers to be 10% greater than that of quadrimers, dimers, and monomers, all of which he assumed to exhibit the same volume per molecule. The mole volume of octomers corresponds to V_1 and that of the remaining species corresponds to V_2.

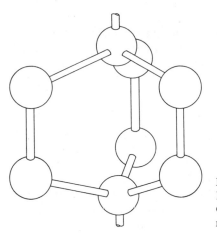

Figure 7. Octomer structure proposed by Eucken[35] as the open-packed structure in water. Compare with Figures 1 and 2 to determine the relation of this structure to the ice lattice.

In order to evaluate the mole fraction of octomers, x_1, Eucken assumed that water behaves normally at 10,000 atm. He fitted an empirical equation of state to the high-pressure data and extrapolated to atmospheric pressure. This procedure resulted in a relatively small value of ΔV.

Smith and Lawson[25] have questioned the validity of the assumption upon which this extrapolation is based. In their investigation of the velocity of sound in water, the anomalous behavior was still observed at 10,000 atm. At each temperature, Eucken equated the difference between the measured and extrapolated curves to the product $x_1 \Delta V$ and calculated x_1. In order to obtain mole fractions for each of the species in the close-packed structure, he first estimated the relaxational contribution to the specific heat versus temperature and then chose the mole fractions of monomers, dimers, and quadrimers as well as the corresponding energy differences (including that between octomers and quadrimers) in order to account for these estimates.

Eucken calculated $C_{p,r}$ by means of a detailed analysis of the various components of $C_{p,0}$. The expression he used was

(32) $$C_{p,0} = (C_{p,0} - C_{v,0}) + C_{\text{tr+rot}} + C_{v,r} + C_{\text{int}}$$

where C_{int} is due to internal degrees of freedom and $C_{\text{tr+rot}}$ is due to translations and rotations of the entire molecule. The values of $C_{p,0}$ and $C_{p,0} - C_{v,0}$ were obtained directly from the literature, C_{int} was obtained from measurements of $C_{v,0}$ in the gas at very low pressures, and $C_{\text{tr+rot}}$ was obtained empirically from a large number of substances. This method of determining $C_{\text{tr+rot}}$ was checked by Eucken in several nonassociated substances where $C_{v,r}$ was considered to be 0 and was found to yield the correct results to within ± 0.7 cal/(mole) (deg). That part of the specific heat which remained was taken to be $C_{v,r}$.

K. Buijs and G. R. Choppin[36] (BC) investigated the near-infrared spectra of water and resolved the broad absorption band in the 7700 to 9090 cm^{-1} region (which they associated with the combination of two oxygen–hydrogen stretching and the bending vibrations) into three components located at 8000, 8330, and 8620 cm^{-1}. They attributed these components to the existence of three species of water molecules corresponding to two-, one-, or zero-hydrogen bonds of type CA (see Figure 5). The mole fractions of the corresponding three species were calculated from the temperature dependence of the intensities of the three component bands. The fraction of CA bonds was found to decrease from 0.52 at 6°C to 0.39 at 72°C. No partition function or structural model was suggested.

Vand and Senior[37] (VS) replaced the discrete energy levels in the work of BC by three energy bands each having a Gaussian spread of energy. Utilizing the approach introduced by NS [see (20)] with the summation replaced by an integration and the factors f_i by a Gaussian distribution of energies, VS obtained a partition function. Using the bandwidths as additional adjustable parameters, they were able to fit the free energy, internal energy, entropy, and specific heat to within approximately 1%.

Jhon, Grosh, Ree, and Eyring[38] (JGRE) have proposed a two-state model for water structure consisting of an equilibrium mixture of clathrate-like clusters of approximately 46 molecules dispersed in an ice-III-like structure. The ice-I-like clusters exhibit the density of ice I and are assumed to almost disappear by 4°C. The liquid is assumed to behave "normally" above this temperature. As pointed out above, the concept of water behaving anomalously only in the vicinity of 4°C is erroneous. Actually it is the temperature dependence of β_0 which is anomalous and this anomaly extends over the entire temperature range (see Figure 6). On the basis of the curve in Figure 6, data obtained near 100°C would lead one to expect a negative value of β_0 in the vicinity of 0°C. It is therefore logical to assume that the structural mechanism that causes the maximum density at 4°C is still active at 100°C.

The significant-structure theory was applied to this model by JGRE, who obtained a partition function containing nine parameters. Five of these parameters normally appear in the significant-structure theory of liquids and are evaluated accordingly. The other four constants are two-state parameters which are adjusted to fit the specific volume versus temperature and the shear viscosity versus pressure. In the latter case three additional adjustable parameters were introduced.

Of historical interest is the theory of Bernal and Fowler.[39] They assumed that below 4°C water is a mixture of a tridymite and a quartzlike structure. The tridymite structure was assumed to disappear at approximately 4°C, after which water was considered to be mainly quartzlike until a solid ammonialike structure appeared at high temperature. This model was primarily qualitative and yielded neither a partition function nor thermodynamic parameters.

Walrafen[15,19,20] made extensive measurements of the Raman spectra due to the intramolecular OH- and OD-stretching modes in water. These measurements were carried out as a function of both temperature and pressure. Under all conditions he observed an asymmetric contour which he decomposes into Gaussian contributions. In pure water[19] both the infrared and Raman intramolecular spectra require four Gaussian contributions in their decompositions, which may be grouped into pairs: two intense low-frequency components that decrease with temperature; and two weak high-frequency components that increase with temperature. The temperature dependence of the high- and low-frequency pairs of components leads to an isosbestic point, that is, a frequency at which the intensity is independent of temperature. The occurrence of such an isosbestic point is consistent with the existence of two species of water molecules. Walrafen has identified these two species as water molecules that are hydrogen bonded with C_{2v} symmetry and those that are nonhydrogen-bonded.

The intramolecular spectra can be greatly simplified by investigating a dilute solution of HDO in either H_2O or D_2O.[15] The simplification arises because of the *inter*molecular decoupling of the *intra*molecular OH- and OD-stretching vibrations in HDO. Unlike the situation in pure water, the decoupled stretching contour decomposes into two (rather than four) Gaussian components: the low-frequency component decreases and the high-frequency component increases with temperature. Thus the occurrence of pairs of components in pure H_2O and D_2O is due to a vibration plus an intermolecular coupling (overtone) that does not show up in dilute solutions. As in the case of pure water, the decoupled stretching vibration in the dilute solutions exhibit an isosbestic point. Falk and Wyss[40] have recently pointed out that isosbestic points occur in solids as well as liquids. In particular the OH-stretching vibration in ice exhibits such an isosbestic point (or, more precisely, points). The nature and origin of the isosbestic points in water and ice appear to be quite different. In the spectra of ice there are two such points, one on either side of the maximum (see Ref. 40). Such points are the result of a single component sharpening up with decreasing temperature. The spectra of water, on the other hand, exhibit a single isosbestic point on one side of the maximum (see Ref. 19). Furthermore, the Gaussian components composing the spectra do not sharpen over the temperature range.

Stimulated Raman OH-extensional spectra in concentrated solutions of HDO have recently been reported by Colles, Walrafen, and Wecht.[18] The results of this investigation provide exceedingly strong evidence for the decomposition into four Gaussian components. A comparison of the stimulated

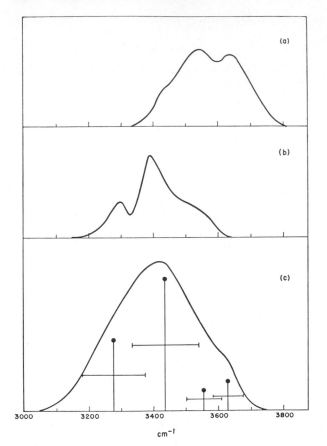

Figure 8. Microdensitometer tracings corresponding to (*a*) stimulated Raman scattering from 50% H_2O, 50% D_2O (by volume); (*b*) stimulated Raman scattering from 60% H_2O, 40% D_2O in the OH-stretching region; (*c*) spontaneous Raman spectrum from 50 mole % H_2O in D_2O (component positions and halfwidths indicated). The spontaneous Raman was decomposed into Gaussian components. The stimulated and spontaneous Raman spectra are considered to arise primarily from HDO. (Furnished by G. E. Walrafen; to appear in Ref. 20.)

and spontaneous Raman spectra are given in Figure 8. The stimulated spectra from a mixture of 50% H_2O and 50% D_2O (by volume) is shown in Figure 8*a*. HDO forms in such a mixture of H_2O and D_2O in the molar ratios of approximately 2:1:1.[41] A pair of components is observed whose maxima agree quite well with the two weak high-frequency nonhydrogen-bonded components in the spontaneous Raman spectra (see Figure 8*c*). On the other hand, the stimulated spectra from a mixture of 60% H_2O and

40% D_2O is given in Figure 8*b*. The two components observed in this mixture agree in frequency with the intense low-frequency hydrogen-bonded components shown in Figure 8*c*. The mutually exclusive stimulation of these two classes of components provides strong evidence for the existence of distinguishable species in water but is in complete disagreement with the continuum model.

Walrafen has compared the temperature dependence of the *inter*molecular Raman intensities arising from hydrogen-bond stretching and librational intensities. To within experimental error the temperature dependencies are found to be the same and yield values of $\Delta H°$ and $\Delta S°$ of -5.1 kcal/mole and -17.0 cal/(deg) (mole). Walrafen points out that only those molecules which are fully hydrogen bonded (i.e., tetrahedrally coordinated exhibiting C_{2v} symmetry) contribute appreciably to the Raman intermolecular intensity. Thus the values of ΔH and ΔS obtained in this manner are assumed to correspond to the difference between the fully bonded and the completely unbonded state. The corresponding values of ΔH and ΔS per hydrogen bond are taken to be one-half of $\Delta H°$ and $\Delta S°$, respectively [i.e. -2.55 kcal/mole and -8.5 cal/(mole) (deg)]. These values agree quite well with the corresponding values obtained by DL: 2.547 kcal/mole and 9.57 cal/(deg) (mole), respectively. Walrafen also measured the Raman stretching intensities versus pressure at 25°C. The value of ΔV (corresponding to the volume change associated with the breaking of a mole of hydrogen bonds) was found to be 7.2 cm^3/mole (actually one-half of 14.4 cm^3/mole as in the case of ΔH and ΔS) compared to 7.00 cm^3/mole obtained by DL.

Walrafen has also obtained estimates of ΔH from the integrated intensities of the two Gaussian components obtained in the decomposition of the *intra*molecular OD extensional mode. The logarithm of the ratio of these integrated intensities versus T^{-1} lies on a straight line whose slope yields a value of -2.5 kcal/mole for ΔH, in good agreement with the value of 2.55 obtained from the Raman intermolecular intensities. Since the Raman *intra*molecular modes are to first order independent of C_{2v} symmetry, division by 2 was unnecessary.

Franck and Lindner[42] have carried out Raman spectra investigations at nearly constant volume (1.0–0.9 g/cm^3) over the temperature range 16 to 400°C. (The corresponding pressure range was 100–3900 bars.) Walrafen[20] also decomposed these data into two Gaussian components and found that, to within experimental error, the temperature dependence of the ratio of integrated intensities were accounted for by the value of ΔH previously obtained (see Figure 9). One quite important observation that can be made on the basis of the work of Franck and Lindner is that the frequency of the nonhydrogen-bonded component remains relatively constant (2644–2649 cm^{-1}) and well below the low-density gas value (2722 cm^{-1}) over the entire

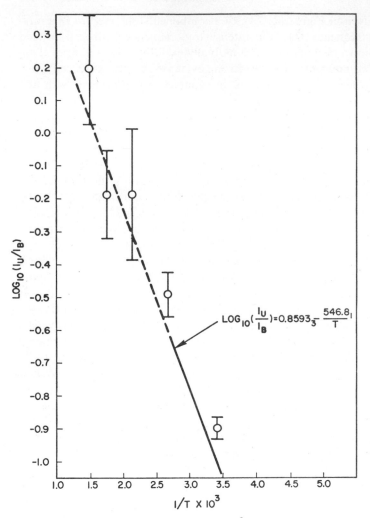

Figure 9. Plot of $\log_{10}(I_U/I_B)$ versus $1/T \times 10^3$ from the high-temperature–high-pressure laser–Raman data of Franck and Lindner.[42] I_U and I_B refer to the integrated Gaussian component intensities of the nonhydrogen-bonded and hydrogen-bonded OD stretching components from a 6.2-M solution of D_2O in H_2O. The solid curve is that obtained previously by Walrafen (see Ref. 15). The dashed curve is the extrapolation to above 400°. The error bars refer to several analog computer trials involving three Gaussian fits of the contours. (Furnished by G. E. Walrafen; to appear in Ref. 20.)

temperature range (16–400°C). This suggests the existence of a temperature-independent nondirectional intermolecular perturbation of the OH extensional vibration. Another significant observation is that the frequency of the hydrogen-bonded component increases from 2516 to 2618 cm^{-1}. This shift is in the direction of hydrogen bond weakening. Such a weakening might occur in several ways: an increase in the OH$\cdots$O bond length, consecutive hydrogen-bond breaking or a bending of the bond. An increase in the OH$\cdots$O bond length at constant density must be accompanied by an increase in the coordination number. Narten and Levy[13] have stated, however, that from 0 to 200°C the coordination number in water remains constant. Thus it would appear that one of the latter two explanations is more likely. Walrafen favors consecutive hydrogen-bond breaking.

Using (4) and the values of ΔH and ΔS obtained previously, Walrafen estimated the mole fraction of tetrabonded molecules: 0.7034 at 0°C and 0.1602 at 100°C. Assuming that $C_{p,\infty}$ for water was 9.0 cal/(C°) (mole), which is approximately the average of that exhibited by ice at 0°C and steam at 110°C (9.24 and 8.66 cal/(C°) (mole), respectively), he estimated $C_{p,r}$ versus temperature. At temperatures above 25°C, these values did not agree with the corresponding values obtained from (11). The discrepancy between the calculated and estimated values was attributed to the presence of two-bonded species. On the basis of a stepwise breaking of hydrogen bonds, Walrafen expressed $C_{p,r}$ in the form

$$(33) \qquad C_{p,r} = \frac{(\Delta H)^2}{RT^2} \left[\frac{x_4'}{1 - x_4'} + \frac{x_3'}{1 - x_3'} \right]$$

where x_4' and x_3' are related to the mole fractions of tetra-, tri-, and dibonded species (x_4, x_3, and x_2, respectively) by the expressions

$$(34) \qquad x_4' = \frac{x_4}{x_4 + x_3}$$

and

$$(35) \qquad x_3' = \frac{x_3}{x_3 + x_2}$$

with $x_4 + x_3 + x_2 = 1$. Using the value of x_4 obtained from (4), Walrafen chose x_3 and x_2 to yield the estimated value of $C_{p,r}$ at 50, 75, and 100°C. The resulting value of x_2 varied from 0.0400 at 50°C to 0.2398 at 100°C.

4 Relaxational Thermodynamic Parameters

Liquids differ from solids primarily in their ability to flow, and structural relaxation is the process by which flow occurs. The added degrees of freedom associated with flow result in structural contributions to the thermodynamic parameters. A theory that fails to consider these relaxational contributions is more nearly applicable to an amorphous solid than to a liquid.

The molecules in a liquid are assumed to occupy positions relative to their neighbors reminiscent of a quasi-crystalline or glassy solid. The energy and volume of the system are characterized not only by the temperature and pressure but also by an ordering parameter that describes the geometric state. In the simplest two-state models of the structure of water the ordering parameter is identified with the mole fraction in the open-packed state. If a step function of pressure were applied to a liquid, the volume would change in two stages: first, by means of a rapid solidlike change (indicated with a subscript, ∞) whose magnitude is approximately the same as would occur in an amorphous solid; and second, by means of a slower relaxational change in the ordering parameter (indicated with a subscript, r). In the latter case the controlling mechanism is characterized by τ_p. In a large number of liquids the volume change due to the relaxational process is approximately equal to the solidlike change (see Tables 1 and 2 of Ref. 4). The individual contributions to $\kappa_{T,0}$, β_0, and $C_{p,0}$ due to these mechanisms may be obtained for a number of viscous liquids by using ultrasonic waves (see Refs. 28 and 49). These measurements are made over a range of frequencies whose periods extend from times long compared to τ_p to times that are short compared to τ_p. Unfortunately, this technique is not applicable to pure water because of its relatively short structural relaxation time ($\sim 10^{-12}$ sec). Alternative approaches for estimating the magnitude of these parameters are discussed later.

Slie et al.[29] have obtained an experimental estimate of $\kappa_{T,r}$ from measurements of structural relaxation in water-glycerol mixtures where, because of the high viscosity, τ_p could be measured ultrasonically. The water content of these mixtures was varied from 0 to 68 mole %. The high-frequency modulus of compression, $K_{T,\infty} = (\kappa_{T,\infty})^{-1}$, was found to be a linear function of water content. By extrapolating to 100% water, an estimate of $K_{T,\infty}$ (and therefore $\kappa_{T,\infty}$) for water was obtained. The corresponding value of $\kappa_{T,r}$ in pure water at 100°C was estimated to be 33.5 × 10^{-12} cm²/dyne. This is approximately 70% of $\kappa_{T,0}$. This method is of considerable interest, since it is now the only experimental technique available for estimating τ_p and $\kappa_{T,r}$ of water. The method cannot be used to estimate β_r and $C_{p,r}$, however.

Using (9) to (11), the relaxational parameters may be calculated from a number of structural models proposed for water. The results of such calculations based on various structural models are considered in some detail later.

The values of ΔV, ΔH, and ΔS obtained by the various investigators are given in Table 1. The estimates of ΔV obtained by DL, Walrafen, and Hall (and therefore of SL and LC, who used Hall's estimate) are from two to

Table 1 Comparison of the Two-State Parameters Obtained by Various Investigators

Investigators	ΔV (cm³/mole)	ΔH (cal/mole)	ΔS (eu)	H_0^{st} (cal/mole)
Davis and Litovitz[27] (DL)	7.24	−2547	−9.57	−1037
Eucken[35]	1.9	−3200[a]	−11.7	−3250[c]
Frank and Quist[31] (FQ)	22.0	−2210	−12.43	
Grjotheim and Krogh-Moe[23] (GK)	2.9	−2600	−8.9	−1456
Hall[22]	7.6 to 9.2			
Jhon, Grosh, Ree, and Eyring[38] (JGRE)	2.0	−482	1.86	
Litovitz and Carnevale[26] (LC)	8.4	−870	−1.4	−592
Némethy and Scheraga[3] (NS)	3.58	−2640[b]		−1246[d]
Smith and Lawson[25] (SL)	8.0[e]	−2245[e]	−8.22[e]	−1120
Wada[24]	2.80	−2510	−9.79	−1456
Walrafen[15]	7.2	−2550	−8.5	−756
Safford[16]		−2600 (at 1°C)		

[a] Enthalpy difference between octomers and quadrimers.
[b] Enthalpy difference between tetrabonded and unbonded molecules, therefore, represents the maximum value.
[c] Sum of enthalpy to obtain monomers, dimers, and quadrimers from octomers; neglects enthalpy to produce octomers initially.
[d] Obtained from $\sum_{i=1}^{4} H_i x_i$, where i represents the number of hydrogen bonds.
[e] Calculated by the present authors on the basis of the assumptions of SL.

four times as large as those of other investigators. The estimates of ΔV are strongly reflected in the values of $\kappa_{T,r}$ [see (10)] and to a lesser extent in the values of β_r [see (9)], both of which are discussed later. With the exception of the estimates of LC and JGRE, the values of ΔH and ΔS are all in the range −2200 to −3200 cal/(mole) and −8.5 to −12.5 eu, respectively. The values of ΔH and ΔS given for LC were calculated in Ref. 27 from their temperature-dependent F on the assumption that both ΔH and ΔS remained constant. The value of ΔH given by LC is too small to account for the value of $C_{p,r}$ in water.

The initial structural breakup at 0°C requires an amount of energy H_0^{st}. This quantity should be less than the heat of fusion, H_f (= 1436 cal/mole), since the heat of fusion cannot be attributed entirely to the appearance of close-packed structure. A fraction must be associated with increased disorder independent of the ordered structural change: in addition, in those models in which the open-packed structure is not merely the extrapolated ice structure, the internal energy can be expected to exceed that of ice. For simple two-state models $H_0^{st} = x_2 \Delta H$. The values corresponding to H_0^{st} are included in Table 1. The values of H_0^{st} obtained by DL, Walrafen, NS, and SL are less than H_f and therefore reasonable. In the case of Wada, and GK (where the open-packed structure is that of extrapolated ice) the value of H_0^{st} was chosen equal to H_f. Finally, Eucken's approach involves four states. Neglecting the initial breakup into octomers, when the remaining structural breakup at 0°C is considered, the enthalpy change involved is approximately 3250 cal/mole, which is already in excess of H_f by a factor of more than 2.

The values of β_r versus T obtained by various investigators are compared in Table 2. The values obtained by DL, Walrafen, and SL are considerably larger than those obtained by the other investigators, and they exceed in magnitude the entire thermal expansion coefficient of most associated liquids. The values attributed to Walrafen were calculated from his parameters by the present authors. Values of x_4, x_3, and x_2 were obtained versus temperature by interpolation. The model becomes three-state at temperatures above 40°C and, since the values of ΔV for the transition from the second to the third state are not given, the values of β_r were only calculated to 40°C.

The value obtained by FQ, based on Pauling's gas hydrate, is much smaller. At 0°C they found the value of x_1 to be 0.82. The very nature of the model makes it impossible for x_1 to drop below approximately 0.79. This represents a serious limitation, leading to small values of $\delta x_1/\delta T$ and consequently

Table 2 Relaxational Thermal Expansion Coefficient ($\times 10^{-3}$/°C)

	Temperature (°C)					
Investigators	0	20	40	60	80	100
Davis and Litovitz[27] (DL)	−1.666	−1.444	−1.258	−1.101	−0.969	−0.858
Eucken[35]	−0.738	−0.515	−0.320	− 0.186	−0.101	−0.053
Frank and Quist[31] (FQ)	−0.33					
Grjotheim and Krogh-Moe[23] (GK)	−0.702	−0.539	−0.379	−0.288	−0.194	−0.120
Némethy and Scheraga[3,a] (NS)	−0.72	−0.56	−0.40	−0.26	−0.15	−0.09
Smith and Lawson[25,b] (SL)	−1.68	−1.43	−1.19	−0.97	−0.79	−0.68
Wada[24]	−0.650	−0.473				
Walrafen[15,b]	−1.434	−1.473	−1.263			

[a] The values in this row were calculated from $\Delta V \Delta x_{01}/V \Delta T$ using data given in Ref. 3.
[b] The values in this row were calculated by the present authors using (9).

β_r. Similarly, the constant value of ΔG in Hall's model leads to small values of $\delta x_1 / \delta T$ and therefore of β_r. The remainder of the estimates are intermediate to these. At 100°C, however, all the values of β_r except that of DL have become quite small—the two-state models of SL and Walrafen are not applicable at 100°C—and therefore those values of β_r cannot account for the large changes in β_0, which continue to occur at even higher temperatures.

In Table 3 the values of $\kappa_{T,r}$ versus T are compared. The estimates of DL, Walrafen, SL, FQ, and LC agree reasonably well with that of Slie et al. In those liquids where $\kappa_{T,r}$ has been measured (see Ref. 28), it is found to be from 20 to 77% of the value of $\kappa_{T,0}$. Over the temperature range considered, the values of $\kappa_{T,r}$ found by DL vary from 62 to 53%, whereas those of Walrafen over a 40°C temperature range vary from 53 to 60% of $\kappa_{T,0}$, exhibiting a maximum at 20°C. Eucken's values over the 100°C temperature range vary from 23 to 1% of $\kappa_{T,0}$. This is consistent with the small value of ΔV estimated by Eucken as discussed previously. In like manner, the estimates of $\kappa_{T,r}$, obtained from the work of GK, NS, and Wada, are small. The estimates of GK vary from 10 to 1% of the total. As in the case of

Table 3 Relaxational Isothermal Compressibility ($\times$ 10^{-12} cm²/dyne)

Investigators	Temperatures (°C)						
	0	20	30	40	60	80	100
Davis and Litovitz[27] (DL)	30.9	27.8	26.8	26.1	25.3	25.2	25.3
Eucken[35]	11.6	7.2		4.2	2.0	1.2	0.6
Frank and Quist[31] (FQ)	22.8		16.4				
Grjotheim and Krogh-Moe[23] (GK)	5.2	3.9		2.6	1.7	1.1	0.6
Litovitz and Carnevale[26] (LC)	36.5		30.1				
Némethy and Scheraga[3] (NS)	7.8	5.8	4.8	4.0	2.5	1.5	0.8
Slie, Donfor, and Litovitz[29]	33.5	26.8	25.8				
Smith and Lawson[25,a] (SL)	39.1	35.7	33.0	31.6	27.6	23.8	21.5
Wada[24]	4.5	3.4					
Walrafen[15,a]	26.4	29.1	28.4	26.7			
Hall[22,b]	32.8	27.8	26.7	26.2	26.3	27.1	

[a] The values in this row were calculated by the present authors using (10).
[b] For $\kappa_{T,\infty} = 18 \times 10^{-12}$ cm²/dyne.

Eucken, their work relies on a drastic extrapolation from high temperatures rather than high pressures.

One can compare the temperature dependence of $\kappa_{T,r}/\kappa_{T,0}$ calculated for water with the values actually measured ultrasonically in other associated liquids. This is done for glycerol (a typical associated liquid) in Table 4. Note that between 0 and 100°C for glycerol,[43] the values of $\kappa_{T,r}/\kappa_{T,0}$ drop only by about 35%. This behavior is far less temperature dependent than Eucken's prediction for water, where a decrease by a factor of 20 is indicated. The data of GK and of NS exhibit a tenfold decrease in $\kappa_{T,r}/\kappa_{T,0}$. Only the results of the work of SL and of DL exhibit a decrease in $\kappa_{T,r}/\kappa_{T,0}$ comparable with that of glycerol and other associated liquids; 50 and 15%, respectively.

The values of $C_{v,r}$ obtained by the various investigators are compared in Table 5. The estimates of Eisenberg and Kauzmann[5] were obtained by subtracting their calculations of the vibrational specific heat from the total. The vibrational component was obtained by assigning to the hydrogen-bonded water structure $6N$ modes of vibration distributed in two Debye spectra having characteristic frequencies estimated from the observed infrared spectra of water and ice. The resulting values of $C_{v,r}$ represented the only theoretically calculated values available. The estimates of Eucken were not obtained from his model, but instead, as discussed in Section 3, from a detailed analysis of the specific heat of water itself. The models of DL, Walrafen, GK, Wada, and NS yield reasonable agreement at low temperatures with the results of Eisenberg and Kauzmann; the model of SL appears somewhat large. The estimates of LC and of FQ are low by an order of magnitude. This is due to the small value of ΔH used in the work of LC. In Ref. 31, FQ discuss in some detail the reason for their low estimate of $C_{v,r}$. Their

Table 4   Temperature Dependence (%) of $\kappa_{T,r}/\kappa_{T,0}$

	Temperature (°C)		
Investigators	0	50	100
Davis and Litovitz[27] (DL)	61.7	58.4	52.7
Eucken[35]	23.2	7.0	1.2
Grjotheim and Krogh-Moe[23] (GK)	10.4	4.9	1.2
Némethy and Scheraga[3] (NS)	15.6	7.4	1.7
Smith and Lawson[25] (SL)	78.0	67.0	44.8
Walrafen[15]	52.4	59.5 (at 40°C)	
Piccirelli and Litovitz[43,a]	41.0	37.4	26.3

[a] Glycerol.

Table 5   Relaxational Specific Heat [cal/(C°) (mole)] at Constant Volume

Investigators	Temperature (°C)					
	0	20	40	60	80	100
Davis and Litovitz[27] (DL)	11.4	10.5	9.4	8.0	6.8	5.8
Eucken[35]	8.6	8.4	8.15	7.75	7.35	
Eisenberg and Kauzmann[5]	8.7	8.3	7.7	7.2	6.5	5.9
Frank and Quist[31] (FQ)	0.6					
Grjotheim and Krogh-Moe[23] (GK)	11.2	9.4	7.6	6.3	5.2	4.0
Litovitz and Carnevale[26] (LC)	1.1	0.8 (at 30°C)				
Némethy and Scheraga[3, a] (NS)	9.7	8.0	6.1	4.3	3.0	
Smith and Lawson[25, b] (SL)	10.2	9.0	7.5	6.0	4.9	4.2
Wada[24]	10.4	8.4				
Walrafen[15]	9.7	10.6	9.3	9.59	9.38	8.7

[a] Estimates of $C_{p,r}$ required in (11′) for Némethy and Scheraga were obtained from their results by means of the expression

$$C_{p,r} - \sum_{i=1}^{4} H \frac{\Delta x_i}{i \Delta T}$$

where i refers to the number of hydrogen-bonded nearest neighbors and $\Delta x_i/\Delta T$ was evaluated over 10 and 20°C intervals. It is assumed that $H_i \approx E_i$ is with respect to (20).
[b] The values in this row were calculated by the present authors.

opinion is based partly on the low value of $\delta x_1/\delta T$ that follows from (19) and partly on the neglect of a third state resulting from a breakdown of the clathrate structures. They note that the "flickering cluster" model requires that a certain percentage of the molecules be unbonded at any instant. The values of $C_{v,r}$ obtained by DL agree most closely with the calculations of Eisenberg and Kauzmann. This is an interesting result, since the estimates of DL were based on a simple two-state approach over the entire temperature range.

5   Hydrogen Bonds in Water

The degree of hydrogen bonding in water has been experimentally determined from infrared measurements by Luck[44] and from the chemical shift associated with nuclear magnetic resonance by Hindman[45] and Muller.[46] In all three cases the existence of two species in water was assumed. Luck's estimates were obtained by associating the intensity of certain lines in the infrared spectra of water with the presence of broken hydrogen bonds. He assumed that all hydrogen bonds were broken at the critical point. The fraction of broken hydrogen bonds at any lower temperature was equated to the ratio of intensity at the given temperature to that at the critical point. The percentage of broken hydrogen bonds calculated in this manner varied from approximately 10% at 0°C to 20% at 100°C.

Nuclear-magnetic-resonance estimates are based on the fact that the alternating frequency required to cause transitions of protons between the two allowed energy levels is a function not only of the applied magnetic field, H_m, but also the local environment. Thus when a hydrogen-bonded liquid condenses from its vapor, the value of H_m required for a given alternating field decreases. The stronger the hydrogen bonds the greater the reduction in H_m. In water such a shift occurs during condensation, but in addition the shift becomes larger as the temperature is reduced. Hindman[45] assumed that water is an equilibrium mixture of a hydrogen-bonded icelike species and a monomer species. The interaction of the latter with its neighbors is considered to arise entirely from London dispersion forces. A semi-empirical approach was taken to calculate the chemical shift and energy associated with the condensation of water vapor into the monomer species. The monomer species was assumed to resemble the condensed states of H_2S and H_2Se (each of which are members of the same isoelectronic sequence as H_2O). The percentage of broken hydrogen bonds obtained in this manner varied from 15.5% at 0°C to 35% at 100°C.

Both the infrared and the NMR methods involve assumptions; however, they represent the most direct estimates available. Taken together, they provide a reasonable range for the percentage of broken hydrogen bonds and may be used to evaluate the various mixture models. The estimates of DL and Walrafen agree most closely with this range of values. Likewise in reasonable agreement are the results of Hall and Narten, Danford, and Levy. Narten, Danford, and Levy[47] assumed an ice I model containing vacancies and interstitial molecules to account for the RDCs of water. This model is described in section 7. The percentage of broken hydrogen bonds equals the percentage of interstitial water molecules. The remaining mixture models predict a much larger percentage of broken hydrogen bonds. The values obtained are compared in Table 6.

Finally, Muller[46] used NMR data to calculate the change in the number of broken bonds versus temperature, but not the absolute number. In this way, while assuming two species of hydrogen bonds, he circumvented the need for naming the two species precisely. He compared his calculations with the temperature dependence of various mixture models. The most satisfactory agreement is obtained with the results of DL.

Table 6   Percentage of Broken Hydrogen Bonds

Investigators	Temperature (°C)					
	0	20	40	60	80	100
Davis and Litovitz[27]						
(DL)	0.204	0.242	0.278	0.306	0.332	0.354
Eucken[35]	0.631	0.677	0.713	0.743	0.767	0.788
Grjotheim and Krogh-						
Moe[23] (GK)	0.385	0.466	0.538	0.600		
Hall[22]	0.285	0.298	0.309	0.320	0.329	0.338
Hindman[45]	0.155					0.350
Litovitz and Carnevale[26]						
(LC)	0.715	0.746 (at 30°C)				
Luck[44]	0.100					0.200
Narten, Danford, and						
Levy[47]	0.18	0.19	0.20	0.21	0.22	0.23
Némethy and Scheraga[3]						
(NS)	0.472	0.538	0.591	0.630	0.656	0.675
Wada[24]	0.575		0.650			
Walrafen[20]	0.07	0.11	0.16	0.19	0.23	0.27

6   Structural Relaxation Times

It is generally recognized that liquids exhibit some resistance to flow and that this resistance is described by a shear viscosity, η_s. (The greater the time required for structural rearrangement to occur as a result of shear forces, the greater the magnitude of η_s.) However, it is less generally recognized that liquids also exhibit a resistance to the structural rearrangements caused by compressional forces. This resistance is described by a bulk (also called volume or compressional) viscosity,[28] η_b. The most striking demonstration of the existence of a bulk viscosity is found in the measurement of ultrasonic absorption in liquids.[49] The observed absorption coefficient, α_{obs}, can be written in terms of both η_s and η_b. At frequencies much smaller than $(\tau_p)^{-1}$ the resulting relation is

$$(36) \qquad \alpha_{obs} = \frac{2\pi^2 f^2}{\rho v^3} \left(\tfrac{4}{3}\eta_s + \eta_b\right)$$

where v is the velocity of sound, f is the frequency, and ρ is the density.

Since there are no viscosimeters available that measure η_b, the measurements of ultrasonic absorption play a unique role in the study of volume viscosity. Using (36), η_b may be calculated from the measured value of η_s and α_{obs}. Furthermore, η_b and τ_p may be shown to be related by the expression[28]

$$(37) \qquad\qquad \eta_b = \frac{\kappa_{T,r}}{\kappa_{T,0}^2}\, \tau_p$$

[This expression is also restricted to a frequency region well below $(\tau_p)^{-1}$.] Thus, if any two of the above-mentioned parameters are known, the third may be calculated. The values of $\kappa_{T,r}$ obtained for the models discussed in Section 3 were combined with experimental values of η_b determined by Pinkerton[48] to calculate τ_p and thus to provide a check on the mixture models just considered.

The structural relaxation time has been measured directly in many associated liquids by means of ultrasonic techniques.[49] Conventional ultrasonic techniques have thus far been limited to frequencies less than 5×10^8 Hz. Therefore, only relaxation times longer than 3×10^{-10} sec can be measured by this process. Using Brillouin scattering of light, it is possible to extend this range by an order of magnitude.[50] Recently the accuracy of this technique has been greatly improved through the use of lasers.[51] However, structural relaxation in water occurs in times much shorter than the periods corresponding to these frequencies. Thus at present it is not possible to measure the structural relaxation time, τ_p, in water by the techniques cited here. Estimates have been obtained by other methods.

Using the relaxation times measured in water-glycerol mixtures, Slie et al.[29] extrapolated to pure water and obtained estimates of τ_p over the range 0 to 30°C. The value of τ_p obtained in this manner varied from 4.3×10^{-12} sec at 0°C to 1.8×10^{-12} sec at 30°C. The inelastic scattering of neutrons provides another means of estimating the structural relaxation time in water. The time required for a single diffusive step to occur was determined by Safford et al.[16] from the width of the neutron inelastic scattering peak. In the temperature range 1 to 25°C, the results of Safford et al. were found to agree quite well with a delayed-jump diffusion model, that is, with a model in which the molecules are assumed to remain in a quasi-lattice site for a time, τ_0 (the residence time), which is much longer than the time, τ_1, during which a diffusive jump takes place. In this case ($\tau_0 \gg \tau_1$), the motions observed correspond to the jumping of individual water molecules. At higher temperature, the approximation $\tau_0 \gg \tau_1$ becomes less valid and the probability increases that, in addition to the jumping of individual water molecules, correlated motions of groups of water molecules may contribute to structural rearrangement. Safford et al. point out that such groups of water molecules should not be confused with "flickering clusters" or "icebergs." The value obtained varied from 2.4×10^{-12} sec at 1°C to 1.7×10^{-12}

sec at 25°C. This agrees quite well with the estimates of Slie et al., discussed previously.

Finally, limits may be placed on the structural relaxation time by means of dielectric relaxation. Litovitz and McDuffie[52] considered the relation between the dielectric relaxation time in a liquid and its structure. They observed that in all hydrogen-bonded liquids where both dielectric and structural relaxation times have been measured, the dielectric time is always found to be the longest (see Table 7). If this is true for water (and there is no evidence to indicate it is not), it places an upper limit on the structural relaxation time.

Table 7   Ratio of Dielectric to Structural (Constant Volume) Relaxation Time

Substance	Tempera-ture (°C)	$\frac{\tau_D}{\tau_v}$
Glycerol	−8	2.1
iso-Butylbromide	140.1	2.2
Propanediol	−38	1.4
Butanediol-1,3	−9.5	6.6
Hexanetriol-1,2,6	−8.0	3.8
2-Methylpentanediol-2,4	−8.0	3.5
n-Propyl alcohol	−130	~1600
n-Butyl alcohol	+2.1	~120

(Adapted from Ref. 52.)

Saxton[53] has measured the dielectric relaxation time, τ_D, in water over a range of temperature from 0 to 50°C. At each temperature the dielectric constant was observed to relax in a single step. At 20°C the value of τ_D was found to be 10.1×10^{-12} sec. This is approximately five to ten times as long as the value of τ_p estimated by ultrasonic relaxation or the time τ_0 determined by neutron diffraction measurements. It thus appears that dipole rotations do not occur when the molecules are bonded to their neighbors (i.e., are in a quasi-crystalline state); instead, such rotations happen during the relatively brief periods when the molecules are free to move (i.e., are not hydrogen bonded to their neighbors). Furthermore, since τ_D is considerably longer than τ_p, it would appear that the structure must break up and reform several times before the dipole is reoriented. Finally, unless $\tau_p \ll \tau_D$, the existence of two or more species in water is inconsistent with a dielectric relaxation process characterized by a single dielectric relaxation time.

Substituting the measured values of η_b[48] and the values of $\kappa_{T,r}$ obtained from the various models of the structure of water into (37), τ_p was calculated as a function of temperature. The results are compared with the extrapolated estimates of Slie et al., the diffusion time obtained from the inelastic scattering of neutrons, and the dielectric relaxation time (see Table 8). The results of DL, FQ, Walrafen, and Hall (as well as the related theories of LC and SL) agree extremely well with the estimates of Slie et al. and with the result obtained from neutron scattering. It is of interest that these estimates of τ_p, which agree so well, are obtained by means of three experimental techniques (ultrasonic absorption, neutron inelastic scattering, and the intermolecular Raman spectra) and the detailed structural model of DL. The values of τ_p obtained from NS, GK, Eucken, and Wada exceed the expected values of τ_p by as much as an order of magnitude. In addition, these estimates exceed the dielectric relaxation time.

It is quite significant that τ_0 and τ_p are approximately equal. The time that an individual molecule remains in a quasi-crystalline site before undergoing a diffusive jump to a new site is τ_0. Since τ_p is about equal to τ_0, the rate-controlling step associated with structural rearrangement in water must

Table 8   Relaxation Times ($\times 10^{-12}$ sec)

	Temperature (°C)						
	0	20	25	30	40	50	60
Constant-Pressure Structural Relaxation Time, τ_p							
Davis and Litovitz[27]							
(DL)	4.4	2.1		1.7	1.3		1.0
Eucken[35]	11.6	8.2			8.2		12.4
Frank and Quist[31] (FQ)	5.9		2.7				
Grjotheim and Krogh-Moe[23] (GK)	25.9	15.1			13.2		14.6
Hall[22]	4.1	2.1		1.6	1.3		0.9
Litovitz and Carnevale[26] (LC)	4.1			1.5			
Némethy and Scheraga[3] (NS)	17.2	10.2		9.3	8.6		9.9
Slie, Donfor, and Litovitz[29]	4.3	2.1		1.8			
Smith and Lawson[25] (SL)	3.5	1.6		1.3	1.1		0.9
Wada[24]	29.9	17.3					
Walrafen[20]	5.1	2.0		1.6	1.3		
Neutron Scattering[16] τ_0	2.4		1.7				
Dielectric Relaxation[52] τ_D	18.7	10.1			5.9	4.7	

also be the jump of individual water molecules and cannot be the result of the melting of a cluster of molecules. Such a flow mechanism is quite inconsistent with the flickering cluster theory of Frank and Wen,[33] the clathrate theory of Pauling, and the models of NS and FQ. The concept of stepwise breaking of hydrogen bonds suggested by Walrafen is entirely consistent with such a flow mechanism.

7 X-Ray Radial Distribution Curves

Thus far our discussions have been concerned with predicting the bulk properties for water on the basis of empirical models. The more rigorous approach is to first predict the radial distribution curve (RDC), and, using it, to calculate the bulk properties.[34] This approach has been successfully applied to argon[6] and to some extent to the molten metals.[54] Using either a Monte Carlo[55] or a molecular dynamics[56] technique, a system of molecules interacting through a known potential is brought to thermodynamic equilibrium and the RDC is calculated. In the case of water, the intermolecular potential is not well known. Recently Del Bene and Pople[57] have applied molecular orbital theory to the interaction of small groups of both open and cyclic polymeric water structures containing up to six molecules. Polymers having OH–OH–OH chains were found to be preferred; cyclic structures were most stable. Hydrogen-bond energies deviated considerably from additivity.

The directional nature of the intermolecular potential leads to further mathematical complications more difficult than any encountered in the case of argon or the liquid metals. Although the RDC of water has not been calculated in this manner, any structural arrangement proposed must be consistent with experimentally determined RDCs. It should be kept in mind that, since the RDCs of a liquid are one-dimensional, it is impossible to determine a unique structure as may be done for the case of crystalline solids.

Morgan and Warren[11] approximated the RDC by means of a smoothed-out ice distribution (see Figure 10). The difference between the experimentally determined RDC of water and the smoothed-out icelike distribution in the region between 3 and 4 Å, indicated by the shaded area, exhibits a broad peak in the region between 3.0 and 4.0 Å. Eucken and GK did not make detailed calculations, but they suggested that the first peak in the RDC might be caused by the presence of two types of bonds: a strong hydrogen bond and a weaker bond. Eucken accounted for the first peak in the RDC by means of two peaks: one with a maximum at approximately 2.75 Å, corresponding to the hydrogen bonds within icelike aggregates, and another maximum at

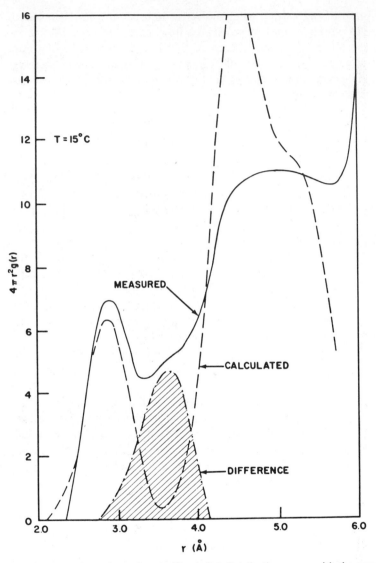

Figure 10. Comparison of an icelike radial distribution curve with the measured data of Morgan and Warren:[11] solid curve, experimental data; dashed curve, smoothed-out icelike distribution; shaded region, difference between experimental and smoothed-out curves in the region between 3 and 4 Å. (Adapted from Ref. 11.)

3.0 to 3.1 Å, corresponding to bonding between these icelike aggregates. The latter bond was considered to be weaker than a hydrogen bond but stronger than those arising entirely from London dispersion forces. Finally, Eucken considered a third peak at approximately 3.7 Å. He suggested that this peak arose from the interaction between single molecules with each other and with the larger aggregates.

NS used their values of the mole fractions, x_i, to calculate the nearest neighbor coordination number as the sum of contributions from several peaks corresponding to several preferred intermolecular distances in the liquid. Each hydrogen-bonded molecule is assumed to have neighbors at a distance equal to the O–H···O bond length. The average number of hydrogen-bonded neighbors is

(38)
$$N_H = \sum_{i=1}^{4} i x_i$$

A second peak, representing the nearest-neighbor distance in the unbonded liquid, is centered in the range 2.8 to 3.9 Å corresponding to the van der Waals contact radii of two water molecules. The average number of neighbors in this case is expressed in the form

(39)
$$N_L = x_u z_u + \sum_{i=1}^{3} x_i z(i)$$

where z_u, the average coordination number of the unbonded species, is taken to be 8, and $z(i)$, the average number of unbonded neighbors of the molecules of species i, are obtained by assuming a linear sequence from $z(4) = 0$ to $z_u = 8$. Narten and Levy[13] have recently reported that when the numbers given by NS are used to obtain an RDC, the predicted area of the first peak is much too large.

The calculation of the RDC plays an essential role in the work of DL. The nonhydrogen-bonded nearest-neighbor distance (see Section 3) was chosen to give the correct position for the "extra" peak observed by Morgan and Warren between 3.0 and 4.0 Å. Taking the location of this peak to be 3.6 Å results in too large a separation between the peaks corresponding to the hydrogen-bonded and nonhydrogen-bonded nearest-neighbor distances. The location of this "extra" peak was therefore assumed to be approximately 3.5 Å. This corresponds to a nonhydrogen-bonded distance of 3.03 Å. Placing the hydrogen-bond nearest-neighbor distance at 2.82 Å gave the best fit to the RDC. These peaks were broadened by means of an error function and summed up in the form

(40)
$$4\pi r^2 g(r) = \sum_{k=1}^{3} \frac{N_k a_k}{(\pi r_k^3)^{1/2}} \exp\left[-\frac{a_k^2(r - r_k)^2}{r_k^3}\right]$$

where $N_1 = 3.28$, $N_2 = 1.80$, $N_3 = 1.80$, $r_1 = 2.82$ Å, $r_2 = 3.03$ Å, $r_3 = 3.50$ Å, $a_1 = 13$, and $a_2 = a_3 = 11$. The small value of a_2 and a_3 corresponds to greater broadening. Since the hydrogen bond is highly directional, it seems reasonable that the corresponding distribution would be narrower. The values of r_k in (40) were chosen to give the proper width to the peaks. Four adjustable parameters are involved in this calculation. These are r_1, r_3, a_1, and a_2. The results are compared with the experimental results of Morgan and Warren in Figure 11. The difference curve shown was obtained by subtracting the calculated curve from the measured curve.

On the basis of this model, the first peak in the RDC of water is made up of two closely spaced peaks corresponding to the hydrogen-bonded nearest neighbors at approximately 2.82 Å and the nonhydrogen-bonded nearest neighbors at approximately 3.03 Å. In addition, an "extra" peak occurs at

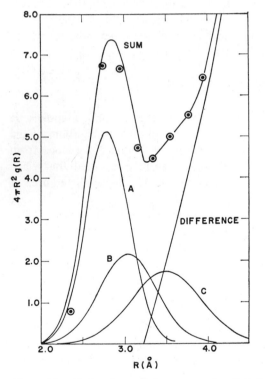

Figure 11. Comparison of the measured radial distribution curve of water[11] with the calculations of Davis and Litovitz:[27] curve A, hydrogen-bonded nearest neighbors; curve B, nonhydrogen-bonded nearest neighbors; curve C, next-nearest neighbors in the close-packed structure.

3.50 Å. The position of this "extra" peak corresponds to the face diagonal in Figure 3, which is assumed to occur in the close-packed structure. The simultaneous presence of both the icelike structure and the close-packed structure leads to a large number of coordination shells in the region beyond 4 Å. In view of the large number of peaks involved and the uncertainty in the number of neighbors at these distances, DL did not extend the calculations beyond 4 Å. The difference curve, however, can be reasonably assigned to neighbors in the region 4 to 5 Å. Cole[58] has used the results of DL to successfully calculate the number of neighbors in the range 4 to 6 Å.

Narten and Levy[13] have recently pointed out that DL did not specify how the structures proposed for the open- and close-packed species were arranged with respect to each other. However, all assumptions and calculations by DL are related simply to individual molecules located either at the center of a tetrahedrally coordinated structure (open-packed species) or at the center of a body-centered cubic structure made up by filling three of the empty cube corners in the tetrahedrally coordinated structure (close-packed species). No specification of the extent of either of the structures was given. In fact, as pointed out in Section 3, the analytical approach taken is inconsistent with the existence of extensive structures of either type. More realistically, the approach used by DL would apply to a random mixture of molecules of the open- and close-packed species. Many arrangements of these species can be expected to account for the RDC (the calculations of Cole[58] are a case in point). The model of DL is thus consistent with the RDC with regard to those intermolecular distances and coordination numbers which are used in the thermodynamic calculations. Furthermore, there appears to be no inconsistency between this model and the RDC beyond 4 Å.

The only detailed agreement with the RDCs of water has been obtained by Danford and Levy[59] at room temperature and subsequently by Narten, Danford, and Levy[47] from 4 to 200°C. The model used consists of an anisotropically expanded ice I lattice containing both vacant lattice sites and interstitial molecules. For the purpose of calculation, the reduced X-ray intensity curves were fit by means of a nonlinear least-square expansion. At a given temperature only three intermolecular distances were treated as independent variables: two, corresponding to different species of hydrogen bonds, characterized the ice I lattice; the third, corresponding to the non-hydrogen-bond distance, characterized the interstitial site. All other distances (out to 10 Å) were expressed in terms of these three distances. Three additional independent variables were introduced: two spreading factors (or temperature factors) associated with the two hydrogen-bonded lattice distances, and the occupancy numbers of the ice I lattice sites. These six independent variables proved sufficient to ensure a good fit to the intensity

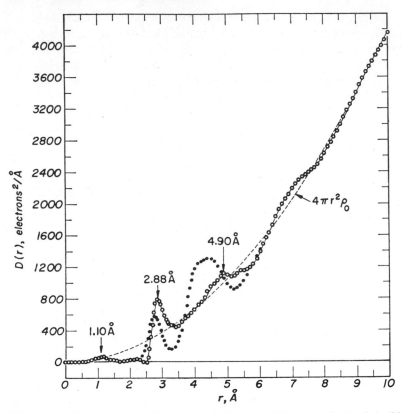

Figure 12. Comparison of the measured (open circles) and calculated (solid circles) radial distribution curves of water obtained by Danford and Levy.[59] Dashed curve indicates calculations based on the chlorine hydrate model proposed by Pauling.[32] $T = 25°C$. (From Ref. 59.)

function except at low angles, where additional independent variables were required. An example of the resulting fit to the RDC appears in Figure 12.

A unique feature of this model is the existence of two distinct species of hydrogen bonds having different bond lengths. Walrafen[20] has recently pointed out that the existence of two such species of hydrogen bonds is inconsistent with the C_{2v} symmetry required to account for spectroscopic data.

Finally, Danford and Levy[59] have calculated the radial distribution for the gas-hydrate structure proposed by Pauling[32] and used by FQ[31] in their model of water. The RDC obtained on the basis of this model is compared with the experimental curve in Figure 12. Large discrepancies exist, and it appears obvious that Pauling's pentagonal dodecahedral model must be ruled out as a possible structure for pure water.

8 Conclusions

Recent evidence (particularly from inter- and intramolecular Raman spectra and from structural relaxation times) points to a comparatively simple model for the structure of water. In this model, near the melting point, water molecules are pictured as tending to be located in one of two nearest-neighbor environments. These are: (a) an open-packed structure characterized by C_{2v} symmetry in which the reference molecule (taken to be located at the center of a cube) is hydrogen bonded to four neighbors located in alternate cube corners and (b) a close-packed structure in which the center molecule, although only partially hydrogen bonded, is surrounded by more than four nearest neighbors. The transport process by which a molecule initially in one environment makes a transition to the other environment appears to be the jumping of an individual molecule with a corresponding increase or decrease in the number of hydrogen bonds it makes with its neighbors. The existence of a single dielectric relaxation time and a value of structural relaxation time approximately the same as the neutron residence time appears to rule out the presence of Frank–Wen clusters in pure water.

Finally, it should be pointed out that the model suggested in this chapter is a uniform (continuum) mixture of two molecular environments whose percentages change with temperature. Thus the "continuum" properties suggested by X-ray spectra would appear to be entirely consistent with the "mixture" properties suggested by structural relaxation and Raman spectra.

Acknowledgments

The authors would like to express their gratitude to T. A. Litovitz and G. E. Walrafen for many helpful discussions and valuable suggestions made during preparation of this chapter.

Appendix 1

DL	Davis and Litovitz (Ref. 27).
LC	Litovitz and Carnevale (Ref. 26).
SL	Smith and Lawson (Ref. 25).
GK	Grjotheim and Krogh-Moe (Ref. 23).
FQ	Frank and Quist (Ref. 31).
NS	Némethy and Scheraga (Ref. 3).
BC	Buijs and Choppin (Ref. 36).
VS	Vand and Senior (Ref. 37).
JGRE	Jhon, Grosh, Ree, and Eyring (Ref. 38).

REFERENCES

1. Bell Telephone Laboratory Bibliography No. 124, *Structure and Physical Properties of Liquid Water*, 1957–1968.
2. W. C. Röntgen, *Ann. Phys. Chem.* (*Wied*), **45**, 91 (1892).
3. G. Némethy and H. Scheraga, *J. Chem. Phys.*, **36**, 3382 (1962).
4. C. Davis and J. Jarzynski, *Advan. Mol. Relax. Proc.*, **1**, 155 (1967–1968).
5. D. Eisenberg and W. Kauzmann, *The Structure and Properties of Water*, Oxford University Press, New York, 1969.
6. S. Rice and P. Gray, *The Statistical Mechanics of Simple Liquids*, Interscience, New York, 1965.
7. J. Bernal, *Nature*, **185**, 68 (1960).
8. K. Furukawa, *Rept. Progr. Phys.*, **25**, 365 (1962).
9. K. Lonsdale, *Proc. Roy. Soc.* (*London*), *Ser. A*, **247**, 424 (1958).
10. J. Lennard-Jones and J. Pople, *Proc. Roy. Soc.* (*London*), *Ser. A*, **205**, 155 (1951).
11. J. Morgan and B. Warren, *J. Chem. Phys.*, **6**, 666 (1938).
12. G. Brady and W. Romanow, *J. Chem. Phys.*, **32**, 306 (1960).
13. A. Narten and H. Levy, *Science*, **165**, 447 (1969).
14. C. J. Pings, *The Physics of Simple Liquids*, Interscience, New York, 1968, Chapter 10.
15. G. Walrafen, *Hydrogen-bonded Solvent Systems*, Taylor and Francis, London, 1968, pp. 9–29.
16. G. Safford, P. Leung, A. Naumann, and P. Schaffer, *J. Chem. Phys.*, **50**, 4444 (1969); G. Safford, P. Schaffer, P. Leung, G. Doebbler, G. Brady, and E. Lyden, *J. Chem. Phys.*, **50**, 2140 (1969).
17. W. Senior and R. Verrall, *J. Phys. Chem.*, **73**, 4242 (1969).
18. M. Colles, G. Walrafen, and K. Wecht, *Chem. Phys. Letters*, **4**, 621 (1970).
19. G. Walrafen, *J. Chem. Phys.*, **47**, 114 (1967).
20. G. Walrafen, in *Water: A Comprehensive Treatise, Vol. I, Physics and Physical Chemistry of Water*, Ed., F. Franks, Plenum Press, New York, to be published.
21. J. Schiffer, *J. Chem. Phys.*, **50**, 566 (1969).
22. L. Hall, *Phys. Rev.*, **73**, 775 (1948).
23. K. Grjotheim and J. Krogh-Moe, *Acta Chem. Scand.*, **8**, 1193 (1954).
24. G. Wada, *Bull. Chem. Soc. Japan*, **34**, 955 (1961).
25. A. Smith and A. Lawson, *J. Chem. Phys.*, **22**, 351 (1954).
26. T. Litovitz and E. Carnevale, *J. Appl. Phys.*, **26**, 816 (1955).
27. C. Davis and T. Litovitz, *J. Chem. Phys.*, **42**, 2563 (1965).
28. T. Litovitz and C. Davis, *Physical Acoustics*, Vol. IIA, Academic Press, New York, 1965, pp. 281–349.
29. W. Slie, A. Donfor, and T. Litovitz, *J. Chem. Phys.*, **44**, 3712 (1966).
30. C. van Panthaleon van Eck, H. Mendel, and W. Boog, *Discussions Faraday Soc.*, **24**, 200 (1957).
31. H. Frank and A. Quist, *J. Chem. Phys.*, **34**, 604 (1961).
32. L. Pauling, *The Nature of the Chemical Bond*, 3rd ed., Cornell University Press, Ithaca, N.Y., 1960, Chapter 12.
33. H. Frank and W. Y. Wen, *Discussions Faraday Soc.*, **24**, 133 (1957).
34. T. Hill, *Statistical Mechanics*, McGraw-Hill, New York, 1956.
35. A. Eucken, *Nachr. Akad. Wiss. Göttingen*, II, *Math-Physik., Kl.*, **38** (1946); *Z. Elektrochem.*, **52**, 255 (1948); *Z. Elektrochem.* **53**, 102 (1949).
36. K. Buijs and G. Choppin, *J. Chem. Phys.*, **39**, 2035 (1963).

37. V. Vand and W. Senior, *J. Chem. Phys.*, **43**, 1878 (1965).
38. M. Jhon, J. Grosh, T. Ree, and H. Eyring, *J. Chem. Phys.*, **44**, 1465 (1966).
39. J. Bernal and R. Fowler, *J. Chem. Phys.*, **1**, 515 (1933).
40. M. Falk and H. Wyss, *J. Chem. Phys.*, **51**, 5727 (1969).
41. J. Bigeleisen, *J. Chem. Phys.*, **23**, 2264 (1955).
42. E. Franck and H. Lindner, doctoral dissertation of H. Lindner, University of Karlsruhe, 1970.
43. R. Piccirelli and T. Litovitz, *J. Acoust. Soc. Amer.*, **29**, 1009 (1957).
44. W. Luck, *Discussions Faraday Soc.*, **43**, 115 (1967); *Discussions Faraday Soc.*, **43**, 132 (1967).
45. J. Hindman, *J. Chem. Phys.*, **44**, 4582 (1966).
46. N. Muller, *J. Chem. Phys.*, **43**, 2555 (1965).
47. A. Narten, M. Danford, and H. Levy, *Discussions Faraday Soc.*, **43**, 97 (1967).
48. J. Pinkerton, *Nature*, **160**, 128 (1947).
49. K. Herzfeld and T. Litovitz, *Absorption and Dispersion of Ultrasonic Waves*, Academic Press, New York, 1959, pp. 353–516.
50. I. Fabelinskii, *Usp. Fiz. Nauk.*, **63**, 355 (1957).
51. R. Mountain, *Rev. Mod. Phys.*, **38**, 205 (1966).
52. T. Litovitz and G. McDuffie, *J. Chem. Phys.*, **39**, 729 (1963).
53. J. Saxton, *Proc. Roy. Soc. (London), Ser. A*, **213**, 473 (1952).
54. N. March, *Liquid Metals*, Pergamon Press, New York, 1968.
55. W. Wood, "Monte Carlo Studies of Simple Liquid Models," in *Physics of Simple Liquids*, Wiley, New York, 1968, Chapter 5.
56. B. Alder and W. Hoover, "Numerical Statistical Mechanics," in *Physics of Simple Liquids*, Wiley, New York, 1968, Chapter 4.
57. J. Del Bene and J. Pople, *J. Chem. Phys.*, **52**, 4858 (1970).
58. L. Cole, "Application of Davis–Litovitz Two-State Theory of the Structure of Water to Calculate the Temperature Dependence of Various Properties of Water," Masters dissertation, American University, 1967, Order No. M1451, Univ. Microfilms, 300 N. Zeeb. Rd., Ann Arbor, Mich., 48106, 1968.
59. M. Danford and H. Levy, *J. Amer. Chem. Soc.*, **84**, 3965 (1962).

I I   Thermodynamics of Dilute Aqueous Solutions of Nonpolar Solutes

A. Ben-Naim, Department of Inorganic and Analytical Chemistry, The Hebrew University of Jerusalem, Jerusalem, Israel

Abstract

Examination of the thermodynamic quantities associated with the dissolution in water of nonpolar solutes, in particular, inert gases, reveals the anomalous nature of liquid water as a solvent for these solutes. Various attempts at explaining these phenomena on a molecular level are surveyed. A possible relation between the properties of such systems and the hydrophobic interaction is discussed.

1   Introduction

The study of dilute aqueous solutions of nonpolar gases has been gaining interest in recent years. These solutions reveal some anomalous properties in comparison with nonaqueous solutions and have therefore presented an attractive challenge in their own right. However, what now seems to be a more fundamental motivation for their investigation is their bearing upon two other important problems, namely, the structure of pure water and the origin of the so-called hydrophobic interaction.

From a formal point of view, the three problems of interest—the structure of water, the properties of dilute aqueous solutions of nonelectrolytes, and the nature of the hydrophobic interaction—can be reduced to the statistical–mechanical study of the three systems, namely, pure water with zero, one, and two solute particles, respectively.

This chapter is mainly about the *thermodynamic* properties of highly *dilute* aqueous solutions of nonelectrolytes, specifically with inert gas solutes. Emphasis is on the theoretical aspects of the problem developed in the past three decades. An extensive review, mainly concerned with experimental aspects of aqueous and nonaqueous solutions has recently been published by Battino and Clever.[1]

Survey of the various theoretical treatments of these systems reveals that

425

most, if not all, of the approaches are rather speculative. Critical assessment of the many assumptions and approximations introduced explicitly, and more often implicitly, in these theories is at present impossible. Indeed, not only is a rigorous and well-founded theory lacking, but also one starting from first fundamentals and proceeding with a well-defined set of approximations is unavailable. This state of affairs is not surprising in view of the high degree of complexity of our system.

The presentation of the material in this chapter compromises between a critical review and a mere compilation of the various opinions. It is critical in the sense that attention is drawn to certain unclearly stated assumptions and their justification is questioned. It is, however, uncritical insofar as no preference for one approach rather than another can be categorically established from this analysis.

The interest in aqueous solutions of nonpolar gases first arose when it was observed that some general regularities in the solubilities of gases in nonpolar liquids were not obeyed in aqueous solutions. These irregularities were systematically analyzed and their explanation first attempted by Eley[2] and later by Frank and Evans.[3] It was already realized by these authors that elucidation of the phenomena is contingent upon an understanding of the peculiarities of pure liquid water. Recent enhanced interest in the structure of water as a medium for most of the chemical activities within biological systems has brought a novel impetus to research in this field. It was also recognized that a systematic study of these systems could lead to significant progress in the comprehension of the nature of the hydrophobic interaction, supposedly responsible for the stabilization of certain conformations of biopolymers.

The next section summarizes the pertinent experimental facts required to establish the anomalous nature of water as a solvent for nonpolar solutes. The theoretical discussion, which begins with an outline of the general framework of a molecular theory in Section 3, continues with more specific treatments in the three sections that follow. Finally, a short discussion of the hydrophobic interaction and its relation to the main contents of this chapter is presented.

2   Experimental Observations

In this section some pertinent experimental facts relating to the system under discussion are presented, primarily to demonstrate the contrast between aqueous and typical nonaqueous solutions. A more extensive compilation of sources of data can be found in the review article by Battino and Clever.[1] The most important thermodynamic anomalous quantities are the following.

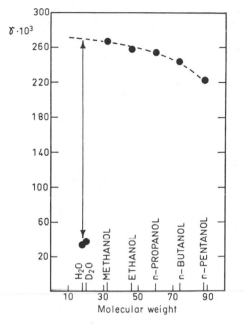

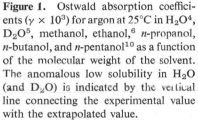

Figure 1. Ostwald absorption coefficients ($\gamma \times 10^3$) for argon at 25°C in H_2O[4], D_2O[5], methanol, ethanol,[6] n-propanol, n-butanol, and n-pentanol[10] as a function of the molecular weight of the solvent. The anomalous low solubility in H_2O (and D_2O) is indicated by the vertical line connecting the experimental value with the extrapolated value.

1. The solubility of a given gas in water is usually lower than in typical organic liquids. Figure 1 shows values of the Ostwald absorption coefficients, γ (defined below), as a function of molecular weight of the solvent for a series of homologous alcohols. The difference between the experimental value for water and the extrapolated value according to molecular weight is demonstrated. It should be noted that the low solubility of gases in water is not a property of this liquid exclusively, and other liquids in which the solubility is of the same order of magnitude are readily found (e.g., the value of γ for argon in ethylene glycol at 25°C is about 0.035).

2. The second striking difference is the temperature dependence of solubility. Figure 2 shows typical curves for the solubility of argon in water and in methanol as a function of temperature.

These differences tend to diminish with rise in temperature: in fact, the plot of the solubility of gases in water as a function of temperature passes through a minimum and then takes on a positive slope as in normal solvents.[8]

The entropy and enthalpy of solution of gases in water are definitely lower than in typical simple liquids. This anomaly is revealed from the difference in the temperature dependence of the solubility, and direct calorimetric measurements of heat of solution have rarely been made.[9] Figures 3 and 4 demonstrate the difference between values of $\Delta \bar{H}_s^\circ$ and $\Delta \bar{S}_s^\circ$ for water and for a typical

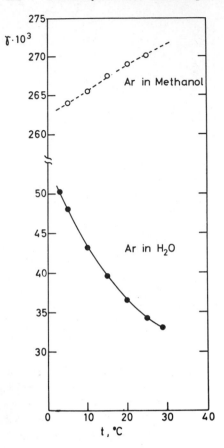

Figure 2. Temperature dependence of the solubility ($\gamma \times 10^3$) of argon in water[4] and in methanol.

simple liquid. (The standard states referred to are discussed in subsequent paragraphs.)

3. The partial molal volume of gases is usually smaller in water than in organic liquids, as illustrated in Table 1.

4. Finally, the partial molal heat capacity of gases in water is much larger[8] than in other liquids, the value for argon in water being about 50 cal/(mole) (deg) and very nearly 0 in other liquids. This quantity is much more difficult to obtain experimentally, since it involves a second derivative of experimental curves. However, the order of magnitude of these quantities seems to be quite established.

Before proceeding to interpret these phenomena, it is appropriate to digress into a brief discussion on the units and standard states employed for expressing solubilities and thermodynamic quantities of solution.

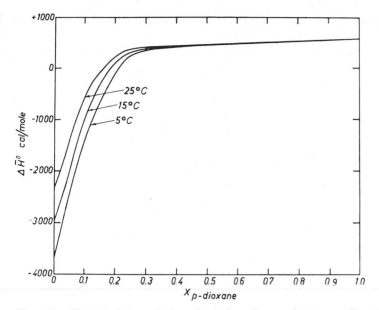

Figure 3. Change of the enthalpy of solution of argon in water–*p*-dioxane system, as a function of the mole fraction of the *p*-dioxane.

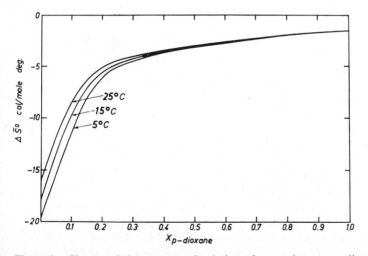

Figure 4. Change of the entropy of solution of argon in water–*p*-dioxane system, as a function of the mole fraction of the *p*-dioxane.

Table 1   Partial Molal Volumes of Gases in Water and in Organic Liquids at 25°C (cm³/mole)

	Ar	N_2	CH_4	C_2H_6
H_2O	32[a]	33[b]	37[c]	51[c]
C_6H_6	43[d]	58[c]	52[d]	73[c]
CCl_4	44[d]	52.5[d]	52.4[d]	
CS_2	45[d]		56.1[d]	

[a] Ref. 45.
[b] Ref. 2.
[c] J. Horiuti, *Sci. Papers Inst. Phys. Chem. Res. (Tokyo)*, **17**, 125 (1931).
[d] E. B. Smith and J. Walkley, *J. Phys. Chem.*, **66**, 597 (1962).

There exist a variety of units from which different authors have made their choice in reporting solubility data.[11, 12] Our choice of the Ostwald absorption coefficient, γ, was made essentially on grounds of convenience. This quantity is directly related to those measured in a solubility apparatus and was also found to be the most convenient for the interpretation of the thermodynamic quantities associated with the dissolution process.

The Ostwald absorption coefficient is defined as the ratio of the volume of gas dissolved, V^g, to the volume of the dissolving liquid, V^l, both phases being at the same temperature, T, and pressure, P.

$$(1) \qquad \gamma = \frac{V^g}{V^l}$$

A second useful form follows immediately. Let C_s^l and C_s^g be the *molar* concentrations of the solute, s, in the liquid, l, and in the gas, g, phases, respectively. Since $C_s^g V^g = C_s^l V^l$

$$(2) \qquad \gamma = \left(\frac{C_s^l}{C_s^g}\right)_{eq.}$$

where the subscript eq stands for equilibrium condition. In this form, γ can be viewed also as a distribution coefficient of s between the two phases.

Since V^g and V^l are measurable quantities, γ can be calculated from (1). However, it is the thermodynamic aspect that is of greater use in form (2). Taking α and β as two phases, and assuming that the solutions are always sufficiently dilute, the chemical potentials of the solute, s, in the two phases may be written as

$$(3) \qquad \mu_s^\alpha = \mu_s^{\circ\alpha} + RT \ln C_s^\alpha$$

$$(4) \qquad \mu_s^\beta = \mu_s^{\circ\beta} + RT \ln C_s^\beta$$

Where the standard chemical potentials are formally defined by

$$\text{(5)} \qquad \mu_s^{\circ\alpha} = \lim_{C_s^\alpha \to 0} [\mu_s^\alpha - RT \ln C_s^\alpha]$$

$$\text{(6)} \qquad \mu_s^{\circ\beta} = \lim_{C_s^\beta \to 0} [\mu_s^\beta - RT \ln C_s^\beta]$$

$\mu_s^{\circ\alpha}$ and $\mu_s^{\circ\beta}$ are traditionally interpreted as the chemical potential of s in a "hypothetical ideal" solution of unit molarity. This follows formally from (3) and (4). However, the physical meaning of these hypothetical states is in general obscure. One does not expect (3) and (4) to hold for high concentrations of the solutes, and extrapolation to one molar concentration therefore generally produces a meaningless situation. The effect of such an extrapolation is more pronounced when using a mole fraction scale, that is, when μ_s^α is expressed in terms of the mole fraction, χ_s^α

$$\mu_s^\alpha = \mu_s^{*\alpha} + RT \ln \chi_s^\alpha$$

Supposing that s is very dissimilar from the solvent, say argon in water, $\mu_s^{*\alpha}$, as defined similarly to (5), depends upon the nature of the interaction of s with its local environment, which is built up *entirely* by solvent molecules. The extrapolation to a unit mole fraction of s presupposes a system of *pure liquid s*, the properties of which are determined as if each solute molecule was surrounded by solvent molecules only—a queer situation indeed!

Moreover, not only are such hypothetical states awkward and perhaps even confusing, but their use is in fact never really necessary. We are usually interested in thermodynamic quantities for the transfer from one phase to another, and for this purpose, differences of standard states can be interpreted without any mention of these hypothetical states. To show that, let us first recall that if equilibrium is maintained

$$\text{(7)} \qquad 0 = \mu_s^\alpha - \mu_s^\beta = \mu_s^{\circ\alpha} - \mu_s^{\circ\beta} + RT \ln \left(\frac{C_s^\alpha}{C_s^\beta}\right)_{eq.}$$

or

$$\text{(8)} \qquad \Delta\mu_s^\circ(\alpha \leftarrow \beta) = \mu_s^{\circ\alpha} - \mu_s^{\circ\beta} = RT \ln \left(\frac{C_s^\beta}{C_s^\alpha}\right)_{eq.} = RT \ln \frac{\gamma^\beta}{\gamma^\alpha}$$

where γ^β and γ^α are the Ostwald absorption coefficients of s in the two phases, respectively. The interpretation of $\Delta\mu_s^\circ(\alpha \leftarrow \beta)$ as defined in (8) can be achieved by subtracting (4) from (3) to get

$$\text{(9)} \qquad \mu_s^\alpha - \mu_s^\beta = \Delta\mu_s^\circ(\alpha \leftarrow \beta) + RT \ln \frac{C_s^\alpha}{C_s^\beta}$$

It can immediately be seen that $\Delta\mu_s^\circ(\alpha \leftarrow \beta)$, which is determined experimentally by (8), is interpreted through (9) as the free energy change resulting from the transfer of one mole of solute s, from β to α (as indicated in the notation) whenever $C_s^\alpha = C_s^\beta$. This is true no matter what the exact concentrations C_s^α and C_s^β may be, provided that they are low enough to ensure the validity of (3) and (4).

There are other units for reporting gas solubilities, and these are discussed in the literature, together with the interconversion formulas.[1,11,12] It is regrettable that no general agreement on a unified system of units has been established, for quick comparison of data from different sources is thus impeded. Since thermodynamic data as reported sometimes lack a clear specification of the process to which they refer, they may be meaningless.

Basically, the choice of standard states may be regarded as simply a question of units. Yet, since these quantities are usually the subject of further interpretation, the question of their usefulness arises. Again, there seems to be no general agreement about the best choice of standard states, and two of the most frequently used are shown schematically in Figure 5. In the first, the difference of the chemical potential of the solute at the same *molar* concentration is observed, and in the second the *molal* (or the mole fraction) concentration is fixed.

As the next chapter reveals, the chemical potential of s in the very dilute system can be written as

$$(10) \qquad \mu_s = \mu_s^{\circ g} + W_{s,w} + RT \ln C_s^w$$

where $\mu_s^{\circ g}$ includes internal properties of the molecule s, and $W_{s,w}$ is the coupling work for introducing an s molecule to the solvent, w. This quantity is determined by the nature of the local environment of the solute (further

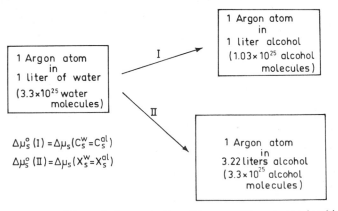

Figure 5. Schematic demonstration of two possible processes in which argon is transferred from one phase to another: I, at constant molar concentration; II, at constant molal (or mole fraction) concentration.

details in next section). This was the basis for our choice of the molar scale standard states. In this case the thermodynamics of transfer involve only the differences in the local properties of the solvent around the solute particles.

As noted above, the choice of a particular unit for reporting solubility data may be made on grounds of convenience. Yet it is recommended that the molar concentration scale be preferred. This choice is particularly and definitely advantageous when the corresponding standard states are constructed. To be more specific, $\Delta\mu_s^\circ$, calculated on the basis of the molar concentration, has a clearcut interpretation in statistical mechanics. As the next section shows, this quantity is related to the work required to transfer an s molecule from a *fixed* position in one phase to a *fixed* position in the second phase. Therefore it reflects the change of the local environment of the solute s in the two phases. This is exactly the kind of information we are striving to get from our measurements. The interpretation given above holds for any kind of solvent—it may be pure water, a mixture of water and alcohol, or an aqueous solution of mixed electrolytes.

We shall now briefly mention the procedure used to construct the standard entropy and enthalpy of solution in the molar concentration scale.

By differentiating (3) and (4) with respect to T and forming the difference in the partial molar entropy at $C_s^\alpha = C_s^\beta$ we get

$$(11) \qquad \Delta\bar{S}_s^\circ(\alpha \leftarrow \beta) = \frac{-\partial}{\partial T}\Delta\mu_s^\circ(\alpha \leftarrow \beta) - RT\frac{\partial}{\partial T}\ln\left(\frac{\rho^\alpha}{\rho^\beta}\right)$$

and

$$(12) \qquad \Delta\bar{H}_s^\circ(\alpha \leftarrow \beta) = \Delta\mu_s^\circ(\alpha \leftarrow \beta) + T\Delta\bar{S}_s^\circ(\alpha \leftarrow \beta)$$

where ρ^α and ρ^β are the densities of the solvents.

Of the variety of experimental information bearing direct relation to our systems, only a part is briefly mentioned. A very wealthy literature on salting-out (and in) of nonelectrolytes by electrolytes in water is available. In most cases, measurements were carried out at one temperature only, so that entropies and enthalpies of transfer from pure water into the ionic solutions cannot be calculated. Only rarely has the temperature dependence been determined.[13-17] Although this kind of data will eventually be important, interpretation is difficult at present, even on qualitative grounds. Here the presence of both the solutes affects the structure of water, so that the thermodynamics is thereby affected. A system of somewhat more immediate importance, is D_2O as a solvent. Known to be a liquid of very similar properties to H_2O, D_2O is presumably more "structured" than the former. Thus any future theory for water or aqueous solutions should be checked against D_2O as a preliminary test of its validity.

Table 2   Thermodynamic Functions for the Transfer of Argon from H_2O to D_2O at Three Temperatures

	Temperature (°C)		
	5	15	25
$\Delta \mu_t^\circ$ (cal/mole)	-63.9	-53.9	-46.2
$\Delta \bar{S}_t^\circ$ (eu)	-1.26	-0.86	-0.64
$\Delta \bar{H}_t^\circ$ (cal/mole)	-414	-302	-237

From Ref. 5.

This has in fact been done on qualitative grounds for many properties of both pure water and its solutions. The relevant thermodynamic quantities for the transfer of argon from H_2O to D_2O are given in Table 2, and the comparison of the solubilities of argon in the solvents is demonstrated in Figure 6. One salient feature that should be noted is that the solubility of argon increases upon replacing H_2O by D_2O[5] (thus "tending" to a more normal

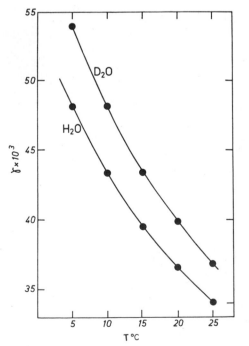

Figure 6. Ostwald absorption coefficients for argon in H_2O and D_2O as a function of temperature.[5]

liquid), whereas the entropy and enthalpy of solution become more negative in D_2O (i.e., the anomaly is increased). This comparison indicates that the molecular reason for the anomalous low value of the solubility, although formally related to the combination of $\Delta\bar{H}_s^\circ - T\Delta\bar{S}_s^\circ$, may not have the same source as the anomalous values of $\Delta\bar{H}_s^\circ$ and $\Delta\bar{S}_s^\circ$ separately.

Mixtures of water and organic liquids provide another source of information when used as solvents for gases.[7,18-21] These mixtures are certainly of interest in their own right.[22] Here we have added a third component and thereby introduced an additional element of complexity. However, by examining the way the thermodynamic property (say entropy or enthalpy of solution) changes from pure water to pure simple liquid, one may obtain some ideas about the sources of its anomalous value in pure water.

Finally, mention must be made of the study of even more complicated systems: solutions of gases in water containing biopolymers and micelles.[23-26] Understanding of these may be important in connection with the problem of the hydrophobic interaction, mentioned in Section 7.

3   General Theoretical Considerations

We shall commence our theoretical discussions with a general description of the system. Although it is recognized that an approach on these lines cannot now be pursued to the point of obtaining numerical results, it has the virtue of establishing the general framework for forthcoming approximations. Our interest being focused on very dilute solutions of gases, it is sufficient to consider a system of N water molecules and one solute molecule, s. The location of the solute, which is presumed not to have internal structure, is specified by the Cartesian coordinates $\mathbf{R}_s = (R_{sx}, R_{sy}, R_{sz})$. A water molecule is specified by both locational $\mathbf{R}_i = (R_{ix}, R_{iy}, R_{iz})$ and orientational coordinates $\mathbf{\Omega}_i = (\phi_i, \theta_i, \psi_i)$, for which the three Eulerian angles[27,28] may be selected. The vibrational partition function is assumed to be separable, so that the canonical partition function takes the form

$$
\begin{aligned}
&Q_{N+1}(V, T, N, N_s = 1) \\
(13) \quad &= \frac{q_w{}^N}{N!\,\Lambda_w{}^{3N}\Lambda_s{}^3(8\pi^2)^N} \int \cdots \int \exp\left(-\beta U_{N+1}\right) d\mathbf{R}_s\, d\mathbf{R}^N\, d\mathbf{\Omega}^N
\end{aligned}
$$

where q_w includes the vibrational and rotational partition function of a water molecule, $\Lambda_w = h(2\pi m_w kT)^{-\frac{1}{2}}$, $\Lambda_s = h(2\pi m_s kT)^{-\frac{1}{2}}$, $\beta = (kT)^{-1}$, with m_w and m_s the masses of water and solute molecules, respectively, k the Boltzmann constant, and h the Planck constant; U_{N+1} is the potential energy of interaction among the $N + 1$ particles at a given configuration,

$\mathbf{R}_s$, $\mathbf{R}^N$, $\mathbf{\Omega}^N$. Where $\mathbf{R}^N = \mathbf{R}_1 \cdots \mathbf{R}_N$, $\mathbf{\Omega}^N = \mathbf{\Omega}_1 \cdots \mathbf{\Omega}_N$. The limits of integration for each set of molecular coordinates are

$$\text{(14)} \qquad \int d\mathbf{R}_i \, d\mathbf{\Omega}_i = \int_V d\mathbf{R}_i \int_0^{2\pi} d\phi_i \int_0^\pi \sin\theta_i \, d\theta_i \int_0^{2\pi} d\psi_i$$

Similarly, the canonical partition function of pure water reads

$$\text{(15)} \qquad \begin{aligned} &Q_N(V, T, N, N_s = 0) \\ &= \frac{q_w^{\ N}}{N! \, \Lambda_w^{\ 3N} (8\pi^2)^N} \int \cdots \int \exp\left(-\beta U_N\right) d\mathbf{R}^N \, d\mathbf{\Omega}^N \end{aligned}$$

The chemical potential of the solute s may be calculated from

$$\text{(16)} \qquad \mu_s = \left(\frac{\partial A}{\partial N_s}\right)_{V,T,N} = A_{N+1} - A_N$$

Thus

$$\text{(17)} \qquad \begin{aligned} -\beta\mu_s &= \ln\left(\frac{Q_{N+1}}{Q_N}\right) \\ &= \ln\left[\frac{\int \cdots \int \exp\left(-\beta U_{N+1}\right) d\mathbf{R}_s \, d\mathbf{R}^N \, d\mathbf{\Omega}^N}{\Lambda_s^{\ 3} \int \cdots \int \exp\left[-\beta U_N\right] d\mathbf{R}^N \, d\mathbf{\Omega}^N}\right] \\ &= \ln\left(\frac{Z_{N+1}}{Z_N \Lambda_s^{\ 3}}\right) \end{aligned}$$

where Z_n stands for the configurational partition function

$$Z_N = \int \cdots \int \exp\left(-\beta U_N\right) d\mathbf{R}^N \, d\mathbf{\Omega}^N$$

The addition of a solute molecule to the system can be accomplished by introducing a coupling parameter, ξ.[29] A modified potential function $U(\xi)$ is defined by

$$\text{(18)} \qquad U(\xi) = \xi \sum_{j=1}^N U_{sj} + U_N$$

where U_{sj} is the intermolecular potential function for the solute water pair, and we have

$$U(\xi = 1) = U_{N+1} \qquad \text{and} \qquad U(\xi = 0) = U_N$$

Similarly, a modified configurational partition function is defined

$$\text{(19)} \qquad Z(\xi) = \int \cdots \int \exp\left[-\beta U(\xi)\right] d\mathbf{R}_s \, d\mathbf{R}^N \, d\mathbf{\Omega}^N$$

where it is observed that

$$Z(\xi = 1) = Z_{N+1} \qquad \text{and} \qquad Z(\xi = 0) = V Z_N$$

Thus (17) can be rewritten in the form

(20)
$$\beta\mu_s = \ln\left(\frac{\Lambda_s^3}{V}\right) - [\ln Z(\xi = 1) - \ln Z(\xi = 0)]$$

$$= \ln \rho_s\Lambda_s^3 - \int_0^1 \frac{\partial \ln Z(\xi)}{\partial \xi} \, d\xi$$

where we put $\rho_s = 1/V$.

We now transform the second term of (20) into a more useful form.

(21)
$$\frac{\partial \ln Z(\xi)}{\partial \xi} = \frac{1}{Z(\xi)} \frac{\partial Z(\xi)}{\partial \xi} = \frac{1}{Z(\xi)} \int \cdots \int \exp\left[-\beta U(\xi)\right]$$

$$\times \left(-\beta \sum_{j=1}^{N} U_{sj}\right) d\mathbf{R}_s \, d\mathbf{R}^N \, d\Omega^N$$

$$= \frac{-\beta N}{Z(\xi)} \int \cdots \int \exp\left[-\beta U(\xi)\right] U_{sw} \, d\mathbf{R}_s \, d\mathbf{R}^N \, d\Omega^N$$

$$= -\beta \int U_{ow} \, d\mathbf{R}_o \, d\mathbf{R}_w \, d\Omega_w$$

$$\times \frac{N \int \cdots \int \exp\left[-\beta U(\xi)\right] d\mathbf{R}^{N-1} \, d\Omega^{N-1}}{Z(\xi)}$$

$$= -\beta \int U_{sw}(\mathbf{R}_s, \mathbf{R}_w, \Omega_w) \rho_{sw}^{(2)}(\mathbf{R}_s, \mathbf{R}_w, \Omega_w, \xi) \, d\mathbf{R}_s \, d\mathbf{R}_w \, d\Omega_w$$

The third form on the right-hand side of (21) follows because the sum over j produces N identical terms. The intermolecular potential function for a solute–water pair is represented by U_{sw}. The fifth form employs the definition of the pair-distribution function $\rho_{sw}^{(2)}$ for solute and water. $\rho_{sw}^{(2)} \, d\mathbf{R}_s \, d\mathbf{R}_w \, d\Omega_w$ is the probability of finding the solute particle at about $\mathbf{R}_s$ and a water molecule (any one of the N) at about $\mathbf{R}_w$ with orientation about Ω_w. We also introduce the pair-correlation function $g_{sw}^{(2)}$ defined by

(22)
$$g_{sw}^{(2)} = \frac{8\pi^2}{\rho_s\rho_w} \rho_{sw}^{(2)}$$

and obtain from (21)

(23)
$$\frac{\partial \ln Z(\xi)}{\partial \xi} = \frac{-\beta\rho_w}{8\pi^2} \int U_{sw}(\mathbf{R}, \Omega) g_{sw}^{(2)}(\mathbf{R}, \Omega) \, d\mathbf{R} \, d\Omega$$

where $\mathbf{R} = \mathbf{R}_w - \mathbf{R}_s$.

The final expression for the chemical potential (per particle) of the solute s is thus

(24)
$$\mu_s = kT \ln \rho_s\Lambda_s^3 + \frac{\rho_w}{8\pi^2} \int_0^1 d\xi \int U_{sw}(\mathbf{R}, \Omega) g_{sw}^{(2)}(\mathbf{R}, \Omega) \, d\mathbf{R} \, d\Omega$$

The first term on the right-hand side of (24) is the ideal part, that is, the chemical potential of very dilute s particles in a gaseous phase. The second term is the work required to couple the solute particle to the solvent through the intermolecular potential energy. In connection with our previous discussion, this term together with $kT \ln \Lambda_s^3$ comprises the standard chemical potential of s in the solution. In particular, when differences of the standard chemical potentials in two phases are constructed, the first terms on the right-hand side of (24) are identical and the only remaining part will be the difference in the coupling terms in the two solvents. Since U_{sw} will be of relatively short range and g_{sw} decays to unity at a distance of the order of a few molecular diameters, one sees that the second term of (24) is determined by the local environment of the solute s.

The coupling term can be further split into what is traditionally known as the "cavity" work and the soft interaction coupling work. The qualitative idea behind this separation has already been expressed by Uhlig[30] and Eley,[2] and further developed recently by Pierotti.[31,32] In order to achieve this split, an effective hard-core diameter, σ, is introduced. It is assumed that U_{sw} can be split into two parts, a hard-core repulsive part $U_{sw}^H(\mathbf{R})$ dependent on the distance only, and a soft attractive tail $U_{sw}^S(\mathbf{R}, \mathbf{\Omega})$. Thus

$$(25) \qquad U_{sw}(\mathbf{R}, \mathbf{\Omega}) = U_{sw}^H(\mathbf{R}) + U_{sw}^S(\mathbf{R}, \mathbf{\Omega})$$

where we require that

$$(26) \qquad \begin{aligned} U_{sw}^H(\mathbf{R}) &= \infty \quad \text{for} \quad 0 \le |\mathbf{R}| \le \sigma \\ &= 0 \quad \text{for} \quad |\mathbf{R}| > \sigma \end{aligned}$$

and

$$(27) \qquad \begin{aligned} U_{sw}^S(\mathbf{R}, \mathbf{\Omega}) &= 0 \quad \text{for} \quad 0 \le |\mathbf{R}| \le \sigma \\ &< 0 \quad \text{for} \quad |\mathbf{R}| > \sigma \end{aligned}$$

Equations 26 and 27 assume that, if σ_s and σ_w are the effective hard-sphere diameters of molecules s and w, then $2\sigma = \sigma_s + \sigma_w$.

Using the separation of the pair potential just shown, one may introduce the solute particle, s, to the solvent in two steps. First we introduce a cavity of radius, σ, at a fixed position in the solvent, and then we couple the soft interaction potential. This procedure may be formally achieved by using two coupling parameters consecutively; that is, instead of $U(\xi)$ in (18) we define

$$U(\xi_1, \xi_2) = \xi_1 \sum_{j=1}^{N} U_{sj}^H + \xi_2 \sum_{j=1}^{N} U_{sj}^S + U_N$$

and similarly

$$Z(\xi_1, \xi_2) = \int \cdots \int \exp\left[-\beta U(\xi_1, \xi_2)\right] d\mathbf{R}_s \, d\mathbf{R}^N \, d\mathbf{\Omega}^N$$

where the state $\xi_1 = \xi_2 = 0$ is equivalent to $\xi = 0$ in (18) and $\xi_1 = \xi_2 = 1$ is equivalent to $\xi = 1$. The analog of (20) will be

$$\beta\mu_s = \ln \rho_s\Lambda_s{}^3 - [\ln Z(\xi_1 = 1, \xi_2 = 1) - \ln Z(\xi_1 = 0, \xi_2 = 0)]$$

$$= \ln \rho_s\Lambda_s{}^3 - [\ln Z(\xi_1 = 1, \xi_2 = 0) - \ln Z(\xi_1 = 0, \xi_2 = 0)]$$

$$- [\ln Z(\xi_1 = 1, \xi_2 = 1) - \ln Z(\xi_1 = 1, \xi_2 = 0)]$$

$$= \ln \rho_s\Lambda_s{}^3 - \int_0^1 \frac{\partial \ln Z(\xi_1, \xi_2 = 0)}{\partial \xi_1} d\xi_1 - \int_0^1 \frac{\partial \ln Z(\xi_1 = 1, \xi_2)}{\partial \xi_2} d\xi_2$$

The next steps are formally identical to (21), the final result of this procedure will lead to

$$\mu_s = kT \ln \rho_s\Lambda_s{}^3 + \frac{\rho_w}{8\pi^2} \int_0^1 d\xi_1 \int_{v_\sigma} U_{sw}{}^H(\mathbf{R}) g_{sw}^{(2)}(\mathbf{R}, \Omega, \xi_1, \xi_2 = 0) \, d\mathbf{R} \, d\Omega$$

(28)

$$+ \frac{\rho_w}{8\pi^2} \int_0^1 d\xi_2 \int_{V-v_\sigma} U_{sw}{}^S(\mathbf{R}, \Omega) g_{sw}^{(2)}(\mathbf{R}, \Omega, \xi_1 = 1, \xi_2) \, d\mathbf{R} \, d\Omega$$

where v_σ indicates that the integration over $\mathbf{R}$ is carried over the volume of a cavity of radius σ around the center of the solute. The term $V - v_\sigma$ includes the region outside this cavity. The first integral in (28) can be identified as the work required to "dig out" a suitable cavity of radius σ, and the second as the coupling work against the soft attractive part of the interaction potential. The behavior of the first integrand is admittedly awkward, since in the range $0 < |\mathbf{R}| < \sigma$, $U_{sw}{}^H = \infty$, and $g_{sw}^{(2)} = 0$. To obtain a more convenient and interesting expression for the cavity work let us rewrite (17)

$$\exp[-\beta\mu_s] = \frac{V\int \cdots \int \exp\left(-\beta \sum_{j=1}^{N} U_{sj}\right) \exp(-\beta U_N) \, d\mathbf{R}^N \, d\Omega^N}{\Lambda_s{}^3 \int \cdots \int \exp(-\beta U_N) \, d\mathbf{R}^N \, d\Omega^N}$$

(29)

$$= \frac{V\left\langle \exp\left(-\beta \sum_{j=1}^{N} U_{sj}\right) \right\rangle}{\Lambda_s{}^3}$$

where the average has been performed over all the configurations of the *solvent* molecules.

Using (25) and assuming $U_{sw}{}^S/kT < 1$, we can expand the exponent with the soft part of the potential

$$\left\langle \exp\left(-\beta \sum_{j=1}^{N} U_{sj}\right) \right\rangle = \left\langle \exp\left(-\beta \sum_j U_{sj}{}^H - \beta \sum_j U_{sj}{}^S\right) \right\rangle$$

$$= \left\langle \exp\left(-\beta \sum_j U_{sj}{}^H\right) \right\rangle + \left\langle \exp\left(-\beta \sum_j U_{sj}{}^H\right)\left(-\beta \sum_j U_{sj}{}^S\right) \right\rangle + \cdots$$

(30)

The first term in this expression is

$$
(31) \quad \left\langle \exp\left(-\beta \sum_{j=1}^{N} U_{sj}{}^{\mathrm{H}}\right) \right\rangle = \frac{\int \cdots \int_V \exp\left(-\beta \sum_j U_{sj}{}^{\mathrm{H}}\right) \exp\left(-\beta U_N\right) d\mathbf{R}^N \, d\Omega^N}{\int \cdots \int_V \exp\left(-\beta U_N\right) d\mathbf{R}^N \, d\Omega^N}
$$

$$
= \frac{\int \cdots \int_{V-v_\sigma} \exp\left(-\beta U_N\right) d\mathbf{R}^N \, d\Omega^N}{\int \cdots \int_V \exp\left(-\beta U_N\right) d\mathbf{R}^N \, d\Omega^N} = P_0(v_\sigma)
$$

The second form of (31) arises from the property of the function $\exp\left[-\beta U_{sj}{}^{\mathrm{H}}\right]$, which behaves as a unit step function, that is,

$$
(32) \quad \exp\left[-\beta U_{sj}{}^{\mathrm{H}}(\mathbf{R}_{sj})\right] \begin{array}{ll} = 0 & |\mathbf{R}_{sj}| < \sigma, j = 1, \ldots, N \\ = 1 & |\mathbf{R}_{sj}| > \sigma, j = 1, \ldots, N \end{array}
$$

Therefore if v_σ denotes the volume of the cavity of radius σ around the center of the solute molecule, then the factor $\exp\left(-\beta \sum U_{sj}{}^{\mathrm{H}}\right)$ effectively reduces the range of integration for all the water molecules from V to $V - v_\sigma$. In view of the interpretation of the quantity

$$
\frac{\exp\left[-\beta U_N(\mathbf{R}^N, \Omega^N)\right] d\mathbf{R}^N \, d\Omega^N}{\int \cdots \int \exp\left[-\beta U_N(\mathbf{R}^N, \Omega^N)\right] d\mathbf{R}^N \, d\Omega^N}
$$

as the probability of observing the configuration $\mathbf{R}^N \, \Omega^N$, it is immediately recognized that $P_0(v_\sigma)$ in (31) is the probability of observing a cavity of radius at least σ around the center of $\mathbf{R}_s$. On the other hand, we can define the quantity W^C as

$$
(33) \quad \begin{aligned} \exp\left(-\beta W^C\right) &= \exp\left[-\beta(A_N(v_\sigma) - A_N)\right] \\ &= \frac{\int \cdots \int_{V-v_\sigma} \exp\left(-\beta U_N\right) d\mathbf{R}^N \, d\Omega^N}{\int \cdots \int_V \exp\left(-\beta U_N\right) d\mathbf{R}^N \, d\Omega^N} = P_0(v_\sigma) \end{aligned}
$$

where $A_N(v_\sigma)$ is the Helmholtz free energy of a system of N molecules that are excluded from v_σ. Hence W^C is the work of creating such a cavity. We thus identify W^C with the first integral of (28). If we denote the second integral of (28) by W^S, we get

$$
(34) \quad \mu_s = W^{\mathrm{ideal}} + W^C + W^S, \qquad W^{\mathrm{ideal}} = kT \ln \rho_s \Lambda_s{}^3
$$

It is important to note that, although a formal split of any partial molar quantity, like the one in (34), can be performed, expression in terms of the pair-correlation function alone is not always possible, and higher order correlation functions must be introduced.

This completes our general discussion. In the next sections we turn to more specific treatments of the problem from various points of view.

4 Nonstructural Approach to the Problem

As pointed out earlier, Eley's idea was to split each thermodynamic quantity associated with the dissolution process into the two parts represented by (35) and (36). It is important, however, to note that the quantity W^C in (34) refers to the formation of a cavity at a *fixed* position in the solvent. One may also discuss the introduction of a *free* cavity (i.e., a free hard-sphere solute of suitable diameter). This distinction is important when dealing with partial molar quantities of the solute. It is less important for differences in the standard partial molar quantities, in which case cancelation of the ideal term occurs.

$$(35) \qquad \Delta \bar{H}_s^\circ = \Delta H^C + \Delta H^S$$

$$(36) \qquad \Delta \bar{S}_s^\circ = \Delta S^C + \Delta S^S$$

The free energy of cavity formation was identified with the work necessary to increase the surface by the amount $4\pi\sigma^2\Gamma$, that is,

$$(37) \qquad \Delta G^C = 4\pi\sigma^2\Gamma$$

where σ is the radius of the cavity and Γ the (macroscopic) surface tension. We note that a solute particle with radius R_s will produce a cavity in the liquid of radius $\sigma = R_s + R_w$. By an "empty" cavity we understand the volume excluded to the *centers* of all particles, as illustrated in Figure 7.

Equation 37 has been used as a basis for correlating solubilities of gases with the surface tension of the solvent through

$$(38) \qquad -RT \ln \gamma = \Delta \bar{G}_s^\circ = \Delta G^C + \Delta G^S = 4\pi\sigma^2\Gamma + \Delta G^S$$

If ΔG^S is approximately insensitive to variation of the solvent, we expect a plot of $RT \ln \gamma$ against surface tension to produce a straight line for a given

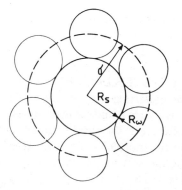

Figure 7. A solute particle with hard-core radius, R_s, produces a cavity of radius, $\sigma = R_s + R_w$, where R_w is the effective hard-core radius of the solvent particles.

solute, s. Such linear relations were observed by Uhlig,[30] and a more complete plot has been presented by Saylor and Battino.[33,34] It is quite surprising that the point for water could be accommodated on the same plot with a large variety of solvents.

Eley's pioneering effort to work out a statistical–mechanical model for these solutions has not met with much success. This may be partially accounted for by his choice of a lattice-type model for liquid water. As for the *sign* of $\Delta \overline{H}_s^\circ$ of gases in water, Eley's argument was quite plausible: in normal liquids ΔH^C is usually large and positive, whereas ΔH^S is small and negative, resulting in a positive value for $\Delta \overline{H}_s^\circ$. For water, Eley assumed, ΔH^C may be small, since "natural" cavities may be found in the network of the relatively open structure. As a result $\Delta \overline{H}_s^\circ$ may acquire a negative value.

As for the entropy of solution, Eley was compelled to assume that the gas molecules are distributed in a restricted number of holes in the liquid, and that entropy and enthalpy of cavity formation was nearly 0 at 4°C. Neither of these assumptions won much acceptance. In the first place, research in this field gave considerable support to the idea that structural changes in the liquid might be of crucial importance. This aspect is discussed in detail later. Second, the development of the scaled particle theory[35-37] provided a powerful means for calculating thermodynamic quantities for cavity formation. This was exploited by Pierotti,[31,32] who employed the method for aqueous solutions of gases.

The essential link between the scaled-particle theory (SPT) and our problem is the SPT prescription for calculating the cavity work. As discussed earlier, it must be assumed that a hard-core diameter can be assigned to each particle. Thus if the solute has a hard-core diameter, σ_s, and a solvent, σ_w, the introduction of the solute to the solvent requires, as a first step, digging a cavity of radius $\sigma = (\sigma_s + \sigma_w)/2$. For this situation the SPT provides the following approximate expression for the free energy of cavity formation

$$(39) \qquad G^C = K_0 + K_1\sigma + K_2\sigma^2 + K_3\sigma^3$$

where the K_i's are functions of T, P, and ρ, as well as the effective diameter of the solvent, σ_w. If we now write for the standard chemical potential

$$(40) \qquad \mu_s^\circ = G^C + G^S = G^C + H^S - TS^S$$

and make an estimate of H^S and S^S, one can in principle calculate all the relevant thermodynamic quantities. This study was undertaken by Pierotti.[31,32] The SPT was utilized for the cavity work and its corresponding derivatives. The soft part of the enthalpy, H^S, was estimated from the total interaction energy of the solute with the surrounding solvent molecules. On the other hand, since there is no simple way of estimating the soft part of the entropy, S^S, it was assumed to be negligible. Although it is difficult to assess the validity of such an approximation, an interesting question, independent of this approximation, remains to be studied. It may be asked whether the anomalous properties of aqueous solutions of inert gases are revealed by hard-sphere gas or equivalently by a solute particle not coupled to the solvent by a soft interaction. Table 3 shows Pierotti's calculations for such a case. The striking difference between water and benzene found in this comparison is that, in water, H^C is much smaller (as follows also from Eley's prediction), and S^C is much more negative. (The standard states employed by Pierotti[31,32] are different from the ones used in this discussion.)

Table 3 also includes a comparison between experimental and theoretical figures, given by Pierotti, showing that the theory succeeds in calculating thermodynamic quantities for both aqueous and nonaqueous solutions of gases. It should be emphasized, however, that although the special structure of water is not featured explicitly in the theory, it does so implicitly.

The SPT is basically not a pure molecular theory in the sense that it does not predict macroscopic properties in terms of molecular parameters only. To be more specific, the aim of a molecular theory should be to give, say the Gibbs free energy, as a function of N, P, and T, as well as the molecular parameters of the system. Given such a function, the density of the system is calculable through thermodynamics (in one component system $\rho = (\partial P/\partial \mu)_T$ where $G = N\mu$). The SPT utilizes the diameters of the particles as molecular parameters, but in addition to the specification of P and T, the density, ρ,

Table 3 Thermodynamic Values for Cavity Formation, G^C, H^C, S^C, Calculated Values[a] of $\Delta \bar{G}^\circ$, $\Delta \bar{H}^\circ$, and $\Delta \bar{S}^\circ$, and the Corresponding Experimental Values for Dissolution of Argon in Water and in Benzene at 25°C

Water		
$G^C = 4430$ (cal/mole)	$H^C = 690$ (cal/mole)	$S^C = -12.5$ (eu)
$\Delta \bar{G}^\circ_{\text{calc}} = 6110$	$\Delta \bar{H}^\circ_{\text{calc}} = -2445$	$\Delta \bar{S}^\circ_{\text{calc}} = -28.7$
$\Delta \bar{G}^\circ_{\text{exp}} = 6270$	$\Delta \bar{H}^\circ_{\text{exp}} = -2820$	$\Delta \bar{S}^\circ_{\text{exp}} = -30.5$
Benzene		
$G^C = 3610$ (cal/mole)	$H^C = 3520$ (cal/mole)	$S^C = -0.3$ (eu)
$\Delta \bar{G}^\circ_{\text{calc}} = 4060$	$\Delta \bar{H}^\circ_{\text{calc}} = 278$	$\Delta \bar{S}^\circ_{\text{calc}} = -12.8$
$\Delta \bar{G}^\circ_{\text{exp}} = 4170$	$\Delta \bar{H}^\circ_{\text{exp}} = 420$	$\Delta \bar{S}^\circ_{\text{exp}} = -16.6$

[a] Based on the scaled-particle theory; from Refs. 31, 32.

should be supplemented as a measurable quantity. This additional information carries with it implicitly the peculiarities of the structure of the solvent.

A comment concerning the use of a constant diameter, σ_w, for water has been raised.[38] Following Mayer's findings[39,40] that the experimental temperature dependence of the surface tension of water could not be predicted by the SPT with constant σ_w, the adequacy of such an assumption for calculating S^C was questioned.[38] It was found that by using the temperature dependence of σ_w as required to fit the experimental surface entropy, and calculating S^C, the agreement with the experiment is destroyed. Pierotti[41] has requestioned the reasonability of using Mayer's temperature dependence of σ_w. Although admitting that some form of temperature dependence of σ_w should be incorporated, he suggested that compressibility data may serve better as a source for such a quantity.

5 Comparison with Gas Hydrates

An interesting comparison between the thermodynamics of aqueous solutions of gases and gas hydrates has been developed since Frank and Evans[3] raised this possibility. The structure of the gas hydrates has been studied by X-rays.[42,43] A systematic comparison of the two processes has been presented by Glew.[44] Table 4 lists the pertinent "reactions", combinations of which were utilized to produce some interesting information on the energetics of water-solute interaction.

From reactions II and IV an estimate can be made of the heat of crystallization of n water molecules, where n is the number of molecules "reacting"

Table 4   Table of Reactions (for °C) Used by Glew

I	$(CH_4)_g$	$\rightarrow (CH_4)_l$	$\Delta H_I =$	-4.621 (kcal/mole)
II	$(CH_4)_g + n(H_2O)_l$	$\rightarrow CH_4 \cdot n(H_2O)_{hyd}$	$\Delta H_{II} =$	-12.830
III	$(CH_4)_l + n(H_2O)_l$	$\rightarrow CH_4 \cdot n(H_2O)_{hyd}$	$\Delta H_{III} =$	-8.228
IV	$(CH_4)_g + n(H_2O)_{ice}$	$\rightarrow CH_4 \cdot n(H_2O)_{hyd}$	$\Delta H_{IV} =$	-4.553
V	$(H_2O)_l$	$\rightarrow (H_2O)_{ice}$	$\Delta H_V =$	-1.436

(From Ref. 44)
I—Transfer of one mole of methane from the gaseous phase, subscript g, into the liquid, subscript l.
II—Formation of solid gas hydrate, subscript hyd, from gaseous methane and liquid water.
III—Formation of solid gas hydrate from liquid water and dissolved methane.
IV—Formation of solid gas hydrate from gaseous methane and solid ice.
V—Crystallization of liquid water into solid ice.

with the solute. The molar heat of crystallization, ΔH_V is, on the other hand, known independently, so that an estimate of n can be derived from the equation

$$(41) \qquad\qquad \Delta H_{IV} - \Delta H_{II} = -n\,\Delta H_V$$

In this way, Glew obtained $n = 5.765$, which agrees very well with the expected water-to-gas-ratio in a crystalline hydrate ($n = 5.75$).

The next observation is that

$$(42) \qquad\qquad \Delta H_{IV} - \Delta H_{II} = 8.277 \simeq -\Delta H_{III} = 8.228$$

which indicates very little difference, energetically, between the ice crystal and water in crystalline hydrate (reaction III is basically a crystallization of water molecules into crystalline hydrate). If this comparison is accepted, one can modify reaction IV into IV′ without affecting the energetics very much. Thus

$$(43) \qquad IV' \quad (CH_4)_g + n(H_2O)_{\text{pure}} \rightarrow (CH_4)\cdot n(H_2O)_{\text{hyd}}$$
$$\text{cryst.}$$
$$\text{hydrate}$$

with $\Delta H_{IV'} \simeq \Delta H_{IV}$.

It is now observed that ΔH_{IV} is also very close to the value of ΔH_I. This means that the dissolution process of gas into liquid and into the hydrate framework, to form the solid crystalline gas hydrate, are similar. It should be emphasized, however, that such an analogy cannot be carried further without any additional information, in particular for the entropy of the two processes. Indeed, this was already indicated by Frank and Evans,[3] who first made this comparison and concluded that the two processes may be similar in their *geometrics* and *energetics*. Later it was shown by Namiot[45] that although ΔH values for the two processes agree, the ΔS values (and hence the $\Delta\mu$ values) are markedly different.

6 Structural Approaches to the Problem

In this section some of the specific structural models elaborated in connection with our present problem are discussed. We first summarize some of their most important common features.

The essential idea of the structural approach is the classification of water molecules into groups of different species. Such classification requires definition of the various species to be distinguished. Of the various methods of definition, the most common in our case depends on the concept of the hydrogen bond. If we agree to call a pair of molecules at a certain relative configuration hydrogen bonded, then for each configuration of the (classically described) system, the water molecules can be classified in groups, so that a molecule in the kth group is linked to the rest of the system by k hydrogen bonds. The average (over the appropriate ensemble) number of molecules in the kth group will be denoted by N_k so that

$$\text{(44)} \qquad \sum_{k=0}^{4} N_k = N$$

where N is the total number of water molecules. We shall call a molecule in the kth group a k-cule. (This nomenclature is borrowed from a similar one suggested in ref. 37.)

One may argue that this kind of discrete classification of molecules contradicts the continuous range of configurations that a pair of molecules may possess. In principle the same argument can be raised against any rule that prescribes a distinction between a pair of free atoms and a molecule. In the present case we split the (continuous) range of all possible configurations for a pair of water molecules into two parts. The separation, inspired by our chemical intuition, is admittedly not free from an element of arbitrariness. However, once we have defined the concept of the hydrogen bond, the classification procedure may be carried out in an exact fashion.[46] In (44) we have also acknowledged the quite acceptable fact that a molecule can be linked, at most, by four hydrogen bonds.

Supposing a system of N water molecules and N_s solute molecules at given temperature, T, and pressure, P (in all subsequent discussion partial derivatives will always be at P, T constant). An extensive thermodynamic quantity, E, can be viewed as a function of either the variables (P, T, N_s, N) or $(P, T, N_s, N_0 \cdots N_4)$. Along with the second set of variables we have a set of four equilibrium conditions, and utilizing these reduces the number of independent components to two. Certainly the thermodynamic properties of the system are not affected by the choice of either point of view. Interest now centers on the partial molar quantities of the solute s, defined by

$$\text{(45)} \qquad \bar{E}_s = \left(\frac{\partial E}{\partial N_s} \right)_N$$

Viewing water as a mixture of five species we write

$$\text{(46)} \qquad dE = \left(\frac{\partial E}{\partial N_s} \right)_{N_0 \cdots N_4} dN_s + \sum_{k=0}^{4} \left(\frac{\partial E}{\partial N_k} \right)_{Nj,\ j \neq k} dN_k$$

Dividing by dN_s and requiring that the total number of water molecules be constant, one obtains

$$\text{(47)} \qquad \bar{E}_s = \left(\frac{\partial E}{\partial N_s} \right)_N = \left(\frac{\partial E}{\partial N_s} \right)_{N_0 \cdots N_4} + \sum_{k=0}^{4} \left(\frac{\partial E}{\partial N_k} \right)_{Nj,\ j \neq k} \left(\frac{\partial N_k}{\partial N_s} \right)_N$$

Using condition (44) and the symbols

$$E_s^* = \left(\frac{\partial E}{\partial N_s}\right)_{N_0 \cdots N_4} \qquad \bar{E}_k = \left(\frac{\partial E}{\partial N_k}\right)_{Nj, \, j \neq k}$$

$$\Delta E_s^r = \sum_{k=1}^{4} (\bar{E}_k - \bar{E}_0)\left(\frac{\partial N_k}{\partial N_s}\right)_N$$

one obtains

(48) $$\bar{F}_s = E_s^* + \Delta E_s^r$$

The significance of the two terms on the right-hand side of (48) is quite obvious. To obtain the experimental variation of E with the addition of s, the effect of s on E when all N_k's are constant is first considered, and then the system is allowed to relax into its final equilibrium state. A more picturesque description, appealing to the chemist's intuition, is the following.

We first invent an imaginary catalyst in whose presence the system of various species of water molecules can attain its equilibrium position. Obviously if this catalyst is absent then the system behaves as if all N_k's are independent variables. Suppose that at the beginning the system that includes the catalyst is at equilibrium with composition $N_s N_0 \cdots N_4$. The catalyst is then removed and dN_s moles of s transferred into the system. Since N_k are kept constant, the measurable partial molar quantity in this step will be E_s^*, to be referred to as the static part of $\bar{E}_s$. Upon reintroduction of the catalyst into the system, the concentrations of the various species may change to their new equilibrium values. This process contributes the part ΔE_s^r to the total partial molar quantity $\bar{E}_s$ and will be referred to as the relaxation part. (The term "structural part," also used in this context, seems less appropriate. In the first place both E_s^* and ΔE_s^r depend on the structure of the solvent for any conceivable definition of the structure; second, the term relaxation fits naturally if we describe the process as taking place in two steps, for then it conforms with similar usages of the term in other fields of physics and chemistry.)

In order to proceed, more must be known about the properties of the various species and their interaction with the solute, s. This is a difficult task, particularly since no explicit definition has been given to the various species. In order to simplify, therefore, we propose to study, as a first step, the two extreme species, 0-cules and 4-cules. This simplification may not be a good approximation, for there is reason to believe that an important role is played by the intermediary species. For the moment their presence will be ignored and we shall be interested in the effect of the addition of s on the two extreme species only. (From the formal point of view, the discussion will still be exact if by 4-cules we mean all four-bonded molecules, and by 0-cules, all other molecules including intermediary species).

For such a system, (48) can be rewritten for some of the relevant partial molar quantities

$$(49) \qquad \bar{H}_s = H_s^* + (\bar{H}_4 - \bar{H}_0)\left(\frac{\partial N_4}{\partial N_s}\right)_N$$

$$(50) \qquad \bar{S}_s = S_s^* + (\bar{S}_4 - \bar{S}_0)\left(\frac{\partial N_4}{\partial N_s}\right)_N$$

$$(51) \qquad \mu_s = \mu_s^* + (\mu_4 - \mu_0)\left(\frac{\partial N_4}{\partial N_s}\right)_N$$

In accordance with our views on the nature of the hydrogen bond, we can assume that a 4-cule forms part of the network of fully bonded molecules. The bonding energy of a 4-cule to the rest of the system will therefore be larger than that of an 0-cule. Thus, even without too detailed a specification of the two species, it can be assumed that $\bar{H}_4 - \bar{H}_0$ and $\bar{S}_4 - \bar{S}_0$ are negative, that is, the transformation

$$(52) \qquad \text{(0-cule)} \to \text{(4-cule)}$$

involves negative enthalpy and entropy of reaction. Because H_s^* is expected, qualitatively, to be of the order of magnitude of $\bar{H}_s$ in simple liquids, the anomalously large negative value of $\bar{H}_s$ in water is attributed to ΔH_s^r.

This explanation is satisfactory if $(\partial N_4/\partial N_s)_N$ can be shown to be positive, producing together with the negative value of $\bar{H}_4 - \bar{H}_0$, a term of the required sign. As for the partial molar entropy, the situation is probably more complicated. The quantity S_s^* is likely to be anomalous because of the special structure of water; also, the relaxation term ΔS_s^r may contribute an additional negative value to $\bar{S}_s$. The important conclusion to be drawn at present is that, although $\bar{H}_s$ and $\bar{S}_s$ may have contributions from both the static and the relaxation parts, only the former contributes to the chemical potential, μ_s. This is a consequence of the condition of chemical equilibrium

$$\mu_4 = \mu_0$$

by which (51) reduces to

$$(53) \qquad \mu_s = \mu_s^*$$

Using this result and also (49) to (51) we get

$$(54) \qquad T\Delta S_s^r = \Delta H_s^r$$

which means that the relaxation part of $\bar{H}_s$ exactly compensates the relaxation part of $T\bar{S}_s$. This is an important conclusion in connection with the general principle of "enthalpy-entropy compensation in aqueous solutions."[47]

It should be noted also that this conclusion is valid for the five-species system or for any other classification of water molecules into various groups. This conclusion also has some implications regarding the effect of structural changes on the solubility of the gas, s.[48] Continuing our pictorial description of the two terms, we conclude that removing a catalyst from a system at equilibrium containing s will affect neither the chemical potential, μ_s, nor the solubility, although other partial molar quantities may well be changed.

The central question left unanswered is the sign of $(\partial N_4/\partial N_s)_N$. If the sign is shown to be positive, corresponding roughly to the concept of structure forming,[3,5] a qualitative explanation for part of the negative contribution to the entropy and enthalpy of solution in water is provided. The original conjecture made by Frank and Evans[3] was suggested in order to explain the experimental data. An attempt to elaborate on this question at the molecular level has been the subject of many studies of particular models, as discussed later. However, before introducing any specific model, the problem of finding the sign of $(\partial N_4/\partial N_s)_N$ may be transformed into an equivalent problem. Such a transformation was suggested[49,50] as a first step toward a molecular treatment of the problem, and follows from the identity

$$(55) \qquad \left(\frac{\partial N_4}{\partial N_s}\right)_N = -(\mu_{4,4} - 2\mu_{4,0} + \mu_{0,0})^{-1}\left[\frac{\partial(\mu_4 - \mu_0)}{\partial N_s}\right]_{N_4 N_0}$$

where $\mu_{ij} = \partial\mu_i/\partial N_j$ and $(\mu_{4,4} - 2\mu_{4,0} + \mu_{0,0})$ is a positive quantity.[51] This identity is a central relation in the present context, since it connects a derivative at equilibrium (left-hand side) with a derivative in the "frozen" system (right-hand side).

Thus a study of the sign of either $\partial(\mu_4 - \mu_0)/\partial N_s$ or $\partial N_4/\partial N_s$ is sufficient for our purpose. The former is, however, more directly amenable to molecular examination. The physical meaning behind this transformation is very simple. To illustrate, using again the imaginary catalyst, and starting with a system at equilibrium (with catalyst present), we have the equality

$$(56) \qquad \mu_4(N_4, N_0, N_s) = \mu_0(N_4, N_0, N_s)$$

Next the catalyst is removed and dN_s moles of s added. In general, at this stage

$$(57) \qquad \mu_4(N_4, N_0, N_s + dN_s) \neq \mu_0(N_4, N_0, N_s + dN_s)$$

For concreteness, suppose that we find

$$(58) \qquad \mu_4(N_4, N_0, N_s + dN_s) < \mu_0(N_4, N_0, N_s + dN_s)$$

then, if the catalyst is reintroduced into the system, it will cause a "flow" of molecules from the 0-state to the 4-state (i.e., the final value of N_4 will be larger than its initial value), so that $\partial N_4/\partial N_s$ is positive. An examination of the change of $\mu_4 - \mu_0$ in the "frozen up" system is therefore equivalent to examination of the change of N_4 in the equilibrated system, which is exactly the meaning of (55).

To see the advantage of using (55), a reasonable guess should be made about the radial distribution functions around a 4-cule and an 0-cule, so that something can be said about the sign of $\partial(\mu_4 - \mu_0)/\partial N_s$ and hence also about $\partial N_4/\partial N_s$. The pertinent chemical potentials of the two species can be written in the form

$$(59) \qquad \mu_4 = kT \ln \rho_4 \Lambda_4{}^3 + W_4$$

$$(60) \qquad \mu_0 = kT \ln \rho_0 \Lambda_0{}^3 + W_0$$

where Λ_i includes internal degrees of freedom of the i-cule. W_i is the total coupling work of the i-cule to the system. The assumption is now made that the introduction of s to the system does not effect Λ_0 and Λ_4. This is a plausible assumption for an inert solute, s. Hence

$$(61) \qquad \left(\frac{\partial(\mu_4 - \mu_0)}{\partial N_s}\right)_{N_4, N_0} = \left(\frac{\partial(W_4 - W_0)}{\partial N_s}\right)_{N_4, N_0}$$

It is conceivable that both Λ_0 and Λ_4 will be affected by the solute, s. This kind of effect must be treated in the quantum mechanical language. Here we shall provide a qualitative argument based on classical arguments only.

Equation 61 reduces the question to an examination of the change of the local properties of the 0-cule and the 4-cule upon the addition of s. An 0-cule is usually viewed as a molecule in a locally closed packed region, whereas the 4-cule as part of a hydrogen-bonded network of water molecules. The essential difference in the response of these two species to the addition of an inert molecule is depicted in Figure 8. The total interaction of the 4-cule with its environment increases by the addition of s, since the original structure of water molecules around it is presumed not to change appreciably. A solute added to the environment of an 0-cule will, on the average, replace water molecules previously surrounding the 0-cule, resulting in a decrease of its total interaction with the rest of the system. Therefore, as far as energetics is concerned, we expect that $\partial W_4/\partial N_s < 0$ and $\partial W_0/\partial N_s > 0$, and hence $\partial(W_4 - W_0)/\partial N_s$ is negative. Therefore, in view of (61) and (55), $(\partial N_4/\partial N_s)_N$, is positive.

Some words of caution are necessary. In the first place, it must be remembered that all intermediary states of the molecules were ignored at the start. The idea was to gain insight into the responses of the two species to solute addition. It can then be conjectured that intermediary species will respond

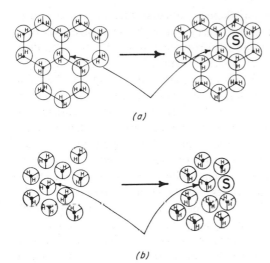

(a)

(b)

Figure 8. Schematic description of the change of the environment of (a) a 4-cule and (b) an 0-cule upon the addition of an inert solute, *s*.

somehow in an intermediary fashion. For an actual calculation of thermo-dynamic quantities, all the species must be included in the discussion, for their role may be dominant in determining the macroscopic properties.

In the second place, it should be noted that the arguments given in this section to show the unsymmetrical response of the 4-cule and the 0-cule were essentially energetic. A full treatment should handle the chemical potential of the two species, including an entropy effect. This can only be done if a good approximation for the various pair-correlation functions involved is available. Finally, the arguments just given, following the demonstration in Figure 8, are based on the fairly stringent requirement that interpenetration of the two species is precluded. In a sense this requirement is closer to a two-phase system than to a mixture of two species. This treatment ignores water molecules on the boundaries of the two species (usually molecules in the intermediary groups), and it may be a reasonably good approximation for very large clusters of hydrogen-bonded molecules. For smaller clusters, sur-face effects must be included as well. The last difficulty may be overcome by changing the tactics of grouping the molecules into species. Instead of looking at different single molecular species, the whole system can be subdivided into clusters of hydrogen-bonded molecules so that C_n is a cluster of n water molecules. The set of chemical reactions is

(62)
$$C_n \rightleftarrows nC_1$$

This approach may be more satisfactory, since surface effects are automatically taken care of. An attempt to carry out such a treatment has been undertaken,[49,50] although not without introducing some drastic simplifications. The effect of the solute, s, on the chemical potential of a cluster containing n molecules was divided into two parts. One comprises the effect produced by s molecules, which penetrate the framework of the clusters; the second operates through the surface of the clusters. It was concluded that each of these effects may lead to a stabilization effect, provided that n is large enough. It is presently believed[52] that the surface effect alone cannot produce a stabilization of the clusters. This conclusion was reached from an exact examination of a one-dimensional model,[52] where it was shown that penetration of solute particles into the "holes" of the cluster is required in order to produce stabilization. Of course such penetration does not completely eliminate different effects that may exist in a more complicated model.

Before turning to a discussion of some specific models that have been pursued to the point where numerical results were derived, comment is required on the possibility of releasing one of the basic assumptions of the two-structure model; namely, that internal degrees of freedom of water molecules are not appreciably affected by the presence of solute. This assumption can be disposed of by further classification of the two (or more) species into groups. A 4-cule adjacent to a solute molecule, and which has suffered a change in internal degrees of freedom will be called a 4'-cule and those that are unperturbed will be called 4"-cules. Similar distinction is made for the 0-cule. Thus

$$(63) \qquad N_4 = N_{4'} + N_{4''}$$

$$(64) \qquad N_0 = N_{0'} + N_{0''}$$

The partial molar quantity, E_s, is now given by

$$(65) \qquad \bar{E}_s = \left(\frac{\partial E}{\partial N_s}\right)_{N_4 \cdot N_{4''} N_0 \cdot N_{0''}} + \sum_k \bar{E}_k \frac{\partial N_k}{\partial N_s}$$

where summation is carried over all species involved. Furthermore, one can use the condition

$$(66) \qquad N_4 + N_0 = N$$

and in a dilute solution, with respect to s, we assume that the total number of affected water molecules is proportional to N_s, that is

$$(67) \qquad N_{4'} + N_{0'} = \alpha N_s$$

Equation (65) can now be written as

$$(68) \quad \bar{E}_s = E_s^* + (\bar{E}_{4'} - \bar{E}_{0'}) \frac{\partial N_{4'}}{\partial N_s} + (\bar{E}_{4''} - \bar{E}_{0''}) \frac{\partial N_{4''}}{\partial N_s} + \alpha(\bar{E}_{0'} - \bar{E}_{0''})$$

Equation 68 shows that three relaxation processes occur. A study of similar systems has been carried out by Vdovenko et al.[53]

We shall now proceed to a brief review of some specific models. In the two-structural domain, Namiot[45] assumed an ideal mixture of icelike water $(H_2O)_i$ and closed-packed water $(H_2O)_c$ in chemical equilibrium

$$(69) \qquad\qquad (H_2O)_i \rightleftarrows (H_2O)_c$$

with the condition

$$(70) \qquad\qquad \mu_c{}^\circ - \mu_i{}^\circ = RT \ln \frac{X_i}{X_c}$$

where X_i and X_o are the mole fractions of the two species. Namiot further estimated $\mu_c{}^\circ - \mu_i{}^\circ \simeq 500$ cal/mole, which gives $X_i = 0.71$ for pure water at 0°C.

The process of dissolution of gas into aqueous solutions has been compared with the process of hydrate formation from gaseous solute and ice. The reactions involved are described in Table 5. The estimation of X_i in pure

Table 5   Table of Reactions Used by Namiot

I
$(H_2O)_i \rightarrow (H_2O)_c$ $\mu_c{}^\circ - \mu_i{}^\circ - RT \ln (X_i/X_c)$ (i = icy, c = close-packed)   Estimation: $\mu_c{}^\circ - \mu_i{}^\circ \sim 500$ cal/mole $\Rightarrow X_i \sim 0.71$ at 0°C

II

$n(H_2O)_i + s^{(g)} \rightarrow (s \cdot nH_2O)_i,$ $\mu_s{}^\circ - \mu_s{}^{og} - n\mu_i{}^{ol} = RT \ln [X_i{}^n P_s / X_{s \cdot nH_2O}]$
 Assumptions:
    $P_s = K_H X_s,$ $X_s = X_{s \cdot nH_2O}$
    P_s is the partial pressure of s over the solution
    K_H = Henry's constant

III
$n(H_2O)$ solid ice $+ s^{(g)} \rightarrow [s \cdot nH_2O)_{hyd}$ $\mu_{s \cdot nH_2O}^{o(hyd)} - \mu_s{}^{og} - n\mu_i{}^{o(ice)} = RT \ln P_H$ P_H = equilibrium pressure of s over the crystalline hydrate   Equating $\Delta\mu^\circ$ of II and III leads to $P_H = K_H X_i{}^n$

From Ref. 45.

water together with the identification of the standard free energy for processes II and III were combined to produce the relation

$$(71) \qquad P_H = K_H X_i^n$$

where K_H is the Henry constant for the gas solubility, P_H is the equilibrium partial pressure of the gas over the crystalline hydrate, and n is the number of water molecules combining with one solute molecule. The last relation may be used to calculate K_H for gases in water.

As pointed out earlier, the analogy between dissolution into liquid and hydrate formation is justified to the extent of comparing the enthalpies of the two processes. There is no direct evidence that the entropies accompanying the two processes are comparable, and therefore the same holds for the free energies. The ideality assumption about the mixture should be viewed as a severe simplification; it holds only in the rare case of two very similar components. In the present case, we have a mixture of two "kinds" of water molecules; therefore, although similar in composition, they differ markedly in their properties. The most relevant property in this context is their interaction with the rest of the system, which by definition is very dissimilar. Thus no kind of ideality should be accepted without strong reservations. Finally, the assumption that the gas molecules actually "react" with one species only is also a restrictive condition and is probably responsible for the numerically low solubilities calculable from this model.

Similar models where the solute interacts with one species have been employed. Examples are the application of Pauling's[54] or Samoilov's[55] models for water to the problem of dilute aqueous solutions.[56-58] Frank and Franks[59] have recently developed a two-structure model for aqueous solutions of nonelectrolytes. In some respects this is an improved version of the Pauling model treated previously by Frank and Quist.[56] The basic improvement is the rejection of the assumption that the icelike species (or the bulky phase)[59] serves as sole medium for the dissolution of the solute. In this model, the solute reacts with the two species. The shift in the concentrations follows from the unsymmetrical response of the two species to the presence of the solute. In this model, too, the distinction is made between a bulky "phase" (in place of the term "icelike," which is frequently misinterpreted) and a dense "phase." The solute is permitted to dissolve in both "phases." The basic assumption of ideality is introduced for pure water. Thus, if f is the mole fraction of the bulky species, one gets for pure water, an equilibrium constant of the form

$$(72) \qquad K = \frac{f}{1-f}$$

For the nonelectrolyte solution Frank and Franks[59] distinguished between the fraction, g, of solute in the bulky "phase" and $1 - g$ in the dense "phase." The free energy of the system is written as

$$(73) \qquad\qquad G = G° - TS^M$$

where $G°$ includes the various standard free energies involved and S^M is the entropy of mixing. The latter quantity is constructed from combinatorial factors that reflect the difference in the solvation properties of the two species. Once G is constructed, all the thermodynamic properties of the system can be derived in a straightforward manner. In spite of the various ideality assumptions introduced in S^M, this model is capable of offering a deeper insight into the molecular origin of the experimentally observed quantities, which may well be valid in real systems. It is likely that some shortcomings of the Frank–Quist model led to this improved version. In the same way, it is hoped that the general features of the present model are sufficiently close to reality that any failure to produce the experimental results can be traced back to the assumptions of the model, and further improvements can be suggested. Work on these lines seems therefore very promising.

An analysis of the statistical mechanical properties of a five-species mixture model has been carried out by Némethy and Scheraga.[60] The molecules were classified according to the number of hydrogen bonds. As described at the beginning of this section, this procedure can be made exact in principle.[46] This classification can most conveniently be accomplished by starting with the Grand partition function for a system of pure water in a volume, V, at a given temperature, T, and chemical potential, μ.

$$(74) \qquad\qquad \Xi(\mu, V, T) = \sum_{N=0}^{\infty} y^N \frac{Z_N(V, T)}{N!}$$

where

$$y = \frac{q \exp (\beta\mu)}{8\pi^2}$$

q is the partition function of a single water molecule, and Z_N the classical configurational partition function.

$$(75) \qquad Z_N = \int \cdots \int \exp [-\beta U_N(\mathbf{X}_1 \cdots \mathbf{X}_N)] \, d\mathbf{X}_1 \cdots d\mathbf{X}_N$$

where $\mathbf{X}_i$ stands for $\mathbf{R}_i$, $\mathbf{\Omega}_i$.

If we are now given a rule by which, for each configuration $\mathbf{X}_1 \cdots \mathbf{X}_N$, a unique classification of all the particles into five groups can be carried out, then we can write Z_N as

$$(76) \qquad Z_N = \sum_{\substack{\{N_k\} \\ \sum N_k = N}} \frac{N!}{\prod_k N_k!} \int \cdots \int_{D(N_k)} \exp (-\beta U_N) \, d\mathbf{X}_1 \cdots d\mathbf{X}_N$$

where the range of integration, denoted by $D\{N_k\}$ is over that part of the phase space for which particles $1 \cdots N_0$ are 0-cules, particles $N_0 + 1 \cdots N_0 + N_1$ are 1-cules, and so forth.

The summation is carried out over all *possible* sets $\{N_K\}$ that conform with some physical requirement. The condition $\sum N_k = N$ is not the only restriction on the possible sets $\{N_k\}$; for example, the sets $\{N_0 = N, N_1 = N_2 = N_3 = N_4 = 0\}$ or $\{N_0 = N - 5, N_1 = 4, N_2 = 0, N_3 = 0, N_4 = 1\}$ are allowable, but not the set $\{N_0 = N - 1, N_1 = N_2 = N_3 = 0, N_4 = 1\}$. All these examples fulfill the condition $\sum N_k = N$. Substituting Z_N into (74) produces an expression resembling a partition function of a mixture of five species. It is difficult to proceed from this stage because there is no obvious way of factorizing this expression into a product of partition functions for each of the species involved. Facing such a deadlock, Némethy and Scheraga suggested using a factorizable partition function chosen in a somewhat heuristic fashion. The main idea of this approach is to assign to each molecule in the k-group a single—apart from internal energy levels, which are treated separately—"energy level." This reduces the infinitely many possible interaction potential energies that each molecule in the kth group may possess to just a single number, E_k. The next step is to treat these parameters, E_k, as if they were real energy levels assigned to independent molecules, in which case the partition function is

$$(77) \qquad Q_N = \sum_{\{N_k\}} N! \prod_{k=0}^{4} \frac{1}{N_k!} \{f_k \exp(-\beta E_k)^{N_k}\}$$

where f_k takes care of the internal as well as rotational and translational partition function of a molecule in the kth group. The summation is carried over a restricted set of $\{N_k\}$ to conform with certain geometrical requirements derived from experimentation with molecular models of compact clusters of water molecules.

Once the partition function has been written, the extraction of thermodynamic quantities is straightforward. The numerical results obtained from this model met with remarkable success for both pure water and for its solutions of nonelectrolytes. In applying this model to aqueous solutions, the "energy levels" were modified. The idea behind this modification is similar to that given earlier in this section; that is, a 4-cule is expected to increase its total interaction energy with the medium, and a 0-cule suffers an opposite change. It is not clear how the "energy level" of an intermediary molecule changes in the presence of a solute, but the model assumes that all energy levels are raised except the lowest, which is lowered. In fact, it is not clear how the "energy level" of even a 4-cule, which is not in the deep interior of a cluster, is changed by the addition of a solute, for example, a four-bonded molecule on the surface of a cluster, which is exposed to interactions with all the species present in the solution. The arguments given on the change in the environments of a 4-cule and a 0-cule may both hold for molecules on the surface of a cluster.

Once this change of energy levels is assumed, the effect on the distribution of water molecules in the various groups is qualitatively evident. This follows from the particular form of the partition function chosen. Here the Boltzmann factors, $\exp(-\beta E_k)$, govern the probabilities of occupying the kth group. Therefore, if E_4 is lowered while all others are raised, the eventual redistribution of water molecules can be expected to favor occupation of the fourth-group.

Thus the starting point of view taken by Némethy and Scheraga is certainly very appealing and in principle can be made formally exact. The replacement of the total interaction energy of a k-cule by a single parameter E_k, can be viewed as an outcome of an averaging process. However, by using these parameters as if they were real energy levels of single independent molecules, a "gaslike" partition function results which can hardly represent the situation in the liquid state.

In concluding this long section, it is appropriate to reflect on the usefulness of the various approaches. Recognizing from the outset the formidable difficulty of an exact treatment, one should not judge any theory by the closeness of its numerical results to the experimental ones. Instead, it would seem more useful at this stage to concentrate on understanding the qualitative sources of the various phenomena, rendered possible by the use of highly simplified models. Such models could be employed in order to gain insight into some unanswered questions.

For instance, the molecular reasons for the low solubility of gases in water are still not understood, even qualitatively. As noted before, the possibility of structural effects on the solvent could give some qualitative arguments to explain the entropy and the enthalpy of solution, but not the solubility. A second quantity of a potential interest is the partial molar heat capacity of gases in water. This quantity may contain valuable information on the structural changes in the solvent.[61] These, as well as other quantities, could be studied in model systems, which have some common features with water yet are simple enough to be amenable for exact analysis. The most primitive system that can be constructed with some features in common with water is a one-dimensional analog of aqueous solutions of nonelectrolyte.[52] The results obtained from this model have of course no real value save the demonstration of some qualitative trends typical to a system of this kind. The examination of some outcomes of these models can shed some further light on the molecular origin of the various phenomena observed.

7 Small Deviations from Ideal Dilute Solutions: A Possible Relation to the Problem of Hydrophobic Interaction

The highly dilute solutions of nonelectrolytes so far discussed are characterized by the absence of any contribution due to solute-solute interaction. The statistical mechanical treatment of such systems is analogous to the one-body problem dealt with in the gaseous phase; but in our case the empty space surrounding the single solute particle is filled with solvent molecules. The next natural system to be examined is the analog of the two-particle system, that is, two solute molecules surrounded by solvent molecules. Phenomenologically, this study corresponds to deviation from ideal dilute solutions in which Henry's law usually fails to be valid. One can raise the purely academic question of whether any particular anomaly will be discovered in such a study, when comparing aqueous with nonaqueous solutions. To the best of my knowledge, such a study has never been undertaken, presumably because of lack of stimulating motivation.

New interest in this subject was aroused only recently,[62-64] when it was recognized that a central problem in biopolymer chemistry—namely, the so-called hydrophobic interaction—may have some features in common with our system. In this section we shall not give details of the experimental information accumulated on the role of hydrophobic interaction in conformational stabilization of biopolymers.

However, mention must be made of existing evidence that, apart from the conventional "bonds" (including hydrogen bonds, disulfide bonds, and ion-ion interactions) known to contribute to the maintenance of a particular "active" conformation, another driving force apparently operates between nonpolar side chains, tending to bring these groups close together. These forces are believed to have their origin in the special structure of water. Figure 9 depicts a schematic "reaction" involving hydrophobic interaction.

A study of the complete set of factors determining the equilibrium constant for such a reaction is very complicated. A detailed definition of the two isomers, without which an equilibrium constant for this reaction may be meaningless, must first be constructed. The various factors to be considered in this "reaction" include many internal changes in the polymer, as well as changes induced on the solvent. All this makes the problem formidably difficult. So far the most successful tactic for such a study has been the investigation of the effect of the isolated factors on model systems, from which inference can be drawn as to their combined action. Focusing attention on one such factor, let us look at two nonpolar groups (say, methyl groups on valine or leucine), whose separation is believed to be an important factor determining the equilibrium constant for this reaction. The original explanation for the existence of such an effect comes from the notion that the two

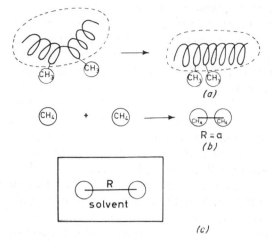

Figure 9. Schematic description of the various stages in the study of hydrophobic inter-action. (*a*) Change of conformation of a biopoylmer involving a separation of two nonpolar groups. (*b*) Dimerization of two simple nonpolar solutes. The pair is identified as dimer whenever the separation, R, is smaller than a chosen parameter, a. (*c*) Study of the change of the free energy of a system of two inert particles in water as a function of their distance, R.

nonpolar groups do not *favor* their water environments and therefore tend to adhere to each other, thus exposing less of their surface to interaction with the water: hence the name, hydrophobic interaction. This picturesque interpretation, although very common, has no theoretical basis however.

In order to isolate this particular effect from the main problem, let us ignore, for the moment, the polymer and watch the reaction between two nonpolar molecules to form a dimer (Figure 9*b*). This in itself requires a definition, and the simplest is the following: a distance, a, can be chosen so that a pair of molecules at distance $R \leq a$ will be called a dimer. This possi-bility has an element of arbitrariness, since we have no a priori information about the distance, a, which defines the dimer. For an exact formulation of the problem, a should preferably be an outcome of the investigation, rather than a chosen parameter. Another possible definition of a dimer may be given in terms of a particular tool of measurement. An interesting example would be the dimerization of carboxylic acids in aqueous solutions.[65] Here one may determine the equilibrium constant for dimerization, and hence the standard free energy of the reaction may be calculated. Similar data may be obtained by other methods.[66] The outcome of these measurements strongly indicates that the free energy of formation of the dimer increases with the increase of the length of the aliphatic radical.

This method has a few shortcomings, however. In the first place, the method is applicable only to solutes that carry some functional groups such as OH, COOH, or NH_2. Therefore, the driving forces for the formation of a dimer is not entirely due to hydrophobic interaction, and the distinction between the various contributions cannot be made unequivocally. Second, the geometry of the dimer is determined primarily by the functional groups (e.g., by the formation of a hydrogen bond between the two solute particles). This geometry restricts the set of configurations accessible for the nonpolar groups, which may not include the typical configuration required for the formation of a strong hydrophobic interaction. Finally, the two nonpolar groups, being attached to a polar functional group, will see each other through an already perturbed medium. Therefore, the tendency to adhere to each other may be different from what would have existed, were the nonpolar groups the only factors that affected the structure of the medium.

From the theoretical point of view, the most direct method of studying this problem should be to investigate the free energy of the system of N water molecules and two nonpolar solutes. If the macroscopic parameters are V, N, and T, the appropriate thermodynamic quantity is the Helmholtz free energy, for which an exact expression can be written. In the system under discussion, depicted in Figure 9c, let the two simple solute particles be at fixed positions $\mathbf{R}_1$ and $\mathbf{R}_2$, then

$$(78) \quad \exp\left[-\beta A(\mathbf{R}_1, \mathbf{R}_2)\right] = \frac{q_w^N}{(8\pi^2)^N N! \Lambda_w^{3N}} \int \cdots \int \exp\left[-\beta U_N(\mathbf{R}^N, \mathbf{\Omega}^N)\right.$$
$$\left. - \beta U_{12}(\mathbf{R}_1, \mathbf{R}_2) - \beta U(\mathbf{R}^N, \mathbf{\Omega}^N/\mathbf{R}_1, \mathbf{R}_2)\right] d\mathbf{R}^N d\mathbf{\Omega}^N$$

where $U_N(\mathbf{R}^N, \mathbf{\Omega}^N)$ includes all the intermolecular potential interactions among the N water molecules, $U_{12}(\mathbf{R}_1, \mathbf{R}_2)$ is the direct pair potential for the two solute particles (assuming orientational independence), and $U(\mathbf{R}^N, \mathbf{\Omega}^N/\mathbf{R}_1, \mathbf{R}_2)$ comprises the interaction between the water and the solute molecules. Integration is carried out over all possible configurations of the solvent particles. The function $A(\mathbf{R}_1, \mathbf{R}_2)$ is in fact a function of the distance $R = |\mathbf{R}_2 - \mathbf{R}_1|$.

The central question is to see if this function has some particular properties originating in the unique structure of water as a solvent. The connection with the solvent properties can be demonstrated more explicitly, taking the gradient of A with respect to say, $\mathbf{R}_1$, we get

$$(79) \quad -\beta \, \nabla_1 A(\mathbf{R}_1, \mathbf{R}_2) =$$
$$\frac{-\beta \int \cdots \int \exp\left[-\beta U_N - \beta U_{12} - \beta U(N/1, 2)\right] \nabla_1 \left[U_{12} + U(N/1, 2)\right] d\mathbf{R}^N d\mathbf{\Omega}^N}{\int \cdots \int \exp\left[-\beta U_N - \beta U_{12} - \beta U(N/1, 2)\right] d\mathbf{R}^N d\mathbf{\Omega}^N}$$

where an obvious shorthand notation has been used. Since

$$(80) \qquad U(N/1, 2) = \sum_{i=3}^{N+2} [U_{i1}(\mathbf{R}_i, \mathbf{\Omega}_i, \mathbf{R}_1) + U_{i2}(\mathbf{R}_i, \mathbf{\Omega}_i, \mathbf{R}_2)]$$

we get from (79)

$$
\begin{aligned}
-\nabla_1 A(\mathbf{R}_1, \mathbf{R}_2) &= -\nabla_1 U_{12}(\mathbf{R}_1, \mathbf{R}_2) \\
&\quad - \frac{\sum_{i=3}^{N+2} \int \cdots \int \exp\{-\beta[U_N + U(N/1, 2)]\} \nabla_1 U_{i1} \, d\mathbf{R}^N \, d\mathbf{\Omega}^N}{\int \cdots \int \exp[-\beta(U_N + U(N/1, 2))] \, d\mathbf{R}^N \, d\mathbf{\Omega}^N} \\
&= -\nabla_1 U_{12}(\mathbf{R}_1, \mathbf{R}_2) - \langle \nabla_1 U_{w1} \rangle \\
&= -\nabla_1 U_{12}(\mathbf{R}_1, \mathbf{R}_2) - \int [\nabla_1 U_{w1}] \rho \, (\mathbf{R}_w, \mathbf{\Omega}_w / \mathbf{R}_1, \mathbf{R}_2) \, d\mathbf{R}_w \, d\mathbf{\Omega}_w
\end{aligned}
$$

(81)

The first form of the right-hand side of (81) utilizes the definition of $U(N/1, 2)$ in (80). The second form shows that the quantity $-\nabla_1 A(\mathbf{R}_1, \mathbf{R}_2)$ is made up of two terms, the direct force, $-\nabla U_{12}$, exerted on particle 1 by the second particle at $\mathbf{R}_2$, and an average force exerted by the solvent. The third form of (81) utilizes the quantity $\rho(\mathbf{R}_w, \mathbf{\Omega}_w / \mathbf{R}_1, \mathbf{R}_2)$, which is the conditional probability density for finding a water molecule at $\mathbf{R}_w$ with configuration $\mathbf{\Omega}_w$, given two solute particles at fixed positions $\mathbf{R}_1$ and $\mathbf{R}_2$.

We are now in a position to examine the following question relevant to the problem of the hydrophobic interaction. Having two solvents, water and some other typical simple liquid, the quantity $\nabla_1 A(\mathbf{R}_1, \mathbf{R}_2)$ can be written for the same two solute particles in the two solvents. The first term on the right-hand side of (81) is the same in both systems and cannot lead to any differences between the solvents (which is true within the assumption made in (78) for the potential energy of the system). The second term will differ between water and other solvents. In particular, it is expected that in water the average interaction with the solvent will tend to push the two particles closer to each other, so that there should be some range of distances in which the function $A(\mathbf{R}_1, \mathbf{R}_2)$ has a particularly low minimum (compared with other solvents). A study along these lines may reveal the optimal distance for hydrophobic interaction. A detailed investigation of the function $A(R)$ in a complicated liquid such as water does not seem to be feasible at present. The difficulties in pursuing an investigation along these lines have been outlined recently.[67] An important contribution is a recent work by Kozak, Knight, and Kauzmann,[68] in which the second "virial coefficient" of the osmotic pressure has been estimated from various sources. If ρ_s is the density of the solute, the osmotic pressure can be expanded in a power series of ρ_s.

$$(82) \qquad \frac{\pi}{kT} = \rho_s + B_2^* \rho_s^2 + B_3^* \rho_s^3 + \cdots$$

where B_k^* is the analog of the kth virial coefficient in the gaseous phase. Using the McMillan–Mayer[69] theory of solution, B_2^* can be written as

$$(83) \qquad B_2^* = -(2V)^{-1} \iint \{\exp[-\beta W^{(2)}(\mathbf{R}_1, \mathbf{R}_2)] - 1\} d\mathbf{R}_1 \, d\mathbf{R}_2$$

where $W^{(2)}$ is the potential of average force[70] for the two particles at $\mathbf{R}_1$ and $\mathbf{R}_2$ immersed in a solvent. We see that B_2^*, which can be determined experimentally by (82), is related through (83) to the two-particles-in-a-solvent problem. In fact, $W^{(2)}(\mathbf{R}_1, \mathbf{R}_2)$ is identical with the quantity $A(\mathbf{R}_1, \mathbf{R}_2)$ up to a quantity whose gradient is 0; that is, we have

$$(84) \qquad \nabla_1 A(\mathbf{R}_1, \mathbf{R}_2) = \nabla_1 W^{(2)}(\mathbf{R}_1, \mathbf{R}_2)$$

The determination of B_2^* does not provide any detailed information on the function $W^{(2)}(\mathbf{R}_1, \mathbf{R}_2)$, but since it gives some measure of the overall behavior of this function, it may contain some useful information on the hydrophobic interaction.

Finally, a further reduction of the problem is suggested which may facilitate the theoretical investigation. In discussing the properties of infinitely dilute solutions of gases in water, we have mentioned the possibility of anomalous properties being revealed for hard-sphere solutes. This is the equivalent of a consideration of the thermodynamics of cavity formation, or in other words, the problem of cavity-solvent interaction. If our first reduction of the present problem from two side-chain groups hung on a skeleton of a polymer to two-solute particles in a solvent proves successful, we may try to investigate two hard particles in a solvent, or equivalently, two cavities in the liquid. The problem will now be to examine the cavity-cavity interaction in water compared with other liquids. In an attempt toward such an investigation,[71] the distribution of pairs of cavities as a function of their distance had been calculated for water and for liquid argon. However, owing to lack of sufficient experimental information, the calculations have been confined to cavities of radius $\sigma_w/2$ where σ_w is the effective "hard-sphere" diameter of a water molecule. As noted in Section 3, this cavity will correspond to a "hard-sphere" solute of radius 0, and may not therefore be a good representation of a real solute. Nevertheless, it may be possible to use the scaled-particle theory to calculate the distribution of pairs of cavities of larger radius. This may be the most elementary test for the existence of hydrophobic interaction in water.

The present discussion certainly offers no simple solution of the problem. It is very likely that the driving forces bringing the two particles close to each other originate from the special interrelations between water and solute and solute and solute. More specifically, the two solute particles at large distances produce independent structural changes; that is, the spheres of influence of the two solutes on the solvents do not overlap. When these particles come close to each other, their combined effect on the structure of water may be different from the sum of their individual effects, possibly providing a significant difference in the energy and the entropy, highly specific to the aqueous environment, for the two states of the pair of solutes.

Therefore, the investigation of the driving forces for hydrophobic interaction is crucially dependent upon our understanding of the local structure around a single solute. This in turn will depend on our knowledge of the molecular structure of pure liquid water.

Acknowledgments

The author is grateful to S. Baer, F. Franks, G. Hertz, and to Y. Marcus for reading the manuscript and kindly offering helpful remarks and suggestions.

REFERENCES

1. R. Battino and H. L. Clever, *Chem. Rev.*, **66**, 395 (1966).
2. D. D. Eley, *Trans. Faraday Soc.*, **35**, 1281, 1421 (1939).
3. H. S. Frank and M. W. Evans, *J. Chem. Phys.*, **13**, 507 (1945).
4. A. Ben-Naim, *J. Phys. Chem.*, **69**, 3245 (1965).
5. A. Ben-Naim, *J. Chem. Phys.*, **42**, 1512 (1965).
6. A. Lannung, *J. Amer. Chem. Soc.*, **52**, 68 (1930).
7. A. Ben-Naim, *J. Phys. Chem.*, **71**, 4002 (1967).
8. D. M. Himmelblau, *J. Phys. Chem.*, **63**, 1803 (1959).
9. D. M. Alexander, *J. Phys. Chem.*, **63**, 994 (1959).
10. J. C. Gjaldbaek and H. Niemann, *Acta Chem. Scand.*, **12**, 1015 (1958).
11. L. Friend and S. B. Adler, *Chem. Eng. Progr.*, **53**, 452 (1957).
12. D. M. Himmelblau, *J. Chem. Eng. Data*, **5**, 10 (1960).
13. A. Eucken and G. Herzberg, *Z. Physik. Chem. (Leipzig)*, **195**, 1 (1950).
14. T. J. Morrison and F. Billett, *J. Chem. Soc.*, 3819 (1952).
15. T. J. Morrison and N. B. B. Johnstone, *J. Chem. Soc.*, 3655 (1955).
16. F. A. Long and W. F. McDevit, *Chem. Rev.*, **51**, 119 (1952).
17. A. Ben-Naim and M. Egel-Thal, *J. Phys. Chem.*, **69**, 3250 (1965).
18. S. A. Shchukarev and T. A. Tolmacheva, *Zh. Strukt. Khim.*, **9**, 21 (1968).
19. A. Ben-Naim and S. Baer, *Trans. Faraday Soc.*, **60**, 1736 (1964).
20. A. Ben-Naim and G. Moran, *Trans. Faraday Soc.*, **61**, 821 (1965).
21. A. Ben-Naim, *J. Phys. Chem.*, **72**, 2998 (1968).
22. F. Franks and D. J. G. Ives, *Quart. Rev. (London)*, **20**, 1 (1966).
23. A. Wishnia, *Proc. Natl. Acad. Sci. (U.S.)*, **48**, 2200 (1962).
24. A. Wishnia, *J. Phys. Chem.*, **67**, 2079 (1963).
25. A. Wishnia and T. Pinder, *Biochemistry*, **3**, 1377 (1964).
26. D. B. Wetlaufer and R. Lovrien, *J. Biol. Chem.*, **239**, 596 (1964).
27. A. Ben-Naim and F. H. Stillinger, Jr., Chapter 8 of this volume.
28. H. Goldstein, *Classical Mechanics*, Addison-Wesley, Reading, Mass., 1959, p. 107.
29. T. L. Hill, *Statistical Mechanics*, McGraw-Hill, New York, 1956, p. 180.

30. H. H. Uhlig, *J. Phys. Chem.*, **41**, 1215 (1937).
31. R. A. Pierotti, *J. Phys. Chem.*, **67**, 1840 (1963).
32. R. A. Pierotti, *J. Phys. Chem.*, **69**, 281 (1965).
33. J. H. Saylor and R. J. Battino, *J. Phys. Chem.*, **62**, 1334 (1958).
34. Ref. 1, p. 415.
35. H. Reiss, H. L. Frisch, and J. L. Lebowitz, *J. Chem. Phys.*, **31**, 369 (1959).
36. H. Reiss, H. L. Frisch, E. Helfand, and J. L. Lebowitz, *J. Chem. Phys.*, **32**, 119 (1960).
37. H. Reiss, *Advan. Chem. Phys.*, **9**, 1 (1966).
38. A. Ben-Naim and H. L. Friedman, *J. Phys. Chem.*, **71**, 448 (1967).
39. S. W. Mayer, *J. Chem. Phys.*, **38**, 1803 (1963).
40. S. W. Mayer, *J. Phys. Chem.*, **67**, 2160 (1963).
41. R. A. Pierotti, *J. Phys. Chem.*, **71**, 2366 (1967).
42. (a) M. V. Stackelberg, *Naturwiss.*, **36**, 327 (1949); (b) M. V. Stackelberg and H. R. Muller, *Z. Elektrochem.*, **58**, 25 (1954); (c) M. V. Stackelberg and B. Meuthen, *Z. Elektrochem.*, **62**, 130 (1958).
43. W. F. Claussen, *J. Chem. Phys.*, **19**, 259, 662, 1425 (1951).
44. D. N. Glew, *J. Phys. Chem.*, **66**, 605 (1962).
45. A. Yu. Namiot, *Zh. Strukt. Khim.*, **2**, 408 (1961).
46. A. Ben-Naim, to be published.
47. R. Lumry and S. Rajender, *Biopolymers* **9**, 1125 (1970).
48. A. Ben-Naim, *J. Chem. Phys.*, **45**, 2706 (1966).
49. A. Ben-Naim, *J. Phys. Chem.*, **69**, 1922 (1965).
50. A. Ben-Naim, *J. Phys. Chem.*, **69**, 3240 (1965).
51. See Appendix I of Ref. 49.
52. A. Ben-Naim and R. A. Lovett, unpublished work on a one-dimensional model for aqueous solutions of gases, and *J. Chem. Phys.*, **51**, 3108 (1969).
53. V. M. Vdovenko, Yu. V. Gurikov, and E. K. Legin, *Dokl. Akad. Nauk SSSR*, **172**, 126 (1967).
54. L. Pauling, *The Nature of the Chemical Bond*, 3rd ed., Cornell University Press, Ithaca, N.Y., 1960, p. 469.
55. O. Ya. Samoilov, *Structure of Aqueous Electrolyte Solutions*, transl. D. J. G. Ives, Consultants Bureau, New York, 1965.
56. H. S. Frank and A. S. Quist, *J. Chem. Phys.*, **34**, 604 (1961).
57. V. A. Mikhailov and L. I. Ponomareva, *Zh. Strukt. Khim.*, **9**, 12 (1968).
58. V. A. Mikhailov, *Zh. Strukt. Khim.*, **8**, 189 (1967).
59. H. S. Frank and F. Franks, *J. Chem. Phys.*, **48**, 4746 (1968).
60. G. Némethy and H. A. Scheraga, *J. Chem. Phys.*, **36**, 3382, 3401 (1962).
61. A. Ben-Naim, *Trans. Faraday Soc.*, **66**, 2749 (1970).
62. W. Kauzmann, *Advan. Protein Chem.*, **14**, 1 (1959).
63. G. Némethy, *Angew. Chem.*, **6**, 195 (1967).
64. G. Némethy and H. A. Scheraga, *J. Phys. Chem.*, **66**, 1773 (1962).
65. E. E. Schrier, M. Pottle, and H. A. Scheraga, *J. Amer. Chem. Soc.*, **86**, 3444 (1964).
66. D. K. Kunimitsu, A. Y. Woody, E. R. Stimson, and H. A. Scheraga, *J. Phys. Chem.*, **72**, 856 (1968).
67. A. Ben-Naim, *J. Chem. Phys.*, **54**, 1387 (1971).
68. J. J. Kozak, W. S. Knight, and W. Kauzmann, *J. Chem. Phys.*, **48**, 675 (1968).
69. W. McMillan and J. Mayer, *J. Chem. Phys.*, **13**, 176 (1945).
70. Ref. 29, p. 193.
71. A. Ben-Naim, *J. Chem. Phys.*, **50**, 404 (1969).

I 2　Thermodynamics of Solutions of Electrolytes

Fred Vaslow, Chemistry Division, Oak Ridge
National Laboratory, Oak Ridge, Tennessee*

1　Introduction

For many years discussion of the thermodynamics of electrolyte solutions has been dominated by considerations of the Debye–Hückel theory and the resultant ionic-strength principles and square-root laws for the excess free energy and its thermodynamic derivatives.[1] Conclusive proof of the validity of the theory on both the experimental and theoretical levels has long been available, and the theory can now be considered as a fundamental law of nature, with further studies having a rounding out rather than an investigative character.

The problem of electrolyte solutions remaining is that the theory is necessarily valid only in the limit of 0 concentration.[2] The validity of some of the assumptions becomes uncertain at finite concentrations and also the theory treats only Coulombic-force-related deviations from the laws of ideal solutions and neglects all other deviations.

Experimentally, except for a few probably fortuitous cases, the limiting law for the excess free energy or derivative is approached rather than reached at any finite concentration, and the deviations from the law must be ascribed either to defects of the basic assumptions of the theory at higher concentrations or to the non-Coulombic interactions of the system. Determining which of these two factors causes the deviations or the degree due to each is a problem that has not yet been solved. The recent theoretical treatments based on Mayer's work[3] have essentially bypassed the problem, since in a rigorous statistical-mechanical treatment the Coulombic and other effects are treated on an equal level and separation is possible only in the limit of 0 concentration.

In analyzing data in the classical way, plausible but often different forms (involving an arbitrary variable or fixed distance parameter) are chosen for

* Operated by Union Carbide Corporation for the U.S. Atomic Energy Commission.

465

the Debye–Hückel component of the excess[4] (nonideal) free energy of the solution, and the difference between this form and the experimental data is taken as being due to one or more non-Coulombic effects. The non-Coulombic effects such as specific interaction of the ions or hydration are physically entirely reasonable and, in fact, are necessary components of any rigorous theory of electrolytes, although they are not usually explicit. The problem is that the parameters cannot be independently calculated but are determined from the difference between the particular Debye–Hückel form assumed and the experimental data.

Since at finite concentrations the true magnitude of the purely Coulombic (Debye–Hückel) term is unknown, the meaning of the new parameters is necessarily uncertain. As a result, few if any of the parameters of the classical theory have an independent or even definite physical meaning. If more accuracy is needed in fitting experimental data, more parameters are added and, also, temperature and pressure derivatives of the excess free energy can be treated with still further additions of parameters.

From the point of view of a rigorous understanding or explanation of the solution properties, the methods are relatively unsatisfactory. Nevertheless, until the recent developments of the Mayer methods, there was no satisfactory theoretical alternative. From the purely experimental point of view, parameters including the distance of closest approach or the hydration number appeared plausible; such parameters often behaved consistently and provided a number of very valuable systematizations of the solution properties. The newer rigorous methods of treatment are very complicated mathematically, and thus far only a few results have appeared. Hence it is probable that the older methods will continue in use until the new results are more widely understood and more results are available.

The spatial and angular interrelations of molecules and the relative stabilities of particular configurations of molecules are of fundamental importance in understanding almost every branch of chemistry. It seems somewhat strange, therefore, that with a few notable exceptions,[5-7] until recently there has been little concern with the structural concepts in treatments of electrolyte solutions. Particularly this is true in light of the work of Bernal and Fowler[8] pointing out the highly structural nature of water.

It would appear obvious that many of the thermodynamic properties of solutions are intimately related to structural effects, although it is true that the specifically structural effects are often small, difficult to calculate, and in the classical treatment unnecessary, since the effects and parameters are usually sufficiently elastic to cover any specifically structural effect. However, the treatment of entropy effects in solution and the more accurate theoretical calculations necessarily must consider structural effects.

The discussion that follows essentially amplifies the topics of the introduction. The basic assumptions and the nature of the uncertainties of the Debye–Hückel theory are considered. Several of the more important effects used to augment the Debye–Hückel theory are discussed, stressing the limitations and the necessity of going over to the statistical-mechanical theories.

An essentially qualitative and descriptive account of the statistical mechanical theory of electrolyte solutions is given, primarily based on Mayer's methods. Some aspects of the thermodynamic properties of solutions of mixed electrolytes are discussed, primarily based on Friedman's work. A discussion of structural aspects of electrolyte solutions is given, showing the relation of the concept of structure as used in solution chemistry to the more orthodox use in crystals and inorganic chemistry. Finally, a short description is given of recent work on solute-induced cooperative changes in solution structure.

2   Debye–Hückel Theory

The basic features of the Debye–Hückel theory have been adequately discussed many times[9] and are not covered here. However, some discussion of the nature of the basic assumptions of the theory would appear to be of interest. As pointed out by Fowler and Guggenheim,[10] the fundamental assumption of the theory is the setting of an average electrostatic potential, $\phi(r)$, in solution at a point, r, fixed with relation to a particular ion, equal to a certain potentiallike quantity, $W(r)$ (potential of average force). The average electrostatic potential, $\phi(r)$, is equal to $\sum Z_i e_i / r_i D$ (Coulomb's law) where e_i is the electronic charge, Z_i the valence of each ion, r_i the distance from any ion to the point of concern, and D the macroscopic dielectric constant of the solvent. The sum is over all ions in solutions and averaged over all the positions the ions can enter. For the moment, the potential of average force, $W(r)$, can simply be taken as the quantity that, placed in Boltzmann's equation—$e^{-W(r)/kT}$—determines the average number of ions at a point fixed with relation to a particular ion, as in the potential.[11]

Returning to ϕ, in the Debye–Hückel theory, the electrostatic potential is calculated from Poisson's partial differential equation, which is equivalent to Coulomb's law

$$(1) \qquad \nabla^2 \phi = \frac{\partial^2 \phi}{\partial x^2} \frac{\partial^2 \phi}{\partial y^2} \frac{\partial^2 \phi}{\partial z^2} = \frac{-4\pi \rho(r)}{D}$$

where ρ is the density of charge at a point and x, y, and z are Cartesian coordinates. It is characteristic of this equation that the charge and potential are linearly related so that multiplying the charge by a given factor multiplies the potential by the same factor. It is important to realize that, if the positions of the charges are fluctuating, then a solution of the equation using the average charge density correctly gives the average potential.

The Boltzmann equation

(2)
$$\frac{\rho(r)}{\rho_0} = e^{-W(r)/kT}$$

in a special form is also a relation between a potential, W, and the charge density, ρ, where ρ_0 is the overall average number of ions at any point in solution.

Although several definitions of W are possible,[2] a satisfactory one is simply that defined by Boltzmann's equation. It is the function that, put into the Boltzmann expression, gives the average number of ions at a fixed point near to a central ion, relative to the macroscopic average number of ions anywhere in the solution. In effect, W represents a potential energylike quantity for two ions when they are held fixed and all other ions are allowed to freely move and their effects averaged.

The relation between charge and potential in the Boltzmann equation is not linear as in the Poisson equation, and as a result the average electrostatic potential found from the Poisson equation or from averaging Coulomb's law, $\phi = e_i/r_i D$, is not equal to W.[10] Fluctuations in W result in exponential movements of the ions rather than linear movements as in ϕ.

If W is sufficiently small that the Boltzmann expression can be represented by a linear expansion

$$e^{-W/kT} \rightarrow 1 - \frac{W}{kT}$$

then this is again a linear relation between charge and potential, and ϕ can be used in place of W to calculate the average charge distribution. Kirkwood and Poirier[2] rigorously showed that in the limit of 0 concentration these two quantities, $Z_i e_i \phi$ and W, were equal. The linearized Boltzmann expression is substituted into Poisson equation, giving the well known Debye–Hückel expression, $\nabla^2 \phi = \kappa^2 \phi$ where κ is the Debye–Hückel reciprocal length and

$$\kappa^2 = \frac{4\pi \sum\limits_i N_i Z_i^2 |e_i|^2}{V D k T}$$

N_i is the number of ions in solution, Z_i the valence of the ion, and V the volume of the solution.

It is not the expansion in series of the Boltzmann exponential and the truncation of the series that leads to the uncertainties of the Debye–Hückel theory. Several treatments that consider the full exponential are discussed later. It is rather the basic difference between the electrostatic potential energy, $Ze_i\phi$, and the potential of average force, W, that is responsible for the eventual failure of the theory. Kirkwood[12] has carefully examined and identified this difference as a term due to microscopic fluctuation in concentration (fluctuation term); however, it has not been possible to evaluate the term quantitatively. As a result there is an inherent and unknown uncertainty in calculations of the excess free energy.

A thermodynamic consequence of the difference between ϕ and W is a symmetry condition for the electrostatic potential ϕ_α in the neighborhood of an α ion. It is shown thermodynamically that[10]

$$(3) \qquad \frac{\partial \phi_\alpha}{\partial Z_\beta} = \frac{\partial \phi_\beta}{\partial Z_\alpha}$$

where Z_α is the charge of an α ion. Solutions for ϕ_z, where the Boltzmann expression is not linearized, do not fit this condition and are thermodynamically inconsistent.

An extension of the Debye–Hückel limiting law usually considered as part of the basic theory is the use of a distance of closest approach parameter. Such a parameter is physically necessary and in principle should be unobjectionable. The basic objection is that the limiting law to this point has not required any arbitrary numbers or parameters but only fundamental constants and simple measurable properties of water. The extension now introduces a parameter that turns out to be vague and unmeasurable in use. The parameter is of importance when the two ions are close together, but it is just under this condition that the linear expansion of the Boltzmann equation is questionable and also where another important approximation becomes doubtful. This is the assumption of a constant dielectric constant, which is almost certainly invalid for the very close approach of two ions.

Estimates of the concentration where the Debye–Hückel approximation becomes seriously wrong range from 10^{-3} to 1 N, and it is doubtful that a conclusive answer can be given.[13–15] Kirkwood[12] has indicated that the difference between $Z_i e_i \phi$ and W might be about the same order of magnitude as the difference between the linear and exponential form of Boltzmann's equation.

There is thus a great deal of uncertainty in the meaning of the Debye–Hückel theory at higher concentrations and, as mentioned in the introduction, there are other non-Coulombic effects to be considered. Consequently, although the utility and importance of the empirical approaches must be emphasized, it appears that many large and fundamental uncertainties cannot be resolved with these older methods.

3   Non-Debye–Hückel Electrical Treatments

Introduction.   Recognizing the difficulties of extending the Debye–Hückel theory to higher concentrations, there have been a number of attempts to obtain theoretically more justifiable treatments of the electrostatic component of the excess free energy. An early suggestion was to modify the Debye–Hückel equation to allow for the changing of the dielectric constant of the solution by the solute.[16] Another method actually antedates the Debye–Hückel theory and considers that above a certain concentration the ions form a quasi-lattice in solution which gives a $c^{\frac{1}{3}}$ rather than $c^{\frac{1}{2}}$ relationship for $\log \gamma_\pm$ ($\gamma_\pm$ = mean molar activity coefficient).[13,17–20] A third method solves the exponential Poisson–Boltzmann equation, implicitly neglecting the thermodynamic inconsistency (3) and the fluctuation effect.[21–24] The fourth method bypasses the fluctuation and linearization problems by considering two ions with a sufficiently strong electrostatic potential as being paired and not contributing electrically to the rest of the system, which follows the Debye–Hückel theory.[25–27] A few details and criticisms of the four methods follow.

Dielectric Constant Variations.   That changes in the dielectric constant should affect the results of Debye–Hückel calculations was pointed out by Hückel[16] in 1925, when he suggested a correction to $\log \gamma_\pm$ proportional to C. The problem can be considered in detail by examining the expressions for the electrostatic component of the excess free energy. The electrostatic excess free energy consists of two terms; one a self-energy term relating to charging the ions in the solvent at infinite dilution, and a second term giving the effect of the interaction of the ions at finite concentrations. The self-energy term has the form[10]

$$(4) \qquad SE = \sum_i \frac{N_i Z_i^2 e_i^2}{a_i D}$$

This quantity is taken relative to the discharged state of the ions.

The interaction part has the form

$$(5) \qquad IE = -\sum_i \frac{N_i Z_i^2 e_i^2 \kappa \tau(\kappa a_i)}{3D}$$

Where the quantity a_i is either the diameter of the ion or the distance of closest approach for an ion pair, and they are assumed the same here. $\tau(\kappa a)$ is a function usually listed in textbooks on electrolyte solutions; it is unity at low concentrations and slowly decreases with concentration. The other quantities were defined earlier.

If D does not change on changing concentration, then the self-energy is part of the standard-state free energy and can be neglected. If D does change, then the self-energy term contributes to the excess free energy and the contribution is opposite in sign to that of the interaction term. The effect of a changed dielectric constant then depends on which of the two terms is larger.

The problem is a difficult one since, first of all, the D's in each quantity are fundamentally different. The dielectric constant associated with the self-energy depends mostly on the properties of water molecules close to the ion and strongly perturbed. The D of the interaction term, for moderate concentrations at least, depends mainly on water molecules far away from the ions, which have normal properties. Also, the proper values of D must be obtained as averages from an integration when the ions are charged from 0 to their final charge.[10]

What can be done for a crude estimate is to neglect the charging process and consider that water molecules adjacent to an ion are unaffected by concentration changes. The changes in self-energy are due only to changes in the dielectric constant far from the ion, and these changes are the same as those seen in the interaction term. To the extent that these considerations are valid and for a solute that decreases the dielectric constant, at low concentrations the self-energy term is the larger and the excess free energy is increased. At high concentrations, the interaction term is the larger, and the excess free energy is decreased with a crossover in the neighborhood of $\kappa a = 3$.

Since much of the self-energy term involves water molecules near to the ion that are assumed to be relatively little perturbed by concentration changes, it is possible that the crossover could occur at much lower concentrations. If so, even at moderate concentrations the effect of a lowered dielectric constant could be one of lowering rather than raising the excess free energy.

In actual fact, so little is known about the ion-solvent interactions that any correction involving the dielectric constant is purely empirical. For high concentrations of ions where the cospheres[7] of the ions are on the average interpenetrating, the concepts of self- and interaction energies as well as dielectric constant lose their meaning. The close approach of two ions necessarily disturbs the self-energy of the ions so that the energy of interaction is not separable from the self-energy. The dielectric constant, which is normally a fixed quantity relating the energy and distance of two charges, now is a complex function of the distance and the structures of the ion cospheres.

In solvents of low dielectric constant and in mixed solvents, both at low salt concentrations, the dielectric constant again has an unambiguous meaning as the macroscopic dielectric constant of the solvent. However, for higher concentrations in mixed solvents, complicated problems of salting out or salting in of solvent components occur, which are not discussed here.[28] Large ions that appreciably decrease the average dielectric constant in their neighborhood can also be thought of as being salted out by other smaller ions. The energetics of the process would probably be best treated by the general methods for the close approach of two ions in water extensively developed by Levine and co-workers.[29]

Quasi-Lattice Treatment.   Before the $c^{1/2}$ behavior of activity coefficients at the lowest concentrations was empirically discovered or the Debye–Hückel theory developed, a $c^{1/3}$ behavior was considered as well representing the experimental data.[17] This is the energy behavior of a crystal lattice and might be considered as evidence of a quasi-lattice in solution. It is still true that for many salt solutions,[8] between about 0.001 and possibly 0.1 N, $\log \gamma_\pm$ is closely proportional to $c^{1/3}$, and some apparent molal heat content curves[30] are also well represented by a $c^{1/3}$ law. In this range of concentrations there is no good empirical reason for preferring the Debye–Hückel expression for $\log \gamma_\pm$ and, if additional terms such as hydration or virial coefficient type terms are added to the cube-root term, very satisfactory data fits are obtained up to relatively high concentrations.[19,20]

Theoretically, there is no definitive answer to whether the Debye–Hückel or the quasi-lattice theory is preferable at higher concentrations; however, some plausible arguments have been made for a quasi-lattice. On qualitative and general grounds it could be argued that the fluctuation term of the electrostatic theory becomes large enough to destroy the validity of the Debye–Hückel theory at 0.01 N[14] or even 0.001 N.[13] In very dilute solutions, the ionic atmospheres cause only slight perturbations of the motions and average energies of the ions. As in an ideal gas or solution these motions are random, and the ions move very nearly independently of one another. As the concentration increases, the number of strong ionic interactions (energy $\gg kT$) increases, and at moderate concentrations each ion is almost constantly exposed to these strong forces from its neighbors. Under these conditions it is at least plausible that some degree of ordering of the ions occurs in order to minimize mutual interference.

More quantitatively, Kirkwood[31] and others[2,32-34] have found in their statistical-mechanical treatments of ionic solutions that above $\kappa a = 1.03$ (Kirkwood), the radial distribution function for oppositely charged ions (i.e., average density of opposite ions at a given distance from a central ion) goes from an exponentially decaying to an oscillating form. The exponentially decaying form is characteristic of the Debye–Hückel theory, and the oscillating form is that of a quasi-lattice.

Many approximations are made in the calculations, and it is not clear either that such a transition does actually occur or, if it does, at what concentration. It is also possible that long-range structure correlations in solution would bring the transition at a much lower concentration. The many different types of calculations that have found this transition suggests that it be taken seriously on an experimental level, although the experimental nature the effect might have is unclear.

Monte Carlo calculations[35] of the radial distribution functions have found oscillations in a 0.001-N solution, although only a very small number of ions (32 cations and 32 anions) were used in the calculation and the result may be an artifact. Other Monte Carlo calculations[35a,35b] did not show oscillations at low concentrations.

Thus the quasi-lattice theories, which are valid only at higher concentrations, have the same advantages and defects as the Debye–Hückel theory at comparable concentrations. Good experimental fits may be obtained, and the ideas are physically plausible, but there is no definitive evidence of their validity, and the parameters tend to be empirical.

Nonlinearized Poisson–Boltzmann Equation. If it is tacitly assumed that the fluctuation term[12] of the potential of average force is small and the thermodynamic inconsistency neglected, then the nonlinearized Poisson–Boltzmann equation,[10] that is

(6)
$$\nabla^2\phi = -\frac{4\pi}{DV}\sum N_i Z_i e_i e^{-Z_i e_i \phi / kT}$$

may be solved to obtain the excess free energy of the system. As in the linear Debye–Hückel equation, an arbitrary distance parameter is used in the solution. The magnitude of the fluctuation term is unknown, and if it is small, then solutions of the equations could give reasonably accurate values of the excess free energy.

A number of different analytical or numerical methods have been used to solve this equation,[21–24] and mathematically they are all somewhat cumbersome. Since the distance parameter is arbitrary and since non-Coulombic effects are neglected, it is difficult to know what the validity of the results is; however, in several respects the results appear better than those from the linear form.

A first improvement is that, whereas the linear form of the Debye–Hückel equation often required inconsistent or negative values for the distance parameter, the nonlinear form uses more reasonable and more consistent values of the parameter. A second related phenomenon concerns the experimental log $\gamma_\pm$ curves for 2:2 electrolytes, which often fall below the Debye–Hückel limiting law. The nonlinear equation has shown such deviations without the use of negative distance parameters or the arbitrary assumption of ion pairs.[22]

Ion-Pair Treatments. The last method of treating the Coulombic interactions was originally developed by Bjerrum,[25] with a number of later modifications. In essence, very strong Coulombic interactions, such as between two small ions at short distances or larger ions in a medium of low dielectric constant, are treated by considering the ions as being associated. The degree of association, α, is given by

$$(7) \qquad \alpha = \frac{N_\pm}{V} \int_a^q e^{-Z_i Z_j e_i e_j / DkTr} 4\pi r^2 \, dr$$

where $N_\pm$ is the number of positive or negative ions in the volume, V. The exponential part of the integrand is the Boltzmann expression for the average density of type-i ions at a distance r from the central j ion, and the total integrand is the average number of ions in the spherical shell of thickness, dr. The parameter $q = Z_i Z_j e_i e_j / 2DkT$ is the distance within which ions are considered associated and a the distance of closest approach; and q is also the distance at which the number of ions in any spherical shell reaches a minimum. A pair of ions within the distance q is considered uncharged, and those outside such pairs obey the normal Debye–Hückel equation.

The treatment uses the complete exponential Boltzmann factor with its much stronger influence on ions rather than the linear term of the expansion, as in the Debye–Hückel approximation. At the close distances where the treatment is applied, the potential of average force between the two ions is strongly dominated by the potential of the central ion alone. In effect, the fluctuation terms from the other ions are minimal, and the thermodynamic inconsistency of the exponential Poisson–Boltzmann equation is avoided. The results obtained with the ion-pair theory are generally similar to those obtained with the exponential Poisson–Boltzmann equation, but the treatment has been of particular value in treating salts in media of low dielectric constant.

As in each of these semiempirical treatments, there are unknown quantities and arbitrary constants.[35] Here the principal new arbitrary constant is q, but apparently the results obtained do not depend strongly on its choice. Other treatments consider only contact ion pairs,[27,36] but basically the concept and results are similar.

4   Non-Coulombic Interactions

Introduction. Using the various treatments of electrostatic interaction mentioned previously, experimental activity or osmotic coefficient data can be fitted only to a very limited extent, the extent depending of course on the complexity and number of disposable parameters used. It is necessary, therefore, to consider such other effects as those that might arise in solutions of nonelectrolytes in water.[1] These nonelectrolyte effects can be quite large, although at very low concentrations (i.e., < 0.01 N) a simple behavior such as Henry's law is observed. This is in contrast to electrolyte solutions, where the Debye–Hückel limiting law (which is in itself a deviation from Henry's law) is only approached and is not reached at the lowest experimentally accessible concentrations.

The strong electric fields of ions interacting with the hydrogen-bonded framework of water will tend to accentuate these non-Coulombic effects to such a degree that the interionic and nonionic effects cannot really be separated except at the very lowest concentrations, where only the square-root behavior remains.

Classically, the problem has been treated by accepting one of the electrostatic treatments as valid at the concentration of interest and considering that the difference between this and the experimental data is due to one or more of the noninterionic effects. Two such effects or treatments are considered here. One treats the non-Coulombic interaction of ions in a series expansion in half or integral powers of the individual ion concentrations analogous to the virial expansion of imperfect gases. The second method attempts to allow for the very strong interactions of ions and solvent or, in other words, hydration. These two are probably the most important and general treatments, and others such as salting out or excluded volume effects can be related to these.

Virial Coefficient Expansions for Electrolyte Solutions.   The justification for virial series expansions for electrolytes lies largely in the Van't Hoff expression for the osmotic pressure.[1] In very dilute solutions $\Pi \overline{V}_1 = X_2 RT$ where Π is the osmotic pressure, $\overline{V}_1$ the molal volume of pure water, and X_2 the mole fraction of solute. For electrolytes, X is replaced by νX where ν is the number of moles of ions per mole of salt. This expression is very analogous to the perfect gas law, $PV = nRT$, and it is therefore tempting to continue the development of the excess osmotic pressure (i.e., the difference between the experimental Π and that given by Van't Hoff's law) in a virial expansion in the same way that imperfect gases are treated. For electrolyte solutions these terms are in addition to the term given by the Debye–Hückel theory.

For the imperfect gas, PV/RT is expanded as a power series in $1/V$ or, equivalently, the gas density ρ in the form

$$(8) \qquad \frac{PV}{RT} = 1 + \frac{B_2}{V} + \frac{B_3}{V^2} + \cdots = 1 + B_2'\rho + B_3'\rho^2 + \cdots$$

The constants B_n are known as the nth virial coefficients. The first term (i.e., 1) represents ideal behavior, and each successive term represents deviations from ideal behavior caused by the simultaneous collisions or interactions of groups of n molecules.

The osmotic pressure of an electrolyte solution can be considered in a very similar way with an ideal term, a Debye–Hückel term, and a series of terms representing the effects of the specific, non-Coulombic interactions of the various groups of ions. The expansion is in terms of powers of the concentration (density) of each type of ion in solution, and there are terms of each power for each of the possible combinations of ion type in solution.[37] Mixtures of salts are very readily treated in this formalism.

The osmotic pressure is, of course, a measure of the partial molal free energy of the solvent. With knowledge of the concentration dependence of π and a suitable extrapolation to 0 concentration, the Gibbs–Duhem equation is used to obtain the partial molal free energy of the solutes and, finally, the total free energy. It is, of course, the excess rather than the ideal part of each of these quantities that is of interest, and these quantities can be obtained in terms of the Debye–Hückel expression plus the virial coefficients and necessary integration constants. By expressing the constants for the total excess free energy as further series functions of the temperature and pressure, very elaborate and complicated algebraic structures can be built up, these can then be differentiated to give excess partial molal free energies, enthalpies, entropies, or volumes for any component of the solution.

From a practical point of view, there can be little question of the utility of these power series expansions in representing experimental data. With the use of electronic computers, the constants can easily be optimized by a least-squares or other procedure, and terms can be added to any power that is statistically justified by the accuracy and detail of the data. The basic question is, Do these terms have anything more than a purely empirical significance in the solution? It should be noted that, in a gas, the importance of the terms is definite, signifying the interaction of a certain number of molecules, although calculation or measurement of the higher coefficients is often very difficult.

The necessary difference between an interaction in a gas and one in a solution should first be noted. In a solution there can be no collision or interaction involving just a pair of solute molecules. Each collision must involve the displacement of several solvent molecules. Since in water there is a substantial degree of positional and angular correlation of the displaced water molecules with nearby water molecules, the collision must involve extensive movements of a great many water molecules. The interaction of the two ions thus involves averaging over the positions of these displaced water molecules, and the energy is not a simple potential energy as in a gas but is a potential of average force. This different nature of the interaction in solution can be recognized and does not involve the validity of the expansion.

A purely mechanical problem is that, if the number of experimental points in a curve relating a thermodynamic property to the concentration is limited, the values of the constants will generally depend on the number of points, their precision, and the number of constants used to represent the work. Assuming that the constants can be accurately evaluated, there is still the fundamental uncertainty of the Coulomb component of the excess free energy. Consequently, an interpretation of the virial coefficient in terms of specific ion interactions must reflect the uncertainty of the Coulomb calculation.

The Mayer[3,38] theory, which is discussed later, provides some definite information about the virial expansion. According to this theory, if the solvent is kept at unit activity (by applying a pressure equal to the osmotic pressure of the solution), then it is possible to express the excess free energy of the solution in a form resembling the virial expansion. The coefficients of this series are complex functions of the ionic strength, however, and so the concept of the virial expansion in the usual sense does not appear to be valid. If a process occurs at constant ionic strength, such as the mixing of two salt solutions, then the coefficients are constant and the theory does suggest that the viriallike treatment is valid.

Hydration.   One of the simplest and most effective ways of augmenting the Debye–Hückel theory has been to use the concept of hydration.[18,25,39] That ions interact very strongly with water is obvious, and that this interaction would perturb the excess free energy behavior at higher concentrations would seem equally obvious. Ions with very strong electric fields would in effect remove water molecules from solution excluding their use as solvent by other ions and, in a sense, would raise the concentration. Another type of hydration effect is based on the fact that the activity coefficient of a salt reflects (in part) the free energy necessary to separate a pair of ions from surrounding water molecules at finite concentrations, relative to that free energy at infinite dilution. It could be expected in the manner of a chemical equilibrium that a lowering of the water activity would make it easier to remove (raise the activity of) the ions from solution.[25]

There is no reason to believe that these effects are not valid, and their validity is reflected in the good fits to experimental data obtained with these methods. As with each of the classical treatments, the problem is that there is no independent way of calculating the parameters. The parameters can only be obtained by fitting the curves, and the hydration parameters are particularly vague in their meaning.

Quantitatively, there have been several treatments of hydration, differing somewhat in their basic concepts. The first quantitative treatment was by Bjerrum,[25] who introduced the water-activity effect and also used the exclusion effect, which had been known before. The Bjerrum equation for the mean activity coefficient is as follows:

$$(9) \qquad \ln \gamma_{\pm} = \ln \gamma_{\pm}{}^1 - \frac{h}{\nu} \ln a_w + \ln \left[1 - \frac{(h - \nu)m}{55.5} \right]$$

where $\gamma_{\pm}$ is the observed activity coefficient; $\gamma_{\pm}{}^1$ is a hypothetical activity coefficient calculated from the assumed electrostatic form (Bjerrum used a $c^{1/3}$ law); h is the hydration number, assumed to be the average number of water molecules bound to the ions of the salt; ν is the number of moles of ions in a mole of salt; and a_w is the activity of water. The last term of the equation represents the change in activity coefficient caused by the hydration-induced change in concentration, and the term in $\ln a_w$ is the change in activity of salt due to a change in water activity. The two terms are numerically of the same magnitude for most salts. The term in $\ln a_w$ comes from an expression for the equilibrium between hydrated and dehydrated ions of the form

$$(10) \qquad \frac{a_d(a_w)^h}{a_h} = 1$$

where a_h is the activity of the hydrated form of the salt, a_d the activity of the dehydrated form and, for lack of information, the equilibrium constant is set equal to unity. It can be noted that in the water-activity part of the equation, the water-molecule–ion relationship is an equilibrium one, whereas in the exclusion part the water molecules are firmly fixed, so that the two parts of the equation are inconsistent. A point of interest is that, although the water activity is a parameter of the equation, it is not fixed by the equation but could be related to the salt activity using the Gibbs–Duhem equation.

A treatment of hydration by Robinson and Stokes[18] resulted in an equation for $\log \gamma_{\pm}$ identical with Bjerrum's equation (except for the Coulomb interaction), but the concepts were different. Robinson and Stokes said that the total excess free energy was uninfluenced by hydration, but the concentration should be redefined so that the solute component included h water molecules and there were h fewer solvent water molecules per mole of ions. The last term of (9) has the same significance as in Bjerrum's equation, but the term $h/\nu \ln a_w$ now signifies a loss in free energy attributed to the water component that has been removed from the solvent. Since the process represents a change in definition of components rather than a physical effect, the change in total excess free energy is necessarily 0.

The treatment of hydration by Glueckauf[39] considered only the concentration-change term and not the water-activity component. Also the exclusion (concentration) term was based on volume-fraction rather than mole-fraction statistics as in the first two treatments. Glueckauf calculates the total excess free energy assuming that a fixed number of water molecules is firmly bound (although an average number is interpreted), and the activities of solvent and hydrated solute each depend on their volume fraction in solution. The hydration-dependent component of the Gibbs excess free energy, G_s, is then given by

$$(11) \qquad \frac{G_s}{kT} = \sum_i n_i \ln \frac{N_i V_i}{V}$$

where N_i is the number of hydrated ions or solvent molecules of each species, V_i the partial molal volume of each, and V the total volume. By differentiating with respect to solvent or solute component, the partial molal excess free energy of each is obtained, which for $\log \gamma_\pm$ of the salt is as follows (the electrical term is excluded)

$$(12) \qquad \begin{aligned} \log \gamma_\pm &= \frac{h - \nu}{\nu} \log (1 + 0.018\ mr) \\ &\quad - \frac{h}{\nu} \log (1 - 0.018\ mh) \\ &\quad + \frac{0.018\ mr(r + h - \nu)}{2.3(1 + 0.018\ mr)} \end{aligned}$$

The quantity r is equal to ϕ_v/V_w°, (i.e., the ratio of the apparent molal volume of the salt to the molal volume of pure water), m is molality, and h is the hydration number; r can be measured so that h is the only empirical parameter introduced.

The Glueckauf equation is quite successful in representing data up to high concentrations, but its main advantages over the other hydration equations are (a) the hydration numbers vary less with concentration in this theory and (b) the hydration number for a given ion does not vary with salt type as in the other equations. Also in favor of this treatment is its emphasis of the importance of the excluded volume of the ions. The significance of this quantity to excess free energies was discussed quite early by Onsager,[40] and in the sense of hard spheres, the excluded volume plays a vital role in the calculations of the Mayer theory.

The defects of these theories are first of all the usual ones that the electrical component is indefinite and the hydration component is obtained as a difference. The meaning of the hydration number itself is very vague. For purposes of calculation, it is necessary to consider a definite number of fixed water molecules; but this has to be considered as representing some sort of average interaction of the ion with all the water molecules it perturbs (i.e., the cosphere). At moderate concentrations the cospheres overlap extensively, and this treatment cannot be strictly correct.

As an example of this, the corrections to the excess free energy from the Glueckauf theory are purely entropic, and the entropy of the solution of a strongly hydrated salt should become increasingly negative with increasing concentration. Actually, the reverse is true, and the excess entropy of a hydrogen chloride or lithium chloride solution increases with concentration, whereas the entropies of the weakly hydrated salts potassium bromide or sodium bromide decrease with concentration.[41] The theory could probably be improved by introducing temperature coefficients for the hydration numbers, but it would be difficult to understand the meaning of such numbers.

Some comments can also be made on the water-activity effect in Bjerrum's equation. In the Mayer treatment it is necessary to correct measured activity coefficients to an osmotic pressure where the solvent has unit activity. Calculations[42] made in this way show only a small correction in the solute activity coefficient rather than the relatively large change predicted by the Bjerrum equation.

5   Statistical-Mechanical Treatment of Ionic Solutions

The treatment of electrolyte solutions by exact statistical-mechanical methods has in the past been rather discouraging—any model that was even remotely interesting physically was extremely complicated to handle mathematically. The treatments by Bogoliubov,[43] Kelbg,[44] Kirkwood,[12] and others, in spite of drastic simplifications, were still extremely complicated, and only the first terms of various expansions were obtained. As a result of the simplifications and truncation, the ranges of validity of the calculations were difficult to assess and probably extended little beyond that of the classical Debye–Hückel work.

Although the complications of the subject have not decreased since the early work, powerful new methods have recently been developed for dense fluids and these have been carried over to the treatment of electrolyte solutions. Using these methods, calculations are now available for models of substantial physical interest to an accuracy that reasonably reflects the properties of the model rather than the uncertainties of the calculation or the theory. The calculations have been carried out to concentrations as high as 1 M and have given significant answers to such problems as the meaning of a distance of closest approach and the thermodynamic nature of hydration.

Although the theories of liquids are highly complex and specialized, it is felt that the importance of the results obtained make some review and description of the methods necessary for any chapter on the thermodynamics of electrolyte solutions. What is attempted here is primarily a nonmathe-matical description and history of the subject, with mathematical formulas used mainly as illustrations of particular points. Much of the mathematics of the theory is difficult, but many of the physical concepts are relatively simple and can be reasonably described.

In the development of the theories, a number of parallel and contributing paths have been taken. The development given here largely follows the methods of Mayer, which have made the most advanced development. Bogoliubov, Kelbg, Kirkwood, and many others have also made outstand-ing contributions and many of these are implicitly or explicitly included here.

Outline of Development. The theoretical developments have taken place over about 30 years and have had several distinct parts, which can be out-lined as follows. The initial developments were in treating the properties of imperfect gases, and the first development was the calculation of the thermo-dynamic properties in terms of the virial expansion, that is, $PV/RT = 1 + B_2/V + B_3/V^2 + \cdots$. Ursell[47] originated and Mayer[40] and co-workers largely developed simple methods for calculating the virial coefficients in terms of the repulsive and attractive forces of mathematical clusters of mole-cules.

Another method of treating imperfect gases was to calculate the pair radial distribution function from certain integral equations that were de-veloped for the purpose. With the radial distribution function, thermo-dynamic properties can be calculated in a rather direct manner. McMillan and Mayer[38] then showed how these calculations, valid for imperfect gases or moderately dense fluids, could be directly related to the properties of nonionic solutions. The osmotic pressure would directly take the place of the gas pressure, and the virial coefficients for the solutions would be cal-culated from potentials of average force rather than true potentials; other-wise they would have identical significance.

The application of these methods to ionic solutions involved some new and difficult problems which were again solved primarily by Mayer.[3] The Mayer methods were first applied using the virial coefficient (cluster expan-sion) method, and in the most recent work combined with some new and powerful methods in the radial distribution function technique.

It should be emphasized that the discussion following is oversimplified, and the original papers and particularly Friedman's[45] monograph should be read for complete and exact treatment.

Cluster Expansion Theory. The Helmholtz free energy per molecule, F, is given by the statistical-mechanical equation[10]

(13) $$F = kT \ln Q$$

The quantity Q, the partition function, in principle can be calculated from the geometry, mass, and the intra- and intermolecular potential energy curves of the molecules of the system. With knowledge of the variation of Q with temperature and density, it is thus possible to calculate any thermodynamic property of the system directly from the individual and collective molecular properties.

The partition function can itself be factored into a product of terms; one involving the kinetic energy of the particles, a second involving the internal motions and energy of the molecules, and a third, the configuration integral, Q_c, which depends on the total intermolecular potential energy of all the molecules of the system. The kinetic and internal energy factors of Q can be calculated fairly easily for most classical (i.e., nonquantum-mechanical) systems, and tables of these are available in standard textbooks on statistical mechanics.

The evaluation of the configuration integral is the basic problem, and this is a very difficult problem, being exactly solved for only a few simple cases. The approximate methods used for calculating Q_c for liquids are the concern of this discussion. The configuration integral is given by the formula[10]

$$(14) \qquad Q_C = \int e^{-U(x_1 y_1 z_1 \ldots z_n)/kT} \, dx_1 \cdots dz_n$$

The quantity U is the total intermolecular potential energy of the system when the n molecules are at the Cartesian coordinates $x_1 y_1 z_1 \cdots z_n$. The integral is taken over the volume, V, of the system for the coordinates of each molecule, so that for a perfect gas where U is 0, the value of the integral is V^n. For a real system where U is not 0 and there are simultaneous interactions of very large numbers of molecules, the integral is evidently an extremely complicated function that cannot be solved without extensive simplification.

The quantity U can be regarded as the sum of the individual interactions between each possible pair of molecules of the system or[46]

$$(15) \qquad U = \tfrac{1}{2} \sum_i^n \sum_{j \neq i}^n u_{ij}$$

where the sum is over each molecule of the system and the $\tfrac{1}{2}$ is to avoid counting each pair twice. The quantities u_{ij} are the potential energies of the pairs of molecules i and j and are, of course, functions of the distance between the pairs. In general, it is true that the potential energy of a given pair will also depend on the positions of other molecules, but the effect is probably small and in any case much more complicated.

The form of the functions u_{ij} will depend on the forces involved which may be among others van der Waals, repulsive, or Coulombic for ion pairs.

For the simpler treatments of ions, a hard-sphere potential is added to the Coulomb component. This type of potential has the property that the potential is $+\infty$ when the radii of the ions overlap and 0 otherwise, so effectively the ions cannot overlap.

As mentioned earlier, evaluating a configuration integral involving a large number of molecules is an extremely complicated matter. The contributions of Ursell[47] and Mayer[46] were to show how this integral could be broken down into a series of separate terms involving the interactions of successively larger groups of molecules. The first terms of the series were quite simple to evaluate, and for gases of moderate density these made up the largest part of the configuration integral. Only a relatively few of the more complicated terms could be calculated; however, the methods of treatment Mayer developed were of basic importance to much of the further developments of gas theory and to the extensions to nonionic and ionic solutions.

Return to the configuration integral given by (14) and substitute the expression for U in terms of the individual pair potentials as follows:

$$(16) \qquad Q_c = \int_V e^{-\frac{1}{2} \Sigma_i \Sigma_j u_{ij}/kt} \, dx_1 \cdots dz_n$$

The exponential of a sum can, of course, be written as a product, that is

$$(17) \qquad e^{\frac{1}{2} \Sigma_i \Sigma_j u_{ij}} = e^{\frac{1}{2}} \prod_{ij} e^{u_{ij}} = e^{\frac{1}{2}} \cdot e^{u_{ij}} \cdot e^{u_{jk}} \cdots$$

of exponentials referring to each possible pair of the system. The function f_{ij} can now be defined as $f_{ij} = e^{-u_{ij}/kt} - 1$, so that each factor in the product $\prod e^{-u_{ij}/kt}$ can be replaced by its equivalent $(1 + f_{ij})$ and the product by $\prod_{ij} (1 + f_{ij})$.

The integrand of the configuration integral $e^{-U/kt}$ now becomes a sum of all the possible products of the factors f_{ij} and 1 or

$$(18) \quad e^{-U/kt} = 1 + \sum_{ij} f_{ij} + \sum_{ijk} f_{ij}f_{ik} + \sum_{ijkl} f_{ij}f_{kl} + \sum_{ijkl} f_{ij}f_{jk}f_{kl} + \cdots$$

and the series finishes with a product involving every possible pairing of the molecules.

The set of molecules in any product of the f_{ij} functions is said to be a cluster, and each f_{ij} function is spoken of as an f bond. It is emphasized that this is a mathematical way of representing and accounting for each of the possible interactions of the system; it does not at all represent a physical bond or cluster.

The clusters have a simple pictorial representation, so that each factor f_{ij} can be represented by a diagram with points i and j and a line between them, that is, $f_{ij} \rightarrow i \overset{\bullet-\bullet}{\quad} j$. Products of the f_{ij} functions can then be represented by points for each molecule in the product and a line for each factor f_{ij}. For three molecules the possible diagrams are

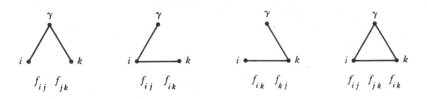

Any number of lines can be drawn from different molecules to one center, although the numerical value of such complex terms would be low and in the higher terms all possible interconnections of all the molecules are to be made.

The potential energy of any pair of molecules for non-Coulombic forces at a large distance is small, as is any term involving this pair as a factor. However, the molecules may be connected by a chain of other molecules so that, although they are far apart, an integral involving the particular pair may still have a significant value. Integrals involving more than a few molecules become increasingly difficult to calculate, and even for the simplest potential functions only a few have been obtained.

The great advantage of the theory for simpler applications is that it allows the very complex relations of interacting systems to be thought of in terms of two-body, three-body, and so on, interactions; and for moderate densities these make the most important contributions to the integrals and the excess free energies. In more advanced developments of gas and liquid theories, such as in the radial distribution function theories, the use of the diagrammatic technique along with topological considerations allows substantial simplification, classification, and factoring of the more complex diagrams. As a result, these methods have become basic to much of the advanced developments in fluid theories.

Before thermodynamic results can be obtained from the cluster integrals, it is necessary to consider the combinatorial problem of counting all the possible combinations of molecules. For any cluster having the same kind and number of molecules, the integrals will have the same value, and the problem is of counting the number of times such a term will occur in the overall sum. This problem is one of moderate difficulty, but it has been solved and the solutions can be found in several references.[39]

It is sometimes more convenient to evaluate the virial expansion coefficients B_n rather than the configuration integral directly, although the expansions are closely related. Generally, the relation between the individual cluster integrals and the virial coefficients is somewhat involved; however, the evaluation of B_2 is very straightforward and is

$$(19) \qquad B_2 = 2\pi \int_0^\infty (1 - e^{-u_{ij}/kT}) r^2 \, dr$$

where $1 - e^{u_{ij}/kT}$ is of course the f_{ij} function.

Radial Distribution Function Methods. The radial distribution function, $g(r)$, gives the probability of finding a molecule of the gas or solution at a specified distance from an arbitrary central molecule. Intuitively, it would appear that if the potential energy curve for the two molecules is known and their average distribution of distances obtained from the distribution curve, it would be possible to calculate many average energetic quantities of the system. This is true and in fact by using the $g(r)$ with various functions, it is possible to calculate the pressure or compressibility of a gas; and from the variations with temperature or volume, excess free energies and other thermodynamic quantities can be obtained. The mathematical expressions for obtaining the average energy, the pressure and the compressibility are [48]

$$(20) \qquad \bar{U} = \frac{\rho}{2} \int^V g_{12} u_{12} \, dr_{12}$$

$$(21) \qquad \frac{P}{\rho kT} = 1 - \frac{\rho}{6kT} \int g_{12} \left(\frac{du_{12}}{dr_{12}}\right) r_{12} \, dr_{12}$$

$$(22) \qquad kT \left(\frac{\partial \rho}{\partial P}\right) = 1 + \rho \int (g_{12} - 1) \, dr_{12}$$

where $\bar{U}$ is the average energy of the molecules of the system, ρ is the density in molecules per cubic centimeter, $\partial \rho / \partial P$ is the compressibility, g_{12} is the radial distribution function, and u_{12} the potential energy corresponding to a distance, r, between molecules 1 and 2.

In (20) the energy of interaction is averaged over the volume of the system, and in (21) the quantity $r_{12} (du_{12}/dr_{12})$ (the virial) is averaged where du_{12}/dr_{12} is of course the force exerted on one molecule by the other. Equations 20 and 21 are valid only if the potential, U_{12}, depends only on the relative positions of molecules 1 and 2 (i.e., pairwise additive); however, (22) is generally valid.

It is apparent that, if the radial distribution function for the molecules of a system is known, then it is relatively easy to calculate the thermodynamic properties of the system, and knowledge of this distribution function is of the highest importance.

In contrast to the virial expansion methods, which depend on the interactions of groups of molecules of all sizes, the properties now seem to depend entirely on the properties of pairs of molecules in the sense that only the pair-distribution function is involved. This dependence is only apparent, however, since g_{12} giving the actual distance distribution depends on the interactions of all the other molecules of the system. Consequently, it is not obvious that calculating g is any simpler than calculating the sum of the virial coefficients and, in fact, one important method of calculating g does use a series of cluster integrals very similar to those for the virial coefficients.[49]

The other methods of calculating g involve integral equations that are obtained by statistical-mechanical methods. In general, the methods of obtaining the equations themselves, and the solutions are all highly complex; however, there are some qualitative aspects that are of interest to the nonspecialist.

These integral equations have been of two different types—an earlier form in which some of the major contributors have been Bogoliubov,[43] Kirkwood,[31] Born, Green, and Yvon,[50] and a later group of equations, with major contributors being Ornstein,[51] Zernicke,[51], Percus,[52] Yevick,[52] von Leeuwen,[53] Groeneveld,[53] De Boer,[53] Meeron,[54] and many others.

The major problem of these calculations can be illustrated with a simple example of the earlier type equation[11]

$$(23) \qquad kT \frac{\partial n_2}{\partial r_1} + \frac{\partial u_{12}}{\partial r_1} n_2 = - \int n_3 \frac{\partial u_{13}}{\partial r_1} dr_3$$

In (23) n_2 is similar to g except that the probability of the first particle being at a particular position is also included. It is the combined probability of a particle 1 being at a particular position and a particle 2 being at a specified position with respect to 1; n_3 is the combined probability of a particle being at a specified position and two other particles having specified positions with respect to the first one. The differentiation is with respect to the position of the first particle and the integration over the coordinates of the third particle.

A set of similar equations can be written in which n_3 can be expressed in terms of n_4, and so forth, on up to the number of molecules of the system. In these equations each of the functions n_i has to be determined in terms of the next higher one (e.g., the two-particle distribution in terms of the three-particle distribution). In order to solve the equation, either a higher n_i must be known, which is not generally possible, or a relation must be obtained between a higher and lower n_i function. The usual method for this type of equation is to use the superposition principle of Kirkwood[12] in which n_3 is set equal to the product of the three possible two-particle distribution functions, that is,

$$(24) \qquad n_3(1, 2, 3) = n_2(1, 2)n_2(2, 3)n_2(1, 3)$$

The parentheses refer to the particular particles in each distribution function.

According to (24), the probability of finding three particles at specified locations is equal to the product of the three separate two-particle probabilities; thus the assumption says that in effect a third particle does not affect the average spatial relation of any of the pairs. The relation for n_3 in (24) can be substituted in the integral equation, a potential function assumed, and the equation solved. Even with the simplifications, the solutions are still difficult and have been relatively unsatisfactory in terms of accuracy.

Although the discussion now concerns imperfect gases, it might be noted here that equations similar to these have been applied to electrolyte solutions, the ions being considered as a gas. These extensions involve considerably greater complications, however. Kirkwood,[12] using a treatment of this type and the superposition principle, showed that the Debye–Hückel equation could be directly obtained from statistical mechanics. Also, Kirkwood demonstrated the nature of the deviations from the Debye–Hückel assumption mentioned earlier and found the transition of the radial distribution function at $\kappa a = 1.03$, also mentioned earlier.

In a later treatment of a similar form, Kirkwood and Poirier[2] showed that the Debye–Hückel limiting law could be rigorously obtained without using the superposition principle in the derivation, but the limiting law necessarily implied the validity of the principle. Although interesting results were obtained, the mathematical difficulties and approximations necessary did not permit quantitative calculations significantly beyond the limiting law, and generally the treatment of fluids and of electrolyte solutions has followed a later type of integral equation method.

Two of the principal newer type equations used for electrolyte solutions are known as the Percus–Yevick[52] and the hypernetted chain[53,54] (HNC) approximations. Each can be obtained in several different ways using different concepts, but neither is remotely simple to describe. One possibly more picturesque way of describing these equations is by way of a relatively old equation given by Ornstein and Zernike[51] in 1914 as

$$(25) \qquad\qquad h_{12} = c_{12} + \int c_{13} h_{23} \, dr_3$$

where $h_{12} = g_{12} - 1$ is known as the total correlation function and g_{12} the usual pair radial distribution function; c_{12} is called the direct correlation function and is defined by the equation. The total correlation function might be conceived as being made up of two parts. In one, the correlation in position between two molecules results from the direct unmediated interaction of the two particles, and this is c_{12}. The second part of the interaction is an indirect part resulting from the intermediary action of particle 3, which acts as a bridge between 1 and 2, with the position of particle 3 averaged over the entire volume. As in the first type of integral equation, there are two types of functions, c_{12} and h_{12}, to be found, and in order to solve the equation, a relation between the two must be obtained.

As indicated, there is no simple way to describe the two relations between c and h—the Percus–Yevick and HNC approximations—that have been found. These approximations may be obtained from certain special types of cluster integral graphs which, although they involve many molecules and contain important contributions to the configuration integral, are also mathematically tractable.[48]

These approximations for $c_{ij}(r)$ are[54]

(26) $$\text{PY} \quad c(r) = [1 - e^{u(r)/kT}][1 + h(r)]$$

(27) $$\text{HNC} \quad c(r) = h(r) - \ln(1 + h(r)) - u(r)/kT$$

The relationships can be substituted in the Ornstein–Zernicke equation and the result numerically solved on a computer to give the radial distribution function.

For nonionic fluids at densities well below the triple-point value, these two equations have given highly satisfactory results. Since in ionic solutions one is also dealing with a solute of moderate density (concentration), these equations would appear to be of value and, in fact, Carley[55] has directly applied these equations and has obtained results similar to those of Rasaiah and Friedman,[56] which are described later.

McMillan–Mayer Theory.   Up to this point the theory discussed has been for a gas or fluid of moderate density, and it is necessary to convert these results to terms of solutions. This is done using the McMillan–Mayer theory.[38] Among other things, the theory treats the problem of calculating for multi-component systems the fugacity of a set of substances at one concentration set in terms of the fugacities at another reference concentration set.

A possible state for the reference concentration is that of infinite dilution, and the activities of substances at finite concentrations can be then calculated in terms of the cluster integrals at infinite dilution. The quantities used in the integrals are, of course, potentials of average force rather than potentials.

A second important point of the theory is that one component (the solvent) can be treated in an unsymmetrical manner so that its standard state is the pure solvent at standard pressure. In addition, the solution for which the calculation is to be made, that is, at finite concentration, is to be placed under sufficient pressure (osmotic pressure) so that the solvent is at unit activity. Under these conditions the treatment for imperfect gases can be carried over to solutions in almost unchanged form, with osmotic pressure replacing gas pressure and, as mentioned, potentials of average force replacing potentials.

The cluster integral in solution for a given group of solute molecules includes the direct potential interactions of the isolated group of solute molecules, which are averaged over the rotational and internal coordinates of the molecules if they are not symmetrical. The positions of the solvent molecules are averaged over all possible positions in calculating the ion interactions, but the positions of ions not in the set are not considered, since the set is at infinite dilution. If infinite dilution were not the reference state, then averaging over the positions of all the other solute molecules would be necessary; this, of course, is a much more complicated problem. In calculating the potential of average force for any configuration, contributions from solvent–isolated-solute interactions, perturbations of intramolecular solute interaction by the solvent, and the perturbation of solvent-solvent interactions by the solute are subtracted, since these are in effect the contributions of the standard state of infinite dilution from which differences are calculated.[11]

A simple example of the nature of the potential of average force for ions is in the dielectric constant. For ions that are far apart, the energy and force depends on the averaging of the positions of all the water molecules that are near or between, and the potential energy is thus a potential of average force. When the ions are sufficiently close to each other that water molecules must be displaced, the dielectric constant loses its meaning, but the potential of average force is still a meaningful and necessary quantity.

The McMillan–Mayer theory is as much concerned with radial distribution functions as with virial coefficients and consequently can be used with either of these approaches.

Mayer Electrolyte Theory.[3]   As a final stage, the theories developed for gases and nonionic solutions must be adapted to solutions of electrolytes. There are two related ways by which it can be seen that the theory in its original form cannot be directly applied to electrolyte solutions. First, the theory is based in large part on the evaluation of integrals of the form

$$(28) \qquad \int^{V} (e^{-u_{ij}/kT} - 1)\, dV$$

where the limit is taken as the volume approaches infinity. For ionic solutions the potential u is equal to $Z^2 e^2 / Dr$, and this together with the volume element $4\pi r^2\, dr$ is substituted in the integral to obtain

$$(29) \qquad \int^{V} (e^{-(Z^2 e^2 / DrkT)} - 1)\, 4\pi r^2\, dr$$

For evaluation, the integral is expanded in series

$$(30) \qquad e^{-(Z^2 e^2 / DrkT)} = 1 - \frac{Z^2 e^2}{DrkT} + \frac{1}{2!}\left(\frac{Z^2 e^2}{DrkT}\right)^2 \cdots$$

The term linear in $1/r$ will cancel, since a summation over each type of pair is taken, but the square terms will give rise to infinite integrals of the form $\int^V T\,dr$ where T is a constant and the volume of integration is infinite. Physically, it is evident that the integrals must converge to a finite value; but mathematically, it appears necessary that some sort of summation must be performed before it is possible to use the cluster integrals.

The second problem is in the experimental fact that the excess free energy of the ionic solutions is proportional to the square root of the concentrations at low concentrations. The virial expansions, however, can give only integral powers and these cannot represent the square-root behavior. Thus the twin problems of nonconvergent integrals and the necessity of a square-root term require additions to the original theory. The methods for treating these problems are as usual complex and mathematical, and the description that follows is again a very limited one.

As for imperfect gases, the configuration integral is expanded in terms of the potential energy of pairs of ions (although the pair restriction can be removed[41]). The potential energy of a pair is taken as

$$(31) \qquad w_{ij} = \frac{e_i e_j}{Dr} + w_{ij}^*$$

where w^* is a short-range non-Coulombic potential that could also have the hard-sphere form. Each f function of the original theory can then be taken as the combination of the nonionic f function $(e^{w_{ij}^*/kT} - 1) = k_{ij}$ and the exponential of the Coulomb force or

$$(32) \qquad \begin{aligned} f_{ij} &= e^{-w_{ij}^*/kT}\, e^{-(e_i e_j/DrkT)} - 1 \\ &= (1 + k_{ij})\, e^{-(e_i e_j/DrkT)} - 1 \end{aligned}$$

Before expanding the configuration integral in terms of products of the f function as before, the exponential of the ionic term is expanded in a power series, that is,

$$(33) \qquad e^{-(e_i e_j/DrkT)} = 1 - \frac{e_i e_j}{DrkT} + \frac{1}{2!}\left(\frac{e_i e_j}{DrkT}\right)^2 \cdots$$

If the short-range term, $e^{-w_{ij}^*/kT} - 1$, is now called k_{ij}^* (i.e., k bond) and $1/r_{ij}$ is represented by g_{ij} (i.e., g bond), then each f_{ij} is given by the expansion

$$(34) \qquad f_{ij} = k_{ij}^* + \frac{[1 + k_{ij}^*]\sum\limits_{\rho \geq 1}(e_i e_j g_{ij}/DkT)^P}{P!}$$

Each f_{ij} bond of the nonionic system is replaced by an infinite number of terms that will have one or k_{ij}^* as a factor and any number of g bonds. The g bonds in each of such terms are all in parallel.

Each cluster integral having products of f_{ij} bonds will give rise to new terms, some of which will be products only of g bonds and no k bonds, and among these there will be some involving only linear closed chains or cycles and no crosslinks.

In graphical form, for any size cluster there will be terms of the form

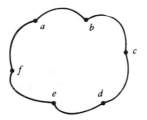

The $\bullet\!\!\frown\!\!\bullet$ symbol represents a $1/r_{ij}$ or g bond interaction, and this particular term could appear in the six-ion cluster integral.

Mayer showed that this particular type of cluster integral (i.e., closed chains of g bonds) gave rise to the most divergent type of integral, and the divergence could be removed by summing this term type over all chain lengths. Rather than first evaluating each cluster integral for a given number of molecules and then obtaining the thermodynamic properties by summing over the different cluster integrals, it is now necessary to pick out a certain type of term in each cluster and sum over this type in all the clusters. The thermodynamic properties are finally obtained by summing these special sums over all their different types.

The integrals for the chains can be evaluated by Fourier transformations, and it is found that the integral over the chain can be represented as an integral over

(35) $$e^{e_i e_j e^{-\kappa r_{ij}/D r_{ij} kT}}$$

where κ is the usual Debye–Hückel quantity. In effect, the Coulomb potential $e_i e_j/Dr$ is replaced by the Debye–Hückel potential of average force

(36) $$q_{ij} = \frac{e_i e_j e^{-\kappa r_{ij}}}{DkTr_{ij}}$$

and the interaction potential between two ions is now dependent on concentration through the molar ionic strength, κ. Integrals involving the Debye–Hückel potential of average force are convergent and do not lead to difficulties in integration. This particular type of interaction is called a q bond. Finally, the combinatorial problems have to be solved and the final sums obtained. There have been serious problems in proving the series convergent, but these now appear to be controlled, and the results are given in the form of a series[41] for the Helmholtz free energy

$$(37) \qquad S = \sum S_u = \frac{F^{\text{ex}}}{kTV}$$

Each S_u is the sum of the particular types of diagrams and F^{ex} is the excess Helmholtz free energy. The first term of the series S_c is obtained by summing the pure g-bond cycles and is

$$(38) \qquad S_c = \frac{\kappa^3}{12\pi}$$

which is identical with the result of the Debye–Hückel theory for F^{ex}/kTV.

The second term of the series corresponds approximately to the second virial coefficient of nonionic solutions and is obtained by integrating over the coordinates of one pair of ions interacting with short-range forces (k bond); but the ionic interaction is the result of summing over chains of all lengths and all numbers of chains between the selected pair of ions. The contributions from each ion pair type must be summed and for a 1:1 electrolyte S_2 has the form

$$(39) \qquad (c_+)^2 B_{++}(\kappa) + c_+ c_- B_{+-}(\kappa) + (c_-)^2 B_{--}(\kappa)$$

where c_+ and c_- are the molar concentrations of positive or negative ions. The form of each B_{ij} is as follows:

$$(40) \qquad B_{ij}(\kappa) = \frac{4\pi}{n} \int_V \left(e^{-w_{ij}^*/kT} e^{-q_{ij}} - 1 - q_{ij} - \frac{q_{ij}^2}{2} \right) dV$$

The quantity n is 2 for like ions and 1 otherwise, and q_{ij} is the Debye–Hückel potential of average force—(36). If the charges are 0, q is 0 and the integral becomes the ordinary second virial coefficient.

The expression for S_2 is explicitly proportional to the square of the concentrations and so is analogous to the virial series expansion mentioned earlier. The coefficients, however, are dependent on the concentration through κ and this dependence may be of the order of either positive or negative powers of κ. As a result, the formal analogy of a virial expansion is not valid, but also it becomes possible for higher terms of S_n to contribute terms to F^{ex} of the order of c^2 or even cause the series to diverge as κ decreases.

Mayer,[3] Haga,[59] and Friedman[60] have each examined these questions, and it is found that for symmetrical electrolytes B_{ij} is of the lowest order κ^0 (kappa to the 0th power) but also may contribute higher order terms. For unsymmetrical electrolytes, B_{ij} is of the order $\ln c$, so that there is a term $c^2 \ln c$ in the excess free energy. The terms S_3 and S_4, involving short-range interactions among 3 and 4 ions, respectively, and explicit c^3 and c^4 factors, contribute terms of the order of $c^{5/2}$ for unsymmetrical electrolytes and c^3 $\ln c$ and c^3, respectively, for symmetrical electrolytes.

The general question of convergence has been examined by Friedman,[60] and it appears reasonably certain that the more complicated integrals do not diverge and are both of higher order and smaller value than those given.

As with the virial expansion for gases, these expansions also converge slowly, and it has been found possible to obtain more accurate and complete results by going to the modern methods of radial distribution function theory. Based on work by Meeron[61] and Allnatt,[62] Rasaiah and Friedman[56] have applied the equivalents of the hypernetted-chain and Percus–Yevick theories to electrolyte solutions.

In these theories for nonionic systems the total and direct correlation functions are calculated using combinations of f_{ij} i.e., (k_{ij}) functions $= (e^{-w^*/kT} - 1)$. Using the Mayer results for ionic solutions, it was found possible to replace the f_{ij} functions of nonionic solutions by combinations of two types of interactions, ϕ^{11} and q bonds, where

(41) $$\phi''_{ij} = [1 + k_{ij}] e^{q_{ij}} - 1 - q_{ij}$$

and $q_{ij} = e_i e_j e^{-\kappa r/DrkT}$ as before. Again, if the charges are 0 these interactions reduce to the nonionic case and also the Percus–Yevick and hypernetted-chain equations derived using them reduce to the nonionic case.

Pair-correlation functions can be obtained for each type of ion pair and the compressibility and pressure (virial) equations extended to calculating the osmotic pressure of the ionic solution. The first paper of this series used the Coulomb plus hard-sphere potential for the ions;[56] however, a recent calculation has used a very interesting and slightly more complicated "square-well" potential for the ions.[63] This potential considered the ion and water molecules as having their crystallographic distances of approach, but a small increment of energy was added for the ions coming closer than the diameter of one water molecule between. The ions in contact were considered as hard spheres. Further interesting types of potential curves are being examined,[64] and results have recently been published.

There are, of course, many unanswered problems in this work, and it is possible that it may be outmoded by newer developments or that inconsistencies and defects may be found. Nevertheless, it now appears that these results, which are discussed in the next section, are among the most interesting and important theoretical developments since the Debye–Hückel theory.

6   Comparisons of Theory with Experiment

Pressure and Volume Corrections.   The theoretical calculations are for a system of constant volume (molarity scale and Helmholtz free energy, F) and constant solvent activity; most experimental data, however, are given in terms of the molality scale (and Gibbs free energy, G), and the solvent activity varies with concentration. It is, therefore, necessary to correct one or the other sets of results in order to make comparisons possible. This was first done by Poirier,[42] who corrected the theoretical results to the experimental and compared his theoretical calculations with the experimental work.

Friedman[41] has made the reverse adjustment, experimental to theoretical, and this appears to be simpler and is given by the following equation:

$$(42) \qquad \frac{G^{ex}}{mRT} = -\frac{S}{C} - R_1(m)$$

and

$$(43) \qquad R_1 = \ln\left(\frac{V(m, o)}{V(o, o)}\right) + \frac{\Pi^2}{2mRT}\frac{\partial V(m, P)}{\partial P}$$

where S is the sum of the Mayer terms given previously and R_1 the correction to the experimental excess Gibbs free energy for the volume and pressure changes; $V(m, o)$ is the volume of an m molal solution having 1 kg of solvent and $V(o, o)$ the volume of 1 kg solvent, both at 0 pressure. The osmotic pressure of the solution is Π, m is the molality, and $V(m, P)$ is the volume of the m molal solution at pressure P (the osmotic pressure). The magnitude of the correction terms is given by a table from Friedman[45] for several salts at 6 m.

	LiCl	NaCl	CsCl	NaI
$\ln\dfrac{V(m, o)}{V(o, o)}$	0.113	0.120	0.234	0.205
$\dfrac{\Pi^2}{2mRT}\dfrac{\partial V(m, P)}{\partial p}$	0.000	0.002	0.004	0.000

These quantities may be compared with values of G^{ex}/mRT for lithium chloride and sodium chloride of 0.26 and -0.34, respectively.[71] The volume correction is quite large at this high concentration, but the pressure correction is still almost negligible.

Poirier[42] gives corrections for $\log \gamma_\pm$ for several salts at 1 N in the form $\log \gamma_{\pm,exp} - \log \gamma_{\pm,theo}$, as -0.031, -0.06, and -0.07 for NaCl, CaCl$_2$, and LaCl$_3$, respectively, where a molar rather than a molal scale is used. These corrections are quite small, as are the corrections to the total free energy at these concentrations.

An immediate result of this correction calculation concerns one part of Bjerrum's hydration theory.[17] According to this theory, a large part of the hydration effect is due to the lowering of the solvent activity—an unlikely supposition from these calculations, although it is very probable that at much higher concentrations the Bjerrum concept would be valid.

Since the corrections up to 1 m are quite small, they can be neglected for qualitative comparisons of experimental and theoretical results, and in any case the forms of the curves are very similar.

Corrections Using S_2. Poirier's theoretical calculations of $\log \gamma_{\pm}$ to the S_2 (i.e., second virial coefficient) term were given in the form of a numerical table containing a large number of components that had to be summed to give values of $\log \gamma_{\pm}$.[42] Each of these terms was given for a range of discrete values of κa. Since the sum S_n discussed earlier is proportional to the total excess free energy of the system, the quantities given by Poirier were derivatives of S_2.

Comparison with experiment was made by considering a as an adjustable parameter and choosing it so that the corrected theoretical and experimental values of $\log \gamma_{\pm}$ for each salt coincided at a given concentration and remained within experimental error below that concentration. The upper limit of the calculation was a point where a distinct divergence of the experimental and calculated curves occurred.

The reference points for $NaCl$, $CaCl_2$, $ZnSO_4$, and $LaCl_3$ solutions were 0.1, 0.05, 0.01, and 0.01 N, respectively, and for these same salts the upper limits of calculations were 0.4, 0.1, 0.05, and 0.033 N. At these concentrations the difference in $\gamma_{\pm}$ between experimental and theoretical results were 1.5, 3.7, 9.2, and 7.8%, with the theoretical results being lower.

In addition to the comparisons, Poirier[42] also gives uncorrected curves for a number of values of a and for different charge types. The set of such curves for 1:1 electrolytes is shown in Figure 1, where f is the uncorrected molar mean activity coefficient and A is equal to $aDkT/e^2$, where a is the usual distance of closest approach parameter and e the electronic charge. At 25°C in water, a is equal to 7.137 A, and the range of values of a, 1.43 to 5.71 Å, is approximately that obtained in Debye–Hückel treatments of 1:1 electrolytes. The experimental activity coefficient curves are quite similar to the theoretical curves of a corresponding a value.

It is seen from these curves that the general features of 1:1 electrolyte activity coefficient curves are described rather well even if only the second virial coefficient (S_2) term is used. The curves for the other charge types are similar with curves for the 2:2 electrolytes close together as found experimentally. For very small values of a_{ij} for 2:2 electrolytes, the theoretical curves came below the limiting-law values, as did the results of the exponential Poisson–Boltzmann equation discussed earlier. Each curve for larger a values shows the characteristic upturn of $\log \gamma_{\pm}$ shown experimentally, without any assumption other than the size of a. The values of a used are, however,

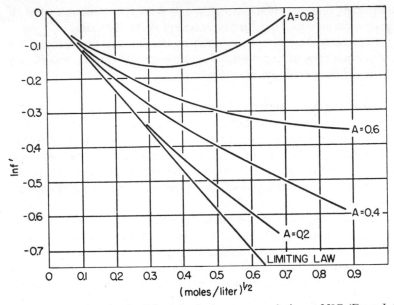

Figure 1.   The function $\ln f'$ for 1:1 salts in aqueous solution at 25°C. (From J. C. Poirier, *J. Chem. Phys.*, **21**, 972 (1953), with permission of the author.)

significantly larger than crystallographic radii and similar to those of the classical Debye–Hückel theory.

The results obtained using the S_2 term are highly interesting and important, but it is also necessary to recognize the deficiencies. Accepting the primitive model (Coulomb law and hard spheres), there remain two basic uncertainties; first the magnitude of the higher terms of the Mayer series, and second the arbitrary nature of the approach distance used. Consequently, it is not certain whether the good results obtained are a result of a valid physical theory or an artifact of using a physically unmeaningful parameter. The validity of the primitive model is, of course, also open to question, but neither question can be answered until a more definite answer can be obtained on the magnitude of the higher terms.

Integral Equation Calculations.   The effect of the higher terms can be obtained, although not in explicit form, by going to the integral equation methods. At the concentration of 1 m the density should be comparable to moderate gas densities, and reasonably accurate results for the particular model should be obtained.

Calculations have been made by Rasaiah and Friedman for the primitive model[56] and also a square-well potential model.[63,66] Calculations were made with both the Percus–Yevick and hypernetted-chain theories as well as some simpler equations, and the radial distribution functions obtained with each method were then used with both the pressure and compressibility equations to obtain the activity coefficients, osmotic coefficients, and excess free energy.

This whole set of calculations, along with those of Carley[55] which did not use the Mayer results, forms a test of the self-consistency and validity of the various sets of calculations. It was found that there were only minor disagreements among the various methods, but the hypernetted-chain method appeared to be somewhat more consistent than the Percus–Yevick. In addition to the internal consistency of the calculations, Monte Carlo calculations[35a] using the primitive model have also given excellent agreement with the hypernetted-chain results.[35b] There are, therefore, grounds for believing that these calculations give a reasonably accurate description of the properties of the model.

Unfortunately, it is not possible to reduce the results to a set of tables or a simple computer program, and so each case must be calculated separately. Consequently, there are relatively few results available; but several of the most interesting features of the results can be seen in Figure 2. The figure shows $F^{ex}/2RTI$ as a function of I (the molal ionic strength) for two types of theoretical calculation compared with the corrected experimental curve for lithium bromide. The theoretical curves include one using the limiting law plus the S_2 (B_2 in the figure) term as in Poirier's calculations (presumably

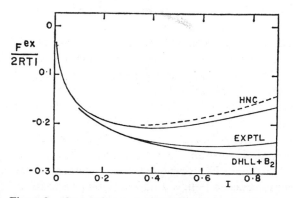

Figure 2. Comparison of $F^{ex}/2RTI$ from the hypernetted-chain and Debye–Hückel limiting-law plus B_2 approximations with the experimental values for lithium bromide in water at 25°C. Solid HNC curves, all $a_{ij} = 4.6$ Å; dashed HNC curves, $a_{++} = 3.6$ Å, $a_{+-} = 4.6$ Å, $a_{--} = 5.6$ Å. The experimental values have been corrected to the McMillan–Mayer standard states. [From J. C. Rasaiah and H. L. Friedman, *J. Chem. Phys.*, **48**, 2472 (1968) with permission of the authors.]

all $a_{ij} = 4.6$ Å) and two HNC calculations. For lithium bromide the value of a used in the empirically extended form of the Debye–Hückel theory is usually about 4.3 Å.[9] As already noted, the S_2 approximation gives the general shape of the experimental curves, but it is seen by comparison with the HNC curves that even at concentrations of 0.2 m or less, the higher terms make significant contributions.

It cannot be said with certainty that the short-range repulsive forces rather than the more complex Coulomb terms are responsible for the upward turn of the curves, but intuitively this does appear reasonable. It therefore seems probable that the high values of the HNC curves relative to the S_2 curve are due to the increased influence of the short-range forces. Also, the large value of the HNC curve relative to the experimental curve would indicate that the choice of a in the empirical or S_2 treatment is not physically reasonable, but is determined by the necessity of compensating for deficiencies of the treatment, and the deficiency is primarily the lack of a treatment of the short-range forces.

The two curves having different like-ion distances but similar unlike distances indicate that there will be a significant dependence of the thermodynamic properties on the interactions of like ions, as shown by the difference in the two curves. Although the interaction is small for a single salt, for mixtures, it could amount to a rather large part of the mixing effect. The amount of interaction is, of course, smaller for smaller ion sizes, since there is an increase in the electrostatic repulsion. The implication of the difference in curves is that like-charged ions can approach each other to distances where the short-range forces are significant; this conclusion does not depend on the use of the primitive model but should be of general validity.

In their next paper, Rasaiah and Friedman[63] made calculations of the osmotic coefficients of sodium chloride solutions for a square-mound potential model using the HNC method. Here the distances of closest approach of the ions were taken as the sum of their crystal radii or $r_+ = 0.95$ Å and $r_- = 1.81$ Å. Within the sums of any of these radii, the interactions were taken as those of hard sphere. For a distance between unlike ions of less than $r_+ + r_- + 2r_w$, where r_w is the radius of a water molecule or 1.36 Å, an additional small, positive, and constant term, d_{ij}, was added to the potential of average force. This could be considered as an average energy necessary to displace a water molecule and allow a transient ion pair to form. The existence of such transient ion pairs has been suggested by Fajans,[65] and Samoilov[66] has suggested that the energy of displacement of a water molecule might be quite small. Consequently, the picture is a plausible one; there are no arbitrary distance parameters in this model, and the only adjustable parameter is the height of the square well.

Results for the experimental osmotic coefficient of a sodium chloride solution compared with the theoretical curves for several different well potentials are shown in Figure 3. It is seen that the curve for the mound potential equal to $\frac{1}{4}kT$ matches the experimental results very well. Since the square-well potential is the only arbitrary parameter and the only assumptions are the very general ones of the HNC theory and the primitive model, these are very encouraging results.

In a subsequent paper,[66] Rasaiah made similar calculations for most of the alkali-metal halides and also calculated the excess energies E^{ex} for the solution. As for sodium chloride, Pauling radii[66] were used for the interionic distances, although, as Rasaiah points out, the sum of the two radii is the same for most of the partitions of radii that have been proposed. In any case, the general character of the curves is little affected by the choice of radii.

The parameter d_{ij}, the height of the square-well or mound potential, varied from $+1.3\ kT$ for lithium iodide to $-0.3\ kT$ for cesium iodide. The negative potential for cesium iodide could signify an increased lability of water molecules in the neighborhood of a cesium ion or negative hydration in Samoilov's[66] sense, although any short-range attraction would provide a valid explanation. In general the results obtained were very similar in quality to those shown in Figure 3 for NaCl. For the parameters giving the best fit, the theoretical results tended to be below experimental values at low concentrations and above at higher concentrations, and it is possible that a concentration-dependent potential would be necessary to improve this fit.

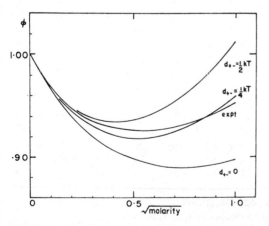

Figure 3. Osmotic coefficients calculated from the HNC equation using hard-sphere plus square-well potential. [From J. C. Rasaiah and H. L. Friedman, *J. Phys. Chem.*, **72**, 3352 (1968) with permission of the American Chemical Society and the authors.]

The E^{ex} for a solution is given by the analog of (20), which is for a gas. The difference is in the use of a potential of average force, which is generally temperature dependent, whereas the potential between two gas molecules is temperature independent. The temperature dependence requires the use of $\partial(w_{ij}/T)/\partial(1/T)$ rather than w_{ij} or u_{ij} in the equation and the equation for a solution is

$$(44) \qquad E^{ex} = \tfrac{1}{2} \sum_{i=1}^{\sigma} \sum_{j=1}^{\sigma} c_i c_j \int_0^{\infty} \frac{\partial(w_{ij}/T)}{\partial(1/T)} g_{ij} 4\pi r^2 \, dr$$

The summations are over the different types of solute ions or molecules in solution.

Calculations in which d_{ij} was considered as temperature independent did not compare well with experiment, and consequently a second parameter, $\partial d_{ij}/\partial T$ had to be introduced. The results of the calculations are shown in Table 1 for a number of salts, along with the values of the parameters and

Table 1   Comparison of E^{ex}/I^a for Aqueous Solutions of Alkali Halides at 25°C with the HNC Results for the Square-Well Model

		LiCl	LiBr	NaCl	NaBr	NaI	KCl	KBr	KI	CsCl
	d_{+-}/kT	0.90	0.90	0.25	0.30	0.35	0	0	0	−0.20
Molarity	$\frac{1}{k}\frac{\partial d_{+-}}{\partial T}$	1.57	1.39	1.03	1.22	1.28	0.65	0.81	0.88	0.60
0.1	Exptl[b]	100	108	84	75	72	80	67	57	55
	HNC	100	108	84	75	72	80	67	57	55
	Diff.	0	0	0	0	0	0	0	0	0
0.4	Exptl[b]	183	193	69	49	32	67	31	−11	−8
	HNC	130	152	79	51	43	66	22	−6	−6
	Diff.	53	41	−9	2	−11	1	9	−5	−2
0.7	Exptl[b]	239	250	26	−5	−33	27	−33	−98	−87
	HNC	148	163	46	4	−10	26	−40	−88	−94
	Diff.	91	87	−20	−9	−23	1	7	−10	−7
1.0	Exptl[b]	284	300	−28	−64	−105	−20	−110	−184	−172
	HNC	118	165	8	−50	−68	−19	−109	−171	−180
	Diff.	166	135	−36	−14	−37	−1	−1	−12	8

From J. C. Rasaiah, J. Chem. Phys., 52, 704, (1970).

[a] The heat of dilution in the McMillan–Mayer system, (cal/mole).

[b] The experimental value at each molarity was interpolated from the heats of dilution compiled by V. B. Parker (U.S. National Bureau of Standards, NJRDS-NBS2, Washington, D.C., 1965).

experimental results. It is seen that, with the exception of the lithium chloride and lithium bromide results, the agreement with experiment is very good, and again in view of the simplicity of the model, a very encouraging result.

It can be noted that the concept of hydration is implicitly contained in the calculation. In effect, the square-well potential allows for an explicit equilibrium between hydrated and unhydrated forms of the ions. Each calculation where the distance between ions is larger than the crystallographic radii is in effect assuming an exclusion process similar to that of the hydration picture. For the square well, however, no fixed hydration sphere is necessary, and the effect is temperature dependent as it should be physically.

These calculations are at a very early stage, and it is probable that they can be made with more realistic potential functions. Potentials depending on the positions of three particles might be included and also more realistic descriptions of the electrostatic forces between two ions at short distances.[29] The correlation functions for water molecules extend out to 7 Å or more,[67] and two ions perturbing these functions might interact with non-Coulombic forces out to relatively large distances. In spite of its simplicity, the square-well model appears to have given a remarkably accurate result, and it would appear that such models can form a framework within which the more complex models can be studied and interpreted.

7 Mixtures

Friedman's Theory.[45,68]   The thermodynamics of the solutions of mixtures of salts has recently been reviewed in detail by Harned and Robinson,[69] and only a few special aspects are discussed here. Mainly, these points concern Friedman's application of the Mayer theory to mixtures, some structural considerations, and some comments on linear relationships such as Harned's[9] rule.

As noted earlier for single-electrolyte solutions, theoretical results depending on the S_2 approximation have been generally superseded by the integral equation results. Recently results for mixtures have, however, appeared using the more complete calculations, much of the available experimental and theoretical work does use the series-expansion method. Although quantitatively the results may not be accurate, there are several conclusions, mainly of a limiting-law nature, that seem to be inherently valid and interesting.

The problem of converting from the molar, constant-solvent activity theoretical system to the molal, constant-pressure experimental system is more complex for mixtures than for single-salt solutions. In general, two different salt solutions at the same molal ionic strength will have different molar ionic strengths, and also the solvent activities in the two will be different. The correction has to be made as a function of the ion or salt fraction, and generally there are insufficient experimental data available to be able to make more than a first approximation.

As given earlier, the excess free energy of a solution of a single salt or mixture is given by

$$(45) \qquad \frac{F^{\text{ex}}}{kTV} = S = \frac{\kappa^3}{12\pi} + \sum_{n=2}^{\infty} e^n B_n$$

where each coefficient B_n is a function only of the molar ionic strength, κ, and each term $e^n B_n$ is a sum of terms for all possible combinations of ions whose sum of exponents is n as in (39) for S_2. Each B_{ij} is specific for that particular combination. The thermodynamic changes of mixing at constant, κ, can then be expressed as the sum of the terms B_{ij} multiplied by the change in concentration or mixing of that particular combination of ions.

The theoretical coefficients must be related to the experimentally measured changes and these are taken in the form (for a two-component system)

$$(46) \quad \Delta_m G^{\text{ex}}(yI) = I^2 RTy(1 - y)[g_0 + g_1(1 - 2y) + g_2(1 - 2y)^2 + \cdots]$$

where the g_i are constants at fixed ionic strength and y and $1 - y$ are the ionic strength fractions of the pure components in the mixture. The excess Gibbs free energy of mixing is defined by

$$(47) \qquad \Delta_m G^{\text{ex}}(y, I) = G^{\text{ex}}(y, I) - y G^{\text{ex}}(1, I) - (1 - y) G^{\text{ex}}(0, I)$$

where $G^{\text{ex}}(y, I)$ is the excess free energy of a quantity of solution containing 1 kg of solvent at mixing fraction, y, at molal ionic strength, I.

Details of the conversion between the standard states and of the relations between the quantities g_i and B_n coefficients are given in Friedman's book and papers and are not given here. It should be noted, however, that higher B_n coefficients can contribute to g_0 and other low terms of the g series, so these terms are not necessarily dependent only on low-order interactions of the ions.

At low concentrations where the difference in standard states is small, the excess free energy of mixing two 1:1 salt solutions with a common ion is given by (47). The salts are designated by the ions 1,3 and 2,3.

$$(48) \quad \Delta_m G^{\text{ex}} = \frac{I^2 RT}{2} y(1 - y) \frac{\sum c_s Z_s^2}{\sum m_s Z_s^2} [B(1, 2) - B(1, 1) - B(2, 2)]$$

The coefficients $B(i, j)$ are identical with the B_{ij} given by (40) of the theoretical section. It can be seen that terms involving the common ion of mixing,

3, do not appear in the S_2 terms, and only terms involving ions of like sign do appear. Another note is that the ratio of molar to molal ionic strengths $\sum c_s Z_s^2 / \sum m_s Z_s^2$ occurs and is an approximation to the standard-state correction term. The $B(i, j)$ can be calculated from a table given by Friedman,[45] or more efficiently with a simple computer program given by Meeron.[70]

As pointed out earlier, it is unlikely that the calculations would be accurate in view of the deficiencies of the S_2 term being used alone. However, as the concentration decreases, this term should have increased importance, and some interesting and very probably valid conclusions can be obtained in the limit of κ approaching 0. For monovalent ions and for like ion pairs only, all the q_{ij} in the expressions for $B(i, j)$ are equal, and the expression in brackets of (48) becomes

$$(49) \quad 4\pi[B(1, 2) - B(1, 1) - B(2, 2)] = \int_0^\infty e^{q''}[2k_{12} - k_{11} - k_{22}]r^2 \, dr$$

where $k_{ij} = e^{-w^*/kT} - 1$ and refers only to the short-range interactions of the ions. It is evident that, for mixtures with a common ion and at low concentrations, the effects arise from short-range forces among like ion pairs. For κ equal to 0 the expression remains finite and indicates that the coefficient g_0 in (43) does not go to 0. For 0 charge, (43) has the same form as the expression used in regular solution[10] theory; although it should be noted that the theory is primarily a lattice rather than a gas theory. In addition to the finite value of g_0 at 0 concentration, Friedman[68] also points out that $1/g_0 \, dg_0/d(I)^{1/2}$ has a finite and definite value at 0 concentration. A second limiting law is derived by Friedman[68] for mixtures involving unsymmetrical electrolytes. According to this law, g_0 goes to $-\infty$ as $\ln I$. This law depends on the long-range Coulomb forces rather than the short-range forces of the first law.

Experimentally, these limiting laws will be extremely difficult to test, although there is evidence from heats of mixing experiments of unsymmetrical electrolytes that h_0 corresponding to g_0 is increasing with decreasing concentration,[71] in accordance with theory.

A conclusion, which was apparent in the S_2 calculation and was made more emphatic by the integral equation work, concerns the relatively large contribution of like-ion short-range interaction to the thermodynamics of mixing. It has been considered for many years that because of the electrostatic repulsion of like ions there would be no specific short-range interaction of such ions. There would only be a general effect in which all ions of like sign and charge would interact with each other in a uniform nonspecific manner. In introducing these principles, Brønsted[72] used coefficients to express specific unlike-charge interactions and general salting-out terms to express like-ion interactions. However, Guggenheim[73] in a very similar treatment did not use the salting-out coefficient. Brønsted and his co-workers[74] did in fact obtain a good deal of data on activity coefficients in mixtures where the principle of nonspecificity of like-ion interactions was apparently valid. According to these principles, g_0 and h_0 for the symmetrical mixing of salts with a common ion should be 0 and this was found true for g_0. More recently, work on the heats of mixing of electrolytes with a common ion[71] has shown that generally h_0 is not 0 but can have in some cases relatively large values.

There is therefore some degree of discrepancy between the theoretical calculations and the heat measurements on the one hand, which both indicate specific like-charge ion interactions, and the Brønsted principle, which denies this point. In part, some of the discrepancy occurs because different systems were studied in the different cases, and also excess free energies of mixing are usually small and difficult to measure accurately. Nevertheless, it is fairly likely that the Brønsted principle is not generally valid but is true only in special cases where the ions have an unusually small influence on the surrounding water organization or where there is compensatory behavior in the ion cospheres. The deviations are often very small, however.

Although the primitive model (i.e., charged hard spheres) is certainly inadequate for the study of mixtures, it would nevertheless appear true that using the integral equations, minimum conditions can be set for like-ion interactions. As the ion size grows smaller, the electrostatic repulsion grows larger, and below certain sizes, changes in the hard-sphere radius of the ions (for like-pair interactions) contribute insignificantly to the excess free energy. Conversely, above certain sizes the short-range forces will contribute regardless of their nature. Estimates of these minimum sizes can be obtained with the primitive model, relative, of course, to the degree that a given size of change in the excess free energy can be considered significant.

In water at 25°C there is some correlation in the position of the molecules out to 7 Å or more.[67] If an ion is taken to perturb the structure of water to a distance somewhere of this magnitude, then it would appear from the calculations that all singly charged ions would have some degree of like-ion interaction, although this is often quite small.

A final point in the discussion of mixtures concerns the interpretation of Harned's rule,[9] namely

$$(50) \qquad \log \gamma_1 = \log \gamma_{1(0)} - \alpha_{12} X_2$$

or other linear relationships between a salt component fraction, X, and a thermodynamic quantity such as the partial molal free energy, $\log \gamma_{\pm}$, or the heat. In the equation γ_1 is the mean activity coefficient of salt 1 in the mixture, $\gamma_{1(0)}$ is the mean activity coefficient of the pure salt 1 solution, X_2 is the fraction of component 2 at constant molality, and α_{12} is an empirical coefficient. This linear behavior would occur if binary short-range interactions of unlike-charge ions only were of importance to the excess free energy and the simultaneous short-range interactions of higher multiples of the ions were negligible. Linear relationships such as Harned's rule do occur rather frequently for mixtures and are often quite accurate, and this has been taken as evidence that the binary interactions are of much higher importance than any others.

There are several reasons for doubting that this is the case, and for believing that the linear behavior is the result of other not necessarily known factors, however. One reason is the substantial size of the higher interaction terms as shown by comparison of the S_2 and integral equation calculations. The S_2 calculations take into account only two-particle short-range interactions, whereas the integral equation accounts for multiple interactions, and the results are substantially different. This must be qualified by the possibility that the more complex Coulomb interactions of the integral equations also could contribute to the upturn of the activity coefficient curves.

A second reason for believing that multiple interactions are important is that the linear behavior appears to hold better at high concentrations[1] (i.e., above 1 m), rather than at lower concentrations where binary encounters would be expected to predominate. A third reason lies in the range of the water correlation function around each ion. If this range in the neighborhood of 7 Å is a valid one, then in a 1-N solution each ion cosphere is on the average interacting simultaneously with a number of its nearest neighbors.[67]

A possibility for an alternate explanation of the linear behavior is in the analogy with regular solutions already made. In its simplest form, the excess free energy for a regular solution can be written in the form[10]

$$(51) \qquad \Delta_m G^{ex} = y(1 - y)w_{AB}$$

w_{AB} is the change in energy on exchanging an A for a B particle in the respective lattices and is equal to

$$(52) \qquad w_{AB} = \int (2k_{AB} - k_{AA} - k_{BB}) \, dr$$

where the k_s can have the same form as for ions as in (49). Equation 52 requires that the distribution of particles be completely random and the equation is derived for a lattice, although the coordination number drops out. For the ionic solution, if the ion-distribution functions are unchanged by mixing, the potentials depend only on two particles, and the water activity is unchanged on mixing, it is possible that a similar theory might apply. It should be noted that the k_{ij} for ions are temperature dependent, which is not true of the w_{AB} of regular solution theory.

The equations are certainly oversimplified, but the point to be made is only that the linear behavior need not imply binary interactions alone.

8   Structural Concepts in Solution

The use of such concepts as hydration or a variable dielectric constant as well as the use of power-series expansions has been criticized earlier in this review as being primarily empirical, and serious doubts about the meaning of the parameters were voiced. In each case it has been pointed out that, without a detailed knowledge of the intermolecular geometrical relationships and forces, it would be extremely difficult to determine the nature of the various parameters. Also, in the development of the Mayer approach to electrolyte solutions, it was pointed out that more accurate developments of the calculations would require a detailed knowledge of the potential of average force between ions. This requires knowing the correlation and instantaneous configurations or structures of the solvent molecules in solution.

The concept of structure has thus been introduced as being fundamental to this study, but the question is whether, as before, there is in principle a physically valid effect that in application turns out to be empirical and qualitative as well as ambiguous. There is no doubt that this is often the case, particularly since, in comparison to electrostatic effects, the energy changes of purely structural effects are often very small. The initial act of placing ions in aqueous solution at infinite dilution involves major structural as well as electrical interactions. As the ion concentrations increase, however, for ions with strong fields it is probable that the effects due to electrical and short-range forces vary much more than those entirely attributable to structure.

As with the other effects, it is very difficult to make a direct calculation from basic principles, and the applications of structure have been also qualitative and empirical. Some semiempirical estimates have been made, however, of entropy effects associated with purely structural processes.[6]

In spite of these defects, an accurate theoretical consideration of electrolyte-solution thermodynamics cannot avoid the experimental evidence of the highly structured nature of water. A quantitative treatment must consider the basic geometry, correlations, and forces of the system, and it is in this sense that structure is fundamental to the discussion.

In the experimental sense that structure is discussed in an aqueous salt solution, what is commonly considered are various effects on the entropy, specific heat, or viscosity of the solution.[6,7,76] The entropy of forming a given ionic solution may be larger or smaller than one could expect from the entropy of solution of other ionic or nonionic substances using reasonable models for comparison. Similarly, increases in the apparent molal specific heat or the viscosity of a solution are taken as evidence of structure in solution. Generally one speaks of increases and decreases in the structure rather than of a specific type of organization. These are concepts that, superficially at least, are quite different from the inorganic chemist's or crystallographer's use of structure, which refers to the fixed angular and spatial relationships of the atoms and bonds in a molecule or crystal.

In an ideal crystal, each unit has a definite and permanent spatial relation to every other unit in the macroscopic crystal. Each vibrational motion centers about a fixed point in an unchanging coordinate system. In a liquid, the relative positions of two particles at macroscopic distances are always random, and even at close range there is at best only a temporary spatial relation between any pair of units in the system. Hence, for a liquid, the concept of structure must inherently be different from that in a solid, even without the added complication that thermodynamic or transport concepts rather than geometric concepts are involved. Since there are such basic differences, it becomes important to understand exactly what is meant by the concept of structure in electrolyte solutions and the legitimacy of using the term structure on the basis of a thermodynamic type of measurement.

If an isolated group of water molecules in close proximity is considered, it is clear that the particles cannot have completely random positions relative to one another, but in large measure the various repulsive and attractive forces will determine a state where on the average there are more or less definite spatial relationships. Both the positions of the centers and orientations of the axes of the molecules will be conditioned by the positions and orientations of adjacent molecules.

One could imagine that different types of configurations could exist for the molecules. One extreme type would be a small group of configurations or a single configuration that had a sharp minimum in the potential energy relative to that of other possible configurations. Since a small group or a single configuration was involved, the entropy of this state would be small, the entropy being proportional to the logarithm of the number of possible configurations of the system. In the second state, a relatively large number of approximately equal energy configurations would exist and would have a high entropy. The first situation would be described as highly structured and the second would have a low degree of structure.

The correlations in position and angle of two molecules can extend well beyond immediate neighbors. One can examine the probability of finding the center of a water molecule (in liquid water) at some given distance from a particular molecule and also the probability of the axis of the second molecule pointing in a particular direction when the reference molecule has a specified direction. For large distances between the molecules, the probability of finding the second molecule is simply the random probability, depending only on the density and not the distance; the directional probability is also random. For short distances between the molecules and at low temperatures, the probability of finding the second molecule depends on the distance. The probability will be 0 if the radii strongly overlap, have a maximum just outside a center-to-center distance of one diameter, and have two or more additional maxima before reaching the long-distance random value. This radial distribution function can be determined by X-ray diffraction and is shown in Figure 4 for water at several temperatures.[67] It is seen that up to about 7 Å or approximately three molecules from the center, the probability is still significantly different from random. The angular correlation cannot be determined from X-rays and is more complicated, but it must exist and knowledge of it would be necessary to calculate the thermodynamic properties of the system.

The positions given by the radial distribution function (or the angular correlation) are average and do not at all relate to the positions of any specified particles. There is, nevertheless, the implication of structure or order rather than randomness. The peaks and valleys of the distribution curve imply that, on the average, the molecules are preferentially constrained to stay out of certain regions and preferentially constrained to stay in other regions.

In the simplest interpretations, these constraints are reflected in a loss of entropy of the system. Any substance or effect that increases the distance where the distribution function is different from random or that will raise the height or narrow the breadth of the peaks increases the degree of order of the system. This increase in order can be associated with a decrease in entropy and an increase in structure. Conversely, substances that flatten or broaden the peaks and decrease the range increase the entropy and decrease the degree of structure.

As the temperature increases, the degree of order decreases and, in a sense, the structure gradually melts. As a consequence of this structure melting, the specific heat of the system is raised for structure-making substances and decreased for structure breakers.

The potential of average force, $W(r)$, is also a useful concept in considering structure and can be defined in terms of the radial distribution function, $g(r)$[11]

$$(53) \qquad g(r) = e^{-W(r)/kT}$$

In an average sense it is a measure of the force between the reference and outer molecules, since the derivative of a potential with respect to distance is a force. If a foreign particle perturbs a water molecule within the correlation range of a second molecule, it is in effect exerting a force on the second molecule. The size and range of this force depend, of course, on the nature of the perturbation and the degree of the correlation. Two foreign particles in water would presumably feel this force if the correlation ranges of the water molecules adjacent to each were to overlap.

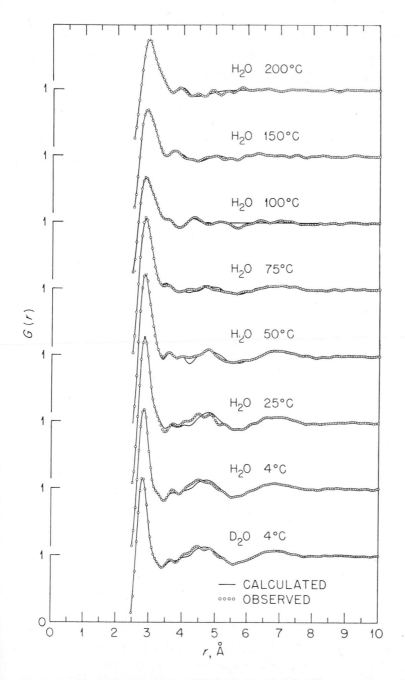

Figure 4. Observed and model radial distribution function for water at several temperatures: circles, observed function; solid curve, as calculated from icelike model. [From A. H. Narten, M. D. Danford, and H. A. Levy, *Discussions Faraday Soc.*, **43**, 97 (1967) with permission of the Faraday Society and the authors.]

These general concepts can be applied specifically to ionic solutions. The configurations of the water molecules adjacent to the ion must be considered, as well as the effect of the adjacent ion on the correlation functions of adjacent with distant water molecules. In the classical view of hydration, the water molecules form a symmetrical tetrahedron or octahedron around the ion with the appropriate end of the dipole pointing at the center of the ion. For large charge-to-radius ratios of the ions, the freedom of motion of each water molecule adjacent to the ion is very limited, and the water molecules surrounding the ions have only a narrow range of configurations available to them. This is therefore a highly structured state of low entropy. Since the positions of the dipoles are narrowly determined in the four or six directions of the polyhedron, it is unlikely that such a configuration will be compatible with the normal hydrogen-bonded arrangement in water. Hence it is probable that the correlations of the inner and outer water molecules will decrease, and there will be an intermediate region of increased entropy and decreased structure.[6]

Another conceivable situation would be one where relatively small ions fit into the water matrix distorting but not necessarily disrupting the structure.[77] The inner shell would not be as ordered as in the classical picture, but more of the other correlation would be preserved. The total degree of structure and entropy of such a system would be similar to the classical case.

If the ions produce electrical fields comparable in strength to those produced by the dipoles of water molecules, other situations can arise. The orienting force on a water molecule adjacent to an ion is due to a summing of the electric fields of the ion and the surrounding water molecules. If these fields are in opposition and approximately cancel, there will be relatively little correlation in position and angle of the adjacent water molecule with either the ion or the surrounding water molecules. Rather than the fixed, sharply oriented configuration of the classical hydration picture, many configurations with a variable number of water molecules in contact with the ion could occur. The entropy in this situation would be high and the degree of structure low. The water molecules adjacent to the ion would diffuse more freely than those in pure water, resulting in the negative hydration of Samoilov.[66]

For very weak fields, such as those in the large quaternary ammonium ions, still another situation is possible. Here the forces due to the ions are so weak that adjacent water molecules are effectively shielded from constraints on the one side and an increased degree of correlation is obtained with the outer water structure on the other side. It is even possible that a clathrate-like structure could be produced under these circumstances, although there is no direct evidence for such a state.[78] The water molecules adjacent to the ion cannot vibrate into an open space as at a gas-water inter-face, and the result is a lowering of the entropy and increased structure.

Direct evidence for this increased degree of correlation can be obtained from the radial distribution function of a tetra-*n*-butylammonium fluoride solution obtained by X-ray diffraction.[78] The distribution function is shown in Figure 5 along with that for pure water. The solution studied had a ratio of 41 water molecules to 1 salt ion pair and was slightly lower in concentration than the crystal hydrate[79] of 32.8 water molecules. The dotted and dashed curves in the figure are calculated distribution functions for two models of the solution which are not of direct concern here although they can be mentioned. The dashed curve is based on the Danford–Levy–Narten icelike structure[67] and the dotted curve is based on the crystalline clathrate model; and it can be noted that the icelike structure is considerably closer to the experimental curve than the clathrate[79] model.

Curves such as these show in essence the electron density probability, and intramolecular electron density correlations will appear, as well as those due to ion-oxygen and oxygen-oxygen correlations, with the protons giving very small contributions.

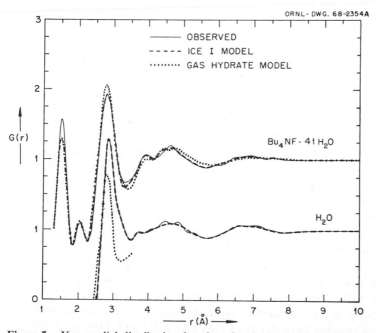

Figure 5. X-ray radial distribution functions for the system Bu$_4$NF–H$_2$O at 25°C. solid curves, observed values; dashed curves, Ice I model; dotted curves, gas-hydrate model. (From A. H. Narten and S. Lindenbaum, *J. Chem. Phys.*, **51**, 1108 (1969) with permission of the authors.)

Since there are so many factors contributing to a distribution function of this type, it is difficult to interpret exactly. Nevertheless, if the extraneous scattering is estimated and corrected for, the scattering does appear remarkably similar to that of pure water. The maximum for the nearest neighbor water distance has decreased from 2.85 to 2.80 Å, and also the number of water molecules in this first peak has decreased. Both of these factors suggest an increase in structure. Thermodynamic evidence also suggests that the Bu_4N^+ ion is one of the strongest structure-promoting cations known.[76]

Radial distribution curves for the ammonium halide[80] solutions do not show the shortening of nearest-neighbor distances or the decrease in coordination number. NH_4I solution does in fact show a lengthening of the nearest neighbor oxygen–oxygen distance, and the I^- ion is thermodynamically considered as a strong structure breaker.[6]

Up to this point the properties of isolated ions have been considered; however, in studying aqueous solutions at finite concentrations, we are mainly interested in the mutual interactions of the ions. The properties of the isolated ions are those of the standard state; the properties usually studied experimentally are the changes from the standard state. From the structural point of view, these changes reflect the increasing degree of intersection of the ion cospheres.[7]

A number of possible ion-solvent interactions have been mentioned, and it would appear that a very large and complex set of cosphere interactions could occur. Only a few of the possibilities are discussed here. Desnoyers[81] and co-workers have recently given an extensive discussion of these interactions and the ideas expressed here are rather similar.

The question can be asked, What happens when two of the zones where the correlation functions have been significantly perturbed by nearby ions overlap? Since these zones might extend 7 Å or more from the center of the ion, the range of this interaction could be quite large and well into the range where the theoretical calculations showed that like-ion interactions could be of significance.

For spheres with a high degree of correlation and low entropy, the overlap will result in an increase in entropy of the system, since the total volume perturbed by the ions is now smaller and the perturbed volume has a relatively low entropy. In a similar way, the overlap of disordered spheres will decrease the entropy, since the total volume of extra disorder is now smaller. A tacit assumption is that there is no proportionate increase or decrease of entropy in the overlapping regions. Generally, in these systems the charges in excess enthalpy and entropy have the same sign and $T\Delta^{ex}$ and ΔH^{ex} generally are substantially larger than the change in ΔG^{ex}.

Another type of structural influence might occur for small ions with cospheres relatively strongly correlated with the external water structure. An oppositely charged ion that could enter this structure without disrupting it could form a relatively stable and low-energy pair under these circumstances. This mechanism has been suggested to explain the low activity coefficients of LiOH or LiAc.[77] Many more situations could be suggested, and several of these are discussed in the Desnoyers paper.[81]

Although not a thermodynamic property, the viscosity of electrolyte solutions has also been important in suggesting the validity of the structural concept.[7] The viscosity relationships of electrolyte solutions are in general highly complicated, but some simple qualitative relationships can be discussed.

The viscosity can be considered as a measure of the friction between adjacent, relatively moving parallel planes of the liquid. Anything that increases the interaction between the planes will raise the friction and therefore increase the viscosity; conversely, effects decreasing the interaction will lower the viscosity.

If large (on the molecular scale) spheres are placed in the liquid, the planes will be keyed together and the viscosity increased. Also, if the average degree of hydrogen bonding between the planes is increased, the friction between the planes will increase. The increase in the average degree of hydrogen bonding in the neighborhood of a structure-promoting ion will have a similarity to the sphere effect and to the pure hydrogen-bonding effect. An ion with a large rigid cosphere will act in the same way as a solid sphere placed in the liquid to increase interplanar friction. Similarly, an ion increasing the degree of hydrogen bonding or, equivalently, the degree of correlation among adjacent water molecules, will also increase the viscosity. Conversely, ions destroying correlation would decrease the viscosity. It might be added that the Coulomb interaction of the ions also keys the planes together and gives a $c^{1/2}$ dependence on concentration;[82] however, this effect is quite small at moderate concentrations.

In considering structural effects it is always necessary to remember that they are one part of a very complex interaction. There are strong electrical forces between the ions and between the ions and water, and it is not really possible to separate them all. Nevertheless, if careful judgment is used, valid conclusions can be drawn from entropy and other thermodynamic measurements relating to the degree of structure and order of the system.

9 Cooperative Effects

A final aspect of structure concerns the possibility of cooperative changes in the solvent structure induced by the solute.[30,85] The physical evidence for such changes consists in relatively abrupt changes in the slope or curvature of various properties such as the apparent molal volume or heat content of the solute.

Such abrupt changes probably cannot be due to a site-filling effect in which the solute molecules enter a lowest energy group of sites and the transition occurs when this lowest set is filled. In order that there be a sharp separation of the two types of sites, a large difference in the free energy per site is necessary. Since no sharp changes in the partial molal free energy (not the slope) are seen, a cooperative effect involving only small energy changes for each ion or molecule is probable.[77]

Substantial evidence for the existence of such transitions has been presented for the case of alcohols,[83] acetone,[84] and other solutes, and the effects appear to be quite large; there is no evidence for a discontinuity in any thermodynamic derivative, however. The explanation offered is that the solute molecules enter spaces in the water structure without essentially destroying the structure. As the concentration of solute increases, the structure becomes unstable and eventually is transformed in a cooperative manner.

For electrolyte solutions, Samoilov[66] has suggested a very similar model. According to Samoilov, for very dilute solutions the structure is determined by the forces of pure water, whereas at high concentrations the structure is similar to that of crystal hydrates. It is suggested that a transition of the critical type can occur at some intermediate point. Samoilov also suggests the eutectic composition as the probable neighborhood of the transition. On this point it could be argued that the transition might take place when significant overlap of the correlation ranges of an ion and its nearest neighbor occurred, and this could be at 1 m or less. Experimentally, it might be expected that a transition involving relatively long-range interactions of the cospheres would show up in extremely small changes, and this is the case.[30,85]

Measurements of the apparent molal volume of several alkali-halide solutions have shown what are apparently discontinuous slopes of ϕ_v versus $\sqrt{c}$ curve.[85] Figure 6 shows $\phi_v/\Delta\sqrt{c}$ for several salts where $\Delta\phi$ and $\Delta\sqrt{c}$ represent the differences of succeeding values of these quantities.

The error bar shown for the sodium iodide curve represents an uncertainty of ± 2 parts in 10^7 in the density, and consequently the effects are extremely small. The curves are reproducible and the effects substantially larger than the estimated uncertainty. Also, earlier density work and curves for other properties such as heat and viscosities appear consistent with the existence of this effect. Consequently, the evidence appears convincing, although the effects require extremely delicate measurements and independent confirmation remains necessary.

Another type of cooperative transition could be one in which the radial

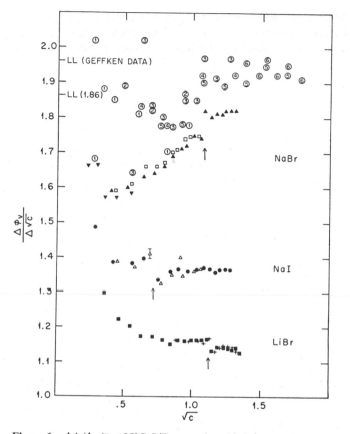

Figure 6. $\Delta\phi_v/\Delta\sqrt{c}$ at 25°C. LiBr, ■ and + ; NaI, $\triangle$ and ● ; NaBr (this work), □, ▲ and ▼; W. Geffcken and D. Price, *Z. Physik. Chem.*, **B26**, 81 (1934); W. Geffcken, A. Kruis, and F. Solana, points (1) through (6). The Geffcken results are displaced upward 0.1 unit, the theoretical limiting law is shown by LL, and a vertical arrow on each curve indicates where an anomaly may be present. [From F. Vaslow, *J. Phys. Chem.*, **73**, 3745 (1969).]

distribution functions for the ions change abruptly, as opposed to a change primarily involving the solvent or ion-solvent relationships, as the Samoilov picture might imply. A transition of the ionic type could be the change in the ion radial distribution function from exponential to oscillating form found by Kirkwood and mentioned earlier.[31] A number of different theoretical calculations[32-34] have suggested the existence of this transition, and the concentrations suggested are in the range found experimentally. There are many approximations in the calculations, and there is no definite evidence that a discontinuity in the excess free energy or any of its derivatives, such as partial molal excess free energy or volume, should occur. Consequently, the applicability of such theories is unknown. That both solvent and ion distribution functions are involved combining both mechanisms is also a good possibility.

Acknowledgments

The author wishes to thank Dr. H. L. Friedman, Dr. S. Lindenbaum, and Dr. M. J. Pikal for critically reviewing the manuscript as well as many earlier discussions in the course of writing this chapter.

REFERENCES

1. G. N. Lewis and M. Randall, *Thermodynamics*, rev., K. S. Pitzer and L. Brewer, McGraw-Hill, New York, 1961.
2. J. G. Kirkwood and J. C. Poirier, *J. Phys. Chem.*, **58**, 591 (1954).
3. J. E. Mayer, *J. Chem. Phys.*, **18**, 1426 (1950).
4. G. Scatchard, *Chem. Rev.*, **8**, 321 (1931).
5. H. S. Frank and A. L. Robinson, *J. Chem. Phys.*, **8**, 933 (1940).
6. H. S. Frank and M. W. Evans, *J. Chem. Phys.*, **13**, 507 (1945).
7. R. W. Gurney, *Ionic Processes in Solution*, McGraw-Hill, New York, 1953.
8. J. D. Bernal and R. H. Fowler, *J. Chem. Phys.*, **1**, 515 (1933).
9. H. S. Harned and B. B. Owen, *The Physical Chemistry of Electrolyte Solutions*, Reinhold, New York, 1958.
10. R. H. Fowler and E. A. Guggenheim, *Statistical Thermodynamics*, Cambridge University Press, London, 1939.
11. T. L. Hill, *Statistical Mechanics*, McGraw-Hill, New York, 1956.
12. J. G. Kirkwood, *J. Chem. Phys.*, **2**, 767 (1934).
13. H. S. Frank and P. T. Thomson, *J. Chem. Phys.*, **31**, 1086 (1959).
14. R. M. Fuoss and L. Onsager, *J. Phys. Chem.*, **61**, 668 (1957).
15. G. A. Martynov, *Sov. Phys. Usp. (Engl. transl.)*, **10**, 171 (1967).
16. E. Hückel, *Physik. Z.*, **26**, 93 (1925).
17. N. Bjerrum, *Z. Anorg. Chem.*, **109**, 275 (1920).
18. R. A. Robinson and R. H. Stokes, *Electrolyte Solutions*, Butterworths, London, rev. ed., 1965.
19. E. Glueckauf, in *The Structures of Electrolytes Solutions*, ed., W. J. Hamer, Wiley, New York, 1959.
20. M. H. Lietzke, R. H. Stoughton, and R. M. Fuoss, *Proc. Natl. Acad. Sci. (U.S.)*, **59**, 39 (1968).
21. H. Müller, *Physik. Z.*, **29**, 78 (1928).
22. E. A. Guggenheim, *Trans. Faraday Soc.*, **56**, 1152 (1960).
23. T. H. Gronwall, V. K. LaMer, and K. Sandved, *Physik. Z.*, **29**, 358 (1929).
24. G. Scatchard, *Z. Physik. Chem. (Leipzig)*, **228**, 354 (1965).
25. N. Bjerrum, *Kgl. Danske Videnskab Selskab, Mat-Fys. Medd.*, **7**, No. 9 (1926).
26. R. M. Fuoss, *Trans. Faraday Soc.*, **30**, 967 (1934).

27. J. T. Denison and J. B. Ramsey, *J. Amer. Chem. Soc.*, **77**, 2615 (1955).
28. B. E. Conway, in *Annual Review of Physical Chemistry*, vol. 17, Annual Reviews, Palo Alto, Calif., 1966, p. 481.
29. S. Levine and D. K. Rosenthal, in *Chemical Physics of Ionic Solutions*, Eds., B. E. Conway and R. G. Barrados, Wiley, New York, 1966.
30. F. Vaslow, *J. Phys. Chem.*, **70**, 2286 (1966).
31. J. G. Kirkwood, *Chem. Rev.*, **19**, 275 (1936).
32. T. H. Berlin and E. W. Montroll, *J. Chem. Phys.*, **20**, 75 (1952).
33. F. H. Stillinger, Jr., and R. Lovett, *J. Chem. Phys.*, **48**, 3858 (1968).
34. C. W. Outhwaite, *J. Chem. Phys.*, **50**, 2277 (1969).
35. J. C. Poirier, in *Chemical Physics of Ionic Solutions*, Eds., B. E. Conway and R. G. Barrados, Wiley, New York, 1966.
35a. D. N. Card and J. P. Valleau, *J. Chem. Phys.*, **52**, 6232 (1970).
35b. P. N. Vorontsov-Vel'yaminov and A. M. El'yoshevich, *Elektrochimia (Engl. transl.)*, **4**, 1430 (1968).
36. R. M. Fuoss, *J. Amer. Chem. Soc.*, **80**, 5059 (1958).
37. G. Scatchard, in *Proteins, Amino Acids and Peptides*, Eds., F. J. Cohn and J. T. Edsall, Reinhold, New York, 1943.
38. W. G. McMillan, Jr., and J. E. Mayer, *J. Chem. Phys.*, **13**, 276 (1945).
39. E. Glueckauf, *Trans. Faraday Soc.*, **51**, 1235 (1955).
40. L. Onsager, *Chem. Rev.*, **13**, 73 (1933).
41. H. L. Friedman, *J. Chem, Phys.*, **32**, 1351 (1960).
42. J. C. Poirier, *J. Chem. Phys.*, **21**, 965, 972 (1953).
43. N. N. Bogoliubov, *Problems of a Dynamical Theory in Statistical Physics*, Goslekhizdot, Moscow, 1946; Engl. transl. in *Studies in Statistical Mechanics*, vol. 1, Eds., J. de Boer and E. E. Uhlenbeck, North Holland Publishing, Amsterdam, 1963.
44. G. Kelbg, in *Chemical Physics of Ionic Solutions*, Eds., B. E. Conway and R. G. Barrados, Wiley, New York, 1966.
45. H. L. Friedman, *Ionic Solution Theory*, Interscience, New York, 1962.
46. J. E. Mayer and M. G. Mayer, *Statistical Mechanics*, Wiley, New York, 1940.
47. H. D. Ursell, *Proc. Cambridge Phil. Soc.*, **23**, 685 (1927).
48. J. P. O'Connell and J. M. Prausnitz, *Ind. Eng. Chem.*, **60**, 36 (1968).
49. J. E. Mayer and E. Montroll, *J. Chem. Phys.*, **9**, 2 (1941).
50. M. Born and H. S. Green, *Proc. Roy. Soc. (London), Ser. A*, **188**, 10 (1946).
51. L. S. Ornstein and F. Zernike, *Proc. Acad. Sci. Amsterdam*, **17**, 793 (1914).
52. J. K. Percus and E. J. Yevick, *Phys. Rev.*, **110**, 1, (1958).
53. J. M. J. von Leeuwen, J. Groeneveld, and J. DeBoer, *Physica*, **25**, 792 (1959).
54. E. Meeron, *J. Math. Phys.*, **1**, 192 (1960).
54a. J. S. Rowlinson, *Contemp. Phys.*, **5**, 359 (1964).
55. D. D. Carley, *J. Chem. Phys.*, **46**, 3783 (1967).
56. J. C. Rasaiah and H. L. Friedman, *J. Chem. Phys.*, **48**, 2742 (1968).
57. J. G. Kirkwood and E. M. Boggs, *J. Chem. Phys.*, **48**, 2742 (1968).
58. B. J. Alder and T. E. Wainwright, *J. Chem. Phys.*, **31**, 459 (1960).
59. E. Haga, *J. Phys. Soc. Japan*, **8**, 714 (1953).
60. H. L. Friedman, *Mol. Phys.*, **2**, 190, 436 (1959).
61. E. Meeron, *J. Chem. Phys.*, **28**, 630 (1958).
62. A. R. Allnatt, *Mol. Phys.*, **8**, 533 (1964).
63. J. C. Rasaiah and H. L. Friedman, *J. Phys. Chem.*, **72**, 3352 (1968).
64. P. S. Kamanathan and H. L. Friedman, *J. Chem. Phys.*, **54**, 1086 (1971).

64a. H. L. Friedman, in *Modern Aspects of Electrochemistry*, vol. 5, Eds., J. O'M. Bockris and B. E. Conway, Plenum Press, New York, 1969.

65. K. Fajans and O. Johnson, *Trans. Electrochem. Soc.*, **82**, 273 (1942).

66. O. Ya. Samoilov, *The Structure of Aqueous Solutions and the Hydration of Ions*, Consultants Bureau, New York, 1965.

66a. J. C. Rasaiah, *J. Chem. Phys.*, **52**, 704 (1970).

66b. L. Pauling, *The Nature of the Chemical Bond*, 3rd ed., Cornell University Press, Ithaca, N.Y., 1960.

67. A. H. Narten, M. D. Danford, and H. A. Levy, *Discussions Faraday Soc.*, **43**, 97 (1967).

68. H. L. Friedman, *J. Chem. Phys.*, **32**, 1134 (1960).

69. H. S. Harned and R. A. Robinson, *Multicomponent Electrolyte Solutions*, Pergamon Press, London, 1968.

70. E. Meeron, *J. Chem. Phys.*, **26**, 804 (1957).

71. R. H. Wood, J. D. Patton, and M. Ghamkhar, *J. Phys. Chem.*, **73**, 346 (1969).

72. J. N. Brønsted, *J. Amer. Chem. Soc.*, **45**, 2898 (1923).

73. E. A. Guggenheim and L. A. Wiseman, *Phil. Mag.* [7], **25**, 45 (1938).

74. J. N. Brønsted, *J. Amer. Chem. Soc.*, **44**, 877 (1922).

75. T. F. Young, Y. C. Wu, and A. A. Krawitz, *Discussions Faraday Soc.*, **24**, 76 (1957).

76. H. S. Frank and W.-Y. Wen, *Discussions Faraday Soc.*, **24**, 133 (1957).

77. F. Vaslow, *J. Phys. Chem.*, **67**, 2773 (1963).

78. A. H. Narten and S. Lindenbaum, *J. Chem. Phys.*, **51**, 1108 (1969).

79. R. K. McMullan, M. Bonamico, and G. A. Jeffrey, *J. Chem. Phys.*, **39**, 3295 (1963).

80. A. H. Narten, submitted to *J. Phys. Chem.*

81. J. E. Desnoyers, M. Aril, G. Perron, and C. Jolicoeur, *J. Phys. Chem.*, **73**, 3346 (1969).

82. H. Falkenhagen and M. Dole, *Z. Physik. Chem.* [*B*], **6**, 159 (1929).

83. F. Franks and D. J. E. Ives, *Quart. Rev.* (*London*), **20**, 1, (1966).

84. V. A. Mikhailov, *Zh. Strukt. Khim.*, **6**, 677 (1961); Engl. transl., p. 625.

85. F. Vaslow, *J. Phys. Chem.*, **73**, 3745 (1969).

13 The Partial Molal Volumes of Electrolytes in Aqueous Solutions[*]

Frank J. Millero, Rosenstiel School of Marine and Atmospheric Sciences, University of Miami, Miami, Florida

Introduction

The thermodynamic quantity known as the partial molal volume has proven to be a very useful tool in elucidating the interactions occurring in aqueous and non-aqueous solutions. Studies of the partial molal volume of electrolytes have been used to examine ion-solvent, ion-ion, and solvent-solvent interactions. They have also proved useful in determining the effect of pressure on ionic equilibria. In this chapter, we examine the partial molal volumes of various electrolytes at infinite dilution as a function of temperature. The partial molal volumes of electrolytes at infinite dilution (where ion-ion interactions vanish) are particularly appropriate to study ion-solvent interactions, since volume properties are easy to visualize (i.e., geometrically) and relatively easy to determine from experimental measurements. We hope to demonstrate the use of partial molal volume data in studying ion-water interactions and to stimulate further use of this property in examining ion-solvent interactions in other solvent systems.

1 Definitions

The partial molal volume of an electrolyte, $\overline{V}_2$, may be visualized by considering a large reservoir of water, so large that the addition of one mole of electrolyte will not alter the concentration. The change in volume of the water upon the addition of one mole of electrolyte, 2, to this large reservoir of volume, V, is the partial molal volume of the electrolyte at the indicated concentration at constant temperature, T, pressure, P, and moles of the other

[*] Contribution Number 1350 from the University of Miami Rosenstiel School of Marine and Atmospheric Sciences. The author would like to acknowledge the support of the Office of Naval Research (Contract NONR 4008-02) and the National Science Foundation (GA-17386) for this study.

519

components, n_1 .... Mathematically, the partial molal volume can be represented by the partial derivative of the total volume with respect to concentration at constant T, P, and n_1 ... (n_1 = moles of water)

(1)
$$\overline{V}_2 = \left(\frac{\partial V}{\partial n_2}\right)_{T, P, n_1 \cdots}$$

The partial molal volume concept is useful because it can be shown mathematically[1] that

(2)
$$V = n_1 \overline{V}_1 + n_2 \overline{V}_2 + \cdots \qquad T, P \text{ constant}$$

The partial molal volume of an electrolyte, $\overline{V}_2$, may be evaluated from density measurements.[2] The graphical methods are described in detail elsewhere;[1] the use of the apparent molal volume, ϕ_V, is more convenient, particularly for binary solutions.

The apparent molal volume, ϕ_V, is defined by the relation

(3)
$$\phi_V = \frac{V - n_1 \overline{V}_1^\circ}{n_2} \qquad T, P \text{ constant}$$

where V is the volume of the solution containing n_1 moles of water and n_2 moles of electrolyte and $\overline{V}_1^\circ$ is the molar volume of pure water at a given T and P ($\overline{V}_1^\circ = M_1/d^\circ$, where M_1 is the molecular weight of water and d° is the density of water). Since $V = n_2 \phi_V + n_1 \overline{V}_1^\circ$, the partial molal volume of the electrolyte, 2, and water, 1, are given by the equations

(4)
$$\overline{V}_2 = \left(\frac{\partial V}{\partial n_2}\right)_{T, P, n_1} = \phi_V + n_2 \left(\frac{\partial \phi_V}{\partial n_2}\right)_{T, P, n_1}$$

and

(5)
$$\overline{V}_1 = \frac{V - n_2 \overline{V}_2}{n_1} = \frac{1}{n_1}\left[n_1 \overline{V}_1^\circ - n_2^2\left(\frac{\partial \phi_V}{\partial n_2}\right)\right]_{T, P, n_1}$$

In terms of the experimentally measured density, d, and the molecular weights, M_1 and M_2, of water and the electrolyte, the apparent molal volume is given by

(6)
$$\phi_V = \frac{1}{n_2}\left(\frac{n_1 M_1 + n_2 M_2}{d} - n_1 \overline{V}_1^\circ\right)$$

When the molal concentration scale is used, $n_2 = m$, the molality, and n_1 is equal to the number of moles of water in 1000 g, so that one has

(7)
$$\phi_V = \frac{1}{m}\left(\frac{1000 + m M_2}{d} - \frac{1000}{d^\circ}\right)$$
$$= \frac{1000(d^\circ - d)}{mdd^\circ} + \frac{M_2}{d}$$

where $d°$ is the density of pure water and d is the density of the solution. When the molar concentration is used, $n_2 = c$, the molarity, and (6) becomes

$$(8) \qquad \phi_V = \frac{1000(d° - d)}{cd°} + \frac{M_2}{d°}$$

When the molal concentration scale is used to determine ϕ_V as a function of concentration, (4) and (5) become

$$(9) \qquad \bar{V}_2 = \phi_V + \frac{m^{1/2}}{2}\left(\frac{\partial \phi_V}{\partial \sqrt{m}}\right)$$

$$(10) \qquad \bar{V}_1 = \bar{V}_1° - \frac{M_1 m^{3/2}}{2000}\left(\frac{\partial \phi_V}{\partial \sqrt{m}}\right)$$

By using the relation

$$(11) \qquad c = \frac{md°1000}{1000 + \phi_V md°}$$

the partial molal volume of electrolyte and water can be determined from the molarity concentration scale by the equations

$$(12) \qquad \bar{V}_2 = \phi_V + \left[\frac{1000 - c\phi_V}{2000 + c^{3/2}(\partial\phi_V/\partial\sqrt{c})}\right]c^{1/2}\left(\frac{\partial\phi_V}{\partial\sqrt{c}}\right)$$

and

$$(13) \qquad \bar{V}_1 = \frac{2000\bar{V}_1°(18.016/d°)}{2000 + c^{3/2}(\partial\phi_V/\partial\sqrt{c})}$$

Molarity is most frequently used to express the concentration dependence of $\bar{V}_2$ and ϕ_V, since the theoretical limiting Debye–Hückel equation uses this scale.

At infinite dilution, the partial molal volume and the apparent molal volume are equal ($\phi_V° = \bar{V}_2°$). To obtain reliable $\phi_V°$ and $\bar{V}_2°$ values, it is necessary to measure the density, d, or the density difference between the solution and pure solvent, $d° - d$, with great precision. This can be demonstrated by examining the probable error in ϕ_V due to uncertainties in density and concentration. By differentiating (8) with respect to c at constant d and with respect to d at constant c, we obtain[3]

$$(14) \qquad \text{Probable error in } \phi_V = \left[\frac{M_2}{d°} - \phi_V\right]\frac{\delta c}{c}$$

$$(15) \qquad \text{Probable error in } \phi_V = \left(\frac{-1000}{c}\right)\frac{\delta d}{d}$$

Equations 14 and 15 demonstrate that in dilute solutions ϕ_V is not seriously influenced by errors in c; however, errors in d do cause a large uncertainty in ϕ_V (e.g., when $c = 0.01$, a 1-ppm error in d causes an error of 0.1 cm^3/mole in ϕ_V).

2 Extrapolation to Infinite Dilution

The extrapolation of the apparent molal volume of electrolytes to infinite dilution and the expression of the concentration dependence of the apparent molal volume have been made by three major equations over the period of years—the Masson equation,[4] the Redlich–Meyer equation,[5] and the Owen–Brinkley equation.[6] Masson[4] found that the apparent molal volumes of electrolytes, ϕ_V, vary with the square root of the molar concentration by the linear equation

$$(16) \qquad \phi_V = \phi_V^\circ + S_V^* \sqrt{c}$$

where ϕ_V° is the apparent molal volume at infinite dilution (equal to the partial molal volume, $\overline{V}_2^\circ$) and S_V^* is the experimental slope (which varies with electrolyte type and charge). Scott[7] and Geffcken[8] have also examined the ϕ_V's of electrolytes by this equation and found that it adequately represents the concentration dependence of ϕ_V over a wide temperature range (0–100°C).

Redlich and Rosenfeld[9] predicted that a constant limiting slope should be obtained for a given electrolyte charge type (at constant temperature and pressure) if the Debye–Hückel limiting law is obeyed. By differentiating the Debye–Hückel limiting law for activity coefficients with respect to pressure, Redlich and Meyer[5] have calculated the theoretical limiting law slope, S_V, using the equation

$$(17) \qquad S_V = kw^{3/2}$$

The two terms for the limiting-law slope are given by

$$(18) \qquad k = N^2 e^3 \left(\frac{8\pi}{1000\,D^3 RT} \right)^{1/2} \left(\frac{\partial \ln D}{\partial P} - \frac{\beta}{3} \right)$$

where e is the electrostatic charge, D is the dielectric constant of the solvent (the other symbols have their usual meaning),[3] and

$$(19) \qquad w = 0.5 \sum_i \gamma_i Z_i^2$$

where γ_i is the number of ions of species i and valency Z_i formed by one molecule of electrolyte. For electrolytes of a fixed valency (w is constant), the limiting slope depends only on temperature and the physical properties of the solvent (D, $\partial \ln D/\partial P$, and β). The results obtained by Redlich and Meyer[5] for the limiting slope, k, for a 1:1 electrolyte (between 0 and 70°C) can be determined from the polynomial equation (t in degrees centigrade)

$$(20) \quad k = 1.444_7 + 1.6799 \times 10^{-2}\,t - 8.4055 \times 10^{-6}\,t^2 + 5.5153 \times 10^{-7}\,t^3$$

[with an rms deviation of 0.0005 cm³ mole$^{-3/2}$ liter$^{1/2}$]. Redlich and Meyer[5] have suggested the use of the following extrapolation equation:

$$(21) \qquad \phi_V = \phi_V^\circ + S_V\sqrt{c} + b_Vc$$

where S_V is the theoretical limiting slope and b_V is an empirical constant determined from the experimental results.

Results of a number of studies[10-29] for various electrolytes in dilute solutions over a temperature range of 0 to 65°C have confirmed the predictions of Redlich and Rosenfeld,[9] that S_V is independent of the nature of the electrolyte except for valency type. Limiting slopes of 1.868, 9.706, 14.944, 27.454, and 59.071 cm³ mol$^{-3/2}$ liter$^{1/2}$ for 1:1, 2:1, 2:2, 3:1, and 4:1 electrolytes have been confirmed at 25°C. It should also be pointed out that recent experimental results[20, 24, 29] show that S_V increases in a regular manner with increasing temperature, as predicted by Redlich and Meyer.[5] The experimental slope, S_V^* decreases with increasing temperature between 0 to 40°C[7,8] and increases with increasing temperature above 50°C;[7,8] thus at low temperatures the deviations from the limiting law are positive (b_V is positive) and at high temperatures the deviations are negative (b_V is negative) for simple electrolytes.

Although the experimental ϕ_V results for dilute solutions have amply proved the predictions of Redlich and co-workers,[9-12] its implications are still being ignored by many authors, who persist in using the Masson equation[4] to extrapolate ϕ_V to infinite dilution.

The Redlich–Meyer[5] extrapolation equation adequately represents the concentration dependence of many 1:1 and 2:1 electrolytes in dilute solutions; however, recent studies[14,18,24] for some 2:1, 3:1, and 4:1 electrolytes show deviations from this equation (i.e., b_V appears to be a function of concentration). Thus, for polyvalent electrolytes, the more complete Owen–Brinkley[6] equation can be used to aid in the extrapolation to infinite dilution and to adequately represent the concentration dependency of ϕ_V.

The Owen–Brinkley[6] equation, derived by including the ion-size parameter, a (in cm) is given by

$$(22) \qquad \phi_V = \phi_V^\circ + S_V\tau(\kappa a)\sqrt{c} + \tfrac{1}{2}W_V\theta(\kappa a)c + \tfrac{1}{2}K_Vc$$

where S_V is the theoretical limiting slope—(20),

$$(23) \qquad W_V = -2.303\,\gamma RTS_fA'\tfrac{1}{2}\left(\frac{\partial \ln D}{P\partial} - \beta - \frac{2\partial \ln a}{\partial P}\right)$$

$$(24) \qquad (\kappa a) = \left(\sum_i \gamma_i Z_i^2\right)^{1/2}\left(\frac{4\pi Ne^2}{1000\,kDT}\right)^{1/2} a\sqrt{c} = A'\sqrt{c}$$

and K_V is an empirical constant. $\tau(X)$ and $\theta(X)$ (which approach unity as the concentration decreases) are given by

$$(25) \qquad \tau(X) = \frac{3}{X^3}\left[\frac{X^2}{2} - X + \ln(1 - X)\right]$$

$$(26) \qquad \theta(X) = \frac{4}{X^4}\left[\frac{X^2}{2} - 2X + 3\ln(1 - X) + \left(\frac{1}{(1 - X)}\right)\right]$$

The distance of closest approach, a, may be evaluated from activity coefficient or conductivity data, and although W_V may in principle be calculated from theory, it must be evaluated from the data because of the appearance of the unknown derivative, $\partial \ln a/\partial P$. The empirical parameter, K_V, must also be evaluated from the experimental data. At 25°C for a 1:1 electrolyte, $A' = 0.3286\ \mathring{a}$, where $\mathring{a}$ is the ion-size parameter (in Å) and $W_V = 0.0615\ \mathring{a}$ when $\partial \ln a/\partial P$ is assumed to be equal to 0.[3,6]

Owen and Brinkley[6] have used these equations to extrapolate the ϕ_V of some simple 1:1 electrolytes to infinite dilution; however, they did not use a correct value for S_V, the limiting slope. More recently, Spedding and co-workers[18] have found it necessary to use (22) to (26) to evaluate the ϕ_V° of some 3:1 (rare earth) electrolytes (they evaluated the ion-size parameter from activity coefficient and conductance data). Because of its complexity, the Owen–Brinkley equation[6] has not been widely used; however, as more complex systems are investigated over a wide range of temperatures, these equations may be needed to aid in the extrapolation to infinite dilution (see Ref. 6b).

In summary, the preferred method of obtaining ϕ_V° from ϕ_V data as a function of concentration is the use of the limiting slope, S_V, with the Redlich–Meyer[5] or Owen–Brinkley[6] equations. The ideal method is to make the measurements in very dilute solutions where the deviations from the limiting law are very small (i.e., b_V and the higher order terms of the Owen–Brinkley[6] equation do not have to be considered).

Since the limiting law is not known above 70°C, the Masson equation must still be used at high temperatures. Also, although the ϕ_V° values determined by the Masson equation[4] are frequently in error by as much as 3 cm³/mole, the additivity and expansibility of the data derived in this manner can be very useful in estimating the ϕ_V° of unknown electrolytes and the S_V^* constants can be related to ion-ion interactions in concentrated solutions. The partial molal volume at infinite dilution, $\phi_V^\circ = \bar{V}_2^\circ$, the b_V constant, and the S_V^* constant for a number of electrolytes from 0 to 200°C have been tabulated in the Appendix.

The densities of aqueous solutions can be determined from this compilation of data by combining (8) with the Masson equation[4] or the Redlich–Meyer equation.[5] Solving for the density of the solution, d, we obtain

(27) $$d = d° + \left[\frac{M_2 - d°\phi_V°}{1000}\right] c - \left[\frac{S_V^* d°}{1000}\right] c^{3/2}$$

(28) $$d = d° + \left[\frac{M_2 - d°\phi_V°}{1000}\right] c - \left[\frac{S_V d°}{1000}\right] c^{3/2} - \left[\frac{b_V d°}{1000}\right] c^2$$

Equation (27) was originally derived by Root.[30] These equations can be used to estimate the density of unknown solutions by using the additivity principle ($\phi_V°$ and S_V are always additive, S_V^* and b_V appear to be additive for simple systems).

3 The Partial Molal Volumes of Ions at Infinite Dilution at Various Temperatures

The utility of the partial molal volumes of electrolytes at infinite dilution to study ion-water interactions lies in the additivity principle.[27,31-34] By additivity we mean that the partial molal volume of an electrolyte, $\bar{V}_{MX}°$, is equal to the sum of its ionic components, $\bar{V}_{M^+}° + \bar{V}_{X^-}°$. The additivity of various alkali-metal and alkaline earth metal halides at 0, 25, and 50°C is demonstrated in Tables 1, 2, and 3. Although the $\bar{V}°$ results at 0 and 50°C used in Tables 1 and 3 may be in error by as much as 0.6 cm³/mole (because the measurements were not made in dilute solutions and the extrapolations to infinite dilution were made by using the Masson equation[6]), the differences between the $\bar{V}°$ of various ions agree very well with the very careful study made by Dunn.[24] For example, from Dunn's data,[24] we calculate $\bar{V}_{Br^-}° - \bar{V}_{Cl^-}° = 6.61$ at 0°C and 7.49 at 50°C; $\bar{V}_{I^-}° - \bar{V}_{Br^-}° = 10.51$ at 0°C and 12.01 at 50°C; $\bar{V}_{K^+}° - \bar{V}_{Na^+}° = 10.68$ at 0°C and 9.86 at 50°C. The results in Tables 1 and 3 for the same ions agree very well with the differences calculated from Dunn's work.[24] Thus the additivity principle appears to be obeyed in dilute solutions as well as at infinite dilution. An examination of the Masson,[4] S_V^*, constants for these same salts (Tables 4-6) shows that the additivity principle often extends to concentrated solutions (i.e., when ion pairing, etc., are not important). This additivity principle can prove to be very useful in estimating reliable $\bar{V}°$'s from less reliable data as well as the density of aqueous solutions, using (27) and (28). It is interesting to note that the differences between the $\bar{V}°$'s of cations ($K^+ - Na^+$) and anions ($Br^- - Cl^-$) change with temperature in an opposite manner; thus the ion-water interactions of cations and anions (i.e., the volume components) appear to behave differently as the size or temperature is increased.

Table 1   The Additivity of the Partial Molal Volume of Salts at Infinite Dilution at 0°C[a]

	Cl⁻		Br⁻		I⁻	
Li⁺	*15.30*	6.64	*21.94*	10.56	*32.50*	
	−2.94		−3.08		−3.16	(−3.06)
Na⁺	*12.36*	6.50	*18.86*	10.48	*29.34*	
	10.64		10.52		10.71	(10.62)
K⁺	*23.00*	6.38[b]	*29.38*	10.67	*40.05*	
	5.11		5.23		5.10	(5.14)
Rb⁺	*28.11*	6.50	*34.61*	10.54	*45.15*	
	7.20		7.34		7.59	(7.37)
Cs⁺	*35.31*	6.64	*41.95*	10.79[b]	*52.74*	
		(6.57)		(10.56)		
	2Cl⁻		**2Br⁻**		**2I⁻**	
Mg²⁺	*11.40*	12.10	*23.60*	21.30	*44.90*	
	2.00		1.90		2.00	(1.97)
Ca²⁺	*13.40*	12.10	*25.50*	21.40	*46.90*	
	1.30		1.20		1.10	(1.20)
Sr²⁺	*12.10*	12.20	*24.30*	21.50	*45.80*	
	4.70		4.70		4.80	(4.73)
Ba²⁺	*16.80*	12.20	*29.00*	21.60	*50.60*	
		(12.15)		(21.45)		

[a] Data for the alkali metal halides from the calculations made by Scott[7] on the density data of Baxter and Wallace.[35] Data for the alkaline earth halides from the Ph.D. thesis of W. C. Root.[34]
[b] Data not used in determining the "best" averaged value of the $\bar{V}°$ differences between various ions (denoted by parentheses).

By examining the $\bar{V}°$ of ions as a function of size, charge, temperature, and so on, it is possible to study the effect of these parameters on ion-water interactions, with the hope of obtaining a better understanding of the interactions in aqueous solutions. The difficulty of using $\bar{V}°$ data (as well as other thermodynamic data at infinite dilution) to study ion-water interactions on an absolute basis springs from the problem of assigning absolute values to the $\bar{V}°$ of individual ions. The division of the $\bar{V}°$ of electrolytes into their ionic components, $\bar{V}°_{ion}$, can normally be made only by using nonthermodynamic methods. Lack of adequate theories or knowledge of such fundamental parameters of ions, such as radius, increase the difficulties of assigning absolute ionic properties. Once the $\bar{V}°$ of one ion is estimated by some method, the $\bar{V}°$ of the other ions are fixed (due to the additivity principle). Since the proton is frequently the ion that is adjusted or estimated by various methods,

Table 2 The Additivity of the Partial Molal Volume of Salts at Infinite Dilution at 25°C[a]

	F^-		Cl^-		Br^-		I^-	
Li^+	—	—	16.95	6.91	23.78	11.58	35.37	
			−0.33		−0.30		−0.33	(−0.32)
Na^+	−2.37	18.99	16.62	6.86	23.48	11.56	35.04	
	10.16		10.23		10.25		10.20	(10.21)
K^+	7.79	19.06	26.85	6.86	33.73	11.51	45.24	
	4.90		5.05		5.04		5.05	(5.01)
Rb^+	12.69	19.21[b]	31.90	6.87	38.77	11.52	50.29	
	7.34		7.27		7.27		7.26	(7.28)
Cs^+	20.03	19.14[b]	39.17	6.87	46.04	11.51	57.55	
		(19.02)		(6.88)		(11.53)		

	$2Cl^-$		$2Br^-$		$2I^-$	
Mg^{2+}	14.20	13.80	28.00	23.20	51.20	
	4.50		4.40		4.40	(4.43)
Ca^{2+}	18.70	13.70	32.40	23.20	55.60	
	−0.60		−0.60		−0.60	(−0.60)
Sr^{2+}	18.10	13.70	31.80	23.20	55.00	
	5.70		5.30[b]		5.50	(5.60)
Ba^{2+}	23.80	13.30[b]	37.10	23.40	60.50	
		(13.73)		(23.25)		

[a] Data for the alkali metal halides from the "best" $\bar{V}^\circ$ results obtained by various workers.[9–29,37] Data for the alkaline earth metal halides from the Ph.D. thesis of W. C. Root.[34]

[b] Data not used in determining the "best" averaged value of the $\bar{V}^\circ$ differences between various ions (denoted by parentheses).

it is convenient to tabulate $\bar{V}^\circ$ data on a conventional basis (as suggested by Owen and Brinkley[36]) by assigning $\bar{V}^\circ_{H^+} = 0$ cm³/mole. The true absolute partial molal volumes of ions, V°_{ion}, of absolute charge, Z (in electronic units) are given by

$$(27) \qquad \bar{V}^\circ_{ion} = \bar{V}^\circ_{conv} + Z\bar{V}^\circ_{H^+} \quad \text{for cations}$$

$$(28) \qquad \bar{V}^\circ_{ion} = \bar{V}^\circ_{conv} - Z\bar{V}^\circ_{H^+} \quad \text{for anions}$$

where $\bar{V}^\circ_{conv}$ is the conventional partial molal volume and $\bar{V}^\circ_{H^+}$ is the absolute partial molal volume of the proton. In Tables 7 to 10 we have tabulated the conventional partial molal volume of ions from 0 to 200°C at 25-degree intervals from what we feel are the most reliable $\bar{V}^\circ$ data.[37] As is readily apparent, more reliable $\bar{V}^\circ$ data[5,10–29] are available at 25°C than at any other

Table 3   The Additivity of the Partial Molal Volume of Salts at Infinite Dilution at 50°C[a]

	Cl^-		Br^-		I^-	
Li^+	*16.96*	7.41	*24.37*	12.44	*36.81*	
	1.00		0.97		0.87	(0.94)
Na^+	*17.96*	7.38	*25.34*	12.34	*37.68*	
	9.70		9.98		9.90	(9.86)
K^+	*27.66*	7.66[b]	*35.32*	12.26	*47.58*	
	5.29		4.96		5.17	(5.14)
Rb^+	*32.95*	7.33	*40.28*	12.47	*52.75*	
	7.44		7.43		7.57	(7.48)
Cs^+	*40.39*	7.42	*47.81*	12.51	*60.32*	
		(7.38)		(12.40)		
	$2Cl^-$		$2Br^-$		$2I^-$	
Mg^{2+}	*14.70*	14.50	*29.20*	24.90	*54.10*	
	4.50		4.30		4.30	(4.37)
Ca^{2+}	*19.20*	14.30	*33.50*	24.90	*58.40*	
	0.60		0.50		0.60	(0.57)
Sr^{2+}	*19.80*	14.20	*34.00*	25.00	*59.00*	
	5.80		6.10		6.10	(6.00)
Ba^{2+}	*25.60*	14.50	*40.10*	25.00	*65.10*	
		(14.38)		(24.96)		

[a] Data for the alkali metal halides from the calculations made by Scott[7] on the density data of Baxter and Wallace.[35] Data for the alkaline earth halides from the Ph.D. thesis of W. C. Root.[34]
[b] Data not used in determining the "best" averaged value of the $\overline{V}°$ difference between various ions (denoted by parentheses).

temperature. The results at high temperature (75 to 200°C) have been taken from the recent work of Ellis and co-workers,[38,39] and most of the reliable results at 0 to 50°C have been taken from the results of Dunn.[24]

Panckhurst[40] has recently critically reviewed the various methods used to estimate the absolute partial molal of ions at 25°C. Some of the basic assumptions and techniques used by different workers are outlined briefly below.

1. Assume that the ratio of the ionic partial molal volumes of the ions of a given electrolyte, MX, is equal to the ratio of the cubes of the crystal radii, $\overline{V}°_{M^+}/\overline{V}°_{X^-} = r_{M^+}{}^3/r_{X^-}{}^3$. Bernal and Fowler,[31] Darmois,[41] and Zen[42] have used this method for the salt CsCl; Kobayazi[43] used this method for the salt KF; Rice[44] and Eucken[45] used this method for the salt CsI. Millero[29c] has recently used this method on the salt Ph_4AsBPh_4 at 0, 25, and 50°C.

Table 4   The Additivity of the Masson S_V^* Constants for Various 1:1 and 2:1 Electrolytes at 0°C[a]

	Cl⁻		Br⁻		I⁻	
Li⁺	1.990	−0.317	1.673	−0.248	1.425	
	1.250		1.303		1.461	(1.34)
Na⁺	3.240	−0.264	2.976	−0.090	2.886	
	0.051[b]		0.243		0.247	(0.24)
K⁺	3.291	−0.072[b]	3.219	−0.086	3.133	
	−0.004		−0.101		0.025	(−0.03)
Rb⁺	3.287	−0.175	3.112	−0.046[b]	3.158	
	0.006		−0.066		−0.273	(−0.11)
Cs⁺	3.293	−0.247	3.046	−0.161	2.885	
		(−0.25)		(−0.15)		

	2Cl⁻		2Br⁻		2I⁻	
Mg²⁺	7.212	−0.424	6.788	−0.848	5.940	
	0.142		0.000[b]		0.141	(0.14)
Ca²⁺	7.354	0.566	6.788	0.707	6.081	
	0.990		0.132[b]		0.697	(0.84)
Sr²⁺	8.344	−0.424	7.920	−0.142[b]	7.778	
	0.283		0.282		0.000[b]	(0.28)
Ba²⁺	8.627	−0.425	8.202	−0.424	7.778	
		(−0.46)		(−0.66)		

[a] Data for the alkali metal halides from the calculations made by Scott [7] on the density data of Baxter and Wallace.[35] Data for the alkaline earth halides from the Ph.D. thesis of W. C. Root.[34]

[b] Data not used in determining the "best" averaged value for the S_V^* differences between various ions (denoted by parentheses).

2. Assume that the partial molar volumes of two ions are equal on the basis of certain theoretical or experimental evidence. For example, Wirth[46] assumed $\overline{V}_{K^+}^\circ = \overline{V}_{F^-}^\circ$ since their Pauling crystal radii[47] are nearly equal; Fajans and Johnson[48] assumed $\overline{V}_{NH_4^+}^\circ = \overline{V}_{Cl^-}^\circ = \overline{V}_{H_2O}^\circ$, since it appears that neither ion has a large influence on the structure of water (based on various experimental evidence); Hepler[49] has also used this method.

3. Assume that the ionic partial molal volume of cations and anions fit the same semiempirical equation. For example, Couture and Laidler[33] found that if they adjusted the $\overline{V}_{H^+}^\circ = -6.0$ cm³/mole, both cations and anions could be represented by the equation, $\overline{V}_{ion}^\circ = 16.0 + 4.9r^3 - 26|Z|$, where r is the Goldschmidt crystal radius[50] and $|Z|$ is absolute charge on the ion. Mukerjee[51] found that both large cations and anions (monatomic monovalent)

Table 5   The Additivity of the Masson S_V^* Constants for Various 1:1 and 2:1 Electrolytes at 25°C

	Cl⁻		Br⁻		I⁻	
Li⁺	1.488	−0.329	1.159	−0.318	0.814	
	0.665		0.601		0.505	(0.59)
Na⁺	2.153	−0.393	1.760	−0.414	1.346	
	0.174		0.179		0.210	(0.19)
K⁺	2.327	−0.488	1.939	−0.383	1.556	
	−0.108[b]		0.099		0.051	(0.07)
Rb⁺	2.219	−0.181	2.038	−0.431	1.607	
	−0.047		−0.137		−0.028	(−0.07)
Cs⁺	2.172	−0.271	1.901	−0.322	1.579	
		(−0.33)		(−0.37)		

	2Cl⁻		2Br⁻		2I⁻	
Mg²⁺	6.222	−0.848	5.374	−0.990	4.384	
	−0.565		−0.566		−0.566	(−0.57)
Ca²⁺	5.657	−0.849	4.808	−0.990	3.818	
	1.414		0.990		1.132	(1.18)
Sr²⁺	7.071	−1.273	5.798	−0.848	4.950	
	−0.566		−0.141		−0.142	(−0.28)
Ba²⁺	6.505	−0.848	5.657	−0.849	4.808	
		(−0.95)		(−0.92)		

[a] Data for the alkali metal halides from the calculations made by Scott[7] on the density data of Baxter and Wallace.[35] Data for the alkaline earth halides from the Ph.D. thesis of W. C. Root.[34]

[b] Data not used in determining the "best" averaged value for the S_V^* differences between various ions (denoted by parentheses).

become a smooth function of the cube of the Pauling crystal radii[47] if $\overline{V}_{H^+}^\circ$ is adjusted to −4.5 cm³/mole.

4. Assume that the ionic partial molal volume of one ion is equal to a fixed value because of a certain theoretical or intuitive notion. For example, Owen and Brinkley[36] have assumed $\overline{V}_H^\circ = 0$ cm³/mole because of its small size or as a matter of convention. Stokes and Robinson[52] have assumed that the large monatomic monovalent anions, Br⁻ and I⁻, are not hydrated and that they contribute to the volume of the system an amount due to "random close packing"; that is, $\overline{V}_{I^-}^\circ = 4.35\, r^3$, where r is the Pauling crystal radius.[47] Padova[53] has also assumed that the I⁻ ion is not hydrated and $\overline{V}_{I^-}^\circ = \overline{V}_{int}^\circ = 37.1$ cm³/mole where $\overline{V}_{int}^\circ$ is the intrinsic partial molal volume. Padova[53] calculated $\overline{V}_{int}^\circ$ for various ions from compressibility data on salt solutions by assuming $\overline{V}_{int}^\circ$ for H⁺ equals 0.

Table 6   The Additivity of the Masson S_V^* Constants for Various 1:1 and 2:1 Electrolytes at 50°C[a]

	Cl⁻		Br⁻		I⁻	
Li⁺	*1.446*	−0.301	*1.145*	−0.576	*0.569*	
	0.358		0.253		0.257	(0.29)
Na⁺	*1.804*	−0.406	*1.398*	−0.572	*0.826*	
	0.283		0.242		0.342	(0.29)
K⁺	*2.087*	−0.447	*1.640*	−0.472	*1.168*	
	−0.040[b]		0.117		0.056	(0.08)
Rb⁺	*2.047*	−0.290	*1.757*	−0.533	*1.224*	
	−0.089		−0.125		−0.140	(−0.12)
Cs⁺	*1.958*	−0.326	*1.632*	−0.548	*1.084*	
		(−0.35)		(−0.54)		
	2Cl⁻		**2Br⁻**		**2I⁻**	
Mg²⁺	*5.657*	−0.849	*4.808*	−1.555	*3.253*	
	−0.142		−0.283		−0.283	(−0.24)
Ca²⁺	*5.515*	−0.990	*4.525*	−1.555	*2.970*	
	0.990		0.990		0.990	(0.99)
Sr²⁺	*6.505*	−0.990	*5.515*	−1.555	*3.960*	
	−0.424		−0.565		−0.425	(−0.47)
Ba²⁺	*6.081*	−1.131	*4.950*	−1.415	*3.535*	
		(−0.99)		(−1.52)		

[a] Data for the alkali metal halides from the calculations made by Scott[7] on the density data of Baxter and Wallace.[35] Data for the alkaline earth halides from the Ph.D. thesis of W. C. Root.[34]

[b] Data not used in determining the "best" averaged value for the S_V^* differences between various ions (denoted by parentheses).

5. Assume the $\bar{V}°$'s for a series of salts, MX, with a common ion, X⁻, are a linear function of some parameter of the uncommon ions, M⁺, such as molecular weight. By extrapolating the $\bar{V}_{MX}°$ salts to 0, one can obtain the $\bar{V}°$ of the common ion, X⁻. For example, Conway, Verrall, and Desnoyers[16] have assumed that the $\bar{V}°$ of the tetraalkylammonium halides, R_4NX, plotted versus the molecular weight of the cation, R_4N^+, is a straight line. They obtain $\bar{V}°$ of X⁻ by extrapolating to 0 molecular weight. Millero and Drost-Hansen[54] have also used this method.

6. Assume that the $\bar{V}°$ may depend upon the sign of the ionic charge and the $\bar{V}_{ion}°$ are the result of two major components

$$(29) \qquad \bar{V}_{ion}° = \bar{V}_{int}° + \bar{V}_{elect}°$$

Table 7   Conventional Partial Molal Volume of Ions at 0°C

Ion	$\overline{V}^{\circ}_{conv}$ (cm³/mole)	Remarks	References
H^+	0	Conventional value	
Li^+	−0.45	From additivity relations	
Na^+	−3.51	From NaCl	24,29(a)
K^+	7.17	From KCl	24
Rb^+	12.31	From additivity relations	
Cs^+	19.68	From additivity relations	
Ag^+	−2.14	From AgNO₃	37
NH_4^+	17.47	From estimated value of NH₄Cl	37
Me_4N^+	88.59	Extrapolated from Me₄NBr	20
Et_4N^+	147.47	Extrapolated from Et₄NBr	20
nPr_4N^+	212.53	Extrapolated from nPrNBr	20
nBu_4N^+	271.10	Extrapolated from nBu₄NBr	20
Ph_4As^+	291.12	From Ph₄AsCl	29(c)
Mg^{2+}	−21.81	From additivity relations	
Ca^{2+}	−19.84	From CaCl₂	24
Sr^{2+}	−20.77	From additivity relations	
Ba^{2+}	−15.79	From BaCl₂	24
La^{3+}	−40.71	From LaCl₃	37
F^-	−2.21	From estimated value of KF	37
Cl^-	16.45	From HCl	37
Br^-	23.06	From KBr	24
I^-	33.57	From KI	24
OH^-	−6.8	From NaOH and KOH	37
NO_3^-	25.6	From NaNO₃ and KNO₃	37
SCN^-	31.3	Extrapolated from KSCN	37
$PhSO_3^-$	93.8	Extrapolated from NaPhSO₃	37
MnO_4^-	38.5	From KMnO₄	37
ClO_4^-	40.7	From estimated value of NH₄ClO₄	37
CO_3^{2-}	−9.8	From Na₂CO₃ and K₂CO₃	37
SO_4^{2-}	11.1	Extrapolated from MgSO₄	20
CrO_4^{2-}	13.6	From K₂CrO₄	37
Ph_4B^-	270.67	From NaBPh₄	29(b)

where $\overline{V}^{\circ}_{int}$ is the intrinsic partial molal volume and $\overline{V}^{\circ}_{elect}$ is the electrostriction partial molal volume. For example, Glueckauf[55] has assumed that $\overline{V}^{\circ}_{int} = 2.52(r + 0.55)^3$ and $\overline{V}^{\circ}_{elect} = -BZ^2/\bar{r}$, where r is the Pauling crystal radius[47] and $\bar{r} = r_{ion} + r_{H_2O}$, $r_{H_2O} = 1.38$ Å. By plotting $\overline{V}^{\circ}$ for the alkali metal chlorides minus $\overline{V}^{\circ}_{int}$ of the cation (Na^+ to Cs^+) versus $1/\bar{r}$ and extrapolating to $1/\bar{r} = 0$, he obtains $\overline{V}^{\circ}_{Cl^-}$. Noyes[56] assumes that when an ion is sufficiently large, $\overline{V}^{\circ}_{elect}$ can be calculated from the Drude–Nernst equation[57]

$$(30) \qquad \overline{V}^{\circ}_{elect} = \frac{Z^2 e^2}{2rD}\left(\frac{\partial \ln D}{\partial P}\right)_T = -\frac{BZ^2}{r}$$

Table 8 Conventional Partial Molal Volume of Ions at 25°C

Ion	$\overline{V}^\circ_{\text{conv}}$ (cm³/mole)	Remarks	References
H^+	0	Conventional value	
Li^+	-0.88	From LiCl	17,27
Na^+	-1.21	From NaCl	17,20,24,27,29
K^+	9.02	From KCl	17,18,20,24,27
Rb^+	14.07	From RbCl	17,27
Cs^+	21.34	From CsCl	17,27
Ag^+	-0.7	From $AgNO_3$	37
Tl^+	10.6	From TlF, $TlNO_3$, $TlClO_4$, and TlO_2CCH_3	37
NH_4^+	17.86	From NH_4Br	19
Me_4N^+	89.57	From Me_4NBr	14,16,20
Et_4N^+	149.12	From Et_4NBr	20
nPr_4N^+	214.44	From nPr_4NBr	20
nBu_4N^+	275.66	From nBu_4NBr	20
Ph_4As^+	300.65	From Ph_4AsCl	29(c)
nAm_4N^+	339.2	From nAm_4NBr	16
$MeNH_4^+$	36.11	From $MeNH_3Br$	19
$EtNH_3^+$	52.94	From $EtNH_3Br$	19
$nPrNH_3^+$	69.44	From $nPrNH_3Br$	19
$nBuNH_3^+$	85.50	From $nBuNH_3Br$	19
$nPenNH_3^+$	101.44	From $nPenNH_3Br$	19
$nHexNH_3^+$	117.33	From $nHexNH_3Br$	19
$nHepNH_3^+$	133.23	From $nHepNH_3Br$	19
$nOctNH_3$	149.15	From $nOctNH_3Br$	19
$(HOEt)_4N^+$	152.0	From $(HOEt)_4NBr$ and $(HOEt)_4NF$	37
$[Bu_3N(CH_2)_8NBu_3]^{2+}$	528.5	From $(Bu_3N(CH_2)_8NBu_3)Br_2$	37
Be^{2+}	-12.0	From $BeCl_2$	37
Mg^{2+}	-21.17	From $MgCl_2$	23
Ni^{2+}	-24.0	From $NiSO_4$	37
Co^{2+}	-24.0	From $CoCl_2$, $CoBr_2$, $Co(NO_3)_2$, and $CoSO_4$	37
Zn^{2+}	-21.6	From $ZnSO_4$	37
Fe^{2+}	-24.7	From $FeCl_2$, $FeBr_2$, $FeSO_4$, and $Fe(NO_3)_2$	37
Mn^{2+}	-17.7	From $MnBr_2$	37
Cu^{2+}	-27.76	From $Cu(SO_3NH_2)_2$	37
Cd^{2+}	-20.0	From $CdBr_2$ and $Cd(O_2CCH_3)_2$	37
Ca^{2+}	-17.85	From $CaCl_2$	23,24
Hg^{2+}	-19.3	From $HgCl_2$	37
Sr^{2+}	-18.16	From additivity relations	
Pb^{2+}	-15.5	From $Pb(NO_3)_2$	37
Ba^{2+}	-12.47	From $BaCl_2$	23,24
Al^{3+}	-42.2	From $AlCl_3$ and $Al(NO_3)_3$	37
Fe^{3+}	-43.7	From $Fe(NO_3)_3$	37
Cr^{3+}	-39.5	From $CrCl_3$	37
Yb^{3+}	-44.22	From $YbCl_3$	18
Er^{3+}	-42.86	From $ErCl_3$	18
Ho^{3+}	-41.76	From $HoCl_3$	18
Dy^{3+}	-40.83	From $DyCl_3$	18
Tb^{3+}	-40.24	From $TbCl_3$	18
Gd^{3+}	-40.41	From $GdCl_3$	18

(continued)

Table 8 (*Continued*)

Ion	$\bar{V}^{\circ}_{conv}$ (cm^3/mole)	Remarks	References
Sm^{3+}	-42.33	From $SmCl_3$	18
Nd^{3+}	-43.31	From $NdCl_3$	18
Pr^{3+}	-42.53	From $PrCl_3$	18
La^{3+}	-39.10	From $LaCl_3$	15,18
Th^{4+}	-53.5	From $ThCl_4$ and $Th(NO_3)_4$	37
F^-	-1.16	From NaF	22,25,27
Cl^-	17.83	From HCl assuming $\bar{V}^{\circ}_{H+} = 0$	10,11,15
Br^-	24.71	From KBr	24,27
I^-	36.22	From KI	24,27
OH^-	-4.04	From $NaOH$	14
Ph_4B^-	277.62	From $NaBPh_4$	29(b)
ReO_4^-	48.18	From $NaReO_4$	37
OCN^-	26.12	From $KOCN$	37
$SeCN^-$	49.68	From $KSeCN$	37
BF_4^-	44.18	From NH_4BF_4	37
SO_3F^-	47.93	From NH_4SO_3F	37
$SO_3NH_2^-$	41.49	From $NH_4SO_3NH_2$	37
NO_3^-	29.00	From KNO_3	15
NO_2^-	26.2	From $NaNO_2$	37
SCN^-	35.7	From $KSCN$	37
CH_2^-	26.27	From NaO_2CH	26
$CH_3CO_2^-$	40.46	From NaO_2CCH_3	10–12,26
$CH_3CH_2CO_2$	54.0	From $NaO_2CH_2CH_3$	37
$CH_3(CH_2)_2CO_2^-$	70.40	From $NaO_2(CH_2)_2CH_3$	26
$PhSO_3^-$	108.9	From $NaPhSO_3$	37
HCO_3^-	23.4	From $NaHCO_3$ and $KHCO_3$	37
MnO_4^-	42.5	From $KMnO_4$	37
ClO_3^-	36.66	From $KClO_3$	10,11
BrO_3^-	35.3	From $NaBrO_3$ and $KBrO_3$	37
IO_3^-	25.3	From KIO_3 and $NaIO_3$	37
ClO_4^-	44.12	From $HClO_4$ and $NaClO_4$	37
HSO_4^-	35.67	From $NaHSO_4$	37
$HSeO_4^-$	31.1	From $NaHSeO_4$	37
$H_2PO_4^-$	29.1	From NaH_2PO_4	37
$H_2AsO_4^-$	35.2	From NaH_2AsO_4	37
$p\text{-}CH_3PhSO_3^-$	119.6	From $Na(p\text{-}CH_3PhSO_3)$	37
PhO^-	68.7	From $NaOPh$	37
S^{2-}	-8.2	From Na_2S	37
SO_4^{2-}	13.98	From Na_2SO_4 and K_2SO_5	15
SeO_4^{2-}	21.0	From Na_2SeO_4 and K_2SeO_4	37
CO_3^{2-}	-4.3	From Na_2CO_3	37
CrO_4^{2-}	19.7	From K_2CrO_4 and Na_2CrO_4	37
WO_4^{2-}	25.7	From Na_2WO_4	37
MoO_4^{2-}	28.9	From Na_2MoO_4	37
$PtCl_6^{2-}$	150.0	From K_2PtCl_6	37
HPO^{2-}	7.7	From $NaHPO_4$	37
$S_2O_3^{2-}$	34.0	From $Na_2S_2O_3$	37
$C_2O_4^{2-}$	16.0	From $Na_2C_2O_4$	37
$Cr_2O_7^{2-}$	73.0	From $Na_2Cr_2O_7$	37
SO_3^{2-}	8.9	From Na_2SO_3	37
AsO_4^{3-}	-15.6	From Na_3AsO_4 and K_3AsO_4	37
$Fe(CN)_6^{3-}$	120.8	From $K_3Fe(CN)_6$	14
$Fe(CN)_6^{2-}$	74.0	From $K_4Fe(CN)_6$	14

Table 9   Conventional Partial Molal Volume of Ions at 50°C

Ion	$\overline{V}^\circ_{\text{conv}}$ (cm³/mole)	Remarks	References
H⁺	0	Conventional value	
Li⁺	−1.24	From additivity relations	
Na⁺	−0.30	From interpolated value of NaCl	24,29
K⁺	9.57	From interpolated value of KCl	24
Rb⁺	14.71	From additivity relations	
Cs⁺	22.22	From additivity relations	
NH₄⁺	19.20	From NH₄Cl	38
Me₄N⁺	91.2	From Me₄NI	37
Et₄N⁺	151.6	From Et₄NI	37
nPr₄N⁺	218.9	From nPr₄NI	37
nBu₄N⁺	285.0	From nBu₄NI	37
Ph₄As⁺	309.82	From Ph₄AsCl	29(c)
Mg²⁺	−20.90	From additivity relations	
Ca²⁺	−18.22	From interpolated value of CaCl₂	24
Sr²⁺	−17.69	From additivity relations	
Ba²⁺	−11.73	From interpolated value of BaCl₂	24
F⁻	−1.4	From NaF	38
Cl⁻	18.00	From HCl, assuming $\overline{V}^\circ_{\text{H}^+} = 0$	39
Br⁻	25.49	From interpolated value of KBr	24
I⁻	37.52	From interpolated value of KI	24
OH⁻	−4.35	From NaOH and KOH	37
NO₃⁻	30.3	From KNO₃	38
ClO₄⁻	45.1	From NH₄ClO₄	38
SO₄²⁻	16.0₃	From Na₂SO₄ and K₂SO₄	38
Ph₄B⁻	283.93	From NaBPh₄	29(b)

where $B = 4.175$ cm³/(Å)(mole) at 25°C (calculated from the results of Owen et al.[58] Noyes[56] used the following equations to estimate $\overline{V}^\circ_{\text{H}^+}$:

$$(31) \qquad \overline{V}^\circ_{\text{int}} - \overline{V}^\circ_{\text{conv}} = \overline{V}^\circ_{\text{H}^+} + \frac{4.175}{r} + \frac{C_2}{r^2} \quad \text{for cations}$$

$$(32) \qquad \overline{V}^\circ_{\text{conv}} - \overline{V}^\circ_{\text{int}} = \overline{V}^\circ_{\text{H}^+} - \frac{4.175}{r} + \frac{A_2}{r^2} \quad \text{for anions}$$

where $\overline{V}^\circ_{\text{conv}}$ is the conventional partial molal volume, C_2 and A_2 are empirical constants. By assuming various forms for $\overline{V}^\circ_{\text{int}}$, $\overline{V}^\circ_{\text{int}} = 2.52\, r^3 + J r^2$ or $\overline{V}^\circ_{\text{int}} = 2.52(r + a)^3$, $\overline{V}^\circ_{\text{H}^+}$, C_2, A_2, and J or a are evaluated using a least-square "best" fit (with the aid of a computer). Panckhurst[40] has also used the Noyes[56] method.

Table 10   Conventional Partial Molal Volume of Ions from 75 to 200°C $(cm^3/mole)^a$

Ion	Temperature (°C)					
	75	100	125	150	175	200
H^+	0	0	0	0	0	0
Li^+	-1.8	-2.7	-3.6	-4.7	-7.0	-8.1
Na^+	0.8	0.8	0.9	1.2	0.5	-0.1
K^+	10.3	9.5	9.4	9.3	7.8	7.3
Cs^+	23.2	23.0	23.2	23.5	22.3	23.1
NH_4^+	19.5	19.5	19.3	19.1	18.8	20.0
Me_4N^+	95.8	—	—	—	—	—
Et_4N^+	159.3	—	—	—	—	—
nPr_4N^+	227.5	—	—	—	—	—
nBu_4N^+	304.1	—	—	—	—	—
Mg^{2+}	-21.7	-23.4	-26.0	-29.0	-34.0	-37.0
Ca^{2+}	-19.1	-20.0	-23.7	-25.7	-27.0	-30.7
Sr^{2+}	-16.8	-17.4	-19.8	-22.4	-25.5	-29.0
Ba^{2+}	-7.7	-8.1	-10.5	-11.9	-15.0	-17.0
F^-	-3.1	-3.7	-6.4	-10.3	-14.4	-21.3
Cl^-	17.4	16.0	14.1	11.2	7.0	0.5
Br^-	25.1	24.9	23.9	21.5	19.2	13.0
I^-	37.2	38.7	37.5	35.8	34.6	30.2
OH^-	-5.2	—	—	—	—	—
NO_3^-	30.7	31.9	31.6	30.0	28.2	22.3
ClO_4^-	46.9	47.9	48.6	48.2	46.5	43.0
SO_4^{2-}	13.8_5	11.5_5	6.2_5	-0.7_5	-10.3_0	-22.9_5

^{a} Data from the $\bar{V}^\circ$ work of Ellis and co-workers[38,39] except for the R_4N^+ ions, which were extrapolated from the R_4NI salts[37] and the OH^- ion, which was extrapolated from NaOH and KOH.[37]

7. Zana and Yeager[59] have determined the absolute $\bar{V}^\circ_{H^+}$, and the other ions, from ionic vibration potential measurements. As pointed out by Mukerjee,[60] Zana and Yeager's results appear to be the only experimentally determined values (i.e., results that do not depend entirely on $\bar{V}^\circ$ data).

8. King[109] has recently developed a method of calculating the $\bar{V}^\circ_{H^+}$ from the tetraalkylammonium salts using the equation

$$(32a) \qquad \bar{V}^\circ_{R_4NX} - \bar{V}^\circ_{HX} = \frac{1}{d} V_{wR_4N^+} - \bar{V}^\circ_{H^+}$$

where d is the packing fraction $(d = V_w/\bar{V}^\circ_{ion})$, $\bar{V}^\circ_w$ is the van der Waals molal volume $(V_w = 2.52 \, r_w^3$ where r_w is the van der Waals radii, corrected for hydrogen bonding). King found a plot of the left-hand side of this equation versus V_w is linear for $\overline{V}^\circ_{R_4N^+}$ ions when V_w is greater than 50 cm³/mole. He finds $\overline{V}^\circ_{H^+} = -4.1$ to -4.9 ± 0.7 cm³/mole depending on the value chosen for the carbon–nitrogen bond length.

9. Millero has recently[29(c)] examined the use of the $\overline{V}^\circ$ of the large electrolyte tetraphenyl arsonium tetraphenyl boron (Ph_4AsBPh_4) to determine ionic partial molal volumes. The first method he used assumes that the ratio of the $\overline{V}^\circ_{Ph_4As^+}/\overline{V}^\circ_{BPh_4^-}$ is equal to the ratio of $r_{Ph_4As^+}{}^3/r_{Ph_4As^+}{}^3$ (based on Bernal and Fowler's method[4]). Using the Stokes radii of $r_{Ph_4As^+} = 4.70$ and $r_{BPh_4^-} = 4.62$ Å, he obtained $\overline{V}^\circ_{H^+} = -4.1$ cm³/mole. The second method he used assumes that the ratio of the $\overline{V}^\circ_{BPh_4As^+}/\overline{V}^\circ_{BPh_4^-}$ is equal to the ratio of the van der Waals volumes $V_{wPh_4As^+}/V_{wBPh_4^-}$ (based on King's method[109]). Using $V_{wPH_4As^+} = 193.2$ and $V_{wBPh_4^-} = 186.9$ cm³/mole, he obtained $\overline{V}^\circ_{H^+} = -6.7$ cm³/mole while $V_{wPh4As^+} = 196.4$ and $V_{wBPh_4^-} = 186.9$ gave $\overline{V}^\circ_{H^+} = -4.4$ cm³/mole. The average value of $\overline{V}^\circ_{H^+} = -5.0$ cm³/mole agrees very well with most of the values obtained by other workers.

To compare the ionic partial molal volume of ions obtained by these methods on a common basis, it is necessary to use the same molal volume data as well as the most reliable molal volume data. This is necessary because earlier workers used inaccurate $\overline{V}^\circ$ data and often various workers have compared their results to others without normalizing the data. For example, Panckhurst[40] calculates $\overline{V}^\circ_{H^+} = -1.14$ cm³/mole using Noyes's method[56] ($J = 4.03$, $C_2 = -3.66$, and $A_2 = -21.73$, using Pauling radii[47]) and unreliable $\overline{V}^\circ$ data. The more reliable $\overline{V}^\circ$ data given in Table 8 yield $\overline{V}^\circ_{H^+} = -2.77$ cm³/mole ($J = 3.63$, $C_2 = -4.56$, and $A_2 = -27.11$, using Pauling radii[47]).

Using the conventional partial molal volume data of various ions (Table 8) as well as the additivity principle (Table 2), we have recalculated the $\overline{V}^\circ_{H^+}$ using the methods outlined above (i.e., when possible). The values of $\overline{V}^\circ_{H^+}$ are based on $\overline{V}^\circ_{HCl} = 17.83$ cm³/mole,[10,11,15] unlike some earlier estimations. The comparison of the estimated values of $\overline{V}^\circ_{H^+}$ at 25°C are given in Table 11, along with the value obtained by Zana and Yeager.[59]

As pointed out by Panckhurst,[40] most of the methods used to estimate the $\overline{V}^\circ_{H^+}$ have assumed certain values for the crystal radii. Pauling[47] or Goldschmidt[50] crystal radii are generally used, and these are very similar for the monovalent monatomic ions with the exception of Li^+. Since many of these methods are quite sensitive to the radii used, it is necessary to consider the effect of using the "experimentally" determined radii of Gourary and Adrian.[61] Table 12 lists the Pauling,[47] Goldschmidt,[50] and Gourary–Adrian[61] crystal radii for the monovalent monatomic alkali metal and halide ions. The calculated values of $\overline{V}^\circ_{H^+}$ using the Gourary–Adrian[61] radii are given in

Table 11 Comparison of the Various Estimates for the Ionic Partial Molal Volume of the Proton at 25°C

$\bar{V}_{H^+}^{\circ}$	Method Used	Investigators	References
−5.4	Ionic vibration-potential measurements[a]	Zana and Yeager	59
−3.8 (1.9)	$\bar{V}_{Cs^+}^{\circ}/\bar{V}_{Cl^-}^{\circ} = r_{Cs^+}^{3}/r_{Cl^-}^{3}$	Bernal and Fowler	31
		Darmois	41
		Zen	42
−2.7 (3.3)	$\bar{V}_{Cs^+}^{\circ}/\bar{V}_{I^-}^{\circ} = r_{Cs^+}^{3}/r_{I^-}^{3}$	Rice	44
		Eucken	45
−5.3 (−3.7)	$\bar{V}_{K^+}^{\circ}/\bar{V}_{F^-}^{\circ} = r_{K^+}^{3}/r_{F^-}^{3}$	Kobayazi	43
−5.1	$\bar{V}_{K^+}^{\circ} = \bar{V}_{F^-}^{\circ}$	Wirth	46
−0.2	$\bar{V}_{NH_4^+}^{\circ} = \bar{V}_{Cl^-}^{\circ}$	Fajans and Johnson	48
		Hepler	49
0.0	$\bar{V}_{H^+}^{\circ} = 0$, conventional value	Owen and Brinkley	36
−6.0	Adjusting $\bar{V}_{H^+}^{\circ}$ so cations and anions fit the same semiempirical equation	Couture and Laidler	33
−7.6 (−1.3)	I^- is not hydrated, $\bar{V}_{I^-}^{\circ} = 4.35\, r^3$	Stokes and Robinson	52
−4.5 (2.4)	Adjusting $\bar{V}_{H^+}^{\circ}$ so that the $\bar{V}^{\circ}$ of cations and anions are a smooth function of r^3	Mukerjee	51
−0.9	I^- is not hydrated, $\bar{V}_{I^-}^{\circ} = \bar{V}_{int}^{\circ} = 37.1$	Padova	53
−2.8 (2.7)	Adjusting $\bar{V}_{H^+}^{\circ}$ so large cations and anions approach theoretical $\bar{V}_{elect}^{\circ}$ [b]	Noyes Panckhurst	56 40
−6.0	Plot of $\bar{V}_{R_4NX}^{\circ}$ versus MW of R_4N^+ extrapolated[c] to 0 molecular weight	Conway, Verrall, and Desnoyers	16
		Millero and Drost-Hansen	54
−2.6 (−2.6)	Plot of $\bar{V}_{MCl}^{\circ} - \bar{V}_{int}^{\circ}$ of M^+ versus $1/r$ extrapolated $1/r = 0$, $r = r_{M^+} + r_{H_2O}$ [d]	Glueckauf	55
−4.5	Plot $\bar{V}_{R_4NX}^{\circ} - \bar{V}_{HX}^{\circ}$ versus $\bar{V}_{wR_4N^+}^{\circ}$ extrapolated to 0	King	109
−4.1	$\bar{V}_{Ph_4As^+}^{\circ}/\bar{V}_{BPh_4^-}^{\circ} = r_{Ph_4As^+}^{3}/r_{BPh_4^-}^{3}$, where r are Stokes radii	Millero	29(c)
−5.5	$\bar{V}_{Ph_4As^+}^{\circ}/\bar{V}_{BPh_4^-}^{\circ} = \bar{V}_{w,Ph_4As^+}^{\circ}/\bar{V}_{w,BPh_4^-}^{\circ}$	Millero	29(c)

[a] Calculated from average value obtained from the chloride omitting LiCl and NH$_4$Cl.
[b] Calculated using (30) and (31) with $\bar{V}_{int}^{\circ} = 2.52\, r^3 + 3.63\, r^2$, $C_2 = -3.66$ and $A_2 = -21.73$.
[c] Average value obtained for Cl$^-$, Br$^-$, and I$^-$ salts, the R_4NBr results of Franks and Smith[20] give $\bar{V}_{H^+}^{\circ} = -6.2$ cm^3/mole.
[d] Calculated from the results given in Table 8 using a least-square best fit.

parentheses in Table 11. The use of the Gourary–Adrian radii[61] yields results for $\overline{V}_{H+}^{\circ}$ that are more positive by about 5.0 cm³/mole. If we use Wirth's method[46] with the Gourary–Adrian radii[61] (i.e., assume the $\overline{V}^{\circ}$ of two ions are equal when their radii are equal), we obtain $\overline{V}_{Na+}^{\circ} = \overline{V}_{F-}^{\circ}$ or $\overline{V}_{H+}^{\circ} = 0$ cm³/mole, which is also about 5.0 cm³/mole higher than assuming $\overline{V}_{K+}^{\circ} = \overline{V}_{F-}^{\circ}$.

Table 12   The Crystal Radii (Å) of Some Simple Monovalent Monatomic Ions

Ion	Pauling Radii (Ref. 47)	Goldschmidt Radii (Ref. 50)	Gourary–Adrian Radii (Ref. 61)
Li⁺	0.60	0.78	0.94
Na⁺	0.95	0.98	1.17
K⁺	1.33	1.33	1.49
Rb⁺	1.48	1.49	1.63
Cs⁺	1.69	1.65	1.86
F⁻	1.36	1.33	1.16
Cl⁻	1.81	1.81	1.64
Br⁻	1.95	1.96	1.80
I⁻	2.16	2.20	2.05

Panckhurst[40] has criticized most of the methods used to estimate $\overline{V}_{H+}^{\circ}$; he feels that the only valid methods are those used by Noyes[56] and by Fajans and Johnson.[48] He selects a value of $\overline{V}_{H+}^{\circ} = 1.5 \pm 2.0$ cm³/mole as the "best" value which is the average value obtained using Pauling[47] and Gourary–Adrian[61] crystal radii by Noyes's method.[56] Using the more reliable $\overline{V}^{\circ}$ data, we obtain $\overline{V}_{H+}^{\circ} = -0.05$ cm³/mole as the "best" averaged value obtained by using the Noyes[56] method with Pauling[47] and Gourary–Adrian[61] radii. Thus, although some of the criticism made by Panckhurst[40]—concerning the various methods used to estimate $\overline{V}_{H+}^{\circ}$—are valid, most of the methods yield results for $\overline{V}_{H+}^{\circ}$ between 0 and -5.0 cm³/mole. We now have an interesting paradox: if $\overline{V}_{H+}^{\circ} = 0$ and the Gourary–Adrian radii[61] are correct or if $\overline{V}_{H+}^{\circ} = -5.0$ and the Pauling radii[47] are correct, cations and anions of the same size have nearly the same volume (both cross relationships yield the result that the $\overline{V}^{\circ}$ of cations and anions of the same size are not equal).

Since much of this paradox is related to the criticism by Panckhurst[40] of the experimental results of Zana and Yeager,[59] we feel it is worthwhile to examine the comments made by Panckhurst.[40] Panckhurst's main criticisms of Zana and Yeager's work are as follows: (a) Zana and Yeager[59] use $\bar{V}°$ data when they should be using apparent molal volume data and (b) Zana and Yeager use ϕ/a values that appear to be a function of concentration; thus the proper choice of ϕ/a may be in error. Since the simple alkali metal halides do not deviate strongly from the Debye–Hückel limiting law in dilute solutions, the error involved in using $\bar{V}°$ instead of ϕ_V can be estimated from the equation $\phi_V - \phi_V° = 1.868\sqrt{c}$.[40] The maximum concentration used by Zana and Yeager[59] to calculate $\bar{V}°$ of various ions was $3 \times 10^{-1}c$; thus the error in using $\bar{V}°$ instead of ϕ_V is about $1.0\ \text{cm}^3/\text{mole}$. It should also be pointed out that the deviations from additivity for the alkali metal halides at $3 \times 10^{-1}c$ should not be larger than about $0.4\ \text{cm}^3/\text{mole}$ (see Table 5). Using the data for the salts RbCl and CsCl, where ϕ/a is constant over the entire concentration range, one obtains $\bar{V}_{H+}° = -5.7\ \text{cm}^3/\text{mole}$, which agrees with the results obtained for the other salts. The errors due to the criticism raised by Panckhurst[40] appear to be well within Zana and Yeager's quoted experimental error of $\pm 2.0\ \text{cm}^3/\text{mole}$. Because of the internal consistency of the many salts investigated by Zana and Yeager,[59] we feel that the $\bar{V}_{H+}°$ should be close to $-5.0\ \text{cm}^3/\text{mole}$ at 25°C.

The choice of radii is not clear at present and we prefer to use the Pauling[47] or Goldschmidt[50] radii because their tabulations are more extensive. It may be accidental that the choice of Zana and Yeager's value[59] for $\bar{V}_{H+}°$ and the use of Pauling radii[47] yield similar values for the $\bar{V}°$ of cations and anions of the same size, however, the similarity does make the comparison as a function of temperature (or solvent system) a lot simpler.

Since the experimental ionic potential measurements have not been made at temperatures other than 25°C, we must use one or all of the methods to estimate $\bar{V}_{H+}°$ at other temperatures. Table 13 contains $\bar{V}_{H+}°$ estimated by some of the methods previously discussed, as well as the method used by Ellis[38(c)] based on the Criss–Cobble correspondence principle,[62] $\bar{V}_{ion}^t = a\bar{V}_{ion^{25°C}}° + b$. Since this method is valid for both cations and anions only if they have similar $\bar{V}°$, the correspondence method gives results in general agreement with the methods of Mukerjee,[51] Wirth,[46] Kobayazi,[43] Bernal and Fowler,[31] and Rice.[44] We feel that $\bar{V}_{H+}°$ should be less than 0 at all temperatures (since $\bar{V}_{int}°$ for the proton should be nearly 0 and $\bar{V}_{elect}°$ is negative); therefore all the methods except Glueckauf's and Padova's give reasonable values for $\bar{V}_{H+}°$ as a function of temperature. Although the $\bar{V}_{H+}°$ values determined by the various methods differ significantly, the partial molal expansibility of the proton, $\bar{E}_{H+}°$ (Figure 1), of all the methods appears to be a smooth function of temperature. The "best" straight line through all these $\bar{E}°$ values as a function of temperature was found to be

$$(33) \qquad \bar{E}_{H+}° = -0.008 - 3.40 \times 10^{-4}\,t$$

Table 13 The Estimation of the Partial Molal Volume of the Proton at Various Temperatures

Temperature (°C)	References Cited in Footnotes to Table												
	1	2	3	4	5	6	7	8	9	10	11	12	13
0	−5.0	−4.1	−4.2	−3.4	−2.4	0.0	−3.0	−3.0	−3.5	−4.9	−10.2	−3.5	−3.9
25.0	−5.7	−5.1	−5.3	−3.8	−2.7	−0.1	−2.6	−2.8	−4.5	−6.0	−7.6	−0.9	−5.0
50.0	−6.0	−5.5	−5.6	−4.2	−2.9	−0.6	−2.0	−2.8	−5.0	−3.7	−6.3	+0.4	−6.4
75.0	−6.6	−6.7	−6.8	−5.0	−3.6	−1.1	−1.8	−3.3	−6.1	—	−6.6	+0.1	—
100.0	−8.0	−6.6	−6.7	−5.5	−3.0	−1.8	+1.5	−2.9	−6.6	—	−5.1	+1.6	—
125.0	−8.9	−7.9	−8.0	−6.5	−3.5	−2.6	+2.3	−3.5	−7.4	—	−6.3	+0.4	—
150.0	−10.8	−9.8	−9.8	−7.9	−4.3	−4.0	+3.3	−4.5	−9.6	—	−8.0	−1.3	—
175.0	−12.5	−11.1	−11.0	−9.2	−3.9	−5.9	+8.6	−4.4	−10.8	—	−9.2	−2.5	—
200.0	−15.5	−14.3	−14.1	−12.5	−5.8	−9.8	+7.2	−7.4	−14.6	—	−13.6	−6.9	—

1. Ellis's correspondence Method.[38(c)]
2. Wirth's method.[46]
3. Kobayazi's method.[43]
4. Bernal and Fowler's method.[31]
5. Rice's method.[44]
6. Fajans and Johnson's method.[48]
7. Glueckauf's method[55] (least-square best fit).
8. Noyes's method[56] (least-square best fit with $\bar{V}^\circ_{int} = 2.52r^3 + Jr^2$).
9. Mukerjee's method.[51]
10. Conway, Verrall, and Desnoyers's method.[16]
11. Stokes and Robinson's method.[52]
12. Padova's method.[53]
13. Millero's methods[29(c)] (average).

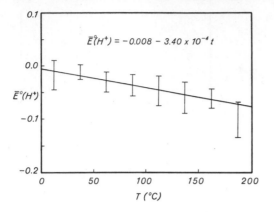

Figure 1. The partial molal expansibility of the proton from 0 to 200°C.

[with a standard deviation of ± 0.02 cm³/(mole)(deg)]. Thus by selecting one value for $\overline{V}_{H^+}^\circ$ at one temperature, we can integrate this equation and determine $\overline{V}_{H^+}^\circ$ as a function of temperature. Using the value of $\overline{V}_{H^+}^\circ = -5.4$ at 25°C obtained by Zana and Yeager,[59] we obtain upon integration of the equation

(34) $$\overline{V}_{H^+}^\circ = -5.1 - 0.008\, t - 1.7 \times 10^{-4}\, t^2$$

The values of $\overline{V}_{H^+}^\circ$ evaluated from 0 to 200°C from this equation are given in Table 14.

Table 14 The Absolute Partial Molal Volume of the Proton at Various Temperatures

Temperature (°C)	$\overline{V}_{H^+}^\circ$ (cm³/mole)
0	−5.1
25.0	−5.4
50.0	−5.9
75.0	−6.6
100.0	−7.6
125.0	−8.8
150.0	−10.1
175.0	−11.7
200.0	−13.5

4 Ion-Water Interactions

Since the multilayer hydration models used by Gurney,[63] Frank and Wen,[64] and Eigen and Wicke[65] have been so successful in describing the thermodynamic properties of aqueous solutions, we shall use this model to examine ion-water interactions. Using this model for ion-water interactions, the partial molal volume of an ion at infinite dilution, $\overline{V}^\circ_{ion}$, can be attributed to the following components:

$$(35) \qquad \overline{V}^\circ_{ion} = \overline{V}^\circ_{cryst} + \overline{V}^\circ_{elect} + \overline{V}^\circ_{disord} + \overline{V}^\circ_{caged}$$

where $\overline{V}^\circ_{cryst}$ is the crystal partial molal volume, $\overline{V}^\circ_{elect}$ is the electrostriction partial molal volume, $\overline{V}^\circ_{disord}$ is the disordered or void-space partial molal volume, and $\overline{V}^\circ_{caged}$ is the caged or structured partial molal volume. This division is made with the realization that the regions may be inseparable because of overlapping effects. For example, void-space effects appear to be caused or related to electrostriction. It is also convenient to divide ions into three classes depending upon which region is predominant; (a) electrostrictive "structure-making" ions when $\overline{V}^\circ_{elect}$ is the dominant region, (b) disordered "structure-breaking" ions when $\overline{V}^\circ_{disord}$ is the dominant region, and (c) hydrophobic "structure-making" ions when $\overline{V}^\circ_{caged}$ is the dominant region.

In the electrostrictive region, the ionic charge on the ion completely orients the surrounding water molecules. The water molecules in this region, although rapidly exchanging, are immobilized to a greater extent compared to bulk water. Samoilov[66] has called this phenomenon positive hydration. Since the water molecules are more firmly packed around an electrostrictive ion than in bulk water, the net volume of the system is decreased in the electrostrictive region. For an ion with a large electrostrictive region (i.e., for ions of high charge and small radii) or when this region is dominant, the ion can be classified as an electrostrictive structure-making[64] or positive-hydrating[66] ion.

Water in the second and third solvation layers will be less immobilized or tightly packed. In the intermediate region between the electrostricted region and the bulk water, the orientation of water molecules is diminished; however, the orientation is still large enough to interfere with the formation of the normal water structure. Water in this region has less structure (or there are fewer hydrogen bonds); thus, if this disordered region is very large, the ion can be classified as a net structure-breaking[64] or negative-hydrating[66] ion.

For ions with a large hydrophobic surface (e.g., the R_4N^+ ions), it is necessary to consider another effect. Water molecules at the surface of these ions are not strongly influenced by the ionic charge or the hydrocarbon portion of the molecule. The water-water interactions next to a hydrophobic ion appear to have a higher degree of hydrogen bonding or structure. This type of ion can be classified as a hydrophobic structure-making ion.[64]

It should be emphasized that the rigid classification of ions into these three classes (although useful), cannot be made, since various physical measurements frequently lead to different classes for the same ion. We also want to make it clear that the terms structure breaker and structure maker refer here to hydration effects (i.e., ion-water interactions) and not water-structure changes (i.e., changes in "icelike" and "non-icelike" forms of water). Consequently, we are neglecting the unique types of structured water that may or may not exist or may or may not be influenced by the ion. We feel that it is possible to discuss ion-water interactions in aqueous solutions by simply recognizing that water is a highly structured solvent and the addition of an ion alters this structure.

The problem of rigidly classifying ions into these three categories can be demonstrated by examining two other methods that can be used to classify ions—viscosity B coefficients[63,67,68] and the partial molal entropies[63,69,70] of various ions. Table 15 gives the viscosity B coefficient,[63,67,68] the partial molal entropy,[63,69,70] and the partial molal volume of some simple ions at 25°C. Positive B coefficients, negative partial molal volumes, and negative partial molal entropies are normally associated with the electrostrictive structure-making ions; whereas negative B coefficients, positive partial molal volumes and positive partial molal entropies are normally associated with structure-breaking ions. To use these methods on a common scale, the intrinsic contribution or the nonelectrostatic terms must be subtracted—that is, the intrinsic or crystal partial molal volume, the nonelectrostatic contribution (Einstein) to the viscosity B coefficient and the intrinsic contribution to the partial molal entropy.

Since the various components of $\overline{V}_{ion}^{\circ}$ may be small when compared to the absolute size of the ion (i.e., the crystal volume), it is difficult to determine the importance of the individual components or to separate overlapping effects. In recent work,[54,71,72] we have tried to solve this problem by examining the effect of temperature on the $\overline{V}_{ions}^{\circ}$ or the partial molal expansibility of ions, $\overline{E}_{ions}^{\circ}$. Other methods that we have used to elucidate the various components include examining the effect of pressure on the $\overline{V}_{ions}^{\circ}$, the partial molal compressibility of ions,[73] and the $\overline{V}_{ions}^{\circ}$ in various solvents.[74-76]

We shall now briefly examine the various components of $\overline{V}_{ions}^{\circ}$ in further detail.

The Intrinsic Partial Molal Volume, $\overline{V}_{int}^{\circ}$. The true "nonhydrated" partial molal volume of an ion in solution cannot be measured directly and is usually evaluated from the crystal volume, $\overline{V}_{cryst}^{\circ}$, by assuming that the ions are perfect hard spheres

$$(36) \qquad \overline{V}_{cryst}^{\circ} = \frac{4\pi}{3} N \times 10^{-24} \times r^3 = 2.52\, r^3$$

Table 15 The Partial Molal Volume, Viscosity B Coefficient, and Partial Molal Entropy of Various Ions at 25°C

Ion	$\bar{V}^\circ_{cryst}{}^a$ (cm³/mole)	$\bar{V}^\circ_{ion}$ (cm³/mole)	$\bar{V}^\circ_{ion} - \bar{V}^\circ_{cryst}$ (cm³/mole)	B^b (liter/mole)	$\bar{S}^{\circ c}$ (e.u.)
H⁺	0	−5.4	−5.4	0.069	−5.5
Li⁺	0.5	−6.3	−6.8	0.149	−2.5
Na⁺	2.1	−6.6	−8.7	0.086	8.6
K⁺	5.9	3.6	−2.3	−0.007	19.0
Rb⁺	8.2	8.7	0.5	−0.030	24.2
Cs⁺	12.2	15.9	3.7	−0.045	26.3
NH₄⁺	8.2	12.4	4.2	−0.007	21.6
Me₄N⁺	105.3	84.2	−21.1	0.12	—
Et₄N⁺	161.3	143.7	−17.6	0.38	—
nPr₄N⁺	232.7	209.0	−23.7	0.86	—
nBu₄N⁺	303.8	270.3	−33.5	1.28	—
Ph₄As⁺	261.6	295.3	33.7	—	—
Mg²⁺	0.7	−32.0	−32.7	0.385	−39.2
Ca²⁺	2.4	−28.6	−31.0	0.285	−24.2
Sr²⁺	3.6	−29.0	−31.6	0.265	−20.4
Ba²⁺	6.2	−23.3	−29.5	0.220	−8.0
F⁻	6.3	4.2	−2.1	0.096	2.2
Cl⁻	14.9	23.2	8.3	−0.007	19.0
Br⁻	18.7	30.1	11.4	−0.042	25.2
I⁻	25.4	41.6	16.2	−0.068	32.1
OH⁻	6.6	1.4	−5.2	0.12	2.9
NO₃⁻	21.1	34.4	13.1	−0.046	40.5
ClO₄⁻	30.3	49.5	19.2	−0.056	49.0
BPh₄⁻	248.5	283.0	34.5	—	—

[a] Calculated from Pauling crystal radii[47] and radii tabulated elsewhere.[75] The radii for BPh₄⁻ and Ph₄As⁺ are Stokes radii.
[b] From Refs. 63, 67, and 68.
[c] From Refs. 63, 69, and 70.

where r is the crystal radii in angstroms (Å). Since there is some doubt about what are the correct crystal radii,[40, 61] one cannot be sure that the $\bar{V}^\circ_{cryst}$ of individual ions determined in this manner are absolute. The crystal volumes for ions appearing in Table 15 were calculated by using Pauling crystal radii.[47] As can be seen from the examination of the difference between the $\bar{V}^\circ_{ion} - \bar{V}^\circ_{cryst}$ given in this table, $\bar{V}^\circ_{cryst}$ is smaller than $\bar{V}^\circ_{ion}$ for some of the large monovalent ions (i.e., $\bar{V}^\circ_{ion} - \bar{V}^\circ_{cryst}$ is positive). For example, $\bar{V}^\circ_{cryst}$ of I⁻ is calculated to be 25.4 cm³/mole, compared to a measured value of 41.6 cm³/mole; it should be noted that neither the use of a lower $\bar{V}^\circ_{I^-}$ obtained by assuming $\bar{V}^\circ_{H^+} = 0$ nor the use of the experimental radii of Gourary–Adrian crystal radii[61] affects these conclusions significantly.

The experimental estimates of the nonhydrated volume from compressibility measurements[77] and those calculated from various semiempirical[33, 49, 51] correlations of $\overline{V}^\circ_{ions}$ are also larger than the crystal volume. In general, the results indicate that the $\overline{V}^\circ_{ion}$ is about 77% larger in water than in the crystal. This discrepancy can be caused by (a) the radius of the ion in solution being 21% greater than the radius in the crystal or (b) the void space around the hydrated ion or some other positive disorder effect.

Since the internal electrostatic pressure in water is comparable to that in the crystal (because both are condensed phases) and pairs of ions enter water with energies of hydration that are almost equal to their lattice energies in the solid salt,[78] it is reasonable to suppose that the radius of an ion in solution is equal to the radius in the crystal.[78–80] Benson and Copeland[80] have estimated the change of the radius of an ion when going from the crystal to water and show that, at best, the radius increases by only 0.01 Å. Thus there is no basis for believing that ions in solution are under less electrostatic pressure than are ions in the crystal, and it is reasonable to equate the intrinsic partial molal volume of an ion in solution to the crystal volume. The apparent increase in the intrinsic volume of an ion in solution is thus caused by void-space packing effects around the nonhydrated ion (i.e., in the $\overline{V}^\circ_{elect}$ region) or around the hydrated ion (i.e., in the $\overline{V}^\circ_{disord}$ region).

The effect of temperature on $\overline{V}^\circ_{cryst}$ is small and positive for normal ions; however, it may be large for ions with hydrocarbon portions (i.e., R_4N^+ ions).

The Disordered Partial Molal Volume, $\overline{V}^\circ_{disord}$.   Various workers[16, 40, 49, 51, 52, 55, 56, 76, 81] have attempted to separate $\overline{V}^\circ_{int}$ into two components—$\overline{V}^\circ_{cryst}$ and $\overline{V}^\circ_{disord}$—by using the semiempirical equations

(37) $$\overline{V}^\circ_{int} = \overline{V}^\circ_{cryst} + \overline{V}^\circ_{disord}$$

(38) $$= Ar^3 = 2.52\,r^3 + (A - 2.52)r^3$$

(39) $$= 2.52r^3 + A'r^2$$

(40) $$= 2.52(r + a)^3 = 2.52\,r^3 + [2.52(r + a)^3 - 2.52\,r^3]$$

Equation 38 assumes that $\overline{V}^\circ_{int}$ is equal to the crystal volume times a constant or that $\overline{V}^\circ_{disord}$ is equal to $(A - 2.52)r^3$. Values for the constant $A = 4.48,$[51] $4.6,$[49] $4.9,$[33] and 5.3[49] have been obtained. These values are nearly twice the theoretical value, 2.52. Using 4.48, the value obtained by Mukerjee,[51] $\overline{V}^\circ_{int}$ is found to be 1.78 times greater than $\overline{V}^\circ_{cryst}$ and $\overline{V}^\circ_{disord} = 1.96\,r^3$. As mentioned earlier, Stokes and Robinson[52] have calculated $\overline{V}^\circ_{int} = 4.35\,r^3$ by assuming that an ion contributes to the volume of the system an amount due to "random close packing." Thus the semiempirical values of $\overline{V}^\circ_{int}$ are nearly equal to the value calculated by including a void-space packing contribution.

Equation 39 assumes that $\overline{V}^{\circ}_{disord}$ is proportional to the surface of the ion, r^2. Various workers have used semiempirical equations to arrive at values of $A' = 4.09,$[56] $4.03,$[40] and 4.0[76] at $25°C$. Conway et al.[16, 81] have calculated the void space or disordered volume by using (39). They calculated $A' = 3.15$ by making the following assumptions: (a) the crystal radius equals the radius of the ion in solution; (b) when r_{ion} is less than r_{H_2O}, (36) holds; (c) when $r_{ion} = r_{H_2O}$, $\overline{V}^{\circ}_{int} = (2r)^3 N$, holds and the ion and water molecules are locally cubically or hexagonally packed; (d) there is a smooth transition between these limiting conditions.

Equation 40 assumes that the radii of all ions increase by a constant amount in solution, or $\overline{V}^{\circ}_{disord} = 2.52(r + a)^3 - 2.52\, r^3$. Values of $a = 0.45,$[76] $0.45,$[56] and 0.436 Å[40] have been found by investigators using different semiempirical equations at $25°C$. Glueckauf[55] postulated that the void space can be represented by a hollow sphere of radius $r + a$ and $\overline{V}^{\circ}_{int}$ can be calculated from (40). He calculates $a = 0.55$ Å at $25°C$ by assuming the void space of an ion with the same radius of a water molecule (1.38 Å) has the same void space as that for pure water

$$(41) \qquad a = \left[\frac{\overline{V}^{\circ}_{H_2O}}{2.52}\right]^{1/3} - r_{H_2O}$$

Although Glueckauf[55] states that void-space effects are the cause of the large $\overline{V}^{\circ}_{int}$ of ions, the addition of the constant, a, to the radius of all ions, in effect, is similar to assuming that all ions expand by a constant amount. Such an assumption, however, does not seem reasonable if one considers the arguments of Benson and Copeland[80] and Stokes.[78] Table 16 compares $\overline{V}^{\circ}_{disord}$ for ions as a function of radius obtained by various semiempirical methods and calculated using the methods of Stokes and Robinson,[52] Conway, Verrall, and Desnoyers,[16, 81] and Glueckauf.[55]

Recently[76] we have examined the $\overline{V}^{\circ}_{ions}$ in the solvents water, methanol, and N-methylpropionamide (NMP) by the following semiempirical equations

$$(42) \qquad \overline{V}^{0}_{ion} = Ar^3 - \frac{BZ^2}{r}$$

$$(43) \qquad = 2.52\, r^3 + A'r^2 - \frac{B'Z^2}{r}$$

$$(44) \qquad = 2.52(r + a)^3 - \frac{B''Z^2}{r}$$

It was found that the constants A, A', and a of (42) to (44) were not proportional to the dielectric constants or the compressibilities of the solvents (as might be expected if the radius of the ion increased in solution). However, the constants do appear to be proportional to the structure of the solvents (e.g.,

Table 16   The Disordered or Void-Space Partial Molal Volume, $\bar{V}^{\circ}_{\text{disord}}$, as a Function of Size at 25°C

Radius (Å)	$\bar{V}^{\circ}_{\text{cryst}}$ (cm³/mole)	$\bar{V}^{\circ}_{\text{disord}}$ (cm³/mole) described in footnotes to Table					
		1	2	3	4	5	6
0	0	0	0	0	0	0	0
0.5	0.31	0.26	1.10	1.85	0.23	0.78	2.61
1.0	2.52	2.06	4.60	5.16	1.83	3.15	6.86
1.5	8.50	6.95	10.35	10.18	6.17	7.08	13.21
2.0	20.16	16.48	18.40	16.90	14.64	12.60	21.62
2.5	39.37	32.19	28.75	25.32	28.59	19.68	32.13
3.0	68.04	55.62	41.40	35.44	49.41	28.35	44.70

1. $\bar{V}^{\circ}_{\text{disord}} = (A - 2.52)r^3$, $A = 4.58$ from semiempirical equation (42)[76] using Pauling crystal radii and data for the monovalent monatomic ions.
2. $\bar{V}^{\circ}_{\text{disord}} = A'r^2$, $A' = 4.60$ from semiempirical equation (43)[76] using Pauling crystal radii and data for the monovalent monatomic ions.
3. $\bar{V}^{\circ}_{\text{disord}} = 2.52(r + a)^3 - 2.52\,r^3$, $a = 0.45$ from semiempirical equation (41)[76] using Pauling crystal radii and data for the monovalent monatomic ions.
4. $\bar{V}^{\circ}_{\text{disord}} = (4.35 - 2.52)r^3$ from Stokes and Robinson's work due to the "random" close packing" of ions and water molecules.[52]
5. $\bar{V}^{\circ}_{\text{disord}} = 3.15\,r^2$ from Conway, Verrall, and Desnoyers's work[16,81] due to the void space for the cubic or hexagonal packing of water molecules around an ion.
6. $\bar{V}^{\circ}_{\text{disord}} = 2.52(r + 0.55)^3 - 2.52\,r^3$ from Glueckauf's work[55] due to void-space packing and represented by a hollow sphere of radius, $r + 0.55$.

$A_{\text{H}_2\text{O}} > A_{\text{NMP}} > A_{\text{MeOH}}$). Thus the $\bar{V}^{\circ}_{\text{disord}}$ appears to be related to the disordered region surrounding the solvated ion. This disordered effect may be visualized as the void space caused by the solvated ion (i.e., including the electrostricted region) rather than improper packing in the electrostricted region. At present it is not possible to state with certainty whether (38), (39), or (40) correctly describes the variation of $\bar{V}^{\circ}_{\text{int}}$ or $\bar{V}^{\circ}_{\text{disord}}$ with radius, since they each appear to "fit" successfully the experimental data.

The effect of temperature on $\bar{V}^{\circ}_{\text{disord}}$ has not been considered by many workers. Ellis[38] has used Glueckauf's method to estimate $\bar{V}^{\circ}_{\text{int}}$ or a from 0 to 200°C. Recently[82] we have examined the $\bar{V}^{\circ}_{\text{ions}}$ by using the semiempirical equations (42) to (44) from 0 to 200°C. In Table 17 we have tabulated the $\bar{V}^{\circ}_{\text{disord}}$ for an ion with a radius equal to 1.0 Å from 0 to 200°C—obtained from the semiempirical equations and from Glueckauf's method.[55] The results indicate that $\bar{V}^{\circ}_{\text{disord}}$ increases slightly with increasing temperature if one assumes that $\bar{V}^{\circ}_{\text{cryst}}$ is independent of temperature.

With the exception of Hepler's work[49] and our recent calculations,[82] most workers have assumed that $\bar{V}^{\circ}_{\text{int}}$ or $\bar{V}^{\circ}_{\text{disord}}$ is the same for cations and anions

Table 17 The Disordered Partial Molal Volume of an Ion with $r = 1.0$ Å at Various Temperatures

Temperature (°C)	$\overline{V}^{\circ}_{\text{disord}}$ (cm³/mole) described in footnotes			
	1	2	3	4
0	1.82	4.10	4.40	6.86
25.0	2.04	4.60	5.16	6.86
50.0	2.25	5.04	5.48	7.04
75.0	2.30	5.15	5.82	7.04
100.0	2.56	5.67	6.16	7.42
125.0	2.68	5.95	6.68	7.61
150.0	2.70	5.97	6.86	7.80
175.0	2.86	6.28	6.68	8.19
200.0	2.77	6.20	6.50	8.80

1. $\overline{V}^{\circ}_{\text{disord}} = (A - 2.52)r^3$ from semiempirical equation (42) using Pauling crystal radii and data for the monovalent monatomic ions Tables 7 to 10; $V^{\circ}_{\text{H}+}$ taken from Table 14.
2. $\overline{V}^{\circ}_{\text{disord}} = A'r^2$ from semiempirical equation (43) using Pauling crystal radii and data for the monovalent monatomic ions Tables 7 to 10; $\overline{V}^{\circ}_{\text{H}+}$ taken from Table 14.
3. $\overline{V}^{\circ}_{\text{disord}} = 2.52(r + a)^3 - 2.52r^3$ from semiempirical equation (44) using Pauling crystal radii and data for the monovalent monatomic ions Tables 7 to 10; $\overline{V}^{\circ}_{\text{H}+}$ taken from Table 14.
4. $\overline{V}^{\circ}_{\text{disord}} = 2.52(r + a)^3 - 2.52r^3$ using values of a calculated by Ellis[38] from Glueckauf's method[55] (41).

of the same size. Hepler[49] found values of $A = 5.3$ for cations and $A = 4.6$ for anions using (42) and the Goldschmidt crystal radii[50] ($\overline{V}^{\circ}_{\text{ions}}$ based on $\overline{V}^{\circ}_{\text{H}+} = -0.1$ cm³/mole). Hepler[49] also pointed out that other workers[83–85] have found the ion cavity for cations to be larger than the ion cavity for anions. The results of our calculations[82] of the semiempirical constants considering cations and anions separately (shown in Table 18) agree with Hepler's results at low temperatures when the Goldschmidt radii are used, that is, for (42). However, most of our results indicate that the $\overline{V}^{\circ}_{\text{int}}$ or $\overline{V}^{\circ}_{\text{disord}}$ is greater for anions than for cations. One could postulate that the difference is due to the different dipole orientations of water molecules around cations and anions. Also, since large $\overline{V}^{\circ}_{\text{disord}}$ effects appear to be related to structure breaking, one might postulate that anions are stronger structure breakers than cations of the same size. However, the observed effect is small and appears to be related to the radii used—and probably to the absolute value of $\overline{V}^{\circ}_{\text{H}+}$—and we therefore feel that it is impossible to state with certainty that $\overline{V}^{\circ}_{\text{disord}}$ is different for cations and anions of the same size.

The effect of temperature on the semiempirical constants obtained by considering cations and anions also appears to show the differences of $\overline{V}^{\circ}_{\text{int}}$ or

Table 18   Constants A, A', and a for Various Semiempirical Equations Considering Cations and Anions Separately at Various Temperatures

Temperature (°C)	Pauling Radii		Goldschmidt Radii		Gourary–Adrian Radii	
	Cations	Anions	Cations	Anions	Cations	Anions
			A			
0	4.18	4.26	4.66	3.93	3.44	4.68
25.0	4.31	4.58	4.80	4.22	3.54	5.03
50.0	4.38	4.78	4.88	4.40	3.60	5.25
75.0	4.48	4.90	5.02	4.51	3.64	5.36
100.0	4.20	5.18	4.98	4.77	3.43	5.69
125.0	4.20	5.31	4.85	4.89	3.54	5.82
150.0	4.20	5.45	4.82	5.01	3.56	5.95
175.0	4.00	5.71	4.51	5.25	3.30	6.22
200.0	4.11	5.78	4.60	5.30	3.43	6.27
			A			
0	2.95	4.23	3.79	3.48	1.90	4.93
25.0	3.20	5.01	4.06	4.21	2.10	5.73
50.0	3.33	5.48	4.20	4.65	2.28	6.22
75.0	3.46	5.78	4.49	4.92	2.31	6.50
100.0	3.48	6.42	3.92	5.53	2.08	7.19
125.0	3.09	6.74	4.15	5.81	2.08	7.48
150.0	3.08	7.08	4.18	6.11	2.13	7.78
175.0	2.70	7.68	3.58	6.67	1.59	8.37
200.0	2.89	7.82	3.66	6.77	1.74	8.44
			a			
0	0.28	0.48	0.36	0.41	0.26	0.55
25.0	0.31	0.55	0.37	0.48	0.29	0.62
50.0	0.33	0.59	0.41	0.52	0.31	0.66
75.0	0.34	0.62	0.45	0.55	0.31	0.68
100.0	0.27	0.66	0.37	0.59	0.25	0.72
125.0	0.30	0.68	0.41	0.62	0.28	0.75
150.0	0.29	0.71	0.41	0.64	0.28	0.77
175.0	0.22	0.75	0.35	0.68	0.22	0.80
200.0	0.21	0.76	0.35	0.69	0.23	0.81

$\overline{V}^\circ_{\text{disord}}$ for cations and anions. The $\overline{V}^\circ_{\text{int}}$ or $\overline{V}^\circ_{\text{disord}}$ for cations appears to go through a maximum as a function of temperature, whereas for anions $\overline{V}^\circ_{\text{int}}$ or $\overline{V}^\circ_{\text{disord}}$ increases smoothly with increasing temperature. Thus the effect of temperature on $\overline{V}^\circ_{\text{int}}$ of cations and anions appears to be different and probably contributes to the differences between the $\overline{E}^\circ$ of cations and anions as a function of size.[71, 72]

The relative contribution of $\overline{V}^\circ_{\text{disord}}$ to $\overline{V}^\circ_{\text{ion}}$ depends upon the magnitude of the electrostricted region, $\overline{V}^\circ_{\text{elect}}$. For ions with a large electrostricted region (large Z^2/r), the disordered volume is small relative to $\overline{V}^\circ_{\text{elect}}$; however, for ions with a small electrostricted region (small Z^2/r) the disordered region is very important. The disordered volume, $\overline{V}^\circ_{\text{disord}}$, appears to be less important at high temperatures because of the increasing significance of $\overline{V}^\circ_{\text{elect}}$ at high temperatures. Thus when $\overline{V}^\circ_{\text{ion}} - \overline{V}^\circ_{\text{cryst}}$ is positive, the $\overline{V}^\circ_{\text{disord}}$ region is greater than the $\overline{V}^\circ_{\text{elect}}$ region and the ion can be classified as a structure-breaking ion or a negative-hydrating ion.

The Electrostriction Partial Molal Volume, $\overline{V}^\circ_{\text{elect}}$. The theoretical prediction of the electrostriction caused by various ions is difficult because of the uncertainty of the form of the interaction of an ion and the water molecule. $\overline{V}^\circ_{\text{elect}}$ cannot be determined simply by subtracting $\overline{V}^\circ_{\text{int}}$ from $\overline{V}^\circ_{\text{ion}}$ because of the problems involved in determining $\overline{V}^\circ_{\text{disord}}$. If one treats the solvent as a continuous dielectric medium, electrostriction can be estimated by differentiating the electrical free energy, $\Delta G^\circ_{\text{elect}}$ of a charged sphere with respect to pressure (i.e., the Born equation)[86]

$$(45) \qquad \Delta G^\circ_{\text{elect}} = N\,\frac{e^2 Z_i^2}{2Dr}$$

Drude and Nernst,[57] (30), were the first to use this method to calculate the electrostriction of an ion. This equation is valid only for large isolated ions and serves only as an approximation. The theoretical values of B in (30) calculated at various temperatures (0–75°C) are given in Table 19 (using D and $\partial \ln D/\partial P$ tabulated by Owen et al.[58]). The values above 75°C have been estimated by extrapolating the B values between 0 to 75°C. Also included in this table are the calculated values of B using the semiempirical equations (42) to (44). Benson and Copeland[80] have used a modified version of the Drude–Nernst[57] equation

$$(46) \qquad \overline{V}^\circ_{\text{elect}} = -\left[\frac{\beta d}{Dd^\circ}\left(\frac{\partial \ln D}{\partial \ln d}\right)_T\right]\frac{e^2 Z^2}{2r}$$

where β is the compressibility of the solvent, d is the density, and the other symbols have their normal meanings. They estimate $B = 6.0$ at 25°C. The semiempirical values are larger than the values calculated from the Drude–Nernst[57] equation (as well as the modified Drude–Nernst equation). This difference is probably due to dielectric saturation effects.

Many authors[53, 87–92] have tried to calculate $\overline{V}^\circ_{\text{elect}}$ by more elaborate methods by considering dielectric saturation effects. The more recent attempts[53, 91, 92] have started by using the equation developed by Frank[93]

$$(47) \qquad \left(\frac{1}{v}\right)\left(\frac{\partial v}{\partial E}\right)_{\mu,\,T} = -\frac{E}{4\pi}\left(\frac{\partial D}{\partial P}\right)_{E,\,T}$$

Table 19   The Partial Molal Electrostriction, $\bar{V}^{\circ}_{elect}$, for a Monovalent Ion with $r = 1.0$ Å at Various Temperatures

Temperature (°C)	$-\bar{V}^{\circ}_{elect}$ (cm³/mole) Described in footnotes					
	1	2	3	4	5	6
0	9.8	13.3	12.0	3.57	—	13.8
25.0	9.2	13.0	11.6	4.18	13.0	13.4
50.0	9.6	13.7	12.1	4.95	—	13.4
75.0	9.8	13.9	12.6	5.92	—	13.8
100.0	11.9	16.0	13.2	7.14*	—	—
125.0	14.0	18.5	16.1	8.63*	—	—
150.0	16.2	20.5	17.4	10.43*	—	—
175.0	20.5	24.8	20.5	12.58*	—	—
200.0	24.9	30.3	27.0	15.12*	—	—

1. $-\bar{V}^{\circ}_{elect} = BZ^2/r$ from semiempirical equation (42) using Pauling crystal radii and data for the monovalent monatomic ions Tables 7 to 10; $V^{\circ}_{H^+}$ taken from Table 14.
2. $-\bar{V}^{\circ}_{elect} = B'Z^2/r$ from semiempirical equation (43) using Pauling crystal radii and data for the monovalent monatomic ions Tables 7 to 10; $V^{\circ}_{H^+}$ taken from Table 14.
3. $-\bar{V}^{\circ}_{elect} = B''Z^2/r$ from semiempirical equation (44) using Pauling crystal radii and data for the monovalent monatomic ions Tables 7 to 10; $V^{\circ}_{H^+}$ taken from Table 14.
4. Calculated from the Drude–Nernst equation[57] using the data of Owen et al.;[58] values estimated by extrapolation are denoted with an asterisk.
5. Calculated from Padova's work.[53]
6. Calculated from work by Desnoyers, Verrall, and Conway[91] and Dunn.[92]

where μ is the chemical potential, v is the volume of the solvent, E is the electrostatic field, D is the dielectric constant, P is the pressure.

Padova[53] has integrated (47) to calculate $\bar{V}^{\circ}_{elect}$ from

$$(48) \qquad \bar{V}^{\circ}_{elect} = \tfrac{1}{2}N \int_{r_e}^{\infty} \int_{0}^{E} \left(\frac{\partial D}{\partial P}\right)_{E,T} d(E^2) r^2 \, dr$$

where N is Avogadro's number, r is the distance from the center of the ion, r_e is the intrinsic radius of the ion, and E is the field strength at distance r from the ion. In the weak-field case, Padova[53] assumes $(\partial D/\partial P)_{E,T}$ is constant and equal to the bulk water value. Thus this equation reduces to the Drude–Nernst equation.[57] For the high-field case, D is a function of field strength and (48) becomes more difficult to solve.

By using graphic integrations, Padova[53] solves (48) for $Z = 1, 2,$ and 3. In the case of $Z = 1$, the dielectric constant is equal to the bulk value when r is greater than 8.0 Å (when $Z = 2$ and 3, r must be greater than 12 and 13 Å, respectively, for D to be equal to the bulk value). For an ion of effective radius equal to 1.0 Å, Padova's results[53] yield $\bar{V}^{\circ}_{elect} \cong -13.0$ cm³/mole at

25°C. Padova[53] has estimated the average theoretical electrostriction per mole of water is about -2.1 cm^3/mole. Thus the theoretical solvation numbers, n, can be calculated from

$$(49) \qquad n = \frac{\bar{V}^\circ_{elect}}{-2.1} = \frac{\bar{V}^\circ_{ion} - \bar{V}^\circ_{int}}{-2.1}$$

Because of the problems involved in estimating $\bar{V}^\circ_{int}$, it is difficult to determine solvation numbers by this method; however, Padova's results[53] are in general agreement with the estimates made by other workers.

Desnoyers, Verrall, and Conway[91] have solved (47) in a different manner. They calculate the effective pressure in the absence of the field that would produce the same volume change that would be produced by E. By substituting the compressibility equation

$$(50) \qquad \frac{dv}{v} = -\beta \, dP$$

into (47), they obtain

$$(51) \qquad dP = \left(\frac{E}{4\pi}\right) \beta \left(\frac{\partial D}{\partial P}\right)_{E,T} dE$$

This equation relates the effective pressure, P, to the field, E. By using (50), the volume change associated with the pressure change, dP, can be calculated. Desnoyers, Verrall, and Conway[91] have calculated the volume change of one mole of water as a function of field, $E = eZ/Dr$, under various conditions at 25°C. The electrostriction for an ion can be calculated from this volume change, Δv, using

$$(52) \qquad \bar{V}^\circ_{elect} = x_1 \, \Delta v + x_2 \, \Delta v$$

where x_1 is equal to the number of water molecules in the first solvation layer and x_2 is equal to the number of water molecules in the second solvation layer. For an ion with a first-solvation-layer solvation number of 4, $Z = 1$, $D = 2.0$, and $r = 1.0$ Å, $\bar{V}^\circ_{elect} \cong -13.4$ cm^3/mole at 25°C. Thus the methods of Desnoyers et al.[91] and Padova[53] yield results for $\bar{V}^\circ_{elect}$ that agree fairly well with the semiempirical value (i.e., for a singly charged ion).

Dunn[92] has recently extended the calculations of Desnoyers, Verrall, and Conway[91] to other temperatures (0–70°C). Although he has not attempted to make a numerical calculation of $\bar{V}^\circ_{elect}$, he has shown that $\bar{V}^\circ_{elect}$ goes through a maximum as a function of temperature and the maximum temperature is a function of field strength. For ions with a high field strength, the maximum occurs at lower temperatures than for ions with a low field strength—which is in agreement with the temperature dependence of $\bar{V}^\circ_{elect}$ found from the semiempirical equations. If one assumes that $-\bar{V}^\circ_{elect}$ is proportional to the compressibility of water, one might expect $\bar{V}^{\circ}_{\text{elect}}$ to go through a maximum, since the compressibility of water goes through a minimum near $40°C$.[94, 95] In Table 19, we have estimated the $\bar{V}^{\circ}_{\text{elect}}$ for a monovalent ion with $r = 1.0$ Å as a function of temperature using the various semiempirical equations and the methods of Padova,[53] Desnoyers, Verrall, and Conway,[91] and Dunn.[92]

The simple continuum model (i.e., the Drude–Nernst equation[57]) predicts the sign and order of magnitude of $\bar{V}^{\circ}_{\text{elect}}$ as a function of size and temperature; however, it does not predict the maximum observed in $\bar{V}^{\circ}_{\text{elect}}$ as a function of temperature. According to the calculations of Dunn,[92] by considering dielectric saturation effects one can account for this maximum in $\bar{V}^{\circ}_{\text{elect}}$ as well as the temperature at which the maximum occurs.

Thus far we have said little about the effect of charge on $\bar{V}^{\circ}_{\text{elect}}$. Couture and Laidler[33] noted that, for ions with nearly the same radius, $\bar{V}^{\circ}_{\text{ion}}$ is lowered by approximately 20 cm³/mole for each unit increase of charge. They also found empirically that $\bar{V}^{\circ}_{\text{elect}} \approx -26|Z|$. The Drude–Nernst equation[57] predicts that $\bar{V}^{\circ}_{\text{elect}}$ should be proportional to Z^2/r; the work of Mukerjee,[51] however, indicates that $\bar{V}^{\circ}_{\text{elect}}$ of divalent and trivalent cations appear to be equal, respectively, to -32.5 and -58.5 cm³/mole (at $25°C$) and independent of radius. Thus $\bar{V}^{\circ}_{\text{elect}}$ is nearly four times greater than the value for a monovalent ion of similar size (i.e., $r = 1.0$ Å), in agreement with the Z^2 relationship; yet $\bar{V}^{\circ}_{\text{elect}}$ for trivalent cations is about seven times greater than a monovalent ion of similar size and not in agreement with the Z^2 relationship.

The effect of temperature on the $\bar{V}^{\circ}$ of divalent cations has been shown to vary with size according to the equation (at $25°C$)[71]

$$(53) \qquad \bar{E}^{\circ}_{\text{ion}} = \frac{-0.031Z^2}{r} + 0.092$$

The value of the slope of this equation is in fair agreement with the theoretical value (0.027) obtained by Noyes[56] by differentiating the Drude–Nernst equation[57] with respect to temperature. As found for the $\bar{E}^{\circ}$ of the monovalent ions, there is a positive contribution (i.e., the intercept of 0.092).[72]

The cause of the decrease in electrostriction with increasing size, the apparent independence of $\bar{V}^{\circ}_{\text{elect}}$ for polyvalent ions as a function of size, and the positive contribution to the $\bar{E}^{\circ}$ of polyvalent ions can be attributed to electrostriction saturation effects (i.e., the inability of the outer water molecules to "see" the true charge on the central ion) or such other positive structural effects as $\bar{V}^{\circ}_{\text{disord}}$. Glueckauf[55] has assigned this positive effect to the long-range structure-forming effect postulated by Frank and co-workers.[64] We prefer to attribute the effect to $\bar{V}^{\circ}_{\text{disord}}$, that is, void-space packing effects of the hydrated ion not being able to fit into the structure of the solvent.[76] It should be noted that the $\bar{V}^{\circ}$ of the Li^+ ion also appears to have a positive

Table 20 The Electrostriction Partial Molal Volume, $\overline{V}^{\circ}_{\text{elect}}$, for Some Divalent Cations at Various Temperatures

	$\overline{V}^{\circ}_{\text{elect}}$[a] $(cm^3/mole)$				
Temperature (°C)	Mg^{2+}	Ca^{2+}	Sr^{2+}	Ba^{2+}	Average
0	33.2	34.2	37.2	36.7	35.3 ± 1.6
25.0	32.3	33.1	35.6	34.5	33.9 ± 1.2
50.0	34.0	34.6	36.4	35.2	35.1 ± 0.5
75.0	36.2	37.0	36.9	35.8	36.2 ± 0.7
100.0	40.0	40.1	39.9	35.8	38.9 ± 1.6
125.0	45.0	46.3	44.9	40.9	44.3 ± 1.7
150.0	50.6	50.9	50.1	44.9	49.1 ± 2.1
175.0	58.9	55.6	56.7	51.6	55.7 ± 2.1
200.0	65.4	62.5	63.5	56.7	62.0 ± 2.7

[a] Calculated from the relation $\overline{V}^{\circ}_{\text{elect}} = \overline{V}^{\circ}_{\text{ion}} - \overline{V}^{\circ}_{\text{int}}$, $\overline{V}^{\circ}_{\text{int}} = Ar^3$, using Pauling crystal radii; values of A from Ref. 82.

component over and above what would be expected (i.e., compared to the other monovalent cations). Further studies on the $\overline{V}^{\circ}$ of polyvalent ions as a function of size, charge, and temperature must be made before we can state with certainty what causes this positive effect.

We have used Mukerjee's method[51] to evaluate $\overline{V}^{\circ}_{\text{elect}}$ for the divalent ions as a function of temperature.[82] The results shown in Table 20 indicate that $\overline{V}^{\circ}_{\text{elect}}$ is nearly independent of radius and goes through a maximum as a function of temperature (in agreement with Dunn's calculations[92]).

With the exception of a few studies,[40, 49, 56] most workers have assumed $\overline{V}^{\circ}_{\text{elect}}$ is equal for cations and anions of the same size. Hepler,[49] using (42), found $B = 4.6$ for cations and 19.0 for anions (using Goldschmidt crystal radii[50]). Noyes[56] and Panckhurst,[40] using (30) and (31) and assuming that $\overline{V}^{\circ}_{\text{int}}$ is the same for cations and anions, also found that $\overline{V}^{\circ}_{\text{elect}}$ is larger for anions than for cations. Our calculations using the Noyes[56] method at various temperatures to evaluate $\overline{V}^{\circ}_{\text{H}+}$ also yield the result that $\overline{V}^{\circ}_{\text{elect}}$ for anions is greater than $\overline{V}^{\circ}_{\text{elect}}$ for cations from 0 to 200°C. Recently we have determined the constant for the semiempirical equations (42) to (44) by considering cations and anions separately from 0 to 200°C. The results for the electrostriction constants (B, B', and B'') or the $\overline{V}^{\circ}_{\text{elect}}$ for a monovalent ion with $r = 1.0$ A are given in Table 21. With a few exceptions, using Pauling[47] or Goldschmidt[50] radii yields the result that $\overline{V}^{\circ}_{\text{elect}}$ for anions is greater than $\overline{V}^{\circ}_{\text{elect}}$ for cations, whereas the Gourary–Adrian radii[61] produce the opposite results. If we assume that $\overline{V}^{\circ}_{\text{int}}$ or $\overline{V}^{\circ}_{\text{disord}}$ is equal for cations and anions of the same size, all the semiempirical results indicate that $\overline{V}^{\circ}_{\text{elect}}$ for anions is greater than

Table 21   Constants B, B', and B'' for Various Semiempirical Equations, Considering Cations and Anions Separately at Various Temperatures

Temperature (°C)	Pauling Radii		Goldschmidt Radii		Gourary–Adrian Radii	
	Cations	Anions	Cations	Anions	Cations	Anions
				B		
0	9.61	7.88	11.32	4.87	13.76	0.96
25.0	8.25	8.65	10.03	5.43	12.20	1.20
50.0	8.30	9.09	10.08	5.74	12.29	1.35
75.0	8.43	10.54	10.12	7.05	12.68	2.38
100.0	8.94	12.35	10.42	8.76	12.97	3.78
125.0	10.41	14.61	12.78	10.83	15.82	5.47
150.0	12.27	18.87	14.24	14.93	17.58	8.97
175.0	14.56	24.28	16.65	20.11	20.40	13.40
200.0	17.11	32.55	19.39	28.24	23.67	20.57
				B'		
0	10.95	13.44	13.06	9.62	15.22	6.08
25.0	9.72	15.21	11.87	11.12	13.83	7.11
50.0	9.78	16.23	11.97	11.99	14.00	7.72
75.0	9.88	18.11	12.13	13.69	14.35	9.06
100.0	10.14	20.56	13.28	16.02	15.21	10.98
125.0	12.16	23.27	14.59	18.50	17.28	13.00
150.0	13.48	27.93	16.05	22.96	19.07	16.76
175.0	15.43	33.98	18.11	28.73	21.47	21.66
200.0	18.00	42.28	20.90	36.84	24.85	28.74
				B''		
0	10.14	15.88	12.61	11.79	12.69	9.39
25.0	9.16	18.15	11.31	13.79	11.72	11.10
50.0	9.33	19.44	11.92	14.90	12.09	12.06
75.0	9.52	21.91	12.46	17.25	12.39	13.51
100.0	10.52	23.72	12.91	18.88	13.18	14.91
125.0	11.43	26.25	14.57	22.06	14.79	17.70
150.0	12.61	31.39	15.84	26.31	16.30	21.48
175.0	14.30	37.27	17.65	31.99	18.40	25.83
200.0	16.67	44.68	20.16	39.36	21.14	32.43

$\overline{V}^{\circ}_{\text{elect}}$ for cations (regardless of what radii are used). Since the $\overline{V}^{\circ}_{\text{ions}}$ used in these correlations are based on $\overline{V}^{\circ}_{\text{H}^+} = -5.4 \, \text{cm}^3/\text{mole}$ at 25°C, it is not surprising that the Gourary–Adrian radii[61] show behaviors different from those of the Pauling and Goldschmidt radii. It should also be pointed out that the use of $\overline{V}^{\circ}_{\text{ions}}$ based on $\overline{V}^{\circ}_{\text{H}^+} = 0 \, \text{cm}^3/\text{mole}$ gives results with the Gourary–Adrian radii that indicate that $\overline{V}^{\circ}_{\text{elect}}$ is greater for anions than for cations of the same size. The problems involved in not knowing what the absolute value of the radii and partial molal volume for ions are make it impossible to state with certainty that $\overline{V}^\circ_{elect}$ is greater for anions than for cations of the same size.

The examination of the effect of temperature on $\overline{V}^\circ_{elect}$ for cations and anions does reveal differences that are independent of the absolute value of $\overline{V}^\circ_{H^+}$, the crystal radii, or the semiempirical equation used. The $\overline{V}^\circ_{elect}$ for cations, like the $\overline{V}^\circ_{disord}$, appears to go through a maximum as a function of temperature, whereas $\overline{V}^\circ_{elect}$ for anions appears to decrease with increasing temperature. Thus the effect of temperature on $\overline{V}^\circ_{elect}$ or $\overline{E}^\circ_{elect}$ appears to be different for the simple monovalent cations and anions. Since the effects of temperature on $\overline{V}^\circ_{disord}$ or $\overline{E}^\circ_{disord}$ also appear to be different, it is not possible at present to decide whether the differences in the $\overline{E}^\circ$ of the simple monovalent cations and anions of the same size are caused by void-space packing effects, by electrostriction effects, or by a combination of two effects.

The relative contribution of $\overline{V}^\circ_{elect}$ to $\overline{V}^\circ_{ion}$ depends upon the charge and radius of the ion. However, it is impossible now to be sure how $\overline{V}^\circ_{elect}$ varies with charge and radius (i.e., whether $\overline{V}^\circ_{elect} \propto Z^2/r, Z/r, Z/r^2$). The effect of temperature on $\overline{V}^\circ_{elect}$ or $\overline{E}^\circ_{elect}$ appears to be predictable by using a sphere in continuum model if one allows for dielectric saturation effects. When $\overline{V}^\circ_{ion} - \overline{V}^\circ_{cryst}$ is negative, the $\overline{V}^\circ_{elect}$ region is greater than $\overline{V}^\circ_{disord}$, and the ion can be classified as a structure-making or positive-hydrating ion.

The Caged or Structured Partial Molal Volume, $\overline{V}^\circ_{caged}$.   The tetraalkylammonium ions, R_4N^+, are a typical example of ions that cause structural effects that are different from those of the simple monovalent ions. One might expect the large R_4N^+ ions to have little or no electrostriction ($\overline{V}^\circ_{elect} \approx 0$); thus $\overline{V}^\circ_{ion} = \overline{V}^\circ_{cryst} + \overline{V}^\circ_{disord}$. An examination of the $\overline{V}^\circ_{R_4N^+}$ ions in Table 15 compared to the crystal volume indicates that $\overline{V}^\circ_{ion} - \overline{V}^\circ_{cryst}$ is negative for these ions. The $\overline{V}^\circ_{R_4N^+}$ ions are similar in this respect to the polyvalent or highly electrostrictive ions. The negative effect decreases in magnitude for the polyvalent ions as the radius increases; however, for the R_4N^+ ions the negative effect increases in magnitude as the radius increases. To explain this type of behavior, we must consider another component of $\overline{V}^\circ_{ion}$, $\overline{V}^\circ_{caged}$—the caged or structural partial molal volume; this behavior is similar to that observed for the aliphatic alcohols.[54]

Measurements of heat capacity,[64] dielectric relaxation times,[96] viscosities,[68, 97, 98] Soret coefficients,[99] NMR relaxation times,[100] conductivities,[101] and heats of dilution and solution,[102, 103] also indicate that the R_4N^+ ions become "hydrophobic structure makers" as the size of the alkyl group increases. The evidence is quite strong indicating that the nBu_4N^+ and nPr_4N^+ ions are strong hydrophobic structure makers. The evidence is not so conclusive about the Me_4N^+ and Et_4N^+ ions. Wen and Saito[104] have pictured the Me_4N^+ ion as slight structure breaker and the Et_4N^+ ion as a slight structure maker. They visualize a competition between the charge effect, $\bar{V}^\circ_{elect}$, and the hydrophobic effect, $\bar{V}^\circ_{caged}$, where the Me_4N^+ ion is just small enough so that the water structure around the ions was still under the influence of its charge and the Et_4N^+ ion is just large enough that the effect of charge on the water structure was slightly overshadowed by the hydrophobic or clathrate effect.

The measurements of conductance[101] and viscosity[68] by Kay and co-workers indicate that the Me_4N^+ ion is a structure breaker, but that the Et_4N^+ ion is in the transition region between a structure maker and a structure breaker.

Since both $\bar{V}^\circ_{elect}$ and $\bar{V}^\circ_{caged}$ appear to be negative, it is difficult to separate the two effects on a volume basis. An examination of $\bar{V}^\circ_{ion} - \bar{V}^\circ_{cryst}$ for the R_4N^+ ions does indicate that the Me_4N^+ ion may behave as an electrostriction structure maker as well as a hydrophobic structure maker (since $\bar{V}^\circ_{ion} - \bar{V}^\circ_{cryst}$ is more negative for Me_4N^+ than for Et_4N^+). The difficulties of calculating $\bar{V}^\circ_{cryst}$ (i.e., calculating the crystal radius) for the R_4N^+ ions make it impossible to separate $\bar{V}^\circ_{ions}$ into all their components. The similarity of the $\bar{V}^\circ$ behavior of the R_4N^+ ions and the aliphatic alcohols has been discussed elsewhere,[54] and it has been postulated that the abnormal volume properties of the R_4N^+ salts may be normal for all solutes able to cause hydrophobic bonding.[54]

Since the large R_4N^+ ions are not hydrated in the normal manner, Conway, Verrall, and Desnoyers[16, 81] (as mentioned earlier) have used the $\bar{V}^\circ$ of the R_4N^+ halides to determine the absolute value of the halide ions. We have also attempted to determine the absolute expansibility for the Cl^- ion by a similar technique.[54] We found, however, that a linear relation is not obtained when $\bar{E}^\circ_{R_4NX}$ is plotted versus the molecular weight of the R_4N^+ ion. At 25°C a smooth curve was obtained with the $\bar{E}^\circ$ of the Pr_4NCl and Bu_4NCl salts appearing to be high compared to the lower molecular weight R_4NCl's. At 30°C a linear relation is obtained for the Me_4N^+, Et_4N^+, and Pr_4N^+ chlorides with the Bu_4N^+ chloride appearing to be high. If we assume that the high values for the Bu_4N^+ salt (and Pr_4N^+ salt) are due to the caged or structural term of $\bar{E}^\circ_{ion}$, one might postulate that $\bar{E}^\circ_{caged}$ or $\bar{V}^\circ_{caged}$ becomes less important as the temperature is increased. Although the linearity of $\bar{E}^\circ_{R_4NCls}$ versus MW of R_4N^+ does not exist for all of the R_4NCl's, it is possible to estimate $\bar{E}^\circ_{H+} = -0.010$ to -0.014 cm³/(mole)(deg) from the results at 25°C in agreement with the value calculated from (32).

The most striking difference between the infinite dilution volume properties of the R_4N^+ salts and those of other electrolytes is that the $\bar{E}^\circ$ of the R_4N^+ salts increases with increasing temperature and the $\bar{E}^\circ$ of other electrolytes decreases with increasing temperature. Hepler[105] has recently developed a method of examining the expansibility of solutes in terms of structure-breaking and structure-making effects using the thermodynamic relation

$$(54) \qquad \left(\frac{\partial \bar{C}_P^\circ}{\partial P}\right)_T = -T\left(\frac{\partial^2 \bar{V}^\circ}{\partial T^2}\right)_P = -T\frac{\partial \bar{E}^\circ}{\partial T}$$

where $\bar{C}_P^\circ$ is the partial molal heat capacity at infinite dilution. The negative $\bar{C}_P^\circ$ values for various electrolytes have been attributed to the ability of the electrolyte to break down the structure of water.[1] Hepler[105] reasons that, since pressure should also break down the structure of water, $\partial \bar{C}_p^\circ/\partial P$ should be positive or $\partial \bar{E}^\circ/\partial T$ should be negative for a structure-breaking solute. Using similar reasoning, Hepler[105] predicts that a positive value for $\partial \bar{E}^\circ/\partial T$ should be associated with structure-making solutes. Thus by Hepler's reasoning most electrolytes would be considered structure breakers and the R_4N^+ salts (as well as most nonelectrolytes) would be considered structure makers. As Hepler[105] and others[106] have pointed out, the classification of solutes into the categories of structure making and structure breaking, although useful in some connections, is only partly satisfactory. This is because we are still unable to define the structure that is being made or broken by the solute or to use one scale to separate or measure structure breaking or making. Nor can we now say whether the unique volume behavior of the R_4N^+ ions (as well as that of nonelectrolytes) is due to void-space effects, $\bar{V}_{disord}^\circ$; changes in the structure of water, $\bar{V}_{caged}^\circ$; or changes in the intrinsic volume, $\bar{V}_{cryst}^\circ$. From the recent values for $\bar{V}^\circ$ of salts in nonaqueous solvent systems,[107, 108] the $\bar{V}^\circ$ of some ions in nonaqueous solvents also appear to go through a maximum as a function of temperature. Thus the model used in this chapter to discuss ion-water interactions appears to hold in other solvents.[110] The recent finding of negative B coefficients for electrolytes in the solvents glycol and glycerol by Crickard and Skinner[111] also support this notion.

5 Summary

A multilayer model has been used to examine the various components of the partial molal volume of some simple monatomic ions as a function of temperature. In Figure 2 the components are given for a monovalent, monatomic ion with a radius equal to 1.0 Å at various temperatures. At low temperatures, $\bar{V}_{disord}^\circ$ is the predominant factor; at high temperatures, $\bar{V}_{elect}^\circ$ is the predominant factor. The maximum observed in $\bar{V}_{ion}^\circ$ as a function of temperature is due to the competition between $\bar{V}_{elect}^\circ$ and $\bar{V}_{disord}^\circ$. For highly charged ions or ions with a small radius, $\bar{V}_{elect}^\circ$ is the dominant factor over the entire temperature range, and the maximum occurs at lower temperatures. We wish to emphasize that this very simple model for ion-water interactions,

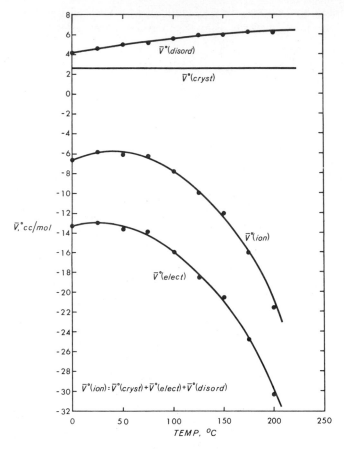

Figure 2. The components for the partial molal volume of a monatomic monovalent ion with $V = 1.0$ Å in water from 0 to 200°C.

although fairly successful for the monovalent ions, represents only the qualitative effects; thus the model cannot be used quantitatively at present. A more careful examination of the effect of temperature, pressure, charge type, and size, does show the different hydration effects between cations and anions as well as unique specific hydration effects (e.g., the Li^+ ion).

More careful studies of the $\overline{V}^\circ_{\mathrm{salts}}$ as a function of temperature, pressure, charge type, and size (as well as solvent) are needed. In the future we hope also to know the exact radius of ions in solution and will thus be able to assign absolute values to $\overline{V}^\circ_{\mathrm{ions}}$, so that more can be said about the differences in hydration between cations and anions.

Acknowledgments

I wish to thank Dr. Desnoyers and his colleagues, Dr. Vaslow, Dr. Dunn, Dr. Hepler, Dr. King, and others for making their manuscripts available prior to publication. I also wish to acknowledge the work of all the authors listed in the references, since much of their work has led to the development of many of my opinions and ideas concerning the use of the partial molal volume in studying the interactions occurring in aqueous solutions.

REFERENCES

1. G. N. Lewis and M. Randall, *Thermodynamics*, 2nd ed., rev. K. S. Pitzer and L. Brewer, McGraw-Hill, New York, 1961.
2. Various methods of measuring the density of solutions are reviewed in the following references; (a) N. Bauer and S. Z. Lewin in *Techniques of Organic Chemistry*, ed., A. Weissberger, 2nd ed., vol. I, part I, Chapter 6, Interscience, New York, 1959; (b) P. Hidnert and E. L. Peffer, "Density of Solids and Liquids," *Nat. Bur. Std. Circ. 487*, Washington, D.C., 1950; (c) F. J. Millero, *Rev. Sci. Instr.*, **38**, 1441 (1967).
3. H. S. Harned and B. B. Owen, *The Physical Chemistry of Electrolytic Solutions*, Amer. Chem. Soc. Monogr. No. 137, Reinhold, New York, 1958.
4. D. O. Masson, *Phil. Mag.* [7], **8**, 218 (1929).
5. (a) O. Redlich and D. M. Meyer, *Chem. Rev.*, **64**, 221 (1964); (b) O. Redlich, *J. Phys. Chem.*, **67**, 496 (1963).
6. (a) B. B. Owen and S. R. Brinkley, *Ann. N.Y. Acad. Sci.*, **51**, 753 (1949); (b) A. Indelli and R. DeSantis, *J. Chem. Phys.*, **51**, 2782 (1969).
7. A. F. Scott, *J. Phys. Chem.*, **35**, 2315 (1931).
8. (a) W. Geffcken, *Z. Physik. Chem.*, **A155**, 1 (1931); (b) *Naturwiss.*, **19**, 321 (1935).
9. (a) O. Redlich and P. Rosenfeld, *Z. Electrochem.*, **37**, 705 (1931); (b) *Z. Physik. Chem.*, **A155**, 65 (1931); (c) O. Redlich, *Naturwiss.*, **19**, 251 (1931).
10. O. Redlich and J. Bigeleisen, *J. Amer. Chem. Soc.*, **64**, 758 (1942); *Chem. Rev.*, **30**, 171 (1942).
11. O. Redlich, *J. Phys. Chem.*, **44**, 619 (1940).
12. O. Redlich and L. E. Nielson, *J. Amer. Chem. Soc.*, **64**, 761 (1942).
13. A. Bodanszky and W. Kauzmann, *J. Phys. Chem.*, **66**, 177 (1962).
14. L. G. Hepler, J. M. Stokes, and R. H. Stokes, *Trans. Faraday Soc.*, **61**, 20 (1965).
15. L. A. Dunn, *Trans. Faraday Soc.*, **62**, 2348 (1966).
16. B. E. Conway, R. E. Verrall, and J. E. Desnoyers, *Trans. Faraday Soc.*, **62**, 2738 (1966).
17. F. Vaslow, *J. Phys. Chem.*, **70**, 2286 (1966).
18. F. H. Spedding, M. J. Pikal, and B. O. Ayers, *J. Phys. Chem.*, **70**, 2440 (1966).
19. J. E. Desnoyers and M. Arel, *Canad. J. Chem.*, **45**, 359 (1967).
20. F. Franks and H. T. Smith, *Trans. Faraday Soc.*, **63**, 2586 (1967).
21. H. E. Wirth, *J. Phys. Chem.*, **71**, 2922 (1967).
22. F. J. Millero, *J. Phys. Chem.*, **71**, 4567 (1967).
23. L. A. Dunn, *Trans. Faraday Soc.*, **64**, 1898 (1968).
24. L. A. Dunn, *Trans. Faraday Soc.*, **64**, 2951 (1968).

25. B. E. Conway and L. H. Laliberté, *J. Phys. Chem.*, **72**, 4317 (1968).
26. E. J. King, *J. Phys. Chem.*, **73**, 1220 (1969).
27. J. E. Desnoyers, M. Arel, G. Perron, and C. Jolicoeur, *J. Phys. Chem.*, **73**, 3346 (1969).
28. F. Vaslow, *J. Phys. Chem.*, **73**, 3745 (1969).
29. (a) F. J. Millero, *J. Phys. Chem.*, **74**, 356 (1970); (b) *J. Chem. Eng. Data*, **15**, 562 (1970); (c) *J. Phys. Chem.*, **75**, 280 1971.
30. W. C. Root, *J. Amer. Chem. Soc.*, **55**, 850 (1933).
31. J. D. Bernal and R. H. Fowler, *J. Chem. Phys.*, **1**, 515 (1933).
32. G. Akerlöf and P. Bender, *J. Amer. Chem. Soc.*, **60**, 1226 (1938).
33. A. M. Couture and K. J. Laidler, *Can. J. Chem.*, **34**, 1209 (1956).
34. W. C. Root, Ph.D. thesis, Harvard University, Cambridge, Massachusetts, 1932.
35. Calculations made by A. F. Scott (Ref. 7) using the density data of G. P. Baxter and C. C. Wallace, *J. Amer. Chem. Soc.*, **38**, 70 (1916).
36. B. B. Owen and S. R. Brinkley, *Chem. Rev.*, **29**, 461 (1941).
37. Taken from the partial molal volume data tabulated in the Appendix.
38. (a) A. J. Ellis, *J. Chem. Soc.* A, 1579 (1966); (b) *J. Chem. Soc.* A, 660 (1967); (c) *J. Chem. Soc.* A, 1138 (1968).
39. A. J. Ellis and I. M. McFadden, *Chem. Commun.*, 516 (1968).
40. M. H. Panckhurst, *Rev. Pure Appl. Chem.*, 45 (1969).
41. E. Darmois, *J. Phys. Radium*, **2**, 2 (1941).
42. E-an Zen, *Geochim, Cosmochim Acta*, **12**, 103 (1957).
43. Y. Kobayazi, *J. Sci. Hiroshima Univ.*, **A9**, 241 (1939).
44. O. Rice, *Electronic Structure and Chemical Binding*, McGraw-Hill, New York, 1940.
45. A. Eucken, *Z. Elektrochem.*, **51**, 6 (1948).
46. H. E. Wirth, *J. Marine Res.*, **3**, 230 (1940).
47. L. Pauling, *The Nature of the Chemical Bond*, 3rd ed., Cornell University Press, Ithaca, N.Y., 1940.
48. K. Fajans and O. Johnson, *J. Amer. Chem. Soc.*, **64**, 668 (1942).
49. L. G. Hepler, *J. Phys. Chem.*, **61**, 1426 (1957).
50. V. M. Goldschmidt, *Skrifter Norske Videnskaps-Akad. Oslo, I: Mat.-Naturv. Kl.*, 1926 (also listed in ref. 47).
51. P. Mukerjee, *J. Phys. Chem.*, **65**, 740 (1961).
52. R. H. Stokes and R. A. Robinson, *Trans. Faraday Soc.*, **53**, 301 (1957).
53. J. Padova (a) *J. Chem. Phys.*, **39**, 1552 (1963); (b) *J. Chem. Phys.*, **40**, 691 (1964).
54. F. J. Millero and W. Drost-Hansen, *J. Phys. Chem.*, **72**, 1758 (1968).
55. E. Glueckauf, *Trans. Faraday Soc.*, **61**, 914 (1965).
56. R. M. Noyes, *J. Amer. Chem. Soc.*, **86**, 971 (1964).
57. P. Drude and W. Nernst, *Z Physik. Chem. (Frankfurt)*, **15**, 79 (1894).
58. B. B. Owen, R. C. Miller, C. E. Milner, and H. L. Cogan, *J. Phys. Chem.*, **65**, 2065 (1961).
59. R. Zana and E. Yeager (a) *J. Phys. Chem.*, **70**, 954 (1966); (b) *J. Phys. Chem.*, **71**, 521 (1967); (c) *J. Phys. Chem.*, **71**, 4241 (1967).
60. P. Mukerjee, *J. Phys. Chem.*, **70**, 2708 (1966).
61. B. S. Gourary and F. J. Adrian, *Solid State Phys.*, **10**, 127 (1960).
62. C. M. Criss and J. W. Cobble, *J. Amer. Chem. Soc.*, **86**, 5385 (1964).
63. R. M. Gurney, *Ionic Processes in Solution*, McGraw-Hill, New York, 1953.
64. H. Frank and W.-Y. Wen, *Discussions Faraday Soc.*, **24**, 133 (1957).
65. M. Eigen and E. Wicke, *J. Phys. Chem.*, **58**, 702 (1954).

66. O. Ya. Samoilov, *Discussions Faraday Soc.*, **24**, 141 (1957); *Structure of Aqueous Electrolyte Solutions and the Hydration of Ions*, trans. D. J. G. Ives, Consultants Bureau, New York, 1965.

67. M. Kaminsky, *Discussions Faraday Soc.*, **24**, 171 (1957).

68. R. L. Kay, T. Vituccio, C. Zawoyski, and D. F. Evans, *J. Phys. Chem.*, **70**, 2336 (1966).

69. K. J. Laidler and C. Pegis, *Proc. Roy. Soc. (London), Ser. A*, **241**, 80 (1957).

70. D. D. Wagman, W. H. Evans, V. B. Parker, I. Halow, S. M. Bailey, and R. H. Schumm, "Selected Values of Chemical Thermodynamic Properties," *N.B.S. Tech.* Note 270-3, Government Printing Office, Washington, D.C., 1968.

71. F. J. Millero, *J. Phys. Chem.*, **72**, 4589 (1968).

72. (a) F. J. Millero and W. Drost-Hansen, *J. Chem. Eng. Data*, **13**, 330 (1968); (b) F. J. Millero, W. Drost-Hansen, and L. Korson, *J. Phys. Chem.*, **72**, 2251 (1968).

73. F. J. Millero and F. Lepple, in preparation.

74. F. J. Millero, *J. Phys. Chem.*, **72**, 3209 (1968).

75. F. J. Millero, *Limnol. Oceanogr.*, **12**, 376 (1969).

76. F. J. Millero, *J. Phys. Chem.*, **73**, 2417 (1969).

77. A. F. Scott, *J. Phys. Chem.*, **35**, 3379 (1931).

78. R. H. Stokes, *J. Amer. Chem. Soc.*, **86**, 979, 982 (1964).

79. J. Burak and A. Treinin, *Trans. Faraday Soc.*, **59**, 1490 (1963).

80. S. W. Benson and C. J. Copeland, *J. Phys. Chem.*, **67**, 1194 (1963).

81. B. E. Conway, R. E. Verrall, and J. E. Desnoyers, *Z. Physik. Chem.*, **230**, 157 (1965).

82. F. J. Millero, paper presented at Mid-Atlantic Regional A.C.S. Meeting, Newark, Delaware, April, 1970; also see, G. Curthoys and T. G. Mathieson, *Trans. Faraday Soc.*, **66**, 43 (1970).

83. W. M. Latimer, *J. Chem. Phys.*, **23**, 90 (1955).

84. R. E. Powell and W. M. Latimer, *J. Chem. Phys.*, **19**, 1139 (1951).

85. R. E. Powell, *J. Phys. Chem.*, **58**, 528 (1954).

86. M. Born, *Z. Physik.*, **1**, 45 (1920).

87. T. J. Webb, *J. Amer. Chem. Soc.*, **48**, 2589 (1926).

88. F. Zwicky, *Physik. Z.*, **26**, 664 (1926); *Proc. Natl. Acad. Sci. (U.S.)*, **12**, 86 (1926).

89. H. M. Evjen and F. Zwicky, *Phys. Rev.*, **33**, 860 (1929).

90. E. Whalley, *J. Chem. Phys.*, **38**, 1400 (1963).

91. J. E. Desnoyers, R. E. Verrall, and B. E. Conway, *J. Chem. Phys.*, **43**, 243 (1965).

92. L. A. Dunn, Ph.D. Thesis, University of New England, Armidale, Australia, 1967.

93. H. S. Frank, *J. Chem. Phys.*, **23**, 2023 (1955).

94. G. S. Kell and E. Whalley, *Phil. Trans. Roy. Soc. (London)*, **258**, 565 (1965).

95. F. J. Millero, R. W. Curry, and W. Drost-Hansen, *J. Chem. Eng. Data*, **14**, 422 (1969).

96. G. H. Haggis, J. B. Hasted, and T. J. Buchanan, *J. Chem. Phys.*, **20**, 1452 (1952).

97. E. Hückel and W. Schaaf, *Z. Physik. Chem. (Frankfurt)*, **21**, 326 (1959).

98. E. R. Nightingale, *J. Phys. Chem.*, **66**, 894 (1964).

99. J. N. Agar, *Advan. Electrochem. Electrochem. Eng.*, **3**, 31 (1963).

100. W. G. Hertz and M. O. Zeidler, *Ber. Bunsenges. Physik. Chem.*, **68**, 821 (1964).

101. R. L. Kay and D. F. Evans, *J. Phys. Chem.*, **69**, 4216 (1965); *J. Phys. Chem.*, **70**, 2325 (1966).

102. S. Lindenbaum, *J. Phys. Chem.*, **70**, 814 (1966).

103. Y. C. Wu and H. L. Friedman, *J. Phys. Chem.*, **70**, 2030 (1966).

104. W.-Y. Wen and S. Saito, *J. Phys. Chem.*, **68**, 2639 (1964).

105. L. G. Hepler, *Can. J. Chem.*, **47**, 4613 (1969).

106. A. Holtzer and M. F. Emerson, *J. Phys. Chem.*, **73**, 26 (1969).

107. G. Gopal and M. A. Siddiqi, *J. Phys. Chem.*, **73**, 3390 (1969).

108. G. Gopal and M. A. Siddiqi, *Z. Physik. Chem.*, **67**, 122 (1969).

109. E. J. King, *J. Phys. Chem.*, to be submitted, 1970.

110. F. J. Millero, *Chem. Rev.*, **70**, 147 (1970).

111. K. Crickard and J. F. Skinner, *J. Phys. Chem.*, **73**, 2060 (1969).

Addendum to Chapter 13

Compilation of the Partial Molal Volumes of Electrolytes at Infinite Dilution, $\bar{V}^{\circ}$, and the Apparent Molal Volume Concentration Dependence Constants, S_V^* and b_V, at Various Temperatures

Frank J. Millero, Rosenstiel School of Marine and Atmospheric Sciences University of Miami, Miami, Florida

Many workers (including Masson,[1] Geffcken,[2] Scott,[3] Bernal and Fowler,[4] Owen and Brinkley,[5] Fajans and Johnson,[6] Couture and Laidler,[7,8] Zen,[9] Harned and Owen,[10] and Mukerjee[11]) have tabulated the partial molal volume of electrolytes at infinite dilution, $\bar{V}^{\circ}$. Since most of the earlier compilations are not complete and in recent years more reliable data have become available, I have recompiled the $\bar{V}^{\circ}$ of various electrolytes from 0 to 200°C. I have also tabulated (when possible the S_V^* constant of the Masson equation[1] ($\phi_V = \phi_V^{\circ} + S_V^* \sqrt{c}$) and the b_V constant of the Redlich–Meyer equation[12] ($\phi_V = \phi_V^{\circ} + S_V \sqrt{c} + b_V c$). Although the ϕ_V° or $\bar{V}^{\circ}$ values determined from the Masson equation[1] are frequently in error by as much as 3.0 cm³/mole (e.g., the $\bar{V}^{\circ}$ of the R_4N^+ halides), the additivity and expansibility of the $\bar{V}^{\circ}$ values determined by this method can be very useful in estimating the $\bar{V}^{\circ}$ of unknown electrolytes. The Masson equation must be used at high temperatures (i.e., greater than 70°C), since the theoretical slope, S_V, is not known. We can demonstrate how the $\bar{V}^{\circ}$ of the electrolyte can be estimated from this compilation by determining the $\bar{V}^{\circ}$ of potassium fluoride at 0°C by two methods.

1. From Ellis's data[13] on potassium fluoride, as well as the $\bar{E}_{KF}^{\circ}$ calculated from my recent work at 25°C,[14,15] we find that the partial molal expansi-

bility, $\bar{E}^\circ$, can be represented from $t = 25$ to $40°C$ by the equation $\bar{E}^\circ_{KF} = 0.146 - 3.56 \times 10^{-3}t$ [with a standard deviation of ± 0.01 cm³/(mole)(deg)]. If we integrate this equation and solve for the constant of integration using $\bar{V}^\circ_{KF} = 7.78$ cm³/mole[16] at $25°C$, we obtain $\bar{V}^\circ_{KF} = 5.04 + 0.146t - 1.78 \times 10^{-3} t^2$; at $0°C$ $\bar{V}^\circ_{KF} = 5.04$ cm³/mole.

2. If we extrapolate the $\bar{V}^\circ$ data[19] for sodium fluoride at 5, 10, 20, and 25°C to 0°C, we obtain $\bar{V}^\circ_{NaF} = -5.8$ cm³/mole. Using $\bar{V}^\circ_{K^+ - Na^+} = 10.68$ cm³/mole at 0°C from Dunn's data,[20] we obtain $\bar{V}^\circ_{KF} = 4.88$ cm³/mole at 0°C. The average value for $\bar{V}^\circ_{KF}$ is 4.96 cm³/mole, which agrees very well with our recent experimental determination for $\bar{V}^\circ_{KF}$.[99] (It is interesting to note that although the $\bar{V}^\circ_{NaF}$ determined by Robertson et al.[19] appears to be in error[16-18] by 1.2 cm³/mole at 25°C, their results at lower temperature appear to be correct.) We have also used this method to estimate the $\bar{V}^\circ$ of other electrolytes—included in Chapter 13.

It is not possible to estimate the uncertainty of all the $\bar{V}^\circ$'s given in this tabulation; however, the values I feel are the most reliable are italicized in Table 1. The $S_{V^\star}$ and the b_V constants given in parentheses have been determined from the original data. The superscripts, a and b, are explained at the end of the table.

Although most workers measure relative densities (i.e., specific gravities assuming the density of water is 1.000000 g/ml at 3.98°C) and report ϕ_V in units of ml/mole; we have used the units cm³/mole throughout Chapter 13 and the Appendix. Since the differences in ϕ_V using either set of units are well within experimental error, the units can be interchanged without serious errors.

Table 1   The Partial Molal Volumes of Electrolytes at Infinite Dilution, $\bar{V}^\circ$, and the Apparent Molal Volume Concentration Dependence Constants, $S_{V^\star}$ and b_V, at Various Temperatures

		0°C			
	Salt	$\bar{V}^\circ$ (cm³/mole)	b_V [cm³ liter mole⁻²]	$S_V^\star$ [cm³ liter½ mole⁻³⁄₂]	References
HCl		16.5			9
		16.40			21
HNO₃		25.6			9
LiCl		15.30		1.990	3
		14.95			111
LiBr		21.94		1.673	3
		21.34			111

Table 1   (cont.)

Salt	$\overline{V}^\circ$ (cm³/mole)	b_V [cm³ liter mole^{-2}]	$S_V{}^*$ [cm³ liter$^{1/2}$ mole$^{-3/2}$]	References
LiI	32.50		1.425	3
	31.83			111
NaCl	*12.86*	1.348		22
	12.95	1.198		20 (0.05°C)
	12.78			23 (0.8°C)
	13.1			9
	12.57			24
	12.4			2
	12.36		3.240	3
	13.45			25
	12.40		3.07	26
	12.59			111
NaBr	18.86		2.976	3
	18.95			2
	18.9			9
	18.98			111
NaI	29.34		2.886	3
	29.47			111
NaOH	−12.9			9
	−10.58		5.334[a]	27
NaNO$_3$	22.4			9
Na$_2$CO$_3$	−15.1			9
Na$_2$SO$_4$	3.7			9
KCl	*23.63*	1.205		20 (0.05°C)
	23.0			2
	23.00		3.291	3
KCl	23.97			25
	23.05			24
	23.4			9
	22.98		3.30	28
	23.66			111
KBr	*30.24*	1.084		20 (0.05°C)
	29.38		3.219	3
	29.34		3.21	29
	30.05			111
KI	*40.75*	1.043		20 (0.05°C)
	40.05		3.133	3
	40.54			111
KOH	−0.50		5.072[a]	24
KMnO$_4$	45.70			30
KNO$_3$	32.5			9
K$_2$CO$_3$	2.7			9
K$_2$SO$_4$	22.7		17.09	31

(*continued*)

Table 1 (cont.)

Salt	$\bar{V}°$ (cm³/mole)	b_V [cm³ liter mole^{-2}]	S_V* [cm³ liter$^{1/2}$ mole$^{-3/2}$]	References
K₂CrO₄	27.98		15.76	31
RbCl	28.11		3.287	3
	28.05		3.25	28
	28.35			111
RbBr	34.61		3.112	3
	34.74			111
RbI	45.15		3.158	3
	45.23			111
CsCl	35.23		3.293	3
	35.40			2
	35.7			9
	35.68			111
CsBr	41.95		3.046	3
	41.95			2
	41.7			9
	42.07			111
CsI	52.74		2.885	3
	52.55		3.05	28
	52.45		3.14	30
	52.56			111
AgNO₃	23.46		3.923	31
MgCl₂	12.7			9
	11.4		7.212	28
MgBr₂	23.60		6.788	28
MgI₂	44.90		5.940	28
MgSO₄	−10.0			9
CaCl₂	13.05	−0.563^b (−0.70)	7.553^b	20 (0.05°C)
	13.8			9
	13.40		7.354	28
CaBr₂	25.50		6.788	28
CaI₂	46.90		6.081	28
Ca₂Fe(CN)₆	25.3		43.4	53
SrCl₂	12.10		8.344	28
SrBr₂	24.30		7.920	28
SrI₂	45.80		7.778	28
BaCl₂	17.11	1.152		20 (0.05°C)
	16.80		8.627	28
BaBr₂	29.00		8.202	28
BaI₂	50.60		7.778	28
LaCl₃	8.56		15.67	29
NaBPh₄	267.16	−7.38		98
Ph₄AsCl	307.57	3.16		99

Table 1 (cont.)

Salt	$\overline{V}^\circ$ (cm³/mole)	b_V [cm³ liter mole^{-2}]	S_V^* [cm³ liter$^{1/2}$ mole$^{3/2}$]	References
		5°C		
LiCl	17.4			33
	16.09	(0.075)		32
LiI	33.39	(−0.318)		32
NaF	−5.08			19
NaCl	14.02	(0.744)		32
	14.02	0.851		20
	14.15			34
	14.17	0.682		22
	15.8			33
NaI	30.37			19
KCl	24.56	0.947		20
	24.65			34
	23.45		3.040[a]	35
	26.8			33
KBr	31.25	0.734		20
	30.40		2.500[a]	35
	31.0			33
KI	42.00	0.597		20
	41.75		2.573[a]	35
	44.6			33
NH$_4$Cl	36.0			33
KNO$_3$	33.50		3.780	35
	36.2			33
KSCN	39.97		10.45	35
Me$_4$NBr	112.20			34
Et$_4$NBr	171.15			34
nPr$_4$NBr	236.30			34
nBu$_4$NBr	295.40			34
NaPhSO$_3$	100.63			19 (8.0°C)
MgCl$_2$	13.4			33
CaCl$_2$	14.49	−0.595[b] (−0.47)	7.292[b]	20
	15.4			33
SrCl$_2$	15.0			33
BaCl$_2$	18.79	−0.181		20
	19.6			33
MgSO$_4$	−9.8			34
		10°C		
HCl	17.2			9
	17.16			21

(continued)

Table 1   (cont.)

Salt	$\bar{V}°$ (cm³/mole)	b_V [cm³ liter mole^{-2}]	S_V* [cm³ liter$^{1/2}$ mole$^{-3/2}$]	References
NaF	−4.34			19
NaCl	14.7			9
	14.45			24
NaBr	21.0			9
KCl	25.0			9
	24.78			24
	24.10		2.800[a]	35
KBr	31.65		2.263[a]	35
KI	42.97		1.875[a]	35
CsCl	37.3			9
CsBr	43.7			9
HNO₃	27.3			9
KNO₃	35.3			9
	35.03		3.310[a]	35
NaOH	−8.16		4.606[a]	27
	−9.1			9
KOH	1.86		4.347[a]	24
NaPhSO₃	101.89			19 (11.0°C)
KSCN	41.48		9.857[a]	35
MgCl₂	14.6			9
CaCl₂	15.7			9
K₂CO₃	7.8			9
Na₂CO₃	−10.1			9
MgSO₄	−6.7			9
Na₂SO₄	8.6			9
[Bu₃N(CH₂)₈NBu₃]Br₂	575.4			36
NaNO₃	25.5			9

	15°C			
NaCl	15.62	0.295		20
	15.58	0.369		22
NaI	33.63			19
KCl	25.98	0.416		20
	24.71		3.612[a]	35
KBr	32.93	0.416		20
	32.65		2.070[a]	35
KI	43.93	0.042		20
	44.00		1.650[a]	35
NaPhSO₃	103.09			19
KNO₃	36.00		2.940[a]	35
KSCN	42.66		9.574[a]	35
K Tartrate	80.47		7.636[a]	35
NH₄Cl	35.4			6

Table 1 (cont.)

Salt	$\bar{V}^\circ$ (cm^3/mole)	b_V [cm^3 liter mole^{-2}]	S_V^* [cm^3 liter$^{1/2}$ mole$^{-3/2}$]	References
NH$_4$CN	38.7			6
Me$_4$NBr	114.29			37
Et$_4$NBr	175.46			37
nPr$_4$NBr	239.19			37
nBu$_4$NBr	299.79			37
	300.0		−9.0	38
CaCl$_2$	16.51	−1.022[b]	7.275[b]	20
		(−2.4)		
BaCl$_2$	21.49	−2.029		20
CrCl$_3$	−21.3			6
(NH$_4$)$_2$SO$_4$	43.4			6
(NH$_4$)$_2$SeO$_4$	57.4			6
(NH$_4$)$_3$MoO$_4$	65.4			6

20°C

Salt	$\bar{V}^\circ$ (cm^3/mole)	b_V	S_V^*	References
HCl	17.8			9
	17.75			21
HBr	23.4			9
LiCl	17.13			39
NaF	−2.22			19
NaCl	16.1			9
	15.98			24
	17.03			4
	15.71			39
NaBr	22.8			9
	23.27			4
NaI	32.65			19
KCl	26.0			9
	26.07			24
	26.97			4
	25.42		2.500[a]	35
	26.10			39
KBr	33.37		1.863[a]	35
KI	44.80		1.700[a]	35
CsCl	38.7			9
	39.03			4
CsBr	45.1			9
	45.32			4
NH$_4$Cl	35.8			6
	34.95			39
HNO$_3$	28.6			9
NaNO$_3$	27.0			9

(*continued*)

Table 1 (cont.)

Salt	$\bar{V}°$ (cm³/mole)	b_V [cm³ liter mole⁻²]	S_V^* [cm³ liter½ mole⁻³⁄²]	References
KNO₃	37.3			9
	37.07		2.640	35
KOH	3.48		3.858[a]	24
NaOH	−6.4			9
	−6.52		4.138[a]	27
	−5.57			4
KSCN	43.77		9.376[a]	35
K Tartrate	81.83		6.717[a]	35
H₂S	19.0			6 (18°C)
NiCl₂	−16.1			6
NH₄N₃	42.9			6
NH₄NO₂	42.9			6
NH₄VO₃	44.9			6
NH₄H₂PO₄	48.9			6
(NH₄)₂WO₄	51.9			6
(NH₄)₂SiO₃	31.8			6
CaCl₂	17.6			9
MgCl₂	15.5			9
K₂CO₃	10.2			9
Na₂CO₃	−6.31			39
	−7.6			9
Na₂SO₄	10.3			9
MgSO₄	−4.8			9
	−9.12			39
ZnSO₄	−11.23			39

25°C

Salt	$\bar{V}°$ (cm³/mole)	b_V [cm³ liter mole⁻²]	S_V^* [cm³ liter½ mole⁻³⁄²]	References
HCl	*17.83*	−1.15		40,41
	17.82	(−0.91)		42
	17.76			43
	17.8			44
	18.0			9
	17.4			9
	18.1			5
	18.07		0.95	45
	17.98			21
	18.20			2
	18.2			6
	18.20		0.83	53
HBr	24.2			9
HO₂CH	*34.69*			73
	34.1			106

572

Table 1 (cont.)

Salt	$\bar{V}^\circ$ (cm³/mole)	b_V [cm³ liter mole^{-2}]	$S_V{}^*$ [cm³ liter$^{1/2}$ mole$^{-3/2}$]	References
HO_2CCH_3	*51.94*			46
	51.93			73
	52.0			106
$HO_2CCH_2CH_3$	67.9			106
$HO_2C(CH_2)_2CH_3$	*69.19*			73
	84.7			106
p-HSO_3Ph–CH_3	119.7			105
H_2CO_3	53.0			106
Phenol	86.0			106
HNO_3	29.0			9
HIO_3	25.17			105
$HClO_4$	44.12			47
	(44.04)			
H_2SO_4	*14.08*			48
	14.9			9
	14.9			49
LiΓ	*−2.18*	1.1		16
LiCl	*16.99*	(−0.367)		50
	16.91	−0.36		16
	17.00		1.488	3
	17.06			2
	17.1			6
	16.85			51
	16.6			52
	17.06		1.42	53
	16.92			111
LiBr	*23.76*	−0.60		16
	23.80	(−0.51)		32
	24.08		1.159	3
	24.07			2
	24.74			111
LiI	35.50		0.841	3
	35.60			2
	35.37	(−0.757)		32
	35.10			111
LiOH	−6.0		3.00	53
$LiNO_3$	28.75		0.84[a]	54
NaCl	*16.61*	0.018		20,55,104
	16.62	0.048		22
	16.62	−0.03		16
	16.63	(0.003)		50
	16.61			56
	16.65			34

(continued)

Table 1   (cont.)

Salt	$\overline{V}^\circ$ (cm^3/mole)	b_V [cm^3 liter mole^{-2}]	$S_V{}^*$ [cm^3 liter$^{1/2}$ mole$^{-3/2}$]	References
	16.64			57
	16.60			58
	16.6			5
	16.67			45
	16.62		1.884, 1.853	26,113
	16.61			51
	16.8			9
	15.9			9
	17.03			59
	16.54			43
	16.8			52
	16.28		2.22	53
	16.28			2
	16.40		2.153	3
	16.35			60
	16.44			61
	16.3			6
	16.56		1.945	62
	16.78			111
	16.45	1.693[a]	0.405[a]	100
NaBr	*23.48*	−0.26		16,63,104
	23.49			58
	23.45			2
	23.64	−3.2		18
	23.45			61
	23.51		1.760	3
	23.4			9
	23.57	(−0.18)		32
	23.60			111
	23.53		1.686	113
NaI	*34.82*	−0.38		16,63
	35.19			19
	34.98	(−0.40)		32
	35.09			64
	35.10		1.346	3
	35.00			2
	35.37		0.80[a]	65
	34.96			111
NaF	*−2.29*	0.52		17
	−2.37	0.46		8
	−2.47	0.64		16
	−1.19			19
NaOH	−5.25	(1.06)		66
	−4.60	0.78		67

Table 1 (cont.)

Salt	$\bar{V}^\circ$ (cm³ mole)	b_V [cm³ liter mole^{-2}]	S_V* [cm³ liter$^{1/2}$ mole$^{-3/2}$]	References
	−5.94		3.98	27 (estimated)
	−6.48		4.00	68
	−6.7		4.18	69,53
	−6.8			5,106
$NaNO_2$	25.0			8
$NaNO_3$	27.8			9
	27.48			69
	28.0		2.18[a]	54
$NaClO_3$	34.9			8
$NaBrO_3$	34.1			8
$NaIO_3$	24.78			105
$NaHSeO_4$	29.9			8
$NaClO_4$	44.9			8
	42.93 (42.85)			47
	43.9			71
	42.7			72
$NaHCO_3$	22			62
	22			8
	22.5			106
$NaHSO_4$	*34.46*			48
	29.9			8
	27		1.8	5
$NaOOCH$	*25.06*	0.013		73
	24.60		2.28	74
	24.8			106
NaO_2CCH_3	*39.23*	0.188		73
	39.24	0.209		41,46
	39.27			75
	39.72		3.42[a]	76
	39.75		2.53	74
	40.1		1.9	5
	39.8		2.22[a]	54
	38.0			106
$NaO_2CCH_2CH_3$	52.82		2.70	74
	52.7			106
$NaO_2C(CH_2)_2CH_3$	*69.19*	0.027		73
	68.41		2.57	74
	69.5			106
$NaPhSO_3$	107.73			19
$NaOPh$	67.5			106
Na Kaolin	179.32	225.071[a]	−32.214[a]	100

(continued)

Table 1   (cont.)

Salt	$\bar{V}^\circ$ (cm³/mole)	b_V [cm³ liter mole^{-2}]	$S_V{}^*$ [cm³ liter$^{1/2}$ mole$^{-3/2}$]	References
NaReO$_4$	46.97		1.99	97
NaBPh$_4$	276.41	−6.94		98
	277.0			102
NaSO$_3$PhCH$_3$	118.4			105
Na$_2$CO$_3$	−6.0			9
	−6.7			5
	−6.74		11.30	58
	−4.7			8
Na$_2$SO$_4$	11.52		12.16	58
	11.62	(2.96)		42
	11.10			60
	11.39		18.27[a]	76
	11.5			5
	10.9			9
	11.64			104
	11.42		12.158	113
Na$_2$S	−10.6			7
NaH$_2$PO$_4$	27.9			8
NaH$_2$AsO$_4$	34.0			8
Na$_2$SO$_3$	6.5			8
Na$_2$HPO$_4$	5.3			8
Na$_2$SeO$_4$	18.5			8
Na$_2$CrO$_4$	17.3			8
Na$_2$MoO$_4$	26.5			8
Na$_2$WO$_4$	23.3			8
Na$_3$AsO$_4$	−19.4			8
Na$_3$P$_3$O$_9$	*66.4*			110
Na$_4$P$_4$O$_{12}$	*73.0*			110
KF	*7.78*	0.52		16
	7.8			13
	6.60		3.35	2
KCl	*26.90*			34
	26.89	(0.099)		50
	25.85			58
	26.81	0.158	1.94	20,55
	26.84	0.08		77
	26.87			16
	26.5			52
	26.23		3.652	78
	26.36		2.41	53
	26.4			9
	26.81			51,104
	26.74			43
	26.74		2.075	62,79
	26.27			70

Table 1 (cont.)

Salt	$\overline{V}^\circ$ (cm³/mole)	b_V [cm³ liter mole⁻²]	S_V* [cm³ liter½ mole⁻³/²]	References
	26.8			5
	26.81			58
	26.36			2
	26.52		2.327	3
	26.25		2.432[a]	35
	27.02			80
	26.50	−2.24[a]	4.89[a]	65
	26.57		2.289	81
	26.87			82
	26.65		3.21[a]	76
	26.4			6
	26.52		2.35	28
	27.04			111
KBr	*33.73*	−0.16		16,63
	33.7			5
	33.73		1.939	3
	33.75	0.032		20,55
	33.73			83
	33.4			9
	33.54			2
	33.89		1.746[a]	35
	33.46		1.847	80
	33.97		1.85	62,79,104
	33.56		2.01	83
	33.88		2.64[a]	76
	33.5			6
	33.86			111
KI	*45.06*	−0.39		16
	45.21	−0.219		20,55
	45.23			2
	45.28			60
	45.48		1.369[a]	35
	45.83			80
	45.55		0.87[a]	65
	45.34		2.28[a]	76
	45.1			6
	45.36		1.556	3
	45.22			111
KOH	4.06			24 (estimated)
	3.6			6
	3.4			5
	2.9		4.35	53

(*continued*)

Table 1 (cont.)

Salt	$\overline{V}°$ (cm³ mole)	b_V [cm³ liter mole⁻²]	S_V^* [cm³ liter½ mole⁻³/²]	References
KNO₃	*38.02*	(0.72)	(4.11)	42
	38.18		2.30	64
	37.98		3.48[a]	76
	38.0		2.31[a]	54
	38.0			5
	38.2			9
	37.77		2.408[a]	35
	37.6			82
	37.8			6
KSCN	44.70		9.250[a]	35
	49.0			6
KOCN	35.14		1.48	97
KSeCN	58.70		1.21	97
KClO₃	*45.68*	0.418		40,41 (24.81°C)
	46.0			82
	43.5			6
KMnO₄	51.54		3.68	30
	51.9			8
	50.4			6
KBrO₃	44			82
	44.8			6
KHCO₃	32.7		2.6	5
	30.7			6
KIO₃	33.5			6
KClO₄	55.2			51
	52.9			6
KHSO₄	34.9			6
KO₂CH	34.7			6
KO₂CCH₃	48.9			6
	47.2		2.48[a]	54
K₂CO₃	14.0			6
	12.2			9
K₂SO₄	*31.99*	(5.17)		42
	31.96		12.53	31
	30.1			9
	31.9			5
	32.28		12.07	62,79
	32.36		12.026	81
	32.30			84
	32.42		11.553[a]	35
	31.7			13
	32.44			104
K₂CrO₄	37.12		11.70	31
K₂PtCl₆	159			51

Table 1 (cont.)

Salt	$\overline{V}°$ (cm³/mole)	b_V [cm³ liter mole^{-2}]	S_V* [cm³ liter$^{1/2}$ mole$^{-3/2}$]	References
K_2SeO_4	39.1			51
K_3AsO_4	11.5			51
K Citrate	112.3		11.723[a]	35
K Tartrate	83.20		6.067[a]	35
$K_3Fe(CN)_6$	147.8	(−12.8)	(12.92)	66
$K_4Fe(CN)_6$	110.0	(−33.1)	(38.71)	66
$K_3P_3O_9$	97.6			110
$K_4P_4O_{12}$	114.0			110
RbCl	31.94	(0.165)		50
	31.87			3
	31.8		2.219	5
	31.71			2
	31.7			6
	32.05			51
	31.94	0.17		16
	31.60		2.40	28
	32.21			111
RbBr	38.84	−0.26		16
	38.71		2.038	3
	38.70			2
	39.03			111
RbI	50.16	−0.05		16
	50.31		1.607	3
	50.40			2
	50.39			111
RbF	12.69	0.55		16
$RbNO_3$	42.85		2.65[a]	54
RbO_2CCH_3	53.5		2.4[a]	54
CsCl	39.15	(0.123)		50
	39.17	0.12		16
	39.02			2
	39.15		2.172	3
	39.3			9,52
	39.2			5
	39.1			6
	39.29			51
	39.44			111
CsBr	46.04	0.09		16
	46.20			2
	46.19		1.901	3
	46.1			5
	45.8			9
	45.26			111

(continued)

Table 1   (cont.)

Salt	$\overline{V}°$ (cm³/mole)	b_V [cm³ liter mole^{-2}]	S_V* [cm³ liter^{½} mole$^{-3/2}$]	References
CsI	57.39	0.11		16
	57.75		1.62	30
	57.74		1.579	3
	57.90			2
	57.65		1.60	28
	57.62			111
CsF	20.03	0.25		16
CsNO$_3$	50.8		2.1[a]	54
	51.38		0.478	81
CsO$_2$CCH$_3$	61.3		2.0[a]	54
TlF	10.5			7
TlCl	27.4			6
TlNO$_3$	39.5		5.9[a]	54
TlClO$_4$	53.55		5.3[a]	54
TiC$_2$H$_3$O$_2$	50.9		4.3[a]	54
AgCl	16.8			6
AgF	−3.1			7
AgNO$_3$	28.02		2.61	31
	28.78		2.46	76
NH$_4$Cl	35.98		1.45	82,83
	36.0			6
	36.26		1.86[a]	76
	36.0			106
NH$_4$Br	42.57	−0.55		63
NH$_4$NO$_3$	47.24			58
	47.4			6
	47.56		0.97	85
NH$_4$ClO$_4$	61.5			13
NH$_4$BF$_4$	62.04		0.67	97
NH$_4$SO$_3$F	65.79		0.279	97
NH$_4$SO$_3$NH$_2$	59.35		2.84	97
Me$_4$NCl	107.4	−2.5		86
	107.0			102
	107.3	−4.60		114
Me$_4$NBr	114.18			87
	114.29	−0.5		86
	114.25	(−1.08)	1.00	66
Me$_4$NBr	114.40			34
	114.8		0.6	38
	115.27			37
	114.25			88
Me$_4$NI	125.8	0		86
	125.7			89
	125.75			88

Table 1 (cont.)

Salt	$\overline{V}^\circ$ (cm^3/mole)	b_V [cm^3 liter mole^{-2}]	S_V* [cm^3 liter$^{1/2}$ mole$^{-3/2}$]	References
MeNH$_3$Cl	53.81	−0.43		114
	55.5			106
Me$_2$NH$_2$Cl	72.47	−1.33		114
	73.1			106
Me$_3$NHCl	90.59	−1.00		114
	91.7			106
MeNH$_3$Br	60.82	−0.63		63
Et$_4$NCl	167.0	−16.0		86
Et$_4$NHCl	138.6	−14.4		114
Et$_4$NBr	173.65			34
	173.60	−3.5		87
	174.3	−10.0		86
	175.3		−3.3	38
	176.5			37
	175.0			102
Et$_4$NI	185.5	−5.6		86
	185.5			89
EtNH$_3$Br	77.65	−0.92		63
nPr$_4$NCl	232.9	−23.8		86
nPr$_3$NHCl	186.8	−22.4		114
nPr$_4$NBr	239.15			34
	239.6	−15.0		86
	240.8		−6.0	38
	240.35			37
nPr$_4$NI	250.9	−9.2		86
	250.9			89
	251.0			102
	250.7	−14.4		114
nPrNH$_3$I	105.7	−2.20		114
nPr$_2$NH$_2$I	156.9	−1.60		114
nPr$_3$NHI	204.8	−1.03		114
nPrNH$_3$Br	94.15	−1.3		63
nBu$_4$NCl	294.3	−35.5		86
nBu$_4$NBr	300.35	(−10.2)	(−5.00)	34
	300.40			55
	301.0	−21.2		86
	301.01			37
	302.9		−8.4	38
nBu$_4$NI	312.4	−18.0		86
	312.2			89
	316.0			102
nBuNH$_3$Br	110.20	−1.7		63
nPen$_4$NBr	363.9	−30.5		86
	365.6		−8.3	38

(*continued*)

Table 1   (cont.)

Salt	$\overline{V}°$ (cm³/mole)	b_V [cm³ liter mole⁻²]	S_V* [cm³ liter½ mole⁻³⁄₂]	References
nPenNH₃Br	*126.15*	−1.9		63
nHexNH₃Br	*142.04*	−2.1		63
nHepNH₃Br	*157.94*	−2.9		63
nOctNH₃Br	*173.86*	−3.1		63
(HOEt)₄NF	150.6		3.0	90
(HOEt)₄NBr	176.9		1.4	90
[Bu₃N(CH₂)₈NBu₃]Br₂	580.9			36
	577.9			36
Ph₄AsCl	318.48	−1.75		99
PhNH₃⁺Cl⁻	101.9			106
Py⁺Cl⁻	90.5			106
Pip⁺Cl⁻	108.2			106
BeCl₂	22			6
	26			51
MgCl₂	*14.49*	(−6.3)		42
	14.5		5.15	62
	14.6			6
	15.3			5
	15.3		4.26	91
	15.6			9,92
	14.20		6.222	28
	14.66		5.404	113
MgBr₂	28.00		5.374	28
MgI₂	51.20		4.384	28
MgSO₄	−7.0			34
	−3.9			9
	−7.16		12.852	113
CaCl₂	*17.78*	−1.8	7.57	42
	17.84	−0.708[b]	6.666[b]	20
	17.0			6,92
	16.86		5.57	62,93
	18.4			9
	18.5			5
	18.26			69
	18.25		6.00	53
	18.54		8.40[a]	76
	18.6			91
	18.70		5.657	28
	17.86		6.253	113
CaBr₂	32.40		4.808	28
CaI₂	55.60		3.818	28
Ca(NO₃)₂	40.164	0.6472[a]	6.3098[a]	112(a)
	37.56	−0.624[a]	9.664[a]	112(b)
SrCl₂	*17.94*			56,104
	17.2			6

Table 1   (cont.)

Salt	$\bar{V}°$ (cm³/mole)	b_V [cm³ liter mole⁻²]	S_V^* [cm³ liter^½ mole^{-3/2}]	References
	18.70			62,91
	18.10		7.071	28
	17.0			92
SrBr₂	31.80		5.798	28
SrI₂	55.00		4.950	28
BaCl₂	23.24	(−3.39)	(6.65)	42
	23.15	−3.027		20
	22.8			6
BaCl₂	23.60		4.83	53
	24.1			94
	23.80		6.505	28
	25.9			92
BaBr₂	37.10		5.657	28
BaI₂	60.50		4.808	28
HgCl₂	16.0			6
CdCl₂	23.24		8.82ᵃ	76
	22.8			6
CdBr₂	33.3		18.9ᵃ	76
CdI₂	67.16		0	78
Cd(O₂CCH₃)₂	71.35		24.25ᵃ	76
Cd(NO₃)₂	38.0			7
	42.38	−0.2479ᵃ	6.460ᵃ	107
CdSO₄	0.80		19.23ᵃ	76
PbCl₂	8.0			6
Pb(NO₃)₂	42.5			7
CuCl	9.4			6
Cu(NH₂SO₃)₂	56.08	1.92	7.07	103(a)
	55.22	0.46		103(b)
ZnCl₂	10.0			6
ZnSO₄	−7.6			7
FeCl₂	13.8			7
	18.0			6
FeBr₂	23.0			7
FeSO₄	−11.9			7
Fe(NO₃)₃	49.7			7
	36.5			7
FeCl₃	24.0			6
COCl₂	10.8			7
	18.0			6
CoBr₂	23.8			7
Co(NO₃)₂	30.6			7
CoSO₄	−4.2			7
	−3.51		14.06	78

(*continued*)

Table 1 (cont.)

Salt	$\overline{V}^\circ$ (cm^3/mole)	b_V [cm^3 liter mole^{-2}]	S_V^* [cm^3 liter$^{1/2}$ mole$^{-3/2}$]	References
	-3.16		13.48[a]	78
NiSO$_4$	-10.1			7
MnCl$_2$	22.0			6
LaCl$_3$	14.28	(-22.9)	(12.68)	42
	14.51	(-18.7)		77
	16.0		11.87	95
	16.06		11.82	29
La(ClO$_4$)$_3$	95.1			109
La(NO$_3$)$_8$	49.37	(-12.9)		77
	47.80			108
PrCl$_3$	10.96	(-23.3)		77
NdCl$_3$	10.18	(-20.5)		77
Nd(NO$_3$)$_3$	44.74	(-32.4)		77
	47.00			108
SmCl$_3$	11.16	(-28.4)		77
GdCl$_3$	13.08	(-27.3)		77
TbCl$_3$	13.25	(-30.0)		77
DyCl$_3$	12.66	(-29.4)		77
HoCl$_3$	11.73	(-24.5)		77
ErCl$_3$	10.63	(-25.2)		77
Er(NO$_3$)$_3$	45.28	(-31.5)		77
YbCl$_3$	9.27	(-20.0)		77
Yb(NO$_3$)$_3$	43.47	(-23.5)		77
AlCl$_3$	11.1			6
	12.9			7
Al(NO$_3$)$_3$	43.0			7
CrCl$_3$	13.9			7
Ce(NO$_3$)$_3$	47.20			108
ThCl$_4$	17.8			7
Th(NO$_3$)$_4$	62.6			7

	30°C			
HCl	18.16			21
NaCl	16.97			24
NaOH	-4.27		3.872[a]	67
	-5.50			27
NaO$_2$CH	24.95		2.27	74
NaO$_2$CCH$_3$	39.29		2.38	74
NaO$_2$CCH$_2$CH$_3$	53.34		2.64	74
NaO$_2$(CH$_2$)$_2$CH$_3$	68.71		2.78	74
KCl	26.98		2.064[a]	35
	26.92			24
KBr	34.30		1.663[a]	35

584

Table 1 (cont.)

Salt	$\overline{V}°$ (cm^3/mole)	b_V [cm^3 liter mole^{-2}]	S_V* [cm^3 liter$^{1/2}$ mole$^{-3/2}$]	References
KI	46.08		1.260[a]	35
KOH	4.50		3.559[a]	24
KNO$_3$	38.45		2.200[a]	35
KSCN	45.53		9.164[a]	35
K Tartrate	84.20		5.444[a]	35
Cd(NO$_3$)$_2$	43.65		5.808	107
Me$_4$NI	126.5			89
Et$_4$NI	186.3			89
nPr$_4$NI	252.4			89
nBu$_4$NI	314.0			89
		35°C		
HCl	18.2			6
LiCl	16.95	(−0.461)		32
	17.10		1.45	96
	17.2			6
LiBr	24.75		0.75	96
NaCl	17.16	−0.095		20
	17.28	−0.127		22
	17.25		1.91	96
	17.3			6
NaBr	24.45		1.55	96
NaI	36.45		1.08	96
NaOH	−5.4			6
NaO$_2$CH	25.24		2.26	74
NaO$_2$CCH$_3$	39.55		2.39	74
NaO$_2$CCH$_2$CH$_3$	53.72		2.64	74
NaO$_2$C(CH$_2$)$_2$CH$_3$	69.08		2.91	74
KCl	27.22		2.16	96
	27.31	0.029		20
KBr	34.57	−0.319		20
	34.27		1.90	96
	34.55		1.85	96
KI	46.28	−0.580		20
	46.25		1.45	96
RbCl	32.4			6
CsCl	39.8			6
NH$_4$Cl	36.2			6
nBu$_4$NBr	304.8		−6.3	38
CaCl$_2$	18.32	−0.719[b]	6.818[b]	20
		(−0.59)		
BaCl$_2$	23.83	−3.542		20

(continued)

Table 1 (cont.)

Salt	$\overline{V}°$ (cm³/mole)	b_V [cm³ liter mole⁻²]	$S_V{}^*$ [cm³ liter½ mole⁻³⁄₂]	References
		40°C		
HCl	18.39			21
NaCl	17.62			24
NaOH	−4.94		3.747[a]	27
NaO₂CH	25.69		2.09	74
NaO₂CCH₃	39.74		2.42	74
NaO₂CCH₂CH₃	54.10		2.64	74
NaO₂C(CH₂)₂CH₃	69.59		2.94	74
KCl	27.53			24
KOH	5.50		3.406[a]	24
Cd(NO₃)₂	45.30		5.160	107
Me₄NI	127.2			89
Et₄NI	187.4			89
nPr₄NI	254.0			89
nBu₄NI	318.5			89
		45°C		
NaCl	17.59	−0.25		22
	17.65	−0.412		20
KCl	27.46	−0.007		20
KBr	34.93	−0.397		20
KI	46.92	−0.645		20
Me₄NBr	117.11			37
Et₄NBr	178.67			37
nPr₄NBr	241.32			37
nBu₄NBr	306.59			37
BaCl₂	24.31	−4.782		20
CaCl₂	18.11	−1.853[b] (−1.73)	8.339[b]	20
		50°C		
HCl	18.0			44
	18.43			21
LiCl	16.96		1.446	3
	16.8		(2.11)	52
	16.82			111
LiBr	24.37		1.145	3
	24.09			111
LiI	36.81		0.569	3
	36.08			111
NaCl	17.96		1.804	3
	18.2		(2.84)	52

586

Table 1   (cont.)

Salt	$\overline{V}°$ (cm³/mole)	b_V [cm³ liter mole^{-2}]	S_V* [cm³ liter$^{1/2}$ mole$^{-3/2}$]	References
	17.97			24
	18.09			111
NaBr	25.34		1.398	3
	25.36			111
NaI	37.68		0.826	3
	37.35			111
NaOH	−4.68		3.705	27
NaBPh$_4$	*283.63*	−4.68		98
Na$_2$SO$_4$	15.5		(10.59)	13
KF	8.2		(2.43)	13
KCl	27.66		2.087	3
	27.4		(4.04)	52
	27.64		2.10	28
	27.87			24
	27.12			111
KBr	35.32		1.640	3
	35.2		(1.64)	13
	35.39			111
KI	47.58		1.168	3
	46.8		(1.85)	13
	47.38			111
KNO$_3$	39.9		(2.58)	13
KOH	5.25		3.353[a]	24
K$_2$SO$_4$	35.1		(8.62)	13
RbCl	32.95		2.047	3
	32.90		2.10	29
	33.18			111
RbBr	40.28		1.757	3
	40.45			111
RbI	52.75		1.224	3
	52.44			111
CsCl	40.39		1.958	3
	40.3		(1.12)	52
	40.64			111
CsBr	47.81		1.632	3
	47.91			111
CsI	60.32		1.084	3
	60.00		1.30	28
	59.90			111
NH$_4$Cl	37.2		(0.92)	13
NH$_4$ClO$_4$	64.3		(1.76)	13
Me$_4$NCl	108.1		2.3	101
				(50.25°C)

(*continued*)

Table 1　(cont.)

Salt	$\bar{V}°$ (cm^3/mole)	b_V [cm^3 liter mole^{-2}]	$S_V{}^*$ [cm^3 liter$^{1/2}$ mole$^{-3/2}$]	References
Me$_4$NI	128.7			89
Et$_4$NCl	170.0		-2.0	101 (50.25°C)
Et$_4$NI	189.1			89
nPr$_4$NCl	237.2		-4.6	101 (50.25°C)
nPr$_4$NI	256.4			89
nBu$_4$NCl	301.2		-5.3	101 (50.25°C)
nBu$_4$NI	322.5			89
Ph$_4$AsCl	*327.94*	-13.242		99
MgCl$_2$	14.70		5.657	28
	15.1		(11.05)	92
MgBr$_2$	29.20		4.808	28
MgI$_2$	54.10		3.253	28
CaCl$_2$	19.20		5.515	28
	17.6		(13.46)	92
CaBr$_2$	33.50		4.525	28
CaI$_2$	58.40		2.970	28
SrCl$_2$	19.80		6.505	28
	19.0		(19.51)	92
SrBr$_2$	34.00		5.515	28
SrI$_2$	59.00		3.960	28
BaCl$_2$	25.60		6.081	28
	27.8		(26.90)	92
BaBr$_2$	40.10		4.950	28
BaI$_2$	65.10		3.535	28
Cd(NO$_3$)$_2$	46.03	-0.0614[a]	5.163[a]	107

55°C

Salt	$\bar{V}°$ (cm^3/mole)	b_V	$S_V{}^*$	References
NaCl	*17.76*	-0.427		20
	17.91	-0.63		22
KCl	*27.68*	-0.234		20
KBr	*35.20*	-0.579		20
KI	*47.24*	-0.660		20
BaCl$_2$	*24.23*	-5.642		20
CaCl$_2$	*17.45*	-3.524[b] (-3.98)	10.523[b]	20

60°C

Salt	$\bar{V}°$ (cm^3/mole)	b_V	$S_V{}^*$	References
HCl	18.30			21
NaCl	18.16			24
NaOH	-4.56		3.685	27
KCl	27.92			24

588

Table 1   (cont.)

Salt	$\overline{V}°$ (cm³/mole)	b_V [cm³ liter mole⁻²]	S_V* [cm³ liter½ mole⁻³⁄₂]	References
KOH	5.25		3.357[a]	24
Cd(NO₃)₂	46.13	−0.209[a]	5.574[a]	107
Me₄NI	130.2			89
Et₄NI	191.1			89
nPr₄NI	258.8			89
nBu₄NI	327.5			89
		65°C		
NaCl	17.69	−0.536		20
KCl	27.54	−0.261		20
KBr	35.17	−0.579		20
KI	47.61	−0.790		20
BaCl₂	23.64	−6.529		20
CaCl₂	16.75	−3.300[b] (−3.33)	10.79[b]	20
Me₄NBr	118.62			37
Et₄NBr	180.96			37
nPr₄NBr	243.35			37
nBu₄NBr	312.49			37
		70°C		
HCl	17.98			21
LiCl	16.14		1.663	3 (70.2°C)
LiBr	23.88		1.312	3 (70.2°C)
LiI	36.85		0.708	3 (70.2°C)
NaCl	18.20			24
NaBr	25.80		1.377	3 (70.2°C)
NaOH	−4.43		3.627[a]	27 (70.2°C)
KCl	27.95		1.982	3 (70.2°C)
	27.85			24
KOH	5.17		3.372[a]	24
CsCl	40.45		2.053	3 (70.2°C)
CsBr	48.24		1.646	3 (70.2°C)
CsI	61.06		1.160	3 (70.2°C)
Me₄NI	131.9			89
Et₄NI	194.3			89
nPr₄NI	262.6			89
nBu₄NI	335.5			89
		75°C		
HCl	17.4			44
LiCl	15.6		(2.64)	52

(*continued*)

Table 1 (cont.)

Salt	$\overline{V}°$ (cm³/mole)	b_V [cm³ liter mole⁻²]	S_V* [cm³ liter½ mole⁻³⁄₂]	References
NaCl	18.2		(3.09)	52
KF	7.2		(3.15)	13
KCl	27.7		(3.96)	52
KBr	35.4		(1.85)	13
KI	47.5		(2.18)	13
CsCl	40.6		(1.15)	52
KNO₃	41.0		(2.58)	13
K₂SO₄	34.1		(9.98)	13
Na₂SO₄	15.8		(10.56)	13
NH₄Cl	36.9		(0.82)	13
NH₄ClO₄	66.4		(1.67)	13
MgCl₂	13.1		(11.84)	92
CaCl₂	15.7		(14.91)	92
SrCl₂	18.0		(20.54)	92
BaCl₂	27.1		(28.08)	92

80°C

NaCl	18.16			24
KCl	27.66			24
Me₄NI	134.0			89
Et₄NI	198.7			89
nPr₄NI	266.8			89
nBu₄NI	347.0			89

100°C

HCl	16.0			44
LiCl	13.3		(3.66)	52
	15.95		1.606	3
NaCl	16.8		(4.00)	52
KF	5.8		(3.38)	13
KCl	25.5		(5.42)	52
	26.49		2.574	3
KBr	34.4		(3.01)	13
KI	48.2		(1.78)	13
CsCl	39.0		(1.30)	52
KNO₃	41.4		(2.61)	13
K₂SO₄	31.4		(11.80)	13
Na₂SO₄	12.3		(12.91)	13
NH₄Cl	35.5		(1.35)	13
NH₄ClO₄	67.4		(1.90)	13
MgCl₂	8.6		(14.10)	92

590

Table 1 (cont.)

Salt	$\overline{V}^\circ$ (cm³/mole)	b_V [cm³ liter mole⁻²]	$S_V{}^*$ [cm³ liter½ mole⁻³⁄₂]	References
$CaCl_2$	12.0		(17.10)	92
$SrCl_2$	14.6		(23.02)	92
$BaCl_2$	23.9		(30.94)	92

125°C

Salt	$\overline{V}^\circ$ (cm³/mole)	b_V [cm³ liter mole⁻²]	$S_V{}^*$ [cm³ liter½ mole⁻³⁄₂]	References
HCl	14.1			44
LiCl	10.5		(4.76)	52
NaCl	15.0		(4.79)	52
KF	3.0		(4.45)	13
KCl	23.5		(6.66)	52
KBr	33.3		(2.83)	13
KI	46.9		(2.74)	13
CsCl	37.3		(1.44)	52
KNO_3	41.0		(2.95)	13
K_2SO_4	25.5		(16.04)	13
Na_2SO_4	7.6		(15.53)	13
NH_4Cl	33.4		(2.20)	13
NH_4ClO_4	67.9		(2.24)	13
$MgCl_2$	2.2		(17.35)	92
$CaCl_2$	4.5		(26.34)	92
$SrCl_2$	8.4		(27.56)	92
$BaCl_2$	17.7		(36.19)	92

150°C

Salt	$\overline{V}^\circ$ (cm³/mole)	b_V [cm³ liter mole⁻²]	$S_V{}^*$ [cm³ liter½ mole⁻³⁄₂]	References
HCl	11.2			44
LiCl	6.5		(6.42)	52
NaCl	12.4		(5.61)	52
KF	−1.0		(5.96)	13
KCl	20.5		(8.27)	52
KBr	30.8		(3.71)	13
KI	45.1		(3.52)	13
CsCl	34.7		(1.62)	52
KNO_3	39.3		(3.83)	13
K_2SO_4	19.0		(19.04)	13
Na_2SO_4	0.5		(19.27)	13
NH_4Cl	30.3		(3.39)	13
NH_4ClO_4	67.3		(2.91)	13
$MgCl_2$	−6.6		(22.03)	92
$CaCl_2$	−3.3		(26.89)	92
$SrCl_2$	0.0		(33.58)	92
$BaCl_2$	10.5		(41.73)	92

(*continued*)

Table 1   (cont.)

Salt	$\bar{V}°$ (cm³/mole)	b_V [cm³ liter⁻²]	S_V* [cm³ liter½ mole⁻³⁄₂]	References
		175°C		
HCl	7.0			44
LiCl	0.0		(9.58)	52
NaCl	7.5		(7.89)	52
KF	−6.6		(8.15)	13
KCl	14.8		(11.59)	52
KBr	27.0		(5.04)	13
KI	42.4		(4.54)	13
CsCl	29.3		(2.04)	52
KNO₃	36.0		(5.50)	13
K₂SO₄	7.0		(27.86)	13
Na₂SO₄	−11.0		(26.20)	13
NH₄Cl	25.8		(5.15)	13
NH₄ClO₄	65.3		(4.47)	13
MgCl₂	−20.0		(30.13)	92
CaCl₂	−13.0		(32.12)	92
SrCl₂	−11.5		(41.72)	92
BaCl₂	−1.0		(51.21)	92
		200°C		
HCl	0.5			44
LiCl	−7.6		(13.84)	52
NaCl	0.4		(11.53)	52
KF	−14.0		(10.73)	13
KCl	7.8		(15.68)	52
KBr	20.3		(8.27)	13
KI	37.5		(6.57)	13
CsCl	23.6		(2.46)	52
KNO₃	29.6		(9.18)	13
K₂SO₄	−8.0		(37.71)	13
Na₂SO₄	−23.5		(32.35)	13
NH₄Cl	20.5		(6.52)	13
NH₄ClO₄	63.0		(5.32)	13
MgCl₂	−36.0		(32.91)	92
CaCl₂	−29.7		(43.48)	92
SrCl₂	−28.0		(54.19)	92
BaCl₂	−16.0		(63.25)	92

[a] Molal concentration, m, was used in the Masson equation instead of molar concentration, c.
[b] Concentration dependence of ϕ_V was expressed by $\phi_V = \phi_V° + S_V*\sqrt{c} + b_V c$.

REFERENCES

1. D. O. Masson, *Phil. Mag.* (7), **8**, 218 (1929).
2. W. Geffcken, *Z. Physik. Chem.*, **A155**, 1, (1931); *Naturwiss.*, **19**, 321 (1930).
3. A. F. Scott, *J. Phys. Chem.*, **35**, 2315 (1931).
4. J. D. Bernal and R. H. Fowler, *J. Chem. Phys.*, **1**, 515 (1933).
5. B. B. Owen and S. R. Brinkley, *Chem. Rev.*, **29**, 461 (1941).
6. K. Fajans and O. Johnson, *J. Amer. Chem. Soc.*, **64**, 668 (1942).
7. A. M. Couture and K. J. Laidler, *Can. J. Chem.*, **34**, 1209 (1956).
8. A. M. Couture and K. J. Laidler, *Can. J. Chem.*, **35**, 207 (1957).
9. E-an. Zen, *Geochim. Cosmochim. Acta*, **12**, 103 (1957).
10. H. S. Harned and B. B. Owen, "The Physical Chemistry of Electrolytic Solutions," 3rd Ed., Amer. Chem. Soc. Monogr. No. 137, Reinhold, New York, 1958.
11. P. Mukerjee, *J. Phys. Chem.*, **65**, 740 (1961).
12. O. Redlich and D. M. Meyer, *Chem. Rev.*, **64**, 221 (1964).
13. A. J. Ellis, *J. Chem. Soc.*, (A) 1138 (1968).
14. F. J. Millero and W. Drost-Hansen, *J. Chem. Eng. Data*, **13**, 330 (1968).
15. F. J. Millero, *J. Phys. Chem.*, **73**, 4589 (1969).
16. J. E. Desnoyers, M. Arel, G. Perron, and C. Jolicoeur, *J. Phys. Chem.*, **73**, 3346 (1969).
17. F. J. Millero, *J. Phys. Chem.*, **71**, 4567 (1967).
18. R. E. Conway and L. H. Laliberté, *J. Phys. Chem.*, **72**, 4317 (1968).
19. R. E. Robertson, S. E. Sugamori, R. Tse, and C.-Y. Wu, *Can. J. Chem.*, **44**, 487 (1966).
20. L. A. Dunn, *Trans. Faraday Soc.*, **64**, 2951 (1968).
21. G. Akerlöf and J. W. Teare, *J. Amer. Chem. Soc.*, **60**, 1226 (1938).
22. F. J. Millero, *J. Phys. Chem.*, **74**, 356 (1970).
23. I. W. Duedall and P. K. Weyl, *Rev. Sci. Instr.*, **36**, 528 (1965).
24. G. Akerlöf and P. Bender, *J. Amer. Chem. Soc.*, **63**, 1085 (1941).
25. T. Batuecus, *Ann. Fis. Quim.* (*Madrid*), **42**, 713 (1946).
26. G. Jones and S. M. Christian, *J. Amer. Chem. Soc.*, **59**, 484 (1937).
27. G. Akerlöf and G. Kegeles, *J. Amer. Chem. Soc.*, **61**, 1027 (1939).
28. C. W. Root, Ph.D. thesis, Harvard University, Cambridge, Mass., 1932.
29. G. Jones and C. F. Bickford, *J. Amer. Chem. Soc.*, **56**, 605 (1934).
30. G. Jones and H. J. Fornwalt, *J. Amer. Chem. Soc.*, **58**, 617 (1936).
31. G. Jones and J. H. Colvin, *J. Amer. Chem. Soc.*, **62**, 338 (1940).
32. F. Vaslow, *J. Phys. Chem.*, **73**, 3745 (1969).
33. T. Isono and R. Tamamushi, *Electrochimi. Acta*, **12**, 1479 (1967).
34. F. Franks and H. T. Smith, *Trans. Faraday Soc.*, **63**, 2586 (1967).
35. Sister Halasey, *J. Phys. Chem.*, **45**, 1252 (1941).
36. T. L. Broadwater and D. F. Evans, *J. Phys. Chem.*, **73**, 164 (1969).
37. S. Schiavo, B. Scrostati, and A. Tommasini, *Ric. Sci.*, **37**, 211 (1967).
38. W.-Y. Wen and S. Saito, *J. Phys. Chem.*, **68**, 2639 (1964).
39. A. B. Lamb and R. E. Lee, *J. Amer. Chem. Soc.*, **35**, 1666 (1913).
40. O. Redlich and J. Bigeleisen, *J. Amer. Chem. Soc.*, **64**, 758 (1942).
41. O. Redlich and J. Bigeleisen, *Chem. Rev.*, **30**, 171 (1942).
42. L. A. Dunn, *Trans. Faraday Soc.*, **62**, 2348 (1966).
43. B. B. Owen and S. R. Brinkley, *Ann. N.Y. Acad. Sci.*, **51**, 753 (1949).
44. A. J. Ellis and I. M. McFadden, *Chem. Commun.*, 516 (1968).
45. H. E. Wirth, *J. Amer. Chem. Soc.*, **62**, 1128 (1940).

46. O. Redlich and L. E. Nielson, *J. Amer. Chem. Soc.*, **64**, 761 (1942).
47. H. E. Wirth and F. N. Collier, Jr., *J. Amer. Chem. Soc.*, **72**, 5292 (1950).
48. R. E. Lindstrom and H. E. Wirth, *J. Phys. Chem.*, **73**, 218 (1969).
49. I. M. Klotz and C. F. Eckert, *J. Amer. Chem. Soc.*, **64**, 1878 (1942).
50. F. Vaslow, *J. Phys. Chem.*, **70**, 2286 (1966).
51. E. Glueckauf, *Trans. Faraday Soc.*, **61**, 914 (1965).
52. A. J. Ellis, *J. Chem. Soc. (A)*, 1579 (1966).
53. F. T. Gucker, *Chem. Rev.*, **13**, 111 (1933).
54. R. A. Robinson, *J. Amer. Chem. Soc.*, **59**, 84 (1937).
55. L. A. Dunn, *Trans. Faraday Soc.*, **64**, 1898 (1968).
56. A. Kruis, *Z. Physik. Chem.*, **B34**, 1 (1936).
57. F. J. Millero, *Rev. Sci. Instr.*, **38**, 1441 (1967).
58. W. Geffcken and D. Price, *Z. Physik. Chem.*, **B26**, 81 (1934).
59. L. H. Adams, *J. Amer. Chem. Soc.*, **53**, 3769 (1931).
60. R. E. Gibson, *J. Phys. Chem.*, **38**, 319 (1934).
61. R. E. Gibson and O. H. Loeffler, *J. Amer. Chem. Soc.*, **63**, 443 (1941).
62. H. E. Wirth, *J. Marine Res.*, **3**, 230 (1940).
63. J. E. Desnoyers and M. Arel, *Can. J. Chem.*, **45**, 359 (1967).
64. R. E. Gibson and J. F. Kincaid, *J. Amer. Chem. Soc.*, **59**, 25 (1937).
65. D. A. MacInnes and M. O. Dayhoff, *J. Amer. Chem. Soc.*, **74**, 1017 (1952).
66. L. G. Hepler, J. M. Stokes, and R. H. Stokes, *Trans. Faraday Soc.*, **61**, 20 (1965).
67. A. Bodanszky and W. Kauzmann, *J. Phys. Chem.*, **66**, 177 (1962).
68. E. H. Lanman and B. J. Mair, *J. Amer. Chem. Soc.*, **56**, 390 (1934); calculated by A. Bodnaszky and W. Kauzmann (see Ref. 67).
69. F. T. Gucker, *J. Phys. Chem.*, **38**, 307 (1934).
70. C. Drucker, *Arkiv. Kemi Mineral. Geol.*, **14A** (15), 1 (1941).
71. J. Bigeleisen, *J. Phys. Colloid Chem.*, **51**, 1369 (1947).
72. H. Kohner, *Z. Physik. Chem.*, **B1**, 427 (1928).
73. E. J. King, *J. Phys. Chem.*, **73**, 1220 (1969).
74. G. M. Watson and W. A. Felsing, *J. Amer. Chem. Soc.*, **63**, 410 (1941).
75. H. E. Wirth, *J. Amer. Chem. Soc.*, **70**, 462 (1948).
76. L. G. Longworth, *J. Amer. Chem. Soc.*, **57**, 1185 (1935).
77. F. H. Spedding, M. J. Pikal, and B. O. Ayres, *J. Phys. Chem.*, **70**, 2440 (1966).
78. R. C. Cantelo and H. E. Phifer, *J. Amer. Chem. Soc.*, **55**, 1333 (1933).
79. H E. Wirth, *J. Amer. Chem. Soc.*, **59**, 2549 (1937).
80. V. K. La Mer and T. H. Gronwall, *J. Phys. Chem.*, **31**, 393 (1927).
81. G. Jones and W. A. Ray, *J. Amer. Chem. Soc.*, **59**, 187 (1937).
82. G. Jones and S. K. Talley, *J. Amer. Chem. Soc.*, **55**, 624 (1933).
83. G. Jones and S. K. Talley, *J. Amer. Chem. Soc.*, **55**, 4124 (1933).
84. L. H. Adams, *J. Amer. Chem. Soc.*, **54**, 2229 (1932).
85. F. T. Gucker and T. R. Rubin, *J. Amer. Chem. Soc.*, **58**, 2118 (1936).
86. B. E. Conway, R. E. Verrall, and J. E. Desnoyers, *Trans. Faraday Soc.*, **62**, 2738 (1966).
87. H. E. Wirth, *J. Phys. Chem.*, **71**, 2922 (1967).
88. B. J. Levien, *Australian J. Chem.*, **18**, 1161 (1965).
89. G. Gopal and M. A. Siddiqi, *J. Phys. Chem.*, **72**, 1814 (1968).
90. W.-Y. Wen and S. Saito, *J. Phys. Chem.*, **69**, 3569 (1965).
91. T. Shedlovsky and A. S. Brown, *J. Amer. Chem. Soc.*, **56**, 1066 (1934).
92. A. J. Ellis, *J. Chem. Soc. (A)*, 660 (1967).
93. G. Pesce, *Z. Physik. Chem.*, **A160**, 295 (1932).

94. G. Jones and M. Dole, *J. Amer. Chem. Soc.*, **52**, 2245 (1930).

95. G. Jones and R. E. Stauffer, *J. Amer. Chem. Soc.*, **62**, 335 (1940).

96. A. F. Scott and R. W. Wilson, *J. Phys. Chem.*, **38**, 951 (1934).

97. J. R. Maurey and J. Wolff, *J. Inorg. Nucl. Chem.*, **25**, 312 (1963).

98. F. J. Millero, *J. Chem. Eng. Data*, **15**, 562 (1970).

99. F. J. Millero, *J. Chem. Eng. Data*, **16**, 229 (1971).

100. A. K. Helmy, F. F. Assaad, M. N. Hassan, and H. Sadek, *J. Phys. Chem.*, **71**, 2358 (1968).

101. I. Lee and J. B. Hyne, *Can. J. Chem.*, **46**, 2333 (1968).

102. W. R. Gilkerson and J. L. Stewart, *J. Phys. Chem.*, **65**, 1465 (1961).

103. (a) E. M. Baker, *J. Amer. Chem. Soc.*, **71**, 3336 (1949); (b) recalculated by Redlich and Meyer, Ref. 12.

104. O. Redlich, *J. Phys. Chem.*, **44**, 619 (1940).

105. O. N. Bonner and R. W. Gable, *J. Chem. Eng. Data*, **15**, 499 (1970).

106. S. D. Hamann and S. C. Lim, *Australian J. Chem.*, **7**, 329 (1954).

107. W. W. Ewing and C. H. Herty, III, *J. Phys. Chem.*, **57**, 245 (1953).

108. F. D. Leipziger and J. E. Roberts, *J. Phys. Chem.*, **62**, 1014 (1958).

109. J. E. Roberts and N. W. Silcox, *J. Amer. Chem. Soc.*, **79**, 1789 (1957).

110. G. Braghetti and A. Indelli, *Ann. Chim. (Rome)*, **59**, 418 (1969).

111. O. Redlich and P. Rosenfeld, *Z. Elektrochem.*, **37**, 705 (1931).

112. (a) W. W. Ewing and R. J. Mikovsky, *J. Amer. Chem. Soc.*, **72**, 1930 (1950); (b) recalculated by Redlich, see Ref. 107.

113. S. Lee, Ph.D. thesis, Yale University, 1966; University Microfilms, Ord. No. 66-4906, Dist. Abst., **B27**, 131 (1966).

114. R. E. Verrall and B. E. Conway, *J. Phys. Chem.*, **70**, 3961 (1966).

I4   Residence Times of Ionic Hydration

*O. Ya. Samoilov, The N. S. Kurnakov Institute of General and
Inorganic Chemistry, The Academy of Science of the U.S.S.R.*

The hydration of ions in aqueous solutions consists of the interaction between the ions of the dissolved substance and the liquid water; the term "solvation" is used when the solvent is any given substance (and not merely water). This interaction has great significance for the properties of the solution and determines its very existence. We have in mind, in this connection, interactions between ions and molecules of water, as well as the alteration of the structure of water in the solution. It should be emphasized that these changes occur in the context of the constant thermal movement of the particles of the solution.

The significance of the structure of water for the properties of aqueous solutions became particularly clear after the appearance of the work of Bernal and Fowler.[1] The works of Frank and his associates,[2-4] in which the effects of the solute on the structure of water are examined, played an especially important role. It is now entirely clear that the problems associated with interactions in solution, the structure of the solution, and the thermal movement of its particles are related so intimately that they constitute a single problem. As a matter of fact, the change in the position of a particle of liquid in space depends on the topography of the potential, and this topography is determined by interaction among particles and their mean positions; the hydration of ions plays an important part here. The structure of the liquid, at the same time, depends on the thermal movement of the particles that occurs, for example, in the temperature change of the coordination number and in the change in this number upon melting.[5] Conway and Bockris have published a survey on the problem of the thermodynamics of solvation.[6] These problems have also been examined in the recent monograph by Mishchenko and Poltoratskii.[7]

According to Bockris,[8] the hydration of ions may be divided into two regions: close hydration associated with the interaction between an ion and the nearest molecule of water, and distant hydration, which involves interaction between an ion and a molecule of water beyond the immediate vicinity of the former.

Distant hydration consists, in the main, of polarization in response to the field of the ions that surround the component volumes of water. The change of the structure of the water is the consequence of this process. According to very convincing models proposed by Mikhailov and Syrnikov,[9] a displacement of the structural equilibrium in the solution water in the direction of molecules with interstitial structures occurs in response to the action of electric fields in the solution. Vdovenko, Guryikov, and Legin[10] have pointed out that the proportion of disordered structures of water is greater in solutions of electrolytes than in pure water. This conclusion agrees with the existence of zonal structure in the model proposed by Frank and Wen.[3]

The specific features of the hydration of this or that ion are principally associated with close hydration. Imagine that the molecule of water nearest an ion is removed from the solution and that the ions dispose themselves in the center of the remaining spaces (actually, of course, such a situation cannot arise). The hydration, then, would depend only upon the charge of the ions and would be independent of its radius or the structure of the electron shell. The specific features of the hydration of the ion would then disappear. The hydration of a series of alkali metals, for example, would remain unchanged. The specific features of hydration, therefore, are produced by the interactions between an ion and the nearest molecule of water. Interactions of this specific type at comparatively large distances probably occur in the solution by the transfer from molecule to adjacent molecule (the thermal movement).

1   The Molecular-Kinetic Close Hydration Configuration

The close hydration of ions in aqueous solutions is frequently represented as the strong binding of nearby molecules. Therefore this type of hydration is characterized by the number of molecules of water bound—the so-called hydration number of the ion. It should be noted that the hydration numbers evaluated for various ions with various properties usually exhibit considerable differences. The error of this view has been indicated more than once.[5,11,12] The concept of a more or less strong bond among molecules of water and ions is justified only in the case of strongly hydrated ions and cannot under any circumstances be treated as a general law of local hydration.

The concept of hydration of ions in solutions according to which the ions are bound to the nearest molecules of water and the liquid water (the solvent) serves merely as a "provider" for these molecules, may be a consequence of the comparison of the total energy of the interaction among the particles of aqueous solutions of electrolytes—that is, the energy necessary to remove the interacting particles to infinite distances. The total interaction energy of a molecule of water in liquid water, as a matter of fact, is approximately 4.5 or 5.0 kcal/mole, whereas the interaction energy of a molecule of water and an ion is tens and even hundreds of kcal/mole greater. It is not necessary, at the same time, to utilize the total energy of interactions in order to separate an ion and a molecule of water.

The distance between particles that had been in contact with each other in a liquid solution does not increase to infinity instantaneously. This increase occurs gradually, to small distances at first. We should assume that the exchange between two adjacent particles becomes possible when the distance between them is of the order of a diameter of a particle. Exchange between adjacent particles must involve energies sufficient only for such an increase in distance. Therefore the exchange between the closest molecules of water in an aqueous solution is not determined by the total interaction energy of the particles, but only by the change in energy at small distances—"local gradients" of interaction energy. A priori we cannot assert that strong bonds between adjacent molecules of water are encountered in all cases, and one cannot investigate close hydration from the standpoint of the concept of the binding of molecules.

It is necessary to use a more general approach to the study of the hydration of ions in aqueous solution. The determination of the binding of a given number of molecules of water must flow from the method chosen only as a result. Such a general approach may be based on a proposed examination of the action of ions on the thermal, and, in the first instance, the translational motion of the closest molecules of water in the aqueous solution. Sufficiently strong bonds among some number of molecules of water would attenuate this movement considerably. Looking at the substitution of other molecules of the aqueous solute for the molecules closest to the ions, we see that close hydration is rather strong if the exchange is not frequent, and it attenuates to the extent that the frequency of exchange increases.

Let t_i (sec) represent the mean time during which a molecule of water that has moved into the immediate neighborhood of an ion occupies its position in the neighborhood. The time, t_i, depends upon the energy, E_i, required by a molecule of water to escape from the immediate neighborhood of the ion. Let the energy in a "free" water of aqueous solution (we have in mind water that does not enter the immediate neighborhood of the surrounding ions) required for a molecule of water to replace its nearest neighbor—a molecule of water—be designated as E. This quantity may be treated as the energy required for the activation of an selfdiffusion molecule in water.[5,13] Let the corresponding time be t sec. The quantity E_i will differ from E in the general case; therefore

$$E_i = E + \Delta E_i$$

The quantities t_i/t and ΔE_i are entirely general properties of the close hydration of the ions in solution. The increase in the values of these quantities is associated with the corresponding increase in the energy of close hydration. Approximately, we can introduce the parameters of the relation

$$(1) \qquad\qquad\qquad \frac{t_i}{t} = e^{\Delta E_i/RT}$$

Equation 1 is approximate largely because the frequency of oscillation of a molecule of water as a whole, near an ion is considered the same as in pure water. This assumption is valid only for small values of ΔE_i. The question of frequencies of oscillation of molecules of water in aqueous solutions, and the values of ΔE_i, have been examined elsewhere.[14]

We developed a method of evaluating ΔE_i for various ions.[5,13] This method is based on experimental data on the temperature dependence of the mobility of ions in solutions and the selfdiffusion coefficient in water. The motion of an individual ion in water and the motions of an ion and the adjacent molecule are taken into consideration. The values of ΔE_i were measured for a group of ions in aqueous solution. These values depend on the properties of the ion—its radius, its charge, and the structure of the electron shell. A number of alkali metal cations were converted from positive to negative ΔE_i as the size of their radii increases. The parameters of ΔE_i were also negative for the halide ions (except F^-). It turned out, then, that two cases are possible:

1. $\Delta E_i > 0$ and $t_i/t > 1$ (e.g., Mg^{2+}, Li^+, Na^+).
2. $\Delta E_i < 0$ and $t_i/t < 1$ (e.g., K^+, Cs^+, Cl^-).

The exchange of molecules nearest the ions occurs less frequently in the first case than the exchange of nearest molecules in water. Therefore it is possible, in this case, to speak of some effective binding of the ions nearest the molecules of aqueous solutions. The first case may be termed positive hydration. In the second case, the exchange of molecules of water nearest the ions occurs more frequently than the exchange of the nearest molecules in water. This phenomenon was termed negative hydration. The terms positive and negative hydration are associated with the sign of ΔE_i. Great interest attaches to the isolated ion case, for which $\Delta E_i = 0$ and $t_i/t = 1$. This case separates positive and negative hydration. The method of evaluating ΔE_i is very approximate. It becomes more precise as ΔE_i decreases, however, the method apparently gives comparatively precise results when $\Delta E_i = 0$; and thus the boundary between positive and negative hydration can be identified in a satisfactory manner.

The approach we have proposed for research on close hydration possesses sufficient generality. It was developed for the special purpose of use in the

Table 1   Values of ΔE_i and t_i/t at 21.5°C

Ion	ΔE_i (kcal/mole)	$\dfrac{t_i}{t}$	r_i (Å)
Li$^+$	0.56	2.60	0.68
Na$^+$	0.14	1.27	0.98
K$^+$	−0.36	0.54	1.33
Cs$^+$	−0.31	0.59	1.65
Cl$^-$	−0.21	0.70	1.81
Br$^-$	−0.89	0.51	1.96
I$^-$	−0.24	0.66	2.20

study of comparatively weak hydration (the alkali cations, the halogen anions). It can, however, be utilized to describe the unbroken transition from comparatively weak (and even negative) hydration to strong hydration, which makes for the appearance of strong aqueous groups in the solutions. The transition to strong hydration is indicated by the ratio t_i/t (in the limiting case $t_i/t = \infty$). Table 1 gives values of ΔE_i for several single-charged ions, measured by the most efficient experimental temperature values of mobility of ions in solution,[13] as well as the corresponding values of t_i/t. The crystal-chemical[15] radii of ions, r_i, are also given. The table shows that t_i and t are of the same order in the alkali metal cations and halogen anions. The ratio that is closest to unity is t_i/t for Na$^+$.

The values of ΔE_i in alkali metal cations are plotted against r_i in Figure 1. The values of ΔE_i are more precise than those published earlier.[5] The more precise value has no effect on the value of the ionic radius, the boundary between positive and negative hydration. The transition is observed at $r_i \approx 1.1$ Å, as previously noted. One might have expected ΔE_i to increase as r_i decreases. This, indeed, is observed in the case of the alkali metal series,

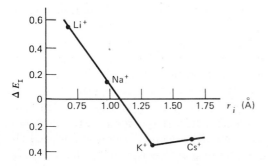

Figure 1.   Values of ΔE_i plotted as a function of alkali metal cation radius.

Li^+, Na^+, K^+. Nevertheless, the value of ΔE_i increases somewhat in going from K^+ to Cs^+. The observed increase appears to be connected with the polarizing propensity of the ion. The disarray of the molecules of water as the distance from the ion increases probably plays a part here.[16] The same may be said of the change in the value of ΔE_i in an entire array of halogen ions (the transition from Br^- to I^-).

Syrnikov's results[17] are of great interest in this connection; that investigator concluded, on the basis of ultrasonic data, that local ions do not enter into the "free" water and that the alkali metal cation may be arranged in the series $K^+ > Cs^+ > Li^+ > Na^+$.

Positive hydration alone is not enough to cause a given ion to move with its hydration shell. Strong hydration (large positive values of ΔE_i) is necessary in order to bring about such movement. Only the Mg^{2+} and La^{3+} ions of those considered earlier[13] move together with their hydration shells. The other cations (Li^+ among them) move through the solution, as a rule, without hydration shells.

The solitary ion for which ΔE_i is 0 has almost no effect on the thermal motion of the nearest molecule of water, which remains the same as in pure water. The presence of a single ion ($\Delta E_i = 0$) is not bound specifically with the structure of water in the event that the specific effect of the ion on the structure results mainly from the thermal motion of the nearest molecule. The close hydration of the single ion may be called nonspecific. Let us note here the result obtained by Zagorets and Bulgakova.[18] These investigators arrived at the conclusion that the minimum change of the structure of water must be observed in the case of ions between 1.1 and 1.15 Å, on the basis of the investigation of the ultraviolet spectrum of the absorption of Cu^{2+} and Fe^{3+} in aqueous solutions to which the alkali metal salts of hypochloric acid have been added.

As we examine the question of various kinds of interactions in the exchange of the molecules of water nearest an ion, we must bear in mind that this value contributes to the exchange with an energy required by the exchange. This contribution, generally speaking, is independent of the total interaction energy (the total potentials are of the molecules) and depends only on the local interaction energy gradient. Exchange often occurs with considerable frequency, however. Therefore one can assert that the exchange of a molecule of water nearest an ion determines the condition of the ion in the solution. It is precisely for this reason that interaction between particles situated considerable distances apart are less important than weaker interactions that attenuate rapidly with distance (short-range forces are important for many properties of aqueous solutions). Peter Debye[19] emphasized the great importance of short-range forces in the properties of electrolytic solutions in his paper.

The distribution of molecules in water is controlled by short-range forces, and short-range forces are of great significance for the properties of electrolytic solutions of electrolytes that have nearly the same structure as water. We begin to comprehend the results of investigations of dilute aqueous electrolytic solutions by thermal methods, for these results permit us to conclude that the condition of an array of ions in a dilute aqueous solution corresponds to the smallest possible change of water structure in the formation of the solution.[5] This conclusion has been confirmed by X-ray studies of solutions.[20]

Lyashchenko[21] recently utilized the data on the density of aqueous solutions of electrolytes as the basis for the conclusion that the inference about the smallest possible change structure of water in the development of a solution is applicable at high concentrations (in many cases up to 6 or 7 mole%). This led him to the conclusion that the structure of water is the matrix in which the entire spontaneous action in aqueous solutions of electrolytes unfolds. In the case of strong hydration, where t_i/t has large values, it appears that entire hydration complexes [e.g., $Mg(H_2O)_6^{2+}$] are altered, and not merely individual ions. It is precisely the major role of the structure of water, and therefore its destruction by ions, that makes the development of negative hydration especially feasible.

The mentioned results (high frequency of water molecules adjacent to ions, the existence of negative hydration, the boundary between positive and negative hydration) is confirmed by a rather large array of experimental studies of aqueous solutions by various methods. We have, at the same time, a confirmation of the basic idea of the entire approach to the study of the close hydration of ions in solutions; this reduces to the conditions of exchange of close particles and, therefore, to the definite role of short-range processes among the particles of a solution. Let us examine these investigations.

The research into the change in the entropy of the water, carried out by Krestov[22-24] is of great interest. This investigator essentially determined the difference between the entropy of a molecule of water adjacent to an ion and the entropy of a molecule of pure water (he determined the sign of this difference, in any event). It turned out that the entropy of water in the nearest adjacent array of ions is less than the entropy of pure water. At the same time, the entropy of water adjacent to some singly charged ion is greater than in the case of pure water. (This case clearly corresponds to negative hydration). The values ΔS_{II} (Figure 2) are plotted along the ordinate, and the signs of these quantities are the opposite of the sign of the difference between the entropy of a molecule of water in pure water and in the nearest surrounding of the ions. The crystal radii of the ions are plotted along the abscissa. The boundary between positive and negative hydration for alkali

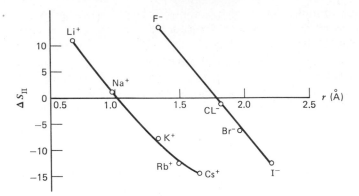

Figure 2. Values of ΔS_{II} plotted as a function of ionic radius for some singly charged ions.

metal cations, according to Krestov, is in good agreement with the data presented previously: Li^+ and Na^+ are hydrated positively; the K^+ and Cs^+ cations hydrate negatively.

Radiospectroscopic methods have yielded valuable results with respect to the hydration of ions in solution. Here one should first recall Valiyev's study, which developed the theory of magnetic nuclear relaxation in solutions of electrolytes,[25,26] and Shcherbakov's work.[27] Valiyev established that I^- and Br^- hydrate negatively. Mazitov[28] investigated temperature as a controlling parameter of the nuclear magnetic relaxation time of protons and deuterons in aqueous solutions of Mn^{2+} and established the occurrence of rapid change from the nearest molecule of water to this cation. It is interesting that the energy of activation was nearly that computed earlier[29] on the assumption that the boundary between positive and negative hydration represents a singly-charged ion with radius equal to 1.1 Å. In their study of the relaxation of deuterons in aqueous solutions, Ionov and Mazitov established that the angular motion of a molecule of water is retarded in response to adjacent Al^{3+}, Mg^{2+}, Li^+, and Na^+ ions but is accelerated in response to the action of K^+ and Cs^+ ions.[30] This demonstrates the occurrence of positive and negative hydration.

The results of recent studies of nuclear magnetic relaxation of Na^+ in aqueous solution[31-33] are extremely interesting. It turns out that t_{Na^+} increases in response to the action of other ions present in the solution. This result is hard to reconcile with the idea of a strong bond between the Na^+ cation and the nearest molecule of water. It is important to note that, in the event of strong hydration (VO^{2+}) t_i decreases[34,35] in response to the action of ions present in the solution. We emphasize that the change in the spin-lattice relaxation time, T_1, measured by nuclear-magnetic-spin relaxation methods, is connected with the processes occurring in the nearest neighborhood of an ion.[31] The nuclear magnetic relaxation method is unique in this sense.

The work of Karyakin, Petrov, Gerlit, and Zubrilina[36] in investigating the action of ions in aqueous solution at supersonic frequency (7300–5000 cm^{-1}) is very interesting inasmuch as it involves the infrared absorption spectrum of water. It is shown that the ions are divided into two groups. One group increases the absorption in the frequency range less than the maximum absorption frequency for pure water. The action of the ions of this group is such that they can be arranged in the series $Al^{3+} > Cr^{3+}$, $Be^{2+} > Cd^{2+} > Zn^{2+} > Mg^{2+}$, $Li^+ > Na^+$ (the effect of the action of the Na^+ ion is almost indiscernible), and $CO_3^{2-} > SO_4^{2-}$, $OH^- > F^-$. The ions of the other group decrease the absorption in the low-frequency range and increase somewhat among the high maxima. The ions in this group may be arranged in the series $Cs^+ > K^+$ and $ReO_4^- > ClO_4^- > I^- > NO_3^- > SNC^- > Cl^-$. The ions of the first group strengthen the bond between the molecules of the solution water and the nearest ions; the ions of the second group attenuate these bonds. The validity of the division of the ions into the two groups—the first including Al^{3+}, Mg^{2+}, Li^+, and Na^+ and the second including K^+ and Cs^+—also follows from Yastremskii's results[37,38] (the dielectric permeability and dielectric losses in aqueous solutions at frequencies of the order of 9400 MHz).

The results of the investigation of the temperature dependence of the coordination numbers of ions in dilute aqueous solutions are closely associated with the concept of close hydration in aqueous solution. It has been shown[39] that

$$\frac{\partial}{\partial T}\left(\frac{n_i}{n}\right) < 0 \quad \text{if} \quad \Delta E_i > 0$$

$$\frac{\partial}{\partial T}\left(\frac{n_i}{n}\right) > 0 \quad \text{if} \quad \Delta E_i < 0$$

where n_i is the coordination number of the ion in the solution, and n is the coordination number of a molecule of water. The temperature dependence of n is known from X-ray data.[5] Experimental studies of the temperature dependence of the coordination number, n_i, for the alkali metals and halogen anions diluted in aqueous solutions has been found, by means of thermochemical measurement, to be Li^+, $\partial/\partial T(n_i/n) < 0$ and $\partial/\partial T(n_i/n) > 0$ in the case of K^+, Rb^+, and Cs^+. The variable n_i behaves in the same manner as n, so that $\partial/\partial T(n_i/n) \approx 0$ in the case of Cl^- and Na^+. One should note, however, that the behavior of the coordination number of the K^+ ion must be studied independently.

Even the data of aqueous solution densities[41] can serve to establish that the close hydration of the Na^+ and K^+ cations differ in sign (the hydration of the Na^+ ion is positive, whereas that of the K^+ ion is negative). Here we have an especially intimate connection between the hydration of ions and the structure of solutions. The difference between the characters of the hydration of the Na^+ and K^+ cations may be readily seen from the pressure dependence of the close hydration of these cations.[42]

The results of magnetochemical investigations of the hydration of ions carried out recently by Yergin and Kostrova[43] are of particular interest. These investigators measured (at room temperature) the magnetic susceptibility of dilute aqueous solutions of the entire group of alkali earth metals and alkali metal halides, as well as the magnetic susceptibility of the salts in the crystalline state. The polarization paramagnetism in the molecules of water nearest the ions was measured and compared with the polarization paramagnetism of pure water—the quantity $\Delta\chi_p^{H_2O}$ was found. It turned out that $\Delta\chi_p^{H_2O}$ can be positive (the earth alkali metal ions, Li^+ and Na^+) or negative (K^+, Rb^+, Cs^+, Cl^-, Br^-, I^-). The first group of cations, as Yergin and Kostrova indicate, undergoes positive hydration, whereas the second undergoes negative hydration. A transition from positive to negative values of $\Delta\chi_p^{H_2O}$ according to their results[43] is observed when the radius of the ion is approximately 1.1 Å.

Thus the experimentally observed data now receive strong confirmation from the existence of negative hydration and from the circumstance that the boundary between positive and negative hydration for alkali metal cations at infinite dilution and temperatures between 20 and 25°C at 1 atm lies between the values for Na^+ and K^+. The fact that the increase in the mobility occurs in the molecules of water nearest the ions is in close agreement with the theoretical results obtained by Valiyev.[25,26] Krestov drew this conclusion, and it also follows from magnetochemical studies of hydration,[43] in particular, from results of nuclear resonance research.[33,34]

The concept of positive and negative hydration is entirely in agreement with Frank's[4] division of the ions that act upon the structure of a solution, into two groups—structure makers and structure breakers. These points of view coincide, according to Engel and Hertz.[44] It should be pointed out, however, that the division of ions into negative- and positive-hydrating complexes presupposes differences in action upon water (the positive and negative values of ΔE_i) and the action of ions on the thermal movement of the *nearest* molecules of water.

Let us note that negative hydration is observed when the contribution of the energy of interaction between an ion and the nearest molecule of water to the threshold energy of translational motion of the molecule is less than the contribution of the interaction of the molecule in water. The presence of "free" water in the solution is not a necessary condition for the initiation of negative hydration. The view of hydration developed here applies only in the realm of dilute solutions which is connected to our limitation of the investigation to an examination of the interaction of a single ion with an adjacent molecule of water. The extension of the theory into the realm of concentrated solutions requires the consideration of other ions. These problems now are being investigated in the works on salting out of electrolytes from aqueous solutions.[45]

Negative solvation is observed only in cases in which both the interaction between the molecules of the solvent and the local ordering are strong (it is worth noting that molecules of water in water generally involve four hydrogen bonds). Krestov established that negative solvation in such liquids as methanol and ethanol is not observed at 25°C. Negative solvation makes its appearance to the extent that the hydrogen bonds in the molecular medium are complex and the solvent becomes more strongly structured (the transition from monoatomic to polyatomic alcohols).[5] The results reported by Nikiforov[47] show that negative solvation may be observed in glycerine solutions of potassium iodide.

2   The Role of the Structural Condition of Solvents

The increase of t_i—the mean time that a molecule of water remains in position adjacent to an ion—indicates a strengthening of close hydration. According to (1), on the other hand, an increase in t involves a weakening of close hydration (ΔE_i decreases in this case). The increase in t corresponds to an increase in the structure of the water. A decrease in t, however, corresponds to a weakening of the structure of the water and, according to (1), leads to a strengthening of close hydration. Therefore we can assert that the strengthening of the structure of the water results in an attenuation and breaking of the aqueous structure—in other words, there is a strengthening of close hydration.

In the event that t_i and t vary in the same direction simultaneously, the change in the close hydration of an ion will depend upon which of these quantities varies more rapidly. In the event that the temperature, T, increases, the values of t_i and t will decrease, but t will, without doubt, decrease faster than t_i. This is a consequence of the strength of the water–water bond in comparison with the ion–water bond, which, in turn, involves a more rapid destruction, since T increases in locally ordered water more rapidly than the molecules of water that surround an ion. Therefore, the close hydration of ions must increase with the increase in temperature. This increase, in turn, is specified by the inequality

$$\left|\frac{\partial t_i}{\partial T}\right| < \frac{t_i}{t}\left|\frac{\partial t}{\partial T}\right|$$

The theoretical conception of the dependence of close hydration upon the structure of water may be compared readily with the results of experimental studies of solutions.

The development of bonds between molecules of water entails the attenuation of the close hydration of ions. The work of Karyakin, Petrov, Gerlit, and Zubrilina,[36] constitutes an interesting illustration of this. The study in question was concerned with the influence of local molecules of water on the infrared absorption spectra of the oxidized anions ClO_4^- and SO_4^{2-} in crystalline hydrates sodium perchlorate and sulfate, as well as the aqueous solutions of these salts. The change in the chlorine–oxygen and sulfur–oxygen lines in the crystalline hydrates due to the influence of the water of crystallization was considerable. Thus the absorption maximum is observed at 1115 cm^{-1} in anhydrous sodium perchlorate, but it is shifted in the direction of low frequencies and is observed at 1085 cm^{-1} in the crystalline hydrate. This shift indicates the development of hydrogen bonds between the water molecules and the oxyanion ClO_4^-. The bands vanish when the solution becomes aqueous, as the absorption maximum is observed at virtually the same frequency as that of anhydrous $NaClO_4$. A similar result is observed in the case of the sulfates. Thus the influence of water on absorption by chlorine–oxygen and sulfur–oxygen becomes much weaker upon the conversion of the crystalline hydrate to an aqueous solution. The investigators have interpreted this interesting fact correctly. There is no doubt that we have here an attenuation of the hydration on anions as a result of the formation of the water–water bonds. The inference about the attenuation of the close hydration because of the strengthening of the water–water bonds has as its origins consideration of the stabilization of the structure of the water by the molecules of methyl alcohol.[48]

The attenuation of close solvation due to the strengthening of bonds between molecules of solvent is seen clearly upon comparison of the behavior of a given salt in H_2O and in D_2O. It is now accepted without qualification that the deuterium bond is stronger than the hydrogen bond.[49] The dipole moments of H_2O and D_2O, however, are almost equal. However, it has been demonstrated at the same time, particularly by nuclear magnetic resonance methods, that an attenuation of close cation solvation may be observed during the transition from H_2O to D_2O.[28,50] Generally the deuterium structure is more regular because D_2O is more endothermic than H_2O. The same endothermicity is associated with the change from H_2O to D_2O.[51] The results obtained by Greyson[52] are also involved here.

The breaking of water–water bonds leads to a strengthening of hydration in solution. Thus when the structure of water is broken because of an increase in temperature, close hydration of ions is reinforced to a certain extent. This is illustrated, in part, by the results obtained by Zagorets, Yermakov, and Grunau,[53] who measured the lattice-spin relaxation of protons in solutions of LiCl in water, and $CoCl_2$ and $CuCl_2$ in methanol at various concentrations and temperatures in the interval between -30 and $+40°C$. The reinforcement of the close hydration of ions with increasing temperature—that is to say, with the breaking of the structure of water—has been demonstrated by Ionov and Mazitov by measuring the spin of deuterium nucleii in solution.[34] This strengthening has also been observed by Krestov and Abrosimov.[55,56] Quantities that were termed limiting temperatures, t_{lim}, are introduced by these authors.[55] Close hydration of ions which is negative at temperatures lower than t_{lim}, becomes positive at high temperatures. The occurrence of limiting temperatures confirms the association of negative hydration with the structure of water.

The data on the structural states of solvents and the phenomenon of the solvation of ions in solution are not applicable to aqueous solutions alone. Interesting results have been obtained in studies of the solubility of sodium chloride in mixtures of ethyl alcohol and methyl alcohol.[57] Small additions of molecules of a second component affect the structure of a liquid differently, depending on the relations among the sizes of the molecules of the dominant and added components, as well as the character of the resulting mixture. Small additions of methyl alcohol to ethyl alcohol lead to the stabilization of the ethyl structure. Small additions of ethyl alcohol to methyl alcohol result in the breaking of the methyl structure. The molecular mechanism of these structural processes has been examined.

The stabilization of the structure of a solvent should produce a decrease in solvation and, therefore, a decrease in solubility of sodium chloride. The breaking of the structure must have the converse result. This has been demonstrated experimentally and can be seen in Figures 3 and 4. The results

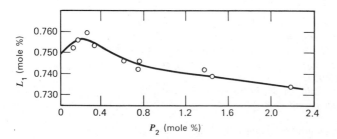

Figure 3.   Solubility (L_1) of sodium chloride in methanol with added ethanol.

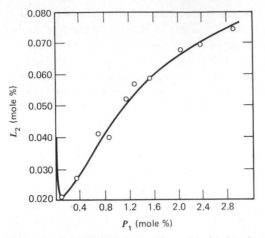

Figure 4.   Solubility (L_2) of sodium chloride in ethanol with added methanol.

obtained are particularly convincing inasmuch as the solubility of sodium chloride in methyl alcohol is greater than it is in ethyl alcohol. The very small additions for which the reversed solubilities are observed point simultaneously to the structural character of the observed effect.

   The relation noted between the close hydration of ions and the structural properties of water constitutes evidence in favor of the approach developed in this chapter and the investigation of the close hydration of ions in aqueous solution based on the residence time of the molecules of water in positions nearest the ion.

REFERENCES

1.   J. D. Bernal and R. H. Fowler, *J. Chem. Phys.*, **1**, 8, 515 (1933).
2.   H. S. Frank and M. W. Evans, *J. Chem. Phys.*, **13**, 2, 507 (1945).
3.   H. S. Frank and W.-Y. Wen, *Discussions Faraday Soc.*, 24, 133 (1957).
4.   H. S. Frank, *Federation Proc.*, **24**, No. 2, Part 3, s-1 (1965).
5.   O. Ya. Samoilov, *Structure of Aqueous Electrolyte Solutions and the Hydration of Ions*, transl. D. J. G. Ives, Consultants Bureau, New York, 1965.
6.   B. E. Conway and J. O'M. Bockris, *Modern Aspects of Electrochemistry*, Butterworths, London, 1954.
7.   K. P. Mishchenko and G. M. Poltoratskii, *Problems in the Thermodynamics and Structure of Aqueous and Anhydrous Solutions of Electrolytes*, "Khimiya" Press, Leningrad, 1968.
8.   J. O'M. Bockris, *Quart. Rev. (London)*, **3**, 173 (1949).
9.   I. G. Mikhailov and Yu. P. Syrnikov, *Zh. Strukt. Khim.*, **1**, 1, 12 (1960).

10. V. M. Vdovenko, Yu. V. Guryikov, and Ye. K. Legin, *Zh. Strukt. Khim.*, **10**, 4, 576 (1969).
11. O. Ya. Samoilov, *Dokl. Akad. Nauk SSSR*, **77**, 633 (1951); *Izv. Akad. Nauk SSSR, Otedel. Khim. Nauk*, 246 (1953).
12. O. Ya. Samoilov, *Discussions Faraday Soc.*, 24, 141 (1957).
13. V. V. Goncharov, I. I. Romanova, O. Ya. Samoilov, and V. I. Yashkichev, *Zh. Strukt. Khim.*, **8**, 4, 613 (1967).
14. O. Ya. Samoilov, *Zh. Strukt. Khim.*, **1**, 1, 36 (1960).
15. G. B. Bokii, *Kristallokhimiya*, Moscow University Press, Moscow, 1960.
16. O. Ya. Samoilov and G. G. Malenkov, *Zh. Strukt. Khim.*, **8**, 4, 618 (1967).
17. Yu. P. Syrnikov, "Szhimayemost Rastvorov Elektrolitov i Nekotorye Voprosy Teorii Ettikh Rastvorov, dissertation, Leningrad, 1958.
18. P. A. Zagorets and G. P. Bulgakova, *Zh. Fiz. Khim.*, **39**, 2, 29 (1965); *Zh. Neorg. Khim.*, **12**, 2, 347 (1967); P. A. Zagorets, "Issledovaniye Solvatasii i Struktury Rastvorov Elektrolitov," dissertation, Moscow, 1969.
19. P. Debye, *Electrolytes*, Pergamon Press, Oxford University Press, London, New York, Paris, 1962.
20. N. M. Shapovalov, I. V. Radchenko, and M. K. Lesovitskaya, *Zh. Strukt. Khim.*; I. V. Radchenko and A. I. Ryss, *Zh. Strukt. Khim.*, **6**, 5, 771 (1965); D. S. Terekhova and I. V. Radchenko, *Zh. Strukt. Khim.*, **10**, 6 (1969).
21. A. K. Lyashchenko, *Zh. Strukt. Khim.*, **9**, 5, 781 (1968).
22. G. A. Krestov, *Zh. Strukt. Khim.*, **8**, 2, 137 (1962).
23. G. A. Krestov, *Izv. Vyschlkh Uchebn. Zavedenii: Khimiya i Khimicheskaya Tekhnologiya*, **8**, 5, 734 (1965).
24. G. A. Krestov, *Zh. Strukt. Khim.*, **8**, 4, 402 (1962).
25. K. A. Valiyev, *Zh. Strukt. Khim.*, **8**, 6, 653 (1962).
26. K. A. Valiyev, *Zh. Strukt. Khim.*, **5**, 4, 517 (1964).
27. V. A. Shcherbakov, *Zh. Strukt. Khim.*, **2**, 4, 484 (1961).
28. R. K. Mazitov, *Dokl. Akad. Nauk SSSR*, **152**, 2, 375 (1963).
29. Ye. S. Lekht and O. Ya. Samoilov, *Zh. Strukt. Khim.*, **3**, 4, 466 (1962).
30. V. I. Ionov and R. K. Mazitov, *Zh. Strukt. Khim.*, **7**, 2, 184 (1966).
31. M. Eisenstadt and H. L. Friedman, *J. Chem. Phys.*, **44**, 4, 1407 (1966).
32. M. Eisenstadt and H. L. Friedman, *J. Chem. Phys.*, **46**, 6, 2182 (1967).
33. V. I. Ionov, R. K. Mazitov, and O. Ya. Samoilov, *Zh. Strukt. Khim.*, **10**, 3, 407 (1969).
34. N. K. Ivoilov, V. I. Ionov, and R. K. Mazitov, *Dokl. Akad. Nauk SSSR*, **183**, 4, 863 (1968).
35. A. I. Rivkind and L. P. Kuznetsova, *Dokl. Akad. Nauk SSSR*, **164**, 4, 860 (1965).
36. A. V. Karyakin, A. V. Petrov, Yu. B. Gerlit, and M. Ye. Zubrilina, *Teor. i Eksp. Khim.*, **2**, 4, 494 (1966).
37. P. S. Yastremskii, *Zh. Strukt. Khim.*, **2**, 3, 268 (1961).
38. P. S. Yastremskii, *Zh. Strukt. Khim.*, **3**, 3, 279 (1962).
39. O. Ya. Samoilov, *Dokl. Akad. Nauk SSSR*, **126**, 2, 880 (1959).
40. M. N. Buslayeva and O. Ya. Samoilov, *Zh. Strukt. Khim.*, **2**, 5, 551 (1961).
41. A. F. Borina and O. Ya. Samoilov, *Zh. Strukt. Khim.*, **8**, 5, 817 (1967).
42. N. A. Nevolina, O. Ya. Samoilov, and A. L. Seifer, *Zh. Strukt. Khim.*, **10**, 2, 208 (1969).
43. Yu. V. Yergin and L. I. Kostrova, *Zh. Strukt. Khim.*, **10**, 6, 971 (1969).
44. G. Engel and H. G. Hertz, *Ber. Buns.*, **72**, 7, 808 (1968).

45. O. Ya. Samoilov, *Zh. Strukt. Khim.*, **7**, 1, 15 (1966); O. Ya. Samoilov, M. N. Buslayeva, K. T. Dudnikova, and N. V. Bryushkova, *Zh. Strukt. Khim.*, **10**, 4, 580 (1969); O. Ya. Samoilov, Yu. P. Aleshko-Ozhevskii, M. N. Buslayeva and P. Ojari, *Zh. Fiz. Khim.* **45**, 4, 974 (1971).
46. G. A. Krestov, *Zh. Strukt. Khim.*, **3**, 5, 516 (1962).
47. Ye. A. Nikiforov, *Zh. Strukt. Khim.*, **10**, 1, 137 (1969).
48. P. S. Yastremskii and O. Ya. Samoilov, *Zh. Strukt. Khim.*, **4**, 6, 844 (1963).
49. I. B. Rabinovich, *Vlyanie Izotopii Na Fiziko-Khimicheskiye Svoistva Zhidkostei*, Izd-Vo "Nauka," Moscow, 1968.
50. J. Bigeleisen, *J. Chem. Phys.*, **32**, 5, 1583 (1960).
51. O. Ya. Samoilov and V. G. Tsvetkov, *Zh. Strukt. Khim.*, **9**, 2, 193 (1968).
52. J. Greyson, *J. Phys. Chem.*, **71**, 7, 2210 (1967).
53. P. A. Zagorets, V. I. Yermakov, and A. P. Grunau, *Zh. Fiz. Khim.*, **39**, 7, 1955 (1965).
54. V. I. Ionov and R. K. Mazitov, *Zh. Strukt. Khim.*, **9**, 5, 895 (1968).
55. G. A. Krestov and V. K. Abrosimov, *Zh. Strukt. Khim.*, **5**, 4, 510 (1964).
56. G. A. Krestov and V. K. Abrosimov, *Zh. Strukt. Khim.*, **8**, 5, 822 (1967).
57. O. Ya. Samoilov, I. B. Rabinovich, and K. T. Dudnikova, *Zh. Strukt. Khim.* **6**, 5, 768 (1965).

I5 Aqueous Solutions of Symmetrical Tetraalkylammonium Salts

Wen-Yang Wen, Department of Chemistry,
Clark University, Worcester, Massachusetts

1 Introduction

Symmetrical tetraalkylammonium salts were first synthesized by Hofmann[1] in 1851 by reacting trialkylamines with the corresponding alkyl halides. These salts were found to be soluble in many polar solvents, including alcohols and water. Although there were some early studies on interesting physical properties of these salts in water, notably those of Lange,[2] they apparently failed to attract much attention of physical chemists until 1940, when Kraus and his co-workers[3] reported the formation of a variety of crystalline hydrates (15 altogether) containing an unusually high proportion of water of crystallization. In 1945 Frank and Evans,[4] in their well-known paper, noted the high viscosity of aqueous solutions of these salts reported by Bingham[5] and proposed the pictorial theory of "iceberg" formation around the hydrocarbon groups of these ions in water. In 1957 Frank and Wen[6] found that the aqueous tetrabutylammonium bromide solution gives large excess heat capacities and attributed them to the "melting" of icelike patches formed around the cations.

Since 1959 McMullan and Jeffrey[7] have carried out a detailed X-ray structural analysis of the clathrate hydrates formed by some of the large members of these salts and have made significant contributions to our understanding of these structures. Their studies have stimulated considerable interest, and the investigations of the aqueous solutions of tetraalkylammonium salts are now very actively pursued in various laboratories. Recently Kay and his co-workers[8] have made extensive studies on the transport process of these salts and have tabulated many conductance data of high precision. Other methods employed in the investigations of these salt solutions include various thermodynamic, spectroscopic, nuclear magnetic resonance, ultrasonic absorption, and dielectric relaxation techniques. The results of these studies are discussed in subsequent sections from a structural viewpoint.

613

The salts of our interest form a class of electrolytes whose properties in water are expected to be somewhere between those of inorganic 1:1 electrolytes (such as alkali halides) and those of surface-active quaternary ammonium salts having long alkyl chains. These "pseudo-surfactants" can be useful as model compounds for studies of hydrophobic interactions. As model compounds for such studies, the symmetrical tetraalkylammonium salts have certain advantages and disadvantages. Among the advantages are the simple shape (tetrahedral) of the cations of the salts, the availability of series of cations with different alkyl chain lengths, and the high solubilities of most of the salts (except certain iodides and perchlorates) in water. One disadvantage for using these salts as model compounds for the study of "hydrophobic bonding" is the presence of counter-ions (anions in this case), which will undoubtedly exert a strong influence on the structure of water. On the other hand, from a different viewpoint, tetraalkylammonium salt series will be useful for the study of anionic properties in water.

Some unsymmetrical quaternary ammonium salts can also be used profitably as model compounds for studies of ion-exchange resins, as demonstrated by the works of Boyd and Lindenbaum and their co-workers.[9] In this chapter the discussion is restricted to the symmetrical tetraalkylammonium salts, that is, tetramethyl-, tetraethyl-, tetrapropyl-, tetrabutyl-, and tetrapentylammonium salts.

2 Clathrate Hydrates

Among many quaternary ammonium salts only tetra-n-butyl- and tetra-i-amyl ammonium salts form crystalline hydrates with high proportion of water of crystallization. These solid hydrates were first prepared by Kraus et al.[3] in 1940, but the hydration numbers (number of water of crystallization) per cation of some of the hydrates reported were later found to be incorrect. The acetate hydrate reported as $(n\text{-}C_4H_9)_4N \cdot CH_3CO_2 \cdot 60H_2O$ turns out to be $(n\text{-}C_4H_9)_4N \cdot CH_3CO_2 \cdot 32H_2O$. The fluoride hydrate reported as $(n\text{-}C_4H_9)_4N \cdot F \cdot 18H_2O$ is actually[7] $(n\text{-}C_4H_9)_4N \cdot F \cdot 32.8H_2O$.

Since the hydration number in the tetra-n-butylammonium series is 32 to 33 (32.8 for an idealized crystal), the stoichiometric formula for these hydrates can be represented by[7]

$$[(n\text{-}C_4H_9)_4N^+]_m \cdot X^{m-} \cdot m \cdot 32.8H_2O \qquad (m = 1 \text{ or } 2)$$

where X = F, Cl, Br, HCO$_3$, CH$_3$COO, C$_2$O$_4$, CrO$_4$, or WO$_4$. They form an isomorphous crystal series which is tetragonal with $a = 23.65 \pm 0.15$ Å and $C = 12.40 \pm 0.15$ Å.

On the other hand, the hydration number in the tetra-i-amylammonium series is 38, and the stoichiometric formula for these hydrates is given by[7]

$$[(i\text{-}C_5H_{11})_4N^+]_m \cdot X^{m-} \cdot m \cdot 38H_2O \qquad m = 1 \text{ or } 2$$

where $X = F$, Cl, CrO_4, or WO_4. (There are hydrates with hydration numbers other than those given here; for details see Beurskens et al.[14]) The hydrates form an orthorhombic isomorphous group with $a = 12.10 \pm 0.10$ Å, $b = 21.50 \pm 0.15$ Å, $c = 12.65 \pm 0.15$ Å. Melting points and densities of these crystals are given in Table 1.

Table 1 Melting Points and Densities of Clathrate Hydrates

Salt	Melting Point ($\pm 0.1°C$)	Density at 5°C (g/cm³), (± 0.005)
$(C_4H_9)_4NF$	24.9	1.032
$(C_4H_9)_4NCl$	15.7	1.026
$(C_4H_9)_4NBr$	12.5	1.080
$(C_4H_9)_4NCH_3COO$	15.1	1.040
$[(C_4H_9)_4N]_2CrO_4$	13.6	1.059
$[(C_4H_9)_4N]_2WO_4$	15.1	1.131
$[(C_4H_9)_4N]_2C_2O_4$	16.8	1.043
$(C_4H_9)_4NHCO_3$	17.8	1.038
$[(C_4H_9)_4N]_2HPO_4$	17.2	1.059
$(i\text{-}C_5H_{11})_4NF$	31.2	1.021
$(i\text{-}C_5H_{11})_4NCl$	29.8	1.021
$[(i\text{-}C_5H_{11})_4N]_2CrO_4$	21.6	1.047
$[(i\text{-}C_5H_{11})_4N]_2WO_4$	22.4	1.109

From Ref. 7.

Crystal structures of some of these hydrates have been analyzed in detail with X-ray by Jeffrey and his co-workers.[10–12] The hydrates studies included $(i\text{-}C_5H_{11})_4N^+F^- \cdot 38H_2O$,[10] $(n\text{-}C_4H_9)_4N^+C_6H_5COO^- \cdot 39H_2O$,[11] and $(n\text{-}C_4H_9)_4N^+F^- \cdot 32.8H_2O$;[12] all were found to be the clathrate type, similar to the gas hydrates.[13] A clathrate is a molecular compound formed by enclosing one or more species of molecules ("guest") in a cage structure formed by another species ("host"). The "host" structure is a polyhedral framework of hydrogen-bonded water molecules that form "cages" and enclose the "guest" molecules or ions. The basic polyhedron in these structures is the pentagonal dodecahedron (20 vertices and 12 faces) formed by the association through hydrogen bonds of 20 water molecules, that is,

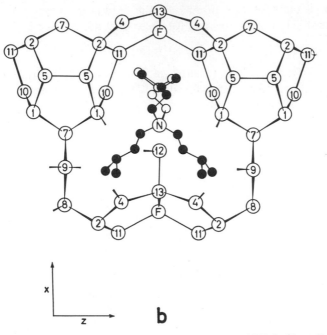

Figure 1.   Crystal structure of $(i\text{-}C_5H_{11})_4N^+F^- \cdot 38H_2O$. From Ref. 10.

$H_{40}O_{20}$. Larger polyhedra that exist in these host structures include tetra-kaidecahedron (24 vertices and 14 faces) and pentakaidecahedron (26 vertices and 15 faces).

In the tetra-n-butyl- or tetra-i-amylammonium salt hydrates, the central N^+ atom lies at a common vertex of four large polyhedra and each of the four alkyl chains extend into one of the polyhedral cages, as shown in Figure 1 for $(i\text{-}C_5H_{11})_4N^+F^- \cdot 38H_2O$. The clathrate host structure of $(n\text{-}C_4H_9)_4N^+F^- \cdot 32.8H_2O$ is similar to that of $(n\text{-}C_4H_9)_4N^+C_6H_5COO^- \cdot 39.5H_2O$.[11] However, in contrast to the benzoate hydrate crystal, in which there are four molecules per unit cell, the fluoride hydrate has five molecules per unit cell. This difference in the number of guests accommodated in essentially the same host structure is possible because the four tetrakaideca-hedra occupied by the benzoate groups are available for the four alkyl groups of the additional cation. In the fluoride hydrate, four of the cations have the central N^+ atom at the fourfold positions and that of the fifth is disordered over the twofold position. Thus the structure has lower symmetry than the benzoate hydrate by reason of the disorder of the four cations over eightfold positions (see Ref. 12).

In addition to tetraalkylammonium salts, some peralkylphosphonium and sulfonium salts form similar hydrate crystals. The crystal data and preparation of these polyhedral clathrate hydrates have been summarized.[14]

3   Thermodynamic Properties

Solubilities of Tetraalkylammonium Salts. Solubility of a tetraalkylammonium salt in water depends sensitively on the sizes of the cation and anion of that salt. Solubilities are generally quite high for salts made up by combination of the following cations and anions:

cation $(CH_3)_4N^+$, $(C_2H_5)_4N^+$, $(n\text{-}C_3H_7)_4N^+$, $(n\text{-}C_4H_9)_4N^+$, $(i\text{-}C_5H_{11})_4N^+$

anion F^-, Cl^-, Br^-, OH^-, HCO_3^-, CH_3COO^-, $C_6H_5COO^-$, CrO_4^{2-}, WO_4^{2-}, HPO_4^{2-}.

Solubilities are low when cations are large and anions are strong water-structure-breaking ions such as I^- or ClO_4^-; solubilities are so low with hydrophobic anions [e.g., $^-B(C_6H_5)_4$], that an accurate gravimetric analysis based on the precipitate can easily be performed.[15]

For a given anion, say Br^-, the solubility changes in a zigzag fashion with cationic size from $(CH_3)_4N^+$ to $(n\text{-}C_4H_9)_4N^+$ but decreases sharply for $(n\text{-}C_5H_{11})_4N^+$. Solubilities of tetra-*i*-amylammonium salts are probably similar to those of tetra-*n*-butylammonium salts. Solubilities of tetraalkylammonium halides given in Table 2 include approximate values. A systematic study on the solubilities of various tetraalkylammonium salts is needed.

Apparent and Partial Molal Volumes. The volumetric behavior of solutes in solution can provide useful information about solute-solvent and solute-solute interactions. (For details of the experimental methods and theories on partial molal volumes of electrolytes, see Chapter 13.) The apparent and partial molal volumes of tetraalkylammonium salts have been investigated rather extensively in aqueous and nonaqueous solvents since 1964. In aqueous solution the volumes of these salts appear to be a sensitive function of concentration, temperature, and pressure.

First we shall examine the concentration dependence of the apparent molal volumes of electrolytes in water at constant temperature and pressure. For this purpose we plot Φ_2, the apparent molal volumes, against $\sqrt{c}$ where c is the molar concentration of the salt solution. The plots for 1:1 electrolytes such as sodium chloride or potassium bromide show that the value of Φ_2 at very low concentration increases with $\sqrt{c}$ in accordance with the Debye–Hückel limiting slope of 1.86 at 25°C.[29] At concentrations greater than about 0.1 M, these plots may deviate from the limiting slope, but the value of Φ_2

Table 2 Solubilities of Tetraalkylammonium Halides in Water at 25°C

Salts	Solubility (m)
Chlorides	
$(CH_3)_4NCl^a$	19.06
$(C_2H_5)_4NCl^a$	9.47
$(n\text{-}C_3H_7)_4NCl^a$	18.66
$(n\text{-}C_4H_9)_4NCl$	Hygroscopic
Bromides	
$(CH_3)_4NBr^b$	6.353
$(C_2H_5)_4NBr^a$	12.65
$(n\text{-}C_3H_7)_4NBr^a$	8.96
$(n\text{-}C_4H_9)_4NBr^a$	26.39
$(n\text{-}C_5H_{11})_4NBr^c$	0.2
Iodides	
$(CH_3)_4NI^b$	0.2715
$(C_2H_5)_4NI^a$	1.93
$(n\text{-}C_3H_7)_4NI^a$	0.508
$(n\text{-}C_4H_9)_4NI^d$	6.22×10^{-2} (at 26°C)
$(n\text{-}C_5H_{11})_4NI^d$	2.50×10^{-3} (at 26°C)
Fluorides	
$n\text{-}(C_3H_7)_4NF$ $(n\text{-}C_4H_9)_4NF$	Extremely high, salts decompose in absence of water

[a] Approximate values according to S. Lindenbaum and G. E. Boyd, *J. Phys. Chem.*, **68**, 911 (1964).
[b] B. J. Levien, *Australian J. Chem.*, **18**, 1161 (1965).
[c] Approximate value according to W.-Y. Wen and S. Saito, *J. Phys. Chem.*, **68**, 2639 (1964).
[d] F. Franks and D. L. Clarke, *J. Phys. Chem.*, **71**, 1155 (1967).

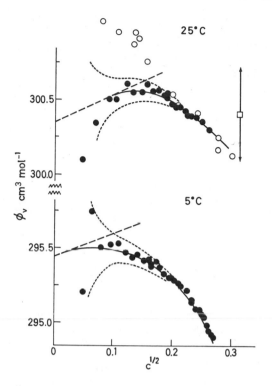

Figure 2. Apparent molal volume of Bu$_4$NBr in aqueous solution plotted against the square root of molality at 5 and 25°C. From Ref. 22.

keeps increasing with the concentration at least up to a concentration of about 4 M.

However, when the same plots are applied to the tetraalkylammonium salts, it has been found[22] that, except for tetramethylammonium salts, the Φ_2 versus $\sqrt{c}$ curve gives negative slope when c is greater than about 0.03. The details of the plots in very dilute range are shown in Figure 2 for the Bu$_4$NBr. At higher concentrations the plot of Φ_2 versus $\sqrt{c}$ for Bu$_4$NBr and Pr$_4$NBr goes through a minimum and then turns upward. The relative apparent molal volumes, $\Phi_2 - \bar{V}_2^\circ$, of four bromides are plotted against $\sqrt{c}$ for an intermediate concentration range in Figure 3.

The partial molal volumes at infinite dilution, $\bar{V}_2^\circ$, for tetraalkylammonium halides have been estimated by several workers,[16–20] but the best values so far seem to be those obtained by Franks and Smith[22] using a magnetic float technique. They are given in Table 3 for the bromides at 5 and 25°C. Since the absolute values of $\bar{V}_-^\circ$ for Br$^-$ has been determined to be 30.2

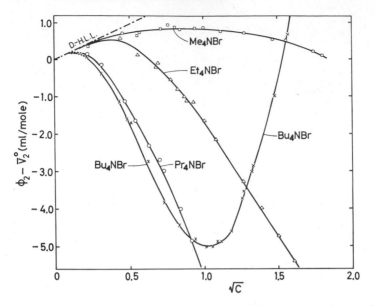

Figure 3. Relative apparent molal volumes of four tetraalkylammonium bromides plotted against $\sqrt{c}$ at 25°C.; D.–H.L.L. curve, Debye–Hückel limiting law. From Ref. 60.

ml/mole at 22°C by Zana and Yeager[30] from ionic vibration potentials, $\bar{V}^{\circ}_{+}$ for R_4N^+ at around 25°C can be estimated. These estimates are given in Table 3.

Table 3   Values of $\bar{V}^{\circ}_{2}$ and $\bar{V}^{\circ}_{+}$ for Tetraalkylammonium Bromides in Water

Temperature (°C)	$\bar{V}^{\circ}_{2}$ (ml/mole)			
	Me_4NBr	Et_4NBr	Pr_4NBr	BU_4NBr
25	114.40	173.65	239.15	300.35
5	112.20	171.15	236.30	295.40
	$\bar{V}^{\circ}_{+}$ (ml/mole)			
	Me_4N^+	Et_4N^+	Pr_4N^+	Bu_4N^+
22–25	84.2	143.5	209.0	270.2

From Refs. 22 and 30.

It has been shown that $\bar{V}_2^{\circ}$ for tetraalkylammonium halides is a linear function of the cation molecular weight,[19] and $\bar{V}_+^{\circ}$ for R_4N^+ is a linear function of the cube of the cationic radius.[31] The partial molal volumes of R_4N^+ in water at infinite dilution may be decomposed to four components

$$\bar{V}_{ion}^{\circ} = \bar{V}_{int}^{\circ} + \bar{V}_{elect}^{\circ} + \bar{V}_{H_2O}^{\circ} + \bar{V}_{caging}^{\circ}$$

where $\bar{V}_{int}^{\circ}$ is the intrinsic volume of the ion; $\bar{V}_{elect}^{\circ}$ the volume change due to electrostriction, $\bar{V}_{H_2O}^{\circ}$ the volume change of water due to its structural changes, and $\bar{V}_{caging}^{\circ}$ the volume change due to "caging effect" or "packing effect."

$\bar{V}_{elect}^{\circ}$ is always negative or nearly 0 for R_4N^+ ions, but $\bar{V}_{H_2O}^{\circ}$ may be either positive or negative depending on whether the bulky structure of water is increased or decreased by the ion. The term $\bar{V}_{caging}^{\circ}$ is negative for large R_4N^+ ions because water molecules tend to form cagelike structures and because the alkyl groups of the cation can be partially fitted inside these cages. In other words, the alkyl chains of R_4N^+ ions are partially hidden in these "holes," resulting in a tighter packing than otherwise would have occurred and causing a volume decrease. The $\bar{V}_{caging}^{\circ}$ term is necessary to explain a considerable increase of $\bar{V}_{ion}^{\circ}$ with the increase of temperature for large R_4N^+ ions. The temperature increase of 20° (see Table 3) would diminish $\bar{V}_{H_2O}^{\circ}$ but can hardly be expected to affect $\bar{V}_{int}^{\circ}$ and $\bar{V}_{elect}^{\circ}$, and therefore $\bar{V}_{caging}^{\circ}$ must increase. With temperature rise, a large negative value of $\bar{V}_{caging}^{\circ}$ at low temperature becomes smaller negative owing to a greater thermal destruction of the cagelike structure of water surrounding R_4N^+ ions at higher temperatures.

The plots of Φ_2 versus $\sqrt{c}$ for $(n\text{-}C_4H_9)_4NX$ where X = F, Cl, Br, and CH_3COO all give a minimum at around 1.6 m,[32] which corresponds to a stoichiometric composition of $(C_4H_9)_4NX \cdot (35 \pm 2)H_2O$. This composition is close to that of the crystalline hydrates $(C_4H_9)_4NX \cdot 32.8H_2O$ investigated by Jeffrey and his co-workers.[7,12] The existence of minima in Φ_2 for these Bu_4NX solutions may be considered as an indication that, among many structures in aqueous solution, there is a clathrate-like component that becomes important at low temperatures. It diminishes with the increase of temperature as observed by the shallowness of minima at higher temperatures.[16] The shape of Φ_2 versus $\sqrt{c}$ curves for Bu_4NX solutions is nearly the same, irrespective of the kind of anions.[32] This seems to indicate the predominance of cation-cation interaction in determining the concentration dependence of the Φ_2 curve. The cation-cation interaction leads to a structural stabilization because of the effect of alkyl groups on the nature of the hydrogen bonding in water. It is probably not hydrophobic bonding in the sense of cation-cation contact, nor micelle formation, since the latter is associated with large positive volume changes.[33]

Next we shall examine more closely the temperature dependence of the partial molal volumes of these ions. Millero[24,34] has determined the partial molal expansibilities, Φ_E°, of R_4N^+ ions in aqueous solution at 25°C. These, and those of halide anions, are given in Table 4. Franks and Smith[22] as well as Millero[24] found that Φ_E° of R_4NX salts increases more or less linearly with the molecular weight of the cation up to Pr_4NX, but the value of Φ_E° for Bu_4NX has been observed to be rather high. In our opinion this observation can again be ascribed to the "caging effect." The concentration dependence of the apparent molal expansibility of aqueous Bu_4NBr shows that Φ_E increases markedly with c, eventually passing through a maximum at the concentration at which Φ_2 undergoes a minimum. Similar but more dramatic change of Φ_E with c has recently been reported for a bolaform electrolyte, $[Bu_3N-(CH_2)_8-NBu_3]Br_2$.[35] These observations also support the concept of cation-cation interaction leading to a structural stabilization that manifests itself as the "caging effect" in the volumetric behavior of large tetraalkylammonium salts in water.

Table 4   Partial Molal Expansibilities of Ions in Aqueous Solution at 25°C

Ion	$r_{Crystal\ Radius}$, (Å)	Φ_E° [ml/(mole)(deg)]
Me_4N^+	3.47	0.033
Et_4N^+	4.00	0.054
Pr_4N^+	4.52	0.095
Bu_4N^+	4.94	0.177
F^-	1.36	0.035
Cl^-	1.81	0.046
Br^-	1.95	0.050
I^-	2.16	0.069

Third, we shall mention briefly on the pressure dependence of the apparent molal volume. Conway and Verrall[36] determined the apparent molal adiabatic compressibilities, Φ_K, of an R_4NX solution by means of differential ultrasonic velocity measurements. Values of Φ_K° obtained are mostly negative, indicating that the salt solutions are relatively incompressible in comparison with pure water. With increasing size of R, Φ_K° becomes more negative, in fact, much more negative than would be expected from the proportionality between Φ_K° and the number of carbon atoms in R_4N^+ ion. For a given cation, the values of Φ_K° increase with increase of the anonic size in the order $Cl^- < Br^- < I^-$.

These observations have been explained[36] qualitatively by assuming that there are three types of water in the salt solution with different compressibilities: (a) unbonded water with high compressibility, (b) icelike water with intermediate compressibility, and (c) electrostricted water with low compressibility. Addition of R_4NX to water is expected to decrease the compressibility by decreasing unbonded water and increasing icelike water and electrostricted water; the effect will be greater, the larger the cation and the smaller the anion. The observations are also consistent with the "caging effect," since a strong caging effect will mean a tight packing of water and ions, leading to a large negative compressibility.

Conway and Laliberté[37] have found that, in very dilute solutions, the tetraalkylammonium salts have larger partial molal volumes in D_2O than H_2O, whereas the smaller, structure-breaking ions have smaller volumes in D_2O.

In contrast to aqueous solution, the concentration dependence of the partial molal volumes of R_4NX in pure alcohols is quite normal;[26,27] It shows, however, some new peculiarities in ethanol–water mixture.[28]

Wen and Saito[38] found that the partial molal volume of the tetrakis(2-hydroxyethyl)ammonium bromide increases with concentration up to high concentrations. This volumetric behavior is similar to that of alkali halides and dissimilar to that of Et_4NBr or Pr_4NBr. The partial molal volume of the $(HOC_2H_4)_4N^+$ ion is only slightly larger than the Et_4N^+ ion in aqueous solutions ($\overline{V}_2 = 151.7$ and 149.6 ml, respectively). The similarity in volumes of these two ions is explained by the high volume of the structural water around the hydrocarbon chains. This structure is destroyed by the terminal hydroxyl group, which fits into the normal hydrogen-bonded structure of the water.

Osmotic and Activity Coefficients. The freezing points of solutions of tetraalkylammonium chlorides and iodides were measured accurately by Ebert and Lange[2] in 1928 and by Lange[2] in 1934. The osmotic coefficients of the dilute solutions at the freezing points showed a behavior characteristic of strong 1:1 electrolytes; the values for the iodides, however, fell below the Debye–Hückel limiting law. The negative deviation from the limiting law has been confirmed for solutions of Et_4NI by Bower and Robinson[39] and Pr_4NI by Lindenbaum and Boyd.[40] Devanathan and Fernando[41] reported extremely high activity coefficients for R_4NI based on measurements of cells with liquid junctions. These high values are unreasonable[42] and are in disagreement with the results of other workers.[2,39,40] In a paper by Frank,[43] the cell measurements of Devanathan and Fernando[41] are reportedly confirmed, but they are interpreted to yield the ratio of iodide single-ion activity coefficients between R_4NI and KI in aqueous solutions.

Lindenbaum and Boyd[40] measured osmotic and activity coefficients of tetramethyl-, tetraethyl-, tetrapropyl-, and tetrabutylammonium chlorides, bromides, and iodides by the isopiestic comparison method. For concentrations below 1 M, the osmotic coefficients for the chlorides are found to increase with the size of the cation, whereas the bromides and iodides showed the reverse order. At higher concentrations, several activity coefficient curves of the chlorides and bromides cross each other in a complicated manner. The systematic studies of Lindenbaum and Boyd have provided much needed data on quaternary ammonium salts. Their interpretation of the low-activity coefficients of large R_4NBr and R_4NI in terms of "water-structure enforced ion pairing"[44] and micelle formation is highly questionable, however, particularly in view of later developments.[45,46] Levien[18] has determined the activity coefficients of Me_4NBr and Me_4NI in aqueous solution up to saturation. Her data seem to be more accurate than those of Lindenbaum and Boyd.[40]

Wen, Saito, and Lee[47] reported osmotic and activity coefficients of tetraalkylammonium fluorides and found them to be strikingly high; in increasing order they are $Me_4NF < Et_4NF < Pr_4NF < Bu_4NF$. The $\gamma_\pm$ values of Bu_4NF are said to be higher than those of any 1:1 electrolytes reported in the literature in the same concentration range at 25°C. One way of explaining the remarkably high activity coefficients of these salts is in terms of the specific way in which the structure of water may be considered to be influenced by the ions. As suggested by Frank,[48] there will be a "medium effect" over and above that caused by changes in dielectric constant due to the specific way in which the structure of water is enhanced or disrupted by the ions. If an ion is a structure maker or a structure breaker, its free energy of hydration will also be influenced by the ease or difficulty with which the surrounding water can accommodate itself to such influences, and this will be changed if other ions are also changing the water structure. There will be, therefore, a "structural salting in" or "structural salting out" of ions by each other, and this will appear as a structural influence on log $\gamma_\pm$ in addition to the influence brought by Hückel's electrostatic salting in or salting out.

Both R_4N^+ and F^- are roughly describable as structure-making ions, but they make different kinds of water structures because they are basically different kinds of ions. The F^- ion is small and tends to hydrate water dipoles in a radial pattern around it, whereas the R_4N^+ ion is large and tends to enhance the "cagelike" structure owing to its hydrophobicity. The cation and anion are thus incompatible in their modes of structure making and, when brought near to each other, will cause a structural repulsion or a structural salting out. Thus each ion raises the escaping tendency of its partner and high osmotic and activity coefficients result. In this connection it is of interest to note that Steigman and Dobrow[49] found the solubility of

$CaSO_4$ at 25°C to rise only slightly and then to decrease markedly with increasing concentrations of Bu_4NBr. The two salts are considered to constitute a pair of antagonistic electrolytes in water; the antagonism is attributed to the different and incompatible water structures surrounding the SO_4^{2-} and Bu_4N^+ ions.

Recently, Desnoyers et al.[50] interpreted the deviations from the limiting Debye–Hückel theory for activity coefficients of alkali halides and tetraalkylammonium halides with a structural interaction model. Essentially, two ions will attract each other if their structural influences, or their tendencies to orient water molecules, are compatible with each other; conversely, an incompatibility in these influences or tendencies will result in repulsive forces. They plotted log γ at 0.2 m (corrected for the Debye–Hückel forces) for these halides as shown in Figure 4, indicating sharp contrast between the fluoride series and the iodide series.

Chen[51] has found from isopiestic measurements that mean molal activity coefficients $\gamma_\pm$ of Me_4NF, Me_4NBr, and Bu_4NBr in D_2O are slightly lower than the corresponding salts in H_2O when compared at equal aquamolal concentrations. On the other hand, $\gamma_\pm$'s of Bu_4NF in D_2O are found to be higher than that in H_2O, indicating the importance of the specific isotope effect on ion-solvent interaction.

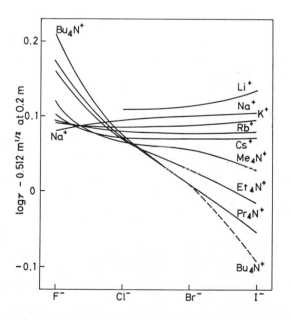

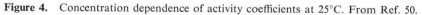

Figure 4.   Concentration dependence of activity coefficients at 25°C. From Ref. 50.

Enthalpies and Entropies of Dilution.   Heats of dilution at 25°C of tetra-alkylammonium chlorides, bromides, and iodides were measured by Lindenbaum,[45] and the corresponding fluorides were measured by Wood et al.[52] These measurements are important, since they have made significant contribution to clarify the nature of water–hydrophobic-ion interaction. Several features of H^{ex} ($= \Phi_L$) versus molality curves shown in Figure 5, reproduced from their results, are of interest. Heats evolved on dilution of the tetrabutyl- and tetrapropylammonium fluorides, chlorides, and bromides are the largest of any 1:1 electrolytes. For a given halide, the order of heat evolved is $Bu_4N^+ > Pr_4N^+ > Et_4N^+ > Me_4N^+$ over the entire concentration range. For a given tetraalkylammonium ion, the order is $F^- > Cl^- > Br^- > I^-$. This regularity is in contrast to the apparent disorder found in the activity coefficient curves of chlorides and bromides at higher concentrations. There is a much larger difference between the activity coefficients of the fluoride and chloride than between the chloride and bromide[40,47] and the same behavior is observed in the values of Φ_L. For instance, at 3 m the values of Φ_L for the tetraethylammonium fluoride, chloride, and bromide are 1302, -502, and -902 cal/mole, respectively.

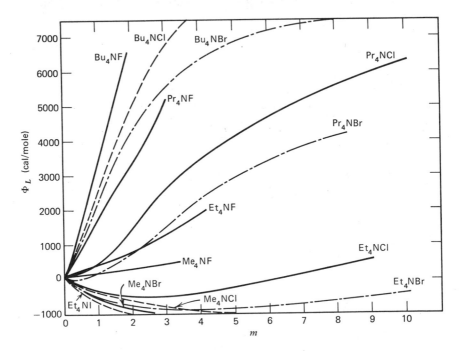

Figure 5.   Relative apparent molal heat content plotted against molality for the tetraalkylammonium halides in aqueous solution. From Refs. 45 and 52.

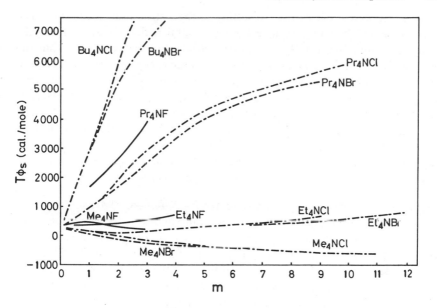

Figure 6. Apparent molal excess entropy plotted against molality for the tetraalkyl-ammonium halides in aqueous solution. From Refs. 45 and 52.

Frank and Robinson[53] have explained the entropies of dilution of aqueous electrolytes using the changes in the overlap of the hydration spheres of oppositely charged ions as the salt is diluted. For R_4N^+ ions, Frank and Wen[6] postulate that larger flickering clusters with longer half lives than "normal" would form around the alkyl groups. Measurements of the heats of dilution of R_4NX showed that the interactions of the larger tetraalkylammonium ions are dominant, since the heats of dilution are relatively independent of the anion. Observation of the entropy of dilution curves given in Figure 6 would reveal that the entropies of dilution of Pr_4NX salts fall very close together, and similarly, the curves for Bu_4NX salts coincide at low concentrations. Wood et al.[52] explained these results in terms of changes in the overlap of the cagelike structures around the tetraalkylammonium ions as the salt is diluted. Then the more highly hydrogen-bonded regions around the tetraalkylammonium ions are completed because there is less overlap and there is more total hydrogen bonding. This increase in structure causes the observed heating of the solution. It also causes an increase in the volume of the solution because of the larger volume of the more highly hydrogen-bonded water.

From their partial molal volume data, Wen and Saito[16] postulated that the water clusters join up at higher concentrations and force the cations and anions to approach each other, resulting in the "structural salting in." The heat-of-dilution data require a lessening of total hydrogen bonding as the clusters join up (overlap). As mentioned earlier, Frank[48] suggested that the cation with a cage has a smaller volume than the cation without a cage, and that the cages link up and stabilize each other as the concentration is increased, thus increasing their half-lives and decreasing the volume. Here is the area that needs further clarification; for the time being, however, these two concepts appear to be quite useful: the concept of overlap of hydration cospheres and the concept of hiding alkyl groups in cages. It appears that the volume is more sensitive to the caging effect than is the entropy and that the entropy is the more sensitive to overlap than is the volume.

Excess Thermodynamic Quantities on Mixing Solutions. Measurements of the changes in thermodynamic quantities on mixing electrolyte solutions provide another interesting way to study the interactions of ions in aqueous solution.[54] If the measurements are made at constant ionic strength, effects of the ionic atmosphere are canceled. In addition, if the measurements are made with a common ion, the effects of oppositely charged ion pairs cancel. Thus the pairwise and triplet interactions of like-charged ions can be conveniently studied. Using Mayer's ionic solution theory,[55] Friedman[56] predicted that like-charged ions should have specific interactions and that these interactions should be more important than triplet interactions for many systems.

We shall follow the formalism of Friedman in discussing the mixing of aqueous 1:1 electrolytes AX and BX, where, for example, A is an R_4N^+ ion, B is an alkali metal ion, and X is a halide ion. The excess free energy of mixing is defined by

$$(1) \qquad \Delta_m G^{ex}(y, I) \equiv G^{ex}(y, I) - yG^{ex}(1, I) + (1 - y)G^{ex}(0, I)$$

where I is the total ionic strength of the mixed electrolyte solution, y is the mole fraction of AX so that $G^{ex}(1, I)$ applies to a solution of pure AX and $G^{ex}(0, I)$ applies to a solution of pure BX. Each term of (1) pertains to a solution of the same ionic strength, temperature, and pressure. The quantity $\Delta_m G^{ex}$ is the increase in excess free energy on forming the mixture from the component solutions at the same I, T, and P and is a measure of the change in ionic interactions in this process. By appropriate differentiation of (1) we may obtain expressions for the corresponding enthalpy and volume changes, $\Delta_m H^{ex}$ and $\Delta_m V^{ex}$, respectively, in this same mixing process.

$$(2) \qquad \Delta_m H^{ex}(y, I) \equiv H^{ex}(y, I) - yH^{ex}(1, I) - (1 - y)H^{ex}(0, I)$$

$$(3) \qquad \Delta_m V^{ex}(y, I) \equiv V^{ex}(y, I) - yV^{ex}(1, I) - (1 - y)V^{ex}(0, I)$$

The excess free energy of mixing can be expanded into the following form[56]

(4) $\Delta_m G^{ex}(y, I) = I^2 RTy(1 - y)[g_0 + g_1(1 - 2y) + \cdots]$

where R is the gas constant, T is the temperature, and $g_0, g_1, \ldots$, are free-energy interaction parameters. Friedman derived expressions for these parameters based on a primitive model consisting of hard spheres in a continuous dielectric as shown for 1:1 electrolytes

$$g_0 = g_{\text{pairs}} + I \cdot g_{\text{triplets}} \qquad g_1 = I \cdot g^1_{\text{triplets}}$$

where

(5) $$g_{\text{pairs}} = 2\bar{g}^*_{AB} - \bar{g}^*_{AA} - \bar{g}^*_{BB}$$

In these equations, however, we shall interpret $\bar{g}^*_{ij}$ as the part of free energy, which changes on mixing because of the structural effects when ion i is in the neighborhood of ion j.

Similarly we may write down corresponding expressions for the enthalpy and volume changes:

(6) $\Delta_m H^{ex}(y, I) = I^2 RTy(1 - y)[h_0 + h_1(1 - 2y) + \cdots]$

where

(7) $$h_0 = 2\bar{h}^*_{AB} - \bar{h}^*_{AA} - \bar{h}^*_{BB} + I \cdot h_{\text{triplets}} + \cdots$$

and

$$h_1 = I \cdot h^1_{\text{triplets}} + \cdots$$

(8) $\Delta_m V^{ex}(y, I) = I^2 y(1 - y)[v_0 + v_1(1 - 2y) + \cdots]$

where

(9) $$v_0 = 2\bar{v}^*_{AB} - \bar{v}^*_{AA} - \bar{v}^*_{BB} + I \cdot v_{\text{triplets}} + \cdots$$

and

$$v_1 = I \cdot v^1_{\text{triplets}}$$

Wood and Anderson[57] have measured the heats of mixing of aqueous solutions of various chloride salts at 0.5 total ionic strength and 25°C. They found that $\Delta_m H^{ex}$ are reasonably symmetrical with respect to mole fraction, where y of maximum $\Delta_m H^{ex}$ ranged from 0.47 to 0.53. A comparison of the results given in Table 5 with measurements on other systems[58,59] with a common anion shows that the tetraalkylammonium ions give somewhat larger heats of mixing than the alkali or alkaline earth metal ions. For the heats of mixing of inorganic salts with common ion, the sign of the heats of mixing follows a simple rule. The mixing of two structure makers or two structure

Table 5   Heats of Mixing at $I = 0.5$ and $25°C$

Salt Pair	RTh_0 (cal/kg $H_2O \cdot m^2$)	RTh_1	$\Delta_m H^{ex}$ at $y = 0.5$ (cal/kg H_2O)
LiCl–Me$_4$NCl	-160.8 ± 0.9	-7.8 ± 1.2	-10.0
KCl–Me$_4$NCl	118.9 ± 0.8	-2.2 ± 1.1	7.4
CsCl–Me$_4$NCl	82.0 ± 1.3	—	5.1
LiCl–Et$_4$NCl	-172.4 ± 1.0	-8.0 ± 1.4	-10.8
KCl–Et$_4$NCl	117.5 ± 0.9	—	7.4
CsCl–Et$_4$NCl	73.1 ± 1.3	-4.4 ± 1.8	4.6
LiCl–Pr$_4$NCl	-693.3 ± 3.8	-33.0 ± 6.1	-43.3
KCl–Pr$_4$NCl	-348 ± 4	-39.0 ± 5.9	-21.8
CsCl–Pr$_4$NCl	-438.6 ± 1.5	12.7 ± 2.1	-27.4
Me$_4$NCl–Et$_4$NCl	-49.8 ± 0.7	1.3 ± 1.5	-3.1
Me$_4$NCl–Pr$_4$NCl	-613.5 ± 2.3	-14.4 ± 3.8	-38.4
Et$_4$NCl–Pr$_4$NCl	-308.4 ± 2.2	-8.3 ± 3.2	-19.3

breakers gave endothermic heats of mixing, whereas a structure maker combined with a structure breaker gave an exothermic heat of mixing. Assuming the same rule to apply, it is noted from the table that Me_4N^+ and Et_4N^+ act as a structure breaker toward the alkali cations. However, mixing of Me_4N^+ and Et_4N^+ is exothermic, indicating that they act as opposites when mixed with each other. The results for Pr_4N^+ ion show that all the mixing of this ion with another ion gives very large negative values for the heat of mixing ($RTh_0 = -300$ to $-600°C$). This suggests that the tetrapropylammonium ions are responsible and that the large interactions of two Pr_4N^+ ions are masking other effects. In any common-ion heat of mixing, say AX with BX, there are three like-charged pair interactions to be considered: (a) the heat of diluting the A^+ ions and therefore the reduction in the overlap of the A^+ hydration sheaths with each other ($-\bar{h}_{AA}^*$), (b) the dilution of the B^+ ions and the reduction of $B^+ - B^+$ overlap ($-\bar{h}_{BB}^*$), and (c) the formation of some $A^+ - B^+$ overlap ($2\bar{h}_{AB}^*$). It is expected that, at any given concentration, Pr_4N^+ ions will have large hydration-sphere overlap with neighbors, and when the ion is mixed with another, the overlap of the hydration sheaths of the Pr_4N^+ ions is reduced. On mixing (diluting Pr_4N^+ ions) more structure is formed per ion and heat is given out (RTh_0 is negative). This interpretation is supported by the heat of dilution of the tetraalkylammonium halides measured by Lindenbaum.[45]

Wen and Nara[60] have measured the volume changes on mixing solutions of potassium halides and tetraalkylammonium halides having common anions at constant ionic strengths at $25°C$. Friedman's theory, when applied to the volume of mixing of AX and BX, gives two relations:

(10)
$$\frac{\Delta_m V^{ex}}{I^2 y(1-y)} = 2\bar{v}^*_{AB} - \bar{v}^*_{AA} - \bar{v}^*_{BB} + \cdots$$

(11)
$$\frac{V^{ex}(AX) - V^{ex}(BX)}{I^2} = \bar{v}^*_{AA} + 2\bar{v}^*_{Ax} - \bar{v}^*_{BB} - 2\bar{v}^*_{Bx} + \cdots$$

In contrast to (10), (11) relates to the difference of excess volumes of two separate solutions at an identical ionic strength, I. In these equations $\bar{v}^*_{ij}$ is the part of the volume which changes on mixing because of the structural effects when ion i is in the neighborhood of ion j. Comparing (10) and (11), we see that if $\bar{v}^*_{AA}$ is much larger than the other terms (comparing absolute values), then both (10) and (11) will be large; if $\bar{v}^*_{Ax}$ dominates, however, (11) will be much larger than (10). A large absolute value of $\bar{v}^*_{AA}$ may be taken to indicate a large contribution of the $R_4N^+-R_4N^+$ interaction to the solution volume. Experimental results obtained are summarized in Table 6.

Table 6   Comparison of Values (ml/kg solvent) Obtained by (10) at $y = 0.5$ and by (11)

	Pr$_4$NBr–KBr		Et$_4$NBr–KBr		Me$_4$NBr–KBr		Bu$_4$NCl–KCl		NaCl–KCl[a]	
I	(10)	(11)	(10)	(11)	(10)	(11)	(10)	(11)	(10)	(11)
0.2	—	—	—	—	—	—	7.22	−11.5	—	—
0.5	4.56	−8.4	2.54	−4.0	0.39	−1.8	5.44	−9.2	−0.045	−0.43
1.0	3.60	−6.8	1.96	−3.4	0.40	−1.2	3.78	−7.3	—	—
2.0	2.54	−5.1	1.60	−2.9	0.42	−1.1	—	—	—	—

[a] H. E. Wirth, *J. Amer. Chem. Soc.*, **59**, 2549 (1937).

These data seem to indicate that $\bar{v}^*_{AA}$ is large for large R_4N^+ ions, but not so large that $\bar{v}^*_{Ax}$ is completely overshadowed. If the assumption is made that the activity coefficients of the ions do not change with concentration, then cation-cation interactions will give a v_0 that does not vary with I, and cation-cation-anion triplet interactions will give a v_0 proportional to I. Thus the value of v_0 at $I = 0$ is a measure of cation-cation pairs only, and by the change in v_0 with I is caused either by changes in activity coefficients or by triplet interactions. The results of Table 6 show that the intercept of a plot of v_0 versus I is quite large, so that the cation-cation interactions are large. Since the volume of mixing is much larger for Pr_4N^+ and Bu_4N^+ than for any other mixtures, the tetraalkylammonium ions are probably causing the effect. The solution volume before mixing is small in the concentration range under discussion owing to the overlap of the "hydration cospheres" and also to the caging effect. If now a KX solution is mixed with an R_4NX

solution, the concentration of the R_4N^+ ion will decrease even though the total ionic strength is kept constant. The decrease of R_4N^+ ion concentration will diminish the overlap and result in the increase of volume. The introduction of K^+ ions will also disrupt the "caging effect" and cause the volume to increase.

The large volume increase of mixing Pr_4NX and KX solutions cannot be attributed entirely to the ion size difference between Pr_4N^+ and K^+ without consideration of the solvent structural changes. This becomes obvious when it is recalled that the mixing of solutions of tetraethanolammonium bromide and KBr gives nearly 0 value for $\Delta_m V^{ex}$, although the size of $(HOC_2H_4)_4N^+$ ion is about the same as that of the Pr_4N^+ ion.[16] The observation is explicable in terms of the ion-water interaction, since in contrast to Pr_4N^+ ion, there can be very little "hydration cosphere" or "cagelike structure" around the $(HOC_2H_4)_4N^+$ ion.

Volume changes on mixing aqueous solutions of KBr and R_4NBr were also measured in D_2O at 25°C and in H_2O at 15 and 25°C.[61] From the experimentally determined values of $\Delta_m V^{ex}$, a quantity, Z, defined by

$$Z = \frac{\Delta_m V^{ex}(y, I)}{I^2 y(1 - y)} = v_0 + v_1(1 - 2y) + \cdots$$

has been calculated and summarized in Tables 7 and 8.

As shown in Table 7, values for the ratio Z_{D_2O}/Z_{H_2O} are 1.09, 0.90, and 0.59 for Pr_4NBr–KBr, Et_4NBr–KBr, and Me_4NBr–KBr, respectively. Since D_2O is known to possess more structure than H_2O, a structure breaker should break more structure of the former than that of the latter, leading to a smaller volume in D_2O than in H_2O. The fact that the ratio Z_{D_2O}/Z_{H_2O} for Me_4NBr–KBr system is considerably less than unity may be taken to imply that both Me_4NBr and KBr are structure breakers. The Z ratio for the Pr_4NBr–KBr system is, on the other hand, very slightly above unity, indicating the structure-promoting effect of Pr_4N^+ ions. The Et_4NBr–KBr system has a value of 0.90 for the Z ratio, which may imply a slightly struc-

Table 7   Comparison of Mean Z Values (ml/kg solvent) Obtained from Mixing R_3NBr and KBr in S_2O and in H_2O at Unit Aquamolality at 25°C

Z	Pr_4NBr–KBr[a]	Et_4NBr–KBr[b]	Me_4NBr–KBr[c]
$Z_{D_2O} - Z_{H_2O}$	0.31	−0.19	−0.15
Z_{D_2O}/Z_{H_2O}	1.09	0.90	0.59

[a] D_2O, 3.75; H_2O, 3.44.
[b] D_2O, 1.69; H_2O, 1.88.
[c] D_2O, 0.22; H_2O, 0.37.

Table 8 Comparison of Mean Z Values (ml/kg H_2O) Obtained from Mixing Pr_4NBr and KBr in H_2O at 15 and 25°C

	$I = 0.50$		$I = 1.0$		$I = 2.0$	
	15°C	25°C	15°C	25°C	15°C	25°C
Z	4.65	4.40	3.77	3.44	3.04	2.62
$Z(15°C)/Z(25°C)$		1.06		1.10		1.16

ture-breaking effect of KBr and very little structural influence of the Et_4N^+ ion. These structural interpretations seem to be in accord with the results of Kay and Evans[8] based on the comparison of the ratio of the Walden product in D_2O to its value in H_2O, which is discussed in detail in Section 4, "Conductance."

In Table 8, mean values of Z obtained from mixing Pr_4NBr and KBr in H_2O at 15 and 25°C, are compared at constant total ionic strengths of 0.5, 1.0, and 2.0. As the temperature is lowered by 10°C, Z values increase in all cases. The increase of Z values with the lowering of temperature for this system clearly indicates the existence of more structure at lower temperature, in agreement with expectation.

The volume changes due to mixing H_2O solutions of Pr_4NBr and $CsBr$ and solutions of Pr_4NBr and Me_4NBr at $I = 1.0$ and at 25°C have also been reported. The Z values for Pr_4NBr–$CsBr$ system are about 8% greater than the corresponding values for the Pr_4NBr–KBr system and are attributed to the greater structure-breaking effect of the Cs^+ ions. On the other hand, the Z value for Pr_4NBr–Me_4NBr system was found to be quite small— only 0.9. This is rather surprising, since the Me_4N^+ ion is a structure breaker by many criteria, including the heat of mixing, and might be expected to act like the Cs^+ ion when mixed with Pr_4N^+.[57] The smallness of the volume change on mixing Pr_4N^+ with Me_4N^+ may be ascribed to the similarity of the hydrophobic surface and to the mutual geometrical compatibility on packing these ions with water molecules. The difference in results of the heat of mixing and the volume of mixing on this system points out the basic fact that $\Delta_m H^{ex}$ is largely a measure of the change in the number of hydrogen bonds of water in the solution, whereas $\Delta_m V^{ex}$ is a measure of the change in the geometrical arrangement of water and ions.

Heat Capacities. Frank and Wen[6] have measured the specific heats of solutions of Bu_4NBr and have reported the concentration dependence of the apparent molal heat capacity, ΦC_p. In dilute solutions the values of ΦC_p are about 270 cal/(deg)(mole). Estimates based on additivity rule found valid for hydrocarbons and other types of compounds indicate that, in the absence of ion-water interaction, ΦC_p must lie somewhere between 120 and 150 cal/(deg)(mole), depending on the assumptions. This requires that something in excess of 120 cal/(deg)(mole) must be regarded as "extra" heat capacity. The straightforward explanation is that this cation causes the water near it to be more structured than normal, and that this extra structure "melts" as the temperature is raised, causing extra heat to be absorbed.

Recently Ackermann et al.[62] reported the apparent heat capacities of dilute Bu_4NBr solutions at three concentrations and temperatures ranging from 10 to 130°C. Their results confirm the values of Frank and Wen[6] at room temperatures, but some of their new data are difficult to explain. For example, the temperature dependence of ΦC_p gives a maximum at about 50 to 90°C for all the alkali halides and alkylammonium salts studied except Bu_4NBr. Values of ΦC_p for this salt are found to be still very large, even at temperatures as high as 130°C.

In sharp contrast to Bu_4NBr, $(HOC_2H_4)_4NBr$ was found to give almost 0 "extra" heat capacity in water at room temperature, as expected.

Enthalpies of Solution and Transfer. Krishnan and Friedman[63] have determined heats of solution in propylene carbonate (PC), dimethyl sulfoxide (DMSO), water, and D_2O. They found that the enthalpy of transfer of the R_4N^+ ions to PC from DMSO comprises additive contributions from the CH_2 groups and that these CH_2 terms are the same as those for the PC ← DMSO transfers of the primary alcohols.[64] This was the basis of their method to obtain single-ion contributions to enthalpies of transfer of the salts. They found that the enthalpy of transfer of R_4N^+ ions to PC from water is dominated by an effect that does not comprise additive contributions from the methylene groups. As shown in Figure 7, ΔH_{tr} reaches a maximum value at tetraamylammonium ion when plotted as a function of chain length. This effect is attributed to structural changes ("icebergs") in the water in the neighborhood of the ions in aqueous solution.

Applying the simple solvation-enthalpy criterion, Krishnan and Friedman concluded that Me_4N^+ is a net structure breaker, structure making and breaking mutually cancel for Et_4N^+, and structure-making effects dominate for Pr_4N^+, Bu_4N^+, and Am_4N^+. These conclusions are in agreement with those reached by other workers on the basis of different experimental probes: Kay and his co-workers[8] from the ratio of Walden products of the ions in H_2O and D_2O, Wood and Anderson[57] from heats of mixing of ionic solutions, and Bunzl[65] from the temperature dependence of the shift of the 0.97-μ infrared band in aqueous solutions of R_4NBr.

Bhatnager and Criss[66] determined the standard heats of solutions at

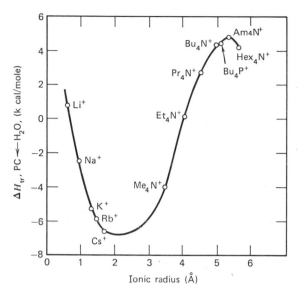

Figure 7. Ionic enthalpies of transfer to propylene carbonate from water. From Ref. 63.

infinite dilution ΔH_{tr}° for Me_4NI, Et_4NI, and Pr_4NI in water and in N,N-dimethylformamide and compared with other ΔH_{tr}° data taken from the literature. They noted surprisingly large changes in ΔH_{tr}° between CsI and Me_4NI for solvent pairs involving water and not involving water, and they cautioned that the large change observed between Cs^+ and Me_4N^+ is not explained by the water structure alone but is the result of special interactions of the tetraalkylammonium ions with solvents in general.

Solubilities of Hydrocarbons in Aqueous R_4NX Solutions. The solubility of benzene in substituted quaternary ammonium bromide solutions has been measured by Desnoyers, Pelletier, and Jolicoeur[67] and by Wirth and LoSurdo.[68] It was found to increase with the length of alkyl chains and the number of alkyl groups on the cation. The results indicate that the increase in structure of water in the vicinity of the large ions plays an important role in salting in by promoting the formation of an association complex between benzene and the organic ions, as previously suggested by Diamond.[44] As in the case of Deno and Spink,[69] Desnoyers et al. found that the equation of McDevit and Long[70] may be used to calculate the solubilities of benzene that are in semiquantitative agreement with the experimental results.

There have been some scattered studies on the solubilities of gases in the aqueous R_4NX solutions, but the results appear to be rather inconclusive.[71,72] Recently Wen and Hung[73] determined the Setchénow constants and ΔF_{tr}°

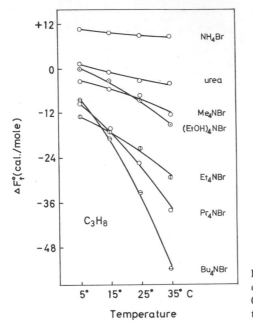

Figure 8. Standard molal free energy of transfer of propane from water to 0.1-*m* aqueous solution at different temperatures.

the standard molal free energy of transfer of hydrocarbons (methane, ethane, propane, and butane), from water to 0.1-*m* R_4NBr solutions at 5, 15, 25, and 35°C. They found the values of ΔF_{tr}° to be all negative except for methane in Me_4NBr solutions at 5°C. At 25 and 35°C, the observed order of ΔF_{tr}° values for each gas is

$$\Delta F_{tr, H_2O \to Et_4NBr}^{\circ} > \Delta F_{tr, H_2O \to Pr_4NBr}^{\circ} > \Delta F_{tr, H_2O \to Bu_4NBr}^{\circ}$$

but at 5°C, a striking change in the order of the salt effect was noticed

$$\Delta F_{tr, H_2O \to Bu_4NBr}^{\circ} > \Delta F_{tr, H_2O \to Pr_4NBr}^{\circ} > \Delta F_{tr, H_2O \to Et_4NBr}^{\circ}$$

As an example, the free energy of transfer for propane is shown in Figure 8. These observations have been interpreted in terms of the important roles played by the structure modifications of water and the hydrophobic bonds between the cations and hydrocarbon molecules.

Temperature of Maximum Density. The shifts, θ, in the temperature of maximum density of water produced by solutes may be expressed by

$$(12) \qquad \Delta\theta = (\theta - 3.98) = \frac{-x\alpha_2 V_2^{\circ}}{2(1-x)\alpha_1 V_1^*} - \frac{1}{2(1-x)\alpha_1 V_1^*} \frac{\partial \Delta V_x^m}{\partial T}$$

where x is the mole fraction of the solute, α_1 and α_2 are the coefficients of expansion of water and the solute, respectively; V_1^* and V_2° refer to the molar volumes of water and solute, respectively; and $\Delta V_x{}^m$ is the excess volume of mixing referred to a mixture of solute mole fraction, x. As pointed out by Wada and Umeda,[74] and also by Franks and Watson,[75] the first term on the right-hand side of (12) is always negative and, for a given solute species, is determined by V_2° and α_2. The second term is identified with solute-structure influence; it is positive for structure makers and negative for structure breakers. Many inorganic salts lower the temperature of maximum density. Two conditions that produce a positive $\Delta\theta$ are that $\partial\Delta V_x{}^m/\partial T$ is negative and that $|\partial\Delta V_x{}^m/\partial T|$ is greater than $x\alpha_2 V_2^\circ$.

Solutes that are found to give positive $\Delta\theta$ are nonelectrolytes having nonpolar groups such as alcohols and amines, particularly with branched chains.[75]

Recently Darnell and Greyson[76] reported negative $\Delta\theta$ for tetraalkylammonium halides and observed that the larger the cation, the greater is the depression of the temperature of maximum density. This is noteworthy, since it has been generally agreed, based on room-temperature measurements, that the R_4N^+ ions are structure makers in aqueous solutions and that their structure-making ability increases with the size of the alkyl group. One would have predicted positive $\Delta\theta$, or, at worst, negative $\Delta\theta$ decreasing with the size of the cation. Darnell and Greyson suggested that, at least in the neighborhood of the temperature of maximum density, ion sizes would play a more important role in influencing water structure than the structure-making or -breaking properties of the ions observed at room temperature. At low temperatures, the quaternary alkylammonium halides disrupt structure because of the incompatibility of the structure of the ion with the structure of the solvent.

Their suggestion seems questionable in view of the near-infrared spectra of Bu_4NBr in H_2O–D_2O mixture at 5°C obtained by Worley and Klotz,[77] which is discussed later. It calls for systematic investigations of various properties of these solutions at low temperatures.

Surface Tension. The surface tension of the tetrabutylammonium halides at the air–aqueous-solution interface and at the hexane–aqueous-solution interface was measured by Tamaki[116] using the drop-volume method. The surface-tension versus concentration curves of these salts are markedly dependent on the kind of their counter-ions (order of surface activity is $Bu_4NI > Bu_4NBr > Bu_4NCl$). The surface activity of Bu_4NI appears to be fairly strong; it corresponds approximately to that of straight-chain electrolyte with an octyl group. However, no sharp change in the surface tension was observed for any of the tetraalkylammonium halide solutions,

indicating the absence of critical micelle concentration. From the thermodynamic parameters of the adsorption of Bu_4NI to the hexane–aqueous-
solution interface (the values for $\Delta F°$, $\Delta H°$, and $\Delta S°$ are -4.85 kcal/mole,
$+1.83$ kcal/mole, and $+22.4$ eu, respectively), Tamaki concluded that the
adsorption is mainly hydrophobic effect[117,118] and not due to the gain of
freedom of internal rotation about the carbon–carbon bonds.[119]

4   Transport Properties

Viscosity.   Aqueous solutions of large tetraalkylammonium salts exhibit
unusually high viscosity. Bingham's data[5] on the viscosity increase
for 1-*m* solutions of some of these salts attracted the attention of Frank
and Evans,[4] who attributed the excess viscosity to "iceberg" formation
around the alkyl groups. Prior to 1966, studies on the viscosities of quaternary ammonium salts in water were reported by Laurence and Wolfenden[78]
for Et_4NBr, Fuoss and Kraus[79] for Bu_4NBr, Wen[80] for Bu_4NBr and
Bu_4NOAc, Hückel and Schaaf[81] for Me_4NI and Et_4NI, Nightingale[82] for
Me_4NBr, and so on. However, the most systematic investigations on the
viscosity to date have been carried out by Kay et al.[83] for tetramethyl-,
tetraethyl-, and tetrapropyl-, and tetrabutylammonium bromides and
iodides in H_2O, D_2O, Ch_3OH, and CH_3CN at concentrations up to 0.2 M
and temperatures between 0 and 65°C.

The relative change in viscosity due to the addition of salt to solvent can
be fitted to the empirical equation of Jones and Dole[84]

$$\frac{\eta - \eta_0}{\eta_0} = Ac^{1/2} + Bc$$

where η_0 is the viscosity of pure solvent and η is the viscosity of solution at
salt concentration of c (mole/liter). The coefficient A (which is quite small
compared to B) is the contribution from interionic forces that tend to interfere with the flow of solution; its value can also be calculated from the
Falkenhagen equation.[85] The B coefficients are related to the size of ions and
their effects on the structure of solvent. Following Kaminsky,[86] Kay et al.
split the B coefficients to the separate contributions of cations and anions
on the basis of additivity rule and of $B_{K^+} = B_{Cl^-}$ at all temperatures. The
ionic B values so obtained in water at various temperatures are given in
Table 9.

The effect of ionic size on B coefficients can be interpreted at least qualitatively, by the Einstein equation,[87] $B = 2.5v/c$, where v is the total volume
occupied by the ions per cubic centimeter of solution. This equation predicts
that the presence of ions should increase the solution viscosity in proportion
to their size, and the increase should be independent of temperature because

Table 9 Ionic *B* values at Various Temperatures in Water

Temperature (°C)	Bu_4N^+	Pr_4N^+	Et_4N^+	Me_4N^+	Br^-
0	1.75	—	—	—	−0.08
10	1.52	1.04	0.44	0.13	−0.07
25	1.28	0.86	0.38	0.12	−0.04
45	1.01	0.66	0.32	0.12	−0.02
65	0.84	0.55	0.28	0.12	−0.01

electrostrictive solvation is independent of temperature change under consideration. Since it is difficult to separate the ionic size effect from the structural effect of *B* coefficients, better criteria for solvent structural influences would be the temperature-dependence values of the *B* coefficients, *dB/dT*. Kay et al. found that, in methanol, values of *dB/dT* for Me_4NBr, Et_4NBr, Pr_4NBr, and Bu_4NBr are about 0—in sharp contrast to the large negative *dB/dT* observed for Pr_4NBr and Bu_4NBr in water. These effects must be due to the large R_4N^+ ions, since the contribution to the *B* values from the bromide ion is very small. Corresponding behavior has been observed in the Walden product for these ions.[8] Thus the viscosity data confirm the conclusions reached from the conductance data that water-structure enforcement about the hydrocarbon side chains of Pr_4N^+ and Bu_4N^+ forms a larger moving entity and, at the same time, increases the bulk viscosity by increasing the degree of hydrogen bonding in their vicinity. On the other hand, the *B* values for Me_4NBr are much smaller in aqueous than in methanol solutions at all temperatures, and they show a slight increase with increasing temperature. The viscosity data seem to indicate the existence of structure-breaking properties of Me_4N^+. The Et_4N^+ ion shows a mixed behavior, since *B* for Et_4NBr is slightly lower in an aqueous than in a methanol solution but still shows a small decrease with increased temperature in aqueous solution. The data can best be interpreted by assuming for this ion that the structure-making and structure-breaking effects cancel each other.

Conductance. As in the case of viscosity there were several reports[2,18] on the conductance of quaternary ammonium salts in water, particularly by Kraus and his co-workers,[86-88] but the most systematic studies were those of Kay and Evans.[8] They measured equivalent conductances of many tetraalkylammonium halides at several temperatures and analyzed the data in terms of the Fuoss–Onsager theory[89]

$$(13) \qquad \Lambda = \Lambda_0 - S(c\gamma)^{1/2} + Ec\gamma \log c\gamma + (J - \theta)c\gamma - K_A c\gamma \Lambda f^2$$

The limiting conductances Λ_0 for the R_4N^+ ions and halide ions at 10 and 25°C are summarized in Table 10.

Table 10   Limiting Cation and Anion Conductances of R_4NX in H_2O

	Temperature (°C)	
	10	25
Me_4N^+	30.93	44.42
Et_4N^+	21.90	32.22
Pr_4N^+	15.33	23.22
Bu_4N^+	12.56	19.31
Cl^-	54.33	76.39
Br^-	56.15	78.22
I^-	55.39	76.98

The ion-size parameter $\mathring{a}$ obtained from J in (13) is found to be much smaller than those obtained for nonaqueous solvents and with increasing cation size; it increases for the chlorides, remains about constant for the bromides, and decreases almost to 0 for the iodides. This variation in the ion-size parameter for iodides is attributed by Kay and Evans to the ion-pair association and was analyzed for an exceedingly small but definite amount of association, K_A, for Pr_4NI and Bu_4NI in dilute solution.

Kay and Evans[8] measured the conductance of R_4NBr series in D_2O and calculated the ratio of the Walden product in D_2O to its value in H_2O both at 25°C; that is

$$R = \frac{(\lambda_0\eta_0)_{D_2O}}{(\lambda_0\eta_0)_{H_2O}}$$

The limiting conductances and Walden products for R_4N^+ ions in D_2O are given in Table 11, which shows systematic and significant deviations from unity. Since D_2O has more structure than H_2O, structure-breaking ions will break more structure in D_2O than in H_2O. The resulting decrease in the local viscosity is greater in the case of D_2O than H_2O and, therefore, it could produce the increase in the conductance solvent viscosity product for the structure breakers in D_2O over that in H_2O. The values of R slightly above unity for Et_4N^+ and Me_4N^+ are taken by Kay and Evans to indicate that these are structure breakers. For Pr_4N^+ and Bu_4N^+, R is less than unity owing to greater enforcement of water structure of these ions in D_2O than in H_2O.

Table 11 Limiting Cation Conductances and Walden Products in D_2O at 25°C

	$\lambda_0{}^+$	$\lambda_0{}^+\eta_0$	R
Me_4N^+	36.61	0.4009	1.0124
Et_4N^+	26.44	0.2895	1.008
Pr_4N^+	18.84	0.2063	0.9981
Bu_4N^+	15.62	0.1711	0.9948

The limiting Walden products for various quaternary ammonium ions are plotted in Figure 9 as a function of the estimated crystallographic radii. It can be seen that, with the exception of Me_4N^1 and Et_4N^1 in hydroxylic solvents, the Walden products for these large ions in nonaqueous solvents are almost identical and fall on the solid line. In aqueous solution, however, the Me_4N^+ ion lies well above this line, suggesting that it is a structure breaker. The Walden products for Am_4N^+, Bu_4N^+, and Pr_4N^+ ions are well below this nonaqueous line, however, suggesting that in aqueous solution these ions are larger than in nonaqueous solvents. This is in keeping with the idea that clathrate-like structure form about the hydrocarbon portions of these ions as the length of the side chain increases. The effect of temperature on the Walden product for these cations is seen more readily in Figure 10. Only the Et_4N^+ ion shows no temperature dependence, owing presumably to a cancellation of the effects of structure breaking and structure making as judged by this criterion. The Me_4N^+ ion has a negative temperature coefficient, typical of structure-breaking ions, as was found with the larger alkali and halide ions. As the temperature increases, there is less structure available to be broken, and the Me_4N^+ ion is less effective in reducing the local viscosity. The Pr_4N^+ and Bu_4N^+ ions both have positive temperature coefficients, as would be expected of ions forming clathrate-like structure around their hydrocarbon side chains. As the temperature increases, these cages of water melt and produce a smaller and therefore faster-moving entity. Thus the temperature dependence, the H_2O–D_2O comparison, and the comparison just stated of Walden products in aqueous and nonaqueous solutions are consistent with the Frank–Wen model[6] for aqueous ionic solutions.

Evans et al.[90] reported the conductance and viscosity of tetraethanolammonium bromide in aqueous solutions at different temperatures. This substitution of a hydroxyl group for a terminal methyl group in the Pr_4N^+ ion is shown to result in the elimination of the effects that have been attributed to water-structure enforcement around the side chains of the large

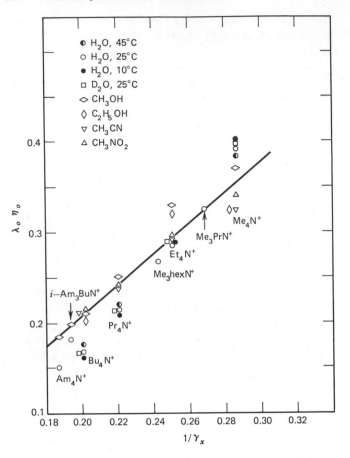

Figure 9. The limiting Walden product for various quaternary ammonium ions as a function of crystallographic radii, temperature, and solvent. From Ref. 8.

tetraalkylammonium ions. This interpretation is consistent with the data of partial molal volumes,[38] activity coefficients,[39] heat capacity,[38] and heat of transport.

Horne and Young[91] have measured the electrical conductivities of $0.1\text{-}M$ R_4NX solutions at 4 and 25°C over the pressure range 1 atm to 4000 kg/cm^2. The variation of the relative specific conductance, $\kappa_P/\kappa_{1\,atm}$, with pressure was found to be smaller than that of alkali halides such as KCl or NH$_4$Cl. The plot of κ_P/κ_1 against pressure goes through a regular sequence Bu$_4$N$^+$ > Pr$_4$N$^+$ > Et$_4$N$^+$ > Me$_4$N$^+$ in the case of the chlorides but goes through a maximum for Pr$_4$N$^+$ in the case of the bromides and for Et$_4$N$^+$ in the case of the iodides. Horne and Birkett[92] postulated the existence of β structure of water at interfaces to distinguish from the α structure of the Frank–Wen clusters and the hydration of small ions; they also proposed that the hydrophobic hydration of the R_4N^+ ions is the β form. Since the β structure is said to be much more stable with respect to the temperature and pressure than the α structure, the hydration of the R_4N^+ ions should be more stable under pressure and, consequently, their electrical conductivity should

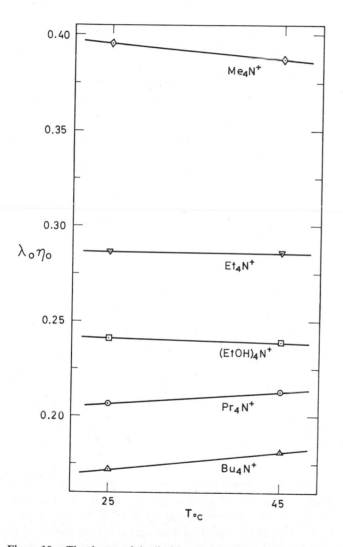

Figure 10. The change of the limiting Walden product for the tetraalkylammonium ions with temperature. From Ref. 90.

exhibit a smaller pressure dependence than that of alkali metal cations. Earlier Wicke[93] postulated the existence of a "third state" of water consisting of dimers to hexamers and proposed that the enhanced water structure around nonpolar groups (i.e., hydration of the second kind) may involve nontetrahedral hydrogen bonds. These postulates are of great interest but rather speculative at the present stage. More work is needed to clarify these structures. We shall discuss Wicke's postulate again in the section on nuclear magnetic resonance.

In the course of an investigation of the electrical conductivity of aqueous Bu_4NBr solutions under high pressure (up to 4470 kg/cm²), Horne and Young[94] found that, as the pressure was increased, the resistance of 0.1-M salt solution abruptly increased by several orders of magnitude. This was attributed to the formation of a new superhigh hydrate of Bu_4NBr stabilized by high pressure.

Heats of Transport.   Agar and Turner[95] have determined heats of transport of electrolytes in 0.01-m solutions. The heat of transport of a solute, Q^* is heat liberated ahead of one of its particles as it moves forward in the solvent. In the case of an ion moving in water this heat may be considered to arise from three sources: (a) statistical-mechanical fluctuations of energy (or entropy) density associated with the motion of a particle relative to its near neighbors, (b) charge effect of an ion, which decreases the entropy of nearby solvent molecules as the ion moves, and (c) structure-making or -breaking effects of the ions upon water.

Table 12   Conventional Heats of Transport (cal/mole) in 0.01 m Aqueous Solution at 25°; $Q^*_{Cl^-} = 0$

Ion	Q^*	Ion	Q^*	Ion	Q^*
Me_4N^+	1746	MeH_3N^+	570	Me_3PhN^+	2210
Et_4N^+	2795	$Me_2H_2N^+$	840	$Me_3(C_2H_4OH)N^+$	1500
Bu_4N^+	4470	Me_3HN^+	1294		

From Table II, p. 96, Ref. 97.

Agar[96,97] found the heats of transport of 1:1 salts to be additive in dilute solutions, and adapted the convention of 0 to Q^* of Cl^- at all concentrations. Table 12 gives Agar's conventional ionic heats of transport at 25°C and 0.01 m. The high heats of transport of the tetraalkylammonium ions shown in the table are correctly attributed to the exceptionally low entropy of water surrounding these hydrophobic ions.

Diffusion. Hertz, Lindman, and Siepe[103] have recently measured the self-diffusion coefficients of tetraalkylammonium chlorides in water by NMR spin-echo technique at 25°C. The self-diffusion coefficients of the cations in D_2O extrapolated to infinite dilution are found to be in satisfactory agreement with those from conductivity measurements: 1.1, 0.725, 0.52, and 0.39×10^{-5} cm²/sec for $(CH_3)_4N^+$, $(CH_3CD_2)_4N^+$, $(C_3H_7)_4N^+$, and $(C_4H_9)_4N^+$, respectively. In addition, they measured the self-diffusion coefficients of H_2O in the solutions of these salts as a function of the concentration and compared them with the self-diffusion coefficients of R_4NCl. Comparison of the corresponding pair of diffusion coefficients led them to conclude that a rigid long-lived hydration sphere of the cations does not exist—or that such an existence is at best very unlikely. Proton and deuteron nuclear-magnetic-relaxation rates of $(CX_3)_4NCl$ and $(C_2X_5)_4NCl$, where $X = H$ or D, in aqueous solution gave evidence for the existence of cation–cation contact for the ion $(C_2H_5)_4N^+$.

5 Spectroscopic Studies

Nuclear Magnetic Resonance. The chemical shift of the proton resonances of H_2O in aqueous salt solutions with respect to pure water was investigated by Hertz and Spalthoff.[98] The chemical shift gives a measure of the extent to which the proton is shielded. The methyl protons of R_4NX salts were used as the internal standard for these measurements. The chemical shift of the H_2O protons in aqueous solutions of tetraalkylammonium bromides and nitrates obtained by them at 25°C are all positive in relation to pure water; that is, a higher external magnetic field is required for resonance of the H_2O protons than in pure water. It indicates that in these solutions the diamagnetic screening of the protons by the electron clouds is stronger than that in pure water. Since a positive chemical shift is observed for pure water with increasing temperature (i.e., the chemical shift increases linearly with temperature) and at 100°C it is 0.7 ppm relative to pure water at 25°C, it appears that the hydrogen bonds between H_2O molecules are weakening and the water structure is breaking down with the increase of salt concentration. Moreover, the chemical shift is seen to increase to greater positive values with the increase of the cation size, which may again appear to indicate the structure-breaking effect of the tetraalkylammonium ions.

Similarly, using methyl protons of the hydrocarbon chain as the internal reference, Clifford and Pethica[99] measured the chemical shifts of H_2O protons in sodium alkyl sulfate solution. Their results also indicate up-field shifts which increase with the length of the alkyl group.

The following three suggestions have emerged so far to account for the chemical shift changes of tetraalkylammonium bromide solutions.

1. Nonpolar groups induce some covalent character to the hydrogen bonds of the water molecules next to them.[99]

2. The up-field shift is caused by the anisotropy imposed on water by the nonpolar groups.[100]

3. According to Wicke's suggestion,[93] the intensified structure of water in the neighborhood of nonpolar groups does not involve icelike structures. Instead, the adjacent H_2O groups may be joined by linkages with stronger electrostatic screening of the protons (i.e., linkages such as the nontetrahedral hydrogen bonds).

Recent study in our laboratory[101] using external reference with bulk susceptibility correction has confirmed the up-field shift. The results clearly indicate the difference which exists between the effects of R_4NX salts on one hand and nonelectrolytes with nonpolar groups on the other in influencing the proton chemical shift of water. Glew et al.[122] have reported down-field shift of water proton resonance with addition of some nonelectrolytes (e.g., t-butanol and tetrahydrofuran) in water at $0°C$. The up-field shifts induced by R_4NX salts are difficult to explain and require further investigation.

Hertz and Zeidler[102] made the nuclear magnetic relaxation rate study of aqueous solutions. When the thermal equilibrium distribution of nuclear spins over the energy states in a magnetic field is temporarily disturbed, the disturbance decays with a time constant, T_1, and the reciprocal, $1/T_1$, is known as the relaxation rate. The relaxation rate for the proton nuclear spins of water molecules provides information about the average time required for them to execute one rotation. These reorientation times are approximately proportional to the measured relaxation rates for the proton nuclear spins. To avoid interference due to the protons of $(C_nH_{2n+1})_4N^+$ ions, Hertz and Zeidler used D_2O as solvent. The reorientation times of the D_2O molecules are then studied by means of the deuteron nuclear quadrupole resonance signal. Their results are shown in Figure 11, in which the relaxation rate, $1/T_1$, of the nuclear quadrupole resonance of the D_2O deuterons in solutions of tetraalkylammonium salts in D_2O is plotted as a function of salt concentration at $25°C$. It can be seen that the reorientation time of D_2O molecules becomes increasingly longer on passing from Me_4N^+ to Et_4N^+ to Pr_4N^+. With Bu_4N^+ ion, the reorientation time is less than Pr_4N^+, indicating a maximum structure-making effect for Pr_4N^+ ion in this series.

Some typical results of nuclear-magnetic-relaxation measurements are given in Figure 12 for an aqueous solution of Et_4NBr. The hydration zone of the ions is clearly indicated. The values beside the curved arrows show the reorientation times of the particles in question; those near the straight arrows are the average times spent by the particles at a given site between two jumps.

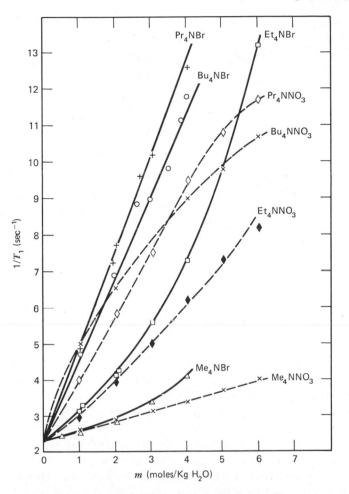

Figure 11. Relaxation rate of the nuclear quadrupole resonance of the D_2O deuterons in solutions of tetraalkylammonium salts in D_2O at 25°C. From Ref. 102.

These residence times were found from the self-diffusion of the particles. Both the reorientation time and the residence time can be taken as 10^{-11} sec in pure water. In the hydration sphere of the bromide ions, the reorientation and transposition take 7×10^{-12} sec (i.e., the H_2O molecules are more mobile than in pure water); in the neighborhood of the Et_4N^+ ion, the molecules take 2×10^{-11} sec for one reorientation (i.e., they are rather more rigidly bound than in pure water).

Lindman, Forsen, and Forslind[100] measured the line width of the [79]Br nuclear magnetic quadrupole resonance signal in aqueous solutions of R_4NBr

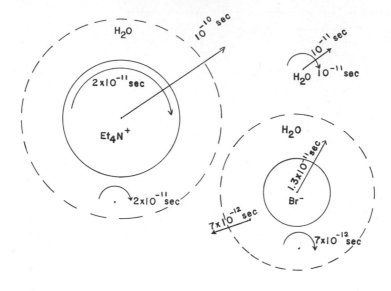

Figure 12.   Reorientation and residence times of the H_2O molecules in various zones and of the ions in an aqueous solution of Et_4NBr at 25°C. From Ref. 102.

series. They found that the relaxation time of the [79]Br nuclei decreases rapidly with increasing length of the alkyl groups on the cation and with increasing concentration of the electrolyte. The signal was found to narrow with increasing temperature. They showed that a simple model in which the Br^- ions are rapidly exchanging between two sites, with different bonding properties for the bromides, provides a description of the observations. The two sites correspond to Br^- ions in the water lattice in bulk and associated with a clathrate-like water lattice in the vicinity of the complex cations. However, as mentioned in the preceding section on diffusion, results of the self-diffusion coefficients using NMR spin–echo technique lead Hertz, Lindman, and Siepe[103] to conclude that a rigid, long-lived hydration sphere of the tetraalkylammonium cations does not exist.

Infrared.   Ever since Suhrmann and Breyer[104] showed that the near-infrared bands of water are characteristically changed by the presence of dissolved electrolytes this method has been used by various workers to study the structure of water in electrolyte solutions. However, there are only few reports in the literature on the study of aqueous quaternary ammonium salt solutions using the near-infrared technique.

Gordon and his co-workers[105] studied the 1.15 to 1.25-μ band of water assigned as $\nu_1 + \nu_2 + \nu_3$ and concluded that a 2-m solution of Me_4NCl showed a structure-breaking effect, whereas Bu_4NBr showed a structure-making effect. However, the choice of the 1.15 to 1.25-μ band to observe these effects seems to be somewhat questionable, since, according to Bunzl,[106] absorption bands of the carbon–hydrogen overtones of the R_4N^+ ions are also present in this wavelength region. Consequently, the 1.15 to 1.25-μ band of water in a Bu_4NBr solution will be altered by the 1.186-μ band of the solute, thus complicating the interpretation. This possibility of an interference of solute and solvent absorption may also be present to some extent in the measurements of Klotz,[107] who selected the 1.45-μ band of water to study the solid hydrate $Bu_4NBr \cdot 32H_2O$.

Bunzl[106] has carried out a systematic near-infrared study of R_4NBr solutions as a function of concentration and temperature. He selected the 0.97-μ band since this band appears in a wavelength region where the salts do not absorb. Temperature dependence of the wavelength of maximum absorbance, λ_{max}, of the 0.97-μ band was analyzed in terms of the structural temperature, t_{str}, following Bernal and Fowler.[108] The value of t_{str} is obtained by selecting a point on one of the solution curves (λ_{max} versus t_{soln}) and finding the corresponding temperature at which pure water has the same λ_{max}. The quantity $\Delta t = t_{soln} - t_{str}$ taken at a given temperature and concentration is then a measure of the change in the association of the water molecules in the solution compared with pure water. When Δt is plotted as a function of t_{soln}, the slope, $\partial \Delta t / \partial t_{soln}$, is found to be large negative for a Bu_4NBr solution, slightly smaller negative for a Pr_4NBr solution, nearly-0 for Et_4NBr, and positive for an Me_4NBr solution. These slopes are taken by Bunzl to indicate the structure-making effect of Bu_4NBr and Pr_4NBr, the structure-breaking effect of Me_4NBr, and the cancellation of two opposing structure effects for Et_4NBr. These notions derived from the near-infrared study of R_4NBr solutions are in good agreement with the conclusion reached by Kay et al.[8,83] from the study of conductance and viscosity.

Worley and Klotz[109] studied the near-infrared spectra for HOD in D_2O solutions containing a variety of solutes. The method takes advantage of the sharpening in band structure caused by uncoupling of fundamental stretching motion. They assigned the absorbance at 1.416 μ to all nonbonded OH species and that at 1.556 μ to all OH groups in a bonded state. From the temperature dependence of the absorbance at the two wavelengths, the enthalpy of the hydrogen bond in water was calculated to be 2.37 kcal/mole. With the addition of 2-m Me_4NBr and Bu_4NBr, the hydrogen-bond strengths of water were observed to be 2.37 and 2.64 kcal/mole, respectively. Up to a concentration of about 2 m, Bu_4NBr was found to lower the structural temperature but it began to act as a structure breaker at higher concentrations.

6 Dielectric Constants and Dielectric Relaxation

In 1952 Haggis, Hasted, and Buchanan[110] carried out the dielectric constant and loss measurements at three microwave frequencies ($\lambda = 9.22$, 3.175, and 1.264 cm) for a wide variety of aqueous solutions including solutions of three tetraalkylammonium salts (Me_4NBr, Me_4NI, and Et_4NCl). These salts were found to lower the static dielectric constant of water, and the extent of the lowering increased with the salt concentration.

When electrolytes or organic substances are dissolved in water up to 1 molar concentration, the single relaxation time of water is shortened or lengthened by an amount proportional to concentration:

$$(14) \qquad \lambda_{s,\,solution} = \lambda_{s,\,water} + c \cdot \delta\lambda_s$$

where λ_s is the wavelength, c is the molarity of the solution, and $\delta\lambda_s$ is the molar lengthening of relaxation wavelength and is proportional to the molar lengthening of the statistical half-period of the relaxation process. It was found that $\delta\lambda_s$ is negative for inorganic salts and positive for organic substances containing nonpolar groups. Since $\delta\lambda_s$ is positive for the three tetraalkylammonium halides studied, the water relaxation time is lengthened by the presence of these salts.

Recently Pottel and Lossen[111] have determined the complex dielectric

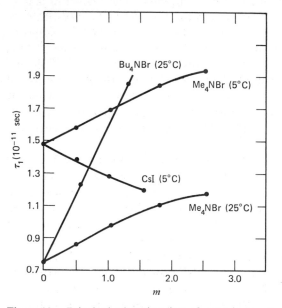

Figure 13. Principal relaxation time of water in aqueous salt solution plotted against the molal concentration. From Ref. 111.

constants of aqueous R_4NBr solutions at about 10 frequencies in the range of 0.5 to 38 $GH_z(\lambda = 0.6 \sim 60$ cm). When the conductance-substracted imaginary part ε'' of the dielectric constant is plotted against the real part ε' for Pr_4NBr or Bu_4NBr solutions, the semicircle of the Cole–Cole plot has a center lying below the ε'-axis. This corresponds to a case of relaxation processes with a continuous distribution of relaxation times. The principal relaxation time, τ_1, of water in solution is plotted against molal concentration, m, in Figure 13; τ_1 changes linearly with m at least up to 1 molal for R_4NBr solutions. The molal changes of the relaxation time of water in solution with respect to the relaxation time, τ_0, of pure water are summarized in Table 13.

Table 13   The Molal Change of the Relaxation Time of Water in Solution, τ_1, with Respect to the Relaxation Time, τ_0, of Pure Water

Salt	Temperature (°C)	$\Delta\tau_1$ $(\tau_0 \Lambda m)$
Me_4NBr	5	+0.12
Me_4NBr	25	+0.14
Et_4NBr	25	+0.36
Pr_4NBr	25	+0.65
Bu_4NBr	25	+0.86
CsI	5	−0.14
CsI	25	−0.078

In the homologous series of R_4NBr, $\Delta\tau_1/(\tau_0\Delta m)$ increases at constant rate with the number of carbons, as shown in Figure 14. The crosses, which indicate the results of NMR measurements by Hertz,[102] agree well with the results of dielectric measurements. Moreover, the relative relaxation time change per mole of CH_2 per kilogram of H_2O, $\Delta\tau_1/(\tau_0 n \Delta m)$ increases with the alkyl chain length. With the increasing chain length for $0 < n < 8$ the influence of the ionic field on the structure of water is greater than that of alkyl group; but for $n > 8$, on the other hand, the ionic influence can be neglected. The relative relaxation time change of R_4N^+ ions with $n > 12$ is 0.054/mole CH_2/kg H_2O at 25°C, which is also common for aliphatic hydrocarbons with arbitrary number of CH_2 groups. The values of $\Delta\tau_1/(\tau_0\Delta m)$ include measurements of Haggis et al.[110] performed on aqueous solutions of alcohols and amines. The additional lengthening of relaxation time for these solutions is probably due to the polar groups (OH, NH_2) that enter into hydrogen bonding with the neighboring water molecules. The reason for

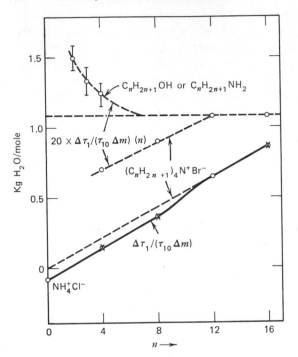

Figure 14. Relative relaxation time changes of water at 25°C, in solutions per mole of salt per kg H_2O for R_4NBr as well as for solutions of aliphatic alcohols and amines, expressed per mole of CH_2 per kg H_2O; n is the number of CH_2 groups per molecule. From Ref. 111.

the alkyl groups in lengthening the relaxation time of water is taken to indicate the presence of the "hydration of second kind."

7   Ultrasonic Absorption

Symons et al.[112] reported in 1966 some ultrasonic absorption of aqueous solutions of tetraalkylammonium halides and found them to show marked relaxations at frequencies of 1.5 to 300 MHz/sec. Their computer analysis indicated that in Bu_4NBr solution there are two or more relaxation processes with relaxation frequencies of about 7.5 and 200 MHz/sec at 25°C. The high-frequency processes increased with decrease in temperature and with increase in concentration. Based on these observations, they suggested that the high-frequency relaxations probably occur because of some sort of conformational change associated with the special hydration properties of the R_4N^+ ions.

Atkinson, Garnsey, and Tait[113] repeated the ultrasonic absorption study of Bu_4NBr solutions in detail and found at least two and possibly three relaxations in the frequency range 10 to 290 MHz, in agreement with the result of Symons et al. However, they appear to reject the explanation concerning the high-frequency relaxations, and they proposed various mechanisms including ion pairing, dimerization of ion pairs,[21] and cation pairing. Results of their kinetic analysis are given in Table 14. They emphasized that

Table 14   Ultrasonic   Absorption   in   Aqueous Bu_4NBr Solutions—Results   of   Kinetic   Analysis   ($M^+ = Bu_4N^+$, $X^- = Br^-$)

Step	Process	$k_1 \times 10^{-8}$	$k_{-1} \times 10^{-8}$	K_1
1	$M^+ + X^- \rightleftharpoons M^+X^-$	—	—	3.9
2	$2(M^+X^-) \rightleftharpoons (M^+X^-)_2$	13.6	4.6	2.9
3	$2M^+ \rightleftharpoons M_2^{2+}$	0.3	0.1	3

other reaction schemes can also be formulated which will give formally correct fits of the experimental data. Since the $R_4N^+X^-$ salts exhibit some of the most complex electrolyte behavior seen to date, further measurements extending to higher frequencies are required for understanding of the relaxation mechanism.

8   X-Ray Diffraction of Aqueous Solution

The diffraction pattern of a concentrated solution of tetrabutylammonium fluoride in water ($Bu_4NF \cdot 41H_2O$) has been measured by Narten and Lindenbaum[114] at 25°C. The effect of Bu_4N^+ ions on the structure of liquid water deduced from the radial distribution function seems to be twofold: the near-neighbor distance between water molecules decreases from 2.85 to 2.80 Å, indicating stronger hydrogen bonding in the solution. At the same time, the average number of nearest neighbors decreases from 4.4 in pure water to 3.8 in the concentrated Bu_4NF solution. The observed shift in the average near-neighbor distances from 2.85 to 2.80 Å is a direct confirmation of the hypothesis[6] that Bu_4N^+ ions promote water structure. Since Danford[115] has observed a significantly smaller decrease in distance (<0.01 Å) in an ammonium fluoride solution of comparable ion concentration ($NH_4F \cdot 45.8H_2O$), Narten and Lindenbaum concluded that the degree of water structure promotion is significantly larger for Bu_4N^+ than for NH_4^+

ions. The change in the average coordination number (not found for the ammonium fluoride solution) may be taken as an indication that not only the average strength but also the average number of hydrogen bonds per water molecule differs in the Bu_4NF solution, in pure water, and in NH_4F solutions.

Narten and Lindenbaum have calculated intensity and radial distribution functions for two different models. The respective models assume modified ice-I and gas-hydrate structures, surrounded by a random distance distribution, as a description for the average short-range order in the solution. In both models the F^- ions and the N^+ atoms of the cations are part of a hydrogen-bonded network, and the butyl chains of the cations are located inside the cavities which, although of different size and shape, are typical of both structures. By comparing the intensity and radial distribution functions calculated for these structures with those obtained from their experiments, they concluded that the ice-I model describes the X-ray data of the solution and pure water. The gas-hydrate model is said to be incompatible with the experimental radial distribution function of liquid water, and the parameters of this model for the solution do not explain the large enthalpy and entropy of fusion of the solid hydrate.

9 Concluding Remarks

Aqueous solutions of symmetrical tetraalkylammonium salts show numerous interesting properties that are now being actively investigated in many laboratories. We have reviewed studies on clathrate hydrates, thermodynamic quantities, transport properties, and applications of various spectroscopic techniques to these solutions.

In dilute solution, evidence is strong that the Pr_4N^+ and Bu_4N^+ ions increase the structure of water, whereas the Me_4N^+ ion diminishes the structure. However, it is not yet clear what kind of water structure is formed around the alkyl groups of the large cations. Is it icelike, clathrate-like, or a structure involving nontetrahedral hydrogen bonds? What is the nature of the so-called hydration of the second kind? Many questions remain to be answered. Can we meaningfully define and assign a hydration number to each cation? For example, for Me_4N^+, Et_4N^+, Pr_4N^+, and Bu_4N^+, Pottel and Lossen[111] assigned the hydration numbers of 16, 21, 27, and 32, respectively; Hertz et al.[103] assumed these numbers to be 25, 30, 35, and 40, respectively, at 25°C.

In 1964 Hertz[120] discussed the structural effect of nonpolar particles in aqueous solution and made an extensive review of their physical properties. He found that certain properties show a structural increase and the others indicate a decrease in the number of hydrogen bonds in solution: he concluded that the structural increase must be one that is attended by rupture or deformation of hydrogen bonds. Wicke[93] suggested that the water structure around nonpolar group may involve nontetrahedral hydrogen bonds that may exist in dimers, trimers, tetramers, pentamers, or hexamers of water. These postulates are of great interest and will stimulate our ideas in search for better understanding. However, the main evidence used by Hertz and Wicke in making these postulates so far seems to be the up-field chemical shifts of H_2O protons in R_4NBr and R_4NNO_3 solutions observed by Hertz and Spalthoff.[98] In view of the recent finding of the down-field shifts of H_2O protons in dilute nonelectrolyte solutions by Glew and others,[122] these postulates are still open to question. This example points out the necessity of further studies on chemical shifts using improved techniques.

On the other hand, the dielectric dispersion studies indicate the presence of a series of relaxation processes in large tetraalkylammonium salt solutions, showing very complex nature of the structures present. Further studies are needed on many fronts to attack the problem. They might include ultrasonic absorption studies, Raman spectral studies, and X-ray diffraction studies of the aqueous solutions. Other studies such as light scattering and measurements on the surface potentials similar to what Jarvis and Scheiman[121] have done on inorganic salt solutions may be useful.

Some of the problems involved are closely related to those for solutions of neutral molecules containing nonpolar group or groups such as hydrocarbons, alcohols, and amines; and the currently active research on these systems will undoubtedly contribute to the better understanding of the aqueous tetraalkylammonium salt solutions.

10   Note Added in Proof

In the period of two years since this manuscript was submitted to the Editor, there have been some rapid developments which must be mentioned.

Friedman and his co-workers[123] used the hypernetted-chain integral equation to calculate the ion–ion correlation functions and thermodynamic properties of models of aqueous alkali halides and tetraalkylammonium halides based on an ion–ion pair potential functions having four terms: the Coulomb term; a core repulsion term; a dielectric repulsion term; and a "Gurney" term. The last represents the effect of the overlap of the structure-modified regions, "cospheres," about the ions when the ions come close together. The only parameter in the potential, which is adjusted to fit excess free-energy data for solutions, is the coefficient A_{ij} of the Gurney term for the interaction of ions of species i and j. The A_{ij} parameters are in the range from 100 to -200 cal/mole, and become more negative with increasing size of halide ion. For a fixed halide ion, the value of A_{+-} is a minimum for Et_4N^+ and becomes less negative with increasing cationic size. In contrast A_{++} is more negative when the cationic size of R_4N^+ is greater. The computed correlation functions, g_{ab}, for models which fit the osmotic coefficient data[125] for aqueous Et_4NCl and Bu_4NCl at 0.3 to 0.4 M show sharp peaks at $++$ distance of about 8 and 10 Å, respectively.[124]

Rupert and Frank measured the transference numbers of Et_4NCl and Bu_4NCl in water at 25°C by the moving-boundary method and obtained the single-ion conductances for Et_4N^+ and Bu_4N^+.[125] They determined the activity coefficients of these salts by the EMF measurements of concentration cells with transference over the molality range of 0.001 to 0.25.

$$Ag, AgCl/R_4NCl(m_R = 0.05)/R_4NCl(m_2)/AgCl, Ag$$

The activity coefficient curves of Et_4NCl and Bu_4NCl are found to cross each other at about 0.14 m and the corresponding osmotic coefficient curves cross each other at about 0.09 m. Lindenbaum et al.[126] measured the osmotic coefficients of some R_4NX salts in water at 65°. Lindenbaum determined the apparent molal heat content of 1-m tetrabutylammonium butyrate (Bu_4NBut) and compared it with corresponding values of $NaBut$, Bu_4NCl, and $NaCl$ and concluded that ion-pairing does not occur in these solutions.[127]

Apparent molal volumes and adiabatic compressibilities of organic nitrogen-containing ions including some R_4NX were measured by Laliberté and Conway,[128] while integral heats of solutions of Bu_4NBr, Pr_4NI, and $NaBP_4$ in water at 3, 5, and 15°C were determined by Sarma and Ahluwalia.[129]

Jolicoeur and Friedman[130] have studied the electron paramagnetic resonance spectra of solutions of hydrophobic EPR probes, 2,2,6,6-tetramethyl-4-piperidine-1-oxyl (A) and bispicolinato vanadyl (B). The rotational correlation time τ is found to be larger for the probes in aqueous solutions (including probe A in 0 to 1.4 m Bu_4NBr) than in nonaqueous solvents when compared at the same viscosity. Since the probe molecule itself has hydrophobic regions, this might be a manifestation of the local influence of hydrophobic groups on water. In the aqueous solutions containing a second hydrophobic species, the composition depence of τ is very nearly given by Walden's rule, $\tau/\eta =$ constant. They found, however, considerable variance with Walden's rule for probe A in $NaBPh_4$ solution and for probe B in Bu_4NBr solution.

Lindman et al.[131] have extended their study of ^{79}Br nuclear quadrupole relaxation to aqueous solutions of mono-, di-, and tri-alkylammonium bromides. The relaxation rate of ^{79}Br was found to increase with increasing number of alkyl groups and with the concentration of salts. They attributed the observed effects to the increased anion-water binding in the vicinity of the cations. This postulate is of interest and worthy of attention.

NMR line shapes for α-methyl protons in Me_4N^+ and Me_3N^+Et salts have been found to vary with concentration, solvent and temperature caused by the changes in ^{14}N quadrupole relaxation rate.[132] The results are interpreted in terms of contact ion pair in DMSO and solvent-separated ion pairs in aqueous solution.

Fister and Hertz[133] investigated ^{17}O nuclear magnetic resonance of aqueous solutions containing Me_4NBr and Bu_4NBr and found that the cation with longer alkyl chains gives a greater up-field shift. This may mean a decrease of hydrogen bonds as the effect of introducing the structure forming alkyl groups into water. Unfortunately, however, the effect is hardly outside the limit of their experimental uncertainties. NMR studies of clathrate hydrates of $(i\text{-}Am)_4NF$ and Bu_4NF have been carried out at a wide range of temperatures.[134,135] They show some types of motion of D_2O as well as guest species and imply a defect structure of the water lattice or a disordered guest structure.

Applying Samoilov's equation, Uedaira has calculated the values of the activation energy of the exchange of the nearest water molecules next to an ion and found them to be positive for all tetraalkylammonium cations.[136] Broadwater and Kay reported conductance measurements on Me_4NBr and Me_4NI for aqueous solutions containing up to 20 mole % t-BuOH and on Bu_4NBr up to 70 mole % t-BuOH.[137]

The ultraviolet absorption spectrum of small amount of iodide (e.g., $5 \times 10^{-5}\ M$) in water shifts with temperature and with addition of solutes.[138] Addition of Me_4NCl and Et_4NCl ($0 < c < 3\ M$) results in a shift of E_{max}, the energy of absorption maximum, to higher energies, while Pr_4NCl shifts the band initially to higher and then to lower energies; in contrast, Pe_4NCl shifts the band to lower energies. These effects do not fit a simple pattern expected from the relative sizes of the cations. A similar contrast occurs for dE_{max}/dT for iodide in these solutions. Addition of Me_4NCl decreases dE_{max}/dT initially but then increases, while Et_4NCl, Pr_4NCl and Pe_4NCl show a trend of increasing dE_{max}/dT.

Acknowledgment

The author wishes to thank the National Science Foundation for a grant (GP-8870) under which this work received support.

REFERENCES

1. A. W. Hofmann, *Ann. Chem.*, **78**, 253 (1851).
2. L. Ebert and J. Lange, *Z. Physik. Chem.*, **139A**, 584 (1928); J. Lange, *Z. Physik. Chem.*, **168A**, 147 (1934).
3. D. L. Fowler, W. V. Loebenstein, D. B. Pall, and C. A. Kraus, *J. Amer. Chem. Soc.*, **62**, 1140 (1940).
4. H. S. Frank and M. W. Evans, *J. Chem. Phys.*, **13**, 507 (1945).
5. E. C. Bingham, *J. Phys. Chem.*, **45**, 885 (1941).
6. H. S. Frank and W.-Y. Wen, *Discussions Faraday Soc.*, **24**, 133 (1957).
7. R. McMullan and G. A. Jeffrey, *J. Chem. Phys.*, **31**, 1231 (1959).
8. R. L. Kay and D. F. Evans, *J. Phys. Chem.*, **69**, 4216 (1965); *J. Phys. Chem.*, **70**, 366, 2325 (1966).
9. G. E. Boyd, A. Schwarz, and S. Lindenbaum, *J. Phys. Chem.*, **70**, 821 (1966); G. E. Boyd, Q. V. Larson, and S. Lindenbaum, *J. Phys. Chem.*, **72**, 2651 (1968).
10. D. Feil and G. A. Jeffrey, *J. Chem. Phys.*, **35**, 1863 (1961).
11. M. Bonamico, R. K. McMullan, and G. A. Jeffrey, *J. Chem. Phys.*, **37**, 2219 (1962).
12. R. K. McMullan, M. Bonamico, and G. A. Jeffrey, *J. Chem. Phys.*, **39**, 3295 (1963).
13. M. von Stackelberg, *Z. Elektrochem.*, **58**, 24, 40, 99, 104, 162 (1954).
14. G. Beurskens, G. A. Jeffrey, and R. K. McMullan, *J. Chem. Phys.*, **39**, 3311 (1963).
15. H. Flaschka and A. J. Barnard, Jr., *Advan. Anal. Chem. Instr.*, **1**, 24 (1960).
16. (a) W.-Y. Wen and S. Saito, *J. Phys. Chem.*, **68**, 2639 (1964); (b), *J. Phys. Chem.*, **69**, 3569 (1965).
17. L. G. Hepler, J. M. Stokes, and R. H. Stokes, *Trans. Faraday Soc.*, **61**, 20 (1965).
18. B. J. Levien, *Australian J. Chem.*, **18**, 1161 (1965).
19. B. E. Conway, R. E. Verrall, and J. E. Desnoyers, *Trans. Faraday Soc.*, **62**, 2738 (1966).
20. R. E. Verrall and B. E. Conway, *J. Phys. Chem.*, **70**, 3961 (1966).
21. H. E. Wirth, *J. Phys. Chem.*, **71**, 2922 (1967).
22. F. Franks and H. T. Smith, *Trans. Faraday Soc.*, **63**, 2586 (1967).
23. R. Gopal and M. A. Siddiqi, *J. Phys. Chem.*, **72**, 1814 (1968).
24. F. J. Millero and W. Drost-Hansen, *J. Phys. Chem.*, **72**, 1758 (1968).
25. T. L. Broadwater and D. F. Evans, *J. Phys. Chem.*, **73**, 164 (1969).
26. J. Padova and I. Abrahamer, *J. Phys. Chem.*, **71**, 2112 (1967).
27. W.-Y. Wen, *Saline Water Conversion Report*, Office of Saline Water, Dept. Interior, Washington, D.C., 1966, p. 13.
28. I. Lee and J. B. Hyne, *Can. J. Chem.*, **46**, 2333 (1968).
29. O. Redlich and D. M. Meyer, *Chem. Rev.*, **64**, 221 (1964).
30. R. Zana and E. Yeager, *J. Phys. Chem.*, **71**, 521 (1967).
31. R. Zana and E. Yeager, *J. Phys. Chem.*, **71**, 4241 (1967).
32. W.-Y. Wen and K. Miyajima, unpublished results.
33. F. Franks, M. J. Quickenden, J. R. Ravenhill, and H. T. Smith, *J. Phys. Chem.*, **72**, 2668 (1968).
34. F. J. Millero, *J. Phys. Chem.*, **72**, 4589 (1968).
35. T. L. Broadwater and D. F. Evans, *J. Phys. Chem.*, **73**, 164 (1969).
36. B. E. Conway and R. E. Verrall, *J. Phys. Chem.*, **70**, 3952 (1966).
37. B. E. Conway and L. H. Laliberté, *J. Phys. Chem.*, **72**, 4317 (1968).

38. W.-Y. Wen and S. Saito, *J. Phys. Chem.*, **69**, 3569 (1965).
39. V. E. Bower and R. A. Robinson, *Trans. Faraday Soc.*, **59**, 1717 (1963).
40. S. Lindenbaum and G. E. Boyd, *J. Phys. Chem.*, **68**, 911 (1964).
41. M. A. V. Devanathan and M. J. Fernando, *Trans. Faraday Soc.*, **58**, 784 (1962).
42. R. H. Stokes, *Trans. Faraday Soc.*, **59**, 761 (1963).
43. H. S. Frank, *J. Phys. Chem.*, **67**, 1554 (1963).
44. R. M. Diamond, *J. Phys. Chem.*, **67**, 2513 (1963).
45. S. Lindenbaum, *J. Phys. Chem.*, **70**, 814 (1966).
46. R. W. Hendricks and S. Lindenbaum, *Abstr. Papers, Div. Colloid and Surface Chemistry*, 157th National Meeting, Amer. Chem. Soc., Minneapolis, April 1969, Paper no. COLL—34.
47. W.-Y. Wen, S. Saito, and C. M. Lee, *J. Phys. Chem.*, **70**, 1244 (1966).
48. H. S. Frank, *Z. Physik. Chem.*, **228**, 364 (1965).
49. J. Steigman and J. Dobrow, *J. Phys. Chem.*, **72**, 3424 (1968).
50. J. E. Desnoyers, M. Arel, G. Perron, and C. Jolicoeur, *J. Phys. Chem.*, **73**, 3346 (1969).
51. C. M. L. Chen, Ph.D. thesis, Clark University, 1968.
52. R. H. Wood, H. L. Anderson, J. D. Beck, J. R. France, W. E. deVry, and L. J. Soltzberg, *J. Phys. Chem.*, **71**, 2149 (1967).
53. H. S. Frank and A. L. Robinson, *J. Chem. Phys.*, **8**, 933 (1940).
54. R. H. Wood and R. W. Smith, *J. Phys. Chem.*, **69**, 2974 (1965).
55. J. E. Mayer, *J. Chem. Phys.*, **18**, 1426 (1950).
56. H. L. Friedman, *J. Chem. Phys.*, **32**, 1134 (1960); *Ionic Solution Theory*, Interscience, New York, 1962.
57. R. H. Wood and H. L. Anderson, *J. Phys. Chem.*, **71**, 1871 (1967).
58. Y. C. Wu, M. B. Smith, and T. F. Young, *J. Phys. Chem.*, **69**, 1868 (1965).
59. R. H. Wood and H. L. Anderson, *J. Phys. Chem.*, **70**, 992 (1966).
60. W.-Y. Wen and K. Nara, *J. Phys. Chem.*, **71**, 3907 (1967); (b) W.-Y. Wen, K. Nara, and R. H. Wood, *J. Phys. Chem.*, **72**, 3048 (1968).
61. W.-Y. Wen and K. Nara, *J. Phys. Chem.*, **72**, 1137 (1968).
62. H. Ruterjans, F. Schreiner, U. Sage, and T. Ackermann, *J. Phys. Chem.*, **73**, 986 (1969).
63. C. V. Krishnan and H. L. Friedman, *J. Phys. Chem.*, **73**, 3934 (1969); see *Abstr. Papers*, 157th National Meeting, Amer. Chem. Soc., Minneapolis, April 1969, paper no. PHYS-121.
64. C. V. Krishnan and H. L. Friedman, *J. Phys. Chem.*, **73**, 1572 (1969).
65. K. W. Bunzl, *J. Phys. Chem.*, **71**, 1358 (1967).
66. Om. N. Bhatnager and C. M. Criss, *J. Phys. Chem.*, **73**, 174 (1969).
67. J. E. Desnoyers, G. E. Pelletier, and C. Jolicoeur, *Can. J. Chem.*, **43**, 3232 (1965).
68. H. E. Wirth and A. LoSurdo, *J. Phys. Chem.*, **72**, 751 (1968).
69. N. C. Deno and C. H. Spink, *J. Phys. Chem.*, **67**, 1347 (1963).
70. W. F. McDevit and F. A. Long, *J. Amer. Chem. Soc.*, **74**, 1773 (1952).
71. T. J. Morrison and N. B. Johnstone, *J. Chem. Soc.*, 3655 (1955).
72. A. Ben-Naim, *J. Phys. Chem.*, **71**, 1137 (1967).
73. W.-Y. Wen and J. H. Hung, submitted to *J. Phys. Chem.*, 1969.
74. G. Wada and S. Umeda, *Bull. Chem. Soc. Japan*, **35**, 646, 1797 (1962).
75. F. Franks and B. Watson, *Trans. Faraday Soc.*, **63**, 329 (1967).
76. A. J. Darnell and J. Greyson, *J. Phys. Chem.*, **72**, 3021 (1968).
77. J. D. Worley and I. M. Klotz, *J. Chem. Phys.*, **45**, 2868 (1966).
78. V. O. Laurence and J. H. Wolfenden, *J. Chem. Soc.*, 1144 (1934).

79. R. M. Fuoss and C. A. Kraus, *J. Amer. Chem. Soc.*, **79**, 3304 (1957).
80. W.-Y. Wen, Ph.D. thesis, University of Pittsburgh, 1957.
81. E. Hückel and H. Schaaf, *Z. Physik. Chem.*, **21**, 326 (1959).
82. E. R. Nightingale, Jr., *J. Phys. Chem.*, **66**, 894 (1962).
83. R. Kay, T. Vituccio, C. Zawoyski, and D. Evans, *J. Phys. Chem.*, **70**, 2336 (1966).
84. G. Jones and M. Dole, *J. Amer. Chem. Soc.*, **51**, 2950 (1929).
85. H. Falkenhagen and E. L. Vernon, *Physik. Z.*, **33**, 140 (1932).
86. H. M. Daggett, E. J. Bair, and C. A. Kraus, *J. Amer. Chem. Soc.*, **73**, 799 (1951).
87. E. L. Swarts and C. A. Kraus, *Proc. Natl. Acad. Sci. (U.S.)*, **40**, 382 (1954).
88. R. W. Martel and C. A. Kraus, *Proc. Natl. Acad. Sci. (U.S.)*, **41**, 9 (1955).
89. R. M. Fuoss and F. Accascina, *Electrolytic Conductance*, Interscience, New York, 1959.
90. D. F. Evans, G. P. Cunningham, and R. L. Kay, *J. Phys. Chem.*, **70**, 2974 (1966).
91. R. A. Horne and R. P. Young, *J. Phys. Chem.*, **72**, 1763 (1968).
92. R. A. Horne and J. D. Birkett, *Electrochim. Acta*, **12**, 1153 (1967).
93. E. Wicke, *Angew. Chem.*, **5**, 106 (1966).
94. R. A. Horne and R. P. Young, *J. Phys. Chem.*, **72**, 376 (1968).
95. J. N. Agar and J. C. R. Turner, *Proc. Roy. Soc. (London)*, *Ser. A*, **255**, 307 (1960).
96. J. N. Agar, *The Structure of Electrolytic Solutions*, Ed., W. J. Hamer, Wiley, New York, 1959, Chapter 13.
97. J. N. Agar, *Advances in Electrochemistry and Electrochemical Engineering*, vol. 3, Interscience, New York, 1963, Chapter 2.
98. H. G. Hertz and W. Spalthoff, *Z. Elektrochem.*, **63**, 1096 (1959); H. G. Hertz, *Ber. Bunsenges. Physik. Chem.*, **67**, 311 (1963).
99. J. Clifford and B. A. Pethica, *Trans. Faraday Soc.*, **60**, 1483 (1964).
100. B. Lindman, S. Forsen, and E. Forslind, *J. Phys. Chem.*, **72**, 2805 (1968).
101. Yn-hwang Lin, M.A. Thesis, Clark University, 1969.
102. H. G. Hertz and M. D. Zeidler, *Ber. Bunsenges. Physik. Chem.*, **68**, 821 (1964).
103. H. G. Hertz, B. Lindman, and V. Siepe, *Ber. Bunsenges. Physik. Chem.*, **73** (1969) in press.
104. R. Suhrmann and F. Breyer, *Z. Physik. Chem. (Leipzig)*, **20**, 17, 23, 193 (1933).
105. H. Yamatera, B. Fitzpatrick, and G. Gordon, *J. Mol. Spectry.*, **14**, 268 (1964).
106. K. W. Bunzl, *J. Phys. Chem.*, **71**, 1358 (1967).
107. I. M. Klotz, *Federation Proc.*, **24**, S-24 (1965).
108. J. D. Bernal and R. H. Fowler, *J. Chem. Phys.*, **1**, 516 (1933).
109. J. D. Worley and I. M. Klotz, *J. Chem. Phys.*, **45**, 2868 (1966).
110. G. H. Haggis, J. B. Hasted, and T. J. Buchanan, *J. Chem. Phys.*, **20**, 1452 (1952).
111. R. Pottel and O. Lossen, *Ber. Bunsenges. Physik. Chem.*, **71**, 135 (1967).
112. M. J. Blandamer, M. J. Foster, N. J. Hidden, and M. C. R. Symons, *Chem. Commun.*, **3**, 62 (1966).
113. G. Atkinson, R. Garnsey, and M. J. Tait, in *Hydrogen-Bonded Solvent Systems*, A. K. Covington and P. Jones, Eds., Taylor and Francis Ltd., London, 1968, p. 161.
114. A. H. Narten and S. Lindenbaum, *J. Chem. Phys.*, **51** (1969) in press.
115. M. D. Danford, "Diffraction Pattern and Structure of Aqueous Ammonium Fluoride Solutions," ORNL-4244 (1968).
116. K. Tamaki, *Bull. Chem. Soc. Japan*, **40**, 38 (1967).
117. W. Kauzmann, *Advan. Protein Chem.*, **14**, 1 (1959).
118. G. Némethy and H. A. Scheraga, *J. Chem. Phys.*, **36**, 3401 (1962).
119. R. H. Aranow and L. Witten, *J. Chem. Phys.*, **28**, 405 (1958); *J. Chem. Phys.*, **35**, 1504 (1961).

120. H. G. Hertz, *Ber. Bunsenges. Physik. Chem.*, **68**, 907 (1964).
121. N. L. Jarvis and M. A. Scheiman, *J. Phys. Chem.*, **72**, 74 (1968).
122. D. N. Glew, H. D. Mak, and N. S. Rath, in *Hydrogen-Bonded Solvent Systems*, A. K. Covington and P. Jones, Eds., Taylor and Francis Ltd., London, 1968, pp. 197–210.
123. P. S. Ramanathan and H. L. Friedman, *J. Chem. Phys.*, **54**, 1086 (1971).
124. P. S. Ramanathan and H. L. Friedman, State University of New York at Stony Brook, 1971, private communication.
125. J. P. Rupert, Ph.D. dissertation, University of Pittsburgh, 1969.
126. S. Lindenbaum, L. Leifer, G. E. Boyd, and J. W. Chase, *J. Phys. Chem.*, **74**, 761 (1970).
127. S. Lindenbaum, *J. Phys. Chem.*, **74**, 3027 (1970).
128. L. H. Laliberté and B. E. Conway, *J. Phys. Chem.*, **74**, 4116 (1970).
129. T. S. Sarma and J. C. Ahluwalia, *J. Phys. Chem.*, **74**, 3547 (1970).
130. C. Jolicoeur and H. L. Friedman, *J. Phys. Chem.*, **75**, 165 (1971).
131. B. Lindman, H. Wennerström, and S. Forśen, *J. Phys. Chem.*, **74**, 754 (1970).
132. D. W. Larsen, *J. Phys. Chem.*, **74**, 3380 (1970).
133. F. Fister and H. G. Hertz, *Ber. Bunsenges. Phys. Chem.*, **71**, 1032 (1967).
134. D. D. Eley, M. J. Hey, K. F. Chew, and W. Derbyshire, *Chem. Commun.*, **1968**, 1474.
135. C. A. McDowell and P. Raghunathan, *Mol. Phys.*, **15**, 259 (1968).
136. H. Uedaira, *Zh. Fiz. Khim.*, **42**, 3024 (1968).
137. T. L. Broadwater and R. L. Kay, *J. Phys. Chem.*, **74**, 3802 (1970).
138. M. J. Blandamer and M. F. Fox, *Chem. Rev.*, **70**, 59 (1970).

16 Hydration of Macromolecules

Gilbert N. Ling, Department of Molecular
Biology, Division of Neurology,
Pennsylvania Hospital, Philadelphia, Pennsylvania

1 Early Hydration Theory and Its Decline

Since the middle of the nineteenth century, biologists have been aware of the immiscibility with water of extruded cytoplasm from plant and animal cells (Nägeli, 1855; Kühne, 1864). This immiscibility, and other evidence as well, led some to conclude that water in the cytoplasm is not free as in dilute salt solutions but, rather, is bound to the macromolecular constituents. One type of experimental evidence cited in support of this view was the demonstration that some water in protein solutions and in living cells refused to freeze even at a temperature as low as $-20°C$ (Rubner, 1922; Thoenes, 1925; Robinson, 1931; Jones and Gortner, 1932). Another type of experimental evidence involved the determination of the freezing point after a known amount of a reference substance such as sucrose had been introduced into the system (the cryoscopic method). Since bound water does not dissolve, or dissolves less sucrose than free water, the concentration of sucrose in the free water within the system is then higher than it is in the case where all water is free. Consequently, the freezing-point lowering observed may also be expected to be lower than in normal water. This type of observation was made by Gortner and his co-workers (Newton and Gortner, 1922; Blüh, 1928; Gortner and Gortner, 1934; Gortner, 1937, 1938; Weissman, 1938).

The view of bound water championed in the 1930s by Gortner and his associates met strong opposition. Hill and his co-workers, on the basis of vapor-pressure measurements, concluded that very little, if any, water in blood or muscle is "bound" (1930). Greenberg and Cohn (1934), using glucose distribution as a criterion, demonstrated that in a solution of 5% gelatin or 6% casein, no more than 0.01 g (casein) or 0.05 g (gelatin) of water could be demonstrated to be bound to 1 g of dry protein; this quantity was considered trivial. MacLeod and Ponder (1936) argued that, if normal living cells possess a substantial amount of bound water, then that portion is unable to dissolve a solute like ethylene glycol. However, they found experimentally

663

that ethylene glycol distributes itself equally between erythrocytes and external solution. They concluded that no bound water existed in these cells.

In a review written in 1940, Blanchard quoted the above-mentioned evidence against the bound water concept. He further argued that, since pure water could be supercooled to $-20°C$, incomplete freezing of water might be attributed to supercooling rather than to water binding. Similarly, he argued that since one could not accurately determine the freezing point in so complex a system as a protein solution, the cryoscopic evidence for bound water that relied on the measurement of freezing point could not be judged valid.

Much of Blanchard's evidence against the bound water concept was eventually proven to be equivocal (see Ling, 1969). Hydration of proteins became established by a large variety of modern techniques, described later. Nevertheless, Blanchard's review was a landmark in the history of cell physiology, which thereafter saw the virtual abandonment of the hydration theory of cell water.

With opposition eliminated, the membrane theory became so generally accepted that textbooks almost universally presented it as undisputed truth. Moreover, bound water was not the only concept that came with the triumph of the membrane theory.

In 1933 Dorothy Jordan Lloyd and H. Phillips pointed out that "the colloidal properties of a protein are dependent on the ease and degree with which it becomes hydrated." This statement clearly brought out the central role of protein hydration in colloidal chemistry. It was not surprising that, with the decline of interest in the cell hydration theory, interest in colloidal chemistry also evaporated.

2   The Rebirth of the Concept of Protein Hydration:
General Theories of Water Sorption on Solid Proteins

The Capillary Condensation Theory.   While the theory of cell hydration waxed and waned, the textile industry had been steadily concerned with the ability of materials such as cotton and wool to sorb water (Alexander and Hudson, 1954). In this connection, the capillary condensation theory prevailed over many years. In the capillary condensation theory, the bulk of the water sorbed was considered to exist as normal liquid filling up pores and narrow channels in the proteins.

Two lines of recent study have shown that, as such, this theory can no longer be considered valid.

Mellon, Korn and Hoover (1949) showed that extensive alteration of the physical structure of wool, egg albumin, and silk fibroin did not alter appreciably the water sorption. Such treatments as they used were certain to destroy or alter any preexisting capillary structures. Benson and Ellis (1948, 1950) in their study of N_2, O_2, and CH_4, sorption on dry proteins, concluded that there was no evidence to indicate the existence of fine pore structures to begin with. Furthermore, they found that the amount of physically adsorbed gases (e.g., N_2, O_2) depends only on the state of subdivision of the proteins. It does not depend on the specific proteins involved. In another paper, Benson, Ellis, and Zwanzig (1950) showed that water sorption, on the contrary, depends only on specific sites of the proteins but not on state of their subdivision.

Theories of Multilayer Adsorption for Inert Gases. The adsorption of gases on the surface of solids has long been of interest to physicists and chemists. In plotting the amount of gas adsorbed against the partial pressure of the gas, a sigmoid-shaped curve has often been observed. In explanation, deBoer and Zwikker offered their "polarization theory" (1929), according to which electric charges on the solid surface induce dipoles in the first layer of gases adsorbed. These dipoles in turn induce dipoles in the gas molecules, forming a second layer, and this process continues until a polarized multilayer is formed. The isotherm they derived is

$$(1) \qquad \log\left(\frac{p}{p_0}\right) = K_2 K_1{}^a + K_3$$

where p is the partial pressure of the gas under study; p_0 is the pressure of the gas at saturation under the same condition; K_1, K_2, and K_3 are all constants under the specified condition of the experiment; a is the amount of gas adsorbed. Equation 1 can be written in the following form:

$$(2) \qquad \log\left(\log\frac{p}{p_0} - K_3\right) = a \log K_1 + \log K_2$$

The same problem was subsequently attacked by Bradley (1936), who introduced an isotherm of the same form as (1) or (2), describing the polarized multilayer adsorption of gases with *permanent dipole moments*, such as water (permanent dipole moment, 1.87×10^{-18} esu). He showed that adsorption of water on CuO crystal surfaces follows the isotherm he derived. However, other gases studied by deBoer and Zwikker and Bradley included noble gases without permanent dipole moments.

The Brunauer–Emmett–Teller (BET) Theory. Brunauer, Emmett, and Teller (1938) criticized the polarization theory of deBoer and Zwikker and Bradley in its application to the adsorption of noble gases such as argon on the ground that the polarizability of the gas is too low to build up more than a single layer on the surface of solid crystals. They suggested instead that the forces holding the multilayers together are the same forces that cause condensation of gases. By generalizing Langmuir's theory of unimolecular adsorption, they derived an isotherm that has been known as the BET isotherm

$$(3) \qquad \frac{p}{v(p_0 - p)} = \frac{1}{v_m c} + \frac{c - 1}{v_m c} \frac{p}{p_0}$$

where v is the total amount of gas adsorbed at vapor pressure, p; v_m is the volume adsorbed in a unimolecular layer, and c is a constant. Thus, in a plot of $p/v(p_0 - p)$ against p/p_0, a straight line should be obtained from which v_m and c may be derived.

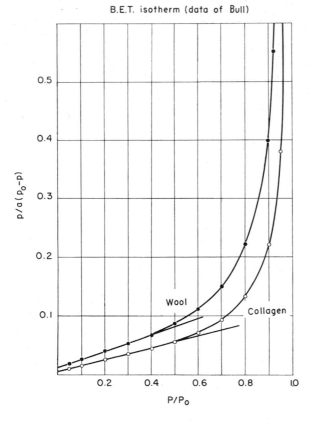

Figure 1. Water sorption on collagen and on wool, plotted according to the BET isotherm using the data of Bull (1944). From Ling (1965) by permission of *Annals of the New York Academy of Sciences.*

The sorption of water vapor by proteinaceous materials at varying vapor pressures had been studied quite early (see Speakman, 1936). However, it was Bull (1944) who first undertook an extensive study of vapor sorptions at different temperatures on a large variety of dry purified proteins. [For a review of water sorption on dry proteins and polymers see McLaren and Rowen (1951).] In years following, these data were often chosen by other authors to test various theories on the sorption of water by proteins. Bull himself analyzed his data with the aid of the BET isotherm. He found that in plots of $p/v(p_0 - p)$ against p/p_0, the curves are linear up to about 40 to 50% saturation. Beyond 50% saturation, however, the departure from the theory is severe, as illustrated in Figure 1.

In 1946 Hill (1946) used statistical–mechanical methods to derive a more generally applicable isotherm having the same form as the BET isotherm. Thus the Hill isotherm can be applied to surface adsorption as well as adsorption in the interior of proteins. It also offers an interpretation of the constant, c, of (3) more explicit than in the BET theory. In Hill's theory, c is equal to $\exp(E_1 - E_L/kT)$, where $E_1 - E_L$ is the heat evolved when one mole of liquid is adsorbed on the localized sites. Also, v_m of (3) is replaced by B, the concentration of localized adsorption sites. The Hill treatment was further extended by White and Eyring (1947) and by Dole (1949).

Solution Theories.   Most of the proteins studied by Bull remain insoluble up to 95% saturation. Salmine, however, becomes an aqueous solution at high vapor pressure. Similarly, the sorption of acetone by cellulose nitrate also results in a solution of the absorbant in the condensed vapor (Rowen and Simha, 1949). These findings led to the development of a theory based on the concept of the formation of solid solutions. Thus, according to the theory of Flory and Huggins (Flory, 1942), the partial molal free energy, ΔF_{sol} for the solution process, is approximately

$$(4) \qquad -\Delta F_{sol} = RT \ln \frac{p}{p_0} = RT (\ln v_1 + v_2 + \mu v_2{}^2)$$

where v_1 and v_2 are the volume fractions of the condensed vapor and polymer, respectively, and μ is a semiempirical parameter determined by the enthalpy and entropy of mixing. This equation does give a better fit than the BET theory to the vapor sorption at high relative vapor pressure ($p/p_0 > 0.5$), but it does not describe the lower region at all.

The Bradley Isotherm for Water Sorption on Proteins.   Hoover and Mellon (1950) found that much of their water sorption data on proteins, polyglycine, and so on, could be described by a simplified version of (1) with K_3 equal to 0. As illustrated in Figure 2, the fit extends all the way from 6 to 93% relative humidity.

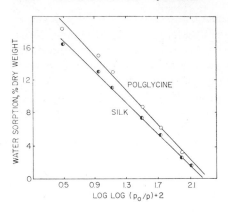

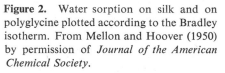

Figure 2. Water sorption on silk and on polyglycine plotted according to the Bradley isotherm. From Mellon and Hoover (1950) by permission of *Journal of the American Chemical Society.*

It should be pointed out that the rejection of the deBoer–Zwikker–Bradley theory by Brunauer et al. was only in reference to the adsorption of noble gases and other molecules possessing no permanent dipole moment. The criticism does not apply to gases with permanent dipole moments. Since water does indeed possess a sizable permanent dipole moment, the application of the deBoer–Zwikker–Bradley isotherm is a far more appropriate choice. That the theory also fits the data much better than the BET isotherm confirms this view.

3   Theories of the Sites of Hydration

Theories of Lloyd, Sponsler, Bath, and Ellis.   Lloyd and her co-workers had long ago considered the protein sites that interact with water (Lloyd, 1933; Lloyd and Phillips, 1933; Lloyd and Shore, 1938). They pointed out that it is through coordination with the oxygen, nitrogen, and hydrogen atoms of the hydroxyl, carboxyl, amino, amido, and, to a lesser extent, the imino and keto groups, that water is bound to proteins.

Sponsler, Bath, and Ellis (1940) noted that, in the dormant period, living protoplasm retains only some 30 to 35% water, instead of the 80% found in active cells. They cited a variety of experimental evidence indicating a restriction of the motional freedom of water in a gelatin-water gel containing less than 35% water. From X-ray diffraction and infrared adsorption evidence, they showed that polar side chains as well as backbone amide groups bind water (Table 1). On the basis of the amino acid analysis of gelatin, they concluded that the polar side chains interact with 438 water molecules and the backbone amide group with 450 to 500. The total of 800 to 850 in fact corresponds well to the 33% water in the gelatin.

Table 1   Water Molecules Coordinated with Hydrophilic Groups

Hydrophilic Groups	Number of Water Molecules (Theoretical)	Number of Water Molecules (Experimental)
H_2O	4	4
—OH	3	3
—COOH	5–4	4
=O	2	2
—NH_2	3	3
$>$NH	2	—
$>$N—	1	—

From Sponsler, Bath, and Ellis, 1940, by permission of *Journal of Physical Chemistry*.

Pauling's Theory.   Pauling (1945), on the basis of data from Bull, suggested that each polar side chain complexes one water molecule except the carboxyl groups that are hydrogen-bonded to an imido group on a glutamine or asparagine side chain (on this issue, see also Green, 1949). Pauling further suggested that the backbone imino and carboxyl groups do not bind water. To support this contention, he quoted Bull's data on the low water sorption of nylon. This synthetic polymer may be represented as

$$\left(\begin{array}{c} -N-C-(CH_2)_3-N-C-(CH_2)_4 \\ \ \ H \ \ \| \ \ \ \ \ \ \ \ \ \ \ \ \ H \ \ \| \\ \ \ \ \ \ O \ \ \ \ \ \ \ \ \ \ \ \ \ \ \ \ O \end{array} \right)_n$$

It resembles protein backbones $\left(\begin{array}{c} -N-C-C \\ \ \ H \ \ \ \ \ \| \\ \ \ \ \ \ \ \ O \end{array} \right)_m$. Table 2 presents a com-

parison of the number of polar groups in each protein with the water molecules in the first adsorbed layer derived from Bull's data using the BET theory. In general, proteins poor in polar side chains, such as silk fibroin, take up fewer water molecules at low relative humidity than such proteins as salmine and gelatin, which are rich in polar side chains. Other proteins fall between these extremes.

Thus, in a general way, Pauling has pointed out an additional significance of Bull's work: a verification of the theory of the sites of hydration as pointed out by Lloyd and her co-workers and later by Sponsler et al. There are, however, two major departures from earlier views.

Table 2   Comparison of the Number of Water Molecules Held by Proteins in Initial Adsorption and the Number of Polar Groups in the Proteins

Protein	Water Adsorbed in First Layer (moles/10^5 g)			Number of Polar Groups[a] (moles/10^5 g)	Total Reported Amino Acids[b]
Silk	226			219–228	107.0[c]
Ovalbumin					
crystallized	329[d]	342	344[e]	277–313	75.7
lyophilized	314				
heat denatured	276				
Wool	366			303–341	71.4[g]
Gelatin, collagen	485	529		328–609	108.8[f]
C-zein, B-zein	210	228		305–390	106.0[h]
Salmine	592[i]			611–707	110.5
Serum albumin	374			424–424[j]	86.8
β-Lactoglobulin				472–508[k]	115.8
crystallized	370				
lyophilized	329				

From Pauling, 1945, by permission of *Journal of the American Chemical Society*.
[a] Of each pair of numbers, the first number represents the polar residues not including the carbonyl groups of proline and hydroxyproline, and the second includes these.
[b] The percentages include the water added to the residues on hydrolysis of the proteins; complete analysis corresponds to 110 to 116%.
[c] Analysis for silk fibroin.
[d] Value found by Shaw.
[e] Value calculated by Shaw from data obtained by Barker.
[f] Analysis for wool keratin.
[g] Analysis for gelatin.
[h] Analysis for zein.
[i] Water adsorbed in first two steps.
[j] Analysis by E. Brand, B. Kassell, and L. J. Saidell, *J. Clin. Invest.*, **23**, 437 (1944); the analysis is for bovine serum albumin, whereas the adsorption data are for horse serum albumin.
[k] E. Brand and B. Kassell, *J. Biol. Chem.*, **145**, 365 (1942), and private communication from Brand.

1. Pauling accepted the basic assumption of the BET theory, namely, that all water taken up by the proteins beyond the first layer exists as normal liquid water. Thus Pauling's conclusion that each polar group combines with one water molecule signifies that this is the case whatever the relative humidity of the environment and whether a protein exists as a solid or as an aqueous solution. This is in direct conflict with the conclusions of Sponsler, Bath, and Ellis (1940) and with the bulk of experimental evidence now available.

2. Pauling departs from the conclusion of Sponsler, Bath and Ellis in his contention that the keto and imino groups of the protein backbone do not combine with water.

In a review of water sorption by swelling high polymers, White and Eyring (1947) reached a conclusion in support of the gist of Pauling's theory of protein hydration, namely, that water sorption occurs on polar side chains and little or no sorption occurs on peptide chains.

The Question of Whether the Backbone Binds Water.   The importance of Pauling's theory is indisputable, as discussed later, but there is considerable evidence to show that the backbone does indeed complex water.

1. Studies of the infrared absorption band of water in gelatin-gel containing 35% water were shown by Ellis and Bath (1938) to indicate water binding.

2. Mellon, Korn, and Hoover (1948) showed that amorphous polyglycine esters containing the same backbone keto and imino groups as normal water adsorb a considerable amount of water (Figure 2). The amount of water sorbed by polyglycine far exceeds that which can be accommodated at the amino and ester ends of the polymer and can only be accounted for by the complexing of water with the backbone amide groups. Similarly, McLaren and Katchman (see McLaren and Rowen, 1951) showed a high degree of hydration of poly-glycine-D,L-alanine (1:1). This subject is discussed again.

3. From investigations of linear synthetic polymers, Dole and Faller (1950) showed that polyamides may or may not adsorb water, depending on their degree of crystallinity; the more crystalline polymers adsorb less water. Since the normal nylon studied by Bull was highly crystalline, its low water sorption can be understood. Dole and Faller also showed that polyvinylpyrrolidone absorbs a considerable amount of water. This polymer has no terminal polar group as in the case of polyglycine esters. Nor does it possess a hydrogen-donating group in its N atom. The only hydrogen-bonding groups are the carbonyl groups, similar to those on protein backbones.

This evidence strongly indicates that the peptide imino and carbonyl groups are inherently capable of binding a considerable amount of water. However, other factors such as intra- or intermolecular hydrogen bonding of the protein chains may prevent its occurrence.

Theory of Bull and Breese.   Very recently Bull and Breese (1968) further analyzed the nature of water-sorption sites in proteins. Their fundamental assumptions are similar to those of Pauling in the following ways.

1. The hydroxyl (on serine, threonine, and tyrosine side chains), carboxyl (aspartic, glutamic, and α-carboxyl), and the basic groups (arginine, histidine, lysine, and α-amino groups) all bind water except when they are in the presence of an amide side chain which, as first suggested by Pauling, was

considered to form a hydrogen bond with the nonamide side-chain polar group and, in so doing, to eliminate the water-binding capability of the latter.

2. The peptide imino and carbonyl groups as water-binding sites are ignored.

Bull and Breese differ from Pauling in two principal ways.

1. They consider the water sorption at a relative humidity of 92%, instead of that corresponding to the initial first layer of water sorption.

2. They assign six water molecules as sorbed on each polar side chain, instead of assigning one molecule to each site.

An equation for the total hydration is

$$(5) \qquad Y = -0.97 \times 10^{-3} + 6.77X_1 - 7.63X_2$$

where Y is the number of moles of water bound per gram of protein at 25°C and at a relative humidity of 0.92; X_1 is the sum of the moles of hydroxyls, carboxyls, and basic groups per gram of protein; X_2 is the moles of amide group (on asparagine and glutamine residues) per gram of protein. The excellent agreement between this theory and the hydration of nine proteins is shown in Table 3. Three other proteins that do not obey (5) because they take up more than the predicted amount of water are silk fibroin, zein, and collagen.

Table 3   Moles of Water Bound per Gram of Protein at a Relative Humidity of 0.92 and at 25°C

	Moles of $H_2O \times 10^3$/g protein	
Protein	Observed	Calculated by (5)
Lysozyme	13.8	14.1
Bovine serum albumin	17.7	(22.3)[a]
Bovine hemoglobin	20.4	21.2
Bovine ribonuclease	19.7	19.4
Horse heart cytochrome c	21.6	22.0
Egg albumin	16.4	16.1
β-Lactoglobulin	17.8	16.8
Sperm whale myoglobin	23.4	23.1
Bovine insulin	13.1	14.6
α-Chymotrypsinogen A	16.1	15.2

From Bull and Breese, 1968, by permission of *Archives of Biochemistry and Biophysics.*
[a] Bovine serum albumin not included in the formulation of (5).

4   Hydration of Proteins in Solution

So far, our review has been concerned with water sorption in solid proteins. The last two decades have also seen the development of a number of methods for the measurement of hydration of proteins in solution. These include calorimetric (Amberg, 1957), osmotic-pressure (Lauffer, 1966), X-ray scattering (Ritland et al., 1950), self-diffusion of solvent (Wang, 1954), sedimentation-velocity (Cox and Schumaker, 1961a), sedimentation-equilibrium (Cox and Schumaker, 1961b), dielectric-dispersion (Oncley, 1943), and nuclear-magnetic-resonance (NMR) measurements (Kuntz et al., 1969). In Table 4

Table 4   Hydration of Proteins in Solution

Protein	Reference Listed at End of Table	Technique Identified at End of Table	Hydration (g H_2O/g protein)
Serum albumin	5	A	0.2
	6	B	0.30–0.42
	10	C	0.19–0.26
	11	D	0.31
	7	E	1.07[a]
	7	F	0.75[a]
	8	G	0.43
	12	F	0.40
	12	H	0.48
	13	A	0.18–0.64
	14	A	0.15
	9	A	0.23
Avg.			0.32
Ovalbumin	2	I	0.18
	5	A	0.1
	6	B	0–0.15
	8	G	0.31
	7	E	0.45[a]
	9	A	0.18
Avg.			0.17
Hemoglobin	6	B	0.20–0.28
	11	D	0.10
	8	G	0.45
	7	E	0.36[a]
	7	F	0.69[a]
	9	A	0.14
	10	A	0.2
Avg.			0.25
β-Lactoglobulin	6	B	0–0.20
	7	E	0.72[a]
	7	F	0.61[a]
	5	A	0.4
	9	A	0.24
Avg.			0.25

(*continued*)

Table 4   (cont).

Protein	Reference Listed at End of Table	Technique Identified at End of Table	Hydration (g H_2O/g protein)
Lysozyme	6	B	0.25–0.32
	11	D	0.24
	8	G	0.36
	7	E	0.89[a]
	8	F	0.33–0.35
	9	A	0.23
Avg.			0.29
Ribonuclease	7	E	0.35
	7	F	0.59
	10	C	0.34
Avg.			0.43
Myoglobin	5	A	0.2–0.3
L-Chymotrypsin	8	G	0.37
Chymotrypsinogen	7	E	0.52
Catalase	7	E	0.70
Urease	7	E	0.53

[a] Maximum hydration value, not used in calculating average for protein.

Techniques
A   Dielectric dispersion.
B   X-ray scattering.
C   Sedimentation velocity.
D   Sedimentation equilibrium.
E   Diffusion coefficient.
F   Intrinsic viscosity.
G   NMR.
H   Frictional coefficient.
I ^{18}O Diffusion.

References
1.   Fisher (1965).
2.   Wang (1954).
3.   Adair and Adair (1963).
4.   Adair and Adair (1947).
5.   Oncley (1943).
6.   Ritland et al. (1950).
7.   Tanford (1961).
8.   Kuntz et al. (1969).
9.   Buchanan et al. (1952).
10.   Cox and Schumaker (1961a).
11.   Cox and Schumaker (1961b).
12.   Anderegg et al. (1955).
13.   Grant et al. (1968).
14.   Haggis et al. (1951).
15.   Miller and Price (1946).

the magnitudes of hydration in protein solution are listed. The theory and techniques involved in a majority of these methods have been well described and need no repetition (Edsall, 1953; Tanford, 1961). We shall briefly describe dielectric-dispersion and NMR measurements, since reports on these have appeared very recently.

Dielectric Measurements.   In 1942 Oncley reviewed his extensive work on the dielectric measurements of aqueous solutions of proteins at megacycle frequencies (Oncley, 1942, 1943). Ten years later Buchanan, Haggis, Hasted, and Robinson (1952) reported their study of the dielectric properties of egg albumin at microwave frequencies. Wide-range studies revealed two frequency regions of dielectric dispersion: the β region, corresponding to relaxation of the protein molecules and the γ region, corresponding to the relaxation of water. Later Schwan (1957) and Grant (1962) observed an extra dispersion, referred to as the δ region, which lies between the β and γ regions (Figure 3). It was suggested (Schwan, 1957, Grant, 1965) that rotation of bound water was responsible for the δ region. Schwan, however, pointed out that the polar side chains might also relax in this frequency region (Schwan, 1957). In a later paper, Pennock and Schwan (1969) concluded that the dielectric dispersion of a hemoglobin solution (7.5–26.6 g/100 c.c.) between 1 and 1200 MHz can be separated into three regions. Below 30 MHz the dispersion is attributed to the dipolar nature of the hemoglobin molecules.

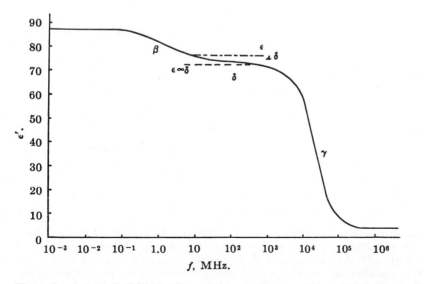

Figure 3.   A typical dielectric dispersion curve for a protein solution. From Grant, Keefe, and Takashima (1968) by permission of *Journal of Physical Chemistry.*

Between 10 and 100 MHz, the dispersion is attributed to the relaxation of polar side chains extending from the surface of water molecules into the solution. Above 100 MHz, the dispersion is attributed to relaxation of a shell of bound water. The estimated amount of bound water most compatible with the data is 0.15 to 0.25 g/g hemoglobin.

NMR Studies of Frozen Protein Solutions. Kuntz, Brassfield, Law, and Purcell (1969) studied the NMR signals of frozen 5 to 10% protein and nucleic acid solution at $-35°C$. At this temperature the signal from ice protons is broadened to the point of indiscernibility; yet the protein solution continues to give relatively narrow signals, as illustrated in Figure 4. The authors found that the signal is directly proportional to the concentration of the macromolecules. On the basis of this and other evidence, they concluded that these signals are due to bound water. The amount of bound water

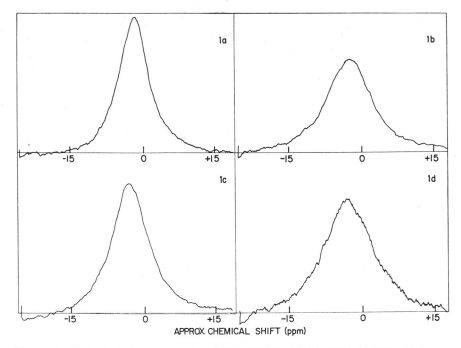

APPROX. CHEMICAL SHIFT (ppm)

Figure 4. Representative spectra at $-35°C$, 60-MHz NMR, 0.001-M KCl: (*a*) lysozyme; (*b*) ovalbumin; (*c*) native serum albumin (pH 5.03); (*d*) pH "denatured" serum albumin (pH 2.45). All solutions were approximately 75 mg/ml protein except (*d*), which was 50 mg/ml but whose spectrum was obtained at twice the spectrum amplitude of the others. Line width (Hz) can be obtained by multiplying the width of the absorption line at half height (ppm) by 60. From Kuntz et al (1969), by permission of *Science*.

measured was between 0.3 and 0.5 g water/g protein, which agrees well with data obtained from hydrodynamic and other methods (Table 4).

Kuntz et al. concluded that their data did not support the idea of an icelike structure for water hydration. The lack of a definite freezing point also indicated that the bound water "is not in a well-organized structure even though it is more restricted in mobility than liquid water." They suggested that part of the water is physically and chemically sorbed on surfaces and that other parts may represent water trapped in small channels or pores. We have already presented the evidence against the capillary water concept. I feel that the polarized multilayer interpretation appears more plausible; this view is developed later.

Many of the data on hydration of protein molecules have been derived on the basis of hydrodynamic measurements. In these studies the validity of the existence of bound water is usually dependent on the validity of the assumptions about the size and shape of the protein molecules. This point is well illustrated by Table 5. The same diffusion coefficient data may be interpreted as indicating hydration or lack of hydration, according to the assumption of a spherical or ellipsoid shape for the protein. It was from this type of occasion that Ogston long ago pointed out the need for extreme caution in viewing hydration data (Ogston, 1956). The NMR study of Kuntz et al. added much more unequivocal evidence that the conventional interpretation of hydration data has proven to be remarkably accurate.

Fisher's Limiting Law. We have already shown how the recent data of Bull and Breese supported Pauling's model of protein hydration in a general way. There is now additional evidence for Pauling's theory, in the sense that the backbone contributes nothing to protein hydration.

In recent years there has been an increasingly popular view that the polar groups of protein molecules in an aqueous solution, are directed outward and nonpolar groups are folded inside the protein molecule (Waugh, 1954; Kendrew, 1963; Fisher and Cross, 1964; Tanford, 1962). By assuming (a) that the size of the hydrated polar groups is more or less constant and (b) that nonpolar groups are not allowed to be on the outside surface, Fisher (1964) suggested that for each protein molecule with a defined amino acid composition, there is a corresponding number of nonpolar amino acid residues that the protein molecule can accommodate. He defined a "polarity ratio," p, as

$$(6) \qquad p = \frac{v_e}{v_i}$$

where v_e and v_i are the volume of the polar outer volume and the volume of nonpolar inner volumes, respectively. Assuming the protein to be a sphere,

Table 5   Solvation and/or Asymmetry Calculated from the Diffusion Coefficients[a]

	Maximum Solvation, $f/f_0 = 1$		Maximum Asymmetry, $\delta_1 = 0$, $f/f_0 = f/f_{min}$	Compromise, $\delta_1 = 0.2$	
	δ_1 (g/g)	R_e (Å)	a/b of Prolate Ellipsoid	$\dfrac{f}{f_0}$	a/b of Prolate Ellipsoid
Ribonuclease	0.35	18.0	3.4	1.05	2.1
Lysozyme	0.89	20.6	6.1	1.21	4.3
Chymotrypsinogen	0.52	22.5	4.2	1.11	3.0
β-Lactoglobulin	0.72	27.4	4.9	1.16	3.7
Ovalbumin	0.45	27.6	3.8	1.08	2.5
Serum albumin	1.07	36.1	6.5	1.25	4.9
Hemoglobin	0.36	31.0	3.4	1.05	2.1
Catalase	0.70	52.2	4.9	1.15	3.6
Urease	0.53	61.9	4.2	1.11	3.0
Tropomyosin	23.0	96	62		
Fibrinogen	8.4	106	31		
Collagen	218	310	300		
Myosin	49	215	100		

From Tanford, 1961, by permission of John Wiley and Sons, Inc.
[a] Alternative interpretations of diffusion data according to the model of solvated spheres (of radius R_e) (column 1) or to that of a nonhydrated protein molecule of large asymmetry (column 3).

and designating p_s as the polarity ratio for a spherical protein molecule, he showed that

$$(7) \qquad\qquad p_s = \frac{r^3}{(r-d)^3} - 1$$

where r is the radius of the molecules. Since the thickness of the outer layer, d, is assumed to be a constant, and r is determined by the total number of amino acid residues in the protein molecule, Fisher suggested that "as the overall size of the protein molecule decreases, the relative number of non-polar groups must decrease, because of the disappearance of interior in which to bury them."

Thus, for each protein with a known total molecular volume, there is a corresponding p_s. If there are more polar side chains, the p value calculated from (6) and the amino acid composition will be lower than p_s. If, on the other hand, there are fewer nonpolar groups, p will be higher than p_s. In the first case, Fisher suggests that the protein will tend to assume other shapes with a larger surface area than spherical area. Fisher examined 35 proteins and found that proteins with p less than p_s, such as insulin, do indeed tend to polymerize. Others, with p much larger than p_s, such as tropomyosin and myosin tend to assume a rodlike structure.

In a subsequent article, Fisher (1965) set forth a corollary to his earlier theory, namely, that there is an upper limit of hydration for each protein molecule. To reach this conclusion he assumed that each polar residue on the outside surface of the protein is 4.0 Å thick, that each protein molecule is covered with a monomolecular layer of water with the same density as water in the bulk solvent, and that in this layer the water molecules have the same random orientation as in the rest of the solvent. Then each water molecule occupies a cubic volume of 33.3 Å³, equivalent to a surface area of 10.3 Å². Fisher represented the weight (in grams) of water bound per gram of protein as H_t, which is related to v_e by

$$(8) \qquad H_t = \frac{18 \text{ g/mole} \times V_e}{4(10.3 \text{ Å/molecule})(6.02 \times 10^{23} \text{ molecules/mole})}$$
$$= 0.725 \times 10^{-24} \, V_e$$

In Table 6 we have presented the H_t values theoretically calculated by Fisher for the seven proteins listed, in comparison with the experimentally determined values of the H_t quoted by Fisher and those from the data collected in Table 4. The overall agreement between theory and experimental

Table 6   Comparison of Theoretical Values of Protein Hydrations According to Fisher and Experimental Data

| Protein | Data (g H_2O/g dry protein) | | |
	Theoretical	Experimental quoted by Fisher[a]	Experimental, from Table 4[b]
Ovalbumin	0.26	0.16	0.17
β-Lactoglobulin	0.28	0.20	0.25
Serum albumin	0.29	0.31	0.32
Hemoglobin	0.25	0.23	0.25
Ribonuclease	0.33	0.44	0.43
Lysozyme	0.32	0.27	0.29
Myoglobin	0.29	0.25	0.25

[a] Averages; from studies of protein crystals and dry proteins.
[b] Averages; from studies of proteins in solution.

results is good, but this accord can by no means be used to justify all the assumptions in the theory. Thus the evidence for internal hydration, in certain proteins at least, is hard to dispute (Edsall, 1953). "Random orientation" of bound water is internally inconsistent, but this does not detract from the theory as a limiting law, as Fisher made clear in his title.

Since in his calculation Fisher counted only arginine, aspartic acid, glutamic acid, histidine, lysine, serine, threonine, and tysosine residues as polar, this model is less precise and discriminating than the later work of Bull and Breese (1968), which took into account the negative correlation between the amide groups (asparagine and glutamine) and hydration. In line with the assumptions of Bull and Breese and Pauling, no consideration was given to hydration of the peptide chains by Fisher.

Possible Reconciliation of Two Apparently Opposing Conclusions. We have just noted two opposing trends regarding the polypeptide chain as sites of hydration. In one view, the peptide imino and carbonyl groups (which far exceed polar side chains in number) are not considered as sites for water binding. Support of this view came from the work of Fisher and Bull and Breese. In the other view, the polypeptide groups are regarded as important sites for water binding. The latter trend began with the work of Sponsler, Bath, and Ellis, and was strengthened by the work of Mellon, Korn, and Hoover, Dole and Faller, and McLaren and Katchman. Indeed, a substantial number of reviewers have expressed the opinion that the peptide bonds cannot be disregarded as sites of water hydration (McLaren and Rowen, 1951, p. 308; Edsall, 1953, p. 561; Gustavson, 1956, p. 150).

A survey of the literature reveals a possible solution to this apparent paradox.

Fisher's theory of hydration is what we may, for convenience, refer to as a nonpeptide participation theory. The seven proteins he cited to agree with his theoretical prediction are all globular proteins. Of the 35 proteins that he examined, the three whose p values depart farthest from the p_s values are tropomyosin, fibrinogen, and myosin. All three are fibrous proteins; the others are globular.

Similarly, of the 13 proteins examined by Bull and Breese, nine agree accurately with the theory, and all nine are globular. Of the three proteins that do not follow (5) but are much more intensely hydrated than predicted, two are fibrous: silk fibroin and collagen. The third, which does not follow (5), is zein. Zein is a seed protein, belonging to the class prolamins (Brohult and Sandegren, 1954). Mellon, Korn, and Hoover (1948) presented evidence to show that 70% of the sorbed water of this protein was due to peptide bonds.

Now the conclusion of Sponsler, Bath, and Ellis that peptide bonds are important sites for water binding was based on the study of one fibrous protein, gelatin. Much of the work of Mellon, Hoover, and their co-workers was done on synthetic fibers. Gustavson, who argued against the concept of nonpeptide participation, based his conclusion primarily on extensive knowledge of another fibrous protein, collagen. Thus it would seem that the polypeptide chain does not contribute hydration in globular proteins, although it may do so in fibrous protein.

The only outstanding departure from this rule is in Pauling's original papers, in which he considered eight proteins, three of which were fibrous and five globular. All seem to agree with Pauling's theory of nonpeptide participation. The answer to this problem is not too far to find, however. Pauling was the only one who considered nothing but the first layer of water.

From this analysis, the following conclusion may be made. In globular proteins, polar groups on the side chains are the primary seats of hydration in an environment of low water activity or high water activity. In fibrous proteins, polar groups are also main sites of hydration of low water activity; at high water activity, a peptide amide bond may make a substantial contribution to total hydration.

In the next section a theoretical basis for this conclusion is presented.

5 Molecular Mechanisms of Hydration

Theoretical Considerations

MONOPOLAR SURFACE. When a charged ion (e.g., Figure 5a) or a plane surface (Figure 5b) is brought into contact with water, the intense electric field surrounding the ion, or charged site, anchors a layer of water molecules around it. In the case of ions, this is what has been referred to as "primary hydration" (Bockris, 1949). The ion-dipole and dipole-dipole interaction does not stop at the first layer of water and is strong enough to influence further layers. The hydration beyond the first layer, referred to as secondary hydration, is known to consist of a relatively loose structure. We suggest that, in part at least, this looseness in secondary hydration is due to the lateral repulsion between water molecules in the same layer.

More or less the same situations occur in positively or negatively charged surfaces. A rough estimate of the energy involved can be made by considering only the nearest-neighbor interaction. Let us consider a water molecule in, say, the ith layer, and let each water molecule be represented as a sphere with a permanent dipole moment of magnitude, μ, and an induced dipole moment, p. Let us further assume that the equilibrium distances between neighboring water molecules are all equal to r. The total dipole-dipole interaction energy

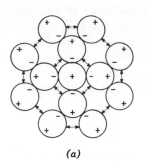

(a)

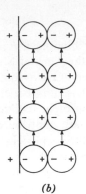

(b)

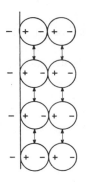

(c)

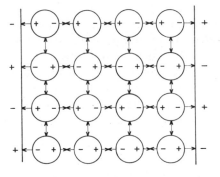

(d)

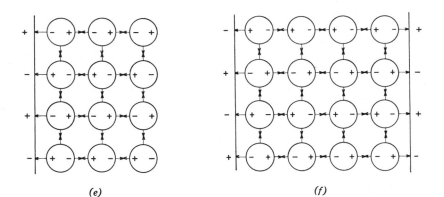

(e)   (f)

Figure 5. Effect of charged site distribution on the stability of polarized multilayers of water molecules: ⟷ repulsion; attraction. →← Unstable multilayers are produced in surfaces because of lateral repulsion between molecules in the same layer: (a) P-type site, charged positive ion. (b) P-type site, uniformly positive; (c) N-type site, uniformly negative. Stable, deep layers of water result from lateral cohesion: (d) N-P type, P-type surface placed face to face with N-type surface; (e) NP-type surfaces with alternating positive and negative sites; (f) two NP-type-surfaces placed face to face (greatest stability).

will consist of attractive components, E_I, and repulsive components, E_{II}, and

$$(9) \qquad E_I = -\frac{1}{r^3}(4\mu^2 + 4\mu p_i + 2p_i p_{i-1} + 2p_i p_{i+1})$$

and

$$(10) \qquad E_{II} = +\frac{1}{r^3}(4\mu^2 + 4\mu p_i + 4p_i p_i)$$

The induced moment, p_i, depends on the electric field, F_i, and the polarizability α, as $p_i = \alpha F_i$. Brunauer, Emmett, and Teller criticized the deBoer–Zwikker–Bradley polarization theory as applied to noble gas adsorption, because without the aid of permanent dipole, the polarizability of the noble gases is not large enough to create adsorption beyond the first layer. Now the permanent dipole moment of water is high (1.87×10^{-18} esu). Multilayers can be formed. But in the specific case under discussion, since $2p_i \simeq p_{i+1} + p_{i-1}$, E_I is virtually equal to E_{II}, but opposite in sign, and they cancel out. The result is that there would be little cohesive force other than that due to the London–van der Waals dispersion energy. Thus we may conclude that if the polar surface has the same type of charge, or if large regions of the surface carry one type of charge predominantly, there is very little reason to anticipate that there will be much more than one layer of bound water or that additional layers of water will have much more cohesion than that of normal water.

This picture is in general agreement with the bulk of data on globular protein hydration. Thus, on the basis of this model, we see that the addition of a very small amount of water to dry globular proteins leads to sorption on polar groups, at a 1:1 ratio (Pauling); further addition of water causes more water to aggregate around each polar group until a total of something like six water molecules for each polar group is reached (Bull and Breese). Water gained thereafter is not much different from normal; in aqueous solution, this water is indistinguishable from the bulk solvent. It should be pointed out that this picture is virtually the same as that deduced by Fuller and Brey (1968) from an NMR study of sorbed water on dry bovine serum albumin. The present model of the mechanism of globular protein hydration is based on the assumption that the polar groups on the protein surface consist of polar groups either isolated or in groups, but carrying charges of predominantly one sign. On the other hand, if there is a pattern of alternately positive and negative charges, a different pattern of hydration emerges, and this is the subject of the next section.

DIPOLAR SURFACE AND BIFACIAL-DIPOLAR SURFACES. In the model just discussed we considered singly-charged ion or monopolar surfaces (negative, N, or positive, P, types) from which we reached the conclusion that little more than one layer of bound water can be anticipated. Let us now examine the case illustrated in Figure 5e, where the charged surface represents a regular array of alternating positively and negatively charged sites (NP type). Making the same assumption, we will have, as before

(11) $$E_I = -\frac{1}{r^3}(4\mu^2 + 4\mu p_i + 2p_i p_{i-1} + 2p_i p_{i+1})$$

But E_{II} is now also attractive

(12) $$E_{II} = -\frac{1}{r^3}(4\mu^2 + 4\mu p_i + 4p_i^2)$$

Thus the total energy is now

(13) $$E_I + E_{II} = -\frac{1}{r^3}(8\mu^3 + 8\mu p_i + 4p_i^2 + 2p_i p_{i-1} + 2p_i p_{i+1})$$

Added together, the dipole-dipole interactions provide both radial and lateral cohesion, and this three-dimensional stabilization will promote the formation of deep polarized multilayers of adsorbed water. The stability of this multilayer will be further strengthened if there are two dipolar surfaces facing each other, which we shall refer to as an NP-NP system. In general, water sorption in an NP or an NP-NP type of system should follow the isotherm of the kind derived by Bradley.

The question of just how many layers of polarized water can be built on a simple NP surface or a double NP-NP surface can best be answered by examining a relatively simple model: a glass surface or two glass surfaces facing each other at a close distance. It is well known that the glass surface represents a two-dimensional array of positive (e.g., Na^+) and negative charges (O^-), and this is a typically NP surface (Zachariasen, 1932).

Model Systems

THE EFFECT OF GLASS SURFACES ON THE PHYSICAL STATE OF WATER.   It has long been known that water confined in narrow spaces between glass surfaces assumes properties different from those of normal water (see Henniker and McBain, 1948). In showing this, nothing could be more dramatic than the demonstration of the powerful influence of glass (and quartz) surface in orienting multiple layers of water. In this respect, the study of Hori (1956) is striking.

As shown in Figure 6, Hori proved that as a layer of water held between two polished glass surfaces becomes thinner and thinner, a point will be reached at about 10,000 Å (1 μ) where the water will not freeze at a temperature as low as $-95°C$. On the other hand, when the plates are more than 20 μ apart, the water freezes above $-40°C$.

Figure 7 demonstrates that water between glass plates about 1300 Å apart

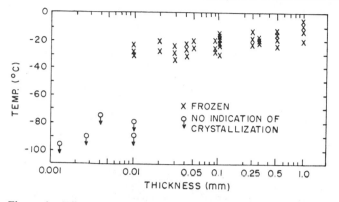

Figure 6. Effect of distance between flat glass plates upon the freezing point of water held between them. Distance between plates given in millimeters on the abscissa. Redrawn after Hori, 1956, by permission of *Low Temperature Science*.

exhibits no measureable vapor pressure and will not evaporate in a vacuum at 300°C. These results are duplicated in water between thin mica sheets.

Hori's experiments gave us an insight into how many layers of water an NP-NP surface can hold under this influence: not less than 10,000 Å, not as many as 3700 layers.

Independent evidence exists that further substantiates these results. Thus Palmer et al. (1952) have shown that at 2 to 3 MHz, thin (2-μ) layers of water held between thin mica sheets give a dielectric constant of 10, in contrast to the dielectric constant of 80 for normal water. Now dielectric constants in essence measure the rotational freedom of the water molecules. Thus the data of lowered dielectric constant would suggest rotational restrictions of deep water layers on account of the polarization on the NP-NP surfaces.

DNA AND THE CLASSIC CONCEPT OF SOLUTE EXCLUSION FROM BOUND WATER. In the long history of the bound-water concept it has frequently been maintained that bound water does not act as solvent for which solute is dissolved in normal water. Qualitatively speaking, this concept is useful. Thus by assuming that a layer of water surrounding desoxyribonucleic acid (DNA) does not dissolve a Cs^+ ion, Hearst and Vinograd (1961) demonstrated that the bound-water sheath is 4-molecules thick. This, of course, is a great deal more water of hydration than occurs with most globular proteins. Like starch and agar, nucleic acids also form gels. It is likely that the large amount of both hydrogen-donating and -accepting groups, particularly those in ribose and phosphate groups, may be contributing factors. We know that, in general, nucleic acid binds more water from other evidence (Langridge et al., 1967).

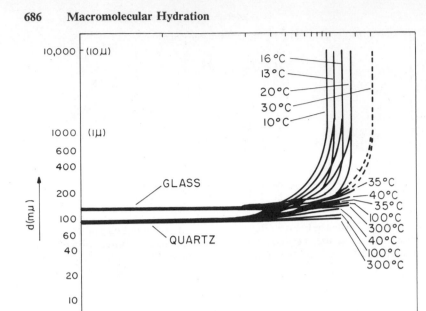

Figure 7. Effect of the distance between flat quartz and glass plates on the vapor pressure of water held between them at various temperatures. Ordinates give the distance between the plates in millimicra; abscissa gives the vapor pressure in mm/Hg. Redrawn after Hori, 1956, by permission of *Low Temperature Science*.

Not the least convincing is the demonstration of Kuntz et al. (1969). Using the low-temperature NMR technique described previously, they showed that *t*-RNA (ribonucleic acid) binds 1.7 g water/g nucleic acid in comparison with the values 0.2 to 0.5 g/g seen for all (globular) proteins studied by Kuntz et al.

6   Solute Exclusion from Water of Hydration

Solute Exclusion from Water in Protein Crystals.   Perutz (1946) studied the distribution of ammonium sulfate in hemoglobin crystals. He found that in the crystals, the salt reaches a concentration that is a constant fraction of that in the external solution with which the protein crystals are in equilibrium (Figure 8). On the basis of these data he concluded that there are two kinds of water in hemoglobin crystals: bound water, which is completely inaccessible to ammonium sulfate and is estimated to amount to 0.3 g/g of protein, and free water (0.39–0.52 g/g protein), which dissolves the same concentration of ammonium sulfate as that in the external solution.

More recently, this interpretation has met with difficulties. In β-lacto globulin crystals, McMeeken and co-workers (1950) found that the fraction of bound water is not a constant (which was the basis of Perutz's theory), but varies with the concentration of sugar studied (Figure 9).

Solute Exclusion from the Water in Copper-Ferrocyanide Gel and in Ion-Exchange Resins. Ion-exchange resins are a three-dimensional matrix richly endowed with polar groups that may reach a total concentraction of 5 moles/kg wet resin. The total water content is roughly 25 moles/kg wet resin. On the basis of this figure, and Bull and Breese's figure of six water molecules oriented by each per polar group, all the water in the resin must be bound. It is then hardly surprising that the equilibrium concentration of sucrose in the resin water (Dowex 50, H_{form}) is only 25% of that in the external medium (Wheaton and Bauman, 1951; Ling, 1965). In the classical view, this would suggest that 76% of the water is nonsolvent and bound and 24% is free. However, another solute, ethylene glycol, reaches an equilibrium distribution in the water

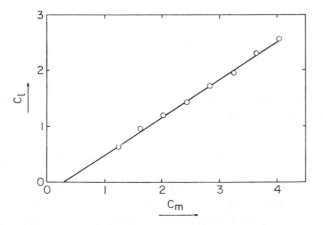

Figure 8. Ammonium sulphate concentration in the liquid of crystallization (C_l) plotted against that in the suspension medium (C_m). No significance need be attached to the fact that the line does not pass through the origin, as the amount of deviation is within the experimental error. From Perutz, 1946, by permission of the *Transactions of the Faraday Society.*

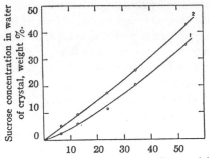

Sucrose concentration in suspending medium, weight %.

Figure 9. Effect of concentration of sucrose in the suspending medium on the sucrose content of β-lactoglobulin crystal. Curve 1 represents data obtained by analysis for sugar, protein, and water. Curve 2 represents concentration of sucrose in the crystal calculated from the equations of Perutz (Perutz, 1946) from data on density. From McMeekin et al., 1950, by permission of the *Journal of the American Chemistry Society*.

of the same resin, 61% of the external solute. Following this reasoning, we would have to postulate that the bound water is only 39% and the free water 61%, which is in conflict with the conclusion made earlier. This conflict is even harder to resolve if one takes into account that the equilibrium distribution ratio of a solute between the resin water and external water (generally referred to as q-value) varies from one solute to another, covering the range from unity to very small fractions (Ling, 1965).

Figure 10 shows the distribution of sucrose between the water of copper ferrocyanide gel and external water. The same type of solute exclusion exhibited in protein crystals and in ion-exchange resins is seen here. These data are significant for two reasons: (a) historically, copper ferrocyanide gel has been an important model of the cell or part of it (Traube, 1867); (b) the data illustrate the fact, well known to colloidal chemists, that hydration and gel formation do not necessarily depend on the presence of long-chain polymers. The first reason is discussed later.

The Entropic Theory of Solute Exclusion from Water Existing in the State of Polarized Multilayers. As part of a general molecular theory of living cells, called the association-induction hypothesis, a theory was offered to explain the general phenomenon of solute exclusion from various water-containing systems (Ling, 1962, 1965). The theory was based on the concept that in those systems water exists as partially immobilized, polarized multilayers. Since, as a rule, water-soluble substances interact with water through hydrogen bonds or ionic bonds, the varying degree of fixation of the water can then impede the normal freedom of motion of the solute. However, since

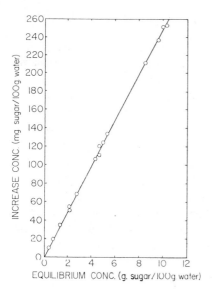

Figure 10. Excess sucrose in suspending medium of copperferrocyanide gel at various sucrose concentrations. Excess sucrose was calculated from the differences between the actually observed sucrose concentration in the suspending medium at equilibrium and theoretical values calculated on the basis of the assumption of equal distribution of sucrose in gel water. From McMahon, Hartung, and Walbran, 1937, by permission of the *Transactions of the Faraday Society.*

the excess energy due to polarization is not very high, the degree of immobilization will depend on how many hydrogen bonds the solute makes with the surrounding medium.

The transfer of a solute from normal water to water existing as polarized multilayers involves the following enthalpy and entropy changes.

1. To accommodate the solute, a hole must be dug in the water. Since the average water-water bond in the multilayer is higher than in normal water, this enthalpy change is unfavorable and thus tends to produce exclusion.

2. To varying degrees this exclusion is offset by the stronger hydrogen bonds the sucrose molecule makes with surrounding water molecules in the polarized multilayers.

3. There is an immobilization and, hence, a decrease in the translational freedom of motion, hence translational entropy. This gain of negative entropy varies with the number of hydrogen bonds the solute forms with the water.

4. There is immobilization that decreases rotational-motional freedom, hence the rotational entropy. This gain of negative entropy should also increase with the increase in the number of hydrogen bonds the solute can form with the surrounding water molecules.

From these considerations one may derive the following theoretical premises.

1. For small molecules with no hydrogen-bonding groups, there will be little or no exclusion ($q \simeq 1$).

2. Large, bulky molecules with few hydrogen-bonding groups on the surface, such as certain globular proteins, may be excluded for both enthalpic and entropic reasons.

3. Molecules that are not too bulky but do possess hydrogen-bonding groups, will be excluded according to the number of hydrogen-bonding groups they possess, and this exclusion occurs primarily for entropic reasons.

From these criteria, the theory predicts that many biologically important solutes such as sucrose and hydrated ions such as Na^+ should be excluded from polarized multilayered water on account of an unfavorable entropy. This can be stated more rigorously in the thermodynamic terms.

Let us refer to the phase containing water in the state of polarized multilayers as phase I and the surrounding normal water solute as phase II and the activities of a solute in question in each of these phases as a_I and a_{II}. At equilibrium

$$(14) \qquad \frac{a_I}{a_{II}} = \exp \frac{\mu_{II}^\circ - \mu_I^\circ}{RT}$$

where μ_I° and μ_{II}° are the standard partial molal free energies of the solute in the two phases. Let us designate $\mu_I^\circ - \mu_{II}^\circ$ as $\Delta\mu^\circ$ and similarly designate $\Delta H^\circ = H_I^\circ - H_{II}^\circ$ and $\Delta S^\circ = S_I^\circ - S_{II}^\circ$, where H_I°, H_{II}°, S_I°, and S_{II}° are the standard partial molal enthalpy and entropy in the two phases. The condition for exclusion of the solute from phase I is that $\mu_I^\circ > \mu_{II}^\circ$ or $\Delta\mu^\circ > 0$. Since

$$(15) \qquad \Delta\mu^\circ = \Delta H^\circ - T\Delta S^\circ$$

the condition under which $\Delta\mu^\circ > 0$ can be satisfied is that ΔH° be positive and/or ΔS° negative.

Now q, which is the ratio of the concentration of solute in phase I and phase II is, to all intents and purposes, equal to a_I/a_{II}. The van't Hoff equation then permits a study of ΔH° by determining how the q value changes with the temperature.

$$(16) \qquad \Delta H^\circ = \frac{R}{4} (T_2 - T_1)^2 \left(\frac{\ln q_{T_2} - \ln q_{T_1}}{T_2 - T_1} \right)$$

where q_{T_1} and q_{T_2} are the q values observed at the temperatures, T_1 and T_2, and R is the gas content. Furthermore, $\Delta\mu^\circ$ can be estimated for this temperature interval by (15) and the relation

$$(17) \qquad \Delta\mu^\circ = RT \ln \left(\frac{q_{T_2} + q_{T_1}}{2} \right)$$

The Molecular Mechanism of Sol-Gel Transformation and the Problem of Fibrous Protein Hydration. We have so far shown how evidence of the existence of water in the form of polarized multilayers can be displayed for a variety of materials. In these cases there is reason to believe that NP-NP surfaces or framework exists. We now suggest a theory of sol-gel transformation of proteins in the following terms.

1. The sol state. A protein solution is in the "sol" state if all or the bulk of its peptide carbonyl and imino groups are forming hydrogen-bonds with other imino or carbonyl groups of the same molecule (e.g., in an α-helical conformation) or of other neighboring proteins, (e.g., in a β-pleated sheet conformation). In this case, the hydration is primarily limited to a thin layer on polar side chains.

2. The gel state. When all or a portion of the polypeptide chain carbonyl and imino groups becomes dissociated from other peptide imino and carbonyl groups and is so distributed that the alternating NH and CO groups become the NP centers of an NP-NP system, water will interact with these charged centers, creating a more or less regular array in polarized multilayers. Under this condition, the protein water system will take on the consistency of a gel. Such water will have a tendency to exclude sucrose and ions such as Na^+.

3. The sol-gel transformation. We next suggest that the sol-gel transformation is cooperative:

$$\overset{\text{Gel}}{\left|\begin{matrix}-NH(H_2O)_nO-C-\\-CO(H_2O)_nNH-\end{matrix}\right|} \underset{\text{transition}}{\overset{\text{cooperative}}{\rightleftarrows}} \overset{\text{Sol}}{\left|\begin{matrix}-NHOC-\\-CONH-\end{matrix}\right|} + 2nH_2O$$

The water in the gel, although not perfectly ordered, is more ordered than normal water. The protein chains in the sol state may be highly ordered and those in the gel state disordered; as far as the water is concerned, however, it is essentially an order-disorder transition from the "ordered" gel state to the "disordered" sol state. Furthermore, since it is a three-dimensional, cooperative assembly, one may anticipate that the sol-gel transformation will have a lambda point (Ling, 1967).

As a corollary to this theory of cooperative sol-gel transformation of protein water systems, we would like to suggest that some proteins do not follow Pauling's theory of nonpolypeptide participation in hydration because they have a part of their polypeptide chain exposed. Not involved in intra- or intermolecular hydrogen bonding with other polypeptide amide groups, they can then interact with more water. Evidence for water sorption on the backbone on collagen (see Berendsen, 1962), gelatin, and zein (Mellon, Korn, and Hoover, 1948) already exists for the three proteins, which Bull and Breese and others have found to sorb more water than was predicted from their polar side chains.

7   The Physical State of Water in Living Cells

The Membrane Theory and Other Views. As mentioned earlier, the colloidal concept of cell protoplasm became all but extinct in the English-speaking world in the years following Blanchard's review of 1940. From that time until now, the majority of cell physiologists have subscribed to the membrane theory. In this view, as pointed out previously: (a) the cell membrane is considered as the universal rate-limiting barrier to the traffic of water and all solutes between the cell interior and the external environment; (b) intracellular K^+ and Na^+ ion are in the free state; (c) the bulk of cell water is normal; and (d) the exclusion of the Na^+ ion and sugars from the cell water is ascribed to hypothetical pumps located in the cell membrane.

Although widely accepted, the membrane theory is not the only one that has been proposed (Neuschloss, 1924; Ernst, 1963; Ling, 1951, 1952; Shaw and Simon, 1955; Troschin, 1958). Thus, according to the association-induction hypothesis (Ling, 1962, 1969), a living cell in essence represents a protein-aqueous fixed-charge system in which a close degree of molecular association exists among water, solutes, and proteins, thereby providing the fundamental coherence of the protoplasm as a functional unit. Coordination as well as energization in work performance is exercised through cardinal adsorbants (e.g., ATP, hormones, drugs), which control the all-or-none type of cooperative changes of the protein water systems.

This view includes the following conditions:

1. The exchange of water and solute may be limited with respect to bulk phase and to surface, according to the nature of the solute. Water is expected to exchange according to a bulk-phase-limited pattern (i.e., diffusion is more or less at the same rate throughout the whole cell).

2. Part of the intracellular Na^+ ion should be free: all the rest of the Na^+ ions and the bulk of the K^+ ions are adsorbed on protein anionic side chains.

3. The bulk of the cell water exists as polarized multilayers.

4. The solute distribution pattern reflects a maintained metastable equilibrium state in which the Na^+ ion, for example, is for entropic reasons excluded from the cell water existing in the state of polarized multilayers. The K^+ ion accumulates because of specific anionic sites which, in a resting state, are maintained at a specified electronic configuration specified by a c value (Ling, 1962). As an equilibrium phenomenon, the maintenance of solute distribution needs no direct energy expenditure.

Experimental Data. The following is a summary of experiments carried out to test these theories:

1. According to the membrane model, to maintain the low intracellular concentration of Na^+ ions in spite of constant inward diffusion, energy has to

be spent to operate the sodium pump as well as the pumps for calcium, magnesium, sugar, and so on. Under special conditions, where the ionic distribution pattern is not changed but the energy input is greatly limited, the minimal energy need of the sodium pump has been shown to be as high as 15 to 30 times the maximum available energy (Ling, 1962, 1965, 1969).

2. It has been shown that the exchange of labeled water with the water in frog ovarian eggs does not follow a surface-membrane-limited pattern (Ling, Ochsenfeld, and Karrman, 1957). Instead, it follows accurately the diffusion pattern if the diffusion coefficient for the labeled water is essentially the same throughout the entire cell (bulk-phase-limited diffusion), and the diffusion coefficient is one-half to one-third the diffusion in normal water. Recently Bunch and Edward (1969) published their study of water and nonelectrolyte diffusion from single muscle fibers from the giant barnacle, *Balanus nubilis*; they interpreted their data on water diffusion as agreeing with the membrane theory, although they presented no experimental data to support this claim. Reisin and Ling (1971) have reinvestigated this problem, using the same single muscle fiber preparation, concluding that Bunch and Edward's data did not include the crucial points made earlier (as judged from their data on non-electrolytes diffusion). When these initial points are included in the data, as they were by Reisin and Ling, perfect agreement is shown with the theoretical curve for bulk-phase-limited diffusion. The diffusion coefficient is also roughly equal to the value obtained from frog eggs.

The conclusion from frog eggs and from barnacle muscles is important not only because it disproves a fundamental tenet of the membrane theory; it has other far-reaching implications as well. Thus merely puncturing a lipid membrane with pores and filling them with water does not change the surface-limited nature of the diffusion. Indeed, if we envisage a cell as consisting of a series of shells (like in an onion), the bulk-phase-limited diffusion demands that the relative cross-sectional area of such aqueous channels be roughly the same in each of these shells. This, in turn, demands that the aqueous channels occupy the same relative proportion in all the shells if they are to be equal to the proportion of water in the whole cell. For frog eggs, this is roughly 50%; for barnacle muscles, it is 80%.

The permeability properties of a surface having 50 to 80% water must be determined primarily by the water and to a far lesser extent by the remaining material. This signifies that the water cannot be normal. We shall come back to this subject in another section.

3. The physical state of intracellular K^+ and Na^+ ions is determined using a wide-line NMR spectroscopy. Cope (1967) demonstrated that a free Na^+ ion in water shows an NMR signal with a height proportional to the concentration of Na^+ ions. This proportionality is not altered by the presence in the solution of 20% serum albumin or of 20% gelatin gel. An Na^+ ion

adsorbed on proteins (e.g., actomyosin), on the other hand, becomes so broadened that it shows no visible signal. Still with the NMR technique, Cope demonstrated that almost half of the intracellular Na^+ ion is in an adsorbed state. Ling and Cope (1969) showed that when intracellular K^+ ion is replaced by Na^+ ion, the extra Na^+ ion is also in the adsorbed state, thereby indirectly showing that the bulk of intracellular K^+ ion is in an adsorbed state.

4. The physical state of water in the living cells is another subject of experimentation. There are at least two reasons for Gortner's failure to convince the majority of cell physiologists that cell water is bound to any significant degree (1938). First, the concept of bound water at that time was largely descriptive, and there was no readily understandable theoretical model to explain what the concept actually meant; second, the experimental techniques then available were not sophisticated enough to elucidate what is fundamentally an elusive and subtle phenomenon.

NMR STUDIES.   The future must determine whether the model of polarized water in multilayers presented here and elsewhere is valid, but there is no question that we now have sophisticated instruments that can give us information about cell water that Gortner could not have dreamed of. Among these, the nuclear-magnetic-resonance spectrometer is outstanding.

We have already discussed the work of Kuntz et al. on the NMR spectrum of frozen protein solution. Brey et al. (1968) and Fuller et al. (1968) also studied the line-width and relaxation times of sorbed water and have derived conclusions in agreement with those on water sorption on globular proteins just discussed.

Within the last two years at least four papers have appeared from different laboratories around the world reporting on the line-width broadening of the water protons in living cells (Chapman and McLauchlan, 1957; Fritz and Swift, 1967; Cope, 1969; Hazelwood et al., 1969). The consensus of these papers seems to be that the bulk of cell water is in a physical state different from normal and that the freedom of motion of cell water molecules is restricted—a finding in direct support or at least in harmony with the theory of polarized multilayers.

Judged as a whole, these NMR studies offer good evidence for the association-induction hypothesis, although it cannot yet be considered decisively proven, since other possible interpretations still exist (see Bratton et al., 1965).

FREEZING EXPERIMENTS.   Blanchard's chief argument against Gortner's evidence for bound water was the supercooling of normal water. Since normal water can supercool to $-40°C$, the failure to freeze at $-20°C$—a criterion Gortner used to distinguish bound water—is not a tenable one. In

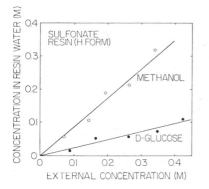

Figure 11. Equilibrium distribution of methanol and D-glucose in the water in a sulfonate ion exchange resin (8% DUB. H form) at 0°C. From Ling, unpublished data.

fact, reliable supercooling of pure water is known to go down to $-40°C$ (Mason, 1958; Hendrix, 1961). However, there is no indication at all that normal water can be supercooled to liquid nitrogen temperature $(-210°C)$; yet Moran (1926) long ago demonstrated the failure of the water in a gelatin gel containing 35% water to freeze even at this temperature. The advent of cryobiology has left no room for doubt that in living cells comparable fractions of water are equally difficult to freeze (Merryman, 1966; McKenzie and Luyet, 1967). Other intracellular freezing experiments supporting the idea that the bulk of intercellular water exists as polarized multilayers have been reviewed elsewhere (Ling, 1967, 1969, 1970).

SOLUTE EXCLUSION. Figure 11 shows the equilibrium distributions of methanol and D-glucose in the water of a sulfonate ion-exchange resin. The q values, measured from the slopes of the curves, are 0.89 for methanol, and 0.25 for D-glucose, respectively. Figure 12 presents the equilibrium

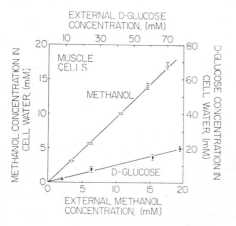

Figure 12. Equilibrium distribution of methanol in freshly isolated frog muscles and of D-glucose in insulin-depleted frog muscles at 0°C. Redrawn after Ling and Will, 1969, by permission of the *Journal of Psychological Chemical Physics*.

distributions of methanol and D-glucose in the water of (insulin-depleted) frog muscle cells. Both lines are rectilinear. The q values are 0.90 for methanol and 0.25 for D-glucose. The equilibrium distributions of methanol and D-glucose in both ion-exchange resin and frog muscle cells follow the theory stated earlier. The data show that there is little or no nonsolvent water in the classical sense (i.e., water impervious to any solute). Instead, the water excludes solutes according to the nature of the solute involved.

The high concentration of polar groups in the ion-exchange resin easily explains why all its water should be polarized. Muscle cells, on the other hand, contain 80% water and only 20% protein. Judging from the low q value for D-glucose, almost all the cell water in the resting muscle cell must be in the polarized multilayer state.

VAPOR SORPTION ISOTHERMS.   Now, if the bulk of the cell proteins is hydrated to the extent of, say, 0.5 g/g protein, the total amount of hydration water in the cell will be no more than $200 \times 0.5 = 100$ g, or one-eighth of the total water. This insufficiency in accounting for the exclusion shown in Figure 12 demands that some other major source or sources of water-binding sites must be utilized in place of and/or in addition to the polar side chains.

In preceding sections we have shown that the polypeptide chains are capable of providing the NP-NP type of fixed sites for long-range water orientation and polarization. We suggest that in resting cells an important fraction of the protein backbones must be in what is conventionally referred to as the "random coil" state, only they are not random but do have their carbonyl and imino groups free to polarize and orient multiple layers of water.

In Figure 13 we have plotted the vapor sorption isotherms of two globular proteins (serum albumin and ovalbumin), one fibrous protein (collagen), and sulfonate ion-exchange resin. The S shape of the curves is typical and well known. Since sulfonate ion-exchange resin contains only polar groups similar to those on protein side chains, the large initial uptake at low humidity and the gentle additional uptake at high humidity seem best explained as characteristics of water sorption on polar groups. This interpretation is in accord with all the other evidence cited previously on the major sites of hydration on globular proteins as well as on collagen.

On the other hand, Figure 14 shows three vapor sorption isotherms of a distinctly different type. All are distinguished by a small initial uptake and a very steep rise of sorbed water between the relative vapor pressures of 0.8 and 0.95, and this steep rise continues beyond this range. I suggest that this steep rise represents cooperative adsorption of water as polarized multilayers. In the case of poly-glycine-D,L-alanine (curve A), the NP-NP type of sites were provided by the carbonyl and imino groups on the polypeptide

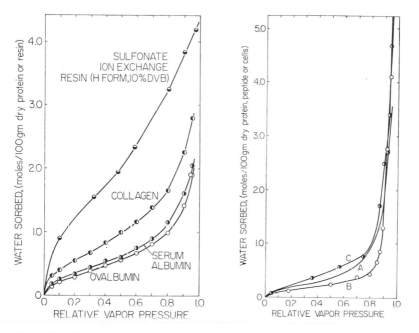

Figure 13.   Water sorption of ovalbumin serum albumin, collagen, and sulfonate ion exchange resin. Data on the 3 proteins from Bull (1944). Data on the ion exchange resin (10% DVB, H form) from Glueckauf (1952). Other data, given by Glueckauf for resin with less DVB, show sharper rise of water sorption at high vapor pressure but not sharper, relatively speaking, than those shown by the proteins.

Figure 14.   Water sorption on cellulose acetate sheets (A), on poly-glycine-D,L-alanine copolymer (B), and on frog muscle (C). Data on cellulose acetato sheet from Palmer and Ling (unpublished); data on poly-glycine-D,L-alanine from McLaren and Katchman (McLaren and Rowne, 1951). Data on frog muscle from Ling and Nagendank (1970).

chains; in the case of cellulose acetate (curve B), by the carbonyl and hydroxyl groups.

Finally, curve C in Figure 14 represents the vapor sorption of intact frog muscles recently reported by Ling and Nagendank (1970). The close similarity of this curve to the other two curves reinforces our earlier conclusion that in resting muscle cells the bulk of cell water exists as NP-NP type of polarized multilayers between protein backbones. The entire water content of the cell may then be under the control of the proteins in a cooperative manner.

Acknowledgments

The preparation of this chapter and the new investigations reported were supported by National Institute of Health Research Grant 2RO1-GM-11422-07, Pennsylvania Hospital General Research Support Grant 5-SO1-FR-05590, and Office of Naval Research Contract Nonr 4371 (00)-105327.

The author is supported by Public Health Service Research Career Development Award K3-GM-19032.

The author thanks Dr. Frank Elliott for his invaluable help and the John A. Hartford Foundation for providing the basic equipment for investigations.

REFERENCES

Adair, G. S., and M. E. Adair, *Proc. Roy. Soc. London, Ser. A*, **190**, 341 (1947).

Adair, G. S., and M. E. Adair, *Proc. Roy. Soc. London, Ser. B*, **120**, 422 (1963).

Alexander, P., and R. F. Hudson, *Wool, Its Chemistry and Physics*, Reinhold, New York, 1954.

Amberg, C. H., *J. Amer. Chem. Soc.*, **79**, 3980 (1957).

Anderegg, J. W., W. Beeman, S. Shulman, and O. Kaesburg, *J. Amer. Chem. Soc.*, **77**, 2927 (1955).

Benson, S. W., and D. A. Ellis, *J. Amer. Chem. Soc.*, **70**, 3563 (1948).

Benson, S. W., and D. A. Ellis, *J. Amer. Chem. Soc.*, **72**, 2095 (1950).

Benson, S. W., D. A. Ellis, and R. W. Zwanzig, *J. Amer. Chem. Soc.*, **72**, 2102 (1950).

Berendsen, H. J. C., *J. Chem. Phys.*, **36**, 3297 (1962).

Blanchard, K. C., *Cold Spring Harbour Symp., Quant. Biol.*, **8**, 1 (1940).

Blüh, O., *Protoplasma*, **3**, 81 (1928).

Bockris, J. O'M., *Quart. Rev. (London)*, **3**, 173 (1949).

Bradley, S., *J. Chem. Soc.*, 1467 (1936).

Bratton, C. B., A. L. Hopkins, and J. W. Weinberg, *Science*, **147**, 738 (1965).

Brey, W. S., T. E. Evans, and L. H. Hitzrot, *J. Collect. Interface Sci.*, **26**, 306 (1968).

Brohult, S., and E. Sandegren, *The Proteins: Chemistry, Biological Activity and Methods*, vol. 2A, Eds., H. Neurath and K. Bailey, Academic Press, New York, p. 487, 1954.

Brunauer, S., P. H. Emmett, and E. Teller, *J. Amer. Chem. Soc.*, **60**, 309 (1938).

Buchanan, T. J., G. H. Haggis, J. B. Hasted, and B. G. Robinson, *Proc. Roy. Soc. (London), Ser. A*, **213**, 379 (1952).

Bull, H., *J. Amer. Chem. Soc.*, **66**, 1499 (1944).

Bull, H. B., and K. Breese, *Arch. Biochem. Biophys.*, **128**, 497 (1968).

Bunch, W., and C. Edwards, *J. Physiol.*, **202**, 683 (1969).

Chapman, G., and K. A. McLauchlan, *Nature*, **215**, 391 (1967).

Cope, F. W., *J. Gen. Physiol.*, **50**, 1353 (1967).

Cope, F. W., *Biophys. J.*, **9**, 303 (1969).

Cox, D. J., and V. N. Schumaker, *J. Amer. Chem. Soc.*, **83**, 2433 (1961a).

Cox, D. J., and V. N. Schumaker, *J. Amer. Chem. Soc.*, **83**, 2439 (1961b).

de Boer, J. H., and C. Zwikker, *Z. Physik. Chem.*, **B3**, 407 (1929).

Dole, M., *Ann. N.Y. Acad. Sci.*, **51**, 705 (1949).

Dole, M., and I. L. Faller, *J. Amer. Chem. Soc.*, **72**, 414 (1950).

Edsall, J. T., H. Neurath, and K. Bailey, *The Proteins*, vol. 113, Academic Press, New York (1953).

Ellis, J. W., and J. D. Bath, *J. Chem. Phys.*, **6**, 723 (1938).

Ernst, E., *Biophysics of the Striated Muscle*, Hungarian Academy of Science, Budapest, 1963.

Fisher, H. F., *Proc. Natl. Acad. Sci. (U.S.)*, **51**, 1285 (1964).

Fisher, H. F., *Biochem. Biophy. Acta*, **109**, 544 (1965).

Fisher, H. F., and D. G. Cross, *Federation Proc.*, **23**, 427 (1964).

Flory, P. J., *J. Chem. Phys.*, **10**, 51 (1942).

Fritz, O. G., and T. J. Swift, *Biophys. J.*, **7**, 675 (1967).

Fuller, M. E., and W. S. Brey, *J. Biol. Chem.*, **243**, 274 (1968).

Glueckauf, E., *Proc. Roy. Soc. (London)*, *Ser. A*, **214**, 207 (1952).

Gortner, R. A., *Selected Topics in Colloid Chemistry*, Cornell University Press, Ithaca, 1937, Chapter 8.

Gortner, R. A., *Outlines of Biochemistry*, 2nd ed., Wiley, New York, 1938.

Gortner, R. A., and W. A. Gortner, *J. Gen. Physiol.*, **17**, 327 (1934).

Grant, E. H., *Nature*, **196**, 1194 (1962).

Grant, E. H., *Ann. N.Y. Acad. Sci.*, **125**, 478 (1965).

Grant, E. H., S. E. Keefe, and S. Takashima, *J. Phys. Chem.*, **72**, 4373 (1968).

Green, R. W., *Trans. Roy. Soc. N.Z.*, **77**, 313 (1949).

Greenberg, D. M., and W. E. Cohn, *J. Gen. Physiol.*, **18**, 93 (1934).

Gustavson, K. H., *The Chemistry and Reactivity of Collagen*, Academic Press, New York, 1956.

Haggis, G. H., T. J. Buchanan, and J. B. Harted, *Nature*, **167**, 607 (1951).

Hazelwood, C. F., B. L. Nichols, and N. F. Chamberlain, *Nature*, **22**, 747 (1969).

Hearst, J. E., and J. Vinograd, *Proc. Natl. Acad. Sci. (U.S.)*, **47**, 1005 (1961).

Hendrix, W. P., K. F. Hurd, and C. Orr, Jr., "A Study of the Alteration of Water Droplets Supercooling by Foreign Vapors," Final Rept; AF Contract 19(604) 4970 Eng. Exptl. Station, Georgia Institute of Technology, Atlanta, 1961.

Henniker, J. C., and J. W. McBain, "The Depth of the Surface Zone of a Liquid," Tech. Rept. No. 5, N60ri-154-T. O. II, Stanford Research Institute, Stanford, Calif., 1948.

Hill, A. V., *Proc. Roy. Soc. (London)*, *Ser. B*, **106**, 477 (1930).

Hill, T. L., *J. Chem. Phys.*, **14**, 263 (1946).

Hoover, S. R., and E. F. Mellon, *J. Amer. Chem. Soc.*, **72**, 2562 (1950).

Hori, T., *Low-Temp. Sci.*, **A15**, 34 (1956) (Engl. transl. no. 62, U.S. Army Snow, Ice and Permafrost Establishment, Wilmette, Ill.).

Jones, I. D., and R. A. Gortner, *J. Phys. Chem.*, **36**, 387 (1932).

Kendrew, J. C., *Science*, **139**, 1259 (1963).

Kühne, W., "Untersuchungen üben das Protoplasma und die Contractilität," *Z. Physik Chem. (Leipzig)*, **1**, 158 (1864).

Kuntz, I. D., Jr., T. S. Brassfeld, G. D. Law, and G. V. Purcell, *Science*, **163**, 1329 (1969).

Langridge, R., H. Wilson, C. Hooper, and M. Wilkins, *J. Mol. Biol.*, **1**, 451 (1967).

Lauffer, M. A., *Biochem.*, **5**, 1952 (1966).

Ling, G. N., *Amer. J. Physiol.*, **167**, 806 (1951).

Ling, G. N., *Symposium on Phosphorus Metabolism*, vol. 2, Johns Hopkins University Press, Baltimore, 1952, p. 748.

Ling, G. N., *A Physical Theory of the Living State*, Ginn (Blaisdell), New York, 1962.

Ling, G. N., *Ann. N.Y. Acad. Sci.*, **125**, 401 (1965).

Ling, G. N., *Thermobiology*, Ed., A. Rose, Academic Press, New York, 1967, Chapter 2.

Ling, G. N., *Intern. Rev. Cytol.*, **26**, 1 (1969)

Ling, G. N., *Intern. J. Neuroscience*, **1**: 129 (1970).

Ling, G. N., and F. W. Cope, *Science*, **163**, 1335 (1969).

Ling, G. N., and W. Nagendank, *Physiol. Chem. Phys.*, **2**, 15 (1970).

Ling, G. N., M. M. Ochsenfeld, and G. Karreman, *J. Gen. Physiol.*, **50**, 1801 (1967).

Lloyd, D. J., *Biol. Rev.*, *Cambridge Philosophical Soc.*, **8**, 463 (1933).

Lloyd, D. J., and H. Phillips, *Trans. Faraday Soc.*, **29**, 132 (1933).

Lloyd, D. J., and A. Shore, *The Chemistry of the Proteins*, Blakiston, Philadelphia, 1938.

McKenzie, A. P. and B. J. Luyet, *Biodynamica*, **10**: 95 (1967).

McLaren, A. D., and B. J. Katchman, as quoted in McLaren and Rowen (1951).

McLaren, A. D., and J. W. Rowen, *J. Polym. Sci.*, **7**, 289 (1951).

MacLeod, J., and E. Ponder, *J. Physiol.*, **86**, 147 (1936).

McMahon, B. C., E. J. Hortung, and W. J. Walbran, *Trans. Faraday Soc.*, **33**, 398 (1937).

McMeekin, T. L., M. L. Groves, and N. J. Hipp, *J. Amer. Chem. Soc.*, **72**, 3662 (1950).

Mason, B. J., *Phil. Mag.*, *Suppl.*, **7**, 221 (1958).

Mellon, E. F., A. H. Korn, and S. R. Hoover, *J. Amer. Chem. Soc.*, **70**, 3040 (1948).

Mellon, E. F., A. H. Korn, and S. R. Hoover, *J. Amer. Chem. Soc.*, **71**, 2761 (1949).

Merryman, H. T., *Cryobiology*, Academic Press, New York, 1966.

Miller, G. L., and W. C. Price, *Arch. Biochem.*, **10**, 467 (1946).

Moran, T., *Proc. Roy. Soc. London (London)*, *Ser. A*, **112**, 30 (1926).

Nägeli, C., *Pflanzenphysiologische Untersuchungen*, vol. 1, Eds., C. von Nägeli and I. Cramer, Nägeli, Zurich, 1855, p. 120.

Neuschloss, S. M., *Pfluegers Arch. Ges. Physiol.*, *Menschen Tiere*, **204**, 374 (1924).

Newton, R., and W. A. Gortner, *Botan. Gaz.*, **74**, 442 (1922).

Ogston, A. G., *Proc. Intern. Wool Text. Res. Conf.*, *1955*, **B**, 92 (1956), Commonwealth Science and Industrial Research Organization, Melbourne, Australia.

Oncley, J. L., *Chem. Rev.*, **30**, 433 (1942).

Oncley, J. L., *Proteins*, *Amino Acids and Peptides*, Eds. E. J. John and J. T. Edsall, Reinhold, New York, 1943, Chapter 22.

Palmer, L. S., A. Cunliffe, and J. M. Hough, *Nature*, **170**, 796 (1952).

Pauling, L., *J. Amer. Chem. Soc.*, **67**, 555 (1945).

Pennock, B. E., and H. P. Schwan, *J. Phys. Chem.*, **73**, 2600 (1969).

Perutz, M. F., *Trans. Faraday Soc.*, **42B**, 187 (1946).

Reisin, I., and G. N. Ling, *Physiol. Chem. Phys.* (1971), in press.

Ritland, H. N., P. Kaesberg, and W. W. Beeman, *J. Chem. Phys.*, **18**, 1237 (1950).

Robinson, W., *J. Biol. Chem.*, **92**, 699 (1931).

Rowen, J. W., and R. J. Simha, *Phys. Colloid. Chem.*, **53**, 921 (1949).

Rubner, M., *Abhandl. Preuss. Akad. Wiss. Phys. Math.*, *Kl.*, *I*, **1** (1922).

Schwan, H. P., *Advan. Biol. Med. Phys.*, **5**, 191 (1957).

Shaw, F. H., and S. E. Simon, *Australian J. Exptl. Biol. Med. Sci.*, **33**, 153 (1955).

Speakman, J. B., *J. Text. Inst.*, **27**, T185 (1936).

Sponsler, O. L., J. D. Bath, and J. W. Ellis, *J. Phys. Chem.*, **44**, 996 (1940).

Tanford, C., *Physical Chemistry of Macromolecules*, Wiley, New York, 1961.

Tanford, C., "Enzyme Models and Enzyme Structure," *Brookhaven Symp. Biol.*, **15**, 227 (1962).

Thoenes, F., *Biochem. Z.*, **157**, 174 (1925).

Traube, M., *Arch. Anat. Physiol. Wiss. Med.* 1867, p. 87.

Troschin, A. S., *Das Problem der Zellermeabilität*, Fischer, Jena, 1958.

Wang, J. H., *J. Amer. Chem. Soc.*, **76**, 4755 (1954).

Waugh, D. F., *Advan. Protein Chem.*, **9**, 325 (1954).

Weissmann, O., *Protoplasma*, **31**, 27 (1938).

Wheaton, R. M., and W. C. Bauman, *Ind. Eng. Chem.*, **H3**, 1088 (1951).

White, H. J., Jr., and H. Eyring, *Text. Res. J.*, **17**, 523 (1947).

Zachariasen, W. H., *J. Amer. Chem. Soc.*, **54**, 3841 (1932).

17   The Viscosity of Water[*]

J. Jarzynski† and C. M. Davis,† Department of Physics,
The American University, Washington, D.C.

1   Introduction

The purpose of this chapter is to summarize the available data on the viscosity coefficients of water. The viscosity of a liquid is a measure of its resistance to flow, or its rate of strain. Since flow takes place by displacements of the equilibrium positions of molecules, studies of the viscosity can lead to information about these displacements.

When the period of the shearing stresses is much longer than the structural relaxation time of the liquid, the principal shearing stresses, P_{ij}, are proportional to the corresponding rates of strain[1]

$$(1) \qquad P_{ij} = P + (\eta_v - \tfrac{2}{3}\eta_s)\, \text{div } \mathbf{u}\, \delta_{ij} + \eta_s\left(\frac{\partial u_i}{\partial x_j} + \frac{\partial u_j}{\partial x_i}\right)$$

where P is the hydrostatic pressure, $\mathbf{u}$ the fluid velocity, η_v the volume viscosity, and η_s the shear viscosity. For a pure shear flow pattern, say flow in the x direction, with a velocity gradient in the z direction, one has in the normal manner

$$(2) \qquad P_{13} = \eta_s \frac{\partial u_1}{\partial z}$$

In the case of a pure expansion, or compression, $P_{ij} = 0$ when $i \neq j$ and

$$(3) \qquad P_{ii} = P + \eta_v\, \text{div } \mathbf{u} \approx P - \eta_v \frac{1}{\rho_o}\frac{\partial \rho}{\partial t}$$

Both terms in P_{ij} are scalar, however the hydrostatic pressure depends only on the momentary value of density, ρ, and temperature, whereas the second term depends on the time variation of the density. It follows that volume

---

[*] The work reported here was supported in part by a grant from the Office of Saline Water, U.S. Department of Interior (Grant #14-30-2518).

† Present address: Physical Acoustics Branch, Naval Research Laboratory, Washington, D.C. 20390.

viscosity is related to the structural changes that occur when a liquid is compressed. It is possible to produce a steady shear flow pattern, and η_s may be measured through steady-state experiments; but it is not possible to produce a steady compressional flow pattern, and η_v cannot be measured directly. However, a longitudinal sound wave can be considered as superposition of pure compression and pure shear. If the period of the sound wave is much longer than the relaxation time of the liquid the absorption coefficient is[1]

(4)
$$\alpha = \frac{2\pi^2 f^2}{\rho c_3} \left(\tfrac{4}{3}\eta_s + \eta_v \right)$$

where c is the velocity of sound, f is the frequency, and ρ is the density. Hence the volume viscosity can be determined by measuring the absorption coefficient of sound waves through the liquid.

When the period of the applied stress becomes comparable to the structural relaxation time of the liquid, it is necessary to make an extension of the hydrodynamic model, (1), and this is done through the viscoelastic theory of liquids.[2] In the hydrodynamic model of liquids it is assumed that there is no elastic resistance to shear forces, and this is true if the forces are applied slowly over a long period of time. However, in the limit of a very-high-frequency alternating stress, the molecules of the liquid do not have time to adjust their positions by inelastic (or viscous) movement over the interval during which the force is applied, and the molecular motion is that of elastic deformation analogous to the behavior of a solid. In general, there are two types of motion: viscous movements in which energy is dissipated and elastic deformations without dissipation of energy. In the simplest version of viscoelastic theory it is assumed the viscous and elastic effects may be treated independently and that terms corresponding to the two effects may be added linearly in the equations of motion. Then each relaxation process occurs with a single relaxation time. For example, if a stepwise shear strain is applied, the corresponding stress will decay to 0 exponentially with a time constant, τ_s, the shear relaxation time. Similarly, if one applies a step function of pressure, the resulting volume change at constant pressure will approach equilibrium value exponentially with a time constant, τ_v, the volume relaxation time at constant pressure. The shear and volume viscosities are related to the corresponding high-frequency parameters through the relaxation times as follows (see the review by Litovitz and Davis[2] for a general discussion of viscoelastic theory):

(5)
$$\eta_s = G_\infty \tau_s \quad \text{and} \quad \eta_v = \frac{\kappa_r}{\kappa_0^2} \tau_P$$

where G_∞ is the high-frequency shear modulus and the relaxational compressibility, $\kappa_r = \kappa_0 - \kappa_\infty$, is the difference between the static compressibility,

κ_0, and the high-frequency compressibility, κ_∞. The volume relaxation time, compressibilities, and viscosity in (5) are either adiabatic or isothermal, depending upon the experiment performed. In Chapter 10 values of κ_r are obtained for the various mixture models of water and combined with experimentally determined values of η_v to calculate τ_P and thus provide a check on these models. Also τ_P is compared with the time required for a single diffusive step as determined by inelastic neutron scattering and with the dielectric relaxation time. Such comparison yields information on the type of molecular motion that determines the rate at which structural rearrangement occurs in water.

2 Experimental

Shear Viscosity. The most reliable absolute values for the shear viscosity at atmospheric pressure and for its temperature dependence come from the National Bureau of Standards[3-5] and the measurements of Korson et al.[6] Similar measurements for heavy water are reported by Hardy and Cottington.[5] The data of these and other workers are summarized in Table 1. Viscosity measurements at higher temperatures are reported by, among others, Moszynski[7] and Dudziak and Franck.[8]

Several workers[9-16] have also studied the pressure dependence of the shear viscosity of water. The data of Stanley and Batten,[9] obtained with a rolling-ball viscosimeter agree consistently with the data of Bett and Cappi,[12] obtained with a falling-body (guided-sinker) viscosimeter, and these data are considered the most accurate measurements of the pressure dependence of η_s. The largest pressure range covered is by Bett and Cappi (up to 10,000 kg/cm^2) and Bridgman (up to 12,000 kg/cm^2). The data are displayed in Figure 1, and it can be seen that there is considerable discrepancy between the results of these investigators. Cappi[17] suggests that Bridgman's data (obtained with a falling-body viscosimeter) are in error partly because of Bridgman's assumption that viscosity was directly proportional to the fall of the sinker at all velocities and partly because of inadequate correction for the inclination of the viscosimeter at the commencement of timing and the acceleration to terminal velocity. Furthermore, Bridgman experienced difficulty in applying his viscosimeter principle to water, and to polar liquids generally, because of the unreliability of his break-and-make contact system of timing. Cappi estimates his η_s values to be accurate to $\pm 0.7\%$.

Cappi[17] also reports measurements of the viscosity of heavy water as a function of pressure at 25°C. The ratio η_{D_2O}/η_{H_2O} remains essentially constant as a function of pressure (1.23 at 0 kg/cm^2 and 1.21 at 10,000 kg/cm^2).

Table 1   Temperature Dependence of the Shear Viscosity of H_2O and D_2O and the Energy of Activation for Viscous Flow, $E_p{}^+$

| Temperature (°C) | Viscosity (cp) | | | $\dfrac{\eta_{s,D_2O}}{\eta_{s,H_2O}}$ | $E_p{}^+$ for H_2O[d] (kcal/mole) |
	Averaged Values[a]	$\eta_{s,H_2O}{}^b$	$\eta_{s,D_2O}{}^c$		
0	1.7865				5.5
5	1.5177		1.9819	1.306	4.8
10	1.3061				4.6
15	1.1381				
20	1.0020[e]		1.2477	1.245	4.2
25	0.8903	0.8903			
30	0.7976				
40	0.6531	0.6527	0.7848	1.202	
50	0.5471				3.4
60	0.4666	0.4665	0.5497	1.178	
80	0.3547		0.4129	1.164	
90	0.3148	0.3143	0.3647	1.159	
95	0.2976		0.3444	1.157	
100	0.2822	0.2820	0.3255	1.153	2.8

[a] Values averaged from data taken at the National Bureau of Standards and quoted in: J. Coe and T. Godfrey, *J. Appl. Phys.*, **15**, 625 (1944); C. Craque, unpublished work quoted by Coe and Godfrey; R. Hardy and R. Cottington, *J. Res. Natl. Bur. Std. (U.S.)*, **42**, 573 (1949).

[b] A. Korosc and B. Fabuss, *Anal. Chem.*, **40**, 157 (1968).

[c] R. Hardy and R. Cottington, *J. Res. Natl. Bur. Std. (U.S.)*, **42**, 573 (1949).

[d] D. Eisenberg and W. Kauzmann, *The Structure and Properties of Water*, Oxford University Press, New York, 1969.

[e] Value assigned at 20°C for calibration.

Volume Viscosity.   As stated in the introduction, values of η_v may be determined by measuring the absorption of sound waves in water. The most reliable measurements at atmospheric pressure were made by Pinkerton.[18] The resulting values of η_v are given in Table 2. Absorption measurements for heavy water have been made by Pancholy,[19] and corresponding values of η_v are included in Table 2.

Measurements of sound absorption as a function of pressure have been made by Tait,[20] Litovitz and Carnevale (LC),[21] Hawley et al. (HAH),[22] and Davis et al. (DFJ).[23] Values of α/f^2, the absorption coefficient divided by the square of the sound wave frequency obtained by these investigators are compared in Figure 2. At 30°C there is good agreement between the various sets of data. Also the data of HAH at 0°C show the same slope as

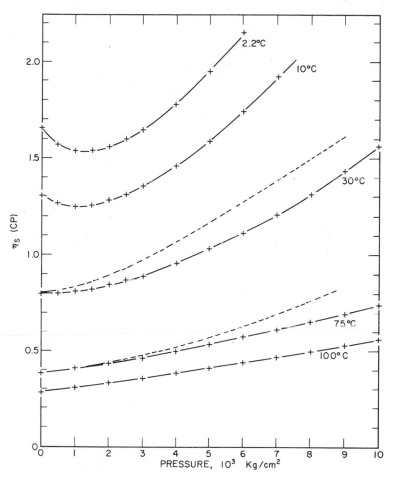

Figure 1. Pressure dependence of the shear viscosity of water according to Bett and Cappi[12,17] (solid curve) and Bridgman[13] (dashed curve).

a function of pressure as the data of DFJ at 1°C, whereas the α/f^2 values of LC decrease more rapidly with pressure. Tait's values of α/f^2 at 16°C are consistently higher than the data of DFJ at 15°C. Since Tait measured sound absorption at the relatively low frequency of 12 MHz, it is possible that his data contain an error due to diffraction effects.[24]

It has been pointed out by Eden and Horne[25] that calculations of the pressure dependence of η_v differ considerably, depending on which values of η_s are used in (4). The values of η_v given in Table 3 have been calculated using the shear viscosity data of Cappi.[17] Since the reproducibility claimed

Table 2   Temperature Dependence of the Volume
Viscosity of H_2O and D_2O

Temperature (°C)	Viscosity (cp)		$\dfrac{\eta_{v,D_2O}}{\eta_{v,H_2O}}$
	η_{v,H_2O}[a]	η_{v,D_2O}[b]	
0	5.61		
5	4.55	5.12	1.13
10	3.75	4.24	1.13
20	2.84	3.10	1.09
30	2.27	2.41	1.06
40	1.81	1.92	1.06
50	1.47		
60	1.22		

[a] Calculated from sound absorption data of J. Pinkerton, *Nature*, **160**, 128 (1947).
[b] Calculated from sound absorption data of M. Pancholy, *J. Acoust. Soc. Amer.*, **25**, 1003 (1953).

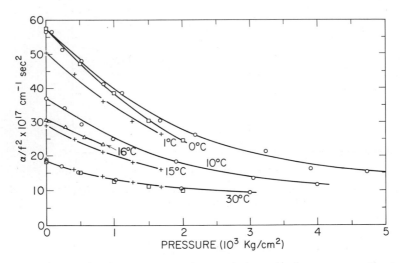

Figure 2.   Pressure dependence of the sound absorption in water according to Hawley, Allegra, and Holton[22] (open circles); Litovitz and Carnevale[21] (squares); Davis, Freeman, and Jarzynski[23] (crosses); and Tait[20] (triangles).

for the sound absorption measurements is between ± 2 and $\pm 3\%$, the resulting uncertainty in the volume viscosity values given in Table 3 can be as high as $\pm 4\%$.

Litovitz and Carnevale claim that the η_v of water exhibits a maximum as a function of pressure at 0°C, and this maximum is absent at 30°C. However, it can be seen from Table 3 that this maximum is a small effect, barely above the uncertainty in the values of η_v. Furthermore, this maximum is not apparent in the η_v calculated from the data of DFJ. Probably the one general conclusion that can be made from the data in Table 3 is that the ratio η_v/η_s for water remains approximately constant as a function of pressure (or density). This is particularly apparent in the data of DFJ and the data of HAH at 10 and 30°C. The data of HAH at 0°C show considerable scatter

Table 3　Volume Viscosity, $P\eta_v$ (cp) versus Temperature, T (°C), and Pressure, (kg/cm²) in H_2O

P	η_v	η_v/η_s	P	η_v	η_v/η_s	P	η_v	η_v/η_s	P	η_v	η_v/η_s
				Hawley, Allegra, and Holton, Ref. 22							
T　0			10			30					
1	5.61	3.13	1	4.00	3.05	1	2.27	2.84			
246	5.53	3.20	278	4.07	3.18	228	2.18	2.72			
522	6.00	3.57	521	3.90	3.07	494	2.20	2.78			
1115	6.23	3.82	978	3.82	3.06	1017	2.21	2.73			
2183	5.94	3.58	1907	3.87	3.02	1978	2.46	2.93			
3283	6.66	3.74	3048	3.90	2.87	3001	2.76	2.79			
3908	6.00	3.21	4008	4.15	2.84						
4724	6.46	3.21									
				Litovitz and Carnevale, Ref. 21							
T　0						30					
1	5.67	3.16				1	2.15	2.69			
500	5.98	3.56				500	2.08	2.60			
1000	5.74	3.50				1000	2.06	2.55			
1500	5.36	3.28				1500	2.01	2.44			
2000	5.19	3.13				2000	2.11	2.50			
				Davis, Freeman, and Jarzynski, Ref. 23							
T　1			15			30			50		
1	5.24	3.03	1	3.12	2.73	1	2.24	2.80	1	1.49	2.71
422	5.18	3.13	422	3.21	2.88	422	2.21	2.76	422	1.46	2.62
844	4.97	3.08	844	3.15	2.84	844	2.26	2.81	844	1.51	2.66
1265	4.84	3.03	1265	3.07	2.76	1265	2.25	2.76	1265	1.50	2.58
1687	4.89	3.05	1687	3.12	2.77	1687	2.23	2.68	1687	1.51	2.57

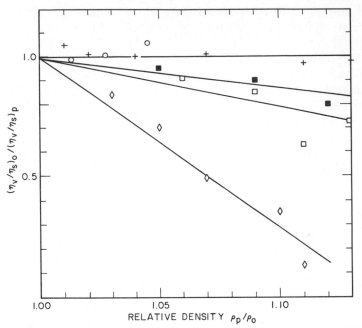

Figure 3. Density dependence of the ratio η_v/η_s: crosses, water at 10°C; open squares, methyl alcohol at 30°C; solid squares, ethyl alcohol at 30°C; diamonds, eugeniol at 30°C, from Hawley, Allegra, and Holton.[22] Circles, glycerol, from Slie and Madigosky[26].

but are also in agreement with this conclusion. It is interesting to note that, inasmuch as the ratio η_v/η_s for water is independent of density, the behavior of water is again different from that of the simple and weakly associated liquids. This is shown clearly in Figure 3, where the density dependence of η_v/η_s for water is compared with several weakly associated liquids[22] and with glycerol.[26] Only in glycerol which, like water, exhibits strong hydrogen bonding, is the ratio η_v/η_s approximately independent of density, as is the case for water.

High-frequency moduli. In many liquids, particularly viscous glass-forming liquids, the relaxation times, τ_s and τ_p, are comparable to the periods attainable with acoustic waves. For such liquids it is possible to determine the high-frequency elastic moduli experimentally by measuring the frequency dependence of the shear and longitudinal sound-wave velocities in these liquids. Conventional ultrasonic techniques have thus far been limited to frequencies less than 5.10^8 Hz, that is, periods greater than 2.10^{-9} sec. Using Brillouin scattering of light, it is possible to extend this range by an order of magnitude.[27] However, structural relaxation in water occurs in times much shorter than these, and at present it is not possible to measure the elastic moduli in water directly by the techniques just mentioned.

Estimates of the high-frequency moduli of water have been obtained by Slie et al.[28] from ultrasonic velocity measurements in water-glycerol mixtures. The water content of these mixtures was varied from 0 up to 68 mole %, which is the maximum water content at which the mixture was still sufficiently viscous that the structural relaxation times were comparable to the period of the ultrasonic wave. The relaxation moduli were found to be linearly dependent on water content and were extrapolated to 100 mole % water. In this manner Slie et al. were able to estimate the temperature dependence of the shear modulus, G_∞, and the high-frequency compressional modulus, K_∞ ($= 1/\kappa_\infty$), for water. Their results are given in Table 4.

Table 4   Temperature Dependence of the Compressional and Shear Moduli of Water Obtained by Extrapolation from Aqueous Glycerol ($\times$ 10^{-10} dyne/cm^2; T in °C)

$$G_\infty = 1.68 - 0.0127\,T$$
$$K_\infty = 5.62 - 0.01\,T$$

3   Theory

Shear Viscosity.   There have been many attempts to construct theories of transport phenomena. On a fundamental level are theories that assume nothing more than the Newtonian equations of motion and the form of the intermolecular force. One such theory is the statistical-mechanical theory leading to kinetic equations.[29] Another approach is the autocorrelation function theory,[30] which considers time-dependent correlation functions of fluctuations of various quantities in the liquid—such as density, velocity, and energy. These theories have so far been applied only to simple liquids where the molecules are assumed to interact through an effective pair potential that depends only on the distance between the molecules.

On a less fundamental level are theories that make use of the similarity between the structure of a liquid and that of a crystalline solid. One approach, applied with some success to a variety of liquids by Eyring,[31] considers the transport mechanism as essentially the movement of a molecule to a nearby vacancy in the quasi-lattice, from which it is separated by an energy barrier. The probability, p_j, of a transition or jump from one site to another is assumed to be the product of the probability, p_E, of attaining sufficient energy to break bonds and the probability, p_v, that there is sufficient local free volume for a jump to occur. It is further assumed that all holes have the same volume, v_h. The shear viscosity is shown to be[2]

$$(6) \qquad \eta_s = \frac{RT}{v_h N_h} \tau_s$$

where τ_s, the shear relaxation time, is the average time between jumps from a site in the quasi-lattice to an adjacent hole. Using reaction-rate theory, the shear relaxation time can be written as

$$(7) \qquad \tau_s = \frac{h}{kT} \exp\left(\frac{h_j - Ts_j}{kT}\right)$$

where h is Planck's constant, k the Boltzmann constant, and s_j and h_j the activation entropy and enthalpy respectively, necessary for a molecule to flow into a hole that is already present.

An approximate expression for the probability p_v was obtained by Hirai and Eyring[32]

$$(8) \qquad p_v = \frac{N_h}{N} = \frac{v}{v_h} \exp\left[\frac{-(e_h + Pv_h)}{kT}\right]$$

where N is the number of molecules, e_h is the energy necessary to create a hole, P is the ambient pressure, and v is the volume of a molecule. Inserting (7) and (8) for τ_s and N_h into (6), we get

$$(9) \qquad \eta_s = \frac{h}{v} \exp\left(\frac{e_h + h_j + Pv_h - Ts_j}{kT}\right)$$

A major success of this theory is that it leads to a viscosity proportional to $\exp(E^+/RT)$, where E^+ is called the activation energy (per mole) for viscous flow. This is the Arrhenius form, which has been observed to hold for simple nonassociated liquids. Values of the activation energy at atmospheric pressure, $E_p^+ = R[\partial \ln \eta / \partial(1/T)]_p$, for water are given in Table 1. It is obvious that the activation energy, E_p^+ is not constant, and the viscosity of water shows large departures from Arrhenius behavior. As pointed out by Jobling and Lawrence,[33] E_p^+ is not of any real theoretical significance because this so-called activation energy takes into account viscosity changes resulting from changes in internal energy, changes in volume, and changes in transport mechanism due to varying degrees of molecular convection. Jobling and Lawrence also argue that the activation energy at constant volume, $E_v^+ = R[\partial \ln \eta / \partial(1/T)]_v$ is a better description of the energy a molecule must possess in order to engage in momentum transport. However, the viscosity of water displays non-Arrhenius behavior even at constant density[17] and E_v^+ is temperature dependent. Macedo and Litovitz[34] point out that this conclusion is not unreasonable if one notes that the degree of hydrogen bonding to near neighbors is temperature dependent and if one makes the further reasonable assumption that $E_v{}^+$ will be dependent upon the degree of hydrogen bonding.

The assumption that all holes have the same volume, v_h, is obviously an approximation. Cohen and Turnbull[35] have developed a free-volume theory of transport processes in liquids by assuming a distribution of hole sizes. They find the probability, $p(v)$, of finding the free volume, v, to be,

$$(10) \qquad p(v) = (\gamma/v_f) \exp(-\gamma v/v_f)$$

where v_f is the average free volume per molecule and γ is a numerical factor, between 0.5 and 1, needed to correct for the overlap of free volume. Cohen and Turnbull further assume that the jump probability, p_j, is determined only by the chance of finding an adjacent local free volume of sufficient size for a molecule to jump into. The resulting expression for viscosity is[34]

$$(11) \qquad \eta = \frac{A}{p_v} = A \exp\left(\frac{\gamma v^*}{v_f}\right)$$

where A is constant and v^* is the minimum local free volume necessary for a jump to occur, usually taken to be equal to v_0, the close-packed molecular volume. The free volume, v_f, is defined as

$$(12) \qquad v_f = v - v_0$$

where v is the volume per molecule.

One method of obtaining v_f is to assume that the free volume is the total thermal expansion at constant pressure. If $v\beta$ is essentially constant over the temperature range considered, then

$$(13) \qquad v_f = \beta v(T - T_0)$$

where β is the coefficient of thermal expansion and T_0 is the apparent temperature at which the free volume becomes 0 (0 mobility temperature or "glass-transition temperature"). For this case (11) becomes

$$(14) \qquad \eta = A \exp\left(\frac{B}{T - T_0}\right)$$

Miller[36] has shown that (14) expresses accurately the viscosity of liquid water between -10 and $+150°C$ ($P_{sat} = 4.7$ atm) with $A = 2.9 \times 10^{-4}$ poise, $B = 510°K$, and $T_0 = 150°K$. Miller also estimates v_f from the viscosity–free-volume relationship $\ln \eta = \ln A + 1/f$, in which $f = v_f/v$, the volume fraction of free volume. There is correlation between the f determined in this manner and the volume fraction of unbonded water molecules as defined by the structural model of Némethy and Scheraga.[37] This correlation is interpreted to mean that the volume of unbonded water molecules appears as "free volume," presumably because the molecular motion required for flow occurs only in the unbonded state. It is emphasized that this is not the same as the true free volume (that is, empty space associated with each molecule) in the Cohen–Turnbull sense.

Bernini et al.[38] obtain the following expression for the factor A in (14):

$$(15) \qquad A = CT^{1/2} = \frac{(mk)^{1/2}}{3\sqrt{3\pi}} gaa^*T^{1/2}$$

where m is the mass of the molecule, g is a geometrical factor, a^* the jump distance, and a the diameter of the sphere to which the molecule is assimilable. It is shown that with $C = 1.2 \times 10^{-5}$ poise $^\circ K^{-1/2}$, $B = 570^\circ K$, and $T_0 = 146^\circ K$; (14) and (15) describe accurately the temperature dependence of the viscosity of water at atmospheric pressure in the range -10 to $100^\circ C$. However, the value for the jump distance, $a^* \approx 7$ Å, estimated from C is in disagreement with the jump distance of 1.8 Å estimated from neutron diffraction experiments.[39]

In common with most properties of water, the viscosity-pressure relationship (Figure 1) is clearly anomalous. The first and most obvious anomaly is the existence of negative viscosity-pressure coefficients and minima in the viscosity isotherms at the lower temperatures. No other compound has been found to exhibit this property. Second, the effect of pressure on the viscosity of water is very much smaller than for other materials. The existence of minima in the viscosity isotherms of water has been attributed by Hamann[40] and Horne and Johnson[16] to a decrease with pressure in the average number of clusters of hydrogen-bonded molecules.

A similar approach, but more quantitative, is taken by Cappi,[17] who bases his explanation of the anomalous behavior of water on the Frank–Wen[41] model for liquid water. According to this model, water is a mixture of two species, an icelike species consisting of clusters of molecules that are hydrogen bonded into a tetrahedral icelike structure with little or no bending of the bonds, and a species consisting of unbonded molecules that have a closer packing than the icelike species.

Cappi assumes that momentum transport in water occurs in the unbonded structure. The failure of water to fit the usual Arrhenius relationship is then attributed to the increase with temperature in the number of molecules available for momentum transport. Application of pressure to water decreases the intermolecular distances and at the same time displaces the equilibrium between the two species in favor of the unbonded species. The latter effect is greatest at low temperatures, and this is reflected in the reduction in the viscosity of water caused by pressure at temperatures below $30^\circ C$. Cappi uses the high-temperature–high-pressure viscosity data for water to estimate the temperature and pressure dependence of the viscosity of "pure" unbonded water. The ratio of the predicted viscosity of the "pure" unbonded species at atmospheric pressure and the measured viscosity of water are assumed to be close to the mole fraction of unbonded water at the same temperature. In this manner Cappi estimates the fraction of the unbonded species to be 0.52 at 0°C and 0.97 at 100°C. However, as discussed in Chapter 10, examination of other properties of water (for example, thermal expansion) suggests that there is still appreciable hydrogen bonding in water even at 100°C.

So far the most extensive fit to the viscosity data for water, as a function of both temperature and pressure, is achieved by Jhon et al.,[42] using a model of water based on the significant-structure theory of liquids. Jhon et al. assume that most of the ice-I-like structure deforms into a denser form (ice-III-like structure) upon melting, still keeping its hydrogen bonding. An equilibrium is established between the ice-I-like species, which consists of clusters of about 46 molecules with the density of ice I, and the ice-III-like species. At the same time fluidized vacancies are introduced. It is assumed that the lifetime of the ice-I-like clusters is long compared to the relaxation time in viscous flow. Therefore, water is considered a solution of the clusters of ice-I-like molecules in the solvent of ice-III-like molecules, and the viscosity of water is then given by the Einstein equation

(16) $$\eta = \eta_{\mathrm{III}}(1.0 + 2.5\phi + \cdots)$$

where ϕ is the volume fraction of the clusters. The expression for the viscosity, η_{III} of the ice-III-like species is taken to be of the form given by (9),

(17) $$\eta_{\mathrm{III}} = C \exp\left(\frac{E^+}{RT}\right) \exp\left(\frac{PV^+}{RT}\right)$$

The parameters in (17) are chosen to fit the high-pressure viscosity where the ice-I-like structure is assumed to have disappeared. The values of these parameters are $C = 2.44 \times 10^{-3}$ poise, $V^+ = 1.5$ c/mole, $E^+ = 3.20$ kcal/mole. The agreement between theory and experimental data is satisfactory over the temperature range 0 to 75°C and pressures up to 5000 kg/cm². In particular, the theory predicts the minima in the viscosity isotherms at the lower temperatures. According to Jhon et al., these minima arise because application of pressure leads to a breakup of the ice-I-like clusters and a decrease in ϕ.

One criticism of the approach of Cappi and Jhon et al. is that in each case the anomalous pressure dependence of the viscosity of water is attributed to the breakup of icelike clusters upon application of pressure. However, as discussed in Chapter 10, the presence of such clusters is difficult to reconcile with the existence, for water, of a single dielectric relaxation time and a value of structural relaxation time approximately the same as the neutron residence time.

Volume Viscosity. Hall[43] was the first to give a molecular theory of the bulk viscosity of water. He was influenced by earlier proposals of Bernal and Fowler, who suggested that the properties of water existed because of the coexistence of two structures in the liquid. If one considers water as a mixture of two structural states of differing mole volumes and energies, then for ideal mixing the volume can be written in the form

$$(18) \qquad V = x_1 V_1 + x_2 V_2 = V_1 + x_2 \Delta V$$

where x_1 and x_2 are the mole fractions of states 1 and 2, V_1 and V_2 are respective mole volumes; and $\Delta V = V_1 - V_2$ is the change in volume is one mole of liquid changes from state 1 to state 2. The relaxational isothermal compressibility, $\kappa_{T,r}$, obtained from (18) (see Chapter 10 for a more detailed discussion) is

$$(19) \qquad \kappa_{T,r} = \frac{(\Delta V)^2}{VRT} x_1 x_2$$

Substituting into (5) one obtains the following expression for the isothermal volume viscosity

$$(20) \qquad \eta_{v,T} = \frac{(\Delta V)^2 x_1 x_2}{VRT(\kappa_{T,0})^2} \tau_{T,P}$$

The expression for $\tau_{T,P}$ obtained by Hall using rate theory is

$$(21) \qquad \tau_{T,P} = \frac{h}{kT} \frac{\exp [G^+/RT]}{1 + \exp [-\Delta G/RT]}$$

where G^+ is the activation free energy for the jump from state 1 to state 2 and ΔG is the difference in the Gibbs free energy between states 1 and 2.

Hall assumes that the activated state is the same for both shear and compressional flow. The reason given for this assumption is that in both shear flow and compressive flow the molecules jump past neighboring molecules into adjoining sites. In both cases energy bonds are broken with the immediate neighbors, and the activation energies for the two flows should be of the same order of magnitude. Molecules in both states 1 and 2 undergo transitions during shear flow, whereas in compressional flow there is a net transition of molecules from one state to the other. Therefore the activation energy for shear flow should be an average over the activation energies for states 1 and 2. Hall, however, makes the approximation of equating the activation energy for shear flow to the activation energy for molecular transitions from the more populated state only. Using Eyring's expression for shear viscosity, Hall finds

(22)
$$\tau_{T,P} = \frac{V\eta_s}{RT[1 + \exp(-\Delta G/RT)]}$$

Substituting into (20) and also using (4) of Chapter 10,

(23)
$$\frac{x_1}{x_2} = \exp\frac{-\Delta G}{RT}$$

one obtains the following expression for isothermal volume viscosity:

(24)
$$\eta_{v,T} = \frac{(\Delta V)^2}{(RT)^2}\frac{\eta_s}{(\kappa_{T,0})^2}x_1 x_2^2$$

where x_1 is the more populated species.

Since ultrasonic propagation is adiabatic, the volume viscosity determined from ultrasonic absorption measurements is the adiabatic volume viscosity, $\eta_{v,s}$. Hall assumed that since over the entire temperature range (0–80°C) $\kappa_{S,0} \simeq \kappa_{T,0}$, no distinction need be made between adiabatic and isothermal loss processes. This assumption is incorrect, since even if the low-frequency adiabatic and isothermal compressibilities are approximately equal the relaxational compressibilities $\kappa_{S,r}$ and $\kappa_{T,r}$ are not necessarily equal. This point is discussed in detail by Litovitz and Davis,[2] who point out that Hall's assumption is equivalent to saying that the difference in enthalpy between the two states, if it does exist, does not play a role. This is inconsistent with the large relaxational specific heat and thermal expansion estimated for water. Litovitz and Davis show how the isothermal and adiabatic volume viscosities are related through the relaxational parameters.

Hall assumed, at 0°C, two trial values (15×10^{-12} and 18×10^{-12} cm^2/dyne) for $\kappa_{T,\infty}$ and estimated the relative change in volume $\Delta V/V$ to be 0.47. He then calculated $\kappa_{T,r}$ from $\kappa_{T,r} = \kappa_{T,0} - \kappa_{T,\infty}$ and ΔG from $\kappa_{T,r}$ and ΔV. Allowing $\kappa_{T,\infty}$ and ΔG to remain constant with temperature, he calculated η_v from (24) as a function of temperature and obtained agreement with experimental measurements to 10% or better over the temperature-range 0 to 80°C. The good agreement with the observed temperature dependence of η_v/η_s arises in Hall's calculations mainly from the assumption that ΔG is independent of temperature. This has the effect that x_1 and x_2 also remain relatively constant, and leads to the prediction that η_v/η_s is approximately independent of temperature, which is in agreement with experimental data. Since both the relaxational expansion coefficient, β_r, and the relaxational specific heat, $C_{V,r}$, are proportional to $(\partial x_1/\partial T)$, a nearly constant value of x_1 leads to values of β_r, and $C_{V,r}$ that are small at all temperatures. This is inconsistent with the relatively large values of these parameters estimated by other methods. (See Chapter 10 for a more detailed discussion.)

The assumptions leading to (24) for the volume viscosity are so general that any mixture theory of water can be used in conjunction with (24) to predict the volume viscosity of water. Therefore, the values of ΔV, x_1, and x_2 obtained by the various investigators and summarized in Chapter 10 are substituted in (24) and the calculated values of η_v/η_s are compared with experimental data in Table 5. In calculating η_v/η_s from (24) it is necessary to identify states 1 and 2 with the species proposed by the various mixture models. The criterion that x_i is the more populated species becomes difficult to apply to those mixture models where the ice-like species is the more populated state in the vicinity of the freezing point and the less populated state in the vicinity of the boiling point. The values of η_v/η_s given in Table 5 were obtained by consistently indentifying state 2 in (24) as the ice-like species. This procedure was followed since it gave the correct temperature dependence for η_v/η_s. The estimates of Walrafen, and Davis and Litovitz, agree reasonably well with experiment. The model of Némethy and Scheraga gives values that are too low, especially at high temperatures, which suggests that this model underestimates the amount of association present at high temperatures and the magnitude of ΔV.

Finally it should be emphasized that the fact that water exhibits a non-zero volume viscosity is not by itself proof of the existence in water of two or more distinguishable species of water molecules, as has sometimes been argued in the literature. What is assumed in Hall's theory is that if water is a mixture of two molecular species then the dominant rate-controlling step

Table 5   Ratio of Volume to Shear Viscosity in Water Calculated from Two-State Relaxation Parameters

Temperature (°C)	Measured[a]	Walrafen[b]	Davis and Litovitz[c]	Némethy and Scheraga[d]
0	3.14	5.89	5.45	1.30
10	2.87		5.00	1.14
20	2.83	5.75	4.82	0.97
30	2.84		4.17	0.78
40	2.77	5.36	3.72	0.61
50	2.69		3.29	
60	2.61		2.89	0.35

[a] J. M. M. Pinkerton, *Nature*, **160**, 128 (1947).
[b] *Hydrogen-Bonded Solvent Systems*, Taylor and Francis, London, 1968, pp. 9–29. Calculated for d = 2.80Å.
[c] *J. Chem. Phys.*, **42**, 2563 (1965).
[d] *J. Chem. Phys.*, **36**, 3382 (1962).

for compressive flow is the jump of a molecule from a state characteristic of one molecular species to a state characteristic of the other species.

The pressure dependence of the volume viscosity of water has been considered by Litovitz and Carnevale,[21] who showed that, in the form presented here, the Hall theory can also account for the η_v as a function of pressure if one assumes that the open-packed state, x_1, is the higher free-energy state. Also it was assumed that ΔV decreased with pressure (a decrease of about 10% at the highest pressure of 2000 kg/cm²). Although Litovitz and Carnevale obtain a remarkably good fit for the pressure dependence of the sound absorption in water, it is difficult to accept their assumption that the open or icelike type of packing is the higher free-energy state. This assumption leads to an estimate of the percentage of broken hydrogen bonds in water much larger than the values determined experimentally from infrared measurements and from the chemical shift associated with nuclear magnetic resonance (see Section 5 of Chapter 10). Most of the other mixture theories of water assign a lower free energy (at least in the vicinity of 0°C) to the open state—one that corresponds to the greatest number of hydrogen bonds per molecule. However, before any of these theories can be applied to calculate the pressure dependence of η_v, it is necessary to arrive at some method of estimating the pressure dependence not only of ΔV but also of the entropy difference, ΔS, and the difference in internal energy, ΔU, between the two states.

Eyring's theory of shear viscosity involves the exchange of positions between a hole and an adjacent molecule. The application of shear stress causes no change in the number of holes present in the liquid. Hirai and Eyring[32] suggest that when an external pressure is applied, the number of holes decreases to an equilibrium number for the new pressure. This is similar to but more sophisticated than an earlier theory of Gierer and Wirtz.[44] Hirai and Eyring develop a hole theory of volume viscosity of liquids as follows. They assume the liquid to consist of two states. State 1 corresponds to the state where the hole has disappeared. The volume and energy of state 1 are larger than those of state 2 by the volume of a hole, v_h, and the energy necessary to create a hole, e_h. This two-state concept is much like that suggested by Hall, with the following changes: (a) the upper-energy state is assumed to have the higher volume and (b) Hall suggests specific structures for his open and closed packing.

In the Hirai–Eyring theory a symmetry parameter, μ, is introduced which predicts the free-energy change in states 1 and 2 due to the applied pressure. The volume viscosity is expressed in terms of e_h, v_h, μ, and $e_j(B)$, $s_j(B)$, the the activation energy and entropy, respectively, of volume viscosity. Experimental data on the pressure dependence of volume and shear viscosities are used to determine v_h and μ for several liquids. In general $v_h \approx (1/5)v$, where v is the volume of a molecule, and μ is about 0.5. For water, however, $v_h \approx (1/9)v$ and $\mu = 0.08$. The physical significance of these results for water is not clear. It is doubtful whether the Hirai–Eyring theory, which does not take specific account of the effects of hydrogen bonding, can be applied to describe the viscosity of water.

Finally it is of interest to compare the viscosity coefficients of water and heavy water. In liquids where momentum transfer involves only convection or vibration of the molecules, isotopic substitution changes the viscosity as follows[45]

$$(25) \qquad\qquad \frac{\eta_i}{\eta} = \left(\frac{m_i}{m}\right)^{\frac{1}{2}}$$

where subscript i refers to an isotopic species. In the case of the H_2O–D_2O system, the masses are such that $(m_i/m)^{\frac{1}{2}} = 1.054$. As can be seen from Table 1, the ratio $\eta_{s,D_2O}/\eta_{s,H_2O}$ falls from 1.306 at 5°C to 1.151 at 100°C. Cappi[17] interprets this result as indicating that, in the case of water, rotation is involved in momentum transfer, as well as convection and/or vibration. When momentum transfer involves only rotation of the molecules, the molecular mass must be replaced in (25) by the moment of inertia, I. For the H_2O–D_2O system, the principal moments of inertia are such that the average value of $(I_i/I)^{\frac{1}{2}}$ is 1.381.

The ratio of the volume viscosities is somewhat different. As shown in Table 2, $\eta_{v,D_2O}/\eta_{v,H_2O}$ falls from 1.13 at 5°C to 1.06 at 40°C. Possibly this means that rotation of water molecules does not contribute appreciably to momentum transfer during compressional flow.

An alternate interpretation of the above results is as follows. First it should be noted that direct application of (25) to the H_2O–D_2O system is open to criticism, since this equation is based on the assumption that isotopic substitution does not change the intermolecular potential. However, Némethy and Scheraga[46] and Davis and Bradley,[47] on applying their respective mixture theories of water to heavy water, find that the hydrogen bond is slightly stronger in D_2O than in H_2O. Both investigators conclude that the fraction of unbroken hydrogen bonds is greater in D_2O than in H_2O at temperatures near the melting point and becomes more nearly equal at higher temperatures. Therefore it is likely that the ratios $\eta_{s,D_2O}/\eta_{s,H_2O}$ and $\eta_{v,D_2O}/\eta_{v,H_2O}$ are determined not only by the ratio of the masses and/or moments of inertia of the D_2O and H_2O molecules, but also by the differences in hydrogen-bond strength and degree of hydrogen bonding.

High-Frequency Moduli.   No theoretical studies of the high-frequency moduli of water have been made to date. Recently Madigosky et al.[48] have presented a phenomenological formulation for the shear compliance (the reciprocal of the shear modulus) of hydrogen-bonded liquids. It is based on the assumption that the compliance of these liquids is a linear function of the reciprocal density of hydrogen bonds. These calculations have not, however, been extended to water.

At present, the main application of the data on the elastic moduli of water is to estimate the structural relaxation time, τ_P. The values for $\kappa_\infty = 1/K_\infty$ obtained by Slie et al. (Table 4) were used together with the low-frequency compressibility, κ_0, obtained from speed-of-sound measurements in water[10] to determine the relaxational compressibility, $\kappa_r = \kappa_0 - \kappa_\infty$. The values of κ_r were used together with the volume viscosity data given in Table 2 to calculate τ_P from (5). The resulting values of τ_P are shown in Table 8 of Chapter 10. The inelastic scattering of neutrons provides another means of estimating the structural relaxation time in water. In the temperature range 1 to 25°C, Safford et al.[39] find that the experimental neutron scattering data are consistent with a model in which the molecules are assumed to remain in a quasi-lattice site for a time, τ_0 (the residence time), which is much longer than the time, τ_1, during which a diffusive jump takes place. In this case the motions observed correspond to the jumping of individual water molecules. Comparison of τ_0 and τ_P shows that these two times are approximately equal. This result is particularly interesting because τ_0 is the time that an individual molecule remains in a quasi-crystalline site before undergoing a diffusive jump to a new site. Since τ_P is about equal to τ_0, the rate-controlling step associated with structural rearrangement in water must also be the jump of individual water molecules and cannot be attributed to the melting of a cluster of molecules. Such a flow mechanism is inconsistent with the flickering cluster theory of Frank and Wen[41] and the mixture model of Némethy and Scheraga.[37]

4 Conclusions

The available data on the viscosity of water are quite accurate; they are also sufficiently extensive to indicate the main features of the pressure and temperature dependence of both the shear and volume viscosity coefficients of water. In common with most properties of water, the viscosity exhibits some anomalous behavior. The unique anomaly is the existence of both negative viscosity-pressure coefficients and minima in the viscosity isotherms at the lower temperatures. The behavior of water is further differentiated from that of the simple and weakly associated liquids because the ratio η_v/η_s, is independent of density. At present there is no theory of the viscosity of water that describes both the shear and volume viscosities and accounts for the anomalies just cited in terms of a single model for water. Further progress lies in developing such a theory, and this work may be of value in determining which, if any, of the present models is a realistic representation of the structure of water.

Measurements of the high-frequency elastic moduli of water are particularly interesting, since they can be used together with volume viscosity data to estimate the structural relaxation time, τ_P. Comparison of τ_P with other relaxation times yields information on the type of molecular motion that determines the rate at which structural rearrangement occurs in water. This information can be used to restrict the number of possible structural models that can be applied to water.

The elastic moduli of water are obtained by extrapolation from water-glycerol mixtures. However, more work is necessary to establish the method as one giving reliable and consistent values of κ_r. In particular, other viscous water mixtures should be studied.

REFERENCES

1. K. Herzfeld and T. Litovitz, *Absorption and Dispersion of Ultrasonic Waves*, Academic Press, New York, 1959.
2. T. Litovitz and C. Davis, *Physical Acoustics*, Vol. IIA, Academic Press, New York, 1965, pp. 281–349.
3. J. Coe and T. Godfrey, *J. Appl. Phys.*, **15**, 625 (1944).
4. C. Craque, unpublished work quoted in ref. 3.
5. R. Hardy and R. Cottington, *J. Res. Natl. Bur. Std.*, **42**, 573 (1949).
6. L. Korson, W. Drost-Hansen, and F. J. Millero, *J. Phys. Chem.*, **73**, 34 (1969).
7. J. R. Moszynski, *Trans. ASME, Ser. C*, **83**, 111 (1961).
8. K. H. Dudziak and E. U. Franck, *Ber. Bunsenges.*, **70**, 1120 (1966).
9. E. Stanley and R. Batten, NAVSHIPSRANDCEN Rept. 2603 (1968).
10. G. Tamman and H. Rabe, *Z. Anorg. Allgem. Chem.*, **168**, 73 (1927).
11. E. Lederer, *Kolloid. Beih.*, **34**, 270 (1932).
12. K. Bett and J. Cappi, *Nature*, **207**, 620 (1965).
13. P. Bridgman, *Proc. Amer. Acad. Arts Sci.*, **61**, 57 (1926).
14. D. Zhuze, *Chem. Abstr.*, **61**, 6417d (1964).
15. R. Cohen, *Ann. Phys.*, **45**, 666 (1892).
16. R. Horne and D. Johnson, *J. Phys. Chem.*, **70**, 2182 (1966).
17. J. Cappi, "The Viscosity of Water at High Pressure," doctoral dissertation, University of London, 1966.
18. J. Pinkerton, *Nature*, **160**, 128 (1947).
19. Pancholy, M., *J. Acoust. Soc. Amer.*, **25**, 1003 (1953).
20. R. Tait, *Acustica*, **7**, 193 (1957).
21. T. Litovitz and E. Carnevale, *J. Appl. Phys.*, **26**, 816 (1955).
22. S. Hawley, J. Allegra, and G. Holton, *J. Acoust. Soc. Amer.*, **47**, 137 (1970).
23. C. Davis, R. Freeman, and J. Jarzynski, Final Rept. Office of Saline Water Grant #14-01-0001-684, 1968.
24. H. Seki, A. Granato, and R. Truell, *J. Acoust. Soc. Amer.*, **28**, 230 (1956).
25. H. F. Eden and R. A. Horne, *J. Acoust. Soc. Amer.*, **40**, 221 (1967).

26. W. Slie and W. Madigosky, *J. Chem. Phys.*, **48**, 2810 (1968).

27. R. Mountain, *Rev. Mod. Phys.*, **38**, 205 (1966).

28. W. Slie, A. Dunfor, and T. Litovitz, *J. Chem. Phys.*, **44**, 3712 (1966).

29. P. Gray, "The Kinetic Theory of Transport Phenomena in Simple Liquids," *Physics of Simple Liquids*, Wiley, New York, 1968, Chapter 12.

30. P. Schofield, "Experimental Knowledge of Correlation Functions in Simple Liquids," *Physics of Simple Liquids*, Wiley, New York, 1968, Chapter 13.

31. S. Glasstone, K. J. Laidler, and H. Eyring, *Theory of Rate Processes*, McGraw-Hill, New York, 1941.

32. N. Hirai and H. Eyring, *J. Appl. Phys.*, **29**, 810 (1958).

33. A. Jobling and A. S. C. Lawrence, *Proc. Roy. Soc. (London), Ser. A*, **206**, 257 (1951).

34. P. Macedo and T. Litovitz, *J. Chem. Phys.*, **42**, 245 (1965).

35. M. H. Cohen and D. Turnbull, *J. Chem. Phys.*, **31**, 1164 (1959).

36. A. A. Miller, *J. Chem. Phys.*, **38**, 1568 (1963).

37. G. Némethy and H. A. Scheraga, *J. Chem. Phys.*, **36**, 3382 (1962).

38. U. Bernini, F. Fittipaldi, and E. Ragozzino, *Nature*, **224**, 910 (1969).

39. G. Safford, P. Leung, A. Naumann, and P. Schaffer, *J. Chem. Phys.*, **50**, 4444 (1969); G. Safford, P. Schatter, P. Leung, G. Doebbler, G. Brady, and E. Lyden, *J. Chem. Phys.*, **50**, 2140 (1969).

40. S. D. Hamann, *Physico-Chemical Effects of Pressure*, Butterworths, London, 1957, p. 82.

41. H. S. Frank and W.-Y. Wen, *Discussions Faraday Soc.*, **24**, 133 (1957); H. S. Frank, *Proc. Roy. Soc. (London), Ser. A*, **247**, 481 (1958).

42. M. Jhon, J. Grosh, T. Ree, and H. Eyring, *J. Chem. Phys.*, **44**, 1465 (1966).

43. L. Hall, *Phys. Rev.*, **73**, 775 (1948).

44. A. Gierer and K. Wirtz, *Z. Naturforsch.*, **5a**, 270 (1950).

45. E. McLaughlin, *Physica*, **26**, 650 (1960).

46. G. Némethy and H. A. Scheraga, *J. Chem. Phys.*, **41**, 680 (1964).

47. C. Davis and D. Bradley, *J. Chem. Phys.*, **45**, 2461 (1966).

48. W. M. Madigosky, G. E. McDuffie, and T. A. Litovitz, *J. Chem. Phys.*, **47**, 753 (1967).

49. V. Rajagopalan and G. Holton, *J. Acoust. Soc. Amer.*, **43**, 108 (1968).

18 Effects of Non-Brownian Motion on Transport Coefficients in Electrolyte Solutions*

Harold L. Friedman, Department of Chemistry, State University of New York at Stony Brook, Stony Brook, New York

1 Introduction

Transport coefficient determinations have played a basic role in the growth of our knowledge of electrolytic systems. This is partly because some of these coefficients, most notably the electrical conductivity, can be measured with great accuracy without much trouble. It is also partly because, for dilute solutions, a very elementary theory is adequate to relate the measured transport coefficients to molecular quantities, such as numbers of solute particles, their size and charge, and a particular measure of the force between them, namely, the association constant. For solutions at higher solute concentration, say over 0.01 M, the experimental advantages of transport coefficients remain with us but the interpretation of the data in terms of molecular properties becomes less certain.

The tools for this interpretation are the theories of the concentration dependence of the transport coefficients invented by Ostwald, Debye, Hückel, and Onsager and developed further by Onsager, Falkenhagen, Fuoss, and their students.[1] It is difficult to extend the range of validity of these theories to higher concentrations or, even at low concentrations, to generalize them to allow refinements of the underlying model. A similar problem, which appears with the Debye–Hückel equilibrium theory and its descendants, has been frequently discussed.[2] Rather recently it has become possible to exploit new developments in the many-body and time-dependent areas of statistical

* This work was supported by a grant from the Office of Saline Water, U.S. Department of the Interior. A substantial portion of this contribution is reproduced, with permission, from a paper appearing in *Journal de Chimie Physique, Numero Special*, p. 75, October 1969.

723

mechanics to formulate the concentration dependence of transport coefficients in a way somewhat different from that of the classical theories, a way that certainly is liberating in terms of the models which can be treated, and which may be more amenable to the region of higher concentrations. To date these new developments have only been carried to the point of calculating the relaxation part of the limiting-law term in the conductivity for solutions of single electrolytes,[3] the electrophoretic part of the same term,[4] the limiting law for conductance for mixed electrolytes,[5] and the limiting law for the Hall effect.[6] It is comforting that the new theories independently lead to the classical results for the limiting law for the electrical conductance, but it may be of more interest to extend them in the directions just mentioned. As with equilibrium theories, there are considerable computational difficulties in doing this, but there is a completely new problem too, and that problem is the subject of this chapter.

2   Requirements of a Model for Calculating Transport Coefficients

A general procedure for calculating solute transport properties of a solution from a model in which only the solute molecules are explicitly represented has been described before, both in detail[3-6] and in a less formal way.[7] Here attention is concentrated on the description of the model required as input to the theory. All the main problems are encountered if we restrict ourselves to the calculation of the conductance of a hypothetical model system chosen to represent an aqueous solution of sodium chloride. Then we require:

1. A description of the fluctuating motion of an Na^+ ion in pure water at equilibrium.
2. The same for the Cl^- ion.
3. A description of the force exerted on a sodium ion by a chloride ion at a specified distance away and moving with a specified velocity, all in water with no other solute ions about.
4. The same for the force on Na^+ due to Na^+, the force on Cl^- due to Cl^-, and the force on Cl^- due to Na^+.

Additional more complicated effects *may* be present, but the functions described are those that *must* be described to enable the calculation of the conductivity. It will be noted that none of them involves the solvent explicitly in the sense that we do not require the force on Na^+ from a water molecule in the neighborhood. On the other hand, all the listed functions involve the solvent in an implicit way.

Of course it is of great importance for our understanding of the molecular interactions in these systems to study also models in which the solvent molecules as well as the ions are explicitly represented. The extensive contributions of P. Resibois and his co-workers to this problem have recently been reviewed.[8] To some extent the studies of the two classes of models are complementary, and further mention of this is made later, but otherwise only models in which the solvent molecules do not explicitly appear are discussed.

At least for simple models, the force that one ion exerts on another comprises three terms. The first is the electrical term, mainly the Coulomb force between the bare ions shielded by the dielectric constant of the solvent. The second is the mutual repulsion at small separations. The specification of these forces for each pair Na^+-Cl^-, Na^+-Na^+, and Cl^--Cl^- is all one requires to calculate the *equilibrium* properties corresponding to the model. The other term of the force is well represented as a hydrodynamic interaction in the limiting case in which the solvent is an incompressible, structureless medium characterized by a dielectric constant and viscosity; in general it results from momentum transfer from a moving ion through the solvent to another ion. It produces the electrophoretic term in the limiting law as well as contributions to higher terms in the concentration dependence of all orders.

The description required in (1) and (2) can be given in many ways that are mathematically equivalent. An example is the differential equation for the motion. The simplest such equation is Langevin's equation

$$(1) \qquad\qquad m\frac{d\mathbf{v}}{dt} = -\zeta\mathbf{v} + \mathbf{R}(t)$$

where $\mathbf{v}$ is the velocity of the ion, ζ its friction coefficient, m its mass, and $\mathbf{R}(t)$ fluctuating force on the ion due to the solvent molecules.

With the assumption that $\mathbf{R}(t)$ is fluctuating at a rate that is fast compared to any process of interest, Langevin's equation becomes the mathematical theory of Brownian motion. This assumption is quite accurate if the solvent molecules are very small (in mass) compared to the ion. Therefore Brownian motion, as described here, is quite consistent for the model in which one assumes that the solvent is an incompressible, structureless medium characterized just by a dielectric constant and a viscosity.

In fact, Langevin's equation is exact in the limit that the solvent/solute mass ratio becomes zero.[9] Also it may be remarked that in Resibois's conductance calculations,[8] Langevin's equation is not assumed; it is derived in the limit that the ion mass is infinite compared to the solvent molecule mass. It does seem that this may be equivalent to using a model with solvent-averaged forces together with the result of Lebowitz and Rubin.[9]

The assumption that Langevin's equation characterizes the fluctuating motion of the *carrier*, the body whose transport gives rise to the flux of interest, is often described by saying that the carrier motion is Brownian, or that it is characterized by small-step diffusion, or (for the analogous rotational motion) that it is Debye-like. In this chapter the assumption of Langevin's equation or any of its equivalent forms is called the Brownian motion assumption.

Some examples of the effects of non-Brownian motion have been studied by Friedman and Ben-Naim[10] and Harris and Friedman,[11] to which reference may be made for information about related work and especially about the extensive theoretical background of the problem to be found in the literature. The purpose of this chapter is to describe some of these results in a more general context, without proofs, and to present some further results.

3   Brownian Motion

Some of the characteristics of Brownian motion, as implied by the theory of the Langevin equation, are outlined in this section.

The motion governed by (1) determines a time-relaxed correlation function given by

$$
(2) \qquad\qquad \langle \mathbf{v}(t) \cdot \mathbf{v}(0) \rangle = \frac{3kT}{m} \exp\left(-\frac{t\zeta}{m} \right)
$$

where $\mathbf{v}(t)$ is the velocity of the ion or other carrier at time t. The quantity on the left is a certain average value of the product $\mathbf{v}(t) \cdot \mathbf{v}(0)$. It may, in thought, be determined by observing the velocity at some time, t_o, and again at a time, t, later. The product of these velocities is $\mathbf{v}(t_o) \cdot \mathbf{v}(t_o + t)$. If the observation is repeated at a great many initial times, t_o, and then averaged, the average is $\langle \mathbf{v}(t) \cdot \mathbf{v}(0) \rangle$. It may also be determined by preparing a great many systems that are replicas of the system of interest and, at some particular instant, t_o, measuring the velocity, $\mathbf{v}(t_o)$, in each system and observing the velocity $\mathbf{v}(t + t_o)$ at a time, t, later. Then the product, $\mathbf{v}(t_o) \cdot \mathbf{v}(t + t_o)$, is formed for each system and the sum of these products is taken and divided by the number of systems. It is, of course, one of the fundamental postulates of statistical mechanics that these two averages, the time average and the ensemble average, are equal.

As these definitions indicate, the time-relaxed correlation function may be calculated for any sort of fluctuating quantity, not just the translational velocity; neither is it limited to Brownian motion. However, the particular exponential time dependence shown in (2) is associated with Langevin's equation. The coefficient in (2) results from the fact that at equilibrium $\frac{1}{2}m\langle v^2 \rangle \equiv \frac{1}{2}m\langle \mathbf{v}(0) \cdot \mathbf{v}(0) \rangle = 3kT/2$. It is obtained for any kind of motion, whereas the simple exponential time dependence is characteristic of Brownian motion.

The correlation function may, in general, be expanded as a power series in the time as follows:

(3) $$\langle \mathbf{v}(t) \cdot \mathbf{v}(0) \rangle = \frac{3kT}{m} (1 + c_1 t + c_2 t^2 + \cdots)$$

In the case of Brownian motion—(2)—we can readily calculate the coefficients in this expansion; we find they are all nonzero. However, for any *physical* case it is possible to show that the c_1 coefficient must vanish, since the change, $\mathbf{v}(t) - \mathbf{v}(0)$, is produced by the forces exerted on the carrier by other molecules. But if t is small enough, then a finite force does not change the velocity of the carrier; during this interval it must travel as though it were in empty space. It may be helpful to note that, for a hypothetical system in which the carrier and the molecules of the medium are hard spheres, the coefficient c_1 is finite just because in this case the forces are infinite (when the spheres collide) and they do produce a finite acceleration in an infinitesimal time interval.

These considerations lead to the well-known conclusion that Brownian motion does not exactly represent the dynamics in any physical system.

It can also be shown by the methods of statistical mechanics that in general the next coefficient, c_2, is determined by a certain average of the force exerted on the carrier by the other molecules and that c_2 is negative. This also is not consistent with (2).

Brownian motion is known to characterize a process that is both Gaussian and Markovian. In our example, where, for simplicity, the probabilities have not been normalized, $\mathbf{v}(t)$ is a "Gaussian process" if the probability to observe a particular velocity is a Gaussian function, for example

(4) $$P_1(\mathbf{v}) = e^{-a\mathbf{v}^2}$$

or if the probability to observe a particular sequence of two velocities is a Gaussian function

(5) $$P_2[\mathbf{v}(t), \mathbf{v}(0)] = \exp\left[-a\mathbf{v}(0)^2 - 2b(t)\mathbf{v}(0) \cdot \mathbf{v}(t) - a\mathbf{v}(t)^2\right]$$

and so on for the series—$P_3[\mathbf{v}(t_2), \mathbf{v}(t_1), \mathbf{v}(0)]$, etc.

Now of course the Gaussian form of P_1 is assured by the Maxwell–Boltzmann distribution law for equilibrium systems in which quantum-mechanical effects are negligible; but there is no corresponding proof even for classical mechanics for the Gaussian form of P_2, P_3, and so on. However, the central-limit theorem, one of the cornerstones of the mathematical theory of probability, requires that any P_n be Gaussian if each of the n velocities is determined by a very large number of independent variables. Therefore the Gaussian form for P_n may be expected to be quite accurate, provided that the intervals between the observations are not too small; that is, $t_n - t_{n-1}$ is long enough so that there are numerous fluctuating contributions to the force, which tends to make $v(t_n)$ different from $v(t_{n-1})$. This would seem to be a much more stringent condition in a hard-sphere fluid, in which the carrier is unlikely to interact with more than one other molecule at a time, than in an aqueous solution, where the carrier must be interacting with many molecules at once owing to the quite long range of the forces and the structure of the medium.

In the same terms used to describe the Gaussian condition on the process, the Markovian condition is

(6) $P_3[v(t_2), v(t_1), v(t_o)]P_1[v(t_1)] = P_2[v(t_2), v(t_1)]P_2[v(t_1), v(t_o)]$

What this requires is simply that the change in velocity from t_1 to t_2 be independent of that from t_o to t_1. That is, there is no memory of the change in the preceding interval. On the other hand, for physical systems memory effects must be expected, since the configuration of all the rest of the molecules, which produce the force that causes the carrier to change to $v(t_1)$ from $v(t_o)$, does not instantly disappear at t_1 because molecular motion is required to change it. Of course if $t_2 - t_1$ and $t_1 - t_o$ are long enough on the time scale of the molecular motion, the Markovian approximation may be reasonably accurate.

One finds again that the Brownian motion representation of the fluctuating motion of the carrier cannot be an exact representation of *any* physical system, but it may be very good if we are only interested in what happens over intervals long compared to the time required for the solvent molecules themselves to move appreciable distances. For normal molecules the latter time is less than 1 psec (10^{-12} sec).

One time scale that is of interest for comparison with this is provided by Brownian motion theory itself, namely, the ratio m/ζ, which appears in the exponent of (2) and obviously has the meaning of a characteristic time for the Brownian motion. In fact, it is the characteristic time for a change in velocity in the sense that the carrier velocity is nearly constant during any interval much shorter than m/ζ, and $v(t)$ becomes nearly independent of $v(0)$ when t is much longer than m/ζ. The friction coefficient, ζ, may be estimated from Stokes's law or obtained from a measured diffusion coefficient or electrical mobility. We find

$m/\zeta \simeq 0.01$ psec for Na_{aq}^+
$m/\zeta \simeq 0.01$ μsec for a 10^3-Å diameter colloid particle in water

It seems reasonable to expect that in the latter case no observation at all can be effected by the error in the Brownian motion approximation. The former case requires further investigation.

4   Vacancy Motion on a Lattice

It might be concluded on the basis of the discussion in the previous section that non-Brownian motion effects can only be detected in experiments that are very fast in an obvious sense. However, this is not the case, as may be illustrated by the following simple example.

In crystals of sodium chloride at equilibrium there are vacancies on some of the Na^+ sites. These vacancies may be filled by a process in which Na^+ hops from a neighboring Na^+ site, leaving a vacancy there. Thus there is a mechanism for a fluctuating motion both of the vacancies and of the Na^+. One can measure the electrical conductance, σ, of the crystal; it is related to the mobility $u = \zeta^{-1}$ of the vacancies by

(7) $$\sigma = ne^2 u$$

where n is the number of vacancies per unit volume and e the electric charge of the sodium ion. One can also measure the self-diffusion coefficient of the Na^+ ion using an isotopic tracer. According to elementary theory this is given by

(8) $$D = \frac{nv}{N} r^2 kT$$

where v is the mean number of hops per second for a vacancy and N is the number of sodium ions per unit volume. Thus nv/N is the mean number of hops per second for an Na^+ ion. The length of a hop, the distance between sodium sites, is r.

We also know that

(9) $$u = vr^2$$

so we expect to find

(10) $$\sigma = \frac{Ne^2 D}{kT}$$

which is the Nernst–Einstein relation of the conductance and diffusion coefficient. However, the experimental data definitely show that the left-hand side of this equation is larger than the right. The explanation of this, due to Bardeen and Herring,[12] is that the assumption on which (8) is based, that successive hops of the Na^+ ion are uncorrelated, is not valid for these systems. When an Na^+ ion hops, say in the x direction, a vacancy hops in the $-x$ direction and is then on the $-x$ side of the Na^+ ion that hopped. The result is that the next hop of the Na^+ ion has a better than even chance of being in the $-x$ direction. This is a very simple example of a non-Markovian process: the direction of a given hop of the Na^+ ion is somewhat affected by the direction of its previous hop. There is no such effect in the motion of the vacancy, which is strictly Markovian in a simple hopping model. The result is to introduce a factor, f, on the left-hand side of (8), f being unity for a Markovian process and less than unity when successive hops are correlated as in this case. This factor has been calculated for various crystal symmetries by Compaan and Haven.[13]

It is instructive to reflect that if one put a single isotopic Na^+ ion in this crystal as a tracer and observed it, he would see trajectories that looked Brownian when watched for intervals that were long compared to the time it takes a vacancy to make several hops but had a distinctly non-Markovian appearance when observed for intervals comparable to the hopping time of the vacancies. Furthermore, the measurements of σ and D require times that are essentially infinite compared to this hopping time; still they are sensitive to the non-Markovian behavior.

It is interesting to note that the model discussed here is somewhat analogous to the model for proton motion in aqueous solutions, where the role of the vacancy is provided by an OH^- ion (i.e., a proton vacancy) and of course, also by an H_3O^+ ion, a defect of another sort. Even when the motions of these defects are considered to be Markovian, they can induce non-Markovian behavior in the proton motion, just as in the Na^+ vacancy case. Further mention of this is made in Section 6.

5 Two-State Brownons[10]

This is a summary of a study undertaken to see whether one could find an example of non-Brownian motion for which transport coefficients in liquid solutions can be calculated. The difficulty is that if one assumes any description of the carrier dynamics other than Langevin's equation, there are problems of enormous difficulty with the calculations of transport coefficients. A trick that seems to give interesting results is to assume that the carrier motion is governed by Langevin's equation but with values of the friction coefficient, ζ, and the mass, m, which change in a random way between two sets of values. This might represent, for example, Na_{aq}^+, which sometimes moves as the unit $Na(H_2O)_4^+$ and sometimes as $Na(H_2O)_6^+$. It is not certain that this behavior describes Na_{aq}^+, but it is of interest to see what the consequences might be. A carrier behaving in this way may be called a two-state brownon.

The trajectory of a two-state brownon drifting in an electric field is shown schematically in Figure 1. The fluctuations in the motion are more noticeable in the state of low ζ, the fast state, than in the slow state.

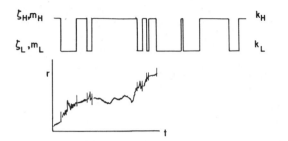

Figure 1. Schematic representation of the trajectory of a two-state brownon drifting in an electric field. The upper part of the figure shows the random change of state as time passes. Subscripts H and L denote states of high friction and low friction, respectively. The lower part of the figure, with the same time scale, is the corresponding trajectory.

In order to describe the two-state brownon completely, six parameters are needed

$$\zeta_s, m_s, k_s \qquad \zeta_f, m_f, k_f$$

where k_i^{-1} is the mean lifetime of state i. The fraction of the carriers in state s at any time is

$$(11) \qquad x_s = \frac{k_f}{(k_s + k_f)}$$

It is expected that if both $k_s \gg \zeta_s/m_s$ and $k_f \gg \zeta_f/m_f$ then the motion cannot be distinguished from the motion of a one-state brownon with properties intermediate to the slow and fast states.

The more interesting extreme is the one in which

$$(12) \qquad k_s \ll \frac{\zeta_s}{m_s} \quad \text{and} \quad k_f \ll \frac{\zeta_f}{m_f}$$

Then it is found that the carrier contributes to the transport properties in the same way as a mixture of carriers with the fraction x_s having the fixed properties, ζ_s, m_s, and the fraction, $1 - x_s$, having fixed properties characteristic of the fast state. In this case the electrical conductance of a system having n of these carriers of charge e per unit volume is

$$(13) \qquad \sigma = ne^2 \left[\frac{x_s}{\zeta_s} + \frac{(1 - x_s)}{\zeta_f} \right]$$

if n is small enough so the mutual interaction of the carriers is negligible.

The corresponding Hall conductance, the coefficient for the current in the $E_\wedge H$ direction when the system of carriers is in both electric and magnetic fields, is

(14)
$$\sigma' = ne^3\left[\frac{x_s}{\zeta_s^2} + \frac{(1 - x_s)}{\zeta_f^2}\right]$$

The two expressions are different because the electrical conductance is a measure of the average velocity of the carrier in the field, whereas the Hall conductance is a measure of the average of the velocity squared. Their comparison provides a way to test for non-Brownian motion in real systems. Thus for motion governed by Langevin's equation one has

$$\sigma = \frac{ne^2}{\zeta} \quad \text{and} \quad \sigma' = \frac{ne^3}{\zeta^2} \quad \text{so} \quad 1 = \frac{ne\sigma'}{\sigma^2}$$

Therefore, for more general motion, the experimentally accessible quantity

(15)
$$R = \frac{ne\sigma'}{\sigma^2}$$

is a measure of the effect of the details of the carrier dynamics upon the transport coefficients; it is unity for Brownian motion.

The calculated R in Figure 2 is obtained for a model system, HCl_{aq}, when it is assumed the hydrogen ion alone is a two-state brownon with average mobility corresponding to the experimental value and mobility of the fast state given by the scale of abscissa.

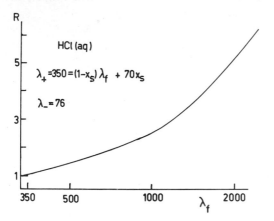

Figure 2.   Hall coefficient of aqueous hydrochloric acid calculated for the two-state brownon model for H^+.

This example shows again that the comparison of two d-c transport coefficients is sensitive to the character of the dynamics of the carriers. Unfortunately the Hall coefficient, which is relatively easy to calculate from a model, has not yet been measured with certainty in ionic solutions. Other transport coefficients that are easier to measure are harder to calculate from a model; this calculation has been carried through for the two-state brownon model only for the "Onsager slope," the coefficient of $\sqrt{c}$ in the expansion of the equivalent conductance

(16) $\Lambda = \Lambda_o + S\sqrt{c} + Ec \ln c + Jc + \cdots$

It may not be obvious that it can be instructive to regard the coefficients S, $E, \ldots$ as transport coefficients, but the situation is in fact quite parallel to the Hall conductance. In the latter the carrier moves under the influence of forces due to the electric and the magnetic fields. To calculate the S and higher coefficients of (16) one must treat the motion of the carrier moving under the influence of forces due to the electric field *and* the forces due to the other ions in the solution (cf. Section 2). In one case the underlying basic fluctuating motion of the carrier in the solvent is perturbed by electric and magnetic fields, in the other by electric and molecular fields.

Since in the limit of small k_s and k_f the carrier acts like a mixture of carriers, one may calculate S for the two-state brownon model just by applying the form of S calculated for mixtures of one-state brownons.[1,5] The appropriate equation is very complicated and not informative about what is going on, so only the results are given here.

These results, shown in Figure 3, have been obtained for the same model used for the Hall conductance: HCl_{aq} in which the Cl^- is a normal brownon, the H^+ is a two-state brownon of varying fast mobility, and the average mobility is fixed at the experimental value. It is clear that the Onsager slope, like the Hall coefficient, is sensitive to the kind of non-Brownian motion introduced by this model. However, in the case of this particular transport coefficient, and no doubt the E coefficient in (16) as well, there is a characteristic feature that interferes with its use to study the underlying dynamics. This is the slow relaxation of the ion atmosphere, which for HCl_{aq} is characterized by a relaxation time, τ_F of 10^{-7} sec at 10^{-4} M and 10^{-10} sec at 0.1 M. If both $k_s\tau_F$ and $k_f\tau_F$ are larger than unity, then this relaxation tends to smooth over the changes in state of the two-state brownon and one sees only its average properties. Therefore, the effects illustrated in Figure 3 may be expected only if k_s and k_f are small compared to τ_F^{-1} and not just small compared to ζ_s/m_s and ζ_f/m_f, respectively. The former is a much more restrictive condition in the range of low concentration, in which S dominates other contributions to $\Lambda - \Lambda_o$, which are less well known.

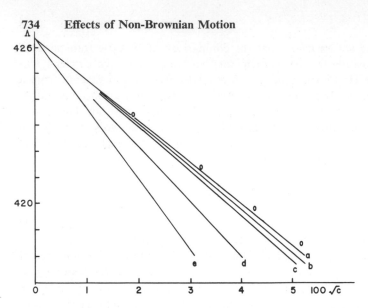

Figure 3. Equivalent conductance calculations for aqueous hydrogen chloride at 25° for the same model as Figure 2. The limiting slopes are calculated from the two-state brownon model for the following cases: A, $\lambda_f = 350$; B, $\lambda_f = 382$; C, $\lambda_f = 420$; D, $\lambda_f = 770$; E, $\lambda_f = 5670$. Case A is the same as for strictly Brownian motion, the Onsager slope. From Stokes and Stokes, *J. Phys. Chem.*, **65**, 1242 (1961).

It is interesting to contrast these results with the behavior of a model for an associating electrolyte

$$A^+ + B^- \underset{k_d}{\overset{k_a}{\rightleftarrows}} AB$$

studied by Onsager and Provencher.[14] If one follows an A particle he sees its mobility change in the manner shown in Figure 1, for the A particle is sometimes in the A^+ state, sometimes in the AB state. In this case it is found[14] that the association process affects the conductivity behavior, provided that k_a is a big enough rate constant (it needs to be near the diffusion-controlled limit) and that the degree of association is more than 0.05 or so. This process is unlike that characteristic of the two-state brownon in that it affects the conductivity when the rates are fast. The differences are that here (a) the change of state is accompanied by a change in charge and (b) the lifetime of an A^+ ion depends on the local concentration of B^- ions. Because of these features, the fluctuations due to the association reaction are coupled to the spatial motions of the ions, an effect which is absent in the two-state brownon model. It is found that this coupling reduces the Debye–Falkenhagen relaxation time, τ_F, so the relaxation term in S in (16) is reduced and the effect increases the conductivity at finite concentration—just the opposite of the effect of the two-state brownon. However, the model for an associated electrolyte does not simply result in a new limiting-law slope because of the concentration dependence of the degree of dissociation and of the diffusion-controlled association rate "constant." In fact, at low enough concentration the effect of the association process vanishes and the standard limiting slope is correct.

6   Electrical Mobility of Protons in Water

Many of the models that have been proposed to explain the large mobility of protons in water may be discussed in terms of the two-state brownon model. An example is given here.

The Grotthuss chain model of proton mobility in water has been made much more precise by Conway, Bockris, and Linton,[15] who have nevertheless retained the central feature that the high mobility is due to the proton exchange reaction

$$(17) \qquad H_3O^+ + H_2O \rightarrow H_2O + H_3O^+$$

More recently the rate of this reaction has been directly determined by Meiboom[16] by means of its effect on the ^{17}O-induced NMR relaxation of protons in the water with the result that at $25°C$ the lifetime of a H_3O^+ ion in water is $\tau = 1.7$ psec, within 40%. This is the mean time that the ion lasts before it reacts according to (17). Later work makes a value nearer to $\tau = 2.2$ psec seem more accurate.[17]

It is interesting to compare this measured value with the estimate $\tau = 0.024$ psec obtained by an a priori calculation by Conway, Bockris, and Linton.[15] This illustrates the difficulty of making a priori calculations of the parameters for dynamic processes in liquids, especially those with complicated structure. These authors assumed the same jump length, r, used here but calculated the electrical mobility associated with the exchange reaction in a way different from that described here, and their method produces a substantial cancellation of the error in τ.

Of course from the point of view of a calculation of the electrical conductivity, the carrier of the current now is neither the proton nor the hydronium ion, but the charge itself, which sometimes rides on one proton or hydronium ion, sometimes on another. It may be noted that this motion of the charge contributes to the self-diffusion coefficient of the protons (i.e., to the tracer self-diffusion coefficient of D in a dilute solution of HOD in H_2O). The situation is analogous to the contribution of the vacancy motion to the self-diffusion coefficient of the sodium tracer in Section 3. However, in the water case this effect turns out to be negligible, since the self-diffusion coefficient of the protons in water is found to be close to that for $^{17}OH_2$, Na_{aq}^+, and so on.

The motion of the charge represented by (17) moves it a distance $r = 2.76$ Å, the average oxygen–oxygen distance in the medium. According to standard Brownian motion theory, the series of random flights of the charge characterized by τ and r corresponds to a self-diffusion coefficient

$$(18) \qquad D_f = \frac{r^2}{6\tau}$$

and an electrical mobility

$$(19) \qquad \lambda_f = \frac{eD_f}{kT}$$

where e is the charge.

Now to calculate the electrical mobility which an experimenter may observe, we must allow for the fact that between exchange steps the H_3O^+ is not fixed, as though it were in a crystal or a solid glass, but is in some kind of random spatial motion. Usually it is assumed that this motion is characterized by an electrical mobility, λ_s, like that of Na^+. Then we have to find how to combine the fast motion characterized by λ_f and the slow motion characterized by λ_s in order to get the observable λ. This problem does not seem to have been carefully studied, but we may compare two approaches that are rather obvious.

It has usually been assumed that one has

$$(20) \qquad \lambda = \lambda_s + \lambda_f$$

which is appropriate for a model in which the slow and fast motions are independent, like the motions of a man walking down the aisle of a moving train. Of course this is not necessarily a realistic view of the motion of the charge in the present problem, when one looks at it at the microscopic level at which the net motion in a field results from the slightly incomplete cancellation of large fluctuating motions. However, assuming (20) with the experimental values of λ for H_{aq}^+ and of λ_s for Na_{aq} together with (18) and (19), we find, for this model for the motion, $\tau = 1.6$ psec in agreement with experiment.

Referring to (13), we see that the model in Section 5 leads to

$$(21) \qquad \lambda = x_s\lambda_s + (1 - x_s)\lambda_f$$

if we assume that the charge is moving by exchange—(17)—for a fraction, $1 - x_s$, of the time and by Na^+-like motion a fraction, x_s, of the time. For example, if the actual exchange process were characterized by a relaxation time, τ_r, during which the electrical polarization around the reaction site adjusted to the final state, and τ is the lifetime of the H_3O^+, then one would have

$$(22) \qquad x_s - \frac{\tau}{\tau + \tau_r}$$

Now it is not clear what to use for λ_f, the mobility during the time that the carrier is moving by the exchange process; but we examine the case in which it is given by (19) with

(23)
$$D_f = \frac{r^2}{6\tau_r y}$$

where $y = \tau_f/\tau_r$ is the ratio of the characteristic time for the Brownian motion in the fast state to the lifetime of the state. [Unless $y \ll 1$, however, (23) does not represent the two-state brownon model accurately and one needs to use the complete theory given in Ref. 10.] Considering that in the present application the motion in the fast state is just a single jump, y is not very well defined, but we only regard it as a parameter to show how it effects the result. Then using (21) and (22) and the experimental data as before, we calculate τ to get

(24)
$$\tau = \frac{1.6}{y} \text{ psec } - 1.2\tau_r$$

The second model can now be fitted to the experimental τ data by adjusting y. This may not be too significant because neither model represents a proper treatment of the dynamical problem; the most that can be concluded is that the two-state brownon model is consistent with the data discussed here.

7   Dielectric Drag

Zwanzig[18] has calculated the force, **F**, on a charged spherical particle moving steadily with constant velocity, **v**, through a dielectric medium characterized by a frequency-dependent dielectric constant, $\varepsilon(\omega)$. The force results because the polarization in the medium is not in equilibrium with the moving charge. Other contributions to the force, mainly the usual viscous drag, are neglected; but even so, Zwanzig's result bears on the question of the applicability of Langevin's equation to ionic motions in solution.

It should be noted that this dielectric drag force, or better, the polaron force, is relatively larger for small ions. For example, it tends to make the mobility of Li^+ less than that of Na^+. Therefore,[19] it tends to "explain" the same mobility data as the hydration model, in which it is assumed that each ion carries a relatively firmly bound number of solvent molecules with it and that smaller ions carry more because of their stronger fields.

We may recall that the hydration model is sometimes invoked to justify the use of Langevin's equation to represent ionic motions in solution. This may seem reasonable because hydration does make the carrier more massive. For example, Li^+ is not heavier than an H_2O molecule, whereas $Li(H_2O)_4^+$ is much heavier. However, if the low mobility of the small ions is due to the polaron force rather than to hydration, then we find in the following way that the use of Langevin's equation is again inaccurate.

Zwanzig's result for the polaron force is not easily reducible to a simple function of the velocity; but for the purpose of this discussion it is sufficient to represent it by the following simple form, which does reduce to his general expression at either small v or large $v (v = |\mathbf{v}|)$

$$(25) \qquad\qquad \mathbf{F} = -\frac{\mathbf{v}A}{1 + (v/B)^2}$$

where A and B depend on the size of the particle and on the frequency-dependent dielectric constant, $\varepsilon(\omega)$, but not on the velocity, v.

Using a simple function for $\varepsilon(\omega)$, Zwanzig has obtained an expression for the largest v for which $\mathbf{F} = -\mathbf{v}A$ within 1%. Using the approximation Equation (25) in Section 8, we may fix B by the condition that this largest v equals 0.1B. Putting in the parameters for an Li^+ ion in water, we find that B is approximately the rms thermal velocity of the carrier, $\sqrt{3kT/m}$.

Finally, in (1) it is required that the random force, $R(t)$, be independent of v. Then if there is a component like the polaron force in the frictional force, it must appear in the systematic term and we have, in place of (1)

$$m\frac{d\mathbf{v}}{dt} = \left(\zeta + \frac{A}{1 + v^2/B^2}\right)\mathbf{v} + \mathbf{R}(t)$$

This is clearly not Langevin's equation, but one might hope that the difference would be negligible because in d-c transport measurements the stationary average velocities in the experimental system are very much less than the rms thermal velocity; but we have seen counterexamples to this type of argument in Section 4.

It is well known that one can also expect such effects corresponding to velocity-dependent friction from hydrodynamics alone. The critical velocity in this case is $v_c = \eta/\rho a$ where a is the radius of the carrier.[20] The force is proportional to the velocity only when the Reynolds number, v/v_c, is small. For a 1A carrier in water, v_c is about 10^6 cm/sec, and so is considerably larger than the thermal rms velocity.

8 Hydrodynamic Interactions

The molecular theory of the momentum transfer from one moving carrier through the medium to another is even more difficult than the molecular theory of the motions of independent carriers. The most that can be done here is to summarize what little is known, especially with regard to the range of validity of the hydrodynamic approximations one is usually compelled to make.

We begin with a hydrodynamic description in which it is supposed that one effect of solute particles (carriers) moving about in a medium is to set up a velocity field, $\mathbf{v}(\mathbf{r})$, in the medium itself. To a certain approximation this is governed by the equations

$$(26) \qquad \mathbf{F}(\mathbf{r}) - \nabla p(\mathbf{r}) + \eta \nabla^2 \mathbf{v}(\mathbf{r}) = 0$$
$$\nabla \cdot \mathbf{v}(\mathbf{r}) = 0$$

where $\mathbf{F}(\mathbf{r})$ is the external force on unit volume of the medium at the point $\mathbf{r}$ (exerted directly by the carriers in the present application), p is the pressure in the medium at point $\mathbf{r}$, and η is the viscosity of the medium; ∇, ∇^2, and $\nabla \cdot$ are, respectively, the gradient, Laplacian, and divergence operators. Equation (26) is Stokes's equation, with the time dependence suppressed, or the Navier–Stokes equation with the time dependence and the kinetic energy of the medium (the "inertial term") suppressed. Thus a number of drastic approximations, in addition to the use of hydrodynamic methods to approximate the effects of molecular interactions are needed to get Eq. (26) but it seems adequate for the discussion of an important question: is the electrophoretic contribution to the limiting law for conductance or other transport coefficients sensitive to the details of the dynamics?

It is of interest to apply (26) to this problem because only the longest-range part of the hydrodynamic interaction can contribute to the limiting law. [There is an interesting paradox here, for it is known that the inertial terms dominate in the velocity field very far from the moving carrier, but still they do not appear in the limiting law (see Ref. 21).] Then we are interested in $\mathbf{v}(\mathbf{r})$ near one carrier particle far from others with which it is interacting; this is expected to depend much more on the velocities of the others than on their accelerations. The long-range requirement brings in a large mass of solvent medium for which a hydrodynamic description ought to be adequate; furthermore, it tends to filter out the effects of the rapidly fluctuating components of the carrier motions, so nothing is lost by looking at stationary-state equations.

To use the hydrodynamic model to calculate the hydrodynamic forces, one expresses the hydrodynamic force $\mathbf{F}_a$ on solute particle, a, at $\mathbf{r}_a$ in terms of the medium velocity, $\mathbf{v}(\mathbf{r}_a)$, due to all the other particles if a were not there, and the velocity, $\mathbf{v}_a$, of particle a itself

$$(27) \qquad \mathbf{F}_a = \zeta_a [\mathbf{v}(\mathbf{r}_a) - \mathbf{v}_a]$$

where ζ_a is its friction coefficient, (1). Depending on the method of calculation, one obtains the electrophoretic term in the transport coefficient by a suitable average of $\mathbf{F}_a$ in the stationary state in which the transport coefficient is measured[1] or by a suitable average of $\mathbf{F}_a$ over the fluctuations of the equilibrium system.[3-8]

Often one begins the calculation of $\mathbf{v}(\mathbf{r})$ for use in (27) by taking the solution to (26) for the case in which there is a single, solid spherical carrier, particle b, moving with a steady velocity such that the hydrodynamic force on it is $\mathbf{F}_b$. In a coordinate system in which b is at $\mathbf{r} = 0$, one finds

$$(28) \qquad\qquad \mathbf{v}(\mathbf{r}) = \frac{1}{8\pi\eta r} T_1(\mathbf{r})\cdot\mathbf{F}_b + \cdots$$

where $r = |\mathbf{r}|$ and the tensor is

$$T_1(\mathbf{r}) = \mathbf{1} + \frac{\mathbf{rr}}{r^2}$$

where $\mathbf{1}$ is the unit tensor and $\mathbf{rr}$ is a dyad. It is easy to see that, with increasing r, $\mathbf{v}(\mathbf{r})$ gets smaller only like $1/r$. The omitted term in (28) falls off like $1/r^3$ as r increases, so it is less long range than the given term and, in fact, does not contribute to the limiting law.

A remarkable feature of hydrodynamics is that, within the range of validity of (26), the long-range term of $\mathbf{v}(\mathbf{r})$ does not depend on the nature of the b particle, nor on whether it is one particle or many, as long as in the latter case the particles are close together compared to r and $\mathbf{F}_b$ is the sum of the forces on all particles. Thus the situation is quite analogous to the electrical potential in the field of a group of charged particles: the term in the potential proportional to $1/r$, where r is the distance from the group, is $(1/\varepsilon r)e$, where e is the sum of the charges of the particles, as long as the particles are close together compared with the length, r.

To illustrate the first point, from equations for the solution of (26) in various cases given by Lamb[22] one finds that (28) is also valid if b is a liquid sphere of any viscosity or an ellipsoid of any axial ratio and any orientation relative to its direction of motion! The friction coefficients are different in the various cases, as are the velocity fields close to the moving particles; but the distant part of the velocity field depends only on the force on the carrier as shown in (28).

It is not hard to obtain a much more general result, which also applies to a medium with many carrier particles in it.[23] One begins by taking the Fourier transform

$$(29) \qquad\qquad f(\mathbf{k}) = \int d^3r\, e^{i\mathbf{k}\cdot\mathbf{r}} f(\mathbf{r})$$

of each term in (26) to get the equivalent equations

$$(30) \qquad\qquad \mathbf{F}(\mathbf{k}) + ikp(\mathbf{k}) - \eta k^2 \mathbf{v}(\mathbf{k}) = 0$$
$$\mathbf{k}\cdot\mathbf{v}(\mathbf{k}) = 0$$

By forming the dot product of **k** with each term of the first equation, one may use the second equation to eliminate $p(\mathbf{k})$ and get

$$(31) \qquad \mathbf{F}(\mathbf{k}) - \frac{\mathbf{k}\mathbf{k}\cdot\mathbf{F}(\mathbf{k})}{k^2} - \eta k^2 \mathbf{v}(\mathbf{k}) = 0$$

Solving for $\mathbf{v}(\mathbf{k})$ we find

$$(32) \qquad \mathbf{v}(\mathbf{k}) = \frac{1}{k^2\eta} T_2(\mathbf{k})\cdot\mathbf{F}(\mathbf{k})$$

$$(33) \qquad T_2(k) = 1 - \frac{\mathbf{k}\mathbf{k}}{k^2}$$

Recalling that we are concerned with the longest-range part of $\mathbf{v}(\mathbf{r})$, we note that this is determined by $\mathbf{v}(\mathbf{k})$ near $k = 0$, as a general property of Fourier transforms. This must be determined by $\mathbf{F}(\mathbf{k})$ near $k = 0$. When $\mathbf{F}(\mathbf{k})$ may be expanded in powers of k, we have

$$(34) \qquad \mathbf{F}(\mathbf{k}) = \mathbf{F}(0) + \mathbf{k}\cdot(\nabla\mathbf{F}(\mathbf{k}))_{k=0} + \cdots$$

so we may write

$$(35) \qquad \mathbf{v}(\mathbf{k}) = \frac{1}{k^2} T_2(\mathbf{k})\cdot\mathbf{F}(0) \qquad k \to 0$$

This is just the Fourier transform of (28) in the case that $\mathbf{F}(0) = \mathbf{F}_b$! Furthermore—refer to (29)—we have

$$(36) \qquad \mathbf{F}(0) = \int d^3r\, \mathbf{F}(\mathbf{r})$$

showing that if there is one particle, then $\mathbf{F}(0)$ is the net force on it, whereas if there are many then $\mathbf{F}(0)$ is the sum of the net forces on them.

This is clear when the carriers generating the velocity field are close together on the scale of r, which is adequate for the present purpose. Otherwise, when the distribution of n carriers is relevant, we may write

$$\mathbf{F}(\mathbf{r}) = \sum_{j=1}^{n} \mathbf{F}_j(\mathbf{r} - \mathbf{r}_j)$$

in (26) so that no information is lost about the locations, $\mathbf{r}_j$, of the carriers in the subsequent steps. Then (32) becomes analogous to the superposition theorem of electrostatic theory.

The result obtained in (36) seems directly applicable to the calculation of the electrophoretic term of the conductance limiting law of real electrolyte solutions. At an ionic concentration of 0.001 M, the mean distance between ions is about 100 Å, so this is a fair estimate of the distance important for the effect of the hydrodynamic interactions. It is of course much larger than the size of simple ions or of an $H_9O_4{}^+$ unit. We may take the latter as the size of the molecular unit, which may be directly involved in the unique transport mechanism in protonic conduction; for example, the charge might be moving from one part of the unit to another.[15] Then, on the basis of the previous discussion, it is expected that at a distance of 100 Å or more from a steadily moving carrier, the $1/r$ part of the local velocity field is determined by the hydrodynamic force on the carrier, but not by the details of its interaction with the medium. As in the comparison of the solid and liquid spheres, the way the force on the carrier is converted into viscous flow of the medium differs from case to case, but this cannot affect the long-range part of the velocity field.

It is of course important to obtain these results from the theory of intermolecular forces by the methods of statistical mechanics. In the only study of this kind, Resibois and Hasselle-Scheurmans[8] have studied both the velocity field in the medium and the hydrodynamic interaction, naturally in the limit of great distance from the perturbing carrier and low frequency of the motion. In this study the carrier is assumed to be infinitely massive compared to the solvent molecules. Then (35) is recovered, except for an undetermined constant factor. Thus a beginning has been made on a difficult problem.

The considerations in this section make it plausible that the hydrodynamic approximation is adequate to get the electrophoretic term of the limiting law for the electrical conductance or other transport coefficient. Presumably this is true for the electrophoretic contribution to the E coefficient in (16) for the same reason. But there seems to be no corresponding basis for expecting the hydrodynamic approximation to be adequate to calculate the electrophoretic contribution to the J (or higher) coefficients of this equation. In such cases, as in the effect on the rate constant of diffusion-controlled reactions,[24] one has to follow the hydrodynamic force between the carriers down to "contact," and if there is much contribution from the parts of the trajectories in which the carriers are close to each other, one may expect these coefficients to be sensitive to details of the dynamics.

REFERENCES

1. Most of this material is reviewed by H. S. Harned and B. B. Owen, *Physical Chemistry of Electrolyte Solutions*, 3rd ed., Reinhold, New York; a recent paper is R. M. Fuoss and K. L. Hsia, *Proc. Natl. Acad. Sci. (U.S.)*, **57**, 1550 (1967).
2. For example, see H. L. Friedman, *Ionic Solution Theory*, Interscience, New York, 1962.
3. P. Resibois, *Electrolyte Theory*, Harper, New York, 1967; H. L. Friedman, *Physica*, **30**, 509, 537 (1964).
4. H. L. Friedman, *J. Chem. Phys.*, **42**, 450 (1965).

5. H. L. Friedman, *J. Chem. Phys.*, **42**, 462 (1965).
6. H. L. Friedman, *J. Phys. Chem.*, **69**, 2617 (1965).
7. H. L. Friedman, in *Chemical Physics of Ionic Solutions*, Eds., B. S. Conway and R. G. Barradas, Wiley, New York, 1966.
8. P. Resibois and N. Hasselle-Schuermans, *Advan. Chem. Phys.*, **16**, 159 (1969).
9. J. L. Lebowitz and E. Rubin, *Phys. Rev.*, **131**, 2381 (1963).
10. H. L. Friedman and A. Ben-Naim, *J. Chem. Phys.*, **48**, 120 (1968).
11. S. Harris and H. L. Friedman, *J. Chem. Phys.*, **50**, 765 (1969).
12. J. Bardeen and C. Herring, *Imperfections in Nearly Perfect Crystals*, Wiley, New York, 1952.
13. K. Compaan and Y. Haven, *Trans. Faraday Soc.*, **52**, 786 (1956).
14. L. Onsager and S. W. Provencher, *J. Amer. Chem. Soc.*, **90**, 3134 (1968).
15. B. E. Conway, J. O'M. Bockris, and H. Linton, *J. Chem. Phys.*, **24**, 834 (1956).
16. S. Meiboom, *J. Chem. Phys.*, **34**, 375 (1961).
17. R. E. Glick and K. C. Tewari, *J. Chem. Phys.*, **44**, 546 (1966); S. W. Rabideau and H. G. Hecht, *J. Chem. Phys.*, **47**, 544 (1967).
18. R. Zwanzig, *J. Chem. Phys.*, **38**, 1603 (1963).
19. H. S. Frank, in *Chemical Physics of Ionic Solutions*, Eds., B. E. Conway and R. G. Barradas, Wiley, 1966.
20. H. Lamb, *Hydrodynamics*, 6th ed., Dover, New York, 1945.
21. H. L. Friedman, *J. Chem. Phys.*, **42**, 459 (1965).
22. H. Lamb, Ref. 20, Art. 337, 339.
23. This result is implicit in a discussion by J. M. Burgers, *Koninkl. Ned. Akad. Wetenschap Verhandel*, sect. 1, **16**, 4, 114 (1938), concerning the work of C. W. Oseen, *Hydrodynamik* (Leipzig, 1927).
24. H. L. Friedman, *J. Phys. Chem.*, **70**, 3931 (1966).

19   Aqueous Solutions under Extreme Conditions

I   HIGH PRESSURES

S. B. Brummer and A. B. Gancy*, Tyco Laboratories, Inc., Waltham, Massachusetts

1   Why High Pressure?

Only recently has a good answer been found to the question, Why high pressure? Correspondingly, although there have been times of great activity, these have been interspersed by long dormant periods. The effect of pressure on the conductance of aqueous solutions was studied as early as the last century by Tammann, Lusanna, and their co-workers; much of this work has been summarized by Cohen[1] and Bridgman.[2] The early interest was purely phenomenological. Since the effects noted were not large and no new qualitative phenomena were found, the subject fell into desuetude until the 1930s. Then, stimulated by the Debye–Hückel revolution, Adams and Hall[3] and Zisman[4] undertook some systematic studies with strong electrolytes. The next important series of investigations was not until the 1950s (Hamann et al.[11]). Since then, the field has been relatively busy. Active groups have involved W. L. Marshall,[55] E. U. Franck,[6] G. J. Hills,[7] R. A. Horne,[8] and ourselves.[9]

The impetus for this renewal of interest in the effects of pressure on the conductance of aqueous solutions has come from the realization that volume effects can give a new kind of information about the interactions between ions and their hydration spheres.[10] Consequently, there has been a real stimulus for the development of techniques for making highly accurate pressure coefficient data and also for producing highly systematized and detailed bodies of data. Intrinsically this work is less phenomenological and more concerned with exacting (quasi-thermodynamic) analysis of the (accurate) data. Because the aim has been to obtain volume data relevant to aqueous solutions at

* Present Address, Industrial Chemicals Division, Allied Chemical Corporation, Solvay, New York.

1 atm, the pressure extremes in this work have not been high (< 3000 atm,[7] ~7000 atm,[8] and ~2200 atm in our own work).

Although reference is made to the earlier studies in appropriate places, the prime intention of this chapter is to summarize the more recent work having relevance to our developing picture of ion–water interactions at ambient temperatures. Thus we emphasize the most accurate data *for dilute solutions of strong electrolytes*; where available, *infinite dilution* data are used.

The bibliography of the early work and the status of the field prior to the early 1960s are described in Section 2. Experimental considerations are described in Section 3. The elimination of ion–ion effects from the pressure coefficients of dilute solutions for the purpose of obtaining infinite dilution parameters are described in Section 4. Recent experimental data for dilute solutions are discussed in Section 5; possible future directions are briefly summarized in Section 6.

2   Early Measurements

Very early work was summarized by Bridgman[2] and Cohen.[1] Hamann[5] has summarized these and other data up to 1957 in a more modern context. Horne[8(j)] has reviewed all the early work and the later studies, with particular emphasis on the "structural" aspects of aqueous solutions. The work from Hamann's group[11] provided the bridge from the first modern studies[3,4] to the more recent work,[7–9,12] that is, from an interest in the overall phenomenology of the effect of pressure to a more detailed interpretation of the results. As we shall see, a most important reason for the delay of this transition was the unexpected difficulty of making accurate measurements.

Very early studies of aqueous solutions (e.g., Cohen,[1]) showed a unique anomaly that has never been found with any other solvent. The conductance of aqueous solutions *increases with increase in pressure*, particularly at lower temperatures. In all other cases, the conductance of solutions decreases with increase in pressure,[18,7(a),7(d),9(a),9(b),19] as might be expected. The reason for this unusual behavior resides in the peculiar "structural" or associative properties of water, both of itself and as it is distorted by ions.[10,20] This peculiarity has stimulated much of the revived interest in high-pressure effects on aqueous solutions.

3   Experimental

We are not concerned in this section with the construction of high-pressure apparatus or with general high-pressure techniques. These are standard and can readily be assimilated from many of the papers referred to throughout this chapter. Rather, we shall deal specifically with problems associated with high-pressure measurements of conductance. Since the resistance of moderately dilute solutions can readily be measured to much better than 0.1%, accurate measurement under pressure would not appear to pose any serious problem. That it has done so is evident from the discord among the reported results. Ellis[12] and Horne et al.[8(a)] have brought this out particularly clearly. Variations of greater than 1% in the measured conductance of quite concentrated solutions ($\sim 0.01 M$) are not uncommon. Problems can arise with electrode seals, with the pressure-transmitting medium, and with contamination from weak electrolytes. We have explored these aspects experimentally recently.[9(d)]

Two intrinsically different methods of solution containment in conductance cells have been made: platinum in glass with mercury pressure transmission[13,7,9] (or open to the pressure medium[4,8]), and platinum in Teflon, with the cell wall acting as the transmitting medium.[14,11,15] Interesting variations in design were made in Refs. 8 and 12.

The platinum-in-glass type has the advantage of good structural integrity and easy calculation—in principle—of the variation of its cell constant with pressure. It was used extensively in the early work on the effect of pressure on solution conductances.[16,17,13] The difficulties with this cell are: (a) breaking of the platinum electrode-cell seals, sometimes under quite moderate pressures, and invariably above ~ 3000 atm; (b) possible solution contamination from the mercury pressure-transmitting fluid or, if none is used, from the (oil) pressure medium. Hamann[5] has noted solution contamination by the pressure-transmitting oil in Zisman's work[4] and by the mercury-transmitting fluid in Buchanan's work.[11(a)]

In a recent study,[9(d)] we have investigated these points in some detail. In agreement with other workers,[7(b)] we noted the tendency of platinum electrodes sealed in Pyrex to come loose in operation. In addition, the seals can "breathe" (i.e., open and close on pressure cycling). This leads to solution contamination even where the electrode structure retains its integrity. Ovenden[7(b)] developed a special platinum-Pyrex seal which was apparently successful with an argon pressure medium. This was modified for use with more contaminating oil-pressure systems.[9(d)] Contamination from mercury with salt solutions was shown to be avoided if the solutions were carefully deoxygenated.[9(d)]

This type of cell can certainly be used successfully for moderately dilute salt solutions (≥ 2 mM) at ambient temperatures (up to 55°C) and moderate pressures (< 3000 atm) with a long-term stability of measured resistances of less than 0.1%. Ovenden[7(b)] claims such use (< 2200 atm) up to 85°C. It cannot be used at higher pressures (3000 atm) because the seals break, nor at higher temperatures because of glass dissolution. With more dilute solutions, the upper temperature of usefulness is lower, but there has been no detailed study.

The Teflon cell[14] has two advantages over the glass cell: use to higher pressures (at least to 12,000 atm) and temperatures (up to 150–200°C with support). These advantages derive from Teflon's greater plasticity and lower solubility than glass. Using Teflon also makes it possible to avoid mercury seals, which are a problem with oxidizing solutions. Teflon undergoes several phase changes in the range 20 to 80°C and 1 to 10,000 atm[21] and is, in any case, highly deformable. The correction for cell constant variation with pressure is difficult to make, and high accuracy cannot be obtained. There have been no detailed studies of the long-term stability of the resistances of solutions in these cells. Pending such investigation, we must withhold judgment about the real abilities of the Teflon to seal accurately around the platinum electrodes. The experiments that have been done[15] indicate a lower level of stability than is acceptable for high-accuracy measurements.

Ellis's cell[12] represented an attempt to combine the advantages of both types. It consisted of a glass former, to hold the electrodes, and a Teflon cell body. No platinum-in-glass seals were made, but the electrodes were strapped to the glass former. This maintained the structural integrity of the electrodes but exposed minimum glass to the solution. The glass and platinum wires were drawn through holes in the wall of the cell—which has the possible disadvantage just mentioned. Another variation of this design, useful substantially above ambient temperatures, has been described.[15]

For the glass cell the correction for the variation of the cell constant with pressure can be accurately made.[9(d)] The cell constant typically changes by 0.1%/1000 atm for glass cells. Special geometric designs are particularly favorable for accurate compensation[7(b),9(d)] ($<0.02\%$ at 2000 atm).

It is necessary in all conductance work to correct measured resistances for the solvent conductance. With 2-mM potassium chloride solutions, the minimum concentration we have found suitable for accurate measurements under pressure, this correction at 1 atm with 5 MΩ water is only $\sim 0.2\%$. If the H^+ and OH^- ions had the same pressure coefficient of conductance as the more normal ions, which is only approximately true,[5] we could (within a desired accuracy of, say, 0.1%) neglect the solvent effect. This is because the solvent effect would be the same (small correction) at pressure, P, as at 1 atm, and would not affect the ratio κ_p/κ_1 ($\kappa \equiv$ specific conductance). Even allowing a small difference ($\sim 20\%$) between the coefficients for H^+ and OH^- and more normal ions, the error in neglecting the solvent conductance is only $\sim 0.04\%$ over a pressure range where the conductance changes by $\sim 10\%$, about 1200 atm at 25°C.

This is not serious but, if one allows for the increase in water dissociation by a factor of 2.11 over 1000 atm,[22] the error increases to $\sim 0.08\%$. For very accurate work with dilute solutions at high pressures, careful correction for water dissociation would have to be made. For this purpose, dissociation of water as a function of pressure may be calculated according to Ref. 23. Some workers[12,7(b)] have attempted to measure the solvent correction as a function of pressure directly. Our own experience has suggested that, in general, this is not a good procedure, since we have had extreme difficulty in maintaining the high resistance of conductivity water in small, high-pressure conductance cells. Such conductivity measurements, then, refer substantially to the pressure behavior of solutions of material leached from the cell. It is clearly not germane to determining the solvent correction to measure the properties of this solution.

These considerations regarding the correction for solvent effects—and increased water dissociation under pressure—are relevant to purification of salts to attain high accuracy. It is well-known (see, e.g., the discussion by Hamann[36]), that all weak electrolytes dissociate under pressure, eventually becoming strong. Correspondingly, the presence of weak electrolyte impurities is to be avoided above all else if high accuracy in the pressure coefficient is to be attained.[9(d)]

Finally, there is the question of the expression of the data. In quite early work (e.g., Ref. 24), it was usually convenient to express results in terms of ratios—R_p/R_1 ($R \equiv$ resistance), for example. Subsequently this method began to see extensive use. Hamann[5] criticizes this procedure on the ground that the data are not fully quantitatively useful. This is not the case, in our view, for if we know κ_p/κ_1 very accurately, we can easily obtain κ_p, since κ_1 is itself accurately known or can be easily determined.

The prime advantage of using a ratio expression (R_p/R_1 or, better, κ_p/κ_1) is that doing so helps to maintain the accuracy of the data in practice, as we shall explain. It also cuts down the labor involved in obtaining accurate measurements.

In our own work, experimental techniques were developed such that after pressure cycles up to 2300 atm, in the range 3 to 55°C, cell resistances reproduced to better than 0.1%.[9(d)] However, occasionally, more marked resistance changes did occur. Also, in the course of an extended experiment, occasionally larger ($\sim 0.3\%$) excursions can occur. It should be noted here that it takes roughly 2 hr for temperature equilibrium after changing the pressure. A full run at one temperature takes three days; setup time is about two days. Runs are inevitably extended, and maintenance of accuracy—and sanity—is tedious. Careful study has shown that such larger excursions do not affect the *ratio* (say, R_p/R_1) to within 0.05% if the most recent value of R_1 is used, or if a new appropriate R_p value is determined. Such accuracy in the pressure ratio over a range of pressure, temperature, *and time* is not, in our experience, to be had any other way—namely, the experiences of Howard, who changed his solutions after each run.[7(a)]

4 Effect of Solution Concentration on the Pressure Coefficient of Conductance

The emphasis in our discussion is on extrapolation of data to infinite dilution—specifically, for 1:1 electrolytes. As with much of the high-pressure conductance field, the early results were confused by a high level of experimental error.

One of the earliest extensive studies of the high-pressure conductance of aqueous solutions was by Körber.[13] He reported that for potassium chloride solutions, κ_p/κ_1 decreased monotonically with increase in concentration in the range 10^{-4} to $3N$. No quantitative or theoretical analysis was made.

Ellis[12] reported that in the range 10^{-3} to $10^{-1} N$ for potassium chloride and hydrogen chloride the effect of concentration was negligible. He recognized that the Debye–Hückel–Onsager theory predicts a large concentration effect and postulated its absence to result from an important artifact: that the ions modify the structure of the water, following the Frank–Evans[25,20,10] description. Hence the variation of viscosity with pressure to be used with the Debye–Hückel–Onsager theory was not that of pure water but should reflect the ".structural temperature" of the water near the ion. The Debye–Hückel terms below 0.1 N should, Ellis argued, lead to a drop in pressure coefficient as the concentration increases. This effect was opposed by a decrease in the structural temperature of the water as the salt concentration increased (hence, we presume he argues, there is a decrease in pressure-destructible order in the solution and an increase in the pressure coefficient at higher concentrations). The two effects appeared approximately to cancel up to 0.1 N. As we show later, Ellis incorrectly interpreted the direction of the Debye–Hückel–Onsager prediction and, in fact, the pressure coefficient of conductance should increase with concentration. This alone is enough to vitiate his conclusion, but we also show that the argument about structural temperatures is not necessary to explain the results; that is, data fit the limiting-law predictions made with normal water properties surprisingly well.

We should say in addition that if Ellis's view about structural temperature were correct, it would have its greatest effect on the concentration dependence of conductance at 1 atm. Deviations from ideal equations attributable to this cause have not been reported in the range up to 0.1 M. In this range of concentration, it does appear appropriate to use the normal viscosity of water in the extended Debye–Hückel–Onsager equations. We do not argue that pressure does not destroy the water structures around the ions—it certainly does; what we say is that, at these concentrations, this effect is purely ion-solvent and that the ion-solvent cospheres of the individual ions are essentially independent. Hence the effects would be concentration independent.

Hamann[5] and Skinner and Fuoss,[26] have also discussed the concentration dependence of the pressure coefficient of conductance. Hamann considered the applicability of the Debye–Hückel–Onsager limiting law to conductance data at high pressure. Large deviations were found even after modifying the equation for the variation of solvent properties with pressure. Fuoss et al.[26] have been much concerned with the effect of concentration on solution conductances under pressure. Usually, however, they have worked with low-dielectric-constant solvents and have mainly been concerned with effects of multiple-ion association.

An explicit treatment of the theoretical effect of concentration on the pressure coefficient of solution conductance for unassociated salt solutions was developed[9(e)] because, as the previous discussion makes clear, there is some uncertainty about even the direction of the effect. This poses difficulties in extrapolation to infinite dilution.

The method of treating the pressure coefficient is formally simple. We consider the equation to describe the concentration dependence at pressure, P, and divide it by the equation at 1 atm, that is

$$
(1) \qquad \Lambda_p = \Lambda_p^\circ - f(c_p)
$$

and

$$
(2) \qquad \frac{\Lambda_p}{\Lambda_1} = \frac{\Lambda_p^\circ - f(c_p)}{\Lambda_1^\circ - f(c_1)}
$$

The problem is in the choice of $f(c_p)$. In principle, this is not a large difficulty. Thus in the concentration range where the Debye–Hückel–Onsager limiting law applies, $f(c_p)$ is given by

$$
(3) \qquad f(c_p) = (\alpha_p \Lambda_p^\circ + \beta_p)(c_p)^{1/2} = S_p(c_p)^{1/2}
$$

where α_p and β_p are the usual constants[27] comprising solvent physical and universal constants.

Normally, deviations from (3) occur at concentrations above about 2 mM in most aqueous solutions. Hence a different form of $f(c_p)$ is required at higher concentrations. The most elaborate and sophisticated of such formulations is that due to Fuoss and Onsager.[26,29] Here, the simplified form suitable for use to ~10 mM for an unassociated salt[30] is

$$
(4) \qquad f(c_p) = S_p(c_p)^{1/2} - E_p' c_p \ln c_p + J_p c_p
$$

An alternative equation, variously attributed but mostly propounded by Robinson and Stokes,[31] is perhaps less accurate (~0.05% at 50 mM) but is easier to handle analytically, namely

$$(5) \qquad f(c_p) = \frac{S_p(c_p)^{1/2}}{1 + a_p \chi_p (c_p)^{1/2}}$$

where χ_p is a collection of universal constants and solvent parameters and a_p is the distance of closest approach of the ions.

The approach was to explore the predictions of the limiting-law equation, (3). It will appear that for simple 1:1 electrolytes, the data to 20 mM never depart from this formulation by more than 0.1%. Data at higher concentrations, where significant deviations occur, were then treated according to (5).

Combining (2) and (3), and recalling that $\Lambda_p/\Lambda_1 = (\kappa_p/\kappa_1)(\rho_p/\rho_p)$, where ρ = density, yields

$$(6) \qquad \frac{\kappa_p}{\kappa_1} = \left(\frac{\kappa_p}{\kappa_1}\right)_{c \to 0} \left\{ 1 + \left[\frac{S_1}{\Lambda_1^\circ} - \frac{S_p}{\Lambda_p^\circ} \left(\frac{\rho_p}{\rho_1} \right)^{1/2} \right] (c_1)^{1/2} + \text{higher terms in } [(c_1)]^n \right\}$$

The variation of water viscosity,[32, 33] η, of dielectric constant,[34] ε, and of density[35] with pressure is known; Λ_1° is also known and, if Λ_p° can be determined, the variation of κ_p/κ_1 with c can be predicted. The method of determining Λ_p° was as follows: κ_p/κ_1 varies approximately as $c^{1/2}$ (see below). Data were graphically extrapolated to $c^{1/2} = 0$ and the value of Λ_p° so derived was used to determine the predicted slope $d(\kappa_p/\kappa_1)/d(c_1)^{1/2}$. This new slope was then used with the data to yield a new value of Λ_p°, which could be used to recalculate the slope, and so on. In practice, the original graphically extrapolated value of Λ_p° was always adequate for slope prediction.

The important result from (6) is that κ_p/κ_1 (unlike Λ itself) should *increase with increase in concentration*. That this can be found in practice is shown in Figure 1 for sodium chloride solutions at 25°C. Similar results were found for a number of salts in the ranges 3 to 55°C and 1 to 2200 atm.[9(e)] Data were taken above 3 mM to maintain accuracy. Despite this high concentration range, above the normal expectation of limiting-law behavior, κ_p/κ_1 is found to increase approximately with $(c_1)^{1/2}$ in the concentration range explored, 3 to 20 mM. We have noted that the limiting-law prediction is that κ_p/κ_1 should increase with $(c_1)^{1/2}$. We see that the deviations from limiting law are small ($\sim 0.1\%$), even at 20 mM. In addition, the preliminary hand-drawn-slopes of the κ_p/κ_1 versus $(c_1)^{1/2}$ plots (constructed without benefit of limiting-law predictions) are more often than not somewhat lower than required by limiting law.

These results are in contrast to those reported by Körber[13] and Ellis.[12] The good agreement below 10 mM with the predictions of limiting law, which itself predicts that κ_p/κ_1 should increase with c, speaks in favor of the recent data. We believe that the previous workers did not find this dependence for the following reasons. Körber found that κ_p/κ_1 decreased as c increased from

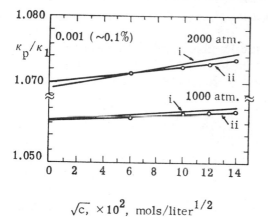

Figure 1. Concentration dependence of the pressure coefficient of conductance for sodium chloride solutions at 25°C: curve i, limiting law; curve ii, preliminary slope. From Ref. 9(e).

very low to high concentrations. The trend at the highest concentrations is certainly correct. The results at low concentrations were too high, almost certainly because of (weak electrolyte?) impurities. Impurities, which affect the results most at low solution concentrations, invariably lead to high κ_p/κ_1 findings. Ellis's data showed much scatter, after large solvent corrections, and this concealed the concentration effect at high dilutions. At higher concentrations ($\sim 0.1\ N$), κ_p/κ_1 deviates considerably below the limiting-law behavior (to be discussed) and takes on values approximately the same as at lower concentrations. This, combined with his scatter, accounts for Ellis's results.

Before we can see the way best to extrapolate data to infinite dilution, we must take account of deviations from limiting-law behavior. We have noted that the data fit limiting-law predictions to within 0.1% out to 20 mM. This is well outside the normal expectation of this theory. Thus at 25°C and 1 atm, 10-mM sodium chloride solutions have a conductance 0.8% greater than the limiting-law predicted value, and the κ_p/κ_1 data follow the limiting law so well at the higher concentrations because of the cancellation of deviations in the ratio. Thus at 10 mM in (2), $f(c_1) \simeq 0.8\%$, and $f(c_{2000}) \sim 0.75\%$; we see the difference (i.e., 0.05% at 2000 atm).

Despite this tendency to cancel deviations, it is evident that data at higher concentrations do deviate systematically below limiting-law behavior. It is of some interest to attempt to describe these deviations. The most sophisticated approach is that of (4). However, this is not applicable above 10 mM. A semiempirical modification of this equation has been proposed[30] but is not suitable for the present purpose.[9(e)] A less rigorous but experimentally reasonably justifiable equation is the Robinson–Stokes form, (5). This equation can be handled analytically and, since we are concerned with examining only small deviations from limiting law ($\sim 0.5\%$ even at 100 mM), its use seems allowable. Then, after ignoring all terms of power higher than c, we obtain

$$
(7) \quad \frac{\kappa_p}{\kappa_1} = \left(\frac{\kappa_p}{\kappa_1}\right)_{c \to 0} \left\{ 1 + \left[\frac{S_1}{\Lambda_1^\circ} - \frac{S_p}{\Lambda_p^\circ}\left(\frac{\rho_p}{\rho_1}\right)\right]^{1/2}(c_1)^{1/2} + \right.
$$
$$
\left. \frac{S_1}{\Lambda_1^\circ}\left\{ \left[\frac{S_1}{\Lambda_1^\circ} - \frac{S_p}{\Lambda_p^\circ}\left(\frac{\rho_p}{\rho_1}\right)^{1/2}\right]\right\}c_1 + a_1\chi_1\left[\frac{a_p}{a_1}\left(\frac{\varepsilon_1}{\varepsilon_p}\right)^{1/2}\left(\frac{\rho_p}{\rho_1}\right) - \frac{S_1}{\Lambda_1^\circ}\right]c_1 \right\}
$$

where the coefficient of $(c_1)^{1/2}$ is the limiting-law term, as before. It is dominant at low c and causes an increase in κ_p/κ_1 with c_1. The second term, within the double braces, is the c_1 term in the power expansion of the limiting law [see (6)]. It also causes an increase in (κ_p/κ_1) with c_1. This term is always small ($\sim 0.04\%$ at 10 mM and 2000 atm). The third term, in c_1, arises from the Robinson–Stokes modification of the limiting law. It should cause a decrease in κ_p/κ_1 with increasing c_1.

To test the applicability of (7), measurements for sodium chloride solutions were extended to over 100 mM. Deviations from the κ_p/κ_1 versus $(c_1)^{1/2}$ limiting-law behavior were found to be proportional to c_1, as required, under all conditions. Detailed testing of the equation, with semiempirical selection of appropriate parameters, was carried out.[9(e)] The effect of ion-pair formation (up to 8% for NaCl at 100 mM[30]) was shown to be negligible and data to 100 mM were fitted within experimental error.

We are, then, in a position confidently to extrapolate data to infinite dilution. Several alternative procedures have been used in the literature for this purpose. One possibility, which has occasionally been followed, is to disregard the relatively small concentration dependence and to take the results at one concentration, say 10 mM.[8(a)] This procedure introduces a systematic pressure- and temperature-dependent error. At 25°C, sodium chloride typically produces errors of 0.35% at 2000 atm and 0.15% at 1000 atm. One can readily work at 3 mM, and similar treatment of these data yield typical errors of 0.25 and 0.07%. These errors are higher at higher temperatures and with smaller ions.

An alternative would be to extrapolate the approximately linear relation between κ_p/κ_1 and $(c_1)^{1/2}$ in the range 3 to 10 mM. This would yield a typical error (NaCl at 25°C) of only 0.10% at 2000 atm. In the worst cases we have examined, the extrapolation error with this procedure would not exceed 0.2%. In cases of unknown physical constants of the solvent under pressure, this method would be adequate.

The best method, however, is to fit the data to a theoretical equation. Since deviations from limiting law, at least insofar as the observed κ_p/κ_1 versus $(c_1)^{1/2}$ slopes are concerned, are appreciable in the accessible concentration range, this is not easy. Consequently, we suggest that data at relatively low concentrations be extrapolated using the calculated limiting-law slope. For aqueous solutions of 1:1 electrolytes, no appreciable error ($<0.1\%$) would be typically incurred if data at 2000 atm were extrapolated in this way from as high as 10 mM. However, extrapolation can easily be made from 3 mM.

The effects of possible salt association need not be considered in any such procedure. We have found that associations as high as 8% do not show up in κ_p/κ_1 ratios if the association constant is not too pressure dependent.[9(e)] For this degree of association to occur at 3 mM, the association constant would have to be >50 liters/mole. For strong electrolytes, this is probable only in solvents for which the dielectric constant is much lower than that of water. Fuoss's treatment of ion association suggests that this method would then be appropriate for solvents with dielectric constants as low as 20.[30]

5   Experimental Data for Pressure Coefficients of Conductance At or Near Infinite Dilution

Having seen that accurate conductance data can be obtained relatively routinely and that the method of extrapolation to infinite dilution is defined, we shall now concern ourselves with the actual data for high-pressure conductance, at or near infinite dilution. Emphasis is put on discussing recent work, and particularly that of Refs. 7 to 9. The work of Franck et al.[6] and Marshall et al.,[55] which is mostly at high temperatures, is discussed in Part II of this chapter.

Horne, Myers, and Frysinger[8(a)] studied the conductance of (0.01-M) KCl, (0.10-M) KOH, and (0.10-M) HCl solutions in the range 5 to 45°C and 1 to ~ 7000 atm. These concentrations are too high to allow one to ignore interionic effects, but they are suitable to indicate the kind of information that can be gained from this type of study. They are also indicative of the general state of discordance in the field: potassium chloride data (25, 35, and 45°C) sharply disagreed with those of Hamann and Strauss.[11(b)] The latter data were described by Ellis[12] as anomalous. In justification of the accuracy, it was noted that sodium chloride data (not reported) were in excellent agreement with those of Adams and Hall,[3] and that Fisher's data[37] appeared to agree with Adams and Hall and with Ellis. Hydrogen chloride data (studied at 5, 15, and 25°C) were only in moderate agreement with Ellis's (25°C).

Activation energies for conductance were calculated as a function of pressure. The absolute values of these are not quantitative because they are based on molal conductances, but comparison between the three electrolytes is instructive. For KCl, E_p (at constant pressure) decreases with increase in P up to about 2000 atm. Thereafter E_p increases. For KOH, the minimum occurs at $\sim$3000 atm. For HCl, there is no minimum, but a break occurs at $\sim$1350 atm. The low P-decrease in E_p is attributed to the same factors that cause a decrease in E_p for water viscosity (i.e., the breakdown of its structure with increasing P). At higher P's, where the water structure is extensively broken down, more normal effects (i.e., the increasing proximity of the species) dominate. That the increase in conductance at low P is the direct result of water structural effects can be seen from Figure 2. We see that the anomalous behavior is sharply accentuated at low temperatures, where structural effects dominate. The data for HCl were interpreted to show that at low P the rate-limiting step in proton migration is solvent rotation rather than proton transfer.

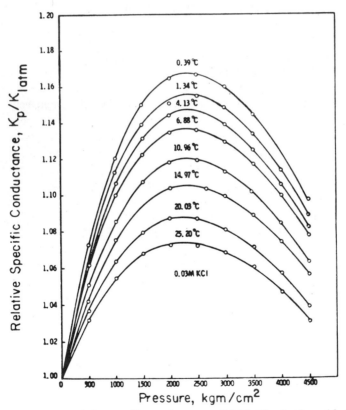

Figure 2. Relative specific conductance of 0.03-M potassium chloride as a function of pressure. From Ref. 8(f).

These results exemplify the kind of information that can be obtained from careful study of accurate high-pressure data. As we shall see, differences in behavior between different electrolytes allow quite subtle description of ion-structure effects. In this vein, Horne[8(b)] analyzed Zisman's data[4] in terms of a modification of Walden's rule. At low P (< 1000 atm), there were substantial differences among the various alkali chlorides, with the pressure coefficient decreasing as the ion size increased. Between ~1000 and 8000 atm very little difference was found among the different salts. The pressure effect in this region was similar to that for the viscosity of water. Applicability of Walden's rule was therefore claimed in this structure-broken region. We believe that the accuracy of the data (both for η and $\Lambda°$) is open to considerable question. We have consistently argued against the futility of the Walden rule approach.[9(a),(b)] At best, when it appears valid, we know that *both* η and $\Lambda°$ depend on exactly the same molecular phenomena. The rule does very little to define those phenomena, however. The important inference in the present context is that increasing P causes dehydration of the solvated ions. It was concluded[8(b)] that above the conductance maximum, where the ions all appeared similar, they were stripped down to their primary solvation layers. Later, it was conjectured that the ions were completely bare at higher P.[8(j)]

In later papers[8(d) (g)] Horne et al. attempt to determine what information on ion–water interactions can be revealed by pressure (and temperature[38]) effects. Studies were also carried out in alcohol–water solvents.[8(l)] Detailed high-pressure data are presented in Refs. 8(f) and (g).

Horne and Young[8(f)] investigated potassium chloride solutions in the ranges 0.03 to 4.0 M, from −8 (not under pressure) to +20°C, and up to ~4000 atm. Potassium chloride solutions were also explored in Ref. 39. It was thought that pressure was even more effective than temperature in destroying water structure. Thus, although increase in temperature tends to destroy mostly the bulk-water structure, high pressure can effectively dehydrate the ions, and Walden's rule was thought to be appropriate at high pressure because the amount of transported solvent is very minor. However, there is a difficulty: the reaction

$$(8) \qquad\qquad M^+ + nH_2O \rightarrow M^+(H_2O)_n$$

should be accompanied by a volume decrease; therefore, pressure should lead to an increase in hydration, not the observed decrease. To account for this discrepancy, a modified interpretation of the Frank–Wen model of water[40] was suggested:[8(e)] the inner, structure-promoted, zone was divided into a dense electrostricted region and a highly (perhaps ice I?) open-structured region. The latter breaks down as P increases and leads to the effective dehydration of the ion. Estimates of the distribution of the inner solvation layer between these two types of close-bound water are made from the concentration dependence of the conductance maximum as a function of pressure. We shall deal with the conductance maximum at infinite dilution and the information it yields on ion–water interactions in discussing our own recent data.

In another important paper,[8(g)] the effect of pressure on 0.10-M solutions of tetraalkyl ammonium salts was investigated. Studies were made at 4 and 25°C up to ~4000 atm with the $(CH_3)_4N^+$, $(C_2H_5)_4N^+$, $(n\text{-}C_4H_9)_4N^+$, and $(n\text{-}C_5H_{11})_4N^+$ halides. The effect of pressure was much closer to that expected from Walden's rule than for smaller ions. This implies that the well-known hydrophobic structured region around these ions is very stable. One might also argue that it is very similar to bulk water. There were irregularities in the order of the cations which were explained on the basis of ion-pair effects and, in particular, cation–cation interactions.

Hills and his co-workers have explored the effect of pressure on the conductance of solutions in methanol,[9(a),7(a)] of solutions in nitrobenzene,[9(a),7(d)] of nitrate melts,[7(c)] and of aqueous solutions.[9(a),7(b),7(c)] They have also looked at the effects of P on double-layer capacity[41] and H_2-evolution.[42] The original interest was stimulated by a desire to measure activation energies at constant volume, E_v.[9(a),43] This procedure greatly simplifies work with nonaqueous solvents. Thus, although E_p is a function of P, ion size, and T, E_v is usually observed to be a function of only V and ion size. It is independent of T,[43,7(a),7(b),9(c)] probably because the potential energy is to a first approximation constant as the temperature is raised. The variation of E_v with V, especially at high densities, is of great interest for examining the physics of the ion-migration problem.[9(a)–(c)] Early work showed that in aqueous solutions E_p decreases with increase in T.[44] Our own studies[9(a)] showed that for ~0.02-N KCl, E_v is just as dependent on T as is E_p. Evidently, for aqueous solutions, constant volume is not even approximately equivalent to constant-potential energy. This is presumably because of structural effects in the solution.

Ovenden's very extensive studies[7(b)] of the effect of pressure on the conductance of aqueous solutions were carried out in this context of measuring E_v's and correlating them with structural effects in the solution. Solutions of LiCl, KCl, CsCl, and $(C_2H_5)_4NCl$ were investigated in the ranges ~2 to 24 mM, 25 to 85°C, and up to 2000 atm, and $\Lambda°$'s were extrapolated with Kohlrausch plots. An uncertainty of ~0.1% in each Λ value was indicated. Agreement with Hensel and Franck[45] (45 and 75°C, LiCl, KCl, CsCl) was only moderate. Based on the great care of Ovenden's experimental procedures, we would favor his data in any such comparison.

The value of E_v was found to decrease approximately linearly with increasing density; it also decreased with increase in T. The variation of E_v with ion was only slight and was inside the (appreciable) experimental scatter, especially at high temperatures.

Volumes of activation for ionic conductance, ΔV^*, were derived graphically from $\ln \Lambda - P$ plots according to

$$(9) \qquad \Delta V^* = -RT\left[\left(\frac{\partial \ln \Lambda}{\partial P}\right)_T - \frac{2}{3}\left(\frac{\partial \ln V}{\partial P}\right)_T\right]$$

where V is the solvent volume. This equation is based on application of transition-state theory[46] to ionic mobility; its derivation is described elsewhere.[9(a)] We should be quite clear at this stage in our understanding that ΔV^* is to be thought of as a convenience to describe the pressure coefficient of conductance. Although ΔV^* is negative at low temperatures and pressures, it becomes increasingly positive (more normal) at higher temperature and pressure, as water's structure breaks down. The experimental scatter was too large to allow reliable discernment of differences among the ions, however. We may say that computer slope evaluation from these excellent data would allow ΔV^*'s to be reliably obtained (see below).

Some of these results, and data for hydrogen chloride, were taken up in Ref. 7(a). Hydrogen chloride data were taken from Ellis[12] and Wall and Gill;[48] transport numbers for H^+ were measured from a concentration cell, and $\lambda_{H^+}^\circ$ was seen to increase more rapidly with P than for other ions. Since increase in P (and T) leads to breakdown of water's hydrogen-bonded structure, the rate-limiting step in proton migration cannot be the transfer of a proton along a Grotthuss chain but must be the reorientation of H_2O molecules in the field of the proton. This is similar to earlier descriptions of the process.[49–51,8(a)]

In our own recent work[47] we have measured the conductances of solutions of LiCl, NaCl, KCl, RbCl, CsCl, NH_4Cl, $(C_2H_5)_4NCl$, KF, KI, KBr, and KNO_3. Studies have been made at 3, 5, 10, 15, 20, 25, 35, 45, and 55°C, usually up to 2200 atm. Extrapolation to infinite dilution was done from above 3 mM, using the calculated limiting-law slope, as indicated previously. An overall accuracy of 0.1% in $(\kappa_p/\kappa_1)_{c\to 0}$ is claimed, except for $(C_2H_5)_4NCl$, where there are extrapolation difficulties and for KF, KI, KBr, and KNO_3, because of limited data. Here, 0.2% in the extrapolated values is more appropriate. Water pressure, volume, and temperature data, to convert κ's to Λ's, were taken from Kell and Whalley[35] (extrapolated above ~ 1000 atm).

Infinite dilution conductances for potassium chloride as a function of P at various temperatures are shown in Figure 3. Comparison is made with the data of Ovenden.[7(b)] For simplicity, the ratio $(\kappa_p/\kappa_1)_{c\to 0}$ rather than $\Lambda_p^\circ/\Lambda_1^\circ$ is shown. Comparison with Ovenden is excellent, with disagreements rarely as great as 0.2% in $(\kappa_p/\kappa_1)_{c\to 0}$. The normalizing effect of high temperature on the pressure effect is clearly seen, as before.

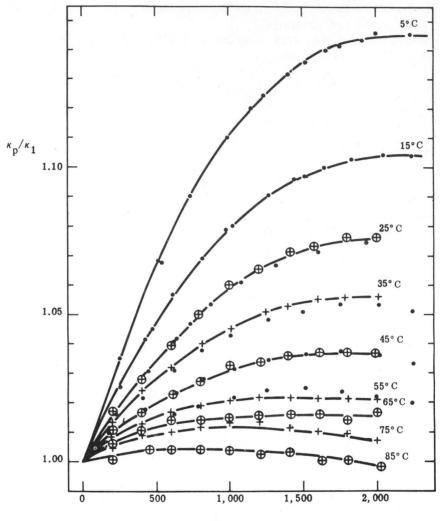

Figure 3. Relative specific conductance of potassium chloride solutions at infinite dilution as a function of pressure: solid circles, data of Ref. 47; crosses, Ovenden's data, from Ref. 7(b).

Values of ΔV^* were calculated from (10) as follows: extrapolated $(\kappa_p/\kappa_1)_{c\to 0}$ data were fitted to third-order polynomials of the form

$$(10) \qquad \ln\left(\frac{\kappa_p}{\kappa_1}\right)_{c\to 0} = \sum_{n,0\cdots 3} D(n)P^n$$

by the method of least squares. Here, P is the gauge pressure. In the worst case (KI), this equation fitted the data to $\sim 0.05\%$, but usually it fitted to within ~ 0.02 to 0.03%. All data manipulations were carried out with the digital computer and ΔV^*'s were estimated to be accurate to 0.02 cm^3/mole at low P and 0.04 cm^3/mole at the highest pressures. The internal consistency was much better than this.

The temperature and density dependence of ΔV^* for potassium chloride solutions are shown in Figure 4. Comparison is again made with Ovenden's calculations and agreement is excellent. The value of ΔV^* is negative at low T and P, but it increases with decrease in specific volume. Usually ΔV^* increases with V,[9(a),(b)] and this opposite dependence in aqueous solutions must result from structural effects. We use the following model of ions in aqueous solution, after Frank and Wen,[40] as a context to interpret the results. The ion is surrounded by an inner sheaf of electrostricted water molecules and an outer layer where the local water is either more (structure-making ion) or less (structure-breaking ion) organized than normal water at the same temperature. The effect of both T and P is to break down the very open (ice-I-like) water structure. We would say that when an ion moves, its local water atmosphere is disturbed and a volume contraction must occur—hence a negative ΔV^*. At high P and T, where there is less structured water to collapse as the ion moves, ΔV^* is less negative.

At high V, ΔV^* increases approximately linearly with decrease in V and the (extrapolated) curves all appear to originate at a common point for each salt, similar to an isosbestic point. Ovenden noted a similar approximation. In our case, the highest temperature (55°C) data are often too high for a perfect fit, but there is a systematic increase of the parameters of the "iso-activation-volume" point with ion size.[47] Such a physical condition, from which the low P data at different temperatures appear to originate, is not easy to understand. An isosbestic point generally implies the conjoint variation of at least two parameters. If ΔV^* decreases (linearly) with increase in total structured water around the ion and if, at low densities, the amount of this (open ice-I-like) structure decreases (linearly) with decrease in V, then we have a condition such that for a given ion, ΔV^* increases as T is varied. However, the regression rate, $\partial \Delta V^*/\partial V$, becomes less. As we raise the temperature, the amount of altered water around the ion decreases until the effect of P is the same on it as on bulk water. At lower temperatures, considerable relative structural rearrangement of the local water around the ion is necessary to

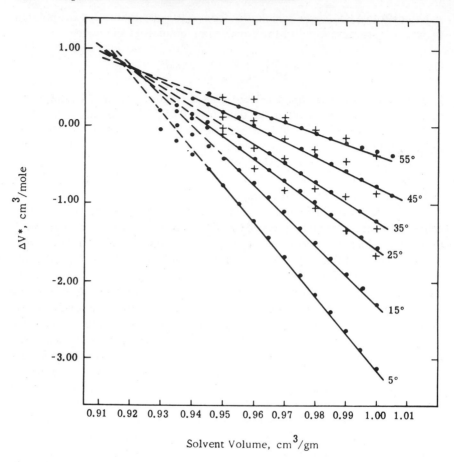

Figure 4. Activation volumes for potassium chloride as a function of density and temperature: solid circles, data of Ref. 47; crosses, Ovenden's data from Ref. 7(b).

achieve the same effect. The crossover point may then correspond to a point at which the structure of water around the ion is the same as in the bulk. We would expect, then, that for the smaller structure-making ions the crossover point would occur at higher P (i.e., lower V), and this is observed.

At high densities, other effects become dominant. The amount of structured water is so small that solvent volume *increases* are necessary to accommodate the translating ion (hole formation). Correspondingly, the straightforward (but not simple) model involving modifications of ambient structure by variations of temperature and pressure is no longer applicable.

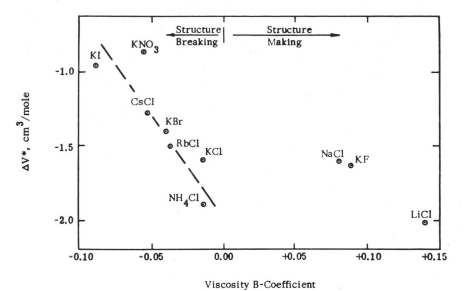

Figure 5. Activation volumes at 25°C ($V = 1.00$ cm³/mole) as a function of viscosity B coefficient. From Ref. 20.

There appears to be no monotonic relation between ΔV^* at low pressure and either ion radius or salt partial molal volumes.[47] Correlation with ion–water structural properties may be obtained, though. Figure 5 shows ΔV^* at low density ($\equiv 70$ atm) at 25°C as a function of viscosity B coefficient. The structure makers and structure breakers separate into two families. There are extremely negative ΔV^*'s associated with both types of ion, but the most negative value is for Li⁺, which is an extremely good structure promoter. There is a well-developed difference among the ions in the structure-breaking family, but rather less ion-property dependence with the structure makers. This is probably because the decrease in ΔV^* with more negative B is opposed by an increase in the size of the transported entity; that is, *transported* Li⁺ carries relatively more H_2O with it than does *transported* Na⁺. There is a systematic change with B on the structure-breaking side because these effects operate in parallel.

At high pressures, we have made comparisons on a constant-ΔV^* basis—specifically at $\Delta V^* = 0$. This state was chosen for two reasons. First, it represents the condition of *easiest* ion transport ($\Delta G_0^* = $ minimum). Second, it *may* correspond to a condition of no net structural change when the ion moves. The solvent volumes corresponding to zero ΔV^* at 25°C appear in Figure 6. Also shown is the relevant point for water self-diffusion.

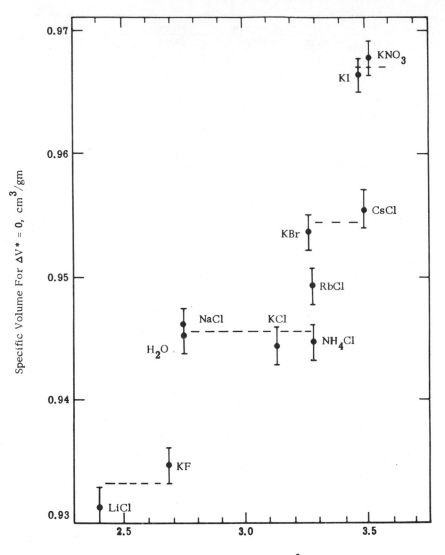

Figure 6. Solvent volumes for $\Delta V^* = 0$ at 25°C as a function of ion crystallographic radius. From Ref. 53.

The volume for zero ΔV^* increases as the ion size increases; this implies a greater disturbance of water for the smaller ions to come into the hole-formation regime from the structure dominated regime. This is expected, since the structure around smaller ions is more well-developed. However, there is not a smooth relation with ion size, and we appear to find a stepped dependence well outside experimental error. We also see this kind of stepped relation between the solvent volume for $\Delta V^* = 0$ and the viscosity B coefficient (Figure 7); of course, we should not expect that high-pressure data will correlate exactly with 1-atm data (e.g., B-coefficients). Such stepped dependences are new to us, and they suggest that *only discrete hydrated ion sizes are stable for transport*. Since it is evident that there is no smooth relation between ΔV^* and other properties, ΔV^* contains new information about the ion and none of the other ion-structural parameters is identical to it. A major problem with more detailed interpretations is the absence (until recently) of high-pressure transport numbers to help separate pressure effects into single-ion quantities.

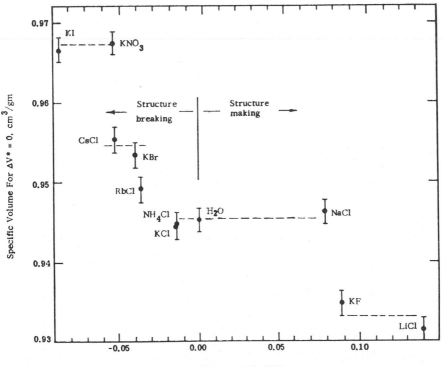

Figure 7.   Solvent volumes for $\Delta V^* = 0$ at 25°C as a function of viscosity B coefficient.

Finally, we shall briefly review some studies of activation energies for ionic mobility in aqueous solutions. Very early on we argued that E_v should theoretically be simpler than E_p.[9(a)] Experimental work showed that in practice this is the case—at least in nonaqueous solvents.[9(a),9(b),7(a),7(d)] In aqueous solution, we are no better off with E_v than with E_p; which is to say we have a temperature-dependent parameter.[44] Some workers have attempted to draw inferences from a study of E_p[38,44,8(a),54,56] both for conductance and for viscous flow.[57] Our aim has been to try to find an equation that would really describe the temperature dependence of $\lambda°$ accurately—and simply.

One approach which appears to have some promise has been revived by Angell:[58]

$$(11) \qquad \lambda°T^{1/2} = A \exp\left[\frac{-K}{(T - T_0)}\right]$$

This "modified" Arrhenius equation has an extra $T^{1/2}$ temperature dependence because of ion-in-cage vibrational considerations;[59] T_0 is the temperature from which liquid fluidity (ion mobility) derives. Angell considers this reference state to be the "ideal glass." Efforts to fit this equation with $\Lambda°$'s either at constant pressure or at constant volume were unsuccessful. With $\lambda°$'s at 1 atm the fit was moderately good, however. The constants for various ions are correlated with the viscosity B coefficient in Figure 8, where T_0 was constant at -135.56 ± 0.09 (mean deviation) °C (also for Ca^{2+}). Angell

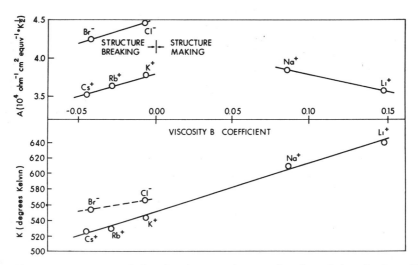

Figure 8.   Constants of glass-forming equation as a function of viscosity B coefficient.

believes that the observed T_0 value should be close to the experimental value for the glass-transition temperature of water, $-133°C$.

The constancy of T_0 over a wide range of ions and the good fit of (11) to the data ($\pm 0.2\%$ from 5 to 55°C), suggest the possible value of this approach. The steady increase of the activation energy parameter, K, with B is what we would expect; the difference between anions and cations is not surprising in view of the known large differences in their solvation characteristics.[10]

In an analogy with the transition state approach, namely

$$(12) \qquad \lambda° = \text{const.} \, L^2 \left(\frac{\Delta S_0^*}{R}\right) \exp \left(\frac{-\Delta H_0^*}{RT}\right)$$

where L is the jump distance, ΔS_0^* is the entropy, and ΔH_0^* the enthalpy of activation,[9(a)] the A term is probably substantially determined by entropy considerations. Then we see that ΔS_0^* varies differently with B for structure-making and structure-breaking ions. This is very exciting. It implies that for the structure breakers, the transition state is more ordered with respect to the initial state the more the ion (in the initial state) tends to destroy structure. This implies that there is some *structure making in the transition state*. For structure-making ions the opposite tendency is found; the transition state is more ordered, vis-à-vis the initial state, the more ordering the ion. This implies that much of this kind of order is substantially maintained when the ion translates. We have already noted that transported Li^+, for example, is probably unexpectedly large.

Further application of this equation, most desirably at constant volume, is indicated.

6 Where Do We Go From Here?

The study of the conductance of solutions under high pressure has been developed such that very accurate data ($\sim 0.1\%$ in $\Lambda°$) can be obtained. Significant inferences about ion-solvent interactions can be made from these data. In our view, further work should proceed in four main directions: (a) extended nonaqueous studies, (b) high-pressure transport number studies in both nonaqueous and aqueous solutions, (c) extension of aqueous studies to higher pressure, and (d) investigation of differing ion types in aqueous solutions.

This is not the place to expand on nonaqueous solutions. The particular interest is in delineating the E_v–V relation [cf. Ref. 9(b)], especially at high pressures (> 2000 atm) and low temperatures ($<$ room temperature down to freezing). From an analysis of Howard's[7(a)] work[60] and from previous studies,[9(a)] the methanol system appears particularly favorable, since the entropy term appears to be volume independent. Transport numbers will soon be required for these extended studies to yield really detailed information.

Such work in nonaqueous media probes the physics of molecular transport in the liquid state. Work with aqueous solutions yields quite different information, since the parameters are dominated by structural effects. The first requirement is the detailed and comprehensive study of transport numbers that will permit the determination of single-ion parameters. The early studies of Wall et al. with HCl[61] and KCl[62] at 25°C (moving-boundary method) have recently been supplemented by Hills et al.[7(c)] (HCl at 25°C to 2000 atm using concentration cells) and by the work of Kay et al.[52] (KCl at 25°C to 2000 atm). The requirement is to generate transport data up to 2000 atm over a range of temperatures. Subsequently, studies at higher pressure seem to be needed. With these data, a detailed evaluation of the properties of ΔV^* and a thorough examination of the application of the glass-forming equation to ionic conductance at constant volume would be possible.

Extension of studies to higher pressures with 1:1 electrolytes would fill in, more accurately, Zisman's studies.[4] The interest is to pursue Horne's thesis that complete ion dehydration occurs.[8(j)]

Finally, we believe that application of our newly increased experimental skills to study of 2:1 and 2:2 electrolytes, and to 1:1 electrolytes of the hydrophobic structure-making ions, would be valuable in developing a quantitative theoretical description of ion mobility.

Acknowledgments

We are pleased to acknowledge support of our high-pressure conductance work by the Office of Saline Water under Contracts OSW-14-01-001-425 and OSW-14-01-0001-966. The support and encouragement of Dr. W. H. McCoy of that Office is gratefully appreciated. We are pleased to thank Helmut Lingertat for his painstaking care in performing much of the experimental work.

REFERENCES

1. (a) E. Cohen, *Piezochemie Kondensierter Systeme*, Akademische Verlagsgesellschaft, Leipzig, 1919; (b) E. Cohen, *Physico-Chemical Metamorphosis and Problems in Piezo-Chemistry*, McGraw-Hill, New York, 1926.
2. P. W. Bridgman, *The Physics of High Pressure*, Bell, London, 1958.
3. L. H. Adams and R. E. Hall, *J. Phys. Chem.*, **35**, 2145 (1931).
4. W. A. Zisman, *Phys. Rev.*, **39**, 151 (1932).
5. S. D. Hamann, *Physico-Chemical Effects of Pressure*, Butterworths, London, 1957.

6. (a) E. U. Franck, *Angew. Chem.*, **73**, 309 (1961); (b) *Endeavour*, **27**, 55 (1961).
7. (a) B. Howard, Ph.D. thesis, University of London, 1963; (b) P. J. Ovenden, Ph.D. thesis, University of Southampton, England, 1965; (c) G. J. Hills, P. J. Ovenden, and D. R. Whitehouse, *Discussions Faraday Soc.*, **39**, 207 (1965); (d) F. Barreira and G. J. Hills, *Trans. Faraday Soc.*, **64**, 1359 (1968); (e) A. F. M. Barton, B. Cleaver, and G. J. Hills, *Trans. Faraday Soc.*, **64**, 208 (1968).
8. (a) R. A. Horne, B. R. Myers, and G. R. Frysinger, *J. Chem. Phys.*, **39**, 2666 (1963); (b) R. A. Horne, *Nature*, **200**, 418 (1963); (c) R. A. Horne, R. A. Courant, and G. R. Frysinger, *J. Chem. Soc.*, **1964**, 1515; (d) R. A. Horne, R. A. Courant, and D. S. Johnson, *Electrochim. Acta*, **11**, 987 (1966); (e) R. A. Horne and J. D. Birkett, *Electrochim. Acta*, **12**, 1153 (1967); (f) R. A. Horne and R. P. Young, *J. Phys. Chem.*, **71**, 3824 (1967); (g) R. A. Horne and R. P. Young, *J. Phys. Chem.*, **72**, 1763 (1968); (h) R. A. Horne and R. P. Young, *J. Phys. Chem.*, **72**, 376 (1968); (i) R. A. Horne, D. S. Johnson, and R. P. Young, *J. Phys. Chem.*, **72**, 866 (1968); (j) *Advances in High Pressure Research*, vol. 2, Ed., R. S. Bradley, Academic Press, London, 1969, Chapter 3.
9. (a) S. B. Brummer and G. J. Hills, *Trans. Faraday Soc.*, **57**, 1816, 1823 (1961); (b) S. B. Brummer, *J. Chem. Phys.*, **42**, 1636 (1965); (c) S. B. Brummer, *J. Chem. Phys.*, **42**, 4317 (1965); (d) A. B. Gancy and S. B. Brummer, *J. Electrochem. Soc.*, **115**, 804 (1968); (c) A. B. Gancy and S. B. Brummer, *J. Phys. Chem.*, **73**, 2429 (1969).
10. This subject has been concisely and elegantly reviewed by J. L. Kavanau, *Water and Solute-Water Interactions*, Holden-Day, San Francisco, 1964.
11. (a) J. Buchanan and S. D. Hamann, *Trans. Faraday Soc.*, **49**, 1425 (1953); (b) S. D. Hamann and W. Strauss, *Trans. Faraday Soc.*, **51**, 1684 (1955); (c) S. D. Hamann and W. Strauss, *Discussions Faraday Soc.*, **22**, 70 (1956).
12. A. J. Ellis, *J. Chem. Soc.*, **1959**, 3689.
13. F. Körber, *Z. Physik. Chem.*, **67**, 212 (1909).
14. J. C. Jamieson, *J. Chem. Phys.*, **21**, 1385 (1953).
15. D. A. Lown and Lord Wynne-Jones, *J. Sci. Instr.*, **44**, 1038 (1967).
16. J. Fink, *Ann. Phys.*, **26**, 481 (1885).
17. A. Bogojawlensky and G. Tamann, *Z. Physik. Chem.*, **27**, 457 (1898).
18. E. W. Schmidt, *Z. Physik. Chem.*, **75**, 305 (1911).
19. (a) W. A. Adams and K. J. Laidler, *Can. J. Chem.*, **45**, 123 (1967); (b) *Can. J. Chem.*, **46**, 1989 (1968); (c) *Can. J. Chem.*, **46**, 2005 (1968).
20. R. W. Gurney, *Ionic Processes in Solution*, McGraw-Hill, New York, 1953.
21. C. E. Weir, *J. Res. Natl. Bur. Std. (U.S.)*, **53**, 245 (1954).
22. S. D. Hamann, *J. Phys. Chem.*, **67**, 2233 (1963).
23. D. A. Lown, H. R. Thirsk, and Lord Wynne-Jones, *Trans. Faraday Soc.*, **64**, 2073 (1968).
24. G. Tammann, *Ann. Phys.*, **69**, 767 (1899).
25. H. S. Frank and M. W. Evans, *J. Chem. Phys.*, **13**, 507 (1945).
26. J. F. Skinner and R. M. Fuoss, *J. Phys. Chem.*, **70**, 1426 (1966).
27. H. S. Harned and B. B. Owen, *Physical Chemistry of Electrolyte Solutions*, 3rd ed., Reinhold, New York, 1957, p. 179.
28. R. M. Fuoss, *J. Amer. Chem. Soc.*, **80**, 5059 (1958).
29. R. M. Fuoss and L. Onsager, *J. Phys. Chem.*, **61**, 668 (1957).
30. R. M. Fuoss and K. L. Hsia, *Proc. Natl. Acad. Sci. (U.S.)*, **57**, 1550 (1967).
31. R. A. Robinson and R. H. Stokes, *J. Amer. Chem. Soc.*, **76**, 1991 (1954).
32. K. G. Bett and J. Cappi, *Nature*, **207**, 620 (1965).
33. R. A. Horne and D. S. Johnson, *J. Phys. Chem.*, **70**, 2182 (1966).

34. B. B. Owen et al., *J. Phys. Chem.*, **65**, 2065 (1961).
35. G. S. Kell and E. Whalley, *Phil. Trans. Roy. Soc. London*, **258**, 565 (1965).
36. S. D. Hamann, *Ann. Rev. Phys. Chem.*, **15**, 358 (1964).
37. F. H. Fisher, *J. Phys. Chem.*, **66**, 1607 (1962).
38. R. A. Horne and D. S. Johnson, *J. Chem. Phys.*, **45**, 21 (1966).
39. R. A. Horne and R. A. Courant, *J. Chem. Soc.*, **1964**, 3548.
40. H. S. Frank and W.-Y. Wen, *Discussions Faraday Soc.*, **24**, 133 (1957).
41. G. J. Hills and R. Payne, *Trans. Faraday Soc.*, **61**, 316 (1965); *Trans. Faraday Soc.*, **61**, 326 (1965).
42. G. J. Hills and D. R. Kinnibrugh, *J. Electrochem. Soc.*, **113**, 1111 (1966).
43. S. B. Brummer, Ph.D. thesis, University of London, 1960.
44. B. B. Owen, *J. Chim. Phys.*, **49**, C-72 (1952).
45. F. Hensel and E. U. Franck, *Z. Naturforsch.*, **199**, 127 (1964).
46. S. Glasstone, K. E. Laidler, and H. Eyring, *The Theory of Rate Processes*, McGraw-Hill, New York, 1941.
47. S. B. Brummer and A. B. Gancy, Final Rep. Contracts OSW-14-01-0001-425 and OSW-14-01-0001-966, October 1969; to be published.
48. F. T. Wall and S. J. Gill, *J. Phys. Chem.*, **58**, 740 (1954).
49. B. E. Conway, J. O'M. Bockris, and H. Linton, *J. Chem. Phys.*, **24**, 834 (1956).
50. M. Eigen and L. De Maeyer, *Proc. Roy. Soc. (London)*, *Ser. A*, **247**, 505 (1958).
51. R. A. Horne and E. H. Axelrod, *J. Chem. Phys.*, **36**, 1518 (1964).
52. R. L. Kay, K. S. Pribadi, and B. Watson, *J. Phys. Chem.*, **74**, 2724 (1970).
53. R. B. Cuddeback, R. C. Koeller, and H. G. Drickamer, *J. Chem. Phys.*, **21**, 589 (1953).
54. R. A. Horne and D. S. Johnson, *J. Chem. Phys.*, **45**, 21 (1966).
55. See, for example, the review by W. L. Marshall, *Rev. Pure Appl. Chem.*, **18**, 167 (1968).
56. R. L. Kay and G. A. Vidulich, *J. Phys. Chem.*, **74**, 2718 (1970).
57. (a) W. Good, *Electrochim. Acta*, **9**, 203 (1964); (b) *Electrochim. Acta*, **10**, 1 (1965); (c) *Electrochim. Acta*, **11**, 759 (1966); (d) *Electrochim. Acta*, **11**, 767 (1966); (e) *Electrochim. Acta*, **12**, 1031 (1967).
58. C. A. Angell and E. J. Sare, *J. Chem. Phys.*, **52**, 1058 (1969); see that paper for other references.
59. M. H. Cohen and D. Turnbull, *J. Chem. Phys.*, **31**, 1164 (1959).
60. S. B. Brummer, unpublished work.
61. F. T. Wall and S. J. Gill, *J. Phys. Chem.*, **59**, 278 (1955).
62. F. T. Wall and J. Berkowitz, *J. Phys. Chem.*, **62**, 87 (1958).

19 Aqueous Solutions under Extreme Conditions

II HIGH TEMPERATURES

A. B. Gancy, Director, Technology Development,
Industrial Chemicals Division, Allied Chemical Corporation

1 Introduction

Reasons for the Interest in Aqueous Phenomena at Higher Temperatures.
For the purpose of this chapter, "high temperature" refers to temperatures
above 100°C. From there up to the critical region for water (374°C), such
studies involve application of enough pressure to prevent boiling. As we
shall see, there has not been a unified purpose, either theoretical or commercial, to act as a context for these investigations. Rather, there is a range
of interests, from geological simulation and mining operations at one extreme
to theoretical models of solutions at the other. It is instructive to take note
of some of these interests before reviewing the data on high-temperature
transport properties.

Until recently, geologists and geochemists had been responsible for the
bulk of the research in the area of high-temperature–high-pressure chemistry.
Thus the early work performed at the Carnegie Institution in Washington[1]
was aimed at an understanding of pure systems, with the recognition that
this was necessary before there could be any hope of understanding the more
complex systems that exist in nature.

Currently the interplay between preparative inorganic chemistry and solid-
state physics has brought about a resurgence of interest in hydrothermal
synthesis. In this case an aqueous solvent under higher temperature and
pressure is used to increase the solubility of a desired substance to a point
at which large single crystals may be readily prepared. A host of crystals
have been prepared in this manner including SiO_2, Al_2O_3, ZnO, V_2O_3,
magnetite, ZnS, PbS, CdS, and many others.[2]

Also, recent research interest in electrolysis at higher temperatures has
resulted from the recognition that ion mobilities increase, to a point, and

overpotential relationships are often radically altered. Thus nickel-electro-deposition rates overtake hydrogen-evolution rates as temperatures are raised beyond 100°C. Ultimately, as temperature continues to rise, there is a maximum in the log $(i) = f(T)$ curves where association of normally strong electrolytes begins to set in and transport of relevant species is limited.[3]

Knowledge of temperature and concentration limits for avoiding scale formation by calcium sulfate is essential in the design of distillation plants for purifying water.[4] Interest is also stimulated by the long-range considera-tion of homogeneous nuclear reactors. The behavior of uranium compounds in conjunction with other aqueous solution components is of vital interest at elevated temperatures. Thus the system $UO_3–CuO–NiO–SO_3–H_2O$ up to 400°C has been studied at Oak Ridge.[5]

From an industrial standpoint, interest lies in areas such as high-tempera-ture hydrolysis reactions and acid stripping of ores.[6] Only recently has there been an investigation of the theoretical aspects of temperatures between 100°C and the critical point of water (374°C); solubilities of 2:2 salts were measured in electrolyte media as a function of ionic strength in order to test the Debye–Hückel theory at higher temperatures.[4]

The region of supercritical water represents a distinctly different environ-ment within which reactions take place, but not until very recently have system-atic investigations been carried out.[7] There is much practical interest in this region with respect to corrosion and materials of construction in connec-tion with nuclear power and, possibly, desalination technology. Deposition of impurities occurs in steam turbines, for example, because solid substances are soluble in, and transported through dense, supercritical steam.

A knowledge of temperatures, pressures, and compositions of coexisting liquids and vapors is necessary for the evaluation of certain proposals for underground power production. One scheme involves the detonation of a fission or fusion device in an underground salt bed, with the subsequent admission of high-pressure water to produce steam.[8]

The problem of the origin of sodium chloride waters in hot springs has received attention, and it has been suggested that the halogens have been transported as the alkali halides dissolved in the magmatic vapor phase.[9]

By far the most interesting aspect of the supercritical region is its structure and behavior as a solution medium. Whereas it had been known for some time that salts were truly soluble in the vapor, the question of the degree of dis-sociation remained, until the last 15 years, unresolved. Pioneering studies of electrolytic conductance showed that indeed strong electrolytes were con-siderably weaker in the supercritical region, and furthermore the formation of ion pairs and triple ions was a marked function of the density (pressure).[10] It became apparent that the ancillary data required for interpretation of conductance data—density, viscosity, and dielectric constant—were not sufficiently known in the supercritical region and beyond. For although there had been no dearth of information in the high-temperature field of aqueous chemistry, the studies performed had diverse objectives. Information for testing and applying theories of electrolyte behavior was limited. Only recently have physical chemists turned their attention to systematic studies above 100°C and into the supercritical region with an accumulation of data on solubility, conductance, PVT properties, acidity, viscosity, dielectric constant, ion association, hydration, and the self-dissociation of water.[11] This work is the prime focus of this part of the chapter.

The organization of the subsequent presentation is as follows. Important articles in the development of the subject are listed in Section 2. Experimental methods and difficulties that have really held the subject back are summarized in Section 3. Conductance data are analyzed in Sections 4 and 5. Ion association at high temperatures and hydration phenomena are discussed in Sections 6 and 7, respectively. High-temperature properties of water, K_w, and of its component ions are presented in Sections 8 and 9. Finally, the work is summarized and new directions suggested in Section 10.

2   Historical

Excellent reviews have been written on high-temperature–high-pressure investigations, which extend back to 1869 when Andrews recognized the continuity between liquid and gaseous phases in the supercritical region of carbon dioxide.[12] We note the early review of Booth and Bidwell in 1949.[13] Ellis and Fyfe in 1957 presented a review of hydrothermal chemistry that covered, among other things, the behavior of water at high temperatures and pressures, solubility relations, high-temperature conductance measurements, and supercritical vapor solutions.[14] Laudise in 1962 wrote a review on the hydrothermal synthesis of single crystals, where high-temperature conditions are invariably employed.[2] In 1961 E. U. Franck presented a discourse on supercritical water as an electrolytic solution medium.[15] He brought this up to date with a similar article in 1968.[16] Marshall in 1968 presented a review of recent conductance and solubility investigations at elevated temperatures.[17] This was followed by his article discussing correlations in aqueous electrolyte behavior to high temperatures and pressure.[11]

The subject of high temperatures has been approached by numerous investigators and from several points of view. Interest has ranged over the fields of magmatic synthesis, hydrothermal laboratory synthesis, geochemistry, geology, metamorphic petrology, mineralogy, electrolysis, corrosion, solubility, macrocrystal growth, and the physical chemistry of electrolytic solutions. A detailed review of these areas and the numerous individual

Table 1 Chronological Summary of High-Temperature Aqueous Conductance Studies

Date	Investigators	Features	Reference
1900–1907	A. A. Noyes and co-workers	Classical studies to 306°C on solutions in equilibrium with vapor ("natural pressures")	1
1940	F. Spillner	NaCl solutions to 400°C and 300 bars; no detectable conductance in the vapor phase	18
1948	A. C. Swinnerton and G. E. Owen	Conductance of dilute solutions increases with increasing temperature, to the critical point; the lower order of conductance in vapor also increases with increasing temperature	19
1951	J. F. Fogo, S. W. Benson, and C. S. Copeland	A pressure-counterbalancing apparatus for measuring conductance of solutions in the supercritical region	20
1954	J. F. Fogo, S. W. Benson, and C. S. Copeland	NaCl, CsCl, and HCl weak electrolytes in near-supercritical region	7
1954	J. F. Corwin and G. E. Owen	Salt–water mixtures are stratified just above critical temperature	21
1956	E. U. Franck	Pioneering work; measured Λ for KCl system to 750°C, 2700 atm	22
	E. U. Franck	Determined Λ_0, K, and α for KCl, 750°C, 2700 atm	10
	E. U. Franck	Determined "extra" conductance for HCl and KOH in supercritical region	23
1959	K. H. Dudziak	Alkali metal chlorides, NaF, and HF in supercritical region	24
1960	J. F. Corwin, R. G. Bayless, and G. E. Owen	NaCl in H_2O to 390°C	25
1961	G. Coulon	NH_3 and NH_4Cl in supercritical water	26
1961	E. U. Franck	Review article; supercritical H_2O as electrolyte solution medium	15
1962	E. U. Franck, J. E. Savolainen, and W. L. Marshall	Cell assembly for use to 800°C, 4000 bars	27
1963	D. Pearson, C. S. Copeland, and S. W. Benson	NaCl in H_2O, 300 to 383°C; no evidence for hydrolysis or triple ions	28

774

Table 1   (cont.)

Date	Investigators	Features	Reference
1963	D. Pearson, C. S. Copeland, and S. W. Benson	HCl in H_2O, 300 to 383°C	29
1963	A. S. Quist, E. U. Franck, H. R. Jolley, and W. L. Marshall	K_2SO_4 in H_2O, 25 to 800°C, 4000 bars	30
1964	F. Hensel and E. U. Franck	LiCl, KCl, CsCl, CsI, K_2SO_4, and Et_4NI to 130°C, 8000 bars	31
1965	E. U. Franck, D. Hartmann, and F. Hensel	Proton mobility at high temperatures and pressures	32
1965	A. S. Quist, W. L. Marshall, and H. R. Jolley	H_2SO_4 in H_2O, 0 to 800°C, to 4000 bars	33
1965	A. S. Quist and W. L. Marshall	Assignment of Λ_0's for single ions to 400°C	34
1966	W. Holzapfel and E. U. Franck	Λ and K_w for H_2O; 1000°C, 100 kb using anvil press	35
1966	A. S. Quist and W. L. Marshall	$KHSO_4$ in H_2O; 0 to 700°C, to 4000 bars	36
1967	W. L. Marshall and A. S. Quist	Ion-pair equilibria independent of changes in dielectric constant	37
1968	A. S. Quist and W. L. Marshall	Independence of isothermal equilibrium on changes in D	38
1968	A. S. Quist and W. L. Marshall	NaCl in H_2O, 0 to 800°C, to 4000 bars; use of the "complete" ion-pair formation constant	39
1968	A. S. Quist and W. L. Marshall	NaBr in H_2O, 0 to 800°C, to 4000 bars	40
1968	A. S. Quist and W. L. Marshall	HBr in H_2O, 0 to 800°C, to 4000 bars	41
1968	G. Ritzert and E. U. Franck	KCl, $BaCl_2$, $Ba(OH)_2$, $MgSO_4$ to 750°C, 6000 bars; improved KCl conductances	42
1968	E. U. Franck	Review article on supercritical water	16
1969	K. Mangold and E. U. Franck	LiCl, KCl, and CsCl, 200 to 1000°C, to 12 kb; application of Fuoss theory for association constants	43
1969	D. Hartmann and E. U. Franck	Conductance of dilute KCl solution in water–argon mixtures at 440°C	44
1969	H. Renkert and E. U. Franck	KCl to 350°C and 8 kb; improved accuracy	45

investigations is not attempted here. For reference, a chronological summary of high-temperature–high-pressure aqueous electrolytic conductance investigations pertinent to the discussions that follow is given in Table 1.

3   Experimental—Apparatus and Methods of Measurement

None of the experiments in this field can be regarded as "easy to do." Indeed, the development of good experimental procedures can, in some ways, be thought of as rate limiting the whole field.

The earliest conductivity cells used to measure higher temperature behavior of aqueous electrolyte solutions were themselves pressure vessels strong enough to withstand large pressure differentials.[46] In an early systematic study of aqueous salt solutions above the critical temperature of water, Fogo, Copeland, and Benson[7] used a pressure-counterbalancing apparatus. The cell was of relatively light platinum-iridium construction, and its internal pressure was counterbalanced by a controlled gas pressure maintained in an outer pressure chamber that enclosed the entire cell assembly. One end of the cylindrical cell was a pressure-sensitive diaphragm. It constituted part of a sensitive differential pressure gauge that indicated both the magnitude and direction of the differential pressure. The required external compensating gas pressure was manually controlled. The primary advantage of this type of cell is the elimination of almost all difficulties in maintaining tight electrode seals. The outer shell served as one electrode, and the inner one consisted of a platinum-iridium rod gold soldered to a platinum-iridium disk that was insulated from the body by means of synthetic sapphire. The cell was filled through a platinum-iridium tube projecting from the side of the cell and spring-sealed against a gold disk. The cell constant corrections were calculated from the known thermal expansion coefficient for platinum-iridium and amounted to ~0.3% over the range 25 to 390°C.

Corwin and Owen[21] have criticized this work on the basis of their observation that mixtures of salt and water are stratified at temperatures just above the critical temperature of water (374°C). Thus the results of Fogo et al. from a shaken cell would represent a mean, and not the conductivity of a homogeneous phase.

In his pioneering work on hydrogen chloride solutions to 750°C and 2700 atm, Franck[22] required an entirely different cell design. His cell was itself a pressure vessel constructed of a special chrome-nickel steel, and sealed at one end by means of a Bridgman closure. The inside surface of this vessel was coated with platinum-black, and served as one electrode. The inner electrode is a platinized platinum cylinder gold soldered to a long platinum rod that extends out of the cell and is insulated from it by means of a sintered alumina cylinder. Here only the cell itself is heated; the pressure seal at the electrode extension is some distance outside the cell and is maintained at room temperature.

Corrosion of the internal parts of the cell and leaching of impurities into the solution make it quite necessary to subtract from the measured solution conductance what is known as a background conductance (*Untergrundleit-fähigkeit*). This background is measured with conductivity water over the entire temperature and pressure range to be studied. In relative terms, this correction only becomes large when the conductance of the aqueous solution is inherently low. For example, the correction is 1.5% for a 0.006-M potassium chloride solution at 600°C (density of 0.6 g/cm³), but it is about 50% for a 0.0023-M solution at a density of 0.23 g/cm³ at the same temperature.

Nevertheless, it is necessary to introduce fresh solution into the cell by pressure slugging and to check these measured resistances against those obtained by isothermal expansion of the same solution. This gives a measure of contamination during the actual measurement, and this quantity was never observed to exceed 4% for the more highly conducting solutions.

Pressures were measured using Bourdon gauges, and were precise to 10 atm up to 1000 atm, and 20 atm up to 2700 atm. The uncertainty in thermocouple readings within the cell was ±1°C. The inherent cell design made thermal cell constant changes negligible.

An improved version of the cell just described was constructed and used by Franck and co-workers.[27] This time the pressure cell had a platinum-iridium lining that was platinized and served as one electrode. The other was a platinum-iridium cylinder at the end of a platinum rod insulated from the cell body by a sintered alumina cylinder. Again, the rod leads out of the cell through a thick-walled metal tube into a cold closure. Around 400°C temperature was controlled to within 1°C, and around 700°C to within 2°C. Pressures were measured using Bourdon gauges and strain gauges and were accurate to 1% between 100 and 4000 bars. Conductances of 0.01-M potassium chloride solution at 800°C showed good reproducibility between stepwise ascending and descending pressure cycles. Background conductances measured with distilled water were determined over the entire pressure and temperature range.

The first extensive measurements of high-temperature conductance of aqueous solutions in this country utilized the improved Franck cell.[30] Potassium sulfate solutions were studied over the range of 25 to 800°C and up to 4000 bars. The corrections for background conductance of distilled water were 2 to 10% for 0.0005-M potassium sulfate and 0.5 to 2% for 0.005-M potassium sulfate. The accuracy of the smoothed values of specific conductance was judged to be ±1 to 2% at the higher concentrations and densities. Later work on hydrogen sulfate solutions[33] increased the uncertainty to ± 4 to 5%, possibly because of corrosion in the tubing and valves. As with potassium hydrosulfate solutions,[36] reproducible measurements on hydrogen sulfate were most difficult to obtain between 100 and 300°C.

In subsequent investigations, a similar cell was designed, but with the elimination of seals that had been subjected to high temperature.[39] The cell had a pressure fitting at each end to make it a flowthrough type that could be flushed thoroughly at the operating temperatures and pressures. Agreement between conductances obtained with this cell and with the nonflow-through type were 1 to 2%, or about the same order as corrections for background conductance. The conditions studied were 0 to 800°C and pressures to 4000 bars.

In another variation of a high-pressure cell that had to withstand only 130°C, Franck[31] used a Teflon-bodied cell in which a large volume of electrolyte was contained within a thin-walled, flexible Teflon region. The entire cell was immersed in n-heptane, which served as the pressure-transmitting fluid. Thus pressure was transmitted to the solution under study by way of the flexible Teflon wall, and the pressure differential was 0. The overall uncertainty in the equivalent conductance was ± 1% at 45°C and ± 1.5% at 130°C. A similar principle was employed in which a bellows of palladium-gold alloy was gold soldered to a metal cell body.[32] In this case kerosene transmitted the pressure to the cell, and measurements were made to 220°C and 8000 bars.

Franck also used a flowthrough-type cell for measurements to 750°C and 6 kb.[42] Pressure was transmitted through oil and thence to the solution under study, which was contained in a flexible, thin-walled Teflon cylinder. The inner and outer coaxial electrodes were platinum. Sintered alumina was the insulator. Induction heating brought temperatures quickly to the desired values, thereby shortening the residence time of solutions in the hot cell. This, along with the continual renewal of solution, gave improved results. At 750°C and 1000 bars, for example, the background conductance correction is only about 8%. The cell constant is known to within 1%.

For his work with supercritical homogeneous water–argon mixtures, Franck[44] used a cell with a platinum inner electrode and a palladium–gold-alloy outer electrode.

Finally, in order to achieve an accuracy of about 1% for 0.01-M potassium chloride solutions up to 350°C and 8 kb, Franck used a cell capped at both ends by deformable Teflon cylinders. Pressure was transmitted to these cooled cell ends by means of liquid water.[45]

Using a static pressure method, the electrolytic conductance of pure water has been measured to 1000°C and at pressures between 20 and 100 kb.[35] Small, flat conductance cells were inserted between the two anvils of a press and contained thermocouples that also served as conductance electrodes.

The statistical error in the measured conductance values was $\pm 40\%$. The results agreed within experimental error with those of David and Hamann,[47] who made conductance measurements in pure water compressed and heated by shock waves. Furthermore, the errors were in the direction to be expected if time were insufficient in the shockwave studies for the attainment of dissociation equilibrium.

Precise conductance measurements at 1 atm and temperatures below 100°C can be accurate to $\pm 0.01\%$. The higher pressure precision studies in the same temperature range have given values that are good to 0.1%.[48] The high-pressure–high-temperature region described previously is less well known than the region just mentioned by at least an order of magnitude. Thus two orders of magnitude in accuracy and precision separate the expanded studies from the highly developed 1-atm data that have been used to test theories of electrolyte solutions over the past 40 years.

4   Conductance Data—Finite Concentrations

Figure 1 shows the behavior of a typical 1:1 electrolyte over a wide range of temperature and pressure;[39] κ, the specific conductance, is plotted against temperature with pressure parametric. Corrections for cell constant variation and for the solvent conductance have been made.

Note the steep ascendancy in κ at all pressures, with an increase in temperature up to about 300°C. A maximum is then reached, and the subsequent drop is very much a function of the applied pressure. From 0 to 300°C, the increase in conductance is caused by the sharp decrease in water viscosity over

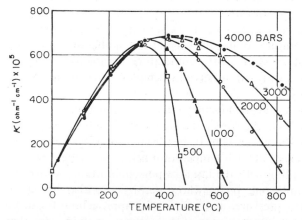

Figure 1. Isobaric variation of specific conductances of 0.01-M sodium chloride solutions as a function of temperature. After Ref. 39.

the same interval, which permits a greater mobility of the ions. This is accompanied by a small decrease in the numbers of ions per unit volume as a result of the thermal expansion of the solution.[49] Although the viscosity continues to drop at still higher temperatures, the sodium chloride is increasingly associated to form ion pairs (discussed later) and the latter effect overtakes the viscosity effect;[11] thus the rapid falling off of the curves in Figure 1.

It has also been proposed that the appearance of a maximum in conductance with increasing temperature depends in some measure on the increase in the effective ion radii (i.e., increased hydration).[50] The marked pressure dependence of the subsequent behavior above the critical temperature may then in part be attributed to the enhanced "stripping" of the ion-hydration atmosphere with increased pressure.[51] This interpretation is not inconsistent with the reporting later on[10] of a constancy of q over this supercritical temperature range, where q is the net change in the amount of hydration water involved in association equilibria.

It is more appropriate in treating conductances to use equivalent conductance, Λ, rather than specific conductance. For 1:1 electrolytes, Λ is calculated from the specific conductance, κ, according to

$$\Lambda = \frac{1000 \, \kappa}{c\rho}$$

where c is the analytical concentration of salt and ρ is the density of the solution at the temperature and pressure of interest. For the dilute solutions under consideration here, the latter is replaced by the density of pure water which is described by PVT data over wide ranges of temperature and pressure.[52] The uncertainties in ρ up to 1000°C and 2500 atm are 1 to 2%, and the errors are compounded with those in κ.

The shape of the Λ versus T curves would be similar to those in Figure 1. Figure 2 gives high-temperature conductance data from a different viewpoint, that is, as a function of pressure, with temperature parametric.[22] Again, there seems to be a sharp distinction between behaviors at subcritical and supercritical temperatures.

It has been customary to plot conductance data as a function of the system density, as in Figure 3. On the right-hand side of the graph the isotherms represent liquid systems for which density has been varied by varying the applied pressure. There exists at each temperature below the critical point a lower density limit coinciding with the liquid-vapor equilibrium. Above the critical point, there is only one phase whose density can also be varied by controlling the applied pressure, and values of ρ approaching 0 can be achieved. It is helpful in the interpretation of these diagrams to imagine that pressure runs generally opposite to density along the abscissa.

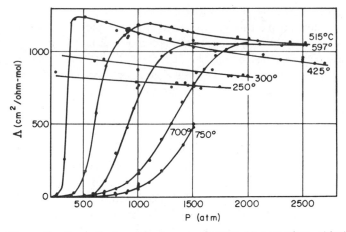

Figure 2. Equivalent conductances of 0.001-M potassium chloride solutions as a function of temperature and pressure. After Ref. 22.

At supercritical temperatures, the sharp rise in equivalent conductance with increasing density corresponds to increased ionization of the electrolyte. The decrease as the highest densities are approached in the subcritical region is related to the increasing viscosity of the solvent, among other things. In this region a 1:1 electrolyte is highly ionized.

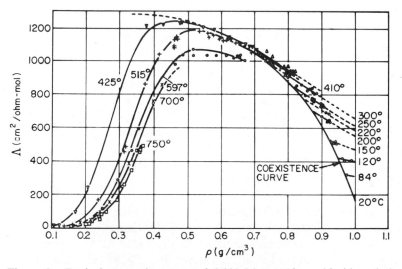

Figure 3. Equivalent conductances of 0.001-M potassium chloride solutions as a function of density and temperature. After Ref. 22.

Inasmuch as strong electrolytes are now assumed to behave as weak electrolytes in the peak region—even though they are highly conducting—their behavior with respect to (analytical) concentration changes should be predictable. Thus as concentration is increased, the degree of ionization decreases, and the maxima are shifted to the right. This must, of course, be accompanied by a fall in the peak value. Such behavior is actually observed.[22]

5   Conductance Data—Infinite Dilution

Thus far conductances have been discussed for aqueous electrolyte solutions in measurable ranges of concentration. A better development of transport theory can only result if ion-ion interactions are eliminated; that is, if infinite dilution values can be obtained. In particular, ion-association effects on Λ would disappear and we can examine ion-solvent interactions, which are of prime interest.

The determination of conductance at infinite dilution, Λ_0, under the circumstances of high pressure and temperature is a formidable, but not insuperable problem. First, the accuracy of extrapolation to 0 concentration depends directly on the accuracy of the data points themselves. We have seen that such data, obtained through ingenious experimental methods though they are, are at least two orders of magnitude less certain than those which one tended to associate with the "classical" electrolyte studies. Second, whenever the electrolyte is appreciably associated, a straightforward extrapolation according to the Debye–Hückel–Onsager equation is not possible.[10] One may resort to a Shedlovsky-type extrapolation.[54] Even this is not feasible however, unless extremely dilute solution data are available and one resorts to the utilization of the Walden rule to determine Λ_0's.[10]

Another difficulty is that, up to the time of the pioneering investigations in high-temperature physical chemistry, the necessary ranges of dielectric constant, density, and viscosity values had not been covered by direct measurement. In earlier stages of the conductance investigations, extrapolations were made based on theoretical and semiempirical expressions.[10] Investigators of high-temperature conductance even resorted to making some of these measurements themselves.[55] The difficult and tedious nature of these efforts made in pursuit of ancillary information attests to the importance of the infinite dilution conductance.

It is worthwhile to examine specific examples illustrating the determination of Λ_0. The following is taken from the work of E. U. Franck[10] for potassium chloride solutions. Over the entire range of densities studied, an attempt was made to extrapolate with the aid of the Debye–Hückel–Onsager limiting law

$$(1) \qquad\qquad \Lambda = \Lambda_0 - a\sqrt{c}$$

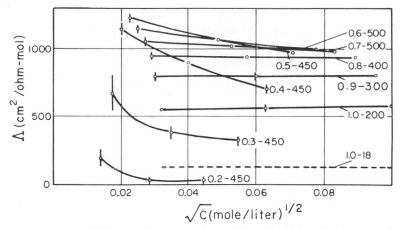

Figure 4. Equivalent conductances of potassium chloride solutions as a function of $\sqrt{c}$. After Ref. 10.

where a is a constant that is independent of c, the concentration. Figure 4 demonstrates that linearity holds only over the density range 0.8 to 1.0 g/cm^3. The other curves are indicative of weak electrolyte behavior, and indeed this was the interpretation given initially to data such as those appearing in Figure 3.

The determination of the Λ_0 and the association constant, K, for weak electrolytes can be determined simultaneously by the use of the Shedlovsky equation

$$(2) \qquad \frac{1}{\Lambda} S(z) = \frac{1}{\Lambda_0} + \frac{c \cdot \Lambda \cdot f_\pm^2 \cdot S(z)}{K(\Lambda_0)^2}$$

where $f_\pm$ is the activity coefficient for the salt and $S(z)$ is defined as

$$(3) \qquad S(z) = 1 + z + \frac{z^2}{2}$$

where

$$(4) \qquad z = a(\Lambda \cdot c)^{1/2}(\Lambda_0)^{-3/2}$$

and a is the constant of (1).

After applying the appropriate values of viscosity and dielectric constant, and determining $f_\pm$ according to the Debye–Hückel approximation, it was found that (2) to (4) provided satisfactory extrapolations only for densities greater than 0.6 g/cm^3 and temperatures below 500°C. The experimental solution concentrations were simply not low enough to allow good extrapolations. The following device was therefore resorted to: (a) satisfactory values of Λ_0 were obtained in the appropriate density and temperature regions using the Shedlovsky extrapolation ($\rho = 0.8$ g/cm³ and 400°C); (b) in the lower-density–higher-temperature regions, appropriate values of Λ_0 were calculated using the Walden rule

$$(5) \qquad \Lambda_0(t, \rho) = \frac{\Lambda_0(400; 0.8) \cdot \eta(400; 0.8)}{\eta(t, \rho)}$$

(c) Λ_0 values derived from (5) were then used in the Shedlovsky expression to calculate K's.

The Λ_0 values determined by these methods are shown in Figure 5, along with the original Λ values. Indeed, the steep slopes of the Λ curves, presumably due to association, become transformed to the monotonic Λ_0 curves, which extend smoothly across the entire density range. The latter can be shown to vary inversely as the viscosity of water in this region.

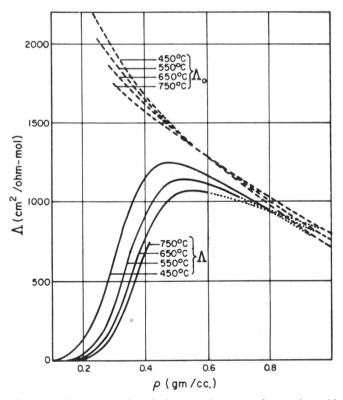

Figure 5. Isotherms of equivalent conductance of potassium chloride solutions: solid curves, 0.001 M; dashed curves, infinite dilution. After Ref. 10.

Table 2   Evaluation of Λ_0

Range of Density (g/cm)³	Standard Deviation (%)
0.70 and above	to 1
0.60–0.65	2
0.5	7
0.45	10
< 0.45	> 10

From Ref. 39.

A slightly different approach was used by Quist and Marshall for sodium chloride solutions.[39] They used both the Robinson–Stokes equation[50]

$$(6) \qquad \Lambda = \Lambda_0 - \frac{Sc^{\frac{1}{2}}}{1 + B\mathring{a}c^{\frac{1}{2}}}$$

and the Fuoss–Onsager equation[57]

$$(7) \qquad \Lambda = \Lambda_0 - Sc^{\frac{1}{2}} + Ec \log c + J(\mathring{a})c$$

to obtain a better fit with the data.

In both expressions $\mathring{a}$ is the mean distance of closest approach of the ions. Equation 7 was found to give a good fit only down to a density of 0.75 g/cm³. Thereafter a Shedlovsky extrapolation was used according to (2).

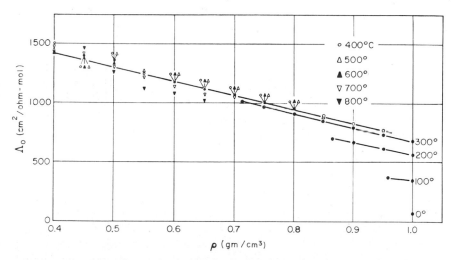

Figure 6.   Limiting equivalent conductances of sodium chloride solutions as a function of density. After Ref. 39.

Table 2 reveals that standard deviation associated with the evaluation of limiting equivalent conductances is a function of the density.

The resulting values of Λ_0 so obtained were plotted against density over the range 0 to 800°C, as illustrated in Figure 6. It is seen that Λ_0 is a linear function of density at higher temperatures; deviations at 700 and 800°C fall within the estimation of error discussed previously. Furthermore, the slopes are all approximately the same at all temperatures. Above about 400°C, at constant density, Λ_0 appears to be independent of temperature; that is, the energy of activation of conductance at constant volume, E_v, described by

$$(8) \qquad\qquad \log \Lambda_0 = A - \frac{E_v}{RT}$$

is 0 at high temperatures and low densities. This is quite different from observations at room temperature.[75] The contrast in results between potassium chloride (Figure 5) and sodium chloride (Figure 6) and the general level of inaccuracy, require us to interpret this conclusion with some caution.

6   Association Equilibria

In studies of sodium chloride solutions over the ranges 300 to 383°C and 0.4 to 0.6 g/cm³, Pearson, Copeland, and Benson[28] applied the Fuoss–Kraus criteria[58] in order to estimate the contribution of triple ions to the overall association. By assuming various values of the triple-ion dissociation constant and comparing the results with actual data, they concluded that they could neglect the contribution of triple ions in this region.

Franck,[10] on the other hand, applied the Fuoss–Kraus criteria to potassium chloride solutions in the range of 600°C and 0.4 to 0.5 g/cm³ to find that indeed triple-ion formation occurred. The existence of triple ions in this case, however, did not affect the determination of Λ_0, inasmuch as the Walden rule was utilized and no assumptions concerning dissociation equilibria were required. In making these estimations it is important to examine the elements that go into the criterion. The Fuoss–Kraus criterion equation is

$$(9) \qquad\qquad D \le \frac{3e^2}{8\,dkT}$$

where e is the electronic charge, d is the triple-ion diameter, and D is the macroscopic dielectric constant. Estimation of d must therefore be made in order to use the test. But more important, D is the macroscopic value and therefore not necessarily the true value of the dielectric constant which should be applied. The "effective" D value near the ions should be lower than the macroscopic value. Hence the use of the latter really sets a criterion that is too high. In the present case, it means that triple-ion formation may be questionable in the intermediate density range but may still apply to low densities and higher temperatures. Notwithstanding these uncertainties, association of some kind unquestionably occurs at intermediate to low densities and at supercritical temperatures.

The ion-pair association constants for potassium chloride solutions studied up to 750°C and 2700 atm were obtained from (2) in Franck's earlier work.[10] The equation can be rearranged to give

$$(10) \qquad K = \frac{\Lambda^2 [S(z)]^2 \cdot f_{\pm} \cdot c}{\Lambda_0 [\Lambda_0 - \Lambda \cdot S(z)]}$$

A degree of dissociation, α, can be defined as

$$(11) \qquad \alpha = \left(\frac{\Lambda}{\Lambda_0}\right) \cdot S(z)$$

Figure 7 shows α plotted against density, with temperature parametric, for a 0.001-M solution of potassium chloride. Although the variations with temperature are small, α is strongly dependent upon density. For densities greater than 0.8 g/cm³, the salt is more than 95% dissociated. But it is

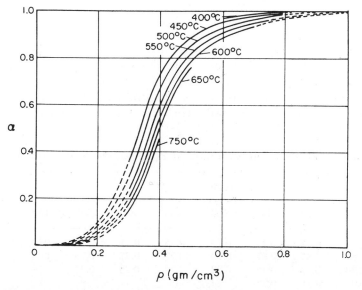

Figure 7. Degree of dissociation of 0.001-M potassium chloride in water at higher temperatures. After Ref. 10.

highly significant that the α's pass from there through all intermediate values and approach 0 at low ρ.

In an expanded investigation, Franck further elaborated upon the region in the upper right-hand region of Figure 7.[43] Conductances of LiCl, KCl, and CsCl were measured over the range 300 to 1000°C, and at pressures up to 12 kb. The question of triple-ion formation was again examined, and estimation of the equilibrium of the salts from the Fuoss equation showed their contribution to be negligible down to densities of 0.7 g/cm³. Thus only ion-pair equilibria need be considered.

The Fuoss–Kraus equation

$$(12) \qquad \alpha = \frac{\Lambda}{\Lambda_0 F(z)}$$

was used to calculate degrees of dissociation, where

$$(13) \qquad z = S(\alpha c)^{1/2} \Lambda_0^{3/2}$$

The required quantities for this calculation are the viscosity, dielectric constant, and the equivalent conductance at infinite dilution. The resultant α values are given in Table 3 for a potassium chloride solution.

At $\rho = 0.7$ g/cm³ the α values are constant over the range 300 to 1000°C. This may be rationalized on the basis of the following argument. The ion-pair constant may be given by

$$(14) \qquad K = \tfrac{4}{3}\pi \frac{N_0 a^3}{1000} \exp\left(\frac{e^2}{akDT}\right)$$

If a, the average ion radius is temperature independent, the entire dependency then rests with the product, DT. At 0.7 g/cm³ this product varies by only 10% from 400 to 1000°C. Thus the constancy of α over the same range.

On the other hand, DT falls with increasing temperature at higher solution densities. This is also reflected in the α values in Table 3. The implication here,

Table 3   Degree of Dissociation for a 0.01-M Potassium Chloride Solution

α (g/cm³)	Temperature (°C)							
	300	400	500	600	700	800	900	1000
0.70	—	0.75	0.75	0.75	0.75	0.75	0.75	0.75
0.80	0.87	0.82	0.81	0.80	0.79	0.77	0.76	0.75
0.90	0.90	0.89	0.85	0.82	0.79	0.75	0.73	—
1.00	0.95	0.91	0.85	0.81	—	—	—	—
1.10	0.96	0.92	—	—	—	—	—	—

of course, is that the interionic attraction theory is valid, and that further-more the effective ion size does not vary with temperature in the supercritical region. The other point of interest is that potassium chloride is not completely dissociated above 300°C, even at high densities.

7   Ion Hydration at High Temperatures

When the K values obtained earlier were converted to logarithms and plotted against log ρ, monotonic curves were obtained which systematically deviated from linearity. However, if average values of straight-line slopes were taken, they were found to be approximately independent of the temperature over the range 450 to 750°C. Considering the direct proportionality between water concentration and water density, Franck was prompted to write the equilibrium

$$(15) \qquad KCl \cdot mH_2O + qH_2O = K^+ \cdot rH_2O + Cl^- \cdot sH_2O$$

where H_2O is assumed to be a reactant and q is the slope of the curves just mentioned. It was found to have a value of 8 to 10. From solubility measure-ments,[60] m was found to have a value of 4. Hence

$$(16) \qquad m + q = r + s = 12 \text{ to } 14$$

The new equilibrium constant, K', is related to the conventional constant according to

$$(17) \qquad K' = \frac{K}{\rho^q \cdot 55.5^q}$$

and experimentally

$$(18) \qquad \log K' \cong \log K - 9.1 \log \rho - 15.8$$

Quist and Marshall have observed this linear relationship for several electrolyte systems over different ranges of temperatures to a maximum tem-perature of 800°C.[39] An example of a plot of log K against log c_{H_2O} is given in Figure 8. From such plots, values can be obtained for q in the expression

$$(19) \qquad K' = \frac{K}{(c_{H_2O})^q}$$

where K' is now termed the "complete" equilibrium constant, and appears to be independent of any isothermal property of the solvent (i.e., dielectric constant or density).

It should be pointed out that such studies extending into the supercritical region allow a change in the molar concentration of water over a rather wide range. This, of course, is impossible for an aqueous solution under normal

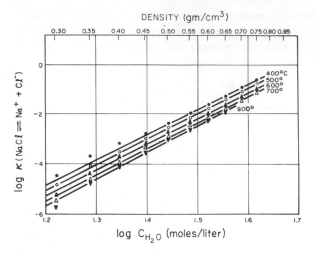

Figure 8.   Log K for the equilibrium $NaCl \rightleftharpoons Na^+ + Cl^-$ as a function of the logarithm of the molar concentration of water at temperatures from 400 to 800°C. After Ref. 39.

conditions. Incidentally, it should also be noted that by changing the pressure of the system at constant temperature it is possible to vary the dielectric constant of the solvent continuously without altering the overall chemical composition.[16] This is another contingency that could not be used in the classical study of electrolyte behavior.

Franck finds a close agreement between the ion-hydration energy obtained from data similar to those in Figure 8 and energy determined for the innermost ion-hydration sheath at ordinary temperatures and pressures.[61] He therefore concludes that, above 400°C, the ions and ion pairs carry with them a relatively tightly bound primary solvation shell.

Quist and Marshall [39] have expanded on the concept of the "complete" equilibrium constant. There are two ways of varying the solvent composition besides varying the pressure in the supercritical region. One is to examine the measurements made in mixed solvent systems wherein one solvent is preponderantly more polar than the other. Water-dioxane mixtures fall into this category. The other is to simply vary the pressure of a system at, say, 25°C. The authors have examined many results of studies on mixed solvent systems, and report a linearity between $\log K$, where K is the conventional equilibrium constant, and $\log c_{H_2O}$.[34] In water-dioxane mixtures, for example, dioxane is considered to be an inert diluent, and water takes part in the equilibrium process as in pure water. An example of this linearity appears in Figure 9, which cites some K_w data taken from Harned and Fallon [62] (left-hand branch). The right-hand branch of the curve in Figure 9 is taken

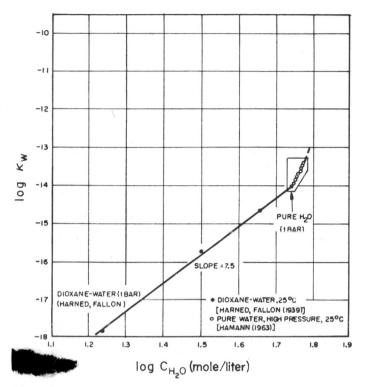

Figure 9. Log K_w for water versus log C_{H_2O} in a dioxane-water mixture[62] and in pure water to 2000 bars at 25°C.[63] After Ref. 11.

from the data of Hamann,[63] who studied the association of pure water to 2000 atm. Note that the slopes of the intersecting curves are not equal. If all the other assumptions going into these determinations are correct, it would appear that dioxane is not essentially inert. Quist and Marshall point out that differences in slope have been observed for acidic and basic electrolytes but not for the few examples of nearly neutral salts that have been studied in both dioxane-water solutions and in water at high pressures.

It was mentioned earlier that the ion-hydration behavior, as reflected in the slope, q, was fairly independent of temperature above the critical temperature. The behavior at lower temperatures is not at all independent of temperature, either in pure water or in water-dioxane mixtures. This is illustrated in Figure 10. At 25°C the data are taken from sodium chloride studies in dioxane-water by Kunze and Fuoss;[64] the 100°C data are those of Dunn and Marshall for sodium chloride in dioxane-water;[65] and the 400 to 800°C data are from Quist and Marshall on sodium chloride in pure water.[39]

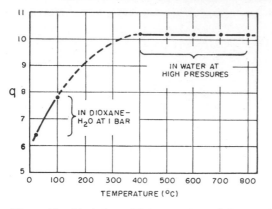

TEMPERATURE (°C)

Figure 10.   Variation with temperature of the net change in waters of solvation, q, for sodium chloride in water and dioxane-water solutions. After Ref. 11.

The constancy of q above 400°C, however, is not universally exhibited. The hydrogen bromide system, for example, shows a variation in q from 12.2 at 400°C to 16.2 at 800°C.[41] The sodium bromide system, in contrast, is characterized by an average q value of 9.85, although the slopes vary randomly in the 400 to 800°C region from 9.76 to 9.94, not a serious amount.[40] The difference between hydrogen bromide and sodium bromide may reflect a unique proton conductance process known to occur in critical regions.

Using the average value of 9.85 for q, (18) was used to calculate values of the complete constant, K'. At 400, 500, 600, 700, and 800°C the plot of log K' versus $1/T$ gives a slight curvature. The points at 500, 600, and 700°C can be represented by a straight line to give an energy change at constant volume, ΔU of -7.5 kcal/mole.

According to (14) the ion-pair formation constant is a function of temperature, dielectric constant of the solvent, and the ion diameter measured from the centers of each ion participating in a stable ion pair. Expressed in its logarithmic form

$$(20) \qquad \log K = -2.598 + 3 \log å + \frac{7.24 \times 10^4}{åDT}$$

Thus, assuming a constant value of $å$, a plot of log K against $1/DT$ (or against $1/D$ at constant temperature) should give a straight line. Quist and Marshall[37] show that such plots are not linear for room-temperature water-dioxane systems, and this is cited as evidence that the dielectric constant is not a controlling factor. Conversely, plots of log K versus log c_{H_2O} are linear, and this is interpreted as a manifestation of a general principle—that correct values of K (K', previously) are independent of dielectric constant not only for water systems at high temperature but for all other solvent-electrolyte systems.

Hartman and Franck[44] have studied the dissociation of potassium chloride in water-argon mixtures in the supercritical region. Argon was selected because of its inert character; water and dioxane, it was pointed out, were miscible through a specific interaction so that dioxane could not function as a true inert diluent. They found that a plot of log K versus $1/DT$ was indeed not a straight line. Two reasons were proposed for this.

1. The value of $\mathring{a}$ obtained from (20) turns out to be a function of D; that is, when the dielectric constant is varied through the use of various water-argon mixtures, the resultant value of $\mathring{a}$ (where $\mathring{a}$ is now treated as a dependent variable) is not constant. This is illustrated in Figure 11.

2. If an ion-diameter term is to be kept constant, then the macroscopic dielectric constant will have to be replaced by an "effective" one. In general, this effective value is lower than the macroscopic one in the close vicinity of the ions. Corollary to this is the assumption that the lower the macroscopic value in the first place, the smaller the discrepancy between the two.

Even if one wished to make no assumptions concerning the individual variations in $\mathring{a}$ and D, it seems clear that the overall effect on K is not a matter of either one alone; both must be considered together. In any further analysis of these phenomena, it would seem that the $(3 \log \mathring{a})$ term in (20) could be used to good advantage.

It is interesting that for high dielectric constants in Figure 11 the value of $\mathring{a}$ is too low (i.e., physically impossible). Yet at the other end, the value of $\mathring{a}$ approaches 3.1 Å. This is close to the sum of the crystallographic radii, 3.14 Å. If one now sets $\mathring{a}$ at 3.1 Å, an effective dielectric constant can be backcalculated. The results are depicted in Figure 12. Included in this figure is a similar determination of effective D based upon the data of Lind and Fuoss[66] for potassium chloride in water-dioxane mixtures at 25°C. The

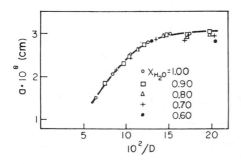

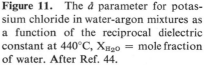

Figure 11. The $\mathring{a}$ parameter for potassium chloride in water-argon mixtures as a function of the reciprocal dielectric constant at 440°C, X_{H_2O} = mole fraction of water. After Ref. 44.

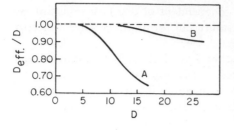

Figure 12. The ratio of effective dielectric constant (in the close vicinity of the ions) to the macroscopic dielectric constant as a function of the macroscopic dielectric constant; curve A, water-argon mixtures at 440°C; curve B, water-dioxane mixtures at 25°C. After Ref. 44.

latter curve is much closer to unity; this is attributable to the greater separation of the ions (5.2 Å) in the first place.

The question of the validity of (20) therefore remains unresolved. But the first matter of importance is to recognize the complex interaction of the variables that comprise it. Although modifications of the equation have been proposed,[64] they do not lead to a linearity between log K and $1/D$. Apparently the most profitable approach at the moment is to introduce the concept of an effective D, which can be obtained by backcalculation.

Quist and Marshall would argue that if the true equilibrium constant, K', were calculated instead of the conventional one, then K' would be independent of changes in D. They draw this conclusion from the observation, for example, that sodium chloride solutions over the range 400 to 800°C are characterized by a constant value of K' which is independent of density over the range 0.30 to 0.75 g/cm³ (and is therefore independent of D, since D does vary under these conditions). It would be interesting to treat the Hartmann–Franck data according to these concepts.

8   Ion Product of Water at High Temperatures

Noyes and co-workers measured the ion product of water under liquid-vapor equilibrium conditions up to 306°C.[1] With increasing temperature, the product rises to a maximum close to 10^{-11} at about 220°C. It is noteworthy that the density of water varies from 1.0 g/cm³ at room temperature to 0.7 g/cm³ at 306°C.

Franck[23] utilizes the K_w value of Noyes at 306°C and 0.7 g/cm³ as a starting point and proceeds from there to the estimation of K_w in the supercritical region and at various constant densities. The primary equations that go into the calculations are

(21) $$\log K' \cong \log K_w - (q + 1) \log \rho - (q + 1) \log 55.5$$

and

(22) $$\log K' \equiv -\frac{\Delta H}{4.57} + \frac{\Delta S}{4.57}$$

Equation 21 is simply the definition of the equilibrium constant, where water is considered as a reactant and appears on the left-hand side of the reaction equation. Equation 22 derives from the relations among equilibrium constant, enthalpy, and entropy changes accompanying the reaction. $q = r + s$, where r and s are the hydration numbers of H^+ and OH^- ions, respectively.

In order to fill in the unknowns in (21) and (22), the conductance data for hydrogen chloride solutions are utilized in conjunction with the following assumptions: (a) the hydration number, s, is equal to that of the chloride ion; (b) the hydration number of the undissociated hydrogen chloride molecule is 0 (this leads to a value for q of 9); (c) the entropy of water dissociation, ΔS, is equal to that for hydrogen chloride dissociation in the supercritical region. The result is given as

(23) $$\log K' = -\frac{20,500}{4.57T} - 20.1 = \log K_w - 10 \log \rho - 17.5$$

From this expression K_w has been calculated and compared with the data of Noyes in Figure 13.

As mentioned earlier, these estimated K_w values do not differ appreciably

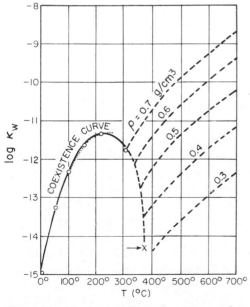

Figure 13. The ion product of water as a function of temperature; arrow indicates critical point. Circles, according to Ref. 1; dashed curves, ρ according to (23). After Ref. 23.

from those obtained in shockwave experiments. The important point to note is that the ion product of water can increase from 10^{-14} at room temperature to $\sim 10^{-9}$ in the realm of 600°C and 0.7 g/cm³, and the latter condition is now routinely accomplished in modern high-temperature–high pressure investigations. It now becomes quite necessary to examine the data on aqueous electrolyte solutions in these regions in terms of hydrolysis. Hitherto in the discussion hydrolysis reactions have been implicitly ignored.

The problem of treating hydrolysis in the supercritical region is complicated because here dissolved salts—even so-called strong electrolytes—behave as weak electrolytes. In the case of potassium chloride, for example, there are at least eight species possible in supercritical water: KCl, KOH, HCl, HOH, K^+, H^+, Cl^-, and OH^-. There are eight corresponding nonlinear equilibrium constant expressions such as

$$(24) \qquad\qquad K_b = \frac{a_{K^+} a_{OH^-}}{a_{KOH}}$$

and four linear equations expressing mass balance, such as

$$(25) \qquad\qquad c_{KCl} + c_{HCl} + c_{Cl^-} = c_0$$

where c_0 is the analytical concentration of KCl. These equations can be solved by introducing reduced variables and utilizing iteration methods on the computer.[15] All equilibrium constants cannot be obtained explicitly, but certain limits of behavior can be described. This exercise is important because it demonstrates that calculations of the degree of hydrolysis in dense supercritical water vapor are not yet possible.

9   Excess Mobility of H^+ and OH^- at High Temperatures

The abnormally high values of the mobility of H^+ and OH^- ions in liquid water have received extensive study. In order to examine separately the dependence of mobility upon temperature, as well as with pressure (or density) Franck[23] has measured the conductances of hydrogen chloride and potassium hydroxide in supercritical water. Because of the corrosive effects of these electrolytes upon the apparatus, solutions more dilute than 0.01 M were not studied. Thus Λ_0, the equivalent conductance at infinite dilution, could not be obtained by the usual methods. Instead, an "excess" conductance attributed to both OH^- and H^+ ions was defined according to the following equations:

$$(26) \qquad\qquad \Lambda_{0,HCl} \simeq \Lambda_{0,KCl} + u^+$$

$$(27) \qquad\qquad \Lambda_{0,KOH} \simeq \Lambda_{0,KCl} + u^-$$

where u represents excess conductance. The values for $\Lambda_{0,KCl}$ are, of course, known in the supercritical region. The approach was therefore to estimate the u's, and thereby determine the Λ_0's. From the latter method the usual degree of dissociation and dissociation constants can be determined.

Below about 200°C, the u values of Noyes[68] were used. They were then extrapolated to the critical point (374°C) along the liquid-vapor coexistence curve. Isochores of u were estimated by the use of the Gierer–Wurtz equation.[69] The Λ_0 values were then calculated using (26) and (27).

At densities at or below 0.6 g/cm³ and at temperatures above 400°C, the values u^+ and u^- are scarcely 10% in excess of $\Lambda_{0,HCl}$ and $\Lambda_{0,KOH}$, respectively. In contrast to room temperature behavior, the values of u_{Cl^-}, u_{OH^-} and of u_{K^+}, u_{H^+} are similar in magnitude.

The degree of dissociation was determined from the above-derived Λ_0 values for KOH and HCl in the supercritical region. It appears that KCl is the strongest electrolyte, KOH is slightly weaker, and HCl is much the weakest in the intermediate density region.

10   Summation and Conclusions

When the temperature is increased from ambient to the critical point, the electrolytic conductance increases sharply almost independent of pressure. Macroscopically, this is owing to the fall in solvent viscosity over this range. The ultimate cause is a disintegration of water clusters and probably an accelerated addition of unbonded water molecules to $H_9O_4^+$ complexes.[32] In addition, the effect of increased pressure is to reduce the effective diameters of the hydrated ions. And, although it has been proposed that a rise in temperature brings about an increase in effective ion size (in concentrated solutions), the evidence for infinitely dilute solutions is that ions are stripped to their most tightly bound primary solvation sheaths, especially in the supercritical region.

Once into the supercritical region, conductances, which may be as much as tenfold greater than at room temperature, begin to drop off, depending on the pressure. Low pressures (low densities) bring about a sharp drop well into the supercritical region; at high pressures the decline is much less severe. In this temperature range, strong electrolytes are now associated to form ion pairs and triple ions and thus behave as weak electrolytes. The strong dependence on density can eventually be attributed to the participation of the solvent in the equilibrium process; that is, solvent density is translatable to water concentration, the latter now appearing in the ion-pair formation equilibrium expression, where it is raised to the appropriate power.

The appropriate determination of equivalent conductances at infinite dilution, where ion-ion interactions are absent, requires ancillary information. Thus PVT data (densities), the dielectric constant, and the viscosity for water are required over broad ranges of temperature and pressure, where measurements become difficult. In spite of the unavoidable inaccuracies in these and in the conductance data themselves, it is clear that "strong" electrolytes weaken in the supercritical region and become weaker yet as the density decreases.

Investigations in the supercritical region have allowed a change in the water concentration through changes in density of the dense supercritical fluid. This was not possible in the classical studies of electrolyte systems. The incorporation of the water concentration into the equilibrium expression for ion association has led investigators to examine room-temperature work in which the water concentration was artificially altered by the use of mixed solvent systems such as water-dioxane. Here too, $\log K$ was found to be a linear function of $\log c_{H_2O}$. This in turn led to the generalized concept of a "complete" equilibrium constant, independent of density (and therefore pressure) in the supercritical region. It also, therefore, appears to be independent of the macroscopic dielectric constant of the solvent.

Doubt is thus cast on the expression for the ion-pair formation equilibrium constant, which is a function of temperature, dielectric constant, and mean ionic diameter. It is true that the derivation of these quantities is based on the concept of charged spheres moving through a dielectric continuum, where the known and recognized structure of water is not taken into account. But this becomes a drawback when one moves away from infinite dilution to higher concentrations of electrolyte, more so than at infinite dilution. The apparent failure to account quantitatively for the behavior of K may be a consequence of the simultaneous variations in a, the ionic diameter, and D', the local, or effective, dielectric constant.

At room temperature and pressure, association (or dissociation) constants are conventionally measured exclusive of the solvent concentration. In the case of aqueous solutions, this concentration is constant at 55.5 M and is lumped into the conventional constant. However, the water concentration is a function of T and P, through the density. The water coefficient, q, in the expression

(28) $$CA + qH_2O \rightleftharpoons C_{aq^+} + A_{aq^-}$$

is also temperature dependent. Thus the complete constant can be expressed as

(29) $$K'(T) = \frac{K(P, T)}{[c_{H_2O}(P, T)]^{q(T)}}$$

It is now clear that conventional constants appearing in handbook tables must be corrected, in general, for the T, P dependence of the concentration of water, and the temperature dependence of $q(T)$. That is, the "constant" that has been lumped into the conventional value of K is not constant at all. Ultimately, it means that the free energy obtained from the conventional constant will have to be corrected as follows:

(30) $$\Delta F^\circ = RT \ln K + \text{correction}$$

(31) $$\text{correction} = q(T)RT \left[\ln 55.5 + \ln \rho(T, P) \right]$$

Inasmuch as $q(T)$ is unknown, such a correction is not readily calculated.

In the supercritical region, $\log K$ was found to be linearly related to $\log \rho_{H_2O}$. But density was then equated to water concentration through the linear equation

(32) $$c_{H_2O}(T, P) = \frac{1000\rho(T, P)}{M}$$

where M is the molecular weight of water. Throughout such treatments, and those extending down to room temperature, M is assumed to be constant with changes in T and P. Yet we know that in general the structure of water is changed by both these variables. Hitherto, when water had been considered the solvent, one only talked about its structure and the effect upon ion mobility; the concentration of electrolyte remained the same regardless of structural changes in water. But now that water is considered a reactant, its effective molecular weight must be reckoned with. Even if we knew little else about water structure, we could suspect that the apparent molecular weight would decrease through thermal dissociation caused by a rise in temperature.[70] The molecular weight has been defined as "the smallest portion of a substance that moves about as a whole."[71] This is clearly unknown for liquid water in the case of "neutral" electrolyte systems, let alone in the special cases of H^+ and OH^- ion mobility.

Thus, although the molecular weight of water will be constant at constant T and P, there is the uncertainty about the actual value that should be used in representing it. In general, its value would be expected to decrease with rising temperature, until in the supercritical region it may be equal to $(18)_q$, where q is a small integer.

It is well known that, in the vicinity of room temperature, increased pressure will shift weak electrolyte equilibria to the right; that is, in a way that increases the degree of dissociation.[72] This is owing to the LeChatelier effect and the electrostriction of water close to the ions. Supercritical temperatures, on the other hand, cause the traditionally strong electrolytes to behave more or less as weak ones, depending on the applied pressure. Thus KCl, HCl, LiCl, NaCl, HBr, NaBr, $BaCl_2$, H_2SO_4, $KHSO_4$, KOH, CsCl, CsI, and K_2SO_4 have been studied in water over wide ranges of temperature and pressure. The 1:1 electrolytes all behave as weak electrolytes at supercritical temperatures. Hydrogen sulfate can be considered a uni-univalent electrolyte above 400°C, and at densities below 0.80 g/cm^3.[33] From 0 to 100°C the specific conductance of potassium hydrosulfate solutions increases rapidly, because of a fall in viscosity. This effect continues to higher temperatures, but the HSO$_4^-$ ion becomes a weaker acid as temperature is increased. Eventually a maximum in the curve of conductance versus temperature is reached. At temperatures of 400°C and above, even at 4000 bars, there is no detectable dissociation of the HSO$_4^-$ ion in this solution.[36] The first and second dissociation constants for BaCl$_2$ at 520°C, however, are only an order of magnitude apart.[42]

Water, an extremely weak electrolyte at room temperature, dissociates to a greater extent as the temperature rises until, at 600°C and 0.7 g/cm^3, for example, K_w has increased by five orders of magnitude. This has been determined by measurement of conductance, and the results agree with shockwave experiments within the limits of probable error. The increased dissociation of water in conjunction with the increased association of electrolytes in the supercritical region can have no other effect than to bring hydrolysis reactions more prominently into play. Complete quantitative treatment is not yet possible.

The dielectric constant of dense, supercritical water can range from 5 to 20 without altering the overall chemical composition, simply upon variation of the applied pressure. This medium allows at least partial dissociation of electrolytes into ions. Inasmuch as ion mobility is also a function of density, however, it would be desirable to alter the dielectric constant by yet another means in order to maintain a constant density. This can be accomplished by using water in admixture with a truly inert diluent. To this end Franck has studied the conductance of potassium chloride in dense, supercritical homogeneous water-argon mixtures.[44] Certain assumptions must then be made concerning the viscosity and dielectric constants of water-argon mixtures in order to estimate the degree of dissociation. This area of study has barely begun to be explored, and holds promise for future investigations.

Also of interest for future study are other polar solvents such as NH$_3$ and HF, both of which possess supercritical regions that are more easily attained than water. For a thoroughgoing physicochemical investigation, however, where ion-ion interactions are always eliminated by extrapolation to infinite dilution, precise viscosity, density, and dielectric constant data are required. On the other hand, studies have been made at higher solution concentrations, where purely empirical correlations are possible.[73]

Whereas electrolytic conductance measurements are a sensitive measure of ion mobility, they do suffer the disadvantage of detecting the large contributions resulting from corrosion of the measurement cell—the so-called background conductance. It is debatable how far improvements in materials will go toward alleviating this situation. In this context, it may be worthwhile to expand our knowledge of the high-temperature–high-pressure region through diffusion measurements. This has the inherent advantage that transport numbers for individual ion migration are not a requirement for a thorough interpretation. Diffusion measurements have been traditionally much less precise than conductance measurements, but it is conceivable that a cell can be designed to reverse the situation in the supercritical region of water.

In recent years solubility measurements carried out by physical chemists have differed from prior hydrothermal chemical research in that an insight into transport and structural aspects has been sought. For example, solubilities can be fitted to an extended Debye–Hückel equation equally well or better than at low temperatures.[74] It is clear that, although we have no definitive theory to predict the behavior of concentrated electrolyte solutions under any conditions, the priority remains for still more research on dilute solutions at high temperature and pressure.

REFERENCES

1. A. A. Noyes, "The Electrical Conductivity of Aqueous Solutions," Carnegie Institution of Washington, D.C., Publ. No. 63, 1907.
2. R. A. Laudise, *Progr. Inorg. Chem.*, Interscience, New York, 3, 1 (1962).
3. S. V. Gorbachev and V. P. Kondrat'ev, *Russ. J. Phys. Chem. (Engl. transl.)*, 35, 10 (1961).
4. W. L. Marshall, Ruth Slusher, and E. V. Jones, *J. Chem. Eng. Data*, 9, 2, 187 (1964).
5. W. L. Marshall et al., "Homogeneous Reactor Program, Quart. Progr. Rept., period ending July 31, 1960, ORNL-2931, UC-4-Chemistry General, and ORNL-3004, UC-81-Reactors-Power; Oak Ridge Natl. Lab., Oak Ridge, Tenn.
6. A. J. Ellis, *J. Appl. Chem.*, 11, 136 (1961).
7. J. K. Fogo, S. W. Benson, and C. S. Copeland, *J. Chem. Phys.*, 22, 2, 212 (1954).
8. S. Sourirajan and G. C. Kennedy, *Amer. J. Sci.*, 260, 115 (1962).
9. D. E. White, *Geol. Soc. Amer. Bull.*, 68, 1637 (1957).
10. E. U. Franck, *Z. Physik. Chem.*, N.F., 8, 107 (1956).
11. W. L. Marshall, *Rec. Chem. Progr.*, 30, 2, 61 (1969).
12. T. Andrews, *Trans. Roy. Soc. (London)*, Ser. A, 159, 547 (1869).
13. H. S. Booth and R. M. Bidwell, *Chem. Rev.*, 44, 477 (1949).
14. A. J. Ellis and W. S. Fyfe, *Rev. Pure Appl. Chem.*, 7, 261 (1957).
15. E. U. Franck, *Angew. Chem.*, 73, 309 (1961).
16. E. U. Franck, *Endeavour*, 27, 55 (1968).
17. W. L. Marshall, *Rev. Pure Appl. Chem.*, 18, 167 (1968).
18. F. Spillner, *Chem. Fabrik.*, 13, 405 (1940).
19. A. C. Swinnerton and G. E. Owen, *Amer. J. Phys.*, 16, 123 (1948).

20. J. K. Fogo, C. S. Copeland, and S. W. Benson, *Rev. Sci. Instr.*, **22**, 765 (1951).
21. J. T. Corwin and G. E. Owen, *J. Chem. Phys.*, **22**, 1254 (1954).
22. E. U. Franck, *Z. Physik. Chem.*, *N.F.*, **8**, 92 (1956).
23. E. U. Franck, *Z. Physik. Chem.*, *N.F.*, **8**, 192 (1956).
24. K. H. Dudziak, thesis, Göttingen, 1959.
25. J. F. Corwin, R. G. Bayless, and G. E. Owen, *J. Phys. Chem.*, **64**, 641 (1960).
26. G. Coulon, thesis, Göttingen, 1961.
27. E. U. Franck, J. E. Savolainen, and W. L. Marshall, *Rev. Sci. Instr.*, **33**, 115 (1962).
28. D. Pearson, C. S. Copeland, and S. W. Benson, *J. Amer. Chem. Soc.*, **85**, 1044 (1963).
29. D. Pearson, C. S. Copeland, and S. W. Benson, *J. Amer. Chem. Soc.*, **85**, 1047 (1963).
30. A. S. Quist, E. U. Franck, H. R. Jolley, and W. L. Marshall, *J. Phys. Chem.*, **67**, 2453 (1963).
31. F. Hensel and E. U. Franck, *Z. Naturforsch.*, **19a**, 127 (1964).
32. E. U. Franck, D. Hartmann, and F. Hensel, *Discussions Faraday Soc.*, **39**, 200 (1965).
33. A. S. Quist, W. L. Marshall, and H. R. Jolley, *J. Phys. Chem.*, **69**, 2726 (1965).
34. A. S. Quist and W. L. Marshall, *J. Phys. Chem.*, **69**, 2984 (1965).
35. W. Holzapfel and E. U. Franck, *Ber. Bunsenges. Phys. Chem.*, **70**, 1105 (1966).
36. A. S. Quist and W. L. Marshall, *J. Phys. Chem.*, **70**, 3714 (1966).
37. W. L. Marshall and A. S. Quist, *Proc. Natl. Acad. Sci. (U.S.)*, **58**, 901 (1967).
38. A. S. Quist and W. L. Marshall, *J. Phys. Chem.*, **72**, 1536 (1968).
39. A. S. Quist and W. L. Marshall, *J. Phys. Chem.*, **72**, 684 (1968).
40. A. S. Quist and W. L. Marshall, *J. Phys. Chem.*, **72**, 2100 (1968).
41. A. S. Quist and W. L. Marshall, *J. Phys. Chem.*, **72**, 1545 (1968).
42. G. Ritzert and E. U. Franck, *Ber. Bunsenges. Phys. Chem.*, **72**, 798 (1968).
43. K. Mangold and E. U. Franck, *Ber. Bunsenges. Phys. Chem.*, **73**, 21 (1969).
44. D. Hartmann and E. U. Franck, *Ber. Bunsenges. Phys. Chem.*, **73**, 514 (1969).
45. H. Renkert and E. U. Franck, *Ber. Bunsenges. Phys. Chem.*, **74**, 40 (1970).
46. A. A. Noyes and W. D. Coolidge, *Z. Physik. Chem.*, **46**, 323 (1903).
47. H. D. David and S. D. Hamann, *Trans. Faraday Soc.*, **55**, 72 (1959).
48. A. B. Gancy and S. B. Brummer, *J. Electrochem. Soc.*, **115**, 804 (1968).
49. S. V. Gorbachev and V. P. Kondrat'ev, *Russ. J. Phys. Chem. (Engl. transl.)*, **36**, 10 (1962).
50. I. M. Rodnyanskii and I. S. Galinker, *Zap. Khar'kov Sel'skokhoz Inst.*, **14**, 43 (1957).
51. R. A. Horne, *Nature*, **200** (4905); *Science*, **418** (1963).
52. G. C. Kennedy, *Amer. J. Sci.*, **248**, 540 (1950); *Amer. J. Sci.*, **252**, 225 (1954).
53. A. D'aprano and R. M. Fuoss, *J. Phys. Chem.*, **67**, 1704, 1722 (1963).
54. T. Shedlovsky, *J. Franklin Inst.*, **225**, 739 (1938).
55. See, for example, J. K. Fogo, S. W. Benson, and C. S. Copeland, *J. Chem. Phys.*, **22**, 209 (1954); K. H. Dudziak and E. U. Franck, *Ber. Bunsenges. Physik. Chem.*, **70**, 1120 (1966); S. Maier and E. U. Franck, *Ber. Bunsenges. Physik. Chem.*, **70**, 639 (1966); A. S. Quist and W. L. Marshall, *J. Phys. Chem.*, **69**, 3165 (1965).
56. R. A. Robinson and R. H. Stokes, *J. Amer. Chem. Soc.*, **76**, 1991 (1954).
57. R. M. Fuoss, L. Onsager, and J. F. Skinner, *J. Phys. Chem.*, **69**, 2581 (1965).
58. R. M. Fuoss and C. A. Kraus, *J. Amer. Chem. Soc.*, **55**, 2387 (1933); *J. Amer. Chem. Soc.*, **57**, 1 (1935).
59. R. M. Fuoss, *J. Amer. Chem. Soc.*, **80**, 5059 (1958).

60. E. U. Franck, *Z. Physik. Chem.,* *N.F.*, **6**, 345 (1956).
61. M. Eigen and E. Wicke, *Z. Electrochem., Angew. Physik. Chem.*, **55**, 354 (1951).
62. H. S. Harned and L. D. Fallon, *J. Amer. Chem. Soc.*, **61**, 2374 (1939).
63. S. D. Hamann, *J. Phys. Chem.*, **67**, 2233 (1963).
64. R. W. Kunze and R. M. Fuoss, *J. Phys. Chem.*, **67**, 911 (1963).
65. L. A. Dunn and W. L. Marshall, "Electrical Conductance and Ionization Behavior of Sodium Chloride in Dioxane-Water Mixtures at 100°C," *J. Phys. Chem.*, in press, 1969.
66. J. E. Lind, Jr., and R. M. Fuoss, *J. Phys. Chem.*, **65**, 999 (1961).
67. See, for example, W. R. Gilkerson, *J. Chem. Phys.*, **25**, 1199 (1956); E. A. Moelwyn-Hughes, *Z. Naturforsch.*, **18a**, 202 (1963).
68. A. A. Noyes, A. C. Melcher, H. C. Cooper, and G. W. Eastman, *Z. Physik. Chem.*, **70**, 335 (1910).
69. A. Gierer and K. Wirtz, *Ann. Physik.*, (6), **6**, 257 (1949).
70. S. Glasstone, *Textbook of Physical Chemistry*, Van Nostrand, New York, 1940, p. 311.
71. *Webster's New International Dictionary*, 2nd ed., Merriam, Springfield, Mass. 1956, p. 1580.
72. S. D. Hamann, *Physico-Chemical Effects of Pressure*, Academic Press, New York, 1957, pp. 137–159.
73. I. N. Maksimova and V. F. Yushkevich, *Elektrokhimiya*, **2**, 5, 577 (1966).
74. W. L. Marshall and E. V. Jones, *J. Phys. Chem.*, **70**, 4028 (1966).
75. S. B. Brummer and G. J. Hills, *Trans. Faraday Soc.*, **57**, 1823 (1961).

Author Index

Subject Index